NEW EXTRAGALACTIC PERSPECTIVES IN THE NEW SOUTH AFRICA

ASTROPHYSICS AND SPACE SCIENCE LIBRARY

VOLUME 209

NEW EXTRAGALACTIC PERSPECTIVES IN THE NEW SOUTH AFRICA

Proceedings of the International Conference on
"Cold Dust and Galaxy Morphology"
held in Johannesburg, South Africa,
January 22–26, 1996

Edited by

DAVID L. BLOCK

Department of Computational and Applied Mathematics,
University of the Witwatersrand, Johannesburg, South Africa

and

J. MAYO GREENBERG

Huygens Astrophysics Laboratory,
University of Leiden, The Netherlands

KLUWER ACADEMIC PUBLISHERS
DORDRECHT / BOSTON / LONDON

A C.I.P. Catalogue record for this book is available from the Library of Congress

ISBN-13: 978-94-010-6637-2   e-ISBN-13: 978-94-009-0335-7
DOI: 10.1007/978-94-009-0335-7

Published by Kluwer Academic Publishers,
P.O. Box 17, 3300 AA Dordrecht, The Netherlands.

Kluwer Academic Publishers incorporates
the publishing programmes of
D. Reidel, Martinus Nijhoff, Dr W. Junk and MTP Press.

Sold and distributed in the U.S.A. and Canada
by Kluwer Academic Publishers,
101 Philip Drive, Norwell, MA 02061, U.S.A.

In all other countries, sold and distributed
by Kluwer Academic Publishers Group,
P.O. Box 322, 3300 AH Dordrecht, The Netherlands.

Printed on acid-free paper

NEW EXTRAGALACTIC PERSPECTIVES IN THE NEW SOUTH AFRICA

Changing Perceptions of the Morphology, Dust Content and Dust-Gas Ratios in Galaxies: January 22 - 26 1996

Scientific Organizing Committee:

R.J. Allen (USA)
G. Bertin (Italy)
D.L. Block (South Africa; SOC Chairperson)
B.W. Burton (Netherlands)
N. Devereux (USA)
J.M. Greenberg (Netherlands)
P. Grosbøl (Germany)
R.D. Joseph (USA)
J. Lequeux (France)
A.N. Witt (USA)

Local Organizing Committee at the Witwatersrand University:
Professors J.P.F. Sellschop (Deputy Vice-Chancellor: Research), P.D. Tyson (Chairman of the National Astronomy Facilities Board: South Africa), R. Crewe (Dean: Science Faculty), C.J. Wright (Head: CAMS) and D.L. Block (CAMS; Chairman of the LOC); G.C. Buric (CAMS), E. Momoniat (CAMS) and M. Nakayama (CAMS). Other LOC Members: Dr R.S. Stobie (Director: SAAO), Dr L. Alberts (former Chairman, South African Academy of Science and Arts); E.K. Block (Vista University), P. van Huyssteen (Price, Waterhouse and Meyernel), T. Hanson (Rennies Conference Management), S. Irish (Sunnyside Residence) and M. Doidge (Wits).
Secretariat:
Mrs Florence Goetze. SOC Secretariat: E. Momoniat

Registration Desk & Camera Ready Copy:
De Wet Ferreira, Siona Bradu, Meg Doidge, Ebrahim Momoniat.

COVER CAPTIONS

Front Cover – (top):
Images secured through H, R and B filters of the famous interacting 'Antennae' galaxy pair NGC 4038/9 (= Arp 244 = VV 245) were assigned to the red, green and blue guns respectively, of a colour digital camera. The difference in the stellar content of the two galaxies is very striking. Approximately 40% of HI in the system is associated with the two galaxies seen here; the remaining 60% is associated with the luminous tails. The observers were K. Sellgren and G. Tiede, using the 1.5m at Cerro Tololo Inter-American Observatory in Chile. Courtesy J. Frogel.

Front Cover – (bottom):
The spiral galaxy NGC 2997. Left: Optical photographs of NGC 2997 (courtesy D. Malin and the Anglo-Australian Observatory) and B CCD images (courtesy S. d'Odorico with the ESO NTT) reveals an optically grand design, m=2 morphology. Middle: A near-infrared 2×3 mosaic at 2.1 microns with the ESO 2.2m telescope, reveals a distinct m=1 stellar component. Observers at ESO, La Silla: D. Block and A. Moneti. Right: After rebinning to a common scale of the B and K′ images, widespread interarm dust (colour coded so that dust=black) is seen at arcsecond resolution in this B-K′ map, produced by D. Block and P. Grosbøl. Reproduced courtesy *Astronomy and Astrophysics*.

Back Cover:
A three dimensional projection of a near-infrared image of the optically *flocculent* spiral galaxy NGC 2841 reveals a remarkable system of long, dark infrared spiral arms. The very regular and smooth *dark, gaseous* spiral arms appear as 'valleys' in this relief projection, and have no stellar counterparts. The arms each span an azimuth of 160°. Image produced using the IBM Data Explorer by B.G. Elmegreen and C. Pickover, from 2.1 micron NICMOS images secured at Mauna Kea, Hawaii by R.J. Wainscoat. Courtesy *Nature*.

CONTENTS

Invited review papers are identified by a † after the surname of the review speaker. A * in a title implies that colour plate(s) from the author(s) are located in the separate Colour Plate Section

Concluding Thoughts

THE PREFACE

The date: September 30, 1880
The place: A private observatory in Hastings-on-Hudson
Profession of the observer: A medical doctor
The instrument: An 11-inch Clark refractor.

The significance of that night marked one of the truly great turning points in the development of astronomical techniques: Dr Henry Draper, a wealthy New York medical doctor, had secured the first photograph of a nebula: a 51-minute exposure on a dry gelatinobromide plate showing the wispy nebulosity of the Orion Nebula. By March 1882, Draper had secured an exposure of 137 minutes, showing far richer detail of both bright *and dark* features.

The rest is history. The photographic era heralded in a universe where hints of the presence of cosmic dust were strongly alluded to: from star-forming regions such as Messier 17, to the Horsehead Nebula in Orion, to the striking *dark finger* in the Cone Nebula, to the magnificent dark bands in the plane of our Milky Way.

"Historically, astromomers from the very beginning have been afraid of dust. But since it had to be accepted, it was hoped that its properties were universal and constant so that corrections for stellar distances could be simply made. But life is not so simple, and passive *'teflon'* dust was an impossible dream. For example, molecular clouds: all the chemical work that was done on molecules in clouds was done using only gas phase interactions such as ion molecule reactions, absolutely ignoring the fact that each cold dust grain is a remarkable chemical factory. Astronomers ignored, or tried to ignore the effects of dust from so many points of view that it is impossible for me to sum them all up here." (JMG)

Professor Michael Disney summed it up eloquently:

"For most astronomers dust is quite simply a nuisance. They want to be confirmed in their fervent hope that there isn't much of it about and they would be happy to see the heretics burned."

What *is* the nature and composition of the dust grains responsible for the visual extinction in our Galaxy and in other galaxies beyond? What are the ranges in temperature of dust grains? Can these be less than 2.7K, the temperature of the cosmic blackbody background? Can cold dust grains be studied optically at unprecedented arcsecond resolution? Might such grains account for dust mass increases by up to one order of magnitude? How does the presence of dust affect the morphology of a galaxy? In particular, can the near-infrared classification of a galaxy be radically different from its optical Hubble type, once the dust veil is lifted? Is this new dust-penetrated view bringing us to the verge of a breakthrough in understanding the connection between galaxy morphology and the underlying physics of galaxies? Are there significant amounts of cold molecular hydrogen gas in galactic disks?

Michael Disney quipped:

"An eminent cosmologist once advised me to forget all about very cold dust .... He sounded plausible, as Cosmologists are apt to sound, but he was in fact totally wrong, as Cosmologists are apt to be... Modest amounts of smoke [dust] can play merry hell with the decomposition of visible light profiles..."

These were some of the key issues which drew approximately 100 astronomers from across the globe, to the University of the Witwatersrand in Johannesburg. The event in January 1996, was a conference entitled *"New Extragalactic Perspectives in the New South Africa: Changing Perceptions of the Morphology, Dust Content and Dust-Gas ratios in Galaxies"*. Much water had passed under the bridge since first-light of the Yale Observatory on our campus, in 1925 (page 578).

The conference was opened by the Deputy Vice-Chancellor for Research at the University, Professor J.P.F. Sellschop (Figure 1). Our Cabinet guest of honour at the conference banquet was Dr. Ben Ngubane, Minister of Arts, Culture, Science and Technology in Parliament. Delegates found Dr Ngubane's banquet address to be most thought provoking and will long be remembered as a conference highlight (Figure 2). A letter received from President Nelson Mandela was read out at the conference, and appears in this volume. Delegates heeded the President's recommendation to *relax* after our packed week of deliberations (colour plate 16)!

The Scientific Organizing Committee (SOC) was formed in late 1994, and in January/February of 1995 – one year in advance of the conference – the SOC members voted for all the speaker categories. Every SOC elected speaker agreed to provide a camera-ready copy of their talk at the Registration Desk. A revised version was permitted to be sent in by late February 1996, one month after the conference.

The ordering of the papers contained in these *Proceedings* essentially follows the *order of presentation* during the conference week. A notable exception is that of Gordon Stacey, who suffered a severe disk-crash at Cornell; with the granting of extra time to prepare his camera-ready copy, Dr Stacey's contribution could then only be accomodated at the end of the oral presentation section. We also very much regret that Dr Michael Fall was unable to submit a camera-ready copy of his talk, in time for the *Proceedings*. We are extremely grateful to Dr E. de Geus and *KLUWER* for the inclusion of a set of sixteen pages of colour plates.

Professor Disney also wisely reminded us as Editors, that conference proceedings are invariably only as valuable as the degree to which conference interactions are reflected. Special steps were taken to ensure that these *Proceedings* could contain over 35 printed pages of the *very lively* and exciting interactions and discussions we all enjoyed together. A television crew filmed the live 'Eyes to the Future' Panel-Audience discussion, for transcription into Latex and finally into KLUWER style format. Our Chairpersons and Panelists are to be thanked for the great effort and care with which they undertook their responsibilities.

• The principal sponsor of the conference was the Anglo-American and de Beers *Chairmans' Fund Educational Trust*. Without the astronomical vision of Chairman Michael O'Dowd (Figure 1), this conference would never have been possible. Mr O'Dowd, we salute you. Mrs. Margaret Keeton and the Board of Trustees, we thank you. We appreciate that cold cosmic dust grains at temperatures of -250°C in galaxies billions of kilometres away may, at first glance, seem rather *far removed* from your gold-mining and diamond activities ... but as Dr. Duccio Macchetto and

Figure 1. Top: *The meeting of two Chairmen.* 'Chairman Mayo' Greenberg (top right) meets Chairman Michael O'Dowd (top left). Mr O'Dowd is the Chairman of the Anglo American & De Beers Chairmans' Fund Educational Trust. Bottom: Professor JPF Sellschop, Deputy Vice-Chancellor for Research at the University (bottom left) in deep discussion with Dr L Alberts (bottom right) of the LOC.

Figure 2. Top: Professor Paul Hodge (right) meets Cabinet Minister Ben Ngubane (left). Professor David Block (center) looks on. Minister Ngubane is the Cabinet Minister of Arts, Culture, Science & Technology. Bottom: Professor Mike Disney (left) emphasizes a point to Minister Ngubane, while Professors David Block, Ron Allen and Brent Tully attentively listen.

... but as Dr. Duccio Macchetto and Dr. Dale Cruikshank reminded us at the conference, without exploding stars there would be no diamonds!

• Our deep gratitude to the University of the Witwatersrand for their financial support. In particular, we gratefully acknowledge financial support from the fund of the Deputy Vice-Chancellor for Research as well as from Professor R. Crewe, Dean of the Faculty of Science, at the University. We also thank members of our Departmental Conference Review Committee, Professor C.J. Wright (Head of the Department of Computational and Applied Mathematics), Professor D.P. Mason and Dr F. Mahomed for their constructive input.

• We are most indebted to the *Foundation for Research Development* for their support.

• As interest in our conference grew beyond all initial expectations, *SASOL* and *TRANSNET* provided a Pentium processor with modem to streamline the editing of the papers in these *Proceedings*.

• Cellular communication with the conference organizing team was facilitated by *ERICSSON* and *VODACOM*.

• To Mr P. Kruger and Mr R. Hugo of SASOL, to Dr M. de Waal of TRANSNET, to Mr M. Levitt of Ericsson and to VODACOM: our words of thanks cannot adequately express our deepest appreciation for the tremendous infrastructure you so graciously provided.

A special word of thanks, too, to:

• Liz Block, wife of David Block, for her pillar of support during the preceeding 15 months of conference preparation. There were some tense moments, such as when Kristen Sellgren arrived with no clothes[1], but Liz stepped into action immediately!

• Ebrahim Momoniat. A note of great appreciation to Ebrahim was recorded at the conference banquet.

• All members of the SOC and LOC.

• De Wet Ferreira for so willingly travelling to the University from Pretoria on innumerable occassions to assist us with the immense task of collation before the Editorial process could begin.

• Traci Hanson of Rennies Travel, and Shirley Irish, Lady Warden of the Sunnyside Residence at Wits University. As Dr Ron Allen emphasized in his vote of thanks, their dedication and professionalism left an indelible impression on us all. Mrs. Irish even delayed taking her retirement in order to assure that the accomodations and arrangements at Sunnyside Residence would be the best possible!

• Mr Mark Irvine (Corporate Communications Department of the Anglo-American Corporation of SA) for the printing of the conference programme (see colour plate 1) and the printing of our conference logo for the conference wine bottles. And to Lauren Wilson and her team for the the streamlining of all presentations at the University's Planetarium.

• Mrs Florence Goetze, secretary to DLB.

• Mr S. Hlubi and Mr G. Lock, computer network specialists.

• Mr P. van Huyssteen and Mr P. Liebenberg.

• Dr Jack Flaks and Mrs Audrey Flaks.

• The staff of our Central Graphics Unit – Colin Emslie, Jo Waltham, Natalie Scrooby, Anna Michaelides, Bernice Fonseca, Gora Jadwat, Joseph Moutloatsi

---

[1] Her luggage had been mislaid enroute from the USA to Africa.

and Vasu Naidoo – your assistance has been invaluable. Our deep gratitude too, to Andy Blake and John Page.

• The South African dried fruit association (SAD) for the provision of South African dried fruits, and SATOUR for providing SA tourism material.

• Photographers Khashifa Jappie and Jo Waltham for securing our conference photographs and, in particular, the informal 'candid shots'. The group photograph was taken by Fidos Kleovoulou. The photograph of President Mandela is copyrighted by Karina Turok and reproduced by permission.

The dust veil is being lifted, and lifted dramatically. We live in a universe filled with New Cosmic Perspectives. From the photographic work of Draper, to unprecedented insights with instruments sensitive to the infrared and the (sub)millimeter. It is with the express aim of trying to captivate the tremendous scientific expertise and thinktank of that great cold dust-morphology conference week at the University of the Witwatersrand, that we submit these *Proceedings* to the greater scientific community.

David L. Block
J. Mayo Greenberg
Johannesburg and Leiden
20 June, 1996.

MESSAGE FROM PRESIDENT MANDELA

I extend a warm welcome to you all on the occasion of the *International Astrophysics Conference.*

Yours is a branch of science which has occupied the mind of every great scientist from Aristotle to Einstein. It is a noble pursuit indeed, and one which fundamentally enriches human knowledge about our habitat and our history.

We are honoured to have had South Africa selected as the venue for the Conference. We hope you will be able to find some time to relax and enjoy our beautiful country.

We wish you every success in your deliberations.

Mandela

Nelson Mandela
President of the Republic of South Africa

January 1996

PHOTOGRAPH: KARINA TUROK

CHANGING PERCEPTIONS OF THE MORPHOLOGY AND DUST CONTENT IN GALAXIES

DAVID L. BLOCK
Department of Computational and Applied Mathematics,
University of the Witwatersrand, Johannesburg
Private Bag 3, Wits 2050, South Africa

Abstract. Optical images of galactic spiral arms invariably show massive, hot and blue stars forming in dense ridges of interstellar gas mixed with dust. Accompanying IR spirals or stellar density waves (if present) in the old stellar disk are invariably bright. Non-interacting galaxies with regular, dark or black IR spiral arms only (i.e those galaxies whose infrared images reveal a clearly defined pattern of pure dust+gas spiral arms only, with little or no star formation in the arms) have hitherto never been encountered. We begin this review by presenting near-infrared imaging of the famous optically flocculent prototype NGC 2841. At K', a remarkable system of four regular dark spirals, each spanning 160° in azimuth, is reported; NGC 2841 is the first *dark IR spiral galaxy* to be discovered. The dark IR spirals are so dense that they are obscured at optical wavelengths by scattered light from the bulge. Not only can spiral arms themselves be dark, but the effective dust albedo at K' in the spiral arms of galaxies can be as high as 0.9 (we derive this result for the M51 arm [dust screen] juxtaposed against the background companion NGC 5195). We conclude the review by discussing the unambiguous detection of dust between the arms of spiral galaxies. Using radiative transfer models (invoking the full effects of multiple scattering by dust grains) on optical/near-infrared images, it is now possible not only to identify, but also to quantify, widespread interarm dust at unprecedented spatial resolution: a resolution two orders of magnitude better than attainable by IRAS and one order of magnitude better than presently attainable with the largest sub-mm/mm telescopes.

'Through space the universe grasps me and swallows me up like a speck; through thought I grasp the universe.' – Blaise Pascal, Pensées

1

D. L. Block and J. M. Greenberg (eds.), New Extragalactic Perspectives in the New South Africa, 1–20.
© 1996 *Kluwer Academic Publishers.*

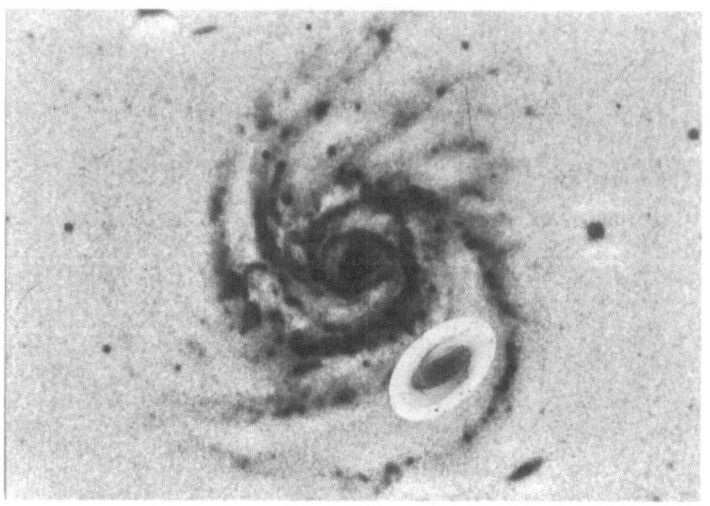

Figure 1. An optical photograph of the ScI spiral galaxy NGC309. In the near-infrared, however, this spiral mimics a galaxy of type SBa with a prominent bar [**13**]. The Population I disk here is therefore not light as may a priori be supposed, but rather is embedded in a heavy stellar disk. NGC 309 is so large that M81 comfortably fits in between its spiral arms: note the inset of M81, printed at the same linear scale as NGC 309.

1. Introduction

The classification of galaxies and the distribution of dust within their disks has traditionally always been inferred from photographs/CCD imaging and colour excesses *shortward* of the 1 micron window, where stellar Population II disks are not yet transparent eg. B(blue; 0.44 μm) minus I (0.83 μm), or R (0.70 μm) minus I. In 1991, we pointed out how two radically different morphologies could actually co-exist in the same galaxy, when that galaxy were imaged beyond 1 micron (Block and Wainscoat [13]). The tracing of dust grains in nature can be very elusive: firstly, high levels of dust *extinction* do *not* necessarily imply that the effects of dust *attenuation* (ie. observed reductions in surface brightness profiles) will also be large, because scattering by dust grains may fill in at least part of the lost surface brightness [82]. Secondly, cold and very cold dust grains in external galaxies were undetectable by IRAS, yet may increase dust masses by nine hundred percent [19]. It has only in recent years been more fully appreciated just *how* dependent galaxy classifications on dust opacities are: the inferences of stellar mass distributions (for example, light vs. heavy galactic disks) from optical imaging can often be summarised in one word: wrong (see Figure 1).

Before proceeding to discuss our technique which carefully probes the full spatial distribution of dust grains of all temperatures, the question I wish to address is:

"Do dark IR spirals exist? Has any galaxy ever been observed where the spiral arms at K′ are only dust+gas (Population I) with no stellar (Population II) counterparts?"

2. The stellar disks of optically flocculent galaxies

2.1. THE BLACK SPIRAL NGC 2841

NGC 2841 is the prototype for a relatively common form of non-barred, non-interacting galaxy in which all of the spiral arms are short, 'flocculent' patches of star formation (Figure 2); optical images show no grand design spiral structure in the underlying stellar disk ([63], [36], [49], [84]). It has a typically large bulge for its Hubble type S(r)b, and a very high rotation rate, 270 km s^{-1}, in most of the disk beyond the bulge where the gas is concentrated in a ring with peak surface mass density ~ 21 M$_\odot$ pc^{-2} ([23], [3], [87], [24]).

In collaboration with Bruce Elmegreen and Richard Wainscoat (see [20]), we report observations of NGC 2841 in the K′ (2.1 micron) regime designed to probe the old disk population and to search for any structures that might be present in the disk at extremely high optical opacities. The extinction at K′ is only 0.1 that in visible light (Martin and Whittet [55]). The distance to NGC 2841 is assumed to be 9.5 Mpc.

No bright stellar waves are seen in our K′ image. Usually such waves are obvious at K′ ([13], [17], [59]). In their pioneering study of azimuthal Fourier decompositions of K′-band light, Rix and Zaritsky [60] showed the arm/interarm contrasts for most spiral arms are a factor 2 or more. The lack of bright K′ spirals in NGC 2841 implies that any stellar waves that may be present are weaker than in grand design spirals by at least a factor of ten.

What is clear in our first K′ image of NGC 2841 (see colour plate 2a) are the remarkable two long and regular *dark* spirals (one outer, one inner) extending from the northeast minor axis to the north, and the other set of dark spirals in the west and south. The dust+gas spirals are seen even more clearly when we illuminate the N-NE section of the disk (colour plate 2b). The dark K′ spiral closest to the bulge in the northeast contains a most unusual amorphous strip of emission in the optical photograph (see arrow in Figure 2). We shall examine that presently. The rest of the K′ disk is fairly smooth. The K′ image also indicates that the position angle of the bulge differs from the position angle of the outer disk by $\sim 10°$, which implies that the bulge is triaxial if the disk is circular.

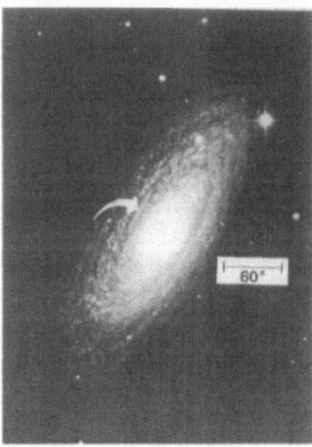

Figure 2.   An optical photograph of NGC 2841: a prototype of the flocculent arm classification. N is up; E to the left. Photo secured by M. Humason; reproduced courtesy the Carnegie Institution of Washington and Caltech. The arrow points to a smooth amorphous strip of optical emission in the inner NE dust spiral.

Colour plate 3a (top) shows a deprojected and enhanced image from our combined V and K′ data. The V band contributes to the blue light, the K′ band to the red light, and the average of the V and K′ bands contributes to the green light in this rgb image. The deprojection was made by assuming a position angle of 150° and an inclination of 68°. The bright vertical band in the center is from the bulge. Dark and bright spiral arms are readily revealed in this image; they are very tightly wound and span azimuthal angles of around 160°.

Galactic spiral arms are usually logarithmic in form, so they are more easily seen as straight lines on plots of log(radius) versus azimuthal angle. Such a plot is shown in Colour Plate 3b (bottom), which covers the radial range from $0.18R_{25}$ to the edge of the image at about $0.7R_{25}$. The spirals are tilted up and to the left, which is in the trailing sense in the original image, as in almost all other spiral galaxies. The tilt also implies that the arms are true spirals and not rings. The pitch angles of the spiral arms are easily read from the slopes of the tilted features; they are around 9°. The 4 vertical bands are highly exaggerated bulge emission.

Colour plate 3b shows both bright and dark tilted features, corresponding to bright and dark spiral arms on the original image. This zebra-stripe pattern leads to the question of whether the real features are bright with a gap between them, or dark with bright background emission on either side of them. The true situation could result from a combination of bright and dark spiral arms, bright from stellar waves and dark from dust in corresponding gaseous waves or shock fronts, but we present evidence that most of the morphology results from dark spirals that are gas and dust.

The arguments in favour of this latter interpretation are the following: (1) Stellar density waves (hereafter SDWs) at K$'$ are invariably regular and smooth (Block and Wainscoat [13], Block et. al. [17], Rix and Rieke [59]) but in colour plate 2, only the *Population I dust+gas* spirals are smooth. The bright K$'$ spirals are not smooth: rather, they are composed of faint patches which correspond to bright star-formation in the optical. The bright K$'$ spirals look like the rest of the flocculent disk (Plate 2b), but outlined or highlighted by the dark spirals. (2) The bright arms on either side of the dark spirals are near to each other and parallel on the log R - θ diagram; usually bright spiral arms are broader and more widely spaced. The inner dark spiral is at the edge of a smooth inner bright disk, i.e., there is no bright spiral that defines its inner edge. (3) The inner dark spiral in the northeast corresponds to a most remarkable smooth, bright strip of visible light in optical images (see arrow in Figure 2). Thus the dark region cannot be a low density interarm between two bright stellar spirals; it must be dust extinction.

The rotation of NGC 2841 is receding on the southern major axis (Bosma [23]), so the near side of the galaxy is the northeast minor axis if all the spirals are trailing. This means that the bright diffuse V-band patch arrowed in Figure 2 in the inner dark NE K$'$-band spiral is on the near side of the galaxy, in front of the bright bulge. There are no other bright diffuse patches of similar extent in this galaxy and this one is exactly on the minor axis. This position suggests that the diffuse optical emission is from bulge light scattered off the top of the gas+dust spiral seen in K$'$. *A prediction: The spectrum of the diffuse patch should contain stellar absorption lines similar to those in the bulge stars.* Long diffuse patches like this are very rare in galaxies; no other cases have been discussed in the literature and we know of no similar examples in Atlas photographs.

The K$'$ intensity profiles can be matched by a radiative transfer calculation that incorporates absorption and phase-dependent scattering (Block, Elmegreen and Wainscoat [20]). The K$'$ band extinctions through the inclined disk viewed with 68° inclination, are equal to 1.32 mag and 0.68 mag for the inner and outer dust spirals, and 0.84 mag and 0.30 mag for the ambient gas at the same radius without the spirals (1 optical depth = 1.086 magnitudes). The extinctions to the midplane are one-half these values; the midplane density in each dust spiral exceeds that in the surrounding regions by factors of 4 and 8, respectively. Because the extinction at V is 10 times larger than the extinction at K$'$, the V-band extinctions through the disk are 13.2 mag and 6.8 mag at the locale of the two spirals, and 8.4 mag and 3.0 mag in the ambient regions. Note that all of these fits were made entirely with gas absorption in an exponential disk, i.e., no bright spirals were allowed to contribute to the features on the minor axis intensity profile. If stellar spirals are also present, then the dark spiral opacities could be smaller.

There is a strong red to blue colour gradient from the near-side to the far-side of the disk [20], which we attribute to dust scattering. The colour gradient with distance is faintly evident in the Wray Colour Atlas [86] where the bulge and far-side minor axis look white and washed out compared to the near-side minor axis. (The effect is much stronger in our data because the wavelength range covered by our V and K' bands is larger).

The red to blue colour gradient be explained by scattered light from dust located high off the plane in the region of the bulge. In such a position, dust will selectively scatter blue bulge light at high angles into our line of sight and leave the direct paths redder; this makes the bulge and far-side *bluer* and more uniform than the near side. The fact that the blue excess occurs at a projected distance from the galaxy center $> 140'' = 6.4$ kpc implies that a substantial amount of dust has to be outside the deep potential well of the bulge. Radiation pressure and a nuclear wind may be involved.

The average V extinctions in the vicinity of the dark spirals can also be estimated also from the HI and H_2 column densities, assuming a galactic gas-dust ratio of $N(HI+H_2) / E(B-V) = 6 \times 10^{21}$ atoms cm^{-2} mag^{-1} as determined by Bohlin, Savage and Drake [21]. Thus, a V-band optical depth of unity corresponds to a combined atomic and molecular hydrogen column density of 2×10^{21} atoms cm^{-2}.

Observations of atomic and molecular gas at arcmin resolution confirm there is a reservoir of interstellar matter in the vicinity of the dark spirals. The long spirals in the K' image are located in the range $0.2-0.5R_{25} = 50'' - 120'' = 2.3 - 5.5$ kpc. The molecular hydrogen column density projected on our line of sight, $N(H_2)$, peaks in a ring at the location of the dust spirals $(100'')$ at a value of 21×10^{20} mol cm^{-2} ([87], [24]). This is a lower limit if there is very cold molecular hydrogen ([1], [51]). The atomic hydrogen column density ([23], [3]) peaks at a larger radius $(120'')$, slightly outside the spirals) with a maximum on the line of sight $\sim 1.1 \times 10^{21}$ atoms cm^{-2}. In the spiral region, the apparent HI column density is $\sim 8 \times 10^{20}$ atoms cm^{-2}. Thus the combined molecular and atomic hydrogen column density at $r \sim 100''$ is $\sim 5.0 \times 10^{21}$ atoms cm^{-2}, so the average V-band optical depth through the whole disk where the dark spirals are located is ~ 2.5. (The CO beam at IRAM is 1 kpc at our adopted distance; compression of the H_2 distribution to smaller annuli to compensate for beam smear could lead to even larger optical depths). To provide an appreciation for such a large opacity value, we recall that the average V optical depth in the very prominent NE dust screen of NGC 4826 is only ~ 2.1 (Block et al [18]).

The gas and dust mass in the spiral region may be calculated by de-projecting the observed surface mass densities and multiplying by the ring areas. The above HI and CO measurements correspond to a peak surface mass density perpendicular to the plane equal to ~ 21 $M_\odot$ pc^{-2} (using a mean mass per H of 1.43, from He and heavy elements). If this is multiplied by the area of a ring around the peak 1 kpc wide and 4.6 kpc ($= 100''$) in radius, the mass enclosed would be $\sim 6.1 \times 10^8$ $M_\odot$. The corresponding dust mass is 4.9×10^6 $M_\odot$ for a dust-to-gas conversion factor of 0.008 (assuming average galactic depletions of gas phase elements with respect to solar abundances (Whittet [80])).

Dust masses – both for objects of low or high redshift z - may be independently computed from 1300 μm fluxes, assuming a distance, a temperature, and a mass absorption coefficient at 1300 μm (see equation 2 in Chini and Krügel [30]). The general formula of Chini and Krügel incorporates the factor 1+z, which describes the ratio of the radius of curvature of space at the present epoch to that at the time of emission (equation 1.7 in Block [8]). For the nearby NGC 2841, the 1+z factor remains at unity. Our NGC 2841 dust masses from (i) atomic HI and molecular H_2 surface mass density considerations (as above) and (ii) from the extrapolated 1300 μm flux, are consistent with IR emission in NGC 2841 described by a Kirchoff-Planck function peaking at a grain temperature of T $\sim$ 20K – or alternatively at a wavelength $\sim 180\mu m$ (section 4c in Disney [33]).

The origin of the long and smooth dust+gas spirals in NGC 2841 is unknown. Numerical studies by Elmegreen and Thomasson [37] of flocculent disks show that if the gas mass column density were to be increased, such gaseous spirals could readily develop. Their numerical models show that when the instability parameter for the stars, $Q_s = \kappa c_s/(3.36 G \sigma_s)$ is high, > 2.5 ($\kappa =$ the epicyclic frequency, $c_s =$ the stellar velocity dispersion, $\sigma_s =$ the stellar mass column density), and when the instability parameter for the gas, $Q_g = \kappa c_g/(\pi G \sigma_g)$ is low, ~ 1.2, long gas+dust spirals can develop in flocculent disks. Now the stellar parameters are not known for NGC 2841, but the gas parameters $c_g = 10$ km s^{-1} and $\sigma_g = 21$ $M_\odot$ pc^{-2} give $Q_g \sim 3$ for a flat rotation curve with speed 270 km s^{-1}. In the absence of large amounts of very cold H_2 to increase the gas mass column density and hence decrease Q_g to ~ 1, the gas would seem to be too weakly self-gravitating in the presence of such strong differential rotation to generate spiral arms on its own. One possibility is that the gas spirals are a response to the *triaxial* bulge potential.

NGC 2841 appears to be the first example of a triaxial bulge in a flocculent galaxy. Most triaxial shapes are in the inner regions of barred galaxies [85], except for M31 ([73], [53]) and NGC 4845 ([6], [38]), which are non-barred galaxies with grand design spirals. Wave excitation by the triaxial bulge in M31 was discussed by Stark [73]. The fact that the spiral response in NGC 2841 is either very weak or confined entirely to the gas illustrates the unusual stability of this stellar disk.

Another intriguing possibility is that the dark IR spirals may be the result of shear [20]. Large, cold clouds with no obvious signs of associated star formation (such as G216-2.5 in our Galaxy [54], [81]) may simply have been subjected to the enormous shear in NGC 2841 of 54 km s^{-1} kpc^{-1}. The column densities at the locale of the dark IR spirals are reminiscent of those for the Rosette GMC. Ofcourse the Rosette GMC extends for only 3.5 degrees (see Blitz and Thaddeus [7] and Block [12] for morphological descriptions). Our dark dust+ gas spirals, spanning 160° in azimuth, *may* be a collection of high mass clouds which have been subjected to very large rates of shear.

2.2. NGC 3223: AN OPTICALLY FLOCCULENT GALAXY SHOWS A BRIGHT, GRAND DESIGN STELLAR DISK

NGC 3223 is a magnificent optically flocculent specimen of Hubble type Sb and van den Bergh speciation I-II. In the near-infrared, however, this galaxy shows bright, grand design SDWs at K' (see Grosbøl & Patsis, this volume). If, in the midst of SDWs, SDW pressures are less than the average turbulent gas pressure, the competition between stars and gas is won by gas: the galaxy would present an *optically* fragmented, flocculent appearance. The star formation history, principally triggered from previous bursts of star formation (rather than SDW pressures), would be stochastic/random. High total gas column densities (and therefore relatively large optical depths at B or V) would be predicted. Indeed, the V optical depth at the radius of the inner ring is ~ 2. The computed dust mass for NGC 3223, assuming a 'dusty' geometry model of Witt, Thronson and Capuano [82] (hereafter WTC) is 6.4×10^7 M$_\odot$. This is consistent with IR emission described by a Kirchoff-Planck function B_λ (T) peaking at a temperature of only $T \sim 17$K.

2.3. NGC 2976: THE OPTICALLY FLOCCULENT COMPANION TO M81

NGC 2976, Sd III-IV (for optical photographs, see Carozzi-Meysonnier [27]), a flocculent companion galaxy to M81, reveals an exceedingly *irregular* morphology in the near-infrared, with no central mass concentration and no coherent stellar morphology at all.

2.4. A REMARK ON DECOUPLING

To conclude this section, we wish to note that the stellar disks within the unbarred flocculent species are not always flocculent/patchy but can be remarkably regular and smooth. Full decoupling can take place. We note 3 possibilities: (i) the stellar disk of the optically flocculent galaxy decouples, is very smooth and contains prominent SDWs (NGC 3223 is a case in point); (ii) the stellar disk of the optically flocculent galaxy is very smooth but with no (or very weak) SDWs (NGC 2841 is a case in point); (iii) the stellar disk may itself be exceedingly flocculent (eg. NGC 2976). Pertaining to the *morphology of dust* in flocculent specimens: the range is from very smooth and *regular* dark dust+ gas spirals– spanning large degrees in azimuth (NGC 2841), to very patchy distributions of dust (eg NGC 2976).

3. The Companion to M51: NGC 5195

3.1. OPTICAL VS. NEAR-INFRARED MORPHOLOGY

We now turn our attention to NGC 5195, the companion of NGC 5194 = M51. Attempts to understand the internal structure of NGC 5195 have produced discordant results (Smith et. al. [70]), given the extremely prominent dust arm of M51 which crosses in front of the east side of NGC 5195. The arm is juxtaposed right up to the nucleus of the companion. Spectroscopic studies give valuable kinematical information (Tully [77]; Schweizer [67]; van Dyck [78]). The properties of the interstellar medium in NGC 5194/5195 are reviewed in Rand and Tilanus [58].

Optically, NGC 5195 has been classified as Ip-Ep (Morgan [56]), Irr II (Sandage [63]), SBa(r) (Spinrad and Harlan [72]), SB (Vorontsov-Velyaminov and Krasnogorskaya [79]) and SB0/a(r) (Thronson, Rubin and Ksir [75]). Our K′ image (see figure 5b in Block et al. [17]) is radically different from the optical and shows *no evidence* for tidal disruption of its old stellar population disk by M 51. *As for NGC 2841, the Hubble classification here clearly does not constrain the morphology of the NGC 5195 stellar disk.* As already noted by Spillar et al. [71], at K′ we see a prominent bulge (displaying a pronounced isophotal twist), a bar, and a disk. The bar does not point *exactly* towards the centre of M51, but slightly eastward (Spinrad and Harlan [72]).

Incipient spiral structure begins from the end of the bar, which is aligned essentially North-South. Are the bar and the incipient spiral structure in NGC 5195 mostly the result of an intrinsic, self-excited mode (Bertin et. al. [4], [5]) which would be observed even in the absence of M51, or is the structure (maybe a damped mode) rather driven by M51?

Now *leading* structures cannot be interpreted as self–excited modes ([4], [5]), so we should consider the geometry of the observed configuration in further detail. We recall that the inclination of NGC 5195 is ∼ 43 degrees (Spillar et al. [71]). The *northern* end of the bar is probably tipped toward the observer (Rand and Lord, private communications). If the near-side of the bar is the northern one, then since the east side of the disk rotates toward us (Schweizer [67]), the incipient spiral structure seen in our near-infrared images is *trailing*.

The symmetry of even the outermost isophotes, when looking 'through' the dust of the foreground M 51 arm, is surprising in view of the spectacular tidal effects that may be responsible for morphologies in interactive QSO systems such as 4C 37.43 (Block and Stockton [14]) and in galaxies such as 'the antennae' NGC 4038/9 (Toomre and Toomre [76]). NGC 5194/5195 has often been described as a typical case of tidal interaction (from the early papers such as Toomre and Toomre [76], to more recent work - e.g., Hernquist [41]). Here the perfect regularity of the infrared arms in NGC 5194 (Rix and Rieke [59]), which can be traced all the way down to the galaxy center (Zaritsky, Rix and Rieke [88]), and of the NGC 5195 Population II disk suggest that tidal interaction is far less dramatic than usually thought, based on the more prominent features of the lighter Population I disks.

3.2. NEAR-INFRARED EFFECTIVE DUST ALBEDO

The distribution of dust in NGC 5195 is highly *asymmetric*, with the most pronounced dust attenuation on the east side where the M51 spiral arm crosses the companion (Figure 3). Such a strongly asymmetric E-W dust distribution provides a unique opportunity to determine the near-infrared albedo of dust grains in the foreground arm of a spiral galaxy directly, subject to three limiting assumptions. Such a technique was first used by Witt and his collaborators [83] to determine the effective K′ albedo of dust grains in NGC 4826 and we follow those precepts here.

(i) We assume that the Galactic extinction curve applies to the analysis ie. that the ratio of extinction opticals depths at K′ compared to that in V is 0.1, (ii) that NGC 5195 would be symmetric with respect to its centre in both surface brightness and its colour distributions in the absence of dust, (iii) that the foreground M51 arm is physically close to (but in front of) its companion [so that the number of photons removed from the line of sight by scattering is equal to the number of photons scattered back into the beam]. The observed velocity difference between M51 and its background companion is only 150 km s^{-1}.

The WTC radiative transfer models [82], which include the full effects

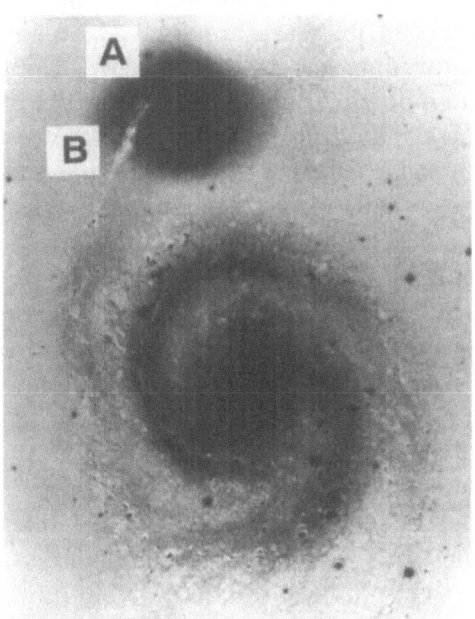

Figure 3. The NGC 5194/5 system, as photographed by Zwicky [89] in this yellow/blue composite with the Palomar 5-m. N is up; E to the left. Field stars A and B permit eye registration with our K′ image in [17]. Reproduced by permission of Palomar/Caltech.

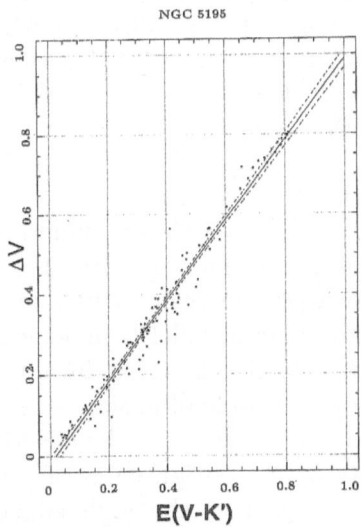

Figure 4. Least squares regression (solid line) for 114 data points to determine the near-infrared effective dust albedo in the M51 arm juxtaposed against the companion NGC5195. 95% confidence limits for the regression (dotted lines) are indicated. The correlation coefficient is 0.98 and the near-infrared albedo of dust grains is ~ 0.9

of scattering as well as of absorption, may be applied to portions of galaxies, where the effects of dust are apparent, with the aim of determining the nature of the star/dust geometry. In the WTC models, differentiations between screen and embedded dusty geometries are prescribed by the ratio of surface brightness reduction in the V-band, ΔV, to the corresponding reddening E(V-K), at unit E(V-K). This ratio has the value 1.18 (or less, if the K' albedo exceeds 0.2) for screen geometries, while embedded geometries have values of 1.30 and greater [82]. Similar contrasts are evident if V is replaced by B surface brightness data.

We systematically surveyed the foreground eastern dust arm in NGC 5195 by producing optical (V) and K' surface brightness profiles oriented through the centre of the M51 companion but sequentially spaced 5 degrees each in position angle. Next, for each position angle, we computed the *differential* colour excess profile E(V-K') = ΔV - $\Delta K'$ and identified those features which exhibited coinciding maxima on the ΔV and E(V-K') profiles. [An important remark is that our determination of colour excesses here does *not* depend on the intrinsic stellar population of NGC 5195 but simply on symmetrically opposite on-screen (ie on-arm, E) and off-screen (W) positions. Because of the (screen) geometry, E(V-K') is simply a differential colour excess produced by the dust screen alone]. The process was repeated for each position angle, and a total of 114 measurement pairs resulted.

Our least squares regression line for the 114 measurement pairs (Figure 4) yields a slope of ΔV / E(V-K') =1.007 $\pm$ 0.018 for the E dusty arm. The WTC models then allow us to identify this value with the geometry of a *foreground screen of dust* where the near-infrared effective albedo of dust grains, a(K'), is very high, $\sim$ 0.97 with a lower limit of 0.90 if the slope of 1.025 is adopted. The best fit for the NGC 4826 data (Witt et al. [83]) was also surprisingly high: effective albedo a(K') =0.86. It must, however, be remembered that the WTC models are spherically symmetric and do not produce the effects of disk brightening [the C_λ (μ) term in Figure 3 of Bruzual, Magris and Calvet [25]. In the case of a disk galaxy approximated, for example, by an infinite plane-parallel homogeneous distribution of stars and dispersive dust grains, light scattering by dust grains is particularly effective in acting as an extra source of radiation at K']. For the NGC 4826 data, Witt et al. (preprint) has shown that the effective near-infrared albedo at K' is 0.6 if the effects of disk brightening are included. Furthermore, as noted by Witt, at least part of the high effective albedo at K' may also result from non-equilibrium near-IR dust emission often observed in galactic reflection nebulae [68].

That the optical and near-infrared albedos may well be identical and that the ISMs of galaxies may therefore contain large grains follows from detailed theoretical studies (for example, Kim, Martin and Hendry [48]) and from observational studies both within our Galaxy (star-forming regions, reflection nebulae) and in the NGC 4826 dust screen ([83]). The high effective albedo of 0.9 at $2.1\mu m$ measured by us for dust grains in the M51 arm may result from scattering by dust grains at least $0.5\mu m$ in radii – *twice as large* as assumed by standard models.

4. Arcsecond Resolution of Warm and Cold Dust

4.1. THE LIMITATIONS OF SUB-MM/MM OBSERVATIONS

Sub-mm/mm observations themselves *cannot* unambigously identify the presence of cold (15K - 25K) and very cold ($< 15K$) dust. The inversion of a spectral energy distribution (SED) into a dust temperature distribution is a badly conditioned problem: Hobson and Padman [46] show that it is possible to find markedly *different* dust temperature distributions from an *identical set* of 30 - $1300\mu m$ flux measurements. The dilemma concerning the detection of (cold) dust in extragalactic systems and the determination of its mass may be summarised as follows:

1. Cold (15K - 25K) dust is not seen by IRAS (see Figure 8 in Sauvage and Thuan [66]). Since the scaling of the total infrared emissive power is $\sim T^6$ where T is the grain temperature (Andriesse [2]), cold grains do not emit much energy and add only a small contribution to the low-frequency end of the emission spectrum, even though these cold grains may constitute by far the bulk of the proportion of the dust mass. IRAS dust masses are usually derived from the measured fluxes at 60 and 100 μm and are not sensitive to the flux at longer wavelengths, where the cold dust would radiate.

2. *Very* cold dust ($< 15K$) will not be seen *even* at sub-mm to mm wavelengths, as beautifully demonstrated by Pajot *et al.* [57]. In some instances has led to claims that very cold dust is simply not present (Clements *et al.* [31]; Carico *et al.* [26]; Eales, Wynn-Williams, and Duncan [35]; Stark *et al.* [74]).

3. Apart from the SED inversion problem being badly conditioned, there are other important reasons why sub-mm observations are far from ideal for determining even the amount of dust which *is detectable* at those wavelengths. Chini and Krügel [29] enumerate these as follows:

    (a) The wavelength dependence of the dust absorption efficiency in the far-IR and sub-mm region is poorly known (Hildebrand [43], Clements *et al.* [31]). Kwan and Xie [50] conclude that the biggest uncertainty in the determination of dust masses is associated with the uncertain dust emissivity law. As elucidated further by Gordon [39], a change of only 20% from $\beta=2$ (i.e. $\beta=1.6$ or 2.4) will cause a 50% error in the ratios of optical depths at 400 and 1300 μm.

(b) Observations at different wavelengths with different beam sizes are difficult to compare.

(c) The 'error beam' (extended side lobes) of sub-mm/mm telescopes operating near their minimum working wavelength can be very important, resulting in large calibration uncertainties and difficulties in comparing point and extended sources (J. Lequeux and C. Purton: private communications).

4. The largest mass fraction of interstellar dust resides in the biggest grains, often found in the densest and coldest environments where accretion and coagulation may be important and where radiative heating is weak (see also Greenberg and Li, this volume).

5. The inference of dust masses from CO observations relies on a *very* uncertain derivation of H_2 column density from the CO line intensities (see especially section 4.3 in Allen and Lequeux [1] and section 3 in Lequeux, Allen and Guilloteau [51]).

6. Furthermore, dust-to-gas ratios themselves still need to be confirmed. A full discussion of this problem for the SMC is presented by Lequeux [52]. It is well known that dust-to-gas mass ratios in spiral galaxies observed by IRAS are exceptionally low [66], by $\sim$ an order of magnitude, compared to the standard value in our Galaxy. The mean [warm dust]-to-gas mass ratio determined by Devereux and Young [32] for 58 spirals is $\sim 1/1080$ (with a similar conclusion reached by Sanders, Scoville and Soifer [65]), compared to the canonical Galactic value of $\sim 1/150$.

At this New South Africa Conference, we further elucidate an independent tool which can be used to map the distribution of both warm *and cold* dust with unprecedented spatial resolution: one order of magnitude better than that possible with the largest sub-mm/mm telescopes, and two orders of magnitude better than with IRAS observations.

Our method provides information pertaining to dust structures which cannot be secured from IRAS or sub-mm observations. Apart from our optical/near-infrared method being independent of the value of β, the availability of near-IR array detectors such as NICMOS (HgCdTe) at major observatories will also give a vastly greater portion of the astronomical community the opportunity to study the spatial distribution of warm and cold dust – compared with, for example, only limited access to, and a restricted number of pointings with, the Kuiper Airborne Observatory (which as of writing, is no longer operational).

4.2. THE METHOD

Full details may be found in Block *et al.* [18] and Block, Witt and Grosbøl [19]. Colour probes such as V-K (or V-K$'$) are almost totally dominated by the extinction at V, allowing V optical depths to be properly and correctly evaluated. Our methodology may be delineated as follows:

1. Optical and near-infrared colour excesses in dusty galaxies are determined.
2. Using the multiple-scattering radiative transfer models of WTC, these colour excesses are used to estimate the V extinction optical depth.
3. These optical depths are converted into dust column densities and hence dust masses.

The quantitative power of the new technique is that the ratio of our computed dust masses to the IRAS dust masses can never be decreased by *a priori* invoking clumpy interstellar media. The effects of a multiphase, hierachical density structure on the penetration of continuum radiation into clumpy molecular clouds has now been carefully studied: from the pioneering paper of Boissé [22] to the more recent investigation of Hobson and Scheuer [44]. The penetration of such radiation can etch (highlight) primordial GMC clumps which can be studied optically (see Block, Dyson and Madsen [15] and Block [12]). Clumpy media with the same amount of dust will always have a higher transmission coefficient than continuous media do. Consequently, by relying on continuous media models, our computed dust masses will always be *lower* limits to the actual amount of dust present. If, in certain (active) galaxies, massive dusty, gaseous clouds are actually undergoing *collapse* in the presence of gravitational and magnetic fields, a full relativistic magnetohydrodynamical approach may be appropriate; as we pointed out in 1974, the geometry of the magnetic field is all important (section 4.3 in Block [8] and the discussion therein).

In our method, a key uncertainty is exactly which unreddened (intrinsic) B-K$'$ or V-K$'$ values to adopt; a complete discussion of uncertainties in the modelling of old stellar populations appears in a paper by Charlot, Worthey and Bressan (preprint). However, for galaxies which we have studied and which have also been observed, for example, in HI and CO, the agreement between our derived optical depths and those derived from the atomic and molecular hydrogen data has been excellent.

We demonstrate the power of subtracting K$'$ images from optical images by presenting a colour map for the galaxy NGC 4736. Of the galaxies I could select to illustrate the method, NGC 4736 remains the most memorable, for it was the first galaxy in which we clearly saw, at arcsecond resolution, embedded warm and *cold* dust optically [18]. Greenberg's far-sighted, 26 year-old prediction was confirmed [40].

4.2.1. *Embedded Interarm Dust in NGC 4736 [18]*

NGC 4736, with its preferentially face-on orientation (i ≈ 35 degrees) and post-starburst mode, is well suited in attempts to probe interarm dust. Extensive amounts of observations, at different wavelengths, of this early type (RSab(s)) spiral, abound in the literature. The galaxy is famous for its inner, 'knotty ring' of HII regions at a radius of ∼ 50 arcseconds (see Figure 1(a) in [18]). Several airborne observations of NGC 4736 have been secured with the Kuiper Airborne Observatory (KAO), the most recent being important new, high spatial resolution 50 and 100 μm measurements reported by Smith *et al.* [69]. Because of the absence of spectral signatures pointing to young massive star formation, Smith *et al.* [69] conclude that the FIR emission from within the inner HII ring, radius ∼ 50″, *is not powered by young massive stars.*

NGC 4736 = IRAS 12485+4123 has been detected in the sub-mm by Chini, Kreysa and Mezger [28] (see their Table 1; note that the listing of NGC 4736 there appears under its UGC designation, which is UGC 7996). Chini *et al.* [28] fitted a 2-component model to the FIR-submm spectrum, yielding two temperatures: 15K and 52K. Sage and Isbell [62] derive an IRAS based temperature of 41.5K – which is intermediate, as expected. A temperature of 41.5K supports our contention of a skewed perception of dust emission as seen by IRAS, which misses the evidence for cold dust in this galaxy entirely.

In colour plate 4, we affirm the existence of widespread *interarm* dust within the inner HII knotty ring. Grain temperature is dependent upon the radiation field (see also item 4 in 4.1): that one can now study the actual distribution (at arcsecond resolution) of warm *and* cold dust grains is because the extinction cross section of a dust grain is *independent* of its temperature: therefore, the presence of such grains in the interarm regions is betrayed through illumination by the interstellar radiation field of old disk stars. The pattern in the distribution of the interarm dust in NGC 4736 is reminiscent of the distribution of discrete dust clouds in M31 (which do not appear to show any affinity to the spiral structure of M31) mapped by Hodge [47]. On the basis of observed B and K′ surface brightness profiles and the B-K′ colour map – which includes the entire HII ring from -50 to +50 arcsec – we conclude that the dust is *embedded* rather than contained in an overlying, foreground screen. This conclusion is based on the widespread presence of filamentary structures in B-K′ colour cuts (eg. Figure 2 in [18]), the ratio of visual extinctions to colour excesses, and the relatively large 0.33 ratio of L(IR)/L(Opt) (a factor two higher than that for the NGC 4826 screen geometry). Typical reddening values are E(B-K′) = 0.2 or more in the interarm regions. That the cold interarm dust is illuminated by the old stellar Population is in excellent accord with Figure 5 of Smith *et al.* [69], who finds that the distribution of the 100 μm emission does *not* trace the spatial distribution of young stars delineated from Hα+[NII] observations but *does* trace the distribution of old stars seen through their broadband red (F) image.

For a 'dusty galaxy' model, this colour excess corresponds to V optical depth values of order 0.75. Adopting a distance to NGC 4736 of 7 Mpc (Sandage and Tammann [64] with H=50 km s^{-1} Mpc^{-1}) and integrating over the inner disk out to a radius of 60 arcseconds, yields a dust mass of $3 \times 10^6 M_\odot$. If the 'cloudy' dust model of WTC is used – where there is a filling factor of 0.33 – the dust mass approaches $6 \times 10^6 M_\odot$. For the radiative transfer in a multiphase, clumpy medium in NGC 4736, all dust estimates will be even *larger* (Hobson and Scheuer [44]; Hobson and Padman [45]). By contrast, the dust mass of NGC 4736 derived from IRAS data is only $7 \times 10^5 M_\odot$ (See Figure 7b of Sage [61]). This clearly illustrates that the bulk of the dust mass is apparently at low enough temperatures to have been undetected by IRAS and that the IRAS dust mass underestimates the total dust mass by $\sim 900\%$. Furthermore, the Galactic dust-gas ratio then becomes representative of that for NGC 4736.

The key point is that embedded warm and cold dust *can* be deconvolved, its existence affirmed shortward of 2.5 μm and quantified by combining appropriate optical and 2.1 μm colour excesses with radiative transfer analyses. The method can also be used for distant ($\sim$ face-on) galaxies provided the demarcation of arm/interarm regions is clear: the range in linear diameter and in linear arm width for spiral galaxies can be very large (Block [9], [10], [11]) and colour images of more distant but intrinsically large specimens (such as NGC 309 in Figure 1, recession velocity $\sim$ 6000 km s^{-1}) still reveal the interarm dust.

5. Conclusions

1. Long and very regular *dark* IR spirals (the first to be encountered) have been detected in the prototype flocculent specimen NGC 2841. The azimuthal angles of the dark IR spirals span 160°. The distribution of dust in that flocculent galaxy follows a coherent, very regular pattern.
2. NGC 2841 appears to be the first example of a flocculent galaxy with a *triaxial* bulge.
3. The strong red to blue colour gradient in NGC 2841 from the near to the far side of the disk is attributed to scattered light from dust located high off the plane in the region of the bulge.
4. The Hubble classification does not constrain the morphology of stellar disks. For example, flocculent galaxies may present (grand design) long arms in the near-infrared with high degrees of symmetry.

5. The near-infrared effective dust albedo for the NGC 5195 dust screen is 0.9.
6. Sub-mm observations of only the central regions of nearby galaxies would seriously underestimate far-infrared to sub-mm ratios of galaxies containing cold and very cold dust grains.
7. Galaxies such as NGC 4736 or NGC 3223 cannot be succesfully modelled without invoking cold dust grains (20K or less) in the interstellar media of these galaxies, necessitating an increase in their computed IRAS dust masses by factors as large as 900%.

6. Acknowledgements.

(a) To my mentors G.C. Buric, A.P. Fairall, M. Sears: *ex amante alio accenditur alius*, said St. Augustine, 'one loving spirit sets another on fire'. (b) In organizing this Conference, my thoughts go back to Kona. In 1985/86, Dr Tully invited me to spend some months at the Institute for Astronomy in Hawaii: it was during that period that he afforded me invaluable experience by asking me to assist with the running of his *Kona Workshop*. Thank you, Brent. (c) To Dr Louis Kroon, Dr Dick Minnitt and Dr Graeme McLean: great thinkers; close friends; true brothers in Christ. This Conference owes so much to your input. (d) The NGC 2841 study is a collaboration with Bruce Elmegreen and Richard Wainscoat; my deep gratitude to you both. (e) Working morphology visits with Professor G. Bertin go back several years. Thank you, Giuseppe. (f) My very special thanks to Preben Grosbøl, Alan Stockton, Nick Suntzeff and Adolf Witt for your invaluable collaborative efforts in the cold dust program. (g) To Professor H. van der Laan – thank you for your invitations to Chile and Garching. (h) To the ESO-OPC for generous allocation of telescope time. (i) Optical images were kindly provided by P. Mulder (NGC 4736), T. Boroson (NGC 5195) and by A. Heller and N. Brosch (NGC 2841). (j) My deep gratitude to S. Courteau, R. Chini, L. Edwards, Rh. Evans, S. Federman, W. Lee, G. Madsen, A. Moneti, P. Patsis, R. Peletier, A. Romeo and M. Thomasson. (k) It is a special pleasure to thank Professor James Lequeux for his incisive comments on our optical/NIR method.

7. Addendum: Astronomy at the University of the Witwatersrand, Johannesburg

For colleagues who could not personally travel to the host University for our Conference and who may be interested in the astronomical activities here, the reader is referred to the contribution by George Buric, which appears in these *Proceedings* on page 578.

References

1. Allen, R.J., Lequeux, J.: 1993, *ApJ.* **410**, L15
2. Andriesse, C.D.: 1977, 'Radiating cosmic dust' in *Vistas in Astronomy* **21**, 107
3. Begeman, K.: 1987, *HI Rotation Curves of Spiral Galaxies*, PhD Thesis, Univ Groningen
4. Bertin, G., Lin, C.C., Lowe, S.A., Thurstans, R.P.: 1989, *ApJ.* **338**, 78
5. Bertin, G., Lin, C.C., Lowe, S.A., Thurstans, R.P.: 1989, *ApJ.* **338**, 104
6. Bertola, F., Rubin, V.C., Zeilinger, W.W.: 1989, *ApJ.* **345**, L29
7. Blitz, L., Thaddeus, P.: 1980, *ApJ.* **241**, 646
8. Block, D.L. 1974.: *Quart. J. Roy. Astron. Soc.* **15**, 264
9. Block, D.L.: 1979, *A & A* **79**, L22
10. Block, D.L. 1982, *A & A* **109**, 336 .
11. Block, D.L.: 1986, 'Spiral Galaxies, Side by Side' in the *1986 Yearbook of Astronomy* (ed. Moore, P.: Sidgwick & Jackson, London), 192
12. Block, D.L.: 1990, *Nature.* **347**, 452
13. Block, D.L., Wainscoat, R.J.: 1991, *Nature.* **353**, 48
14. Block, D.L., Stockton, A.: 1991, *Astron. J.* **102**, 1928
15. Block, D.L., Dyson, J.E., Madsen,C.: 1992, *ApJ.* **390**, L13
16. Block, D.L., Grosbøl, P., Moneti, A., Patsis, P.: 1993, *The Messenger*, **71** 41
17. Block, D.L., Bertin, G., Stockton, A., Grosbøl, P., Moorwood, A.F.M., Peletier, R.F.: 1994, *A & A* **288**, 365 (Paper I)
18. Block, D.L., Witt, A.N., Grosbøl, P., Stockton, A., Moneti, A.: 1994, *A & A* **288**, 383 (Paper II)
19. Block, D.L., Witt, A.N. & Grosbøl, P.: 1995, in *The Opacity of Spiral Disks* (ed J.I. Davies and D. Burstein, Kluwer) p. 227
20. Block, D.L., Elmegreen, B.G., Wainscoat, R.J.: 1996, *Nature* **381**, 674
21. Bohlin, R.C., Savage, B.D., Drake, J.F.: 1978, *ApJ.* **224**, 132
22. Boissé, P.: 1990, *A & A.* **228**, 483
23. Bosma, A.: 1981, *Astr. J.* **86**, 1825
24. Braine, J., Combes, F.: 1992, *A & A. Suppl.* **264**, 433
25. Bruzual, A.G., Magris, A., Calvet, N.: 1988, *ApJ.* **333**, 673
26. Carico, D.P., Keene, J., Soifer, B.T., Neugebauer, G.: 1992, *PASP.* **104**, 1086
27. Carozzi-Meysonnier, N.: 1980, *A & A*, **92**, 189 1193
28. Chini, R., Kreysa, E., Mezger, P.G.: 1986, *A & A.* **166**, L8
29. Chini, R., Krügel, E.: 1993, *A & A.* **279**, 385
30. Chini, R., Krügel, E.: 1994, *A & A.* **288**, L33
31. Clements, D.L., Andreani, P., Chase, S.T.: 1993, *MNRAS.* **261**, 299
32. Devereux, N.A., Young, J.S.: 1990, *ApJ.* **359**, 42
33. Disney, M: 1995, in *The Opacity of Spiral Disks* (ed J.I. Davies and D. Burstein, Kluwer), p. 11
34. Devereux, N.A., Young, J.S.: 1992, *AJ.* **103**, 1536
35. Eales, S.A., Wynn-Williams, C.G., Duncan, W.D.: 1989, *ApJ.* **339**, 859
36. Elmegreen, D.M., & Elmegreen, B.G.: 1984, *Astrophys. J. Suppl*, **54**, 127
37. Elmegreen, B.G. & Thomasson. M.: 1993, *Astron. Astrophys.* **272** 37
38. Gerhard, O.E., Vietri, M., Kent, S.M.: 1989, *ApJ.* **345**, L33
39. Gordon, M.A.: 1995, *A & A* **301**, 853
40. Greenberg, J.M.: 1970, *Interstellar grains and spiral structure* in *Interstellar Gas Dynamics* (ed. H.J. Habing, Reidel, Dordrecht), 305
41. Hernquist, L.: 1990, in "Dynamics and Interactions of Galaxies", ed. R. Wielen, Springer-Verlag, Heidelberg, 108
42. Hildebrand, R.H. et. al.: 1977, *Astrophys. J.* **216** 698
43. Hildebrand, R.H.: 1983, *QJRAS.* **24**, 267
44. Hobson, M.P., Scheuer, P.A.G.: 1993, *MNRAS.* **264**, 145
45. Hobson, M.P., Padman, R.: 1993, *MNRAS.* **264**, 161

46. Hobson, M.P., Padman, R.: 1994, *MNRAS.* **266**, 752
47. Hodge, P.W.: 1980, *AJ.* **85**, 376
48. Kim, S.-H., Martin, P.G., Hendry, P.D.: 1994, *ApJ.* **422**, 164
49. Kormendy, J.: 1977, in *The Evolution of Galaxies and Stellar Populations* (ed. B.M. Tinsley & R.B. Larson, Yale Univ: New Haven), 131
50. Kwan, J., Xie, S.: 1992, *ApJ.* **398**, 105
51. Lequeux, J., Allen, R.J., Guilloteau, S.: 1993, *A & A.* **280**, L23
52. Lequeux, J.: 1994, *A & A.* **287**, 368
53. Lindblad, B.: 1956, *Stockholm Obs. Ann.* **19**, No. 2
54. Maddalena, R., Thaddeus, P.: 1985, *ApJ.* **294**, 231
55. Martin, P.G., Whittet, D.G.B.: 1990, *ApJ.* **357**, 113
56. Morgan, W.W.: 1958, *PASP.* **70**, 364
57. Pajot, F., Boissé, P., Gisbert, A., Lamarre, J.M., Puget, J.L., Serra, G.: 1986, *A & A.* **157**, 393
58. Rand, R.J., Tilanus, R.P.J.: 1990, in "The interstellar medium in galaxies", eds. H.A. Thronson, Jr., J.M. Shull, Kluwer, 525
59. Rix, H-W., & Rieke, M.J.: 1993, *Astrophys. J.*, **418**, 123
60. Rix, H-W & Zaritsky, D.: 1994, in *Infrared Astronomy with Arrays: The Next Generation* (ed. I.S. McLean; Kluwer), 151
61. Sage, L.J.: 1993, *A & A.* **272**, 123
62. Sage, L.J., Isbell, D.W.: 1991, *A & A.* **247**, 320
63. Sandage, A.: 1961, *Hubble Atlas of Galaxies.* Washington DC: Carnegie Institution
64. Sandage, A., Tammann, G.A.: 1981, *A Revised Shapley-Ames Catalogue of Bright Galaxies* Washington DC: Carnegie Institution
65. Sanders, D.B., Scoville, N.Z., Soifer, B.T.: 1991, *ApJ.* **370**, 158
66. Sauvage, M., Thuan, T.X.: 1994, *ApJ.* **429**, 153
67. Schweizer, F.: 1977, *ApJ.* **211**, 324
68. Sellgren, K., Werner, M.W., Dinerstein, H.L.: 1992, *ApJ.* **400** 238
69. Smith, B.J., Harvey, P.M., Colome, C., Zhang, C.Y., DiFrancesco, J., Pogge, R.W.: 1994, *ApJ.* **425**, 91
70. Smith, J., Gehrz, R.D., Grasdalen, G.L., Hackwell, J.A., Dietz, R.D., Friedman, S.D.: 1990, *ApJ.* **362**, 455
71. Spillar, E.J., Oh, S.P., Johnson, P.E., Wenz, M.: 1992, *AJ.* **103**, 793
72. Spinrad, H., Harlan, E.: 1972, *PASP.* **85**, 815
73. Stark, A.A.: 1977, *ApJ.* **213**, 368
74. Stark, A.A., Davidson, J.A., Harper, D.A., Pernic, R., Loewenstein, R., Platt, S., Engargiola, G., Casey, S.: 1989, *ApJ.* **337**, 650
75. Thronson, H.A., Rubin, H., Ksir, A.: 1991, *MNRAS.* **252**, 550
76. Toomre, A., Toomre, J.: 1972, *ApJ.* **178**, 623
77. Tully, R.B. 1974, *ApJS.* **27**, 415
78. van Dyck, S.D.: 1987, *PASP.* **99**, 467
79. Vorontsov-Velyaminov, B.A., Krasnogorskaya, A.A.: 1962, "Morphological Catalogue of Galaxies" (Moscow: Sternberg Institute), **1**
80. Whittet, D.C.B.: 1984, *MNRAS* **210** 479
81. Williams, J.P., de Geus, E.J., Blitz, L.: 1994, *ApJ.* **428**, 693
82. Witt, A.N., Thronson, H.A., Capuano, J.M.: 1992, *ApJ.* **393**, 611
83. Witt, A.N., Lindell, R.S., Block, D.L., Evans, R.: 1994, *ApJ.* **427**, 227
84. Woltjer, L.: 1965, in Stars and Stellar Systems, **5** *Galactic Structure* (ed. A. Blaauw & M. Schmidt, Univ Chicago Press) 531
85. Wozniak, H., Friedli, D., Martinet, L., Martin, P., Bratschi, P.: 1995, *A & A Suppl*, **111**, 115
86. Wray, J.D.: 1988, *Colour Atlas of Galaxies* Cambridge University Press.
87. Young, J.S. & Scoville, N.: 1982, *Astrophys.J.* **260**, L41
88. Zaritsky, D., Rix, H-W., Rieke, M.J.: 1993, *Nature* **364**, 313
89. Zwicky, F.: 1957, 'Morphological Astronomy', Springer, Berlin

HOW COLD COULD GALAXIES BE?

MIKE DISNEY

Physics and Astronomy,
University of Wales at Cardiff,
PO Box 913
Cardiff CF2 3YB, UK
email: mjd@astro.cf.ac.uk

1. INTRODUCTION

Technology is affording us new insights into spiral galaxies. Near Infra Red arrays are peering through the smoky gas at the stellar populations hidden inside them. Sub-millimeter detectors will soon allow us to map the coldest obscuring material directly. Ever more powerful computers permit modelling of the escape of the radiation in progressively more realistic ways. In the process some of our comfortable old ideas, such as the transparency of spiral discs and the notion of a permanent, wavelength-independent morphological classification for a galaxy may have to be abandoned, though probably not without a fight.

Like most recent recruits to this ancient minefield I didn't volunteer, but backed into it by accident. In my case multi-wavelength observations of galactic nuclei, actually made in 1983 in South Africa with the then new CCDs, revealed that the spirals around them often looked strikingly different through different filters. It seemed natural to assume that one was seeing the effects of wavelength-dependent opacity acting within fairly obscured galaxies, and simple calculations seemed to support that notion. Imagine my surprise therefore when I subsequently looked at the literature to find that Holmberg had long since proved that spiral galaxies were transparent and that the latest IRAS observations strongly backed him up.

I mention this personal experience only because it seems typical for many of us now working in this field. The acquisition of a new kind of data, possibly taken for quite another purpose, brings one unintentionally into

21

D. L. Block and J. M. Greenberg (eds.), New Extragalactic Perspectives in the New South Africa, 21–28.
© 1996 *Kluwer Academic Publishers.*

conflict with well established tribes. A vociferous battle ensues, which ends, to the confusion of the onlookers, with both sides proclaiming total victory.

To understand what has been going on one needs to recognise two tendencies, one particular to this field, the other more general. The particular tendency is for most people to enter the dust field with strong prejudices. For most astronomers dust, or rather smoke as it should be called, seeing that it has condensed and not disintegrated, is quite simply a nuisance. They want to be confirmed in their fervent hope that there isn't much of it about and they would be happy to see the heretics burned. The second tendency is to confuse Astronomy with Physics; Astronomy is a much softer science which means that the discovery of a conflict between a theory and an observation should not, on its own, be used to dismiss that theory - if only because the 'observations' themselves will often turn out to contain embedded working hypotheses which shouldn't pass unquestioned.

This is a plea for all of us working in this area to keep our minds open for now. The dogmatic statements and the personal virulence which have recently marred some of the debate betray a lack of understanding of scientific history. We are likely to arrive at the truth more quickly and more enjoyably if we are willing to acknowledge our own prejudices and to entertain the notion that they are probably wrong.

2. WIDER SIGNIFICANCE OF THIS FIELD

David Block has asked me to to review the main reasons why the subject of this conference ought to be of interest to a much wider audience. Since there is a more detailed review in the Proceedings of the recent Cardiff NATO ASW [1] - which incidentally is a mine of recent data, insights and interesting references, these remarks will be fairly brief.

This subject matters to Astrophysics in general because:

(a) We need to unearth the correlations which we suspect may exist between the global properties of galaxies [20]. Such correlations, which proved to be the Rosetta Stone of stellar astrophysics, have so far remained elusive, and what correlations we do find such as Freeman's Law [2] or the Tully-Fisher effect [3], may owe a lot to observational selection [4]. Before searching for such global correlations it is vitally important to know how to correct the measured luminosities of galaxies for their internal extinctions [5], and we need to know how to do it for a wide range of luminosities and morphologies. It is the most luminous, most massive galaxies which are likely to contain the most interstellar material and which will therefore need the largest corrections [6].

(b) Then we come to the Hidden Mass which supposedly dominates the dynamics of gassy galaxies. Before we ascribe it all to exotic agencies we need to exactly estimate how much baryonic material is still shrouded in obscuring smoke. Modest amounts of smoke can play merry hell with the decomposition of visible light-profiles into bulge and disc components [7]. This whole area needs urgent re-examination in the Near IR, particularly in those spirals with well determined rotation curves.

Eighteen years after Bosma's discovery [8] of flat rotation curves it is scandalous that we do not have some independent confirmatory evidence for the huge masses of inferred invisible material in spiral galaxies. Dynamicists have been telling us for years that they understand spiral structure, even before we knew of the hidden mass. Surely some clues to its existence and nature ought to emerge from the comparison of dynamical theory with the cornucopia of recent observations? Here again NIR imaging is looking through the smoke to reveal the massive stellar components far more reliably than we were able to do in the optical. This puts us in a good position to re-examine the whole field of spiral dynamics.

(c) Now that HST is delivering such startling pictures of the distant Universe we realize we cannot interpret them without first knowing the K-corrections quite accurately. But the luminosities, apparent morphologies and the spectral evolution of distant galaxies will be heavily affected by smoke; its amount, its distribution, its microscopic properties and its evolution[9].

(d) Again HST is taking us towards an era where we see the far distant sky crowded with, even confused by innumerable faint galaxies. If they are to some extent opaque their combined opacity may seriously obscure our picture of the high redshift universe. Some authors ([9], [10], [11]) have already speculated that the dearth of QSOs beyond redshift 4 could be owed to this cause. However there are plenty of free parameters to play with and I know that Mike Fall will be expressing a contrary point of view.

(e) We need to identify the main sources of the large Far Infra Red [FIR] outputs found in many galaxies. Some are undoubtedly obscured AGNs, some are starbursts, some contain heavily obscured normal stellar populations, and some may be a combination of all three. The advocacy of one or other, to the exclusion of the rest, will not yield the true story.

(f) Various cosmological measurements eg. the Tully-Fisher correlation as a measure of galaxy distance, supernovae as standard candles, even Cepheid pulsation distances, all require accurate estimates of foreground obscuration.

(g) Arguments about the dynamical evolution of galaxies will need to be re-examined in the light of NIR imaging. The prevalence and size of the bars and bulges may prove to be very different in the NIR [12]. For instance Dressler [13] has argued that the optical bulges in late type spirals are too underluminous for them to be the precursors of SO's. But what if they are obscured?

3. HOW COLD COULD GALAXIES BE?

An eminent cosmologist once advised me to forget all about very cold dust because the T^4 law ensures that it cannot emit, and therefore by implication cannot absorb, much radiation. He sounded plausible, as Cosmologists are apt to sound, but he was in fact totally wrong, as Cosmologists are apt to be.

Imagine pouring a mixture of cold gas and smoke into a spiral galaxy until virtually all its stars are obscured in the optical. Its visible surface brightness would be extinguished but it would glow in the Far IR. Its Black Body temperature T_{BB} would be such as to emit as much FIR as the naked spiral, which had a mean surface-brightness of 24 B mags per sq arc sec $[24B\mu]$, would have done in the optical. Thus:

$$\sigma T_{BB}^4 \sim 24B\mu \sim \frac{L_B}{\pi R_G^2} \sim \left[\frac{10^{33} \times 10^{10}}{(3 \times 10^{22})^2}\right]_{cgs} \tag{1}$$

or:

$$T_{BB} \sim 10^{3/4} \sim 6°K$$

which low temperature reflects the very low surface-brightness of a typical spiral galaxy.

The next question to ask is whether ridiculous amounts of such smoke would be needed to so extinguish a galaxy. If spread out the total cross-section of the smoke would be:

$$A_s \sim Q_o \pi a^2 \left(\frac{M_s}{m_g}\right)$$

where a is the particle-size, M_s the total mass of smoke and Q_o the extinction efficiency of a grain in the optical-UV where most of the absorbing is going on. Then the fraction of the total galaxy area πR_G^2 extinguished by grains wil l be:

$$\frac{A_s}{\pi R_G^2} \sim Q_o \left(\frac{\pi a^2}{\rho_g a^3}\right)\left(\frac{M_s}{M_{gas}}\right)\left(\frac{M_{gas}}{M_\star}\right)\left(\frac{M_\star}{\pi R_G^2}\right) \tag{2}$$

Taking typical values for spirals:

$$\frac{A_s}{\pi R_G^2} \sim \frac{Q_o}{a(\mu)} \tag{3}$$

where $a(\mu)$ is the grain size in microns. If, as Mayo Greenberg [14] is going to tell us, $Q_o \sim 1$ and $a(\mu) \sim 10^{-1}$ then sufficient smoke exists, in principle, to entirely black out a normal galaxy. Indeed one wonders, with Jura [15], whether the rough equality found between A_s and πR_G^2 isn't more than a coincidence. Galaxies having "normal" amounts of smoothly distributed smoke and gas could not attain surface brightnesses much above what is actually observed [16].

Could such an extinguished galaxy be detected? I beg leave to doubt it. Its emission would peak at 600 microns - almost the most difficult region to observe in the entire spectrum. The contrast between the galaxy at 6K, and the surrounding Cosmic background at 2.7K, would be very low, while side-lobe problems in the virtually isotropic radiation field would be horrendous.

It is usually said, however that the emissivity Q_{FIR} of cold smoke particles in the FIR is very low [19], in which case (1) must be amended to:

$$Q_{FIR}(\sigma T^4) \sim L_B/R_G^2 \qquad (4)$$

Treating such particles classically then Mie theory yields, in the $\lambda \gg a$ approximation:

$$Q_{FIR} \sim \left(\frac{\lambda}{a}\right)^{-\beta}$$

where β lies between 1 and 2, depending on the grain material. For instance with $\beta \sim 1$ the smoke temperature rises to $\sim 20K$, and the peak flux [in $\lambda F(\lambda)$ terms] will then emerge at 180 microns, rendering it accessible with the KAO, with COBE in our Galaxy, and with ISO.

The point I am trying to make is that our estimate of what a smoky galaxy ought to look like, were we able to see all its radiations, is entirely dictated by our assumed knowledge of the detailed microscopic properties of its interstellar smoke particles. If we are entirely confident about their size distribution and their emissivities, particularily at long wavelength, and if we could in addition know how they are distributed within a galaxy, only then can we predict with confidence what that galaxy would look like at various wavelengths.

I don't know how confident astronomers can afford to be about this smoke physics. We rely on a tiny handful of specialists whose difficult work is mostly theoretical. Working in the Classical approximation they model grains as dielectric spheres and tell us that at long wavelengths [$\lambda \gg a$] particles are naturally poor emitters. But what about Quantum Mechanics? After all the Hydrogen atom has its resonant transition wavelength 1000 times larger than its diameter. Can we really say with certainty that a complex grain, possibly composed of many dislocated crystals, and coated with various dirty ices, will so lack low energy transitions that its emission cross-section falls to about one thousandth of its geometrical value? If we can't be so certain then we have to keep our minds open to the possible existence of significant amounts of very cold ($< 10K$) material in spiral galaxies.

4. SOME VERY HESITANT CONCLUSIONS

(1) We shouldn't forget that a mere 70 years ago we didn't know whether spiral nebulae were 10 light years away, or 10 million, while it is only 17 years since we were forced to accept that they must be dominated by Hidden Mass. Our knowledge of them remains too fragmentary and fragile for any of us to be dogmatic, or unaware of our variously motivated prejudices.

(2) On the other hand technology in this field is accelerating on an almost monthly basis. We'll probably learn more about galaxies in the next 5 years than we have over the past 50. There is even a chance that we will find out how they formed. In the distance the Hubble Deep Field apparently shows spirals merging out of pre-existing structures [o'heads shown] while closer by Michael Rich and his colleagues have started the exciting business, using HST colour-magnitude diagrams, of disentangling the hi story of star formation in our neighbouring galaxies. We are lucky to be active right now!

(3) Our field depends critically on the symbiosis between galaxy observers on the one side and grain-physicists on the other. To what extent each of the sides understands what it can take from the other as reliably understood, and what is still no more than a necessary working hypothesis, faute de mieux, I don't know. Certainly we must try to avoid the comical situation that prevails in cosmology where astronomers who don't appear to understand particle physics, but think they do, insist that we have to believe in some astronomical theory simply because it is, so they conceive, an unavoidable consequence of particle physics[or vice-versa]. That way lies folly.

(4) In particular we have to be very suspicious of any theoretical argument which tells us NOT to look somewhere because such and such [eg. very cold gas or dust] cannot exist. Theory is at its best when it suggests we turn over some previously neglected stone; at its worst when it tells us not to bother. The early history of theory in X-ray astronomy [17] should be sufficient warning.

In particular Astronomy will always face a dilemma over Occams razor. If you can't detect some agency then Occam sensibly advises you to ignore it. But ignoring it in a theoretical synthesis should never discourage us from going back to look for it again and again with ever more sensitive instruments.

(5) Finally one returns to the Hidden Mass in spiral galaxies. There is no denying the rotation curves, but we must also face the uncomfortable possibility that we are staring at the Emperor's New Clothes. If we are not to become a laughing stock one day, we desperately need some independent confirmatory evidence. If it cannot be found amongst all the newly incoming data then I think we should be very worried.

5. WORLD ASTRONOMY AND SOUTH AFRICA

Being the privileged first non-South-African to speak at this conference I would like to express the hope that South Africa will once again play a key role in world astronomy. It has a long and honourable astronomical tradition and at least one excellent observing site, which I have been lucky to use, at Sutherland. Australia does not possess good photometric conditions, which leaves the Southern Sky to Chile and South Africa. For climatic, seismic, economic and political reasons it would be very unwise to put all our eggs in the Chilean basket. Optical astronomy is at the mercy of oceanic currents whose significance and stability we are only beginning to comprehend.

The New South Africa will have many calls on its purse and pure science may not be its highest priority. Nevertheless South Africa is, by virtue of both geography and history, in a very priveledged position to explore the wider universe. I hope it will take pride in the major discoveries which could be made here, as they have been made in the past. Astronomy doesn't have to be an extravagant science. In my opinion it is the silliest mistake to equate astronomy with particle physics and so to argue that only gigantic new telescopes can make significant discoveries. All history is against such nonsense [18], even if it currently seems to be the fashionable view. Consider only the list of landmark discoveries recently made with modestly sized but well instrumented telescopes: Gravitational Lenses (Kitt Peak 84 inch) , Large Scale Flows (Smithsonian 60 inch), Quasar Absorption Lines (Kitt Peak 84 again), Extra-Solar Planets (Haute Provence 1.9 meter), MACHOs (Mount Stromlo 50 inch): and don't forget that the expansion of the universe was really discovered by Slipher's wooden 24 inch.

The key phrase here is "well instrumented". We cannot expect South Africa to develope state-of-the-art instrumentation without help and indeed support from outside. Until recently the British used to supply such instrumentation in return for observing privileges. It was a wonderful bargain for both sides. Although it wasn't admitted at the time, this arrangement was terminated, so I believe, for political reasons which are now, thankfully, irrelevant. Britain should now resume its long standing partnership with the South African Observatory. But if it isn't smart enough to do so then I sincerely hope that other Northern nations will seize this wonderful opportunity, on behalf of us all, instead. Astronomy needs South Africa just as I believe the New South Africa needs to make the exploration of the Universe part of its emerging dream.

References

1. "The Opacity of Spiral Discs" , 1994, proc. NATO ARW Cardiff July 1994, Ed. J.I. Davies and D. Burstein.
2. Freeman K.C., 1970, ApJ, 160, 811.
3. Tully R.B. & Fisher J.R., 1977, A&A, 54, 661.
4. Disney M.J. & Phillips S., 1983, MNRAS, 205, 1253.
5. de Jong R.S., 1995, Ph.D. thesis, University of Groningen.
6. Wang B., 1995, "Opacity of Spiral Disks", eds. Davies and Burstein,, p345.
7. Davies J.I.,1991, "Dynamics of Disc galaxies", ed Sundelius, Goteborg, p65.
8. Bosma A., 1978, Ph.D. thesis, University of Groningen.
9. Bruzual G., 1995, "Opacity of Spiral Disks", eds. Davies and Burstein,, p 33.
10. Ostriker J.P. & Heisler J., 1984, Ap J., 278, 1.
11. Wright E.L., 1990, Ap.J., 353, 411.
12. Peletier et al., 1995, "Opacity of Spiral Disks", eds. Davies and Burstein,, p243.
13. Dressler A., 1980, Ap.J. 236, 531.
14. Greenberg M. & Li A., 1995, "Opacity of Spiral Disks", eds. Davies and Burstein, p19.
15. Jura M., 1980, Ap.J., 238, 499.
16. Disney M.J., Davies J.I. & Phillips S., 1989, MNRAS,239 ,939.
17. Tucker W & Giacconi R., 1985, "The X-Ray Universe", Harvard UP, p9.
18. Harwit M., 1981 "Cosmic Discovery", Harvester press.
19. Whittet D.C.B, 1992, "Dust in the Galactic Environment", IOP pub.
20. Roberts M., & Haynes M. 1994, Ann. Rev. Astron. Astrophys.

TEMPERATURE FLUCTUATIONS AND VERY COLD DUST

W.W. DULEY
Physics Department
University of Waterloo
Waterloo, Ontario
Canada N2L 3G1

Abstract. The small heat capacity of very small interstellar grains results in temperature spiking on absorption of ultraviolet photons from the interstellar radiation field. These grains cool by emission of infrared radiation at wavelengths that are out of equilibrium with thermal emission at the average grain temperature. One effect of this is the appearance of excess emission at short ($\sim 3\mu m$) and middle ($12 - 25\mu m$) infrared wavelengths which signals the presence of grains that have been transiently heated to high temperatures. Such grains will, however, cool to temperatures that are considerably lower than the average temperature between high temperature spikes and will spend the majority of the time at these low temperatures. As a result, they will contribute excess emission at very long infrared wavelengths and this emission will be seen in IRAS sources at mm wavelengths.

Radiative equilibrium established between the absorption of short wavelength photons by small interstellar dust particles and the re-emission of energy by thermal radiation in the infrared can be used to define an equilibrium temperature, T_e, for the particle (Whittet 1992). The concept of such an equilibrium dust temperature is valid only when the energy deposited by an incident photon is much less than that stored in the grain (Greenberg 1968) exclusive of zero point energy. When the absorbed photon energy is comparable to that stored in the grain, then the grain temperature can be raised significantly in a transient temperature spike.

Temperature spiking in small interstellar particles in response to absorption of photons from the interstellar radiation field (ISRF) has been discussed in some detail (Duley 1973a,b, Greenberg and Hong 1974, Purcell 1976, Draine and Anderson 1985, Guhathakurta and Draine 1989). This effect is most noticeable for particles with sizes $\leq 10nm$ and can be-

D. L. Block and J. M. Greenberg (eds.), New Extragalactic Perspectives in the New South Africa, 29–33.
© 1996 *Kluwer Academic Publishers.*

come very large for particles whose sizes approach the dimensions of large molecules. In such cases, the temperature spike associated with absorption of a 10 eV photon, may approach $10^3 K$ (Draine and Anderson 1985).

Emission of IR radiation as these transiently heated particles cool leads to the possibility of emission excesses at wavelengths extending from the far IR (Duley 1973a) to the near infrared (Sellgren 1984). In particular, it has been suggested that thermal spiking of very small grains (VSG's) is the source of the excess near infrared continuuum emission observed from reflection nebulae (Sellgren, Werner and Dinerstein 1983) and IR excesses in IRAS sources (Puget, Leger and Boulanger 1985, Draine and Anderson 1985). There has also been some discussion concerning the possibility that temperature spiking can lead to enhanced reaction rates for adsorbed molecules on dust grains together with desorption of the products of these reactions (Allen and Robinson 1975, Duley 1976).

When temperature spiking is important, the concept of a time-averaged temperature $< T >$ is still valid. This temperature will not, in general, be identical to that calculated under the assumption of equilibrium heating and cooling (i.e. the temperature T_e). At any temperature T, the cooling rate, R_c, of a small particle is given by

$$R_c = 4 < Q_a(a,T) > \sigma T^4 \qquad (1)$$

where a is the particle radius, σ is the Stefan-Boltzmann constant and $< Q_a(a,T) >$ is the Planck-averaged emissivity

$$< Q_a(a,T) >= 15 \left[\frac{hc}{\pi \kappa T}\right]^4 \int_0^\infty Q_{abs}(a,\lambda)\lambda^{-5} \left[\exp(\frac{hc}{\kappa T}) - 1\right]^{-1} d\lambda \qquad (2)$$

where h is Planck's constant, c is the speed of light, λ is wavelength, κ is the Boltzmann constant and $Q_{abs}(a,\lambda)$ is the Mie absorption efficiency for a spherical particle of radius a.

The strong dependence of R_c on T ensures that a VSG heated to a high temperature by absorption of an energetic photon will cool rapidly to a temperature near T_e. The timescale, tc, for this cooling is approximately

$$t_c \sim \frac{h\nu_{abs}}{4\pi a^2 R_c} \qquad (3)$$

where is the energy of the absorbed photon. t_c is typically ~ 1 sec when $T > 100K$ which is much less than the time interval, t_p, between the absorption of photons from the ISRF. As most of the grain energy is removed over the time t_c and $t_p \gg t_c$ a grain which has experienced transient heating will have the opportunity to cool to temperatures $T \ll T_e$ in the interval between photon absorptions. The effect of this cooling can be seen by calculating the ratio $dP/d(\ln T)$ which expresses the fraction P of the time a grain spends at a temperature greater than T. Plots of $dP/d(\ln T)$ for graphite and amorphous carbon particles are shown in figures 1 and 2, respectively. Refractive indices for graphite were obtained from Draine (1985) while those for amorphous carbon are from Rouleau and Martin (1991). It is apparent from these data that grain temperatures can fall to very low values ($\leq 3K$) in the interval between temperature spikes. As expected from thermodynamic considerations, those particles which initially reach the highest transient temperatures, cool to the lowest final temperature over the interval t_p. For a $1nm$ particle in an unattenuated ISRF, $t_p \sim 10^5 - 10^6 sec$. Thus, if temperature spiking in VSGs is responsible for IRAS excesses at wavelengths of 12 and 25 μm, then these VSGs must be at very low temperature. Figures 1 and 2 show that such VSGs will spend the large majority of their time at a temperature which is lower than the cosmic black body background temperature of 2.7 K. In approaching these temperatures, VSGs will contribute excess IR emission at 12, 25 and $60\mu m$. Radiation at temperatures $T < 2.7K$ would be noticeable as an excess at wavelengths longer than 1 mm. However, these grains lose most of their energy through radiation at higher temperatures.

An excess emission from VSGs at long wavelengths will be detectable against the continuum emitted during high temperature spiking when the ratio $\eta > 1$. where

$$\eta = \frac{\left[e^{h\nu/\kappa T_1} - 1\right]}{\left[e^{h\nu/\kappa T_2} - 1\right]} \left[\frac{t_p}{t_c}\right] \qquad (4)$$

and T_1 and T_2 are the dust temperatures during the 'hot' and 'cold' phases of temperature spiking. For a wavelength of 1 mm, and with $T_1 = 300K$, $T_2 = 3K$ this becomes $\eta = 4.2 \times 10^{-4}(t_p/t_c)$. Since $t_p/t_c \geq 10^5, \eta > 1$ and VSGs will show both long wavelength and short wavelength excesses.

The possible presence of a cold dust component coexisting with the warm dust responsible for IRAS emission in interstellar clouds has been noted by Andreani et al. (1991). The ratio of excess fluxes at $12\mu m$ and 1 mm was observed to be typically $\sim 10^2 - 10^3$. The ratio of fluxes by VSGs at 12 and 1000 μm is

$$\frac{I_{12}}{I_{1000}} = \left[\frac{1000}{12}\right]^3 \frac{\left[e^{14.35/T_2} - 1\right]}{\left[e^{1195/T_1} - 1\right]} \frac{Q_{12}}{Q_{1000}} \frac{t_c}{t_p} \qquad (5)$$

With $T_1 = 300K$, $T_2 = 3K$ and assuming $Q_\lambda \propto \lambda^{-1}$, $I_{12}/I_{1000} \sim 10^8 t_c/t_p$. This result is consistent with the observational data of Andreani et al. (1991) since $t_c/t_p \sim 10^{-5} - 10^{-6}$ in interstellar clouds (see above). To summarize these conclusions: a) VSGs that exhibit high temperature

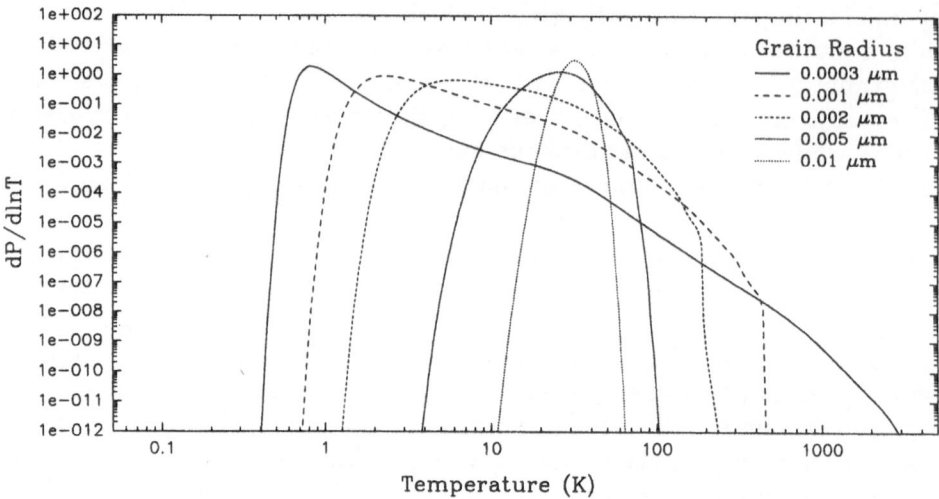

Figure 1. Temperature distribution functions $dP/d(\ln T)$ for graphite grains with various radii in the ISRF.

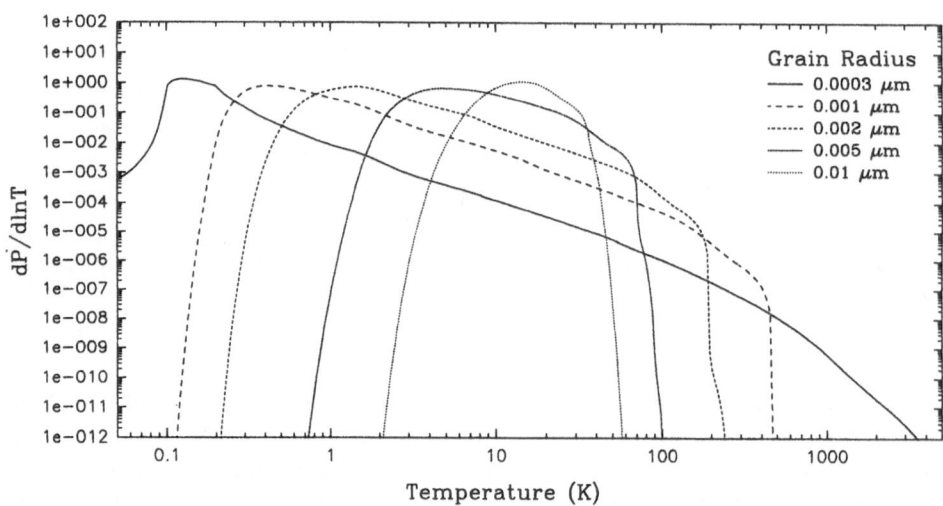

Figure 2. Temperature distribution functions $dP/d(\ln T)$ for amorphous carbon grains with various radii in the ISRF.

transients due to stochastic heating in the ISRF will spend the majority of their time at very low temperatures. b) Temperatures may be $< 2.7K$. c) IRAS excesses at short wavelength (e.g. $12\mu m$) will be accompanied by another excess at long wavelengths (e.g. 1 mm). d) Excess emission at mm wavelengths should be detectable in IRAS sources.

Acknowledgement

This research was supported by grants from the NSERC of Canada.

References

Allen, M. and Robinson, G.W., 1975. Ap. J., 195, 81.
Andreani, P., et al. 1991. Ap. J., 375, 148.
Draine, B.T. 1985. Ap. J. Suppl. 57, 587.
Draine, B.T. and Anderson, N. 1985. Ap. J., 292, 494.
Duley, W.W. 1973a. Nature: Phys. Sci., 244, 57.
Duley, W.W. 1973b. Ap. Space Sci., 23, 43.
Duley,W.W. 1976. Ap. Space Sci., 45, 261.
Greenberg, J.M., 1968 in "Stars and Stellar Systems", vol. 7, eds. B.M. Middlehurst and L.H. Aller (Chicago: Univ. Chicago Press), p. 221.
Greenberg, J.M., and Hong, S.S., 1974. in "IAU Symposium 60, Galactic and Radio Astronomy" ed. F. Kerr and S.C. Simonson (Dordrecht: Reidel), p. 155.
Guhathakurta, P., and Draine, B.T., 1989. Ap. J. 345, 230.
Puget, J.L., Leger, A. and Boulanger, F. 1985. Astron. Ap. 142, L19.
Purcell, E.M., 1976. Ap. J., 206, 685.
Rouleau, F. and Martin, P.G., 1991. Ap. J., 377, 526.
Sellgren, K., 1984. Ap. J., 277, 623.
Sellgren, K., Werner, M.W. and Dinerstein, H.L., 1983. Ap. J., 271, L13.
Whittet, D.C.B. 1992. "Dust in the Galactic Environment" (Bristol: IOP).

THE INTERSTELLAR MEDIUM AS OBSERVED BY COBE

J. C. MATHER

Infrared Astrophysics Branch, Code 685
NASA Goddard Space Flight Center
Greenbelt, MD 20771 USA

Abstract. The Cosmic Background Explorer (COBE) was developed by NASA Goddard Space Flight Center to measure the diffuse infrared and microwave radiation from the early universe. It also measured emission from nearby sources such as the stars, dust, molecules, atoms, ions, and electrons in the Milky Way, and dust and comets in the Solar System. It was launched 18 November 1989 on a Delta rocket, carrying one microwave instrument and two cryogenically cooled infrared instruments. The Differential Microwave Radiometers (DMR) measured the fluctuations in the Cosmic Microwave Background Radiation (CMBR) originating in the Big Bang, with a total amplitude of 11 parts per million on a 10° scale. It also measured the synchrotron and free-free emission of the Galaxy. The Far Infrared Absolute Spectrophotometer (FIRAS) mapped the sky at wavelengths from 0.01 to 1 cm, and compared the CMBR to a precise blackbody. The spectrum of the CMBR differs from a blackbody by less than 0.03%. The FIRAS observed interstellar dust, showing that up to three different temperatures of dust are needed to explain the spectra. It also measured the emission of CO, [C I], [C II], and [N II], tracing several distinct phases of the gas. The Diffuse Infrared Background Experiment (DIRBE) spanned the wavelength range from 1.2 to 240 μm and mapped the sky at a wide range of solar elongation angles to distinguish foreground sources from a possible extragalactic Cosmic Infrared Background Radiation (CIBR). It measured foreground emission from stars, interstellar and interplanetary dust, the polarization of starlight from the Galactic Center, and the reddening of the Galactic plane by the dust absorption.

D. L. Block and J. M. Greenberg (eds.), New Extragalactic Perspectives in the New South Africa, 34–49.
© 1996 *Kluwer Academic Publishers.*

1. Introduction

The COBE satellite was designed to measure the radiation from the early and distant universe, the cosmic microwave and infrared background radiation, but has also provided some of the best measurements of diffuse foreground radiation at wavelengths from 1 μm to 1 cm. The foreground originates in starlight, the interstellar medium, and the interplanetary dust. The data from the mission are now public, available from the National Space Science Data Center, and are very valuable for comparison with data at other wavelengths.

The interstellar medium emits by a variety of processes observed by the COBE. At millimeter wavelengths, synchrotron emission traces the magnetic fields and the high energy tail of the electron distribution. Free-free emission (bremsstrahlung) traces the square of the electron density, as do the Hα emission and the radio recombination lines of hydrogen. The electron emission is very strong in the Galactic plane relative to cosmic background fluctuations, but at the poles it is relatively smooth and not so easy to measure. At millimeter and submillimeter wavelengths, interstellar dust emission is well correlated with the local visible and ultraviolet interstellar radiation field, but the details depend in interesting ways on dust properties. The COBE data have sufficient resolution to distinguish multiple properties of the dust, such as temperature distributions and spectral indices. The dust radiates of order 1/4 of the total luminosity of the Galaxy. At shorter wavelengths, the interstellar dust grains radiate in a nonequilibrium way, transiently heated by individual photons or electrons, so that they can emit strongly at 12 μm. At even shorter wavelengths, the smallest such grains (macromolecules) may be responsible for the diffuse interstellar absorption bands. The millimeter and submillimeter spectral range also includes the dominant cooling lines for the interstellar gas, from the CO rotational ladder, the fine structure doublet of [C I] for cool gas, the much brighter line of [C II] at 158 μm, which is the strongest emission line of the whole Galaxy, and the fine structure doublet of [N II], which can only come from the ionized portion of the gas.

The three COBE instruments cover the entire wavelength range from 1.2 μm to 1 cm. The FIRAS (Far Infrared Absolute Spectrophotometer) spans 105 μm to 1 cm in two bands with a 7° beam. The DMR (Differential Microwave Radiometer) covers 31.5, 53, and 90 GHz (9.5, 5.6, and 3.3 mm), also with a 7° beam. The DIRBE (Diffuse Infrared Background Experiment) spans 1.2 to 240 μm with 10 bands, and also measures polarization in its three shorter bands. Previous works have described the COBE and early results (Mather *et al.* 1990; Mather *et al.* 1991a,b; Janssen and Gulkis 1991; Wright 1990, 1991; Hauser 1991a,b; Smoot *et al.* 1991; Smoot

1991; Bennett 1991; Boggess 1991; Mather 1982; Gulkis *et al.* 1990; Boggess 1992), and the engineering of the spacecraft has also been published (Barney 1991; Milam 1991; Hopkins and Castles 1985; Hopkins and Payne 1987; Volz and Ryschkewitsch 1990; Volz *et al.* 1990, 1991; Volz and DiPirro 1992; Mosier 1991; Coladonato *et al.* 1990; Sampler 1990; Bromberg and Croft 1985).

A liquid helium cryostat cooled the FIRAS and DIRBE to about 1.5 K for 10 months, and after the evaporation of the helium the temperature slowly rose to about 60 K. The DMR radiometers are radiatively cooled, with the 31.5 GHz receivers at 290 K and the others at 140 K. A conical shield protects them from the Sun and Earth, which do not illuminate the instruments directly. The orbit is a sun-synchronous polar orbit, at 900 km altitude and 99° inclination, with a 6 PM ascending node. The Earth's gravitational quadrupole precesses the orbit plane to maintain a 6 PM ascending node, so that the plane is always approximately perpendicular to the Sun. The spacecraft is oriented with its spin axis away from the Earth and about 94° to the Sun. It spins at 0.8 rpm and the DMR and DIRBE beams therefore scan the range from 64° to 124° from the Sun.

2. FIRAS

2.1. FIRAS INSTRUMENT

The instrument is a polarizing Michelson interferometer used as a spectrometer (Mather, Shafer, and Fixsen 1994). It has a beamwidth of 7°, defined by a compound parabolic concentrator. It is symmetrical, with differential inputs and differential outputs. It is a two-beam interferometer, whose output signal (called an interferogram) depends on the path difference between the two beams. A monochromatic input produces an output proportional to the cosine of the phase difference between the two beams. A general input produces an interferogram which is the Fourier transform of the input spectrum. These interferograms are detected, amplified, filtered, digitized, filtered and averaged again, and transmitted to the computers on the ground. They are then Fourier transformed numerically and calibrated to determine the input spectra. The mirror mechanism can scan at two different scan speeds and over two different stroke lengths, to obtain low and high spectral resolution.

2.2. FIRAS COSMOLOGY

The COBE FIRAS (Far Infrared Absolute Spectrophotometer) was designed to compare the CMBR spectrum to a blackbody with great precision. The most recent results show that the CMBR deviates from blackbody form by less than 0.03% of the peak intensity over the wavelength range from 0.5 to 5 mm (Mather et al. 1994). Our interpretation of these results has also been given (Wright et al. 1994a). The FIRAS also showed that the cosmic dipole has the expected shape of the derivative of the Planck function with respect to temperature (Fixsen et al. 1994a). The primary basis of the comparison is a full beam external blackbody calibrator that can be adjusted to match the temperature of the sky. This calibration is described in Fixsen et al. (1994b). The FIRAS is the first instrument to have this opportunity in the protected space environment. The instrument is designed with multiple modes of operation and multiple detectors to test for possible systematic errors.

To derive cosmological results from the FIRAS sky maps, two terms were subtracted from the spectrum of each pixel. A dipole term compensated for the Doppler shift of the spectrum, due to the motion of the Earth. The mean spectrum of the dust in the Galaxy was also derived from the map, and a fraction of this mean spectrum was fitted to each pixel spectrum and subtracted. The resulting map of the CMBR temperature showed almost no trace of the strong Galactic emission, so that a single dust brightness map times a single characteristic spectrum is a good representation of the far infrared Galaxy.

2.3. FIRAS GALACTIC ASTRONOMY

The FIRAS produced the first nearly all-sky far IR survey of Galactic emission at wavelengths greater than 120 μm (Wright et al. 1991). The general features seen in the maps are very similar to those found by the IRAS satellite at 60 and 100 μm. Dust is concentrated in the Galactic plane and is heated by starlight to a temperature ranging from 15 to 30 K, depending on the intensity of the starlight incident on the dust. This dust controls radiative transfer of energy through the Galaxy, and thus the heating and cooling of the interstellar gas. The dust has a mean spectrum which is approximated well by the form $\nu^\alpha B_\nu(T)$, where the index $\alpha = 1.65$ and $T = 23.3$ K.

The mean dust spectrum of the Galaxy can also be well represented by the sum of two such terms with the index $\alpha = 2$, with cold dust with a temperature of about 4.8 K and warmer dust with a temperature of 20.4 K. In this representation, the optical depth of the cold dust is 6.7 times that of the warmer dust component. There are many possible explanations for this kind of dust spectrum. It is possible that there are multiple types of dust grains, so that some can have a very different equilibrium temperature than others. Dust grains could have unusual shapes, like needles (Wright 1982) or fractals (Wright 1987). Small grains with low heat capacities could have variable temperatures, as they are excited by discrete UV photons or cosmic rays. Some dust is protected within optically thick clouds and must reach much lower temperatures than that exposed to the full intensity of the interstellar radiation field, and other dust is very close to stars and must be much warmer than average. The total dust luminosity of the Galaxy is $(1.8 \pm 0.6) \times 10^{10} L_\odot$ (Wright et al. 1991).

This work was greatly extended by Reach et al. (1995). They found that the FIRAS spectra can be represented by a sum of several components with $\alpha = 2$ and different temperatures. The dominant component ranges in temperature from 16 to 21 K. There is also a weaker component with a temperature from 10 to 14 K found in the Galactic plane, where some dust is partially shielded from starlight. The much colder component found by Wright et al. (1991) is found to be very widespread, with a temperature ranging from 4 to 7 K, and an optical depth that is correlated with the dominant component. Shielded dust is not a plausible source of this coldest component, which is more likely to arise from special dust grains like needles or fractals, or grains with a spectral index that differs from 2 and varies with direction. The authors present maps of the distribution of temperature and optical depth of the various components.

The many questions about the nature and emission of the dust make it very difficult to derive cosmological results at the wavelengths where the dust is dominant. The dust emission is correlated to the H I column density, and an extrapolation of the FIRAS data to zero H I suggests that there might be a significant far infrared background radiation field. The extrapolated excess is obvious in the data, but the FIRAS team believes that the variable properties of the interstellar medium make the interpretation as a far IR background quite uncertain. Therefore the full understanding of the dust emission is very important to cosmology.

The FIRAS also provided the first panoramic and unbiased spectral survey of the far IR line emissions from the interstellar medium, including CO, [C I], [C II], and [N II] lines. These results were reported by Wright et al. (1991) and by Bennett et al. (1994a). The FIRAS map of the emission of [C II] at 158 μm shows it to be extremely widespread, both in and out of the Galactic plane. This line is the dominant cooling mechanism for most of the interstellar gas, and is the brightest emission line of the Galaxy, accounting for 0.3% of the dust luminosity. The strength of the line is so great that it can be seen by eye in the raw interferograms as an undamped sinusoid, and with more analysis the Galactic rotation curve can be seen in the line center frequency. The [C II] line is very strongly correlated with the dust emission, as expected from the theory of photodissociation regions. The heating of the gas is provided mostly by photoelectrons emitted from dust grains, and must be in balance with the cooling provided mostly by the [C II] line. Deviations from this correlation are expected where the dust properties change, where the ratio of ionized carbon to dust changes, where the interstellar radiation field changes color, or where other heating and cooling mechanisms become important at extremes of temperature or density. In particular, the ratio of [C II] to dust continuum decreases to nearly zero at the Galactic poles. The reason for this is not known.

The FIRAS also detected the emission of ionized nitrogen [N II] at 205 μm for the first time and measured its frequency well enough for further astronomical (Petuchowski and Bennett 1992) and laboratory (Brown et al. 1994) observations. This is the second brightest far IR line from the interstellar gas and probes a different phase of the interstellar medium because nitrogen ionizes less easily than hydrogen and carbon. The nitrogen and carbon line intensities are also strongly correlated but show the relation [N II] $\propto$ [C II]$^{3/2}$. The meaning of this correlation was explored by Petuchowski and Bennett (1993).

Carbon monoxide is the standard tracer for interstellar molecular gas. The FIRAS has mapped the brightness of the lines of CO from the 2-1 up to the 6-5 transitions, showing that the line ratios are highly dependent on location in the Galaxy. In particular, the CO lines are extremely bright relative to the others in the Galactic Center, where the gas temperature and density are very high and the UV field is shielded by dust. The two lines of neutral carbon [C I] are also seen and mapped by the FIRAS.

Weaker lines observed by the FIRAS include oxygen and water. The water line at 269 μm has never been observed before in astrophysical sources because of the extremely high optical depth of the terrestrial absorption line. The line is seen in absorption by FIRAS (Bennett et al. 1994a) against the continuum emission of the Galactic Center, but at a low level of significance relative to possible systematic errors. The observed absorption is comparable to that expected from the concentration of interstellar water vapor.

3. DMR

3.1. DMR INSTRUMENT

The Differential Microwave Radiometer (DMR) instrument has made the first detection of primeval anisotropy of the CMBR. It operates with two radiometers at each of three frequencies, 31.5, 53, and 90 GHz. These frequencies were chosen to minimize the effects of Galactic free-free, synchrotron, and dust emission on the cosmic measurements. The beamwidth is 7°, with two horns pointing 60° apart and 30° away from the spin axis of the spacecraft. The output of the DMR is a signal proportional to the difference of the brightnesses observed by the two antennas. As the spacecraft spins and orbits, and the orbit precesses through the year, all possible pairs of pixels 60° apart are compared by the DMR. The fundamental data analysis algorithm is a least squares fit of all the difference measurements to a map of the sky. The map has 6144 pixels, with 6 square faces of 32×32 pixels.

To achieve high sensitivity, the DMR uses mixer–preamplifier receivers with Gunn diode local oscillators. The input to each receiver is switched between the two horn antennas at 100 Hz by a ferrite switch. The output from each receiver is detected by a square law diode, and amplified again. The part of the output that is synchronous with the input switch is further amplified, integrated, and telemetered every half second.

The analysis of the data begins with inspection to ensure that no bad data are included in the final products. Corrections are made for known instrument errors, including drift of the DC output of the receiver, and sensitivity to the Earth's magnetic field. The instrument performance was monitored throughout the mission, and calibration with noise diodes was performed every two hours. The CMBR dipole was determined every day to verify the stability of the instrument. The Moon was used whenever it was visible to confirm the calibration stability and verify that the spacecraft orientation was correctly calculated. Detailed analysis of the Moon observations also yield an in-flight map of the antenna beam patterns.

3.2. DMR COSMOLOGICAL RESULTS

The initial results from the DMR were reported by Smoot et al. (1992), Bennett et al. (1992b), and Wright et al. (1992), based on the error analysis of Kogut et al. (1992) and the calibration reported by Bennett et al. (1992a). Recent papers by Bennett et al. (1994b), Wright et al. (1994b), Smoot et al. (1994), Kogut et al. (1994) and Gorski et al. (1994) have reported the results from analysis of two years of flight data. The results are consistent with a far IR balloon survey of the CMBR anisotropy, as reported by Ganga et al. (1993). They showed that there is a good statistical correlation between their maps and the one-year DMR maps. As we reported previously, the angular correlation function of the maps is also consistent with the scale-invariant $n = 1$ Peebles-Harrison-Zeldovich power law spectrum of primordial density fluctuations.

The second year results (Bennett et al. 1994b) are consistent with the first year, and averaging them together improves the precision and accuracy. The two-year 53 GHz data show RMS temperature fluctuations of $(\Delta T)_{rms}(7°) = 44 \pm 7 \, \mu K$ and $(\Delta T)_{rms}(10°) = 30.5 \pm 2.7 \, \mu K$ at 7° and 10° angular resolution respectively. The 53 × 90 GHz cross-correlation amplitude at zero lag is $C(0)^{1/2} = 36 \pm 5 \, \mu K$ (68% CL) for the unsmoothed (7° resolution) DMR data.

3.3. DMR GALACTIC ASTRONOMY

The DMR cosmological results depend on removing the Galactic emissions and demonstrating that the remaining effects are negligible. There are three primary effects at the DMR wavelengths: free-free, synchrotron, and dust emission. The methods and arguments used for the DMR analysis are thoroughly described by Bennett et al. (1992b). They used three different methods for removing the foreground emissions: subtraction, fitting, and combination. In the subtraction method, they used models of dust and synchrotron emission to extrapolate from measurements made by other instruments. Then they made a cosmic background map from the DMR data at 53 and 90 GHz, and subtracted it from the 31.4 GHz map to obtain a measurement of the free-free emission. In the fitting method, they again removed extrapolated models of the synchrotron and dust emission, but then fitted a model of the free-free emission and cosmic background fluctuations to all three DMR maps. In the combination method, they made a weighted sum of all three DMR maps with weights chosen to minimize the Galactic contributions and remove the free-free spectrum entirely. All three of these methods gave very similar cosmological results.

They considered three models of the synchrotron radiation. Their Model A scales the 408 MHz maps with a fixed power law of index $\beta_{synch} \sim 2.75$. Model B scales the 408 MHz data by the spatially variable spectral index between the 408 and 1420 MHz surveys. Model C uses an integral over the local energy spectrum of the electrons in conjunction with the 408 and 1420 MHz data. Deviations from a pure power law energy spectrum should produce a radio spectral index that is a function of magnetic field strength. They provide a table of the theoretical spectral index, showing that it increases with frequency and decreases with magnetic field. For example, at 1 μG field strength, the spectral index ranges from 2.79 between 408 and 1420 MHz, to 3.14 between 53 and 90 GHz. These authors conclude that Model C is superior to the others, as measured by the correlation coefficient between the model and the balloon measurements at 19.2 GHz by Boughn et al. (1992). The analysis was further expanded by Kogut et al. (1995), who correlated the DMR maps with low frequency radio surveys and far IR maps from COBE DIRBE. They found a limit on the synchrotron spectral index $\beta_{synch} < -2.9$ between 408 MHz and 31.5 GHz.

The free-free emission model is much simpler because the theory shows that the antenna temperature is $T_A \propto Z^2 \ EM \ \nu^{-2} T^{-1/2} F(\nu, T, Z)$, where Z is the ionic charge, EM is the emission measure, ν is the frequency, T is the temperature, and F is a logarithmic function of ν, T, Z. The recombination Hα radiation also follows a simple formula, with intensity $I \propto EM \ T^{-1/2} H(T)$, where H is another logarithmic function. As a practical result, one expects that the free-free emission should be very nearly proportional to the Hα emission, with a proportionality constant of 2 μK at 53 GHz per Rayleigh. Reynolds (1984, 1992) has reported measurements of the Hα at high latitudes that may be summarized as I (in Rayleighs) $= 1.2 \csc |b|$ for $|b| > 15°$.

There is another indicator of the presence of the free-free emitting gases, the N$^+$ line at 205 μm measured by the FIRAS. The observed intensity at high latitudes is a factor of 3 greater than expected from the Hα brightness. This may be the result of selective sampling, in that Reynolds initial observations were chosen in relatively dark regions away from bright objects, while the FIRAS beam is very wide and includes bright objects.

Kogut et al. (1995) studied the free-free emission by correlation with the DIRBE maps as well. They find that on scales larger than the 7° DMR beam the rms variation is 5.3 ± 1.8 μK at 53 GHz, with an angular power spectrum $P \propto \ell^{-3}$. If this persists to high ℓ then the free-free emission is not a significant limitation on measurement of the cosmic anisotropy.

The dust emission of the Galaxy is not readily recognized in the DMR maps, because it is not dominant over the other Galactic emissions. The DMR maps were corrected slightly based on extrapolations from the FIRAS spectra, but the correction was small. Kogut et al. (1995) found that the dust could be recognized by correlation with the DIRBE maps, finding that the dust at latitudes $|b| > 30°$ is well fitted by a single component with $T = 18^{+3}_{-7}$ K and an emissivity law $\epsilon \propto (\nu/\nu_0)^\alpha$ with $\alpha = 1.9^{+3.0}_{-0.5}$. The large upper limit on α indicates that the extrapolation to long wavelengths is very uncertain because the emission is very weak.

4. Diffuse Infrared Background Experiment

4.1. DIRBE INSTRUMENT AND OBSERVATIONAL STRATEGY

The DIRBE was designed to search for the Cosmic Infrared Background Radiation (CIBR), recognizable as an isotropic flux after all foreground contributions have been removed. The predicted CIBR is faint relative to the foregrounds, so this is an exceptionally difficult task. The DIRBE accomplishes this by making absolutely calibrated maps of the full sky in 10 photometric bands at 1.25, 2.2, 3.5, 4.9, 12, 25, 60, 100, 140, and 240 μm. To characterize the light of the interplanetary dust, the DIRBE also measures linear polarization at 1.25, 2.2, and 3.5 μm.

The interplanetary dust is the brightest of these foregrounds at most of the wavelengths observed by DIRBE. Its brightness is a function of many variables: the position of the Earth in the dust cloud, the direction of observation, the wavelength, and the polarization of the light. To first order, the dust cloud is symmetrical above and below a plane which is close to the

ecliptic plane, and rotationally symmetrical about an axis nearly perpendicular to that ecliptic plane. The dust grains gradually spiral in toward the
Sun under the Poynting-Robertson drag, which produces a number density
roughly proportional to $1/r$, where r is the radius of the orbit. However, the
perturbations of Jupiter and the other planets have systematic effects that
spoil the simple symmetry of the cloud. Moreover, there are significant
discrete sources of dust, such as individual comets and asteroid families.
These produce bands of dust which are not fully symmetrized nor stable
in time. There are intergrain collisions which grind down the particles into
larger numbers of smaller ones, and there are electromagnetic forces on
the grains. Finally, there are two roughly equal sources of dust particles:
asteroid collisions and comets, which have quite different orbit parameters.

To observe this dust as well as possible, the DIRBE line of sight is
offset from the COBE spacecraft spin axis by 30°. The observing pattern is
a spherical cycloid, sweeping the DIRBE beam across half the sky in a few
orbits. Over the course of a year, the DIRBE observes each direction many
times at all accessible angles from the Sun, with a maximum range from
64° to 124°. The DIRBE achieves true color measurements of each pixel
since all the detectors view the same field of view. The design sensitivity
for each 0.7° square field of view is $\nu I_\nu = 1$ nW m^{-2} sr^{-1} (1 σ, 1 year),
but this is not achieved at 140 and 240 μm. This is much less than the sky
brightness.

The DIRBE instrument contains the only telescope on the COBE, and
is optimized for radiometry of a diffuse source. The telescope is in an off-
axis Gregorian configuration for strong rejection of stray light (Magner
1987, Evans 1983). It utilizes a 32 Hz tuning fork chopper to modulate
the sky signal and compare it continuously with the cold interior of the
instrument. The DIRBE is cooled to less than 2 K by the COBE cryostat.
Its photometric zero point is calibrated by a cold shutter at the prime focus,
and its responsivity is calibrated against known celestial point sources.
The calibration is verified frequently by an internal reference source, and
is reproducible within $\sim 1\%$ or better. The DIRBE instrument has been
described in more detail by Silverberg et al. (1993).

The initial results show the expected features on the maps, published
by Hauser (1993) in four all-sky false color images. The Galactic plane is
bright at all wavelengths, and discrete stars are numerous at wavelengths
from 1.2 to 4.9 μm. At longer wavelengths, diffuse emission from cirrus dust
clouds and dust in star forming regions is bright. The interplanetary dust
is brightest at 12 and 25 μm and clearly detectable at all wavelengths. By
coincidence, the ecliptic plane passes near the Galactic Center.

4.2. DIRBE INFRARED ASTRONOMY

The DIRBE has revealed new features of the Galaxy and the Solar System. Since infrared penetrates the dust toward the Galactic Center, a fresh view of the inner regions is quite revealing. The disk is composed largely of K and M giants. The dust obscuring this region produces a reddening curve that is consistent with other observations (Arendt et al. 1993, 1994), and is detectable in absorption out to 3.5 μm wavelength. The same dust is correlated with the polarized light observed by DIRBE near the Galactic Center (Berriman et al. 1993). The polarization is less than about 2%, and is produced by preferential extinction by dust grains oriented in a magnetic field. After correction for extinction and emission from the disk, the central stellar bulge is found to be asymmetrical (Weiland et al. 1993, 1994), consistent with a bar as described by Blitz and Spergel (1991) and suggested by Liszt and Burton (1980) and Sinha (1979). The DIRBE data allow detailed modeling of the shape and orientation of this bar, and estimates of its luminosity and mass (Dwek et al. 1995).

The low latitude ($|b| < 10°$ far IR maps (140 and 240 μm) were correlated with H_2, H I, and H II maps by Sodroski et al. (1994). They interpret the data in terms of three phases of the interstellar medium. About 60-75% of the far IR luminosity comes from cold (17-22 K) dust associated with diffuse H I clouds, 15-30% comes from cold ($\sim$ 19 K) dust associated with molecular clouds, and less than 10% comes from warm ($\sim$ 29 K) dust in extended low-density H II regions. The derived dust temperatures differ significantly from those obtained from the IRAS observations, suggesting that small transiently heated dust grains (0.5 to 20 nm radius) contribute significantly to Galactic 60 μm emission. The correlation methods were extended to carbon monoxide in the Galactic Center by Sodroski et al. (1995a). They found that the ratio of H_2 to CO in the Galactic Center is a factor of 3 to 10 lower than in molecular cloud complexes in the inner Galactic ring. This suggests great caution in using CO to estimate H_2. Further work was presented by Sodroski et al. (1995b), who constructed a three dimensional model of the Galaxy to fit the IR, CO, H I, and 5 and 19 GHz radio surveys. The results include trends of the dust temperature, gas to dust ratio, and far IR luminosity per H atom with distance from the Galactic center.

The DIRBE also confirms the IRAS observation that "cirrus" dust clouds are prevalent throughout the Galaxy. These clouds are prominent at 12 μm wavelength, even though the equilibrium temperature of dust grains in clouds is far too low to allow emission at 12 μm. The emission is explained by transient heating of small dust grains by energetic UV photons or collisions with cosmic rays. The grains apparently reach temperatures of the order of 300 K and even higher. There is little doubt of the reality of this phenomenon, although there are many possible versions of the theory. The grain materials and optical properties are known to differ in different sources. Grains observed in particular IR and UV sources include graphite, hydrogenated sheets of hexagonal carbon rings (polycyclic aromatic hydrocarbons), silicates, carbonates, sulfides. The structure of the grains is also unknown. It is likely that some grains are needle-shaped (Wright 1982, Hawkins and Wright 1988), others are fractal fluff balls (Wright 1987), and others have dense cores with ice mantles. Given the wide variety of materials, shapes, and size distributions, it is impossible to derive detailed descriptions of the dust from observations that average over these properties. However, it is possible to show from the DIRBE maps and colors that some simplified pictures are not adequate.

The Galactic plane is also seen to be warped in both its stellar and interstellar dust components, particularly in the outer galaxy (Freudenreich et al. 1993, 1994). The dust warp is similar to that observed in interstellar gas. Modeling is needed to determine whether the physical displacement of the stars is similar to that of the gas.

A detailed analysis of the Orion region is presented by Wall et al. (1996). They are able to allocate a part of the emission to heating by the general interstellar radiation field, and a part to local heating by the OB associations. They find that the 100, 140, and 240 μm data are consistent with a single dust component with temperature 18-20 K and emissivity index of 2. However, the 60 to 100 μm color temperature is higher by 5-6 K, suggesting a warmer component of stochastically heated grains. Consequently, dust column densities estimated from the 60 and 100 μm maps underestimate the dust-to-gas ratio by factors of 5 to 10. Assuming that the Orion star forming region is typical of those seen in extragalactic objects, there may be important reinterpretations of observations with the IRAS.

The DIRBE far infrared maps of the Galaxy are strikingly similar to maps of the gamma ray sky obtained by the EGRET. The gamma rays trace the interstellar medium, through decay of π^0 particles created by cosmic ray impacts on atomic nuclei, from bremsstrahlung, and from inverse Compton scattering effects. The detailed comparison of the maps will be of great interest in understanding the origin of the gamma rays and mapping the population of relativistic electrons.

Closer to home, the DIRBE has observed 4 comets. Considering that cometary dust may resemble interstellar dust, the comets are a very useful laboratory. They are very close and bright, and the dynamic processes allow measurements of particle size and albedo. The analysis was reported by Lisse (1993) and Lisse et al. (1994). Comet Austin was observed to be extremely dark (very low albedo), so that the infrared emission is far brighter than the visible scattered sunlight. A favorable orbit geometry enabled the mapping of the time history of the dust emission and the determination of the particle size spectrum. Large dust grains ejected by the comet stay near the comet nucleus, while small ones are blown rapidly away from the Sun by radiation pressure. An individual spot in the map of the comet corresponds to dust emitted at a particular time and with a particular ratio of solar radiation pressure to gravitational force. This ratio is primarily controlled by the particle size.

5. Data Delivery

A full set of data products from the Cosmic Background Explorer Satellite is now available for analysis, and further data releases are planned based on improved calibrations. The data and images now available, and the associated documentation, may be obtained by anonymous ftp to nssdca.gsfc.nasa.gov (equivalent IP address 128.183.36.23). Please provide your e-mail address as the password. Change to directory (cd) anon_dir: [000000.cobe] (" cd cobe" should work) and download AAREADME.DOC for further instructions. The COBE data and images are accessible on the World Wide Web at the following Uniform Resource Locator (URL) address: http://www.gsfc.nasa.gov/astro/cobe_home.html

6. Summary and Conclusions

The COBE was designed to answer three questions: what is the spectrum of the CMBR, what are its anisotropies, and what is the brightness of the CIBR? In the process of answering these, it has also collected extensive data on the interstellar medium and structure of the Milky Way galaxy. These data include maps of the interstellar dust emission at wavelengths from 12 μm to 1 mm, maps of interstellar line emissions from 100 μm to 3 mm, and maps of the infrared polarization of the Galactic Center. All three instruments have given new views of the Galaxy, showing large scale features due to the interstellar medium and the general distribution of material. The Galactic bulge at the center is revealed at wavelengths where it is less obscured by dust absorption, and the Galaxy appears to have a central bar. Line emission from neutral and ionized carbon, ionized nitrogen, and carbon monoxide are widespread, and water may have been seen in absorption against the Galactic Center.

7. Acknowledgements

The National Aeronautics and Space Administration/Goddard Space Flight Center (NASA /GSFC) is responsible for the design, development, and operation of the Cosmic Background Explorer (COBE). GSFC is also responsible for the software development and the final processing of the mission data. The COBE SWG is responsible for the definition, integrity, and delivery of the public data products, and includes the following members:

NAME	AFFILIATION	SPECIAL ROLE
J. C. Mather	GSFC	Project Scientist and FIRAS Principal Investigato
M. G. Hauser	GSFC	DIRBE Principal Investigat<
G. F. Smoot	UC Berkeley	DMR Principal Investigator
C. L. Bennett	GSFC	DMR Deputy PI
N. W. Boggess	GSFC–ret	Deputy Proj. Scientist
E. S. Cheng	GSFC	Deputy Proj. Scientist
E. Dwek	GSFC	
S. Gulkis	JPL	
M. A. Janssen	JPL	
T. Kelsall	GSFC	DIRBE Deputy PI
P. M. Lubin	UCSB	
S. S. Meyer	MIT	
S. H. Moseley	GSFC	
T. L. Murdock	Gen. Res. Corp.	
R. A. Shafer	GSFC	FIRAS Deputy PI
R. E. Silverberg	GSFC	
R. Weiss	MIT	Chairman of SWG
D. T. Wilkinson	Princeton	
E. L. Wright	UCLA	Data Team Leader

Many people have made essential contributions to the success of COBE in all its stages, from conception and approval through hardware and software development, launch, and flight operations. To all these people, in government agencies, universities, and industry, the science team and I extend my thanks and gratitude. In particular, I thank the large number of people at the GSFC who brought this challenging in-house project to fruition.

References

Arendt, R. G., et al., *Back to the Galaxy*, eds. S. S. Holt & F. Verter, AIP Conf. Proc 278, 222, New York (1993)

Arendt, R. G., et al., *ApJL* 425, L85-L88 (1994)

Barney, R. D., *Illuminating Eng. Soc. J.*, 34, 34 (1991)

Bennett, C. L. *Highlights Astron*, 335 (1991)

Bennett, C. L. et al., *ApJ*, 391, 466 (1992a)

Bennett, C. L. et al., ApJ, 396, L7 (1992b)

Bennett, C. L., et al., ApJ, 434, 587-598 (1994a)

Bennett, C. L., et al., ApJ, 436, 423-442 (1994b)

Berriman, G. B., et al., Back to the Galaxy, eds. S. S. Holt & F. Verter, AIP Conf. Proc, 278, 214, New York (1993)

Blitz, L., & Spergel, D., ApJ, 379, 631-638 (1991)

Boggess, N. W. Highlights Astron, 273 (1991)

Boggess, N. et al., ApJ, 397, 420 (1992)

Boughn, S. P. et al., it ApJ, 391, L49 (1992)

Bromberg, B. W. & Croft, J. Adv. Astron. Sci., 57, 217 (1985)

Brown, J. M., et al., ApJL, 428, L37 (1994)

Coladonato, R. J. et al., Proc. Third Air Force/NASA Symp. on Recent Advances in Multidisciplinary Analysis and Optimization, 370, Anamet, Hayward, CA (1990)

Dwek, E., et al., ApJ, 445, 716-730 (1995)

Evans, D. C., SPIE Proc., 384, 82 (1983)

Fixsen, D. J. et al., ApJ 420, 445-449 (1994)

Fixsen, D. J. et al., ApJ 420, 457-473 (1994)

Freudenreich, H. T., et al., Back to the Galaxy, eds. S. S. Holt & F. Verter, AIP Conf. Proc, New York, 278, 485 (1993)

Freudenreich, H.T., et al., ApJL, 429, L69-L72 (1994)

Ganga, K., Cheng, E., Meyer, S., & Page, L., ApJ, 410, L57-L60 (1993)

Gorski, K. et al., ApJL, 430, L89-L92 (1994)

Gulkis, S., Lubin, P. M., Meyer, S. S., & Silverberg, R. F., Sci. Amer., 262, 132 (1990)

Hauser, M. G. Highlights Astron., 291 (1991a)

Hauser, M. G., Proc. Conf. Infrared Astronomy and ISO, Les Houches, 479 (1991b)

Hauser, M. G., Back to the Galaxy, eds. S. S. Holt & F. Verter, AIP Conf. Proc, New York, 278, 201 (1993)

Hawkins, I., & Wright, E., ApJ, 324, 46-59 (1988)

Hopkins, R. A., & Castles, S. H., Proc. SPIE, 509, 207 (1985)

Hopkins, R. A., & Payne, D. A. Adv. Cryogenic Engineering, 33, 925 (1987)

Janssen, M. A. & Gulkis, S., Proc. The Infrared and Submillimetre Sky After COBE, Les Houches, 391, ed. M. Signore & C. Dupraz, Kluwer, Dordrecht (1991)

Kogut, A., et al., ApJ, 433, 435-439 (1994)

Kogut, A., et al., ApJ, 439, L29 (1995)

Lisse, C. M., PhD thesis, University of Maryland, Astronomy (1993)

Lisse, C. M., et al., ApJL, 432, L71-L74 (1994)

Liszt, H. S., & Burton, W. B., ApJ, 236, 779-797 (1980)

Magner, T. J., Opt. Eng., 26, 264 (1987)

Mather, J. C., Opt. Eng., 21, 769 (1982)

Mather, J. C. et al., IAU Colloq. 123, Observatories in Earth Orbit and Beyond, Proc., ed. Y. Kondo, 9, Kluwer, Boston (1990)

Mather, J. C. et al., AIP Conf. Proc. After the First Three Minutes, 222, 43, ed. S. S. Holt, C. L. Bennett, & V. Trimble, AIP, New York (1991a)

Mather, J. C., Highlights Astron., 275 (1991b)

Mather, J. C., Shafer, R. A., & Fixsen, D. J., Proc. SPIE, 2019, 146-157, 1993.

Mather, J. C. et al., ApJ 420, 439-444 (1994)

Milam, L. J., Illuminating Eng. Soc. J., 34, 27 (1991)

Mosier, C. L. AIAA, 91-361 (1991)

Petuchowski, S. J. & Bennett, C. L., ApJ, 391, 137-140 (1992)

Petuchowski, S. J., & Bennett, C. L., ApJ, 405, 595-598 (1993)

Reach, W. T., et al., ApJ, 451, 188-199 (1995)

Reynolds, R. J., ApJ, 282, 191 (1984)

Reynolds, R. J., ApJ, 396, L1 (1992)

Sampler, H. P. Proc. SPIE, 1340, 417 (1990)

Silverberg, R. F. et al., Proc. SPIE, 2019, 180-189 (1993)

Sinha, R. P., Kinematics of H I near the Galactic Center, PhD Thesis, University of Maryland (1979)

Smoot, G. F. *Highlights Astron*, 281 (1991)

Smoot, G. F. et al., *ApJ*, 371, L1 (1991)

Smoot, G. F. et al., *ApJ*, 396, L1 (1992)

Smoot, G. F., et al., *ApJ*, 437, 1 (1994)

Sodroski, T. J., et al., *ApJ*, 428, 638-646 (1994)

Sodroski, T. J., et al., *ApJ*, 452, 262-268 (1995a)

Sodroski, T. J., et al., *Unveiling the Cosmic Infrared Background*, ed. E. Dwek, AIP Conf. Proc, New York, Conf. at U Md., April 1995.

Volz, S. M. & DiPirro, M. J., *Cryogenics*, 32, 77 (1992)

Volz, S. M. & Ryschkewitsch, M. G., *Superfluid Helium Heat Transfer*, *HTD*, 134, 23, ed. J. P. Kelly & W. J. Schneider AME, New York (1990)

Volz, S. M., Dipirro, M. J., Castles, S. H., Rhee, M. S., Ryschkewitsch, M. G., & Hopkins, R., *Proc. Internat. Symp. Optical and Opto-electronic Applied Sci. and Eng.*, 268, SPIE, San Diego (1990)

Wall, W. F., et al., *ApJ*, 456, 566 (1996)

Weiland, J. L., et al., *Back to the Galaxy*, eds. S. S. Holt & F. Verter, AIP Conf. Proc, New York, 278, 137 (1993)

Weiland, J. L, et al., *ApJL* 425, L81-L84 (1994)

Wright, E. L., *ApJ*, 255, 401-407 (1982)

Wright, E. L., *ApJ*, 320, 818-824 (1987)

Wright, E. L. *Ann. NY Acad. Sci., Proc. Texas-ESO-CERN Sym*, 647, 190 (1990)

Wright, E. L. et al., *ApJ*, 381, 200 (1991)

Wright, E. L., et al., *ApJ*, 420, 450-456 (1994a)

Wright, E. L., *ApJ*, 436, 443 (1994b)

MOLECULAR GAS IN SPIRAL GALAXIES

The "X-factor", and evidence for the presence of
a very cold component of molecular gas in disk galaxies

RONALD J. ALLEN
Space Telescope Science Institute
3700 San Martin Drive
Baltimore, MD 21218, USA

1. Introduction

Our current knowledge of the distribution and total mass of molecular gas in galaxies comes almost exclusively from surveys carried out with filled-aperture millimeter radio telescopes in the $\lambda 2.6$ mm line of the ^{12}CO molecule; an excellent example is the comprehensive FCRAO survey by Young *et al.* (1995). The measurement actually made is the surface brightness of the ^{12}CO(1-0) line emission, diluted by the beam of the radio telescope, from which the beam-smoothed and velocity-integrated line intensity I_{CO} is obtained in units of K km s^{-1}. In order to obtain the average column density of H_2, it has been customary for many years to simply multiply the observed CO intensity by a constant factor, the so-called "X-factor". This has led to our present "conventional" view of the distribution and total content of molecular gas in galaxy disks, and to many papers on the variations of the rate and efficiency of star formation in these disks. However, over the past few years evidence has been accumulating that the X-factor is neither constant with position in a given galaxy, nor is it the same from one galaxy to another. A radial gradient of more than a factor of 10 has been reported in our Galaxy, and the values of X for the central regions of starburst galaxies differ by as much as two orders of magnitude from the X for dust clouds in the inner disk of M31.

In this paper I will first briefly review the conventional view of the X-factor, and then summarize the evidence against it. My conclusion is that, with the possible exception of the warm component of the molecular gas in our own Galaxy, we actually know very little about the true distribution and amount of molecular gas in galaxy disks from existing observations of the ^{12}CO(1-0) line.

D. L. Block and J. M. Greenberg (eds.), New Extragalactic Perspectives in the New South Africa, 50–60.
© *1996 Kluwer Academic Publishers.*

In the second part of this paper I will present several observational results, some quite new, which provide evidence for the existence of significant amounts of molecular gas in disk galaxies, both in their inner and in their outer parts. This gas does not appear (or is very faint) in ^{12}CO(1-0) emission, and it is likely to be very cold.

2. The I_{CO} - N(H$_2$) Connection

In the conventional view of the neutral component of the ISM, most of the mass is in atomic and molecular hydrogen, with other species of atoms and molecules present in trace amounts. For example, for every CO molecule in the ISM, there are on the average more than 14,000 H$_2$ molecules. Unfortunately, Nature has made it quite difficult to observe the H$_2$ molecules directly; it is much easier to observe tracers like the (1-0) rotational line of the ^{12}CO molecule which is excited by collisions with other molecules, primarily H$_2$. But, if what we are really after is some idea of the amount of H$_2$ present, then we have to invent an argument to connect the observation of e.g. the velocity-integrated brightness I_{CO} of the ^{12}CO(1-0) line to the column density N(H$_2$) of molecular hydrogen.

2.1. THE CONVENTIONAL VIEW

For more than a decade it has been customary to express the I_{CO} - N(H$_2$) connection as a factor $X = N(H_2)/I_{CO}$ in units of molecules cm^{-2}/(K km s^{-1}), and to assert that this ratio is a constant of Nature (e.g. Young 1990 and references given in §2 there). A fundamental calibration of X has been done by comparing the millimeter CO emission directly with the UV absorption by H$_2$ and the reddening by dust along lines of sight through low-optical-depth "translucent" Galactic molecular clouds in the vicinity of the Sun; this has been summarized in the excellent review on CO in the Galaxy by Combes (1991, §2). However, the optical depths in "normal" molecular clouds are very large, so that in general we see only the skins of the clouds in the ^{12}CO emission. Because of the large accompanying dust obscuration, there is no data either on the UV absorption lines of H$_2$ or the reddening of distant stars. In this more normal case, three other methods have been used to arrive at a calibration of X for giant molecular clouds (GMCs) in the Galaxy (Combes 1991, Young & Scoville 1991). These methods are: an analysis of the excitation conditions in GMCs from observations of several rotational transitions of different molecules, e.g. of ^{12}CO and ^{13}CO; a comparison of CO brightness with virial masses determined from CO cloud sizes and line widths; and comparisons of the CO emission to the gamma-ray emission along the same lines of sight. Of these three methods, the "virial method" is the most widely used.

Young & Scoville (1991) discuss the options and adopt a value for X = 3.0×10^{20} cm^{-2}/(K km s^{-1}) for CO brightnesses measured in the T_R^* scale, and offer several arguments in support of this value and its universal applicability:

- The CO luminosities L_{CO} of Galactic GMCs are linearly correlated with their virial masses M_{VIR}, and from the constant of proportionality one can derive the adopted value of X.

- Several CO clouds observed in the star-forming ring of M31, in the inner part of M33, and in IC 10 fall within a factor of about 2 of the L_{CO} - M_{VIR} correlation, justifying the use of the adopted value of X for external galaxies.

- A comparison of the CO surveys of the Galaxy with the gamma-ray surveys yields an X which is consistent with the adopted value over the range of Galactocentric radius from 2 to 10 kpc, suggesting that variations in metallicity and other cloud properties with radius do not strongly affect the calculated H_2 masses.

2.2. PROBLEMS WITH THE CONVENTIONAL VIEW

The use of the virial theorem and the apparent correlation of the CO luminosities of Galactic GMCs with their virial masses are centerpieces of the argument in favor of adopting a constant value for the X-factor. In attaching physical significance to this correlation, the underlying assumption is that the two quantities L_{CO} and M_{VIR} can be determined observationally and independently of each other. However, Maloney (1990a) has shown that, because of the way L_{CO} and M_{VIR} are computed from the data, there is a built-in correlation between them; it's just another way of writing the size-linewidth relation, itself a purely observational correlation connecting the measured sizes of Galactic GMCs with their measured ^{12}CO(1-0) linewidths. This relation neither requires nor implies virial equilibrium; it may instead be related to a universal underlying spectrum of turbulence (cf. Combes 1991, §4.2, and references cited there). This size-linewidth relation is also found for GMCs in the Small Magellanic Cloud (Rubio *et al.* 1993). If this relation is universal it would also explain why the CO luminosities and virial masses of warm GMCs in other galaxies also appear to fall on the L_{CO} - M_{VIR} correlation.

The apparent constancy of X with Galactocentric radius mentioned earlier as evidence in favor of the conventional view is somewhat surprising, since changes in the metallicity and excitation of GMCs with radius in a galaxy are to be expected and ought to have a measurable effect on cloud properties (e.g. Maloney & Black 1988), and the complicated chemistry of dust grains gives every reason to expect that the dust-to-gas ratio is also highly variable in galaxies (e.g. Greenberg 1991). Recently, Sodroski *et al.* (1995) appear to have found the expected radial dependence in our Galaxy; they combined DIRBE data from COBE with CO survey data to show that the value of X within about 400 pc of the Galactic center is about 3 to 10 times lower than in the solar neighborhood. This also resolves an outstanding discrepancy in the interpretation of the gamma-ray data in the Galactic center region (the "gamma-ray deficit"). Sodroski *et al.* combine their estimates of X with existing virial analyses of GMCs at larger radii to show that, at a Galactocentric radius of 13 kpc, X has increased by a factor of 36 from its value in the Galactic center region! In fact, X increases exponentially with R_{Gal} according to $\log X = -0.34 + 0.12 \times R_{Gal}(kpc)$. The consequences of using this radially-increasing X on the cumulative H_2 mass in the Galactic disk are shown in Figure 1. The correction seems modest enough in this graph, but the effects in the outer Galaxy are actually quite pronounced; the total H_2 mass no longer appears to be converging! The standard value of X predicts that less than 18% of the total molecular gas mass of 1.5×10^9 $M_\odot$ is left outside of $R_\odot$ in the range $8.5 \lesssim R_{Gal} \lesssim 14.5$ kpc, whereas the increasing X factor predicts that more than 44% of the total of 2.0×10^9 $M_\odot$ is in this outer annulus. And we must remember that *this only refers to the warm molecular gas which is detected in current sensitivity-limited surveys for CO.*

2.3. EXCITATION EFFECTS

As mentioned earlier, we see in general only the skins of the molecular clouds in the emission line of $^{12}CO(1-0)$; the observed brightnesses therefore depend directly on the excitation temperatures and beam filling factors of the clouds. In this case it is clear that any sensitivity-limited survey of a galaxy will be biased towards clusters of the warmest GMCs because they will be the brightest. The value of X has been determined for the warm clouds in the Galaxy, and it may also be applicable to clusters of similarly-warm clouds in the nearby galaxies, but there is no evidence that it is appropriate for clouds (or clusters of clouds) which are significantly colder or significantly warmer than the population of observed Galactic GMCs. As has been reviewed by Young & Scoville (1991; see also Maloney 1990b), taking X to be constant is equivalent to assuming that the ratio $\sqrt{n}/T_{CO}$ is a constant, where n is the total gas density in the cloud and T_{CO} is the emergent CO brightness temperature. This dependence is quite different from that expected for a population of warm, optically-thick molecular clouds of varying densities in approximate pressure equilibrium ($nT_{CO} \approx nT_K = $ const., e.g. Turner 1988, p. 158) or molecular clouds which are heated by cosmic rays and cooled by molecular line emission ($T_{CO} \approx T_K$

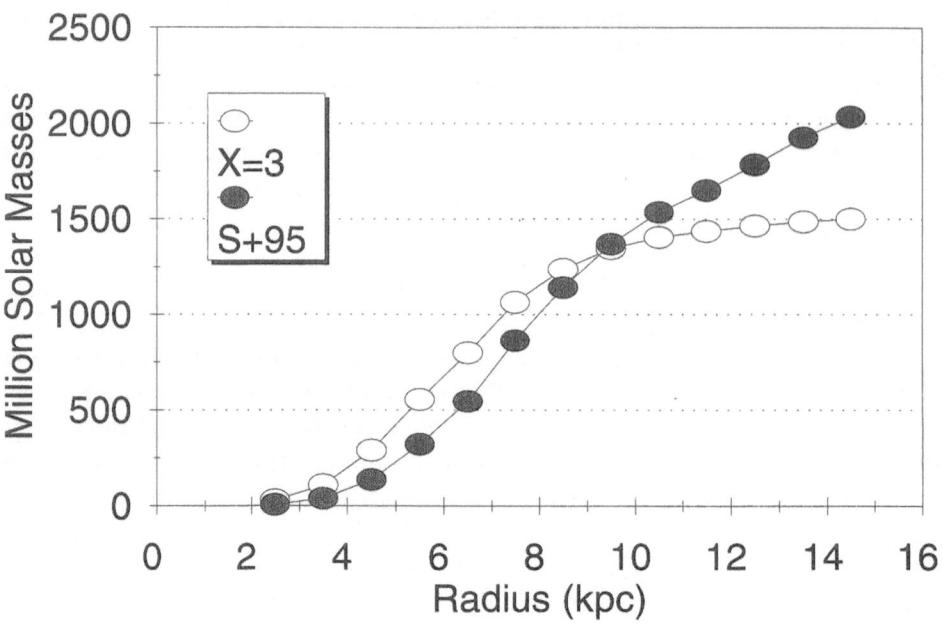

Figure 1.   Radial dependence of the cumulative molecular mass in the Galaxy as it would be measured by an external distant observer, first using the conventional constant X-factor (open ovals, for X = 3.0×10^{20} cm^{-2}/(K km s^{-1})), and second using the increasing X-factor of Sodroski *et al.* (1995) (filled ovals). CO data from Bronfman *et al.* (1988) out to 9.5 kpc; data from 9.5 to 14.5 kpc from Clemens *et al.* (1988) scaled by a constant factor of 0.61 for consistency with Bronfman *et al.* The ^{12}CO volume emissivity has been integrated in Z according to Allen (1992).

= const., roughly independant of density for $10^2 \lesssim n \lesssim 10^5$ cm^{-3}; Suchkov *et al.* 1993).

Observations of the [C II] emission from galaxies at $\lambda = 158\mu$ provides evidence for changes in the ^{12}CO excitation conditions, especially in galaxies with active star formation in their nucleii and disks. The physics of the [C II] line formation in photodissociation regions (PDRs) is reasonably well understood (e.g. Crawford *et al.* 1985, Tielens & Hollenbach 1985). The line is in the far IR, so it is not affected by extinction, and it is optically thin in the cases of interest. The [C II] emission comes from warm ($\sim 100 - 300$ K), dense ($n_H \gtrsim 1000$ cm^{-3}) gas evaporating from the surfaces of molecular clouds which are being heated and dissociated by the UV photons from nearby OB stars or from the general ambient interstellar UV radiation field. The [C II] brightness is mostly a measure of the UV energy density. Crawford *et al.* have found that the [C II] brightness of objects ranging from 10 pc-sized Galactic star-forming regions to many-kpc-sized pieces of entire galaxies correlates well with the measured values of I_{CO} at the same positions over a factor of 100 in surface brightness; they concluded that I_{CO} is apparently also a measure of UV energy density and may not be a reliable indicator of the H_2 mass. Crawford *et al.* also compared in detail their [C II] data on six gas-rich galaxies, including the starburst systems M82, NGC 1068, and M83, with $^{12}CO(1-0)$ data on the same systems and concluded that the ^{12}CO data overestimate the amount of molecular mass present. Stacey *et al.* (1991) extended these results to another dozen gas-rich galaxies, and concluded that the excitation temperatures of the $^{12}CO(1-0)$ line in starburst galaxy molecular clouds is significantly higher than in Galactic disk GMCs. From their [C II] data the overestimate in the molecular gas content using the standard X-factor would be a factor of 3 or more. This is in line with the results of Adler *et al.* (1992), who found that the standard X-factor masses for 20 clusters of GMCs (300–800 pc in size) spread over the inner disk of the bright spiral M51 exceed the virial masses by a factor of 2.5.

Further evidence that the CO brightness in galaxy disks is largely an excitation effect comes from the observed correlation between I_{CO} and T_{NT}, the surface brightness in nonthermal radio continuum emission. I_{CO} itself decreases roughly exponentially with radius in the luminous late-type spirals such as M83, N6946, and IC342 (cf. the review by Young & Scoville 1991, their Fig. 2a); this behaviour is at least qualitatively similar to the distribution of light in the disks of these galaxies. But this similarity ends with the earlier types such as M31, N7331, N2841, and M81; these galaxies show broad central depressions in their I_{CO} distributions which do not resemble the morphologies of their disk light. However, there is another component of the radiation from these galaxy disks which *does* follow I_{CO}, broad central depressions and all; it is T_{NT}, the nonthermal radio continuum emission. Adler *et al.* (1991) have presented the radial dependences of the ratio I_{CO}/T_{NT} for eight nearby spirals and shown that this ratio is surprisingly constant within a given galaxy on scales of $\sim 1 - 2$ kpc, and furthermore it is approximately the same value from one galaxy to another. The total range in average surface brightness spanned by the galaxies in their sample is more than a factor of 100, but the ratio remains at 1.3 ± 0.6. This situation is strongly reminiscent of that for the [C II] emission discussed in the previous paragraph. Adler *et al.* suggest that the two apparently disparate quantities, I_{CO} and T_{NT}, in fact may have a common antecedent in the cosmic ray content of the galaxy disks. The relativistic electron component provides the nonthermal radio continuum emission, while the 10-100 MeV primary component is the main source of heat for the general ISM in regions more than a few hundred parsecs away from local heating sources such as H II regions and SNR. In disks with a high content in cosmic rays, we may therefore expect to see warm molecular gas radiating in the CO lines, as well as bright nonthermal radio continuum. This explanation is admittedly qualitative, and it ignores possible variations in magnetic field strength which will be important for the nonthermal radio emission.

The [C II] and radio continuum data described above provide two different explanations for the main source of heating in the ISM, but in both cases the central theme is clear; *the $^{12}CO(1\text{-}0)$ line in galaxy disks is bright where the ISM is warm, and not necessarily where the H_2 column densities are high.*

3. Evidence for Cold Molecular Gas in Galaxy Disks

The examples reviewed above concern cases where the amount of H_2 is usually *overestimated* by the standard X-factor owing to high heating rates in the ISM. Conversely, it is possible that the molecular gas content is *underestimated* in regions of galaxies where the heating rates are substantially lower than the average in the solar neighborhood. Such regions can be found in the inner disks of the intermediate-type spirals like NGC 7331, M31, and M81, and in the outer parts of most galaxies. There is indeed some evidence for the existence of substantial quantitities of cold molecular gas which has have been missed in sensitivity-limited $^{12}CO(1\text{-}0)$ surveys; this evidence is both direct (in M31), and indirect (in M31 and M81).

3.1. COLD MOLECULAR GAS IN THE INNER DISK OF M31

Allen & Lequeux (1993) have reported observations of emission in the $^{12}CO(1\text{-}0)$ and $^{12}CO(2\text{-}1)$ transitions from two dust clouds in the inner disk of M31. These clouds are located about $10'$ (~ 2 kpc) from the center of the galaxy. The line widths appear to obey the same relation with cloud size as that of Galactic GMCs, suggesting that one is dealing with the same type of massive (few $\times 10^7$ $M_\odot$) objects. However, the luminosity of the $^{12}CO(1\text{-}0)$ line is lower by an order of magnitude compared to that of similar Galactic clouds, and the $^{12}CO(2\text{-}1)/^{12}CO(1\text{-}0)$ line intensity ratio is unusually small ($\sim 0.2 - 0.4$). This suggests low temperatures and thus probably little heating, presumably due to a low intensity of far-UV radiation and a low flux of cosmic rays. Allen *et al.* (1995) have modelled these and other line data from the same clouds and concluded that the emission of the ^{12}CO lines is dominated by a low-density (~ 100 cm^{-3}), very cold gas spread out in extended clouds, while that of the ^{13}CO lines comes in large part from higher-density clumps (~ 3000 cm^{-3}) inside these clouds. The properties of these molecular clouds can be understood as resulting mainly from a very small rate of photodissociation due to a very low UV radiation field, together with a low cosmic-ray density. Typical kinetic temperatures inside the clouds can drop to values less than 5 K.

Loinard *et al.* (1995) have recently conducted a blind search for $^{12}CO(1-0)$ and (2-1) emission over the inner disk of M31 and concluded that cold molecular clouds are not at all rare there; more than 50% of the surface area of the inner disk (inside $\approx$ 8 kpc radius) is covered with such cold clouds. An estimate of the mass using the virial theorem leads to the conclusion that the molecular gas mass surface density near 2 - 6 kpc radius is at least 10 $M_\odot pc^{-2}$, *more than a factor of 50 higher than the Galactic X-factor would predict.*

3.2. REDDENING OF BACKGROUND GALAXIES IN THE OUTER PARTS OF M31

An indirect indicator for the presence of molecular gas comes from the very recent detection of the reddening of background galaxies seen through H I clouds in the outer parts of M31 at radial distances of 23 kpc from the center of the galaxy (Cuillandre *et al.* 1996). A preliminary report on these results is given in the paper by Lequeux & Guélin in this volume. The reddening provides unequivocal evidence for dust mixed with the H I at these large galactocentric distances. The gas here is clearly not primordial, but has an appreciable content in heavy elements, perhaps as much as 0.4 of the value in the solar neighborhood. If dust is there, can molecular gas be far behind? Cuillandre *et al.* have also found evidence for the presence of young main-sequence stars (B1–B9) belonging to M31 in the general area where the H I is located.

3.3. H I AS A PRODUCT OF PHOTODISSOCIATION OF H_2

Large-scale conversion of H_2 into H I by star-formation activity in the spiral arms of galaxies was first proposed to explain the separation of the H I and the dust lanes in M83 (Allen, Atherton, & Tilanus 1986), and later the displacement of the H I from the nonthermal radio continuum in M51 (Tilanus & Allen 1989). Knapen & Beckman (1994) have summarized further evidence for this proposal in their Hα and H I comparisons on 1 kpc scales in M51 and NGC 4321, and concluded that it is a likely explanation for their results at least over the inner parts of these bright galaxies.

Here again the 158μ [C II] data provides new and important support for this picture. Stacey *et al.* (1991) have interpreted their [C II] results in 14 gas-rich galaxies as supporting the suggestion that much of the H I in galaxies is a photodissociation product. Madden *et al.* (1993) have mapped the [C II] emission from the disk of NGC 6946 at 55″ and modelled the PDR

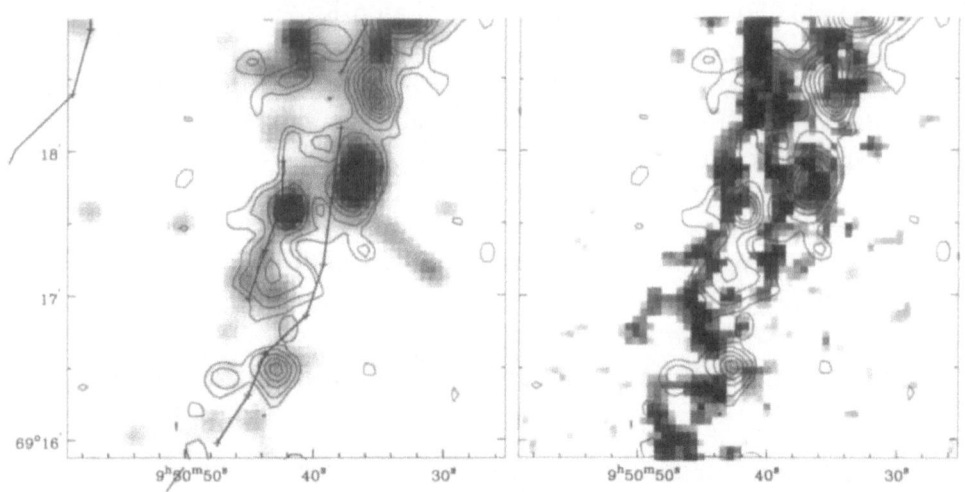

Figure 2. Left Panel: Contours of the far-UV emission from a small area covering the western spiral arm of M81 at $\lambda \approx 150$ nm as observed by the Ultraviolet Imaging Telescope (Stecher *et al.* 1992), overlaid on a gray-scale representation of the Hα from Devereux *et al.* (1995). The line segments indicate roughly where prominent dust lanes are to be found. Right Panel: The same far-UV contours overlaid on a gray-scale representation of the H I from VLA observations by B. Hine & A.H. Rots (see Kaufman *et al.* 1989a). The resolution is $\approx 9'' = 150$ pc for an assumed distance to M81 of 3.7 Mpc.

heating/cooling balance; for the most likely choice of parameters, they find that as much as 70–80% of the observed H I may be photodissociated H$_2$.

For the bright galaxies M51, M83, NGC 4321, and NGC 6946 mentioned above the ISM is generally warm, a result of the high fluxes of UV photons and cosmic rays produced by star-formation activity. There is thus direct evidence for the H$_2$ antecedent of the photodissociated H I, since the ISM is warm and the CO emission is easily detectable. However, there are indications that photodissociation occurs even when the CO emission is very faint. My colleagues Johan Knapen, Ralph Bohlin, and I are doing a detailed comparison of the Hα, H I, and UV morphology of M81 at a linear resolution of about 150 pc. The H I is often found alone in the spiral arms, without associated UV or Hα (at least at the present level of sensitivity). However, in the neighborhood of a far-UV concentration, the H I tends to be brightest at the periphery of the UV. Some extreme cases are quite striking, with the H I forming a "halo" or "skin" around a large UV-emitting region, parts of which may have little or no Hα (e.g. Figure 2). The Hα is generally poorly correlated with the H I at this resolution; the far-UV is much more closely linked with the H I. This suggests two conclusions: First, the H I morphology, at least in the vicinity of the detected far-UV regions, is qualitatively consistent with the H I being a product of dissociation of H_2 by UV photons. Second, there are more regions of "B-stars" capable of photodissociating H_2 into H I than there are regions of "O-stars" capable of ionizing the photodissociated H I into H II. A corollary of this picture is that the disk of M81 must harbor a large reservoir of molecular gas which has so far escaped detection in the $^{12}CO(1-0)$ line; this is consistent with the findings of Kaufman *et al.* (1989b) who concluded that there is more extinction in the dust lanes of M81 than the observed H I + H_2 (the latter from the ^{12}CO data) would predict, although those authors presumed that the molecular gas would be confined mostly to the dust lanes. Our UV/H I comparison points rather to a more extended distribution of molecular gas; this gas would then have to be very cold. It may be visible after long observations as very faint, broad emission lines of $^{12}CO(1-0)$, such as those found in the inner disk of M31. A more quantitative analysis is needed; a model assuming balance between H_2 formation on grains and dissociation by UV ought to permit us to estimate the density of the underlying cold molecular gas in the immediate neighbourhood of the UV sources in the spiral arms.

4. Conclusions

I think we can safely conclude that:

- With the possible exception of the solar neighborhood in the Galaxy, the X-factor is not a reliable way to compute the quantity of molecular gas present in galaxy disks.
- There is good evidence for the presence of large amounts of H_2 in the disks of spiral galaxies, especially among the intermediate morphological types. The amount of H_2 actually present is quite uncertain at the moment, but it may exceed the quantity of H I by a sizeable factor.

References

Adler, D.S., Allen, R.J., & Lo, K.Y. (1991), *ApJ*, **382**, pp. 475–482
Adler, D.S., Lo, K.Y., Wright, M.C.H., Rydbeck, G., Plante, R.L., & Allen, R.J. (1992), *ApJ*, **392**, pp.497–508
Allen, R.J. (1992), *ApJ*, **399**, pp. 573–575
Allen, R.J., Atherton, P.D., & Tilanus, R.P.J. (1986), *Nature*, **319**, pp. 296–298
Allen, R.J., & Lequeux, J. (1993), *ApJ*, **410**, pp. L15–L18
Allen, R.J., Le Bourlot, J., Lequeux, J., Pineau des Forêts, & Roueff, E. (1995) *ApJ*, **444**, pp. 157–164
Bronfman, L., Cohen, R.S., Alvarez, H., May, J., & Thaddeus, P. (1988), *ApJ*, **324**, pp. 248–266
Clemens, D.P., Sanders, D.B., & Scoville, N.Z. (1988), *ApJ*, **327**, pp. 139–155
Combes. F. (1991), *ARAA*, **29**, pp. 195–237

Crawford, M.K., Genzel, R., Townes, C.H., & Watson, D.M. (1985), *ApJ*, **291**, pp. 755–771

Cuillandre, J.-C., Mellier, Y., Fort, B., Allen, R.J., & Lequeux, J. (1996), *A&A*, in preparation.

Devereux, N.A., Jacoby, G., & Ciardullo, R. (1995), *AJ*, **110**, pp. 1115–1128

Greenberg, J.M. (1991), in *Cosmic Rays, Supernovae, and the Insterstellar Medium*, ed. M.M. Shapiro *et al.* (Kluwer; Dordrecht), pp. 57–68

Knapen, J.H., & Beckman, J.E. (1994), in "Physics of the Gaseous and Stellar Disks of the Galaxy", ed. I. King (ASP Conference Series V. 66), pp. 329–336

Kaufman, M. *et al.* (1989a), *ApJ*, **345**, pp. 674–696

Kaufman, M., Elmegreen, D.M., & Bash, F. (1989b), *ApJ*, **345**, pp. 697–706

Loinard, L., Allen, R.J., & Lequeux, J. (1995), *A&A*, **301**, pp. 68–74

Madden, S.C., Geis, N., Genzel, R., Herrmann, F., Jackson, J., Poglitsch, A., Stacey, G.J., & Townes, C.H. (1993), *ApJ*, **407**, pp. 579–587

Maloney, P. (1990a), *ApJ*, **348**, pp. L9–L12

Maloney, P. (1990b), in *The Interstellar Medium in Galaxies*, ed. H.A. Thronson & J.M. Shull (Kluwer; Dordrecht), pp. 493–523

Maloney, P., & Black, D. (1988), *ApJ*, **325**, pp. 389–401

Rubio, M., Lequeux, J., & Boulanger, F. (1993), *A&A*, **271**, pp. 9–17

Sodroski, T.J. *et al.* (1995), *ApJ*, **452**, pp. 262–268

Stacey, G.J., Geis, N., Genzel, R., Lugten, J.B., Poglitsch, A., Sternberg, A., & Townes, C.H. (1991), *ApJ*, **373**, pp. 423–444

Stecher, T.P. *et al.* (1992), *ApJ*, **395**, pp. L1–L4

Suchkov, A., Allen, R.J., & Heckman, T.M. (1993), *ApJ*, **413**, pp. 542–547

Tielens, A.G.G.M., & Hollenbach, D. (1985), *ApJ*, **291**, pp. 722–746

Tilanus, R.P.J., & Allen, R.J. (1989), *ApJ*, **339**, pp. L57–L61

Turner, B.E. (1988), in *Galactic and Extragalactic Radio Astronomy*, ed. G.L. Verschuur and K.I. Kellermann (Springer-Verlag; New York), pp. 154–199

Young, J.S. (1990), in *The Interstellar Medium in Galaxies*, ed. H.A. Thronson & J.M. Shull (Kluwer; Dordrecht), pp. 67–120

Young, J.S. *et al.* (1995), *ApJS*, **98**, pp. 219–257

Young, J.S., & Scoville, N.Z. (1991) *ARAA*, **29**, pp. 581–625

COLD DUST SIGNATURES ON SNR GAMMA RAY SPECTRA

O.C. DE JAGER
Space Research Unit, PU vir CHO,
Potchefstroom 2520, South Africa

1. Introduction

Arendt (1989) found several high confidence associations of IR emission (IRAS) with supernova remnants (SNR), and the dust seems to trace the radio synchrotron emission in many cases. Whereas the dust temperature and magnetic field strength associated with young remnants are relatively high, we find that the older remnants are usually associated with colder dust and lower field strengths (Dwek & Arendt 1992; Reynolds & Ellison 1992).

The EGRET instrument on the Compton Gamma Ray Observatory discovered at least 130 high energy γ-ray sources above 100 MeV, of which 71 are unidentified (Thompson et al. 1996). Sturner & Dermer (1995) found that up to 13 of these unidentified sources may be associated with SNR, but the lack of TeV γ-rays from these remnants (e.g. Lessard et al. 1995) indicates that they are probably not due to cosmic ray (nucleonic) acceleration in the SNR, and the γ-ray spectra must steepen or cut off at $E_\gamma > 10$ GeV.

In this paper we show that the observed γ-ray flux from SNR W44 may be explained by the inverse Compton (IC) scattering of radio emitting electrons on IR photons ($T \sim 25$K) from dust grains swept up by the SNR, provided that the field strength is not too large. Whereas the synchrotron cutoff above $\sim 10^{11}$ Hz is expected to be hidden by the presence of dust emission, the associated IC cutoff would reveal an observable feature in high energy γ-rays.

2. The inverse Compton γ-ray spectrum

The power law radio spectra of shell-type SNR are relatively flat with spectral indices α between 0.3 and ~ 0.8 (Green 1988), and the IC scattering of soft photons (with a blackbody or gray body spectrum) by the same radio emitting electrons, results in a γ-ray spectrum with the same index α for energies well below the cutoff. We generalise the expression of Blumenthal

D. L. Block and J. M. Greenberg (eds.), New Extragalactic Perspectives in the New South Africa, 61–64.
© 1996 *Kluwer Academic Publishers.*

& Gould (1970) (BG) for scattering on a blackbody, to the case for a gray body with temperature T, given the observed IR energy flux F_{IR} (see Table 3 of Arendt 1989). The electron spectrum is obtained by inverting the observed power law synchrotron spectrum for a given magnetic field strength $B_\perp$. The expression for the monochromatic γ-ray flux at 100 MeV is given by (after normalising with $B = 10^{-5}B_{-5}$ G, $F_{IR} = 10^{-9}F_{-9}$ ergs/cm^2/s, the SNR diameter θ' in arcmin, and the dust temperature $T = 25T_{25}$K)

$$F_\gamma = 1.4 \times 10^{-8}(6.6 \times 10^{-4}\frac{T_{25}}{B_{-5}\epsilon_{100}})^\alpha \frac{F_{-9}S_9}{T_{25}B_{-5}\theta'^2} \text{ MeV.cm}^{-2}.\text{s}^{-1}.\text{MeV}^{-1} \quad (1)$$

Using this expression on all SNRs detected by IRAS with a confidence index of 1 (as defined by Arendt 1989), and associated radio spectral information as given by Green (1988), we find that the remnant W44 predicts the largest γ-ray flux, provided that we account for the indication that younger remnants have larger field strengths.

3. Thermal signatures in γ-ray cutoffs

Radio synchrotron emission in shell-type remnants are explained in terms of shock wave acceleration (first order Fermi acceleration) of electrons to relativistic energies (Reynolds & Ellison 1992). Since the acceleration process competes with loss processes, the electron energy is limited to a maximum $\gamma_b mc^2$ (Reynolds 1995), resulting in a synchrotron break frequency ν_b, which is typically larger than 10^{11} Hz. Since the average γ-ray energy associated with the scattering of a soft photon with energy ϵ is $\sim \epsilon\gamma^2$ (the Thomson limit applies in our case), we also expect to see a γ-ray cutoff at an energy of $\sim 3kT\gamma_b^2$ resulting from the scattering of photons from a blackbody with temperature T by the highest energy electrons. By combining the standard expression for the characteristic synchrotron frequency $\nu_c = \nu_b$ (with the magnetic field strength normalised to $B = 10^{-5}B_{-5}$ G), and the abovementioned expression for the average γ-ray energy (normalised to $T = 25T_{25}$K), we may express the expected γ-ray cutoff energy as

$$E_b = 300\, T_{25}B_{-5}^{-1}(\frac{\nu_b}{1.5 \times 10^{12} \text{ Hz}})\text{ MeV}. \quad (2)$$

We find that this energy is in the 50 MeV to 30,000 MeV energy range of the EGRET instrument for typical SNR parameters ($B_{-5} > 0.3$, whereas $B_{-5} \sim 100$ for young remnants such as Kepler and Tycho - Reynolds & Ellison 1992).

Suppose the ambient soft photon density consists of N blackbodies and/or grey bodies with temperatures T_i, $i = 1, ..., N$. If the associated energy densities are comparable, we would expect to see several γ-ray bumps at energies

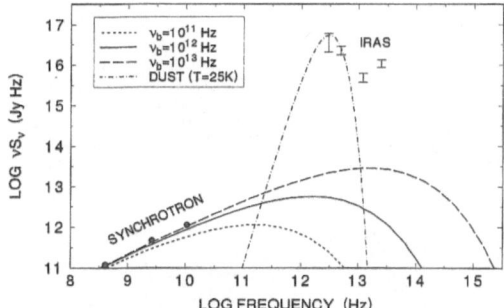

Figure 1. The synchrotron spectrum of W44 (Green 1988) corresponding to three possible cutoff frequencies. The error bars are the IRAS corrected flux values (Arendt 1989). A modified blackbody with $T = 25$K is also indicated.

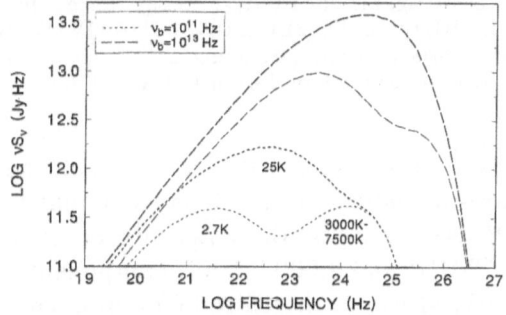

Figure 2. The IC spectrum of W44 corresponding to the two cutoff frequencies as indicated. *Thin lines*: the two bumps result from the scattering of the 2.7K CMBR and optical photon fields only. *Thick solid lines*: inclusion of the dominant 25K component.

$3kT_i\gamma_b^2$ associated with the synchrotron cutoff at ν_b. In our case, $T_1 = 2.76$K (the CMBR), $T_2 \sim 25$K (due to cold dust in older SNR, as well as the average galactic dust temperature - Skibo 1993), $T_3 = 3,000$K, $T_4 = 4,000$K, and $T_5 = 7,500$K (the latter three representing the various galactic optical photon fields - Skibo 1993).

4. The expected cold dust signature of W44

Fig. 1 shows three possible synchrotron spectra, corresponding to the cutoff frequencies as indicated. It is clear that observations above 10^{11} Hz will be dominated by thermal emission from swept-up dust grains. Apart from the cold 25K component, a warmer component may also be needed to explain the IRAS 12μm and 25μm emission.

Fig. 2 shows the expected IC spectra of W44 corresponding to the two cutoff frequencies as indicated. These spectra were calculated from the exact cross section given by BG, assuming a realistic field strength of $B = 10^{-5.2}$

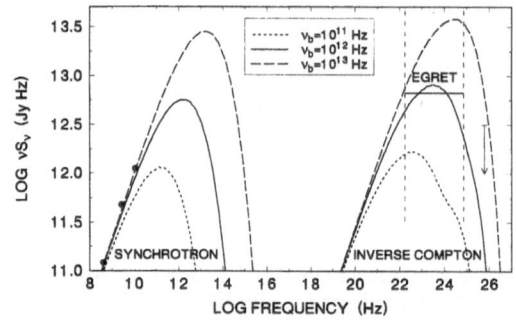

Figure 3. The synchrotron (as in Fig. 1) and corresponding IC spectra (assuming $B = 10^{-5.2}$ G) of W44 (as in Fig. 2). The EGRET energy range is indicated by two vertical dashed lines, and the EGRET flux for 2EG J1857+0118 (W44) given by Thompson et al. (1996) is shown by a horisontal line (assuming a E^{-2} photon spectrum). The arrow indicates the Whipple TeV upper limit by Lessard et al. (1995).

G. The γ-ray bumps are the result of IC scattering on thermal photon spectra with temperatures as indicated, and it is clear that the 25K component (galactic and SNR W44) dominates the γ-ray flux, compared to the contribution from the 2.7K CMBR and 3,000-7,500K components. A γ-ray detection of IC emission from W44 would therefore be dominated by the cold dust as shown by the thick lines.

Fig. 3 shows the synchrotron and the IC spectra as in Fig. 2, but including all the sources of soft photons as shown by the thick lines in Fig. 2. The spectra shown here correspond to three possible values of ν_b as indicated, and it is clear that the average EGRET flux from W44 may be explained if $B = 10^{-5.2}$ G, and $\nu_b \sim 10^{12}$ Hz. However, a detailed spectral fit is needed to derive the two free parameters B and ν_b, and to see if the spectral fits are unique compared to other competitive mechanisms such as relativistic bremsstrahlung (BG).

References

Arendt, R.G. (1989) *Astrophys. J. Suppl.* **70**, 181.

Blumenthal, G.R. & Gould, R.J. (1970) *Rev. Mod. Phys.* **42**, 237 (BG).

Dwek, E. & Arendt, R.G. (1992) *Ann. Rev. Astron. Astrophys.*, **30**, 11.

Green, D.A. (1988) *Astrophys. Space Science* **148**, 3.

Lessard, R.W. (1995) *Proc. 24th Int. Cosmic Ray Conf.* (Rome), **2**, 475.

Reynolds, S.P. & Ellison, D.C. (1992) *Astrophys. J.* **399**, L75.

Reynolds, S.P. *Proc. 24th Int. Cosmic Ray Conf.* (Rome), **2**, 17.

Skibo, J.G. (1993), *Diffuse galactic positron annihilation radiation and the underlying continuum*, Ph.D. Thesis, Univ. of Maryland.

Sturner, S.J. & Dermer, C.D. (1995) *Astron. Astrophys.* **293**, L17.

Thompson, D.J. et al. (1996), *Astrophys. J.*, in press.

OPTICAL AND INFRARED IMAGES OF GALAXIES: WHAT'S TO BE LEARNED?

JAY A. FROGEL, A. C. QUILLEN AND R. W. POGGE

*Department of Astronomy, The Ohio State University
Columbus, Ohio U.S.A.*

Abstract.

Multiwavelength imaging surveys of galaxies are of great value for investigations of their structure and stellar content. The advent of large format near-IR (NIR) arrays means that such surveys need no longer be confined to the optical. There are major advantages in concurrently carrying out a survey in the optical and NIR. While optical light traces the young blue stellar content of a galaxy, the NIR images will be effective tracers of the older stars and hence, as we will argue, of the stellar mass distribution. In the NIR the effects of dust extinction are minimized. The variation in appearance of structures such as spiral arms, bars, and disks as they are viewed from 0.4 to 2.2μm can set constraints on theories of their formation and evolution. The effects of internal dust absorption and scattering on the observed properties of a galaxy can be better studied. The effects of stellar and dynamical evolution on observations of distant galaxies can be better predicted. This review will give examples of applications of multicolor galaxy surveys and show some results from a partially completed OSU survey of ~230 spiral galaxies in *BVRJHK*. A new technique developed to detect dynamical structures in disk galaxies based on observed color gradients will be discussed in some detail.

1. Introduction: The Technical and Scientific Background

"Since the time of the Herschels, surveys of bright galaxies have provided the foundations upon which much of observational cosmology rests."
(Sandage & Tammann 1981)

D. L. Block and J. M. Greenberg (eds.), New Extragalactic Perspectives in the New South Africa, 65–83.
© 1996 *Kluwer Academic Publishers.*

1.1. THE TECHNICAL BACKGROUND

In astronomy our understanding is often limited by the quality of the data; for example insufficient signal to noise, or inability to resolve structures at critical physical scales. Getting around these limitations has been the driving force behind revolutionary advances in the tools of observational astronomy. Such advances take place in two ways: construction of new, large telescopes and innovative instrument design, particularly the development of new types of detectors. The development of the capability to image the sky at infrared wavelengths is an advance of the latter kind.

The past 25 years have seen the rapid development of increasingly sensitive infrared instrumentation, particularly in the NIR, 1–5μm. NIR and optical instrumentation have come to resemble one another more and more, but an important difference still sets special requirements for maximizing instrumental sensitivity in the infrared. Most astronomical observations are an attempt to measure an object against a background. In the visible this background is almost always due entirely to sources beyond the telescope – a few night sky lines, light scattering in the atmosphere, and the zodiacal light. In the NIR night sky emission lines are the dominant source of background shortward of 2.2μm, but they are so numerous and strong that their integrated effect produces a background surface brightness of $\geq 10^4$ times that in the optical. Longward of 2.2μm things get worse exponentially as thermal emission from both the sky and the telescope quickly begin to dominate. the telescope and instrument.

Until recently nearly all infrared observations were seriously limited by the near absence of array detectors. To get more than just the integrated brightness of an extended source at a particular wavelength, it was necessary to painstakingly construct a map one pixel at a time via some scanning technique (e.g. Hackwell & Schweizer 1983; Wainscoat *et al.* 1990). Since the beginning of the use of photographic plates more than 100 years ago, such a laborious approach to observing has been almost unheard of at optical wavelengths. About 8 years ago detector arrays finally became commercially available for use in the NIR. The first arrays were typically 58$\times$62 pixels. Now, 256$\times$256 arrays are "standard" and 1024$\times$1024 arrays should soon become generally available. Early examples of the great leaps forward made in IR imaging are illustrated in Gatley *et al.* (1988). Rix (1994) gives a quantitative comparison of the difficulties and advantages associated with optical and NIR observations of galaxies.

1.2. THE SCIENTIFIC BACKGROUND

There are two key arguments for the importance of obtaining images from 0.9 to 2.5μm of nearby spiral galaxies. The first has to do with their dust content; the second with their stellar mass content.

Observations in the Near-IR minimize the effects of dust. Light at NIR wavelengths emitted by a galaxy will be minimally affected by dust within the galaxy. Dust scattering and absorption can not only severely reduce the intensity and alter the spectral energy distribution of the light that is observed, but can also drastically alter a galaxy's observed morphology. However, extinction at K is only $\sim$10% that at V and $\sim$15% that at I. Images in the NIR of many galaxies have revealed major structural features, such as bars and broad, large amplitude arms that are invisible or only suggested by optical images. For example, although Spinrad and Harlan's (1973) I band photograph was the first detection of the bar in NGC 5195, NGC 5194's peculiar companion, it was not until the NIR observations by Spillar *et al.* (1992, see also Block *et al.* 1994) that one came to realize that NGC 5195's peculiar appearance is due entirely to the effects of the thick overlying dust lane and that in fact it is is a rather ordinary looking SBa or SB0 galaxy. Finally, the effects of dust extinction can be especially bad in regions of star formation such as spiral arms and galactic nuclei. As an extrememe case, consider the fact that from the center of the Milky Way 10^{10} more NIR than optical photons are able to make it to the earth and be detected.

Near-IR light traces the true mass distribution. With few exceptions the visible stellar mass of galaxies is dominated by intermediate ($\sim$1 Gyr) to old ($\geq$10 Gyr) stars. In such systems nearly all of the mass is in main sequence dwarfs while most of the luminosity, bolometric and NIR, comes from the first ascent giant branch and, in the case of galaxies with a sigificant intermediate age population, from asymptotic branch giants (Aaronson 1986; Frogel 1988; Frogel *et al.* 1990). On the other hand, features that tend to dominate optical images of spirals are delineated by young objects that contribute negligibly to the mass *Since first and second ascent giants are direct descendents of the dwarfs at the main sequence turnoff, mapping out the location of the giants will be equivalent to mapping out the mass distribution.* The fact that these giants dominate the NIR light of galaxies accounts for why this light is such a good tracer of the mass. The contribution of red supergiants (and other young stars) to the K light, which if large could render these arguments invalid, has been shown to be <20% (Rix & Zaritsky 1995; see also Rix & Rieke 1993).

Since NIR light is an effective tracer of the mass distribution in galaxies, it can often reveal important new information about their internal structure and dynamics. For example, de Jong (1995, 1996a) uses K band images to derive the fundamental parameters that characerize the bulk of the luminous matter in the bulges and disks of 86 spirals. His work is particularly relevant to Freeman's law for the central surface brightness of their disks.

We are developing new analytic techniques that compare isophotes and isochromes to map out the gravitational potential field in the presence of nonaxisymmetric mass distributions and to explore the internal dynamics of galaxies (Quillen *et al.* 1995a,b; 1996).

In addition to the great value of NIR observations in the study of nearby galaxies, they play a critical role in work on distant galaxies (*cf.* McCarthy 1993) The optical light from galaxies at large z probably comes from an unrepresentative population of young hot stars. In contrast, Lilly (1989) has shown that even radio galaxies at $z \sim 2$ are dominated by an underlying, old stellar population. By analogy with their presumed counterparts in nearby galaxies, the longer the wavelength of the obsevations, the greater will be the relative contribution from this old population. Thus the NIR will be best suited for observing light from this old population and determining the morphology and luminosity of these distant galaxies. The global properties of nearby galaxies will then be a valuable guide in interpreting the observations of the galaxies at large redshift.

Given the importance of NIR imagery for the study of spiral galaxies and the recent advent of appropriate technology, it is not surprising that there are a number of on-going or recently completed NIR surveys of these and other types of galaxies (eg. Thronson *et al.* (1989), Peletier (1989) Wainscoat *et al.* (1990), Peletier *et al.* (1994), Terndrup *et al.* (1994 – initial results from the OSU survey), Rix & Zaritsky (1995), Grauer *et al.* (1995), deJong (1995), Heraudeau *et al.* (1996)). In this review we will discuss topics that bear directly on the two key arguments for the value of of NIR observing mentioned above. We will treat dust not just as a hinderance in studing galaxies, but as an objective of several research projects. Then we will summarize one of the new approaches we are taking to explore galaxy dynamics based on the OSU survey. The review will conclude with a brief overview of the OSU galaxy survey itself together with our opinion on what directions we think further work of this sort should take.

2. Spiral Arms, Bars, and a Galaxy's Mass Distribution

A spiral galaxy's appearance in the NIR can be radically different than in the optical. NIR images reveal structures such as broad arms and bars that must contain significant amounts of matter (Figures 1–3 here). Such structures may be more common than previously thought (*cf.* Franx 1994). Block *et al.* (1994) give examples that illustrate the presence of such features. One reason for these differences in appearance is extinction which is still substantial even at I. Rix & Rieke (1993) pointed out, τ_I is 6 times greater than τ_K. Since there is considerably more dust in the arms than in the inter-arm regions, observations made at I such as are described below

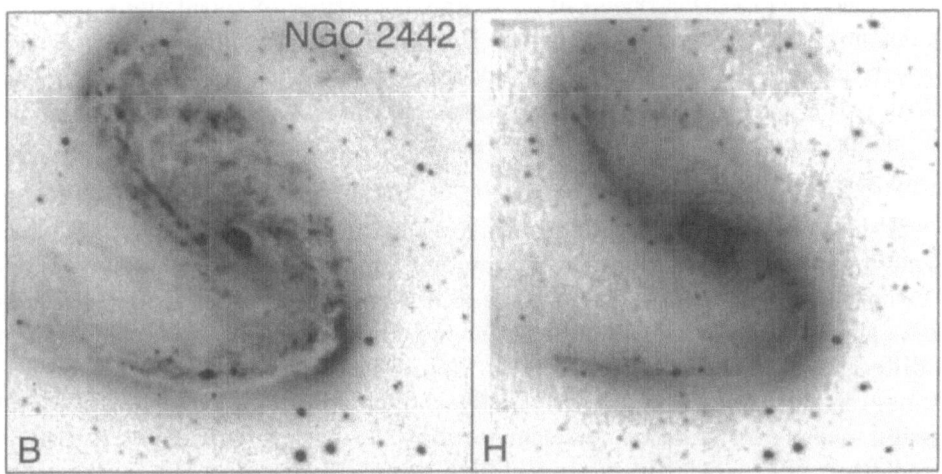

Figure 1. NGC 2442 is clasified SBbc(rs)II in the *RSA*. This pair of *B* and *H* images have the same logarithmic stretch. The *B* image has been precisely shifted, binned, and smoothed to match it to the *H* image. These images should be viewed together with their ratio image in Figure 4. On the *H* image note the greatly increased smoothness, the reduced evidence for dust, the much more prominent nuclear bulge and spiral arms, the sharply defined, long, nearly straight arm, and the near absence of the bright clumps of material above the central bulge visible on the *B* image. See also color plate 6.

will often significantly underestimate the arm to disk contrast (see also Rix & Zaritsky 1995). The second reason is that since the stars that are the dominant source of NIR light are relatively old, they will have been both dynamically well mixed and heated to a relatively large scale height. Combined with differential rotation, the net effect is that a galaxy's appearance in the NIR will be much smoother than in the optical where young stars, HII regions, and dust – the classical tracers of spiral arms but certainly not necessarily representative of the overall mass distribtion – are important contributors. The remainder of this section disucsses these issues further.

2.1. SPIRAL ARMS

Perhaps the first published evidence for a spiral pattern in the red, old disk population is by Zwicky (1955). He showed with photographic subtraction techniques that although the yellow-red and blue stellar populations of NGC 5194 both lay in the disk, they had radically different distributions. The yellow-red stars have, according to Zwicky, "achieved a much higher degree of organization than the blue stars." In modern parlance, they are more relaxed. As is evident from his photographs, the yellow-red stars define a much smoother and thicker spiral pattern with considerably less detailed structure than do the blue stars (Figure 3).

Schweizer (1976) carried out an exhaustive study of spiral structure as a function of wavelength in 6 galaxies. He found a broad spiral pattern in red light from the old disks and attributed the pattern to a 20–30% enhancement in the surface mass density of the old stars, the main constituent of the disks. He further concluded that the red light arises primarily from the giant component of this population.

Elmegreen (1981) classified spirals as "grand-design" which have two-armed symmetry with smooth, continuous, long arms or "flocculent" which have multiple short, broken-up arms. She suggests that the former are due to a spiral density wave, and the latter to self-propagating star formation in a differentially rotating disk (see, $eg.$ Toomre 1981 and references therein). Elmegreen & Elmegreen's (1984) analysis of 34 spirals showed that in grand design ones, strong arms are seen at both B and I, consistent with the prediction that both the young stellar population seen at B, a tracer of recent star formation, and the old population observed at I, a tracer of total stellar mass, should have their surface density distributions perturbed by a spiral density wave. On the other hand, they found that flocculent spirals in I show very little of the structure detected at B implying that in these galaxies the arm structure is primarily due to young stars not to variations in the surface mass density; old stars are largely unaffected by self-propagating star formation patterns in the differentially rotating disk.

NIR observations have revealed the complexity of the issue of three armed spirals (eg. Elmegreen et $al.$ 1992). Rix & Rieke demonstrated via a Fourier analysis of the azimuthal distribution of light in M51 that while an m=3 component is prominent in the visible it is insignificant in the K band. Images of NGC 309 (Block & Wainscoat 1991) in the optical and at K give another particularly striking example of the "disappearance" of a third arm at K. Thus, while for these two galaxies the "third arm" is due primarily to dust and young stars, not to mass enhancement, Rix & Zaritsky (1995) do find K band third arms in 3 of their sample of 18 spirals. Perhaps related to this issue is the finding by Block et $al.$ of a number of galaxies with prominent single arms or m=1 components in the NIR.

Finally, an interesting result of relevance from the OGLE Project's search for micro-lensing events towards Baade's Window is the detection of a narrow, well populated, main sequence $\sim$ 2 kpc from the sun (Paczynski et $al.$ 1994) corresponding in location to the Sagittarius arm. Although knowledge of this arm has been based almost exclusively on young objects (e.g. supergiants, HII regions, etc.) the results of Paczynski et $al.$ indicate a strong concentration of old stars with a density enhancment of at least twice that predicted in the absence of a spiral arm, or as strong as the arm, inter-arm contrast in M51 at K (Rix & Rieke (1993).

2.2. BARS

Pfenniger & Norman (1990) have suggested that the dynamical interaction of a bar and the central mass concentration in a galaxy could cause vertical heating of the disk that in turn would cause the addition of a relatively young, metal-rich, and rotationally supported component to the galaxy's bulge. Dynamically, then, bulges of barred galaxies would be quite different from bulges of non-barred galaxies (Kormendy 1982a). Numerical simulations by Friedli et al. (1994) show that the presence of a bar will result in a significant flattening of the abundance gradient in a galaxy over only a few dynamical timescales. They suggest that non-barred galaxies with little or no abundance gradient may contain a hidden bar. Obviously, these would be good candidates for imaging in the NIR.

Determination of the true frequency of bars and other nonaxisymmetric structures is important since they could be one way of initiating a spiral density wave (Toomre 1969; Sanders & Huntley 1976; Kormendy & Norman 1979; see also Kormendy 1982a) or stellar rings (Buta 1986). Most of the disks in Rix & Zaritsky's (1995) sample show some form of nonaxisymmetric structure. Optical surveys show that a relatively high percentage of Seyfert and starburst galaxies have bars (Adams 1977; Simkin et al. 1980; MacKenty 1990; Jackson et al. 1987). Bars may provide a means to channel gas into the nuclear regions of galaxies and cause enhanced star formation or an AGN (Kormendy 1982b; Combes & Gerin 1985; Shlosman et al. 1989; Pfenniger & Norman 1990; Sakamoto et al. 1995). While Pompea and Rieke (1990) find that bars are not a necessary requirement for nuclear activity as measured by high far-IR luminosity, Moles et al.'s (1995) survey of ~200 active galaxies shows that some sort of nonaxisymmetric structure is necessary.

One of the first bars discovered in the NIR was in NGC 1566 by Hackwell & Schweizer (1983) who mapped this galaxy by laboriously scanning a single element detector. The bar, which is barely visible in the blue, produces "azimuthal brightness variations of up to ~ 25% at H." They concluded, in agreement with Schweizer's (1976) earlier study, that these brightness variations reflected true mass variations since the H light is sampling the dominant old stellar population. Block & Wainscoat (1991) identified a prominent bar in NGC 309. Figure 2 of NGC 5161 shows that this galaxy too has an infrared bar. Much more commonly discovered via NIR imaging are *small* central bars; examples are NGC 1068 (Scoville et al. 1988; Thronson et al. 1989), M100 (NGC 4321; Sakamoto et al. 1995; see also Knapen et al. 1995), NGC 4314 (Benedict et al. 1992; see also Quillen et al. 1994a), M81 (Elmegreen et al. [1995] confirmed its presence after the suggestion by Reichen et al. [1994] based on R band images), and NGC 1097 (Quillen et

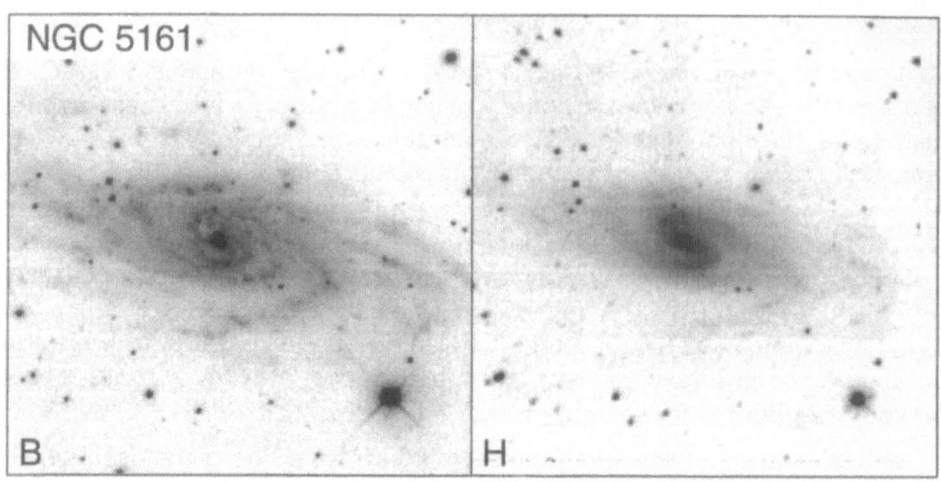

Figure 2.   NGC 5161 is a new example of a NIR bar. is classified as Sc(s)I in the *RSA*. Technical details are the same as for Figure 1. The disk at *H* is much smoother than it is in *B*. Instead of multiple, narrow arms, the *H* image is dominated by two rather broad arms emerging from the ends of a short bar that is masked by the central dust lanes on the *B* image. With considerably less dust than in NGC 2442 (Figure 1), the nuclear bulge in *H* is only slightly stronger than in the optical.

al. 1995). When these small inner bars are found in galaxies with large outer bars, there is no preferred angle of alignment of the two. The consequences of this are discussed by Quillen *et al.* (1995).

3. Dust in Galaxies and Its Effects

From the simple fact that at low galactic latitudes the Milky Way is nearly opaque in some directions while at high latitudes it is nearly transparent, one can conclude that the extinction and reddening observed in other spirals will be a strong function of inclination angle. This can be a major problem if you want to study the stellar content, but a major advantage if you are interested in a galaxy's dust content.

The disks of spiral galaxies are generally assumed to have low internal visual extinction perpendicular to their planes based primarily on the dependence of visual surface brightness on inclination angle (Holmberg 1958). Although this assumption has been challenged by Disney *et al.* (1989) based on models of the dust distribution and by Valentijn (1990) based on a statistical analysis similar to Holmberg's but with significantly more galaxies, most all other studies agree with Holmberg's original conclusion (but see Burstein *et al.* 1991 and Peletier *et al.* 1994). For example, with dynamical arguments Bosma *et al.* (1992) shows that the outer parts of the disks of

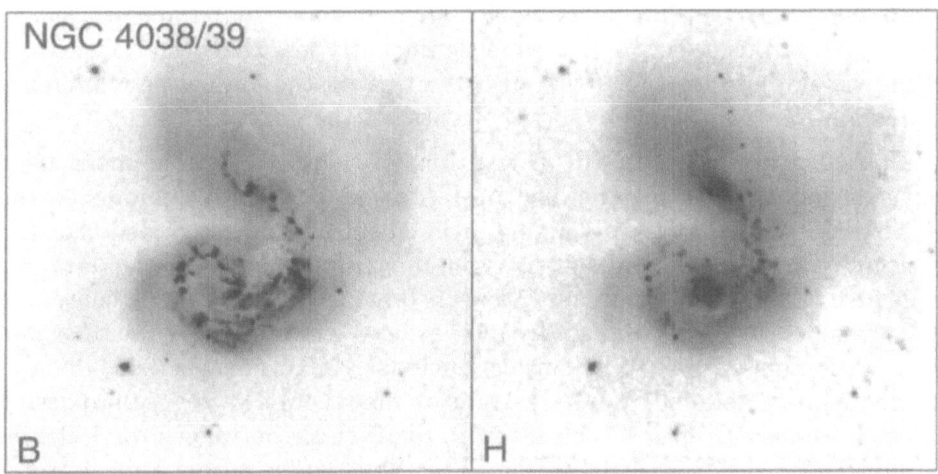

NGC 4038/39

B

H

Figure 3. The famous "Antennae" galaxies; technical details as for Figure 1. This pair of images should be viewed together with the ratio of the two in Figure 4. These galaxies show extreme effects from dust absorption even in the *H* band. Note the following: The smooth outer envelope at both wavelengths. The upper envelope, though, is clearly redder. Both nuclei are heavily extincted, particularly the bottom one which is crossed by nearly opaque dust lanes in the *B* image. There is a very extensive dust cloud between the two nuclei. Many intense, compact, blue regions of star formation are clearly visible. Finally, note that the short arm (?) curving to the left and upward from the upper nucleus is very blue. For a 3-color HRB rendition, see color plate 5 in these *Proceedings*.

spirals are transparent as are the inner parts with a few exceptions. Peletier & Willner (1992) have analyzed H band surface brightness vs. inclination angle and find that τ_H is 0.1 through a typical face-on spiral while τ_B is 1. Rix (1994) summarizes estimates for τ_V in face on luminous spiral disks and finds a mean value of 0.5 to 1.

Determination of the light distribution perpendicular to the disks of galaxies almost always needs to be based upon observations of edge-on galaxies. If these studies are carried out at optical wavelengths, they usually are confined to dust free galaxies, (e.g. extreme lenticulars by Hambe & Wakamatsu 1989), or exclude regions closest to the dusty central plane (e.g., van der Kruit & Searle 1981). Hambe & Wakamatsu speculated that the extremely thin disks they found could be edge-on bars. Such structures would be undetectable in the presence of a central dust layer, which is of course the case for nearly all edge-on disk galaxies. With NIR data one can better sample the true underlying light distribution. For example Wainscoat *et al.* (1990) made NIR maps of three edge-on S0 galaxies with prominent dust lanes and found that all three disks are much more important than expected from optical images alone, even with reasonable allowance for extinction. Barnaby & Thronson (1992) applied van der Kruit & Searle's technique to H band images of NGC 5907 and found that the scale height and length of this galaxy's disk were significantly less than when measured in the visible because the effects of extinction on the optical light increase as one approaches the plane and the center of the galaxy.

Together, optical and NIR observations of nearly edge-on galaxies are a powerful tool in studying the dust itself. Jansen et $al.$ (1994) modelled the O/NIR light distribution in 4 highly inclined spirals and concluded that the dust and stars were well mixed. A screen model for the dust fit the data less well. Also although τ_V was found to be as large as 10 towards the centers of their galaxies, it decreased rapidly outwards. With models for the dust and light in 15 edge-on Sab to Sc spirals Kuchinski & Terndrup (1996) demonstrated that a detailed radiative transfer model for the starlight passing through a galaxy's disk is necessary to distinguish between broad classes of models for the dust distribution. They find τ_V to be between 4 and 7 for nearly edge-on disks and as high as 15 for exactly edge-on cases and that scattering cannot be significant in these galaxies as it does not produce enough reddening. Rix & Rieke (1993) and Jansen et $al.$ pointed out that scattering can be neglected when analyzing light from nearly edge-on systems because nearly all scattered photons that escape from the galaxy are moving perpendicular to the plane. An interesting consequence of this is that there is little change in the color of light that emerges perpendicular to the disk (Witt et $al.$ 1992). This effect may contribute to the high albedo deduced for dust in the K band (Witt & et al. 1994). Giovanelli et $al.$ (1994) find a maximum τ_I of 5 at the centers of the disks of a sample of ~2000 spirals with somewhat smaller inclination angles than those examined by Kuchinski & Terndrup. Consistent with Jansen et $al.$, Giovanelli et $al.$ find that outside of 2–3 scale lengths the disks are almost completely transparent. They also deduce that the scale height of the absorbing layer is half that of the stellar component, similar to Rix & Rieke's result for M51. Wainscoat et $al.$ (1990), on the other hand, found the scale heights of dust and stars to be very similar in 3 galaxies.

4. Stellar Populations of Galaxies

Radial gradients in stellar absorption lines are difficult to detect because of the rapid drop of surface brightness with radius in galaxies. Their detection optically by Davies & Sadler (1987) and Davies et $al.$ (1993) in early-type galaxies demonstrated that line gradients closely follow color gradients. Although colors alone cannot furnish the detailed information that spectroscopy can, they can be measured even in the NIR over the entire face of a galaxy – including ones of low surface brightness and great distance – to high precision. Indeed, information on stellar populations that can be

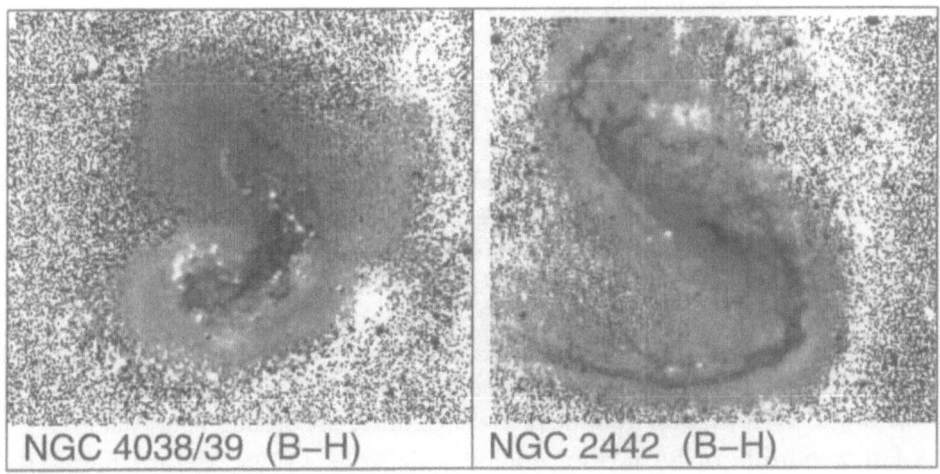

Figure 4. Each of these images is the log of the ratio of the *B* and *H* images from Figures 1 and 3. Both have the same scaling so that the degrees of redness can be compared directly. Dark is red, white is blue. The dust lanes in NGC 2442 are sharply delineated which suggests that they are associated with a shock front. The strong bifrucation of the lower arm may be due to two pattern speeds. The scimtar-like dust feature above and to the right of the central bulge probably indicates that that whole side of the galaxy is closer to us than the lower left part. On the antennae, note how uniform the color is of the outer envelope of the two galaxies. The very blue nearly unresolved clumps stand out quite strongly.

gleaned from imaging data has a big advantage since the few constraints set by such data are generally well defined empirically and easily understood in terms of the gross physical properties of the stellar, gaseous, and dust content (e.g., Aaronson 1977; Frogel *et al.* 1978; Frogel 1985; Impey *et al.* 1986; de Jong 1995, 1996b). There is the further advantage that if a spectral synthesis model is incapable of meeting the "simple" constraints set by colors and extinction alone, there is little sense in further elaboration of such a model. One can use different types of building blocks, particularly star clusters, to decompose the integrated colors of galaxies into more narrowly defined components (e.g. Frogel 1985; Frogel *et al.* 1990; Bica 1988; Bica *et al. 1988*).

With the addition of NIR data to optical images, constraints that can be set on the component stellar populations become significantly stronger. Limits can be set on the initial mass function from observations of luminosity sensitive NIR indices such as CO and H_2O (Tinsley 1972; Aaronson 1977; Frogel *et al.* 1978; Frogel and Whitford 1987). Of particular value is the sensitivity of NIR colors to changes in the cool giant and supergiant populations that result from changes in age and chemical composition. Persson, *et al.* (1983) showed that colors involving the *H* band are especially sensitive to the presence of red AGB stars.

Franx (1994) argued that the age-metallicity degeneracy is too great to be sorted out by broad band color observations alone. Indeed, with optical colors alone Balcells & Peletier (1994) could not easily tell if the bluer colors of Sa and Sb spiral bulges relative to ellipticals of similar luminosity meant that the bulges were younger or more metal poor. However, multicomponent analysis (*eg.* Faber 1973 and Deeming 1964) of blue and NIR colors together for late-type spirals shows that some decoupling can be achieved between reddening, age, and metallicity effects (cf. Frogel 1985). With such data Frogel (1985) convincingly demonstrated that the relatively blue colors of the central regions of Sc bulges were due primarily to an intermediate age or young stellar component rather than to a metal poor one and that the optical and NIR colors were each tracing different stellar components.

For a sample of 86 face-on galaxies de Jong (1995, 1996b) finds that color gradients "are best explained by a combined stellar age and metallicity gradient" with dust playing only a minor role; outer parts of the galaxies tend to be younger and more metal poor. Terndrup *et al.* (1994) also interpret radial color changes in their sample of 43 galaxies in terms of metallicity or age gradients as do Rix & Rieke (1993) for M51.

5. Galaxy Dynamics With Infrared Images and Color Maps

Large NIR arrays have opened a new window into the study of galaxy dynamics because NIR light is a better tracer of a galaxy's luminous matter than is optical light. As stars age they are heated by spiral density waves and collisions with molecular clouds and are mixed into the galaxy by differential rotation. Thus, not only is the stellar population which dominates the infrared images an old one, but it also is primarily a dynamically relaxed one which has been smoothly distributed azimuthally about the galaxy. This is in contrast to optical images in which young objects such as star clusters and HII regions often dominate. These young objects can give a galaxy a blotchy appearance as they will be observed close to their place of birth and will not have been mixed into the general background.

Since near-IR images are a good tracer of mass they can be used to estimate the galaxy gravitational potential which controls the large scale dynamics of both stars and gas. This is despite potential problems such as a varying IR mass-to-light ratio from unknown amounts of dark matter or from changes in stellar population including recent star formation, uncertainty in the 3-dimensional structure of the galaxy, and the effects of dust in regions of high dust column depth. A comparison of the predicted shapes of stellar orbits to observed quantities such as isophotes and isochromes can be used to constrain and study these uncertainties. A preliminary example of such a study can be found in Quillen *et al.* (1994).

Because of the large optical to infrared color gradients that exist in some and perhaps most spiral galaxies, dust models that predict colors over a large wavelength range at different positions in a galaxy must take into account variations in stellar population. For example, even though dust extinction is probably low in the dust lanes in NGC 1097 (which are practically invisible in the near-IR images), large color gradients are observed in this galaxy which are most likely due to a variation in the stellar population (Quillen *et al.* 1995).

Color gradients in galaxies due to population changes can be used to study galaxy dynamics (Quillen *et al.* 1996). For example when there is a bar or other non-axisymmetric shape to the gravitational potential, the closed stellar orbits are no longer circular; a star's speed will depend upon where it is in its orbit. Generally, it will be slowest near the ends of the major axis of the orbit. When the galaxy is dynamically relaxed and has a small stellar velocity dispersion, the morphology of the galaxy is primarily determined by the structure of these closed orbits. However, the varying speed about the non-circular orbits causes the stellar density to vary about the orbit as well. This implies that in a non-axisymmetric galaxy the isophotes – equivalent to contours of constant stellar density – are not coincident with the closed orbits, as is true in an axisymmetric galaxy.

Consider two idealized cases. First, a galaxy with an axisymmetric potential. For a small velocity dispersion all closed orbits will be circular. The speed associated with an orbit will depend only on the orbit's radius, not on position angle. If we assume that the defining characteristics of the stellar population are a function only of radius and the color at any point in the galaxy is a function only of stellar population (neglecting dust), then isochromes will be circular and identical to the isophotes. For the second case consider a galaxy with a non-axisymmetric potential, again with a small velocity dispersion. The only closed orbits will be ellipses. In interests of simplicity, take orbits of different ellipticities to be concentric, never crossing one another. The degree of ellipticity will be a function of radius. Differential rotation will have smoothed out any azimuthal variation in the stellar population, at least in an old one such as NIR light comes from. Thus, a unique stellar population can be associated with orbits of a give ellipticity, i.e. the stellar population will be constant about each closed orbit. Contours of constant color, the isochromes, will then be coincident with the closed orbits. Because of the varying speed over each orbit, the stellar density around the orbit will vary as well. Thus the isophotes will have to connect points on different orbits rather than on the same orbit and their shapes will differ from those of the isochromes. The magnitude and shape of the nonaxisymmetric part of the gravitational potential will determine

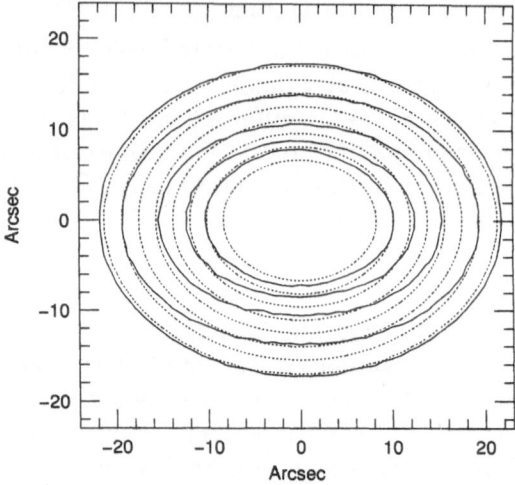

Figure 5. Isophotes (solid lines) and isochromes (dotted lines) for a model galaxy with a flat rotation curve and a gravitational potential with ellipticity $\epsilon = 0.1$. The disk is exponential with a scale length that varies as a function of wavelength because of a population gradient that causes a color gradient. The isophotes were constructed from closed stellar orbits found by numerical integration. The isophotes differ from the isochromes because the speed of stars along the closed orbits varies.

the difference in shape between the isophotes and isochromes at any give radius. Detection of this effect can be used to study the shape of the gravitational potential in non-axisymmetric systems (see figure for illustration).

Barred ring galaxies are a particularly promising place to look for such an effect in the NIR. In such a galaxy typically the contribution from dark matter is significant at the location of the ring. NIR images of these rings do show density variations about the ring (e.g. NGC 3351). Measuring the shape (or ellipticity) of the potential as a function of distance from the nucleus will allow us to set strong constraints on the distribution of dark matter.

6. The OSU Galaxy Survey

We have selected about 230 non-E and non-S0 galaxies from the *Third Reference Catalog of Bright Galaxies* as the primary targets of the OSU galaxy survey. Aside from excluding Es and S0s, we had only two selection criteria: diameters had to be $\leq 6'$ (to fit on our detectors) and have total B magnitudes brighter than 12. According to the RC3, the catalog is at least 95% complete for galaxies this bright. When necessary we have added some additional, generally fainter galaxies of the latest types so as to get a reasonable sample of them. We have also added a few edge-on and face-on galaxies bigger than our size limit for special projects. *JHK* images are being obtained with OSIRIS (the Ohio State Infra Red Imager and Spectrometer) primarily on our 1.8 m at Lowell Observatory, and with CIRIM at CTIO. Additional NIR data have been obtained with IRIM at Las Campanas Observatory. All 3 cameras use 256×256 NICMOS chips. *BVR* images are being obtained with our IFPS (Imaging Fabry-Perot Spectrometer) on the 1.8 m and at CTIO. As of 1996 January we have obtained 6-color images for about half of our sample. At the conclusion of our survey (by mid-1997) we expect to make the final photometrically calibrated images available to the astronomical community in a digital form.

The exceptionally large field of view of OSIRIS on the 1.8-m, 6.6', is particularly valuable for the survey because of the ability to record signal in the faint outer regions of the galaxies. In these regions, with 2×2 pixel binning we achieve a S/N of 4 in 1 second of integrating at brightness levels of 21.6 and 20.1 at J and K, respectively. The value for H is intermediate. Our exposures typically are 600 seconds on source at J and H and 1200 seconds at K. With CIRIM the S/N is about 0.3 mag less so we expose about 1.5× longer to achieve comparable sensitivity.

7. Future Directions

Many important areas of extragalactic research would benefit from a concurrent optical and NIR imaging survey of late type galaxies. Although the OSU survey and several others that are currently being carried out have specifically targeted some of these research areas, there is no shortage of exciting topics to work on and large telescopes are not needed. In what follows when we talk about "a survey" we will mean a combined optical, NIR survey.

Observations in the NIR have revealed structures that are invisible or at best only hinted at in I band images. It should now be possible to set up a simple but completely quantitative classification sequence for galaxies based on what is seen in the NIR images. Such a scheme would be based on the true mass distribution in spiral and other late type systems. The sequence developed would be multi-dimensional as it should measure or test for the presence of things such as strength (amplitude) of arms, number of arms, tightness of winding, non-axisymmetric features such as bars, luminosity profile, color gradients, and central concentration of light, i.e. something like a bulge to disk ratio but without assuming the existence of either. It should also contain a parameter to measure the structural differences between the optical and NIR; such a parameter would in effect be a global measure of star formation activity. Other global parameters of obvious importance would be integrated luminosity and color. Selection of galaxies for the sample must be done carefully and in an unbiased fashion as possible. We think that it would be a sobering experience to match up such a scheme with the Hubble sequence.

Questions that could be addressed with the type of classification scheme we have sketched out would include: How does the density enhancement in spiral arms and bars relate to star formation activity, degree of interaction with neighboring galaxy, morphological type, and luminosity? What is the most meaningful definition of galaxy type? Do population gradients as indicated by color variations, depend on type? How does the "average" stellar population vary with type? How does the surface brightness or mass distribution, vary with galaxy type and luminosity? Can the concept of components such as bulge and disk be more quantitatively defined and how does their stellar content and relative masses vary with type?

Finally, such a quantitative description of properties of galaxies should contribute to the development of numerical simulations of spiral structure and to the resolution of discrepancies between such simulations and real galaxies, e.g. the prediction that Sc galaxies should have many closely spaced arms compared with observations that show few distantly spaced arms (Toomre 1990).

We have already discussed the importance of an accurate determination of the frequency of occurance of bars in galaxies of different types. Here we point out one other interesting experiement concerning gas in bars. If gas transport along the bar has resulted in star formation over the past Gyr or so, the resulting stellar population could be easily detected via *VJHK* colors (Persson *et al.* 1983) becuase of luminous AGB stars.

The attempt to study dust properties in galaxies by observing a sample with various inclination angles is not without serious problems (eg. Burstein *et al.* 1991), not the least of which has to do with how the sample is chosen. However, we think a potentially useful avenue to explore via a joint survey would be based on a sample with all inclination angles where the angle itself did not enter into the selection criteria. In addition we are suggesting that the sample be selected so that population differences between galaxies are minimized. In this way, one could assume that changes in color and optical luminosity with inclination angle was do to the dust itself rather than to a change in the population. Selection criteria such as size or luminosity would have to be based on NIR properties alone so as to minimize the possibility of introducing biases into the sample.

Finally, we consider two technques that are applicable to individual cases. Many highly inclined galaxies show a distinct asymmetrical appearance. Commonly, the dust lanes on one side of the bulge appear more prominent than those on the other (Figs. 1 and 2). This is a projection effect because the dust that is in front of the galaxy will be more obvious due to the larger number of stars that are behind it. In this case, screening becomes important. It should be possible to use this fact to address issues such as dust properties and distribution (eg. Witt *et al.* 1994). A second technique reqires two galaxies to be overapping in such a way that that one of them is observed through the other. NGC 2207 is an excellent example (*cf.* Elmegreen *et al.* 1995): images taken during the course of the OSU survey clearly show the dust lanes in the arms of NGC 2207 crossing the face of its companion IC 2613. Cases like this, particularly if only part of the background galaxy is obscured, should prove to be a powerful tool in studying the dust content in the foreground galaxy.

The OSU galaxy survey is supported by NSF grant AST 92-17716. JAF's research on stellar populations is supported in part by NSF grant AST 92-18281. OSIRIS was built and supported by NSF grants AST 90-16112 and AST 92-18449. We are grateful to the other members of the OSU galaxy survey team for their participation: R. Davies, D. DePoy, K. Sellgren, and D. Terndrup. We are particularly grateful to M. Houdashelt, S. V. Ramirez, G. Tiede, & R. Aviles for substantial contributions to data gathering and reduction, to R. S. deJong and R. Davies for comments on a draft of this paper, and to L Kuchinski for discussions. JAF acknowledges the hospitality of the Department of Physics, University of Durham, England and the support of a Senior Visiting Research Fellowship during completion of this paper.

References

Aaronson, M. 1977, Ph.D. thesis, Harvard University.

Aaronson, M. 1986, in "Stellar Populations", ed. C. A. Norman, A. Renzini, & M. Tosi (Cambridge: Cambridge Univ. Press), 45.

Adams, T. F. 1977, *ApJS*, **33**, 19.

Balcells, M., & Peletier, R. F. 1994, *AJ*, **107**, 135

Barnaby, D., & Thronson, Jr., J. A. 1992, *AJ*, **103**, 41.

Benedict, G. F., Higdon, J. L., Tollestrup, E. V., Hahn, J. M., & Harvey, P. M. 1992 *AJ*, **103**, 757.

Bica, E. 1988, *A&A*, **195**, 76.

Bica, E., Arimoto, N., & Alloin D. 1988, *A&A*, **202**, 8.

Block, D. L., Bertin, G., Stockton, A., Grosbol, P., Moorwood, A. F. M., & Peletier, R. F. 1994, *A&A*, **288**, 365.

Block, D. L., & Wainscoat, R. J. 1991, it Nature, **353**, 48.

urstein, D., Haynes, M. P., & Faber, S. M. 1991, *Nature*, **353**, 515

Buta, R. 1986, *ApJS*, **61**, 609.

Combes, F., & Gerin, M. 1985, *A&A*, **150**, 327.

Davies, R. L., Sadler, E., & Peletier, R. F. 1993, *MNRAS*, **262**, 650.

Deeming, T. J. 1964, *MNRAS*, **127**, 35.

de Jong, R. S. 1995, Ph.D. Thesis, Rijksuniversiteit Groningen.

de Jong, R. S. 1996a, *A&A*, in press.

de Jong, R. S. 1996b, *A&A*, submitted.

DePoy, D. L. 1996, this conference.

Disney, M. J., Davies, J., & Phillips, S. 1989, *MNRAS*, **239**, 939.

Elmegreen, B. G., Sundin, M., Kaufman, M., Brinks, E., & Elmegreen, D. M. 1995, *ApJ*, **453**, 139.

Elmegreen, B. G., Elmegreen, D. M., & Montenegro, L. 1992, *ApJS*, **79**, 37.

Elmegreen, D. M. 1981, *ApJS*, **47**, 229.

Elmegreen, D. M., Chromey, F. R., & Johnson, C. O. 1995, *AJ*, **110**, 2102

Elmegreen, D. M., & Elmegreen, B. 1984, *ApJS*, **54**, 127.

Faber, S. M. 1973, *ApJ*, **179**, 731.

Franx, M. 1994, in IAU Symposium 153, "Galactic Bulges", eds. H. DeJonghe & H. Habing (Dordrecht: Reidel)

Friedli, D., Benz, W., & Kennicutt, R. 1994, *ApJL*, **430**, L105.

Frogel, J. A. 1985, *ApJ*, **298**, 528.

Frogel, J. A. 1988, *ARA&A*, **26**, 51.

Frogel, J. A., Mould, J., & Blanco, V. M. 1990, *ApJ*, **352**, 96.

Frogel, J. A., Persson, S. E., Aaronson, M., & Matthews, K. 1978, *ApJ*, **220**, 75.

Frogel, J. A., Terndrup, D. M., Blanco, V. M., & Whitford, A. E. 1990, *ApJ*, **353**, 494.

Frogel, J. A., & Whitford, A. E. 1987, *ApJ*, **320**, 199.

Gatley, I., DePoy, D. L., & Fowler, A. M. 1988, *Science*, **242**, 1264..

Giovanelli, R., Haynes, M. P., Salzer, J. J., Wegner, G., Da Costa, L. N., & Freudling, W. 1994, *AJ*, **107**, 2036.

Grauer, A. D., Rieke, M. J., & McLeod, K. K. 1995, *ASP Conf. Proc*, **73**, 195.

Hackwell, J. A., & Schweizer, F. 1983, *ApJ*, **265**, 643.

Hambe, M., & Wakamatsu, K. 1989, *ApJ*, **339**, 783.

Heraudeau, P., Simien, F., & Mamon, G. A. 1996, *A&AS*, in press.

Holmberg, E. 1958, *Medn. Lunds Astr. Obs.*, **2**, 136.

Impey, C. D., Wynn-Williams, G., & Becklin, E. E. 1986, *ApJ*, **309**, 572.

Jackson, J. M., Barrett, A. H., Armstrong, J. T., & Ho, P. T. P. 1987, *AJ*, **93**, 531.

Jansen, R. A., Knapen, J. H., Beckman, J. E., & Peletier, R. F., & Hes, R. 1994, *MNRAS*, **270**, 373

Kaufman, M., Bash, F. N., Hine, B., Rots, A. H., Elmegreen, D. M., & Hodge, P. W. 1989, *ApJ*, **345**, 674.

Knapen, J. H., Beckman, J. E., Shlosman, I., Peletier, R. F., Heller, C. H., & de Jong, R. S. 1995, *ApJ*, **443**, L73.

Kormendy, J. 1982a, in "Morphology & Dynamics of Galaxies", eds. J. Binney, J. Kormendy, & S. D. M. White (Sauverny: Geneva Obs.), p. 115.

Kormendy, J. 1982b, *ApJ*, **257**, 75.

Kormendy, J., & Norman, C. A. 1979, *ApJ*, **233**, 539.

Kuchinski, L. E., & Terndrup, D. M. 1996, *ApJ*, in press.

Lilly, S. J. 1989, *ApJ*, **340**, 77.

MacKenty, J. W. 1990, *ApJS*, **72**, 231.

McCarthy, P. J. *ARA&A*, **31**, 639.

Moles, M., Marquez, I., & Perez, E. 1995, *ApJ*, **438**, 604.

Paczynski, B., Stanek, K. Z., Udalski, A., Szymanski, M., Kaluzny, J., Kubiak, M., & Mateo, M. 1994, *AJ*, **107**, 2060

Peletier, R. 1989, Ph.D. Thesis, Rijksuniversiteit Groningen.

Peletier, R. F., Valentijn, E. A., Moorwood, A. F. M., & Freudling, W. 1994, *A&AS*, **108**, 621.

Peletier, R. F., & Willner, S. P. 1992, *AJ*, **103**, 1761.

Persson, S. E., Aaronson, M., Cohen, J. G., Frogel, J. A., and Matthews, K. 1983, *ApJ*, **266**, 105

Pfenniger, D., & Norman, C. 1990, *ApJ*, **363**, 391.

Pierce, M. J. 1986, *AJ*, **92**, 285.

Pompea, S. M., & Rieke, G. H. 1990, *ApJ*, **356**, 416.

Quillen, A. C., Frogel, J. A., & Gonzalez, R. A. 1994a, *ApJ*, **437**, 162.

Quillen, A. C., Frogel, J. A., Kenney, J. D, Pogge, R. W., & DePoy, D. L. 1994b, *ApJ*, **441**, 549.

Quillen, A. C., Frogel, J. A., Kuchinski, L. E., & Terndrup, D. M. 1995, *AJ*, **110**, 156.

Quillen, A. C., Ramirez, S. V., & Frogel, J. A. 1996, *AJ*, in press.

Reichen, M., Kaufman, M., Blecha, A., Golay, M., & Huguenin, D. 1994, *A&AS*, **106**, 523

Rix, H.-W. 1994, in NATO Workshop "Proc. Opacity of Spiral Disks", ed. J. Davies

Rix, H.-W., & Rieke, M. J. 1993, *ApJ*, **418**, 123.

Rix, H.-W., & Zaritsky, D. 1995, *ApJ*, **447**, 82.

Sakamoto, K., Okumura, S., Minezaki, T., Kobayashi, Y., & Wada K. 1995, *AJ*, **110**, 2075

Sandage, A., & Tammann, G. A. 1981, *A Revised Shapley-Ames Catalog of Bright Galaxies*, Carnegie Inst. of Wash. Publication 635.

Sanders, R. H., & Huntley, J. M. 1976, *ApJ*, **209**, 53.

Schweizer, F. 1976, *ApJS*, **31**, 313.

Scoville, N. Z., Matthews, K., Carico, D. P., & Sanders, D. B. 1988, *ApJL*, **327**, L61.

Shlosman, I., Frank, J., & Begelman, M. C. 1989, *Nature*, **338**, 45.

Simkin, S., Su, H., & Schwarz, M. P. 1980, *ApJ*, **237**, 404.

Spillar, E. J., Oh, S. P., et al. 1992, *AJ*, **103**, 793.

Spinrad, H., & Harland, E. 1973, PASP, **85**, 815.

Terndrup, D. M., Davies, R. L., Frogel, J. A., DePoy, D. L., and L. A. Wells 1994, *ApJ*, **432**, 518.

Thronson, Jr., H. A., Hereld, M., Majewski, S., Greenhouse, M., Johnson, P., Spillar, E., Woodward, C. E., Harper, D. A., & Rauscher, B. J. 1989, *ApJ*, **343**, 158.

Tinsley, B. M. 1972, *ApJ*, **178**, 319.

Toomre, A. 1969, *ApJ*, **158**, 899.

Toomre, A. 1981, in "The Structure & Evolution of Normal Galaxies", eds. S. M. Fall & D. Lynden-Bell (Cambridge: Cambridge Univ. Press), p. 111.

Toomre, A. 1990, in "Dynamics & Interactions of Galaxies", ed. R. Wielen (Heidelberg: Springer-Verlag), p. 292.

Valentijn, E. A. 1990, *Nature*, **346**, 153.

van der Kruit, P. C., & Searle, L. 1981, *A&A* **95**, 105.

Wainscoat, R. J., Hyland, A. R., & Freeman, K. C. 1990, *ApJ*, **348**, 85.

Witt, A. N., Thronson, Jr., H. A., & Capuano, Jr., J. N. 1992, *ApJ*, **393**, 611.

Witt, A. N., Lindell, R. S., Block, D. L., & Evans, R. 1994, *ApJ*, **427**, 227.

Zwicky, F. 1955, *PASP*, **66**, 232.

OPTICAL, IR, AND HI OBSERVATIONS OF A LARGE COMPLET CLUSTER SAMPLE

R. BRENT TULLY
Institute for Astronomy
University of Hawaii, Honolulu, Hawaii, USA

AND

MARC A.W. VERHEIJEN
Kapteyn Astronomical Institute
Postbus 800, 9700 AV Groningen, The Netherlands

1. Introduction

It is a chronic problem to achieve completeness in galaxy samples. Also, distance uncertainties confuse the comparison of properties between galaxies. These difficulties can be overcome by studying the members of clusters. The objects are essentially at the same distance and all objects that meet specified selection criteria within the volume of the cluster can be included. For example, a flux threshold might be set in some spectral band and this threshold corresponds to a specific *absolute* threshold for a complete sample.

However, the cluster environment might affect the properties of member galaxies. Our complete sample may only be representative of an unusual location. Perhaps our interest is in the characteristics of galaxies in low density environments, the "field", where most galaxies find themselves. There is the prospect that the Ursa Major Cluster provides us with an assemblage of galaxies with field-like properties within the confines of a cluster.

2. The Ursa Major Cluster

The Ursa Major Cluster is large and nearby but poorly known for precisely the reasons that make it interesting for us: it is irregular with no central concentrations. The cluster is at a distance of 15.5 Mpc, essentially the same distance as the Virgo and Fornax clusters. It lies in the supergalactic equa-

84

D. L. Block and J. M. Greenberg (eds.), New Extragalactic Perspectives in the New South Africa, 84–97.

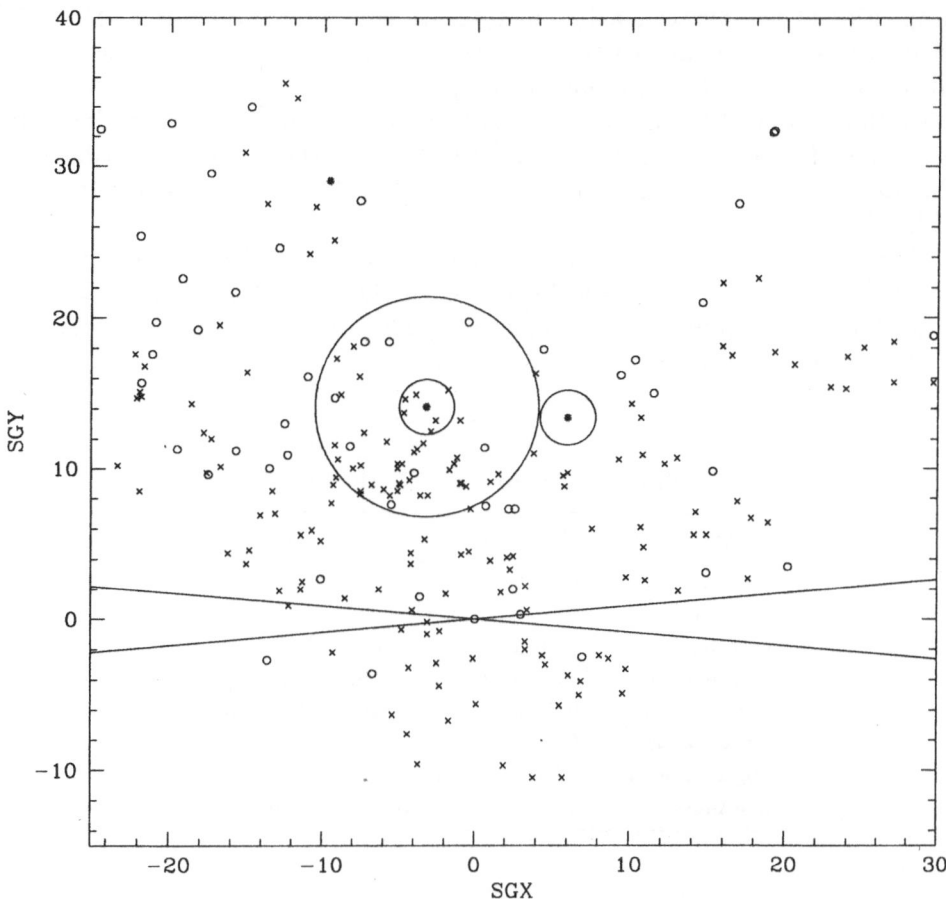

Figure 1. Groups and non-group galaxies in the plane of the Local Supercluster. The double concentric circles are centered on the Virgo Cluster; the inner circle defines the dimension of the cluster and the outer circle indicates the current turnaround radius. The Ursa Major Cluster lies within the the other circle near the tangent with the Virgo infall sphere. The Local Group lies at the origin of the plot and the 10° wedge locates the zone of obscuration where there is incompleteness. The three stars identify Virgo, Ursa Major, and Virgo W, the three clusters with $\log L_B \geq 11.5$. Open circles identify groups with $10.5 \leq \log L_B < 11.5$ and crosses indicate groups/galaxies that are fainter.

torial plane, like Virgo, at an angle of 37° from Virgo and just marginally outside the turnaround radius for infall into Virgo. See Figure 1. The Ursa Major Cluster will merge with the Virgo Cluster in about a Hubble time.

It is interesting to compare the three clusters, Virgo, Fornax, and Ursa Major, that accidently lie at roughly the same distance from us. Incidently, to find a fourth cluster with dozens of luminous galaxies one has to go to twice these distances, to Virgo W, Antlia, Centaurus, and Hydra I. The properties of the three nearby clusters are compared in Table 1. Virgo is easily the most populous and massive. It is a solid Abell richness class 0 cluster and 42% of the systems are HI-poor (E or S0). The Fornax Cluster has a similar density of E-S0 galaxies at the core but lacks the extended halo of spirals. The HI-poor galaxies make up 54% of the total population and the dimension of the cluster is small. The Ursa Major Cluster has the size of the Virgo Cluster and contains almost as many spiral galaxies but only 15% of the galaxies are early-type and this cluster lacks a concentration to any sort of core. The velocity dispersion of the members of Ursa Major is exceedingly low, indicative of a mass $\sim 1/20^{th}$ Virgo and the characteristic crossing time is $0.5H_0$.

TABLE 1. Properties of 3 Nearby Clusters

Properties	Virgo	Fornax	Ursa Major
No. E/S0	66	19	9
No. S/Ir	91	16	53
Distance	15.6 Mpc	14.5 Mpc	15.5 Mpc
Vel. dispersion	715 km/s	434 km/s	148 km/s
Virial radius	0.73 Mpc	0.27 Mpc	0.88 Mpc
Crossing time	0.08 H_0	0.07 H_0	0.5 H_0
Log luminosity	12.15	11.40	11.62
Log mass	14.94	14.10	13.64

Our formal requirement for membership in the Ursa Major Cluster (called group 12-1 in the *Nearby Galaxies Catalog* of Tully 1988a) is that a galaxy lie within 7.5° of $\alpha = 11^h56.9^m$, $\delta = 49°22'$; $SGL = 66.03$, $SGB = 3.04$ and have a velocity $700 < V_{helio} + 300\sin\ell\cos b < 1210$ km/s. The fussy high-end velocity cutoff is required because of the proximity of the group called 12-3 in the *Nearby Galaxies Catalog*. A detailed justification of this formal definition will be given in another paper. It is to be clarified that our present interest is to study the statistical properties of cluster members so we want a list of *high probabilty* members. For our purposes, it is better to loose a few true members from the sample, as long as the reasons are not correlated with the properties of the individual systems, than to add scatter to the sample with spurious candidates. If our goal were to study the dynamics of the cluster, for example, then our group definition would not be appropriate because outliers in position and velocity should be taken into account. Figure 2 shows the projected positions of galaxies we associate with the cluster.

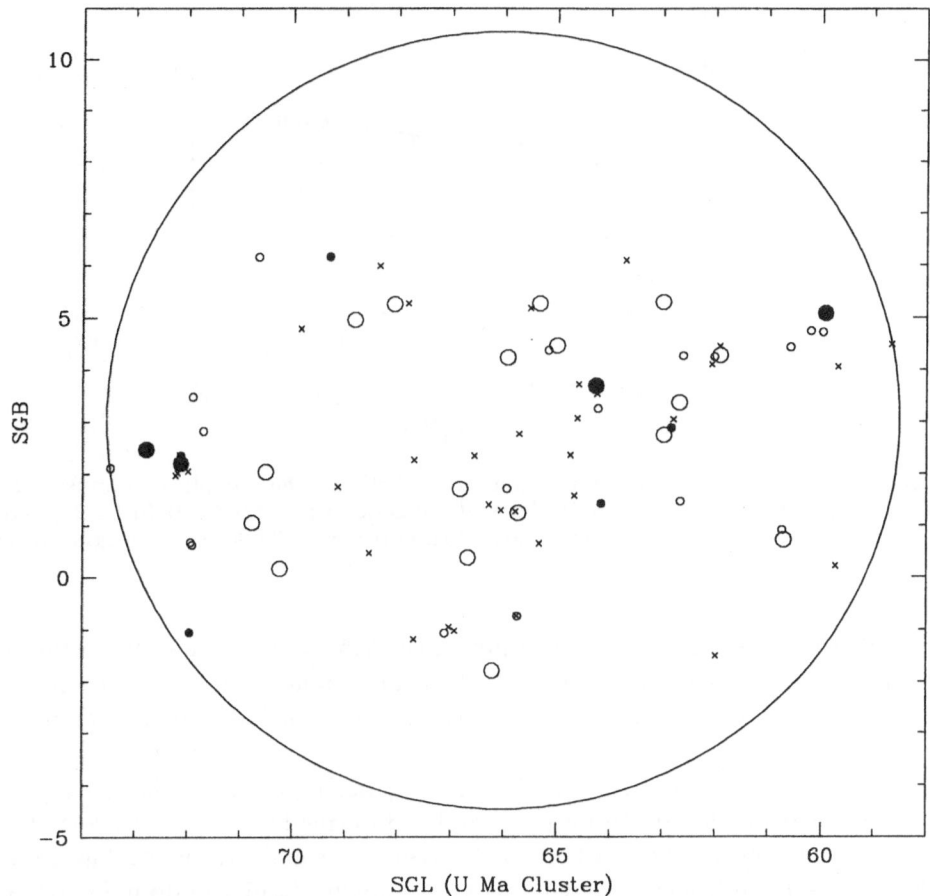

Figure 2.   Projected positions of 79 galaxies in our Ursa Major sample in supergalactic coordinates. The cluster is contained within a 7.5° radius. Filled circles denote types E-S0; open circles, types Sa-Scd; crosses, types Sd-Im. Larger symbols denote galaxies with $M_B < -19$.

The B-band luminosity function of the 79 identified Ursa Major galaxies is displayed in Figure 3. A distance modulus of $(m-M)_0 = 30.95$ is assumed and magnitudes have been adjusted for obscuration effects. The sample is complete brighter than -16.5. Information is retained in our database about any galaxy that falls within our spatial and velocity windows, but the sample is grossly incomplete at $M_B^{bi} > -16$.

Concerning the overall normalization, it was a surprise to us that Ursa Major *almost* qualifies as Abell (1958) richness class 0. The formal requirement is that there be 30-50 galaxies within two magnitudes of the third brightest member within the Abell radius. It turns out that there are 31 galaxies within our spatial-velocity windows that meet the magnitude crite-

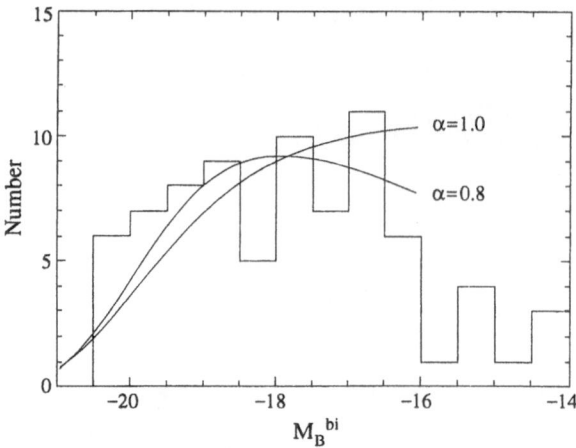

Figure 3.   Ursa Major Cluster luminosity function. The sample is complete for $M_B^{bi} < -16.5$ only. The curves are the Schechter function with $M_B^{\star} = -19.91 + 1.25(1+\alpha)$ where α is -1.0 for the monotonically increasing curve and -0.8 for the curve that turns over.

rion. With our distance scale zero-point, the $7.5° = 2.0$ Mpc radius window is 10% greater than the formal Abell radius. Hence, the luminous population of the Ursa Major Cluster falls just a couple of galaxies short of richness 0.

The Schechter function (1976) curves superimposed on the histogram were taken from the fits to larger samples described by Tully (1988b). The only free parameter allowed here is the overall normalization $\phi^{\star}$. The curve with $\alpha = -1.0$ is consistent with what is known about the domain $M_B^{bi} < -16$ with larger samples (Loveday *et al.* 1992; Marzke *et al.* 1994). The curve with $\alpha = -0.8$ actually gives a better rms fit to the present data over the domain of completeness.

There have been claims that the luminosity function turns up, possibly dramatically, at the faint end (Sandage *et al.* 1985: Driver *et al.* 1994; Marzke *et al.* 1994). These claims are usually made for denser regions than Ursa Major so we are motivated to test the faint end function in this "field-like" environment. Our planned approach for the future is to observe a significant fraction of the cluster at R and the same areas with an HI synthesis telescope. A new 8k × 8k CCD imager allows us to cover 0.5 square degrees with 0.2" per pixel scale. The Westerbork Synthesis Radio Telescope (WSRT) provides a similar coverage within the half-power radius. We have already observed ~ 15% of the area of the cluster with the WSRT although the field centers are chosen on galaxy targets rather than randomly. The CCD observations to date have also targeted known members and provide fields only modestly larger than the sizes of the galaxies.

The new program of wide field photometry and WSRT observations of the same fields will provide interesting constraints on both the HI-rich and HI-poor dwarf populations. The photometry should identify candidates as faint as $M_R = -9$ and the HI observations will provide HI fluxes and redshifts for all galaxies with normal gas properties as faint as $M_B = -12$. We can already draw two preliminary conclusions. The first, from our current coverage of the cluster, is that *the luminosity function for HI-rich systems is at best flat through the dwarf regime, and probably turns down by $M_B \sim -14$.* If the luminosity function goes up at faint magnitudes, as claimed in other clusters, then it has to arise out of an HI-poor population that remains unidentified in Ursa Major. The luminosity function may or may not go up at faint magnitudes but it certainly goes flat, may even drop, at $-18 < M_B < -16$. Hence, the second preliminary conclusion is that *overwhelmingly the mass associated with galaxies in Ursa Major are in the bright ($\sim L^*$) systems.*

3. Individual Galaxies

NGC 3718 is probably the best known peculiar galaxy in the cluster (Schwarz 1985). It is by far the biggest object though not quite the brightest. The interest in this paper is in the statistical properties of "normal" galaxies so not much attention will be given to this unusual system.

NGC 4051 is a Seyfert 1.2 galaxy that is not too extreme but receives attention because it is relatively close. Not surprisingly, among cluster members, it has the brightest nucleus at the resolution limit (although NGC 4111 is close and may be brightest if it were not viewed edge-on).

There are no outstandingly luminous galaxies in this cluster. The brightest galaxy at optical bands is NGC 3992 ($M_B^{bi} = -20.33$) and this galaxy also has the broadest HI linewidth, after correction for inclination ($W_R^i = 519$ km/s). NGC 3953 is slightly brighter at K'. Andromeda is brighter and rotates faster than any galaxy in the Ursa Major Cluster.

There is *no* certified elliptical in the cluster. The few HI-depleted systems seem to have disks and are classified S0. The brightest is not that bright, NGC 4111 at $M_B^{bi} = -19.51$, $B - I = 1.99$. Still, even the moderately small S0's have high central surface brightnesses at the resolution limit. There is no particular tendency for the S0 galaxies to congregate and certainly they do not define any cluster core.

At the faint end, there is a wide range of morphologies from high surface brightness – small exponential scalelengths to low surface brightness – moderate scalelengths. There is one Markarian galaxy (otherwise anonymous), Mkn 1460, which is compact with a high surface brightness, presumably attributable to star formation.

There are some systems that are adjacent in projection and some are probably interacting (NGC 3769 and companion, NGC 3893-96, NGC 3990-98, NGC 4117-18, UGC 6962-73), but the numbers are small and the cases are not dramatic (except perhaps for what is going on with NGC 3718). The spirals are generally normal in their HI flux properties. This aspect of the study will receive a lot more attention in a Ph.D. dissertation by Verheijen (1996). As the full HI synthesis dataset becomes available there will be more that can be said about the binary pairs. Overall in this cluster with a crossing time of half the Hubble time, there are no red flags cautioning against the proposition that the galaxies have typical "field" properties.

4. Optical, IR, and HI Information

We have observed almost all the galaxies in our sample with optical and infrared imagers. The larger galaxies have been observed using an 0.6 cm telescope in order to assure that the sky background is always adequately determined. The University of Hawaii 2.2 m telescope was used if the galaxies were small or the detector had a sufficiently large format to get enough sky. Texas Instrument and Tektronics CCDs have been used at B, R_c, and I_c bands and a 256×256 $HgCdTe$ detector has been used at K'. Details of the data reductions and access to the data will be provided in a separate paper. Here are the global photometric parameters that we record in each of the four passbands: ellipticities, total magnitudes, radii containing 20%, 50%, and 80% of the light, exponential scalelengths, extrapolated central surface brightness of the exponential disks, and the peak central surface brightness at the resolution limit of $\sim 1''$. Colors between bands are recorded. We use HI fluxes and linewidths from the literature for the current discussion but our own WSRT observations will be used in future studies (Verheijen 1994). Absolute properties are calculated assuming the cluster distance is 15.5 Mpc. At this distance, $1'' = 75$ pc, $1' = 4.5$ kpc.

5. Reddening Properties

The availability of photometric information over a wide wavelength range provides leverage on the problem of internal reddening in galaxies. The second important advantage with the cluster sample is the elimination of differential distances. We are able to look for inclination dependencies in two correlations: luminosities vs. HI linewidths (discussed further in the next section) and luminosities vs. optical to infrared color. We take advantage of the fact that there is very little obscuration in the K' band. The spirit of our modeling is the same as that by Tully & Fouqué (1985) except that now the analysis can be done independently in each of the available passbands, rather than just at B. In that earlier work, reddening was modeled with two free parameters: a characteristic optical depth at the wavelength in question, τ_λ, and a geometric term that describes the fraction of light mixed with the obscuring dust vs. the fraction of light above the obscuring layer. The interplay between the different passbands allows the geometric term to be much more reliably defined now. The details will be described in a separate publication. The analysis also involves the use of data from two other clusters. We reach the following conclusions. First, with regard to the geometric term, *we find that* $30\%\pm10\%$ *of the light typically lies outside the obscuring layer and* $70\%\pm10\%$ *lies mixed with the dust.* Second, with regard to the characteristic opacities, *we find* $\tau_B = 0.60$, $\tau_R = 0.30$, $\tau_I = 0.22$, *and* $\tau_K = 0.025$. Our reddening corrections may appear to be smaller than others are finding (Peletier & Willner 1992; Han 1992; Giovanelli *et al.* 1994) but most of the differences are in parameterizations that do not quite have the same meaning.

6. Luminosity – Linewidth Correlations

Part of our motivation in this work is to have a very complete sample so we can study the properties of the luminosity–linewidth correlations for distance measurement applications (Tully & Fisher 1977). Roughly half the full sample can be used for this purpose. Galaxies are *excluded* by type (E/S0) or inclination ($i < 45°$), or pathology (eg, NGC 3718), or HI profile confusion, or low HI signal/noise. The latter problems are in the process of rectification. Figure 4 provides plots of the present observational status for the I and K' bands. Linewidths are in the convention given by Tully & Fouqué (1985). The scatter is similar in these two bands and at R. The K' band has advantages with respect to absorption corrections. However, it is much easier to work at optical wavelengths, especially with low surface brightness targets, because of the competition from sky in the infrared. The distance of 15.5 Mpc that we use for Ursa Major is derived from the zero-point calibration of the apparent magnitude versions of these plots.

7. Central Surface Brightnesses

Here is some speculation. There has been a long-term controversy over the remarkable claim by Freeman (1970) that the extrapolated central surface brightness of exponential disks has a very small dispersion. It has been suggested that the small dispersion might be an artifact of a selection effect (Disney 1976; Davies *et al.* 1994) or internal opacity (Disney *et al.* 1989), although van der Kruit (1987) has provided evidence that the effect is real. It has been known, for example by van der Kruit, that dwarf galaxies do not achieve the same high surface brightnesses as early types. There may

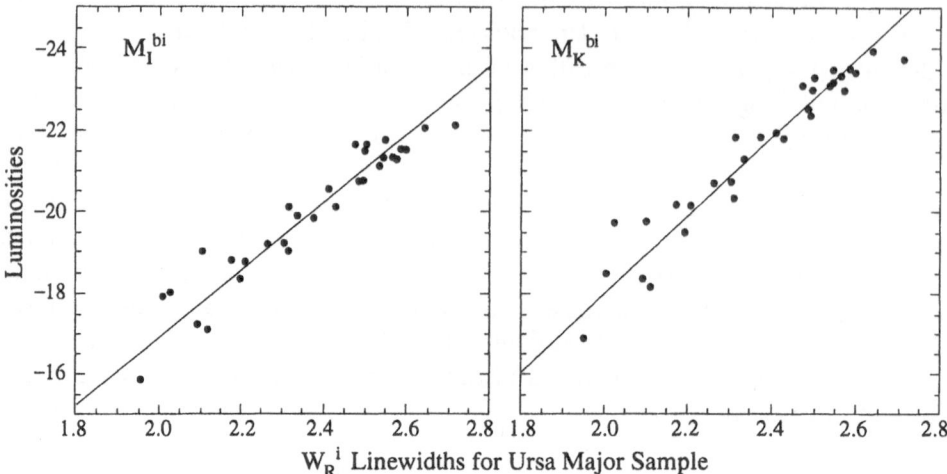

Figure 4. Luminosity–linewidth relations for the Ursa Major Cluster. Excluded are E-S0 types, pathological and confused galaxies, those with unsatisfactory HI information, and inclinations < 45°. The *I* relation is shown in the left panel and the *K'* relation is shown at the right.

be a continuum of properties that is capped at a high surface brightness limit, possibly by obscuration.

We are able to test these alternatives with our multi-band, complete sample. The results are presented for the *I* band data in Figure 5. Although the statistics are small, it is our impression that the distribution is *bimodal.* If the sample is split at the division in morphological type between S*cd* and S*d* then those that are earlier nicely fill the high surface brightness hump ($\mu_0^I = 18.79 \pm 0.80$) and those later fill the low surface brightness hump ($\mu_0^I = 20.58 \pm 0.95$). Of course, the break between spiral and late morphological types are largely an expression of the differences in surface brightness of these two populations. That is, the early-late typing and the central surface brightnesses are highly correlated. The interesting speculative possibility, though, is that a complete volume-limited sample really has this bimodality in the surface brightness distribution.

An intruiging possibility is that we are seeing a manifestation of an old idea by Mestel (1963). He suggested that the two natural radial equilibrium states for disks were *angular velocity* ~ *constant* or *linear velocity* ~ *constant*. The latter requires a strong central concentration and the former implies a very low central concentration. The rotation curves of spirals and dwarfs are in general agreement with these two alternatives.

It is evident enough that the Mestel picture is a good description of the extremes between the large and small galaxies. At issue is whether there is a

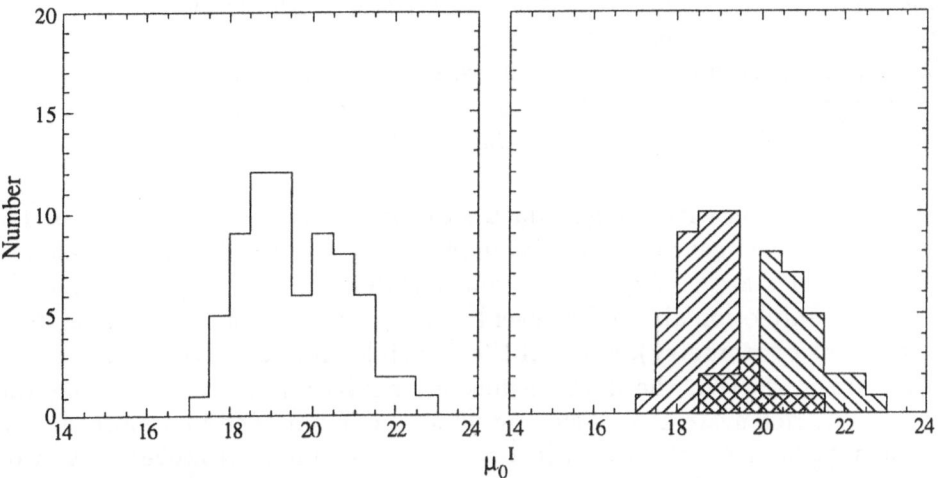

Figure 5. Exponential disk central surface brightnesses at I band. The distribution for the full sample is shown in the left panel. In the right panel, the separate distributions are shown for types Scd and earlier (41 galaxies) and for types Sd and later (32 galaxies).

continuum of properties between the extremes with no discernable break or whether galaxies really have a preference for the extremes of either linear or angular velocity $\sim$ constant. *Figure 5 may be evidence that galaxies prefer to organize themselves in one or the other state and avoid the middle ground.*

By contrast, there is no discontinuity in the luminosity-linewidth relations between the earlier and later spiral types. There is the continuing curiosity that a galaxy with a given luminosity has a well specified linewidth, whether it be compact with a high central surface brightness or extended with a low central surface brightness.

8. Color–Magnitude Correlations and Galaxy Evolution

While speculating, why stop now? The four-band photometry on a complete sample allows us to revisit the optical-infrared correlation found by Tully, Mould, & Aaronson (1982). The earlier discussion involved $B - H$ colors of an incomplete sample drawn from the Virgo Cluster. The magnitudes were from aperture photometry and referred only to the inner parts of galaxies for the H measurements. It had been found that spirals lie along a well-defined track and a sample of S0 galaxies from the cluster lie along a redder, steeper track. There were few galaxies in between. Crude composite stellar models were developed that provided a description of the locus of the spirals. The star-formation rate was monotonically dependent on the galaxy mass, so massive systems converted most of their HI resources to stars quickly and have now reddened. To explain the least massive, bluest systems the model required some combination of low metallicities and a flat initial mass function for the stars. If star formation is halted, then our models would evolve to positions on the S0 track in 4 Gyr. The slight slope of the S0 track – brighter systems are redder – is taken to be a metallicity effect.

The situation with the new data set is demonstrated with Figure 6. The left panel is the new color-magnitude plot, now with the $B-K'$ baseline and the quantities are now global, obtained with electronic imaging detectors. There is completion above the lower line (except for a couple of galaxies with unreduced photometry). Some of the principle characteristics of the earlier work are preserved: the domain filled by galaxies is bounded on the blue side by HI-rich systems in a sloping band and is bounded in a more sharply defined fashion on the red side by HI-poor systems. However now, with the complete sample, galaxies are found at intermediate positions. Toward the dwarf regime a broad range of colors is found. This range could be the consequences of stochiastic star formation episodes that can dominate the light output in small galaxies.

The panel at the right of Figure 6 shows a much tighter correlation. What is happening here? What is this 'correction for HI' term? If we look back at the left panel, there is a dashed line near the blue edge of the distribution of points. We became aware that galaxies with a given total HI flux, F_H, lie in strips parallel to this line. This particular line is the locus for $F_H = 10^7 M_\odot/\text{Mpc}^2$, corresponding to a mass in HI gas of $M_H = 2.4 \times 10^9 M_\odot$ at the distance of Ursa Major. Galaxies with progressively less neutral hydrogen lie progressively to the right. We found a linear relationship between the $\Delta(B-K')$ color offset from the dashed line and the value of $\log M_H$ for a galaxy. The observed HI flux gives us an offset and it is the color minus this offset that is plotted on the horizontal axis of the right panel.

This rather tight correlation after the offset is introduced reveals that there is a 'fundamental plane' in the three-dimensional space of the coordinates M_K (number of old stars or total mass), $B-K'$ (ratio of young to old stars, with a metallicity factor), and M_H (the reservoir for star formation). We are managing to view this plane edge-on in the right panel of Figure 6.

In the spirit of stellar color-magnitude diagrams, this snap-shot in time can be used as a window on cluster evolution. It should be possible to plot trajectories of how the galaxies move about with time in the color-magnitude space. There was a beginning of this enterprise by Tully *et al.* (1982). Returning to speculations, we propose that the natural locus for isolated galaxies is along the blue edge of scatter in the left panel or roughly

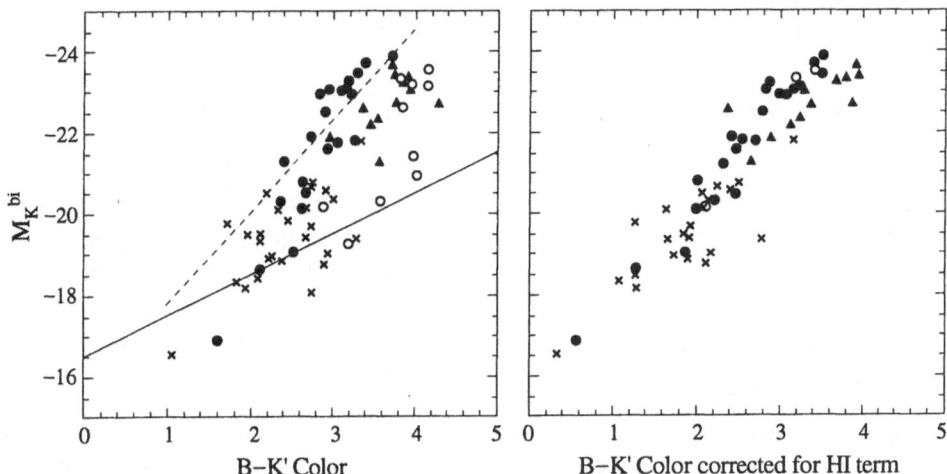

Figure 6. *Left panel:* Color–magnitude diagram for the Ursa Major sample. The vertical axis, M_K^{bi}, is a measure of the populations of old stars. The horizontal axis, $B - K'$, is a measure of the ratio of young to old stars and is affected by metallicity and stellar mass function variations. The solid line marks the completion limit. Open circles are labels for types E-S0; filled triangles, for types Sa-Sb; filled circles, for types Sbc-Scd; crosses, for types Sd-Im. *Right panel:* Color–magnitude diagram adjusted for a third parameter. Colors are adjusted by a factor that depends on the HI flux of a galaxy. The dispersion in the color-magnitude diagram is largely determined by one parameter: the residual HI reservoir.

where galaxies are seen to lie in the right panel. Their location along that locus would depend on a single parameter: total mass. Galaxies with more mass were more effective at star formation at early times, built up more heavy elements, and are redder today. If the galaxies become gas-depleted then they move to the red. The distinct morphologies of the reddest galaxies (S0's with high central surface brightnesses and modest dimensions) suggest that the evolution was not passive. These galaxies are not simply reddened versions of the spirals of comparable mass. Since these S0 span a wide range of luminosities, something other than total mass was driving the evolution. Perhaps this line of reasoning is in line with the proposition that S0's form from interactions.

It is a curiosity to us that the blue edge of the color-magnitude distribution corresponds to $M_H \sim constant$. Accordingly, there are dwarf galaxies with HI reservoirs as large as those found in giant galaxies. Of course, this reservoir is a lot larger fraction of the total mass in the dwarf than in the giant. The giant *started* with a lot more gas. The star formation rates are such that the global HI content depletion curves are now intersecting.

This investigation will not be finished until there are satisfactory models of the stellar and gas populations of the galaxies and it becomes possible to plot evolutionary tracks in the three-dimensional space, both representing the past and the future.

9. Acknowledgements

We thank our collaborators with the extensive observations. Much of the optical imaging was done with M.J. Pierce and the infrared imaging was obtained with J.S. Huang and R. Wainscoat. R. Sancisi is participating in the WSRT HI program. This research is supported by NATO Collaborative Research Grant 940271.

10. References

Abell, G.O. 1958, *Astrophys. J. Suppl.*, **3**, 211.
Davies, J., Phillipps, S., Disney, M.J., Boyce, P., & Evans, R. 1994, *M.N.R.A.S.*, **268**, 984.
Disney, M.J. 1976, *Nature*, **263**, 573.
Disney, M.J., Davies, J., & Phillipps, S. 1989, *M.N.R.A.S.*, **239**, 939.
Driver, S.P., Phillipps, S., Davies, J.I., Morgan, I., & Disney, M.J. 1994, *M.N.R.A.S.*, **268**, 393.
Freeman, K.C. 1970, *Astrophys. J..* **160**, 811.
Giovanelli, R., Haynes, M.P., Salzer, J.J., Wegner, G., Da Costa, L.N., & Freudling, W. 1994, *Astron. J.*, **107**, 2036.
Han, M.S. 1992, *Astrophys. J.*, **391**, 617.
Loveday, J., Peterson, B.A., Efstathiou, G., & Maddox, S.J. 1992, *Astrophys. J.*, **390**, 338.
Marzke, R.O., Geller, M.J., Huchra, J.P., & Corwin, H.G. Jr. 1994, *Astron. J.*, **108**, 437.
Mestel, L. 1963, *M.N.R.A.S.*, **126**, 553.
Peletier, R.F., & Willner, S.P. 1992, *Astron. J.*, **103**, 1761.
Sandage, A., Binggeli, B., & Tammann, G.A. 1985, *Astron. J.*, **90**, 1759.
Schechter, P.L. 1976, *Astrophys. J.*, **203**, 297.
Schwarz, U.J. 1985, *Astron. Astrophys.*, **142**, 273.
Tully, R.B. 1988a, *Nearby Galaxies Catalog*, (New York: Cambridge University Press).
Tully, R.B. 1988b *Astron. J.*, **96**, 73.
Tully, R.B., & Fisher, J.R. 1977, *Astron. Astrophys.*, **54**, 661.
Tully, R.B., & Fouqué, P. 1985, *Astrophys. J. Suppl.*, **58**, 67.
Tully, R.B., Mould, J.R., & Aaronson, M. 1982, *Astrophys. J.*, **257**, 527.
van der Kruit, P.C. 1987, *Astron. Astrophys.*, **173**, 59.
Verheijen, M.A.W. 1994, in *The World of Galaxies II*, Lyon, France, 5-7 Sept.

Verheijen, M.A.W. 1996, Ph.D. Dissertation, University of Groningen, in preperation.

THE RELATIONSHIP BETWEEN NEAR IR EXTINCTION AND CO EMISSION

MICHAEL W. REGAN

Department of Astronomy
University of Maryland
College Park, MD 20742
mregan@astro.umd.edu

Abstract. The relationship between dust optical depth and integrated CO emission is explored for a sample of four galaxies. The dust optical depth is determined by using a radiative transfer model to match the observed optical and infrared colors at different positions in each galaxy. With assumptions of dense dust fractions and a constant dust to gas ratio, the comparision of the ratio of dust optical depth to integrated CO emission can be used to look at the variation in X, the conversion factor between integrated CO emission and H_2 mass.

1. Introduction

Millimeter observations of molecular gas in galaxies are limited to secondary tracers of the molecular gas since the H_2 that dominates the mass lacks a dipole moment and thus does not radiate at millimeter wavelengths. The strongest tracer of H_2 is CO whose $J=(1-0)$ occurs at a wavelength of 2.6cm. Several studies in the local region of the Milky Way have shown that the conversion between CO $J=(1-0)$ integrated emission and H_2 number surface density is fairly well defined as being around 3×10^{20} cm^{-2} T^{-1} km^{-1} s.

There are other observations in different regions of the Milky Way that show variations in the conversion factor (Sodroski et al 1995, Digel et al 1990). Also, observations of molecular clouds in Local Group galaxies show a range of conversion factors (Wilson 1995).

D. L. Block and J. M. Greenberg (eds.), New Extragalactic Perspectives in the New South Africa, 98–104.
© 1996 *Kluwer Academic Publishers.*

Despite these variations in the conversion factor (X), most CO observations of external galaxies outside of the Local Group convert the CO to molecular mass using the "standard" Galactic value. An independent method of determining the conversion value is needed.

In this paper I will discuss how to determine the dust optical depth and thus mass using as input to a radiative transfer model a set of optical and near infrared colors. Then under the assumption of a constant dust to gas ratio we can convert the dust optical depth to an H_2 mass which can be compared to the CO integrated intensity to obtain the X.

2. Dust Model

Using multiple colors to model the dust in a galaxy has been done before in M51 (Elmegreen 1980; Rix & Rieke 1993). In those previous studies only selected regions were modeled. Our current dust model uses optical-infrared and infrared-infrared colors to determine the content and geometry of the dust. The galaxy is modeled as an infinite plane with the stars and dust centered on the midplane of the galaxy. The details of the model are described in Regan, Vogel, and Teuben (1995).

This method of determining the dust content has several advantages. One is that the resolution of the resulting dust map is approximately 2 arcseconds, an order of magnitude better than what can be done observing dust in emission in the far infrared. Also, since we are observing the dust in absorption, the temperature of the dust doesn't matter. Finally, the model has a low ratio of free parameters to observables meaning that the chi^2 value output from the least squares fit can be used as a criterion to reject pixels where the model does not fit well.

An underlying assumption of the model is that the stellar population does not vary between the dust free region and the modeled regions. There are two possibilities where this assumption may be false. One would be when there is a radial color gradient in the galaxy. This effect can be minimized by picking a dust free region near the region of interest. The second possibility is that there is recent star formation in the region being modeled. In this case the intrinsic color of the underlying population are different than the dust free region and the model does not work. Empirically this results in a high chi^2 from the model fitting and we can ignore these regions.

3. Observations

3.1. NEAR INFRARED

The near infrared observations were obtained on three different observing runs. The first set were obtained on the 1.3m at Kitt Peak using the Simultaneous Quad Infrared Imaging Device (SQIID) in January 1994. The second set were obtained on the 1.3m at Kitt Peak using the Cryogenic Optical Bench in March 1994. The third set were obtained on the 1.5m at Cerro Tololo using the CTIO IR Imager (CIRIM).

3.2. OPTICAL

All the optical observations were made on the Isaac Newton Telescope at La Palma. The observations of NGC 1068, NGC 6946, and NGC 2903 were retrieved from the archive. The observations of NGC 1530 are described in Regan et al (1996).

3.3. CO INTERFEROMETRY

The millimeter observations were obtained during 1994 and 1995 using the Berkeley- Illinois - Maryland Array (BIMA) observing the CO J=1−0 transition. All the observations were made in three configurations of six antennas each yielding beams of approximately 3-5 arcseconds.

4. Results

Figure 1(a-c) shows the input data for the ratio determination for the barred spiral galaxy NGC 1530. The CO is primary concentrated in two peaks of emission centered on the inner terminus of the dust lanes. The dust follows this general morphology, but the dust lanes along the leading edge of the bar are visible in the dust map unlike, in the CO map.

By dividing the dust optical depth map by the CO integrated intensity map and scaling to the correct units we can create a map of the X for NGC 1530 (Figure 2). This figure shows that the X is much lower than the local galactic value, implying that there is more CO per unit mass of H_2 in the central region of NGC 1530 than there is locally in the Milky Way.

NGC 1530 has a range of calculated X values which are actually a little higher than what is found for the other four galaxies in the sample. Overall a range of $0.1 \times 10^{20} < X < 0.8 \times 10^{20}$ is found for all the galaxies in the sample.

It is possible to make an estimate of X in regions where we detect dust but do not detect CO above the noise level in the maps by averaging the emission over larger regions. The best example of this is in NGC 1530 where we see dust in the bar dust lanes but do not detect significant CO. If we average over about 15 arcseconds^2 we find that X is approximately 2.4×10^{20}, very close to the local galactic value.

Given the low values of the X found for all the galaxies we need to look at possible reasons for the dust mass to be underestimated. One possible

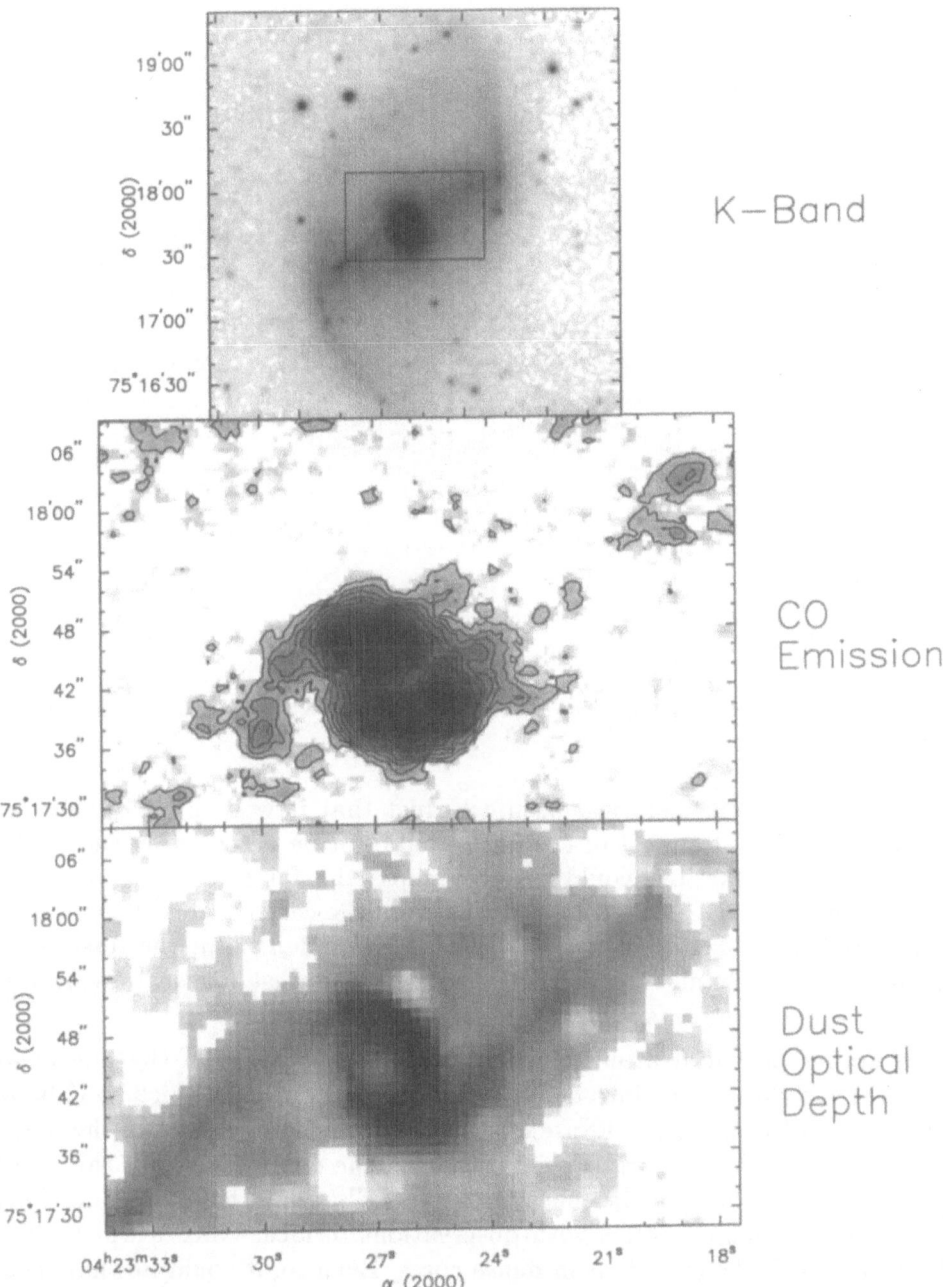

Figure 1. NGC 1530 a) K-band image b) BIMA CO map c) Dust optical depth output from the model.

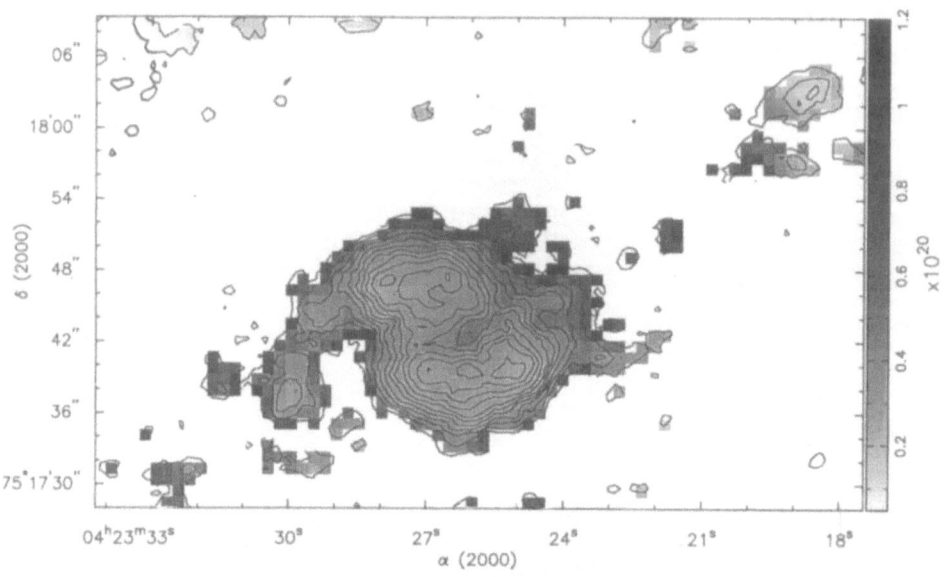

Figure 2.   NGC 1530 X ratio (gray) overlaid with CO emission contours

reason would be if there was so much dust that it was optically thick at K-band. In this case most of the dust would not be detected by the model because the extinction would be gray since the far side stars would be totally obscured in all the colors. If this were the case the dust lanes would suffer at least a factor of 2 extinction in K-band making the dust lanes prominent in K-band images. Since none of the dust lanes are near this level in the K-band images, we can rule out saturation.

Another potential problem would be if the majority of the molecular gas where concentrated into dense cores that are optically thick to K-band emission. In this case, the diffuse dust between the cores generates the colors while the CO emission arises primarily from the dense cores. For this to be true more than 90% of the mass of the GMC's would have to be in these dense cores. This contrasts with observations of local GMC's which show that only 20% of the mass is in dense cores. Even so, it could be that these GMC's are different than the local GMC's and have a large component of dense gas. A constraint on these GMC's being vastly different than local GMC's come from HCN observations of NGC 1068 (Helfer & Blitz 1995) and NGC 6946 (Helfer 1995). There studies show that the HCN/CO ratio for the non-nuclear regions of these two galaxies is very close to the local galactic value. Since HCN traces gas at higher densities than CO, the HCN to CO ratio is a good indicator of the dense gas fraction. The low HCN/CO ratio in these galaxies makes it unlikely that there is a significant difference between the amount of dense gas in these external GMC's and local GMC's. Thus, this cannot explain the low value of X.

Another possibility is that the dust to gas ratio is vastly different in all of these galaxies. We would need there to be a lower dust to gas ratio in these regions than the local galactic value. This is not what you would expect given the higher metallicity found near the centers of galaxies. It could be that there is enough star formation to destroy a significant fraction of the dust grains leading to a lower dust to gas ratio.

5. Discussion

It is not really unexpected to come up with low X values for the central regions of external galaxies given the low value for the central region of the Milky Way determined by Sodroski (1995). Therefore, it is most likely that the central regions of all galaxies have low X values. If so, there are several implications.

The biggest implication is that the molecular mass estimates for the nuclear regions of galaxies are too high, possibly by an order of magnitude. This means that when calculating the fuel reservoir for starbursts there is much less fuel available and the lifetime of the starburst comes down significantly. Thus, starbursts are extremely short lived phenomena. The lower molecular mass in the nuclear regions of galaxies also implies that the mass inflow rate needed to generate the central concentration is much less then previously thought. Finally, the lower gas mass in the nuclear regions means that the gas is not as important dynamically as previously thought and that gas self-gravity is not as important.

Since the Milky Way has a low X near the center and a high value outside of the solar circle there is a gradient in the X. This would be consistent with the relationship found in the Local Group that the X changes with metallicity (Wilson 1995) since most galaxies have a metallicity gradient.

If there were a gradient in the X in galaxies this would affect the interpretation of measurements of the CO scale length. Measurements of the CO scale length in external galaxies have assumed that CO traces the molecular mass, but if there is a gradient in the X this is not true. This could mean that correlations between CO scale length and blue light scale lengths are not between molecular mass and stars. Even so CO could still be associated with star formation since it is the dominant method of cooling the molecular gas and can thus be thought of as a measure of the rate at which gas cools enough to form stars. Thus, CO could still trace star formation even if it doesn't correlate with molecular mass.

6. Conclusions

The most important conclusion is that the conversion factor between CO integrated intensity and molecular mass is not constant. In the central regions of galaxies the correct conversion yields a much lower molecular mass than if the local Milky Way conversion is used. This change cannot be completely due to metallicity variations since we see changes in X over regions that have the same metallicity.

This lower molecular mass is found for all four studied galaxies and is also seen in the Milky Way. Thus, we can conclude that it holds for all galaxies. This means that determinations of starburst lifetimes could be off by an order of magnitude since there is now much less fuel available for star formation.

I would like to acknowledge helpful discussions with Stuart Vogel, Leo Blitz, and David Spergel.

References

Digel, S, Bally, J. and Thaddeus, P (1990), ApJ, 357, L29
Elmegreen, D. (1980), APJS, 43, 37.
Helfer, T.T. (1995), PhD Thesis, University of Maryland
Helfer, T.T. and Blitz, L. (1995), ApJ, 450, 90
Rix, H.W. and Rieke, M.J. (1993), ApJ, 418, 123.
Regan, M.W., Vogel, S.N. and Teuben, P.J. (1995), ApJ, 449, 576.
Regan, M.W., Teuben, P.J., Vogel, S.N. and van der Hulst (1996), in prep
Sodroski et al (1995), ApJ, 452, 262.
Wilson, C.D. (1995), ApJ, 448, L97.

EXTINCTION AND DUST COLUMN DENSITY IN SPIRAL DISKS FROM FIR VS. UV-OPTICAL COMPARISON

C. XU

Max-Planck-Institut für Kernphysik
Postfach 103980, D69117 Heidelberg, Germany

AND

V. BUAT

Laboratoire d'Astronomie spatiale du CNRS
BP 8, 13376 Marseille Cedex 12, France

1. Introduction

A new method has been proposed by us (Xu & Buat 1995; Xu & Helou 1996) to constrain the extinction and the dust column density in spiral disks through comparisons between the FIR emission (dust re-radiation) and the UV and optical emission (escaped starlight). It differs from other similar methods (e.g. Disney et al. 1989) by including explicitly the vacuum UV data, which became available only recently, into the solution of the problem.

2. The sample

Our sample contains 135 spiral galaxies selected from nearby spiral galaxies detected in two recent vacuum UV experiments: FAUST (Deharveng et al. 1994) and SCAP (Donas et al. 1987). See Xu & Buat (1995) for details of the sample selection and sources of data. The monochromatic UV fluxes at 2030Å have been extrapolated to the entire non-ionizing UV domain (912 — 3650Å) using the average broad band UV spectra by Pence (1976) and Coleman et al. (1980) for different Hubble types. The optical (UBV) fluxes have been extrapolated to cover the optical and NIR wavelength range from 3650Å to $3.4\mu m$ using the spectra taken from Yoshii & Takahara (1988). The dust emission $(8 — 1000\mu m)$ has been estimated from the IRAS FIR flux $(40 — 120\ \mu m)$ using an empirical formula (Xu & Buat 1995).

D. L. Block and J. M. Greenberg (eds.), New Extragalactic Perspectives in the New South Africa, 105–108.
© *1996 Kluwer Academic Publishers.*

3. The model & results

In a simple sense, our model treats dust grains as 'frequency-converters' which convert the UV and optical radiation to the IR radiation via the absorption-reradiation (heating-cooling) process, as is shown by the following equation

$$L_{IR}^{dust} = \int_{NIR}^{UV} L_{\nu}^{ob} \times G_{\nu} \, d\nu \qquad (1)$$

where G_{ν}, the 'conversion function', is the ratio between the radiation absorbed by dust grains to the radiation escaping the galaxy and eventually observed:

$$G_{\nu} = \frac{L_{\nu}^{ab}}{L_{\nu}^{ob}} = 10^{0.4A_{\nu}} - 1 \qquad (2)$$

where A_{ν} is the extinction at frequency ν.

In the simplest case of 'gray extinction', $A_{\nu} = A_{gray}$ is constant with the frequency, and

$$A_{gray} = -2.5 \times \log(1 - \frac{L_{IR}^{dust}}{L_{bol}}) \qquad (3)$$

where L_{bol} is the bolometric luminosity, namely the sum of radiations in all bands. We find a mean dust luminosity to bolometric luminosity ratio for galaxies in our sample of 0.31 ± 0.01, which translates to $A_{gray} = 0.40\pm0.01$. So, in the sense of this 'gray extinction', galaxies are not opaque.

In reality the extinction is of course not gray. Nevertheless, if the extinction curve (i.e. A_{ν}/A_B) is known, one can determine the blue extinction A_B by solving the equation

$$L_{IR}^{dust} = \int_{NIR}^{UV} L_{\nu}^{ob} \times (10^{0.4A_B \times \left(\frac{A_{\nu}}{A_B}\right)} - 1) \, d\nu \qquad (4)$$

Taking the empirical 'effective' extinction curve obtained by Calzetti et al (1994) using IUE spectra of a sample of starburst galaxies, we calculate A_B (Fig.1) for galaxies in our sample. The mean is $< A_B >= 0.29 \pm 0.01$ mag.

However, there are a lot of uncertainties with the empirical extinction curve. Furthermore it was derived for the starburst galaxies and therefore may not be applicable to our sample of normal spiral galaxies. Hence we tried another approach, namely developing a detailed radiative transfer model to determine the conversion function G_{ν} (Eq(2)), with the following assumptions:

- The same wavelength dependences for the dust extinction coefficient, albedo and phase function as measured in Solar Neighborhood (Mathis et al. 1983; Herwitz et al. 1991).

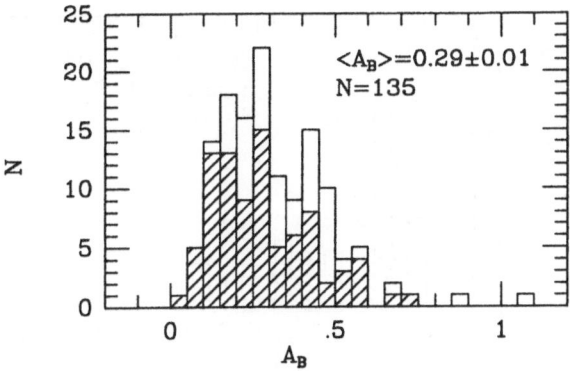

Figure 1. Histogram of the blue extinction A_B calculated by assuming the 'effective' extinction curve of Calzetti et al. (1994). Hatched: late spirals (Sc–Sm).

– A Sandwich configuration for the star and dust distribution, with dust and the UV sources having the same scale height, and the scale height of the optical sources being a factor of 2 larger.

The radiative transfer code, based on the algorithm by van de Hulst & de Jong (1969) which takes fully into account the effects of scattering (see Xu & Helou (1996) for details), specifies G_ν as a function of ν, τ_B (face-on blue optical depth) and the view-angle i (known for each galaxy from axis ratio). Solving Eq(1) one can calculate τ_B. Figure 2 is the histogram of τ_B for galaxies in our sample. The median value is 0.49, and the mean 0.6±0.04. This confirms our suggestion that spiral disks are not opaque. Using the same radiative transfer model and taking into account the view angle, we can translate τ_B to blue extinction A_B. A mean of $A_B = 0.22\pm0.02$ is found, not very different from the one derived using the empirical extinction curve of Calzetti et al. (1994).

It is clear that our radiative transfer model is base on a much simplified picture for a real galaxy disk. We (Buat & Xu 1996) made some tests to estimate the uncertainties introduced by various assumptions made in the model, and found that our results on optical depth τ_B vary within a factor of ~ 2 when we change these assumptions in realistic ranges. On the other hand, the results on the extinction A_B are insensitive to the changes in the radiative transfer model, and are basically constrained by the infrared luminosity to the UV plus optical luminosity ratios.

We have also compared the optical depth τ_B so computed with the gas *surface density* for another sample of 95 spiral galaxies with HI and CO data. Assuming that the H_2 (estimated from CO) is confined in a disk only half as large, while the HI disk is twice larger, as the optical disk, the tentative results indicate a mean τ_B–to–total-gas-surface-density ratio

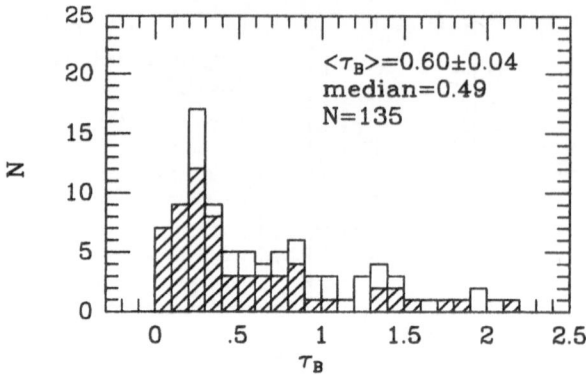

Figure 2. Histogram of the face-on blue optical depth τ_B. Hatched: late spirals (Sc–Sm).

consistent with the Galactic dust-to-gas ratio (Buat et al. in preparation).

4. Conclusion and future work

Our results, which show that the mean blue extinction is about 0.2 to 0.3 magnitude, and face-on blue optical depth is about 0.6, suggest that galaxy disks are semi-transparent for the blue light. However, we cannot rule out that some part of the disks may be opaque. Since we don't resolve galaxies, the τ_B calculated using our model is obviously an average value. The regions where the radiation density is high get larger weight in this average. Our final conclusion is that **on average the bright part of a standard galaxy disk is semi-transparent for the blue light.**

There is an on-going project to study the distribution of τ_B within disks of nearby galaxies which can be resolved in FIR (e.g. through ISO observations) and UV, using the same model presented here. The first example is M31 (Xu & Helou 1996).

References

Buat, V., Xu, C. 1996, A&A, in press
Calzetti, D., Kinney, A.L., Storchi-Bergmann, T. 1994, ApJ 429, 582
Coleman, G.D., Wu, C.C., Weedman, D.W. 1980, ApJS 43, 393
Deharveng, J.M., Sasseen, T.P., Buat, V., Bowyer, S., Wu ,X. 1994, A&A 289, 7
Disney, M., Davies, J., Phillipps, S. 1989, MNRAS 239, 939
Donas, J., Deharveng, J.M., Milliard, B., Laget, M., Huguenin, D. 1987, A&A 180, 12
Hurwitz, M., Bowyer, S., Martin, C. 1991, ApJ 372, 167
Mathis, J.S., Mezger, P.G., Panagia, N. 1983, A&A 128, 212
Pence, W. 1976, ApJ 203, 39
van de Hulst, H.C., de Jong, T. 1969, Physica, 41, 151
Yoshii, Y., Takahara, F. 1988, ApJ 326, 1
Xu, C., Buat, V. 1995, A&A 293, L65
Xu, C., Helou, G. 1996, ApJ 456, 163

THE EFFECTS OF SUPERGIANTS ON THE INFRARED LIGHT DISTRIBUTION IN GALAXIES

D. L. DEPOY, A. C. QUILLEN, A. BERLIND AND S. V. RAMIREZ

Astronomy Department
Ohio State University

1. Introduction

Dynamical models require good estimates of the stellar surface mass density in galaxies. Surface brightness measurements in principle can trace the locations of the stars, but optical wavelength images are often confused by dust extinction and dominated by the light from a relatively few, bright young stars. Near–infrared images are much less affected by dust extinction $(A_K \approx 0.1A_V)$ and, for a normal stellar population, relatively unaffected by the presence of young stars. For example, in a population of stars similar to that found in the disk of the Milky Way $\sim 92\%$ of the total light at K is due to giant stars, $\sim 7\%$ due to main sequence stars, and only $\sim 1\%$ due to supergiants. Since main sequence stars, which dominate the mass, and giants are well–mixed in older populations (see Aaronson 1977 and Frogel 1988), K images should trace the mass effectively.

Sufficiently high rates of star formation and subsequent evolution of the high mass stars to red supergiants could affect the K light. If high mass stars form ~ 30 times more often than the average within the Milky Way's disk, then (ignoring extinction) supergiants would contribute $\sim 20\%$ of the total K light. Such high rates of star formation are observed in galaxies and could exist in isolated regions within a galaxy disk (e.g. Rieke et al. 1980). Note that the estimate of the contribution of high mass stars to the total K light from an evolving star formation event depends strongly on the fraction of their lifetimes massive stars spend as red supergiants. Empirically, this fraction depends strongly on metallicity; something present day massive star models do not adequately predict (see Langer and Maeder 1995).

D. L. Block and J. M. Greenberg (eds.), New Extragalactic Perspectives in the New South Africa, 109–112.
© 1996 *Kluwer Academic Publishers.*

2. Estimating the Contribution of Supergiants

Estimating the contribution of red supergiants to the K light in a galaxy is difficult. Emission is dominated by younger populations and extinction effects at short wavelengths; infrared colors are not sufficiently diagnostic. In some cases, however, a galaxy is near enough to resolve individual supergiants or OB associations. For example, Regan and Vogel (1994) obtained JHK images of M33 and demonstrated that in regions corresponding to the optical positions of OB associations there is a slight increase in the K emission. The overall effect was relatively small, however, only about 5% the total K light.

Fortunately, there are reasonably strong molecular and neutral atomic absorption features that can potentially serve as luminosity indicators in galaxies at greater distance. For example, absorption due to $CO(2,0)$ overtone bandheads are the strongest features near–infrared spectra of late–type stars. The equivalent width of this feature around 2.3 μm is sensitive to surface gravity and effective temperature (see Baldwin et al. 1973); for a given temperature the equivalent width is largest in supergiants and smallest in main sequence stars (see Kleinmann and Hall 1986). There is also some dependence on metallicity (see Bell and Briley 1991). Thus, quantitative interpretation based on CO measurements alone is difficult. Differential measurements may indicate regions of enhanced supergiant emission, however, as shown by Rix and Rieke (1993).

We have imaged several galaxies in narrow–band filters centered on the CO absorption ($\sim$2.36 μm), similar to those of Rix and Rieke (1993), and a filter centered on a relatively featureless region ($\sim$2.22 μm). These two filters are the pair traditionally used to measure a "CO Index" (see Baldwin et al. 1973). The "continuum" image can be compared to the broad–band K image to assess instrumental effects that can affect the results. We find results consistent with those of Rix and Rieke: there is no evidence in the disks of normal galaxies of any supergiant contribution greater than 20% the total K light.

We have also obtained moderate–resolution, long–slit spectra of the same galaxies. These spectra are of sufficient sensitivity to detect neutral metal (CaI, NaI, and, perhaps, MgI and FeI) absorption in addition to several CO bandheads. Figure 1 shows an example of a galaxy spectrum along with a spectrum of an M2III star for comparison. The strengths of the neutral metal lines are very similar to those measured in the giant spectrum. Figure 2 shows a comparison of the sum of the equivalent widths of the CaI and NaI features with $^{12}CO(2,0)$; again the galaxy spectra are most similar to giant star spectra. Unfortunately, there are few adequate spectra of supergiants available for comparison. The results for two are shown in figure

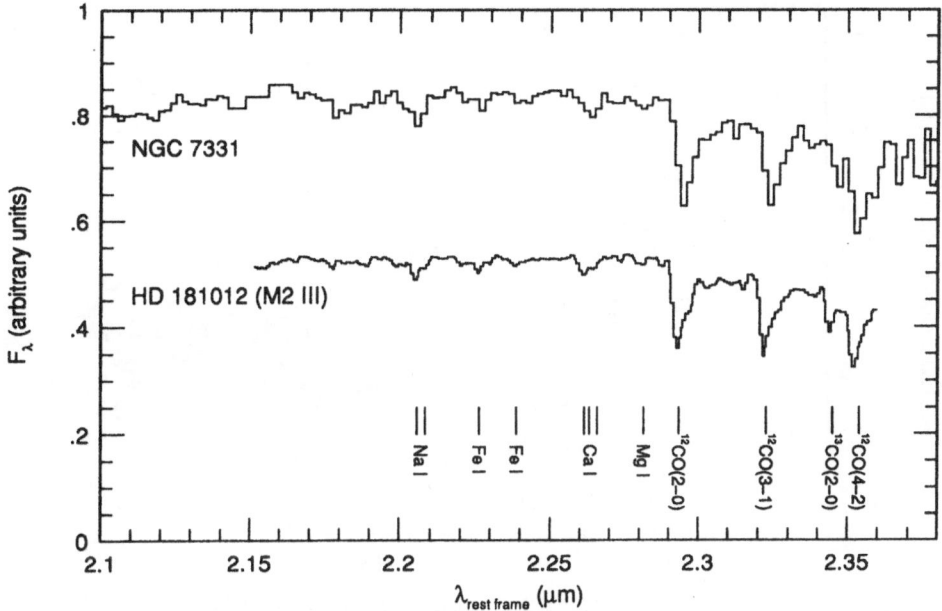

Figure 1. Comparison of K–band spectrum of the spiral galaxy NGC 7331 and an M2III star

2 (taken from Kleinmann and Hall). The difference between supergiants and giants is relatively small, suggesting that further work is required to properly interpret the results.

3. Conclusions

All CO measurements are consistent with a limit of $\sim 20\%$ for the contribution of supergiants to the total K light in a normal galaxy from the absence of significant variations. Spectroscopic measurements can detect other stellar features, but more observations are required to assess the contribution of supergiants.

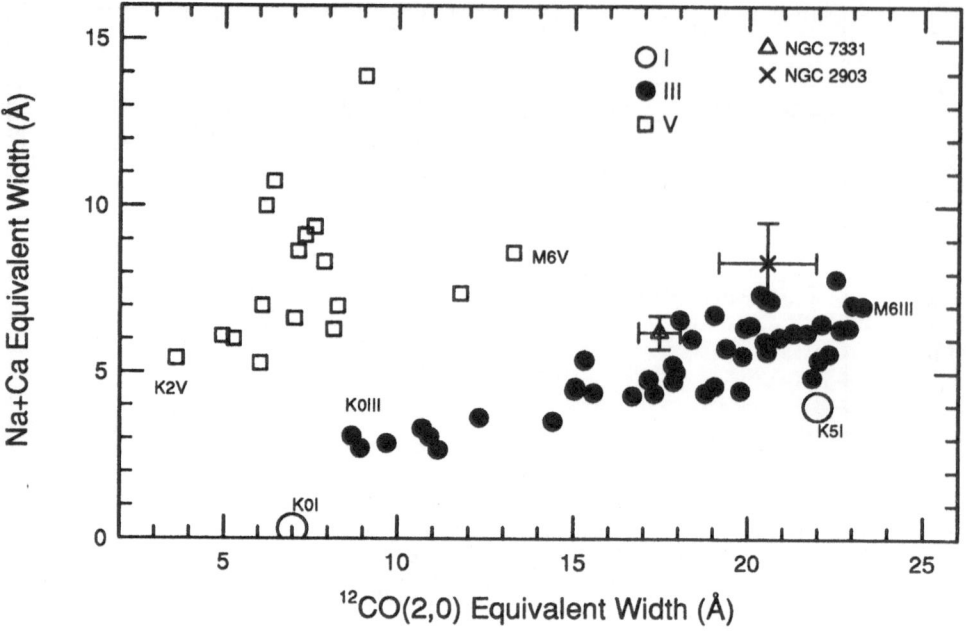

Figure 2. Equivalent widths of NaI and CaI absorption versus a CO bandhead for two spiral galaxies and various types of stars

References

Aaronson, M. 1977, Ph.D. Thesis, Harvard University

Baldwin, J., Frogel, J. A., & Persson, S. E. 1973, ApJ, 184, 427

Bell, R. A.,& Briley, M. M. 1991, AJ, 102, 763

Frogel, J. A. 1988, ARA&A, 26, 51

Kleinmann, S. B., & Hall, D. N. B. 1986, ApJS, 62, 501

Langer, N., & Maeder, A. 1995, A&A, 295, 685

Regan, M., & Vogel, S. 1994, ApJ, 434, 536

Rieke, G. H, Lebofsky, M. J., Thompson, R. I., Low, F. J., & Tokunaga, A. T. 1980, ApJ, 238, 24

Rix, H.-W., & Rieke M. J. 1993, ApJ, 418, 123

REFLECTIONS AT THE REGISTRATION DESK: RAY WHITE

[Editorial Comment: Every delegate delivering an invited talk at the Conference and/or presenting a poster agreed (nearly 12 months in advance) to hand in a camera-ready copy of their work, at the Registration desk. Together with the camera-ready hard copy, a 3.5-inch computer diskette was also required to be handed in, for editorial purposes. In South Africa, a 3.5-inch diskette is called a stiffy. Furthermore, apart from its usual computer connotations, a buzzword in the New South Africa is *PC*: 'Politically Correct'.]

Delegate **Ray White** reflects...

When I came to South Africa I was expecting to find a social revolution, but I had not anticipated how extensive it was.

For example, last night, when I handed in my camera-ready copy to David's assistant, I was asked:

"Do you have a stiffy?"

*After I recovered from the shock, I looked down and realized I **did** have a stiffy.*

So, with stiffy in hand, I offered it to the assistant, who recoiled and blurted:

"That's not PC!"

I was rather embarrassed.

He was right, of course. I had formatted my diskette with a Mac, not a PC!

D. L. Block and J. M. Greenberg (eds.), New Extragalactic Perspectives in the New South Africa, 113.
© 1996 *Kluwer Academic Publishers.*

DISTRIBUTION AND CONTENT OF DUST IN OVERLAPPING GALAXY SYSTEMS

R.E. WHITE III AND W.C. KEEL
University of Alabama
Department of Physics & Astronomy
Tuscaloosa, AL 35487-0324 USA

AND

C.J. CONSELICE
University of Chicago
Department of Astronomy & Astrophysics
5640 S. Ellis Avenue
Chicago, IL 60637 USA

1. Introduction

We present a progress report on our program to determine the opacity of spiral disks *directly*, rather than statistically, by imaging foreground spirals partially projected against background galaxies. The non-overlapping regions of partially overlapping galaxies can be used to reconstruct, using purely differential photometry, how much light from the background galaxy is lost in passing through the foreground galaxy in the overlap region.

There are several benefits to the direct, differential photometric approach we have adopted, including: 1) it is not subject to the selection effects influencing statistical studies; 2) there is no selection against high opacity regions; 3) the imaging technique involves only differential photometry; 4) large, contiguous areas can be analyzed, allowing average values of the opacity to be estimated; 5) there is no need to correct for the internal extinction of the background galaxy or the Milky Way; 6) scattering corrections are also differential, which can keep them slight.

This technique also has some disadvantages relative to others: 1) there are rather few tractable objects nearby enough for spatially well-resolved analysis; and 2) the success of the technique hinges on the degree of symmetry in both the foreground and background galaxies

114

D. L. Block and J. M. Greenberg (eds.), New Extragalactic Perspectives in the New South Africa, 114–117.
© 1996 *Kluwer Academic Publishers.*

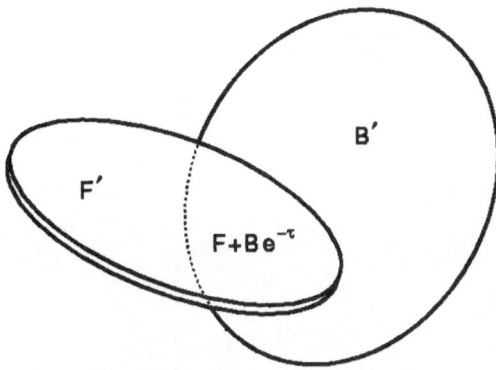

Figure 1.   The ideal overlapping galaxy pair for the direct opacity measures.

Figure 1 depicts the ideal case for constructing maps of opacity using differential photometry: a foreground disk (spiral) galaxy is half-projected against half of a similarly-sized background elliptical galaxy. For the sake of illustration, the (unobscured) surface brightness of each galaxy is taken to be constant, with F and B being the actual surface brightness values of the foreground disk and background elliptical in the overlap region, and τ is the optical depth in the disk. The observed surface brightness in the overlap region is then $\langle F + Be^{-\tau} \rangle$, where brackets are used to emphasize that this whole quantity is the observable in the overlap region and cannot be directly decomposed into its constituent components. We use symmetric counterparts from the non-overlapping regions of the two galaxies to *estimate* F and B and denote the estimates as F' and B'. We can then construct an estimate of the optical depth, denoted τ':

$$e^{-\tau'} = \frac{\langle F + Be^{-\tau} \rangle - F'}{B'}. \tag{1}$$

Here the estimate of the foreground spiral's surface brightness, F', is first subtracted from the surface brightness of the overlap region, $\langle F + Be^{-\tau} \rangle$; this result is then divided by the estimate of the background elliptical's surface brightness, E'. This creates a map of $e^{-\tau'}$ in the overlap region.

2. Sample Selection and Results

Our observing sample is largely drawn from numerical searches for overlapping neighbors in the *RC3, RSA, ESO-Uppsala, UGC, RNGC, NGC2000, MCG*, Karachentsev, and Zhenlong et al (1989) catalogues. We also inspected individual catalog entries in the *UGC, ESO-LV*, and *NGC* listings which were typed as inherently multiple systems. In addition, we visually inspected all pairs in the Arp-Madore catalog, the Arp *Atlas of Peculiar*

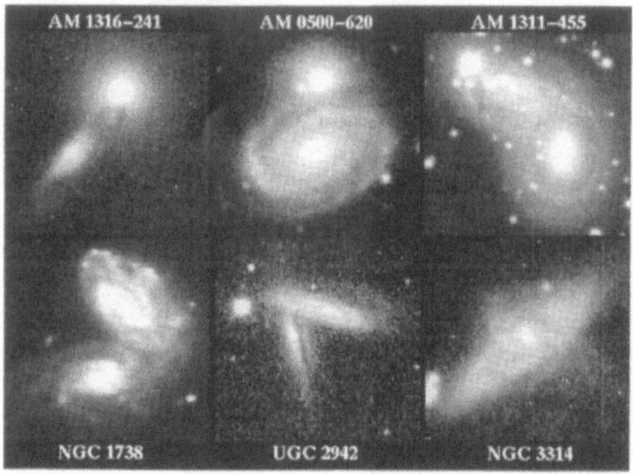

Figure 2. Six galaxy pairs from which extinctions have been deduced.

Galaxies, and the Reduzzi & Rampazzo (1995) catalog of southern pairs. Figure 2 shows six of the dozen or so galaxy pairs that we have successfully analyzed after imaging them in the B and I bands.

3. Results

Table 1 lists the extinctions determined in arm and interarm regions in various backlit galaxies. For each measurement, we list its radial distance from the center of the foreground galaxy in units of R_{25}^B. Typical errors in the extinctions are ~ 0.15 mag. Figure 3 plots the contents of Table 1.

4. Conclusions

We generally find that opacity is concentrated in spiral arms, while interarm regions are nearly transparent, particularly beyond $\sim 0.5R_{25}^B$. This confirms and extends our previous results (White & Keel 1992, Keel & White 1995). Since most disk light comes from spiral arms, the dust opacity is correlated with the emission. This correlation may explain why the surface brightness of spirals is found to be independent of inclination in statistical tests (Valentijn 1991), erroneously implying that spiral disks are generally opaque.

This project was supported by the NSF and the State of Alabama through the $EPSCoR$ program, by the *NSF Research Experience for Undergraduates* program, and by Lowell Observatory and NOAO.

TABLE 1. Face-on-corrected Magnitudes of Extinction

Galaxy Pair	arm R/R_{25}^{B}	A_B	A_I	interarm R/R_{25}^{B}	A_B	A_I
AM 0500-620	0.6	> 2.3	1.64	0.5	0.1-0.47	0.0-0.55
AM 1311-455	1.18	0.73	0.24	0.95	0	0
AM 1316-241	0.75	0.38	0.16	0.4-0.75	0.08	0.05
ESO 0320-51	0.65	0.27	0.17	0.50	0	0
NGC 450				0.95-1.0	< 0.1	< 0.1
NGC 1739	0.65	0.3-0.4	0.24-0.3	0.55	0.2-0.26	0.16
NGC 4568	0.5-0.9	1.1	0.69			
NGC 3314	0.16	1.6	1.24	0.19	1.11	1.60
"	0.34	1.64	0.82	0.28	0.77	0.59
"	0.42	1.11	0.82	0.39	1.75	0.63

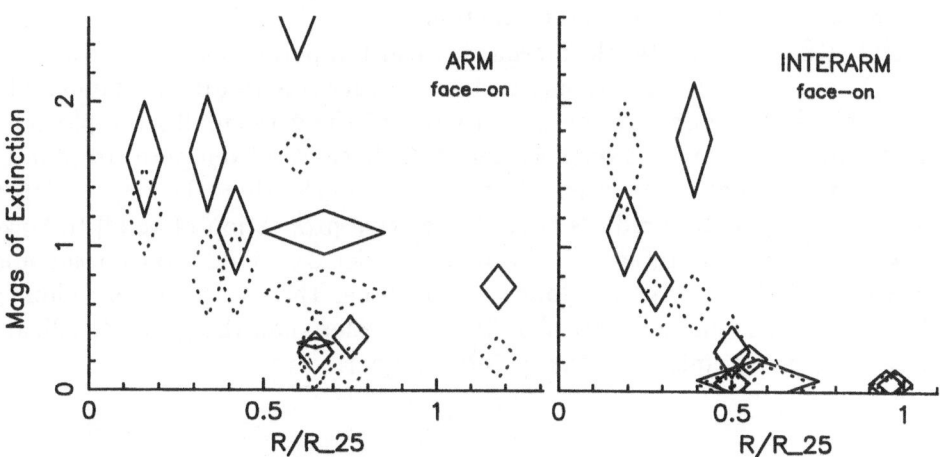

Figure 3. Extinction measures in arm and interarm regions. Values of A_B and A_I are plotted as solid and dotted diamonds, respectively.

References

Keel, W.C. & White, R.E. III 1995, in *The Opacity of Spiral Disks*, eds. J.I Davies & D. Burstein (Kluwer, Dordrecht), p.167

Reduzzi, L. & Rampazzo, R. 1995, ApLComm, 30, 1

Valentijn, E.A. 1990, Nature, 346, 153

White, R.E. III & Keel, W.C. 1992, Nature, 359, 129

Zhenlong et al 1989, Publ. Beijing Astron Obs. 12, 8

EVOLUTION AND EMISSION OF COLD, WARM AND HOT DUST POPULATIONS IN DIFFUSE AND MOLECULAR CLOUD

J.MAYO GREENBERG & AIGEN LI

Laboratory Astrophysics, University of Leiden, Postbus 9504, 2300 RA Leiden, The Netherlands
email: mayo@rulhl1.LeidenUniv.nl; agli@strw.LeidenUniv.nl

Abstract. The cyclic evolution of a multimodal interstellar grain population and its chemical, morphological and optical properties are described
in terms of dust in molecular and diffuse clouds. The temperatures characteristic of the "large" tenth micron core-mantle particles, which dominate
the mass of the dust and the extinction in the visual, are calculated. For
typical situations the tenth micron particle temperatures range from 6 K
to 15 K in cool molecular clouds and are about 16 K in diffuse clouds. The
small (hundredth micron) hump particles and the very small FUV particles
(large molecules) are not only warmer but, because of temperature fluctuations, emit radiation at much shorter wavelengths than the tenth micron
particles. The model predicts that the relative proportion of small to large
particles is larger in diffuse clouds than in molecular clouds so that the mid
infrared emission in diffuse clouds is far higher than in molecular clouds.
The core-mantle particles are significantly cooler than the graphite-silicate
mixture and the emission peak is $70\mu m$ higher.

1. Introduction

The interstellar particles in spiral galaxies encompass a wide range of
sizes and physical properties. The sizes range from the equivalent of large
molecules up to tenth micron. Most of the visible and ultraviolet radiation in the galaxy from stars passes through clouds of particles and heats
them. This heating leads to reradiation at much longer wavelengths extending to the millimeter. The converted radiation is a probe of the particles
and the physical environments in which they find themselves. The particle
properties and the environment are coupled via the stages of evolution of
the interstellar medium and the local stellar population. With the advent

D. L. Block and J. M. Greenberg (eds.), New Extragalactic Perspectives in the New South Africa, 118–134.
© 1996 *Kluwer Academic Publishers.*

of new space telescopes the thermal emission by dust at the lowest temperature is becoming accessible and it should be possible to follow both variations in local environment and in dust populations.

The temperatures of the interstellar dust depend on their optical properties and sizes; i.e., on the way they absorb and emit radiation. The temperatures published by many authors (Mezger et al 1982; Draine & Lee 1984; Siebenmorgen & Krügel 1992) have been calculated based essentially on the graphite-silicate model as originally proposed by Mathis, Rumpl & Nordsieck (1977) with an added component of PAH's (polycyclic aromatic hydrocarbon). An exception to this is the work of Désert et al (1990) where the large particles are considered to be silicate cores with organic mantles as suggested by Greenberg (1973). The graphite-silicate model does not readily lead itself to being considered in terms of variation with environment. On the other hand, the evolutionary model consisting of silicate core-organic refractory and ice mantle grains with small carbonaceous particles and PAH's provides a consistent basis for following the variation of grain properties in different astrophysical regions (Greenberg 1978; Jenniskens & Greenberg 1993;, de Groot et al 1988; Hong & Greenberg 1980; Chlewicki & Greenberg 1990; etc). The consistency of this model with observation has recently been accorded additional confirmation from studies of the silicate spectral features at $9.7\mu m$ and $18\mu m$ (Greenberg & Li 1996) and the $3.4\mu m$ organic feature in the diffuse interstellar medium (Greenberg et al 1995). As a result of these studies it has been possible to obtain a more complete set of grain material optical properties of the core-mantle particles required for predicting their grain temperatures and emission in the infrared and submillimeter.

The situation for the optical properties of the small carbonaceous particles and the PAH's is quite different. In essence these are obtained semi-empirically, that is, by less direct methods.

In this paper we describe in Section 2 a picture of the evolutionary model of interstellar dust in terms of populations and optical properties and how they may be expected to differ in diffuse and molecular clouds. We summarize the optical properties of the different grain populations. We shall not consider here the immediate neighbourhood of stars deferring that to a later paper. Grain temperatures are presented in Section 3 and in Section 4 the resulting grain emissions in the infrared and submillimeter are shown for typical diffuse cloud interstellar dust and for a typical quiescent molecular cloud.

2. Interstellar grain model

2.1. POPULATIONS

The most consistent key to the grain size populations is provided by the interstellar extinction curve as shown in Fig. 1 for the average diffuse regions. This curve may be deconvolved to consist of a component of large tenth micron particles which produce the visual and ultraviolet extinction, hundredth micron particles — most probably carbonaceous — which produce the 2200 Å hump, and very small particles — probably PAH's — which produce the far ultraviolet (hereafter FUV) extinction. There appears to be an additional component producing a linear contribution but, as yet, this has not been ascribed to be an extra component in the grain populations but rather as a possible effect produced by small surface irregularities on the large grains. In the following we shall neglect its contribution to grain temperatures and emissivities.

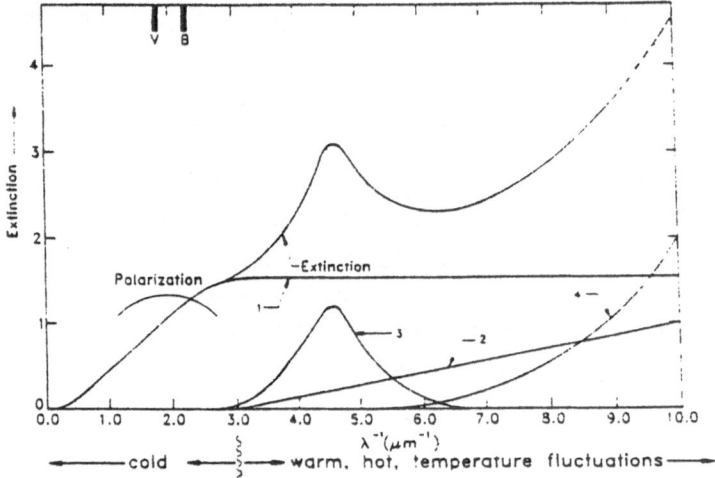

Figure 1. Decomposition of the average interstellar extinction curve into 4 components. Part 1 corresponds to a population of particles with mean size $\sim 0.1\mu m$. Part 2 may be due to small surface perturbation on Population 1 particles. Part 3 is due to a population of small ($\leq 0.01\mu m$) carbonaceous particles. Part 4 is suggested as due to a population of large molecules — PAH's. See Greenberg & Li (1995).

2.2. MATERIAL AND OPTICAL PROPERTIES

We generally define the optical properties as given by the complex index of refraction $m(\lambda) = m'(\lambda) - im''(\lambda)$. We shall assume that the large particles consist of silicate cores with organic refractory mantles. In recent work (Greenberg & Li 1996) we have described how the best match to observations of the $9.7\mu m$ and $18\mu m$ polarization in the Becklin-Neugebauer object can be obtained by using the optical properties of a laboratory silicate (Dorschner et al 1995) mantled by organic refractory material as it has evolved from organics photoproduced from ice mantles in molecular clouds and then further photoprocessed in diffuse clouds. Insofar as the far infrared is concerned, we model the silicates (following Henning & Stognienko 1996) by letting the absorptivity fall off as λ^{-1} beginning at $\lambda = 100.0\mu m$ according to $m'' \propto 0.3\,(100.0/\lambda)$. For the organics we use the data derived from the visible and near ultraviolet by Jenniskens (1993), and for the infrared by Greenberg & Li (1996). In all cases the real and imaginary parts of the indices of refraction are related via the Kramers-Kronig relation so that they are physically consistent. In Fig.2 are shown the optical constants for the silicate and organic refractory materials, respectively used here to calculate extinction and temperatures. Some of the details of these curves are subject to change but these will not affect the derived temperatures in Section 3 significantly.

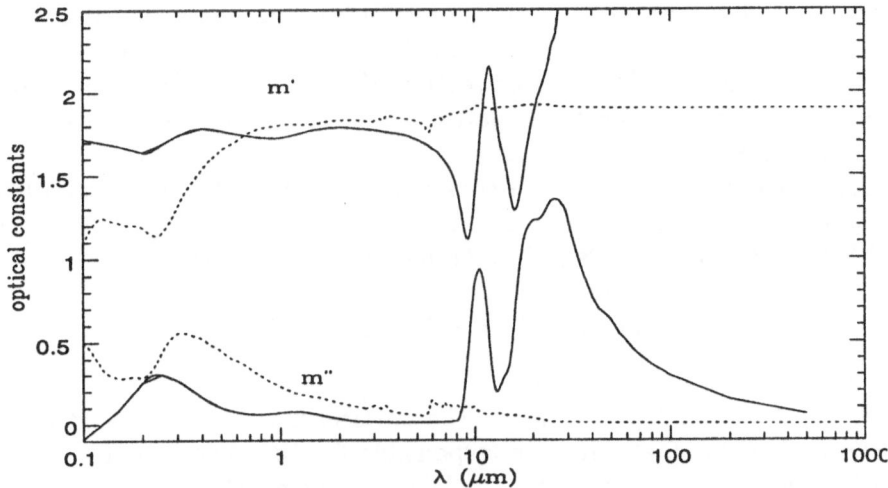

Figure 2.   The optical constants of silicate (solid line) and organic refractory (dotted line).

The particles responsible for the 2200 Å hump were originally thought to be graphite but this has been strongly disputed for numerous reasons. Thus while they are not yet fully identified (Bussoletti et al 1987; Sakata et al 1984; de Groot et al 1988; Draine 1988; etc) it is generally believed that they are carbonaceous. In the hump region ($\lambda \leq 0.5\mu m$) we empirically assume that their absorption cross section has an underlying continuum which is linear in λ^{-1} and, following Desert et al (1990) and Siebenmorgen

& Krügel (1992), we describe them by a Drude profile

$$\frac{Q_{abs}(\lambda,a)}{a} = (6.6 \times 10^4 x + \frac{1.92 \times 10^6 x^2}{(x^2 - x_0^2)^2 + \gamma^2 x^2}) \, cm^{-1} \tag{1}$$

where $x_0 = 4.6 \mu m^{-1}$, $\gamma = 1.0 \mu m^{-1}$ is the peak position and width of the Drude profile, respectively (Fitzpatrick & Massa 1986); x is $\lambda^{-1} \mu m^{-1}$; a is the size of the hump particle; and the absorption efficiency $Q_{abs}(\lambda,a)$ is normalized to match that of small graphite particles (Draine & Lee 1984) at the hump. For the rest of the wavelength region we use the graphite optical constants (Draine & Lee 1984) to calculate the absorption efficiencies by Mie theory (van de Hulst 1957).

The FUV particles, represented by PAH's, can not be described in terms of their complex indices of refraction, although some effort has been put into determining the optical constants of several PAH molecules in the laboratory (see, e.g., Cherchneff et al 1991). The absorption cross sections of PAH's are still uncertain, but some approximate results have been obtained both through laboratory and theoretical investigations (Leger et al 1989a; Leger et al 1989b; Schutte et al 1993; etc.). Following Desert et al (1990), we adopt the absorption cross section of PAH's which is taken empirically by subtraction of the large particle and the hump component from the curvature of the extinction curve in the FUV. This curvature in the diffuse cloud medium appears to be constant (Greenberg & Chlewicki 1983). We shall use a typical mean size defined by the number of carbon atoms per PAH, $N_c \approx 80$, which is ~ 7 Å (Puget & Leger 1989).

2.3. THE EXTINCTION CURVE AND GRAIN SIZES

2.3.1. *Diffuse cloud grains*

Given the optical properties of the particle material described above one can derive the dust particles responsible for the average extinction curve. Although we know that the particles producing the visual extinction are nonspherical, as indicated by interstellar polarization, this will have little effect on their temperatures so that we shall hereafter consider only spherical core-mantle particles. In order to get a reasonable match to the long wavelength part of the extinction, we use particles with silicate cores of radius $a_c = 0.07 \mu m$ and organic refractory mantles with a size distribution $n(a_m) \propto e^{-5(\frac{a_m - a_c}{a_i})^3}$ as in Greenberg (1968) where a_i is an effective cut-off size parameter which produces a mean radius $a_m = 0.10 \mu m$.

The contribution by the hump particles is given by the dashed curve in Fig.3 where we have chosen a mean hump particle size $a_{hump} = 0.004 \mu m$ which is small enough not to produce any scattering contribution in the hump or FUV regions — only absorption.

The contribution by the PAH's is then taken, as prescribed in the previous section, to provide the extra extinction curvature in the FUV. The remaining linear portion is left to be determined.

The combined result for the three particle populations is shown by the thick solid line in Fig.3, to give a rather good match to the full average diffuse cloud extinction (Savage & Mathis 1979).

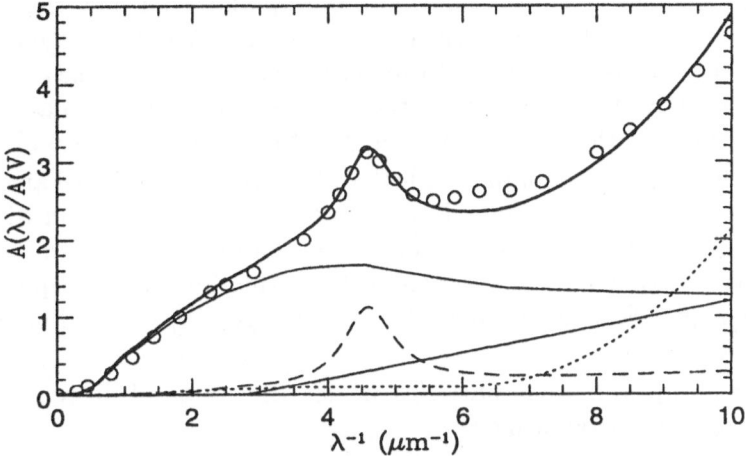

Figure 3. The observational and theoretical mean diffuse medium extinction curve. The circles are the observed average extinction curve from Savage & Mathis (1979). The thick solid line refers to our model prediction. Also shown are the individual contributions of the three dust components : large core-mantle particles (solid line + the linear part); hump particles (dashed line); PAH's (dotted line).

2.3.2. *Molecular cloud grains*

In molecular clouds the grains have mantles in the form of ice which contains many molecules as produced by accretion, surface reactions and photoprocessing (Greenberg et al 1972; Schutte 1996; etc). We are not concerned here with the detail of these mantles but only with the fact that they increase the size of the large particles. The small hump particles do not accrete such mantles from the gas for well defined physical reasons as described earlier (Greenberg & Hong 1974a; Greenberg 1980; Aannestad & Kenyon 1979). On the contrary, as a result of different relative velocities between the large particles and the hump particles one expects the hump particles to hit and stick within the icy mantles. There are, in fact, well known instances where the hump contribution is greatly reduced not only relative to the visual extinction but also relative to the cloud density (Mathis 1990; Laureijs, Clark

& Prusti 1991). Although some of the hump particles may be returned as a result of icy mantle desorption by explosion or other processes (Schutte & Greenberg 1992) we expect that a major fraction remain because they are too massive to be ejected; i.e., they act like very large molecules. Note that we are not considering here the desorption which occurs in the vicinity of hot stars where evaporation may be significant.

For the PAH's, it has been shown that if they accrete in an icy mantle they are likely to be destroyed (Mendoza-Gómez, de Groot & Greenberg 1995) and probably incorporate in the mantle as large molecules.

To treat the decrease in population of hump and PAH particles relative to the large particles along with the increase in the size of the large particles is beyond the scope of this paper. However we have tried to make reasonable estimates to show to what extent the changes in grain populations can produce changes in the overall dust temperature and resulting emission.

For the large particles in one typical molecular cloud we have arbitrarily assumed that they have an added mantle of H_2O ice of 0.005 μm thickness (Schutte 1996) within the cloud. This provides an average 3.1 μm ice band strength. We further let the total to visual extinction ratio be quite high, $R = A_v/E_{B-V} = 5$, as compared with its average diffuse cloud value $R \simeq 3.1$.

For the hump and PAH particles we let their rate of decrease be described by accretion following

$$n(r,t) = n_0 e^{-\alpha(r)t} \tag{2}$$

where $\alpha(r)$ is the depletion time scale given by

$$\alpha(r) = \frac{1}{n_d(r)\sigma_d \delta v} \tag{3}$$

where $n_d(r)$, σ_d are the number density and mean area of the dust grains and where δv is the mean small particle—large particle collision speed. Note $n(r,t)$, $n_d(r)$ and $\alpha(r)$ are functions of the position r within the cloud. For the hump particles, δv is probably greater than would be given by kinetic motion alone and is possibly as large as $\sim 0.1\,km\,s^{-1}$ which would result from turbulent motion (Völk et al 1980). For the PAH's the kinetic speed at $T_{gas} = 30$ K is $\delta v = 0.03\,km\,s^{-1}$ and for the hump particle is even smaller. For the hump particles, δv_{hump} is less than turbulent produced motion.

We use the decrease of hump and PAH particles given by Equation 2 where we consider a typical molecular cloud age of $\approx 10^7$ years and where we assume $\delta v_{PAH} \simeq 0.1\,km\,s^{-1}$ and $\delta v_{hump} \simeq 0.05\,km\,s^{-1}$.

3. Dust temperatures

3.1. THE TEMPERATURES OF CORE MANTLE PARTICLES

Dust grains absorb energy from the interstellar radiation field (hereafter ISRF) and emit the same amount of energy at longer wavelengths. This energy balance defines the steady-state temperature of dust particles as follows

$$\int\limits_{0}^{\infty} \pi a^2 Q_{abs}(a,\lambda) 4\pi J_\lambda d\lambda = \int\limits_{0}^{\infty} \pi a^2 Q_{abs}(a,\lambda) 4\pi B(T_d,\lambda) d\lambda \qquad (4)$$

where a is the size of the dust which is $\sim 0.1\mu m$ for core-mantle particles; $4\pi J_\lambda$ is the flux density of the interstellar radiation field; $B(T_d,\lambda)$ is the Planck function for the grain temperature T_d; $Q_{abs}(a,\lambda)$ is the absorption efficiency. For simplicity, we take the particle as a sphere. Although we are sure that the interstellar dust particles have a size distribution and are nonspherical on the basis of the observed polarization, it is sufficiently accurate to represent the particle as a single $\sim 0.1\mu m$ sphere when one calculates the dust temperatures for which the shape effect is negligible (Greenberg & Shah 1971).

As shown in Equation 4, the ISRF is quite essential to calculate the dust temperature. Historically, the ISRF in the solar vicinity was represented by a 10^4 K black-body radiation diluted by a factor of 10^{-14} (Eddington 1926) (in this paper we confine ourselves to the solar neighbourhood). A more reasonable representation was given analytically by Greenberg (1978). More recently, Mathis, Mezger & Panagia (1983, hereafter MMP)'s extensive investigation described the ISRF by

$$4\pi J_\lambda = 4\pi J_\lambda^{uv} + \sum_{i=2}^{4} W_i 4\pi B_\lambda(T_i) + 4\pi J_\lambda^{dust} \qquad (5)$$

where $4\pi J_\lambda^{uv}$ is the ultraviolet component and $(W_2, W_3, W_4) = (10^{-14}, 10^{-13}, 4 \times 10^{-13})$ and $(T_2, T_3, T_4) = (7500, 4000, 3000)$ K are dilution factors and corresponding black-body temperatures. The first and second part of the right hand side of the equation are the stellar contributions to the ISRF ($\lambda \leq 8\mu m$) and the third part is the dust contribution ($\lambda \geq 8\mu m$). Compared with the MMP field, the Greenberg field (1978) neglects the dust contribution to the ISRF. But this would not make much difference in the dust temperature evaluation because the dust particles absorb mainly at short wavelengths and the ISRF is dominated by the short wavelength part ($\lambda \leq 8\mu m$) which is attributed to stellar emission. Actually, the derived dust temperatures are similar whichever of the two kinds of ISRF approximation is used (Greenberg & Li 1995). In this work, we shall adopt the MMP ISRF.

For dust grains in the diffuse medium, with the optical constants discussed in Section 2.2, the absorption efficiencies $Q_{abs}(\lambda, a)$ of the 0.1 μm silicate core-organic refractory mantle particle can be calculated by Mie theory (Van de Hulst 1957). The dust temperature derived from energy balancing with the MMP ISRF is $\sim$ 16 K.

The case for the dust grains in molecular clouds is much more complicated. In order to obtain the exact radiation field within a molecular cloud one should solve the radiative transfer equations. For simplicity we crudely assume that the ISRF at the position r within a molecular cloud is attenuated by a factor $e^{\tau(\lambda, r)}$ and the lower limit is set to be $\sim 10^4 Ly\alpha\, photons\, cm^{-2}s^{-1}$ which is produced by cosmic rays. $\tau(\lambda, r)$ is the optical depth at position r. The cloud center corresponds to $r = 0$ while $r = R$ corresponds to the cloud surface.

$$\tau(\lambda, r) \propto \frac{\int_r^R n(r)dr}{\int_0^R n(r)dr} \frac{A_v(0)}{1.086} A(\lambda) \qquad (6)$$

where $A_v(0)$ is the visual extinction from the center to the surface of a (spherical) molecular cloud; $n(r)$ is the dust number density distribution within the cloud; $A(\lambda)$ is the average extinction curve within the cloud ($R_v = 5$; Mathis 1990). It is worth noting that our attenuation approximation (simply multiplying the ISRF by $e^{-\tau(\lambda, r)}$) is not unreasonable because, for a moderately obscured molecular cloud, the main heating source of dust is the penetrating UV photons of the ISRF, while the contribution by the infrared reradiation of the dust in the cloud is not significant. Including the heating of the dust reradiation leads to a very small (less than several per cent) increase of the dust temperature and corresponding small extra submillimeter emission (Greenberg 1971; Bernard et al 1992). This effect was originally semi-quantitatively pointed out by Greenberg (1971) and was confirmed in detailed calculations by Bernard et al (1992).

The core-mantle dust particles in a molecular cloud are mantled by a thin additional ice layer. We use Maxwell-Garnett effective medium theory (Maxwell-Garnett 1904; Bohren & Huffman 1983) to calculate the effective optical properties of the silicate core-organic refractory mantle plus ice outer mantle particles. The total dust size with the ice mantle is $0.105\mu m$.

In this work we only consider a molecular cloud with visual extinction $A_v(0) = 5$ mag, as an example. We assume the hydrogen density in the cloud center is $n_H(0) = 2 \times 10^4\, cm^{-3}$ and constant to $\frac{r}{R} \sim 0.2$, thereafter following an $n_H(r) \propto r^{-2}$ distribution to the outer radius of 0.3 pc where it is $n_H(R) = 10^3 cm^{-3}$. It is reasonable to assume that the dust number distribution within the cloud behaves similarly to that of hydrogen. Based on these assumptions, the wavelength dependent optical depth $\tau(\lambda, r)$ can be derived at each point within the molecular cloud and the radiation field in the molecular cloud $4\pi J_\lambda(r)$ follows from this. Therefore we can derive the dust temperature distribution within the cloud and the results are shown in Fig.4. It is clear that the dust in molecular clouds is much colder than in a diffuse cloud; e.g., $T_d \sim 6.5$ K in the molecular cloud center, $T_d \sim 12$ K in the middle of the cloud and $T_d \sim 15$ K at the cloud surface.

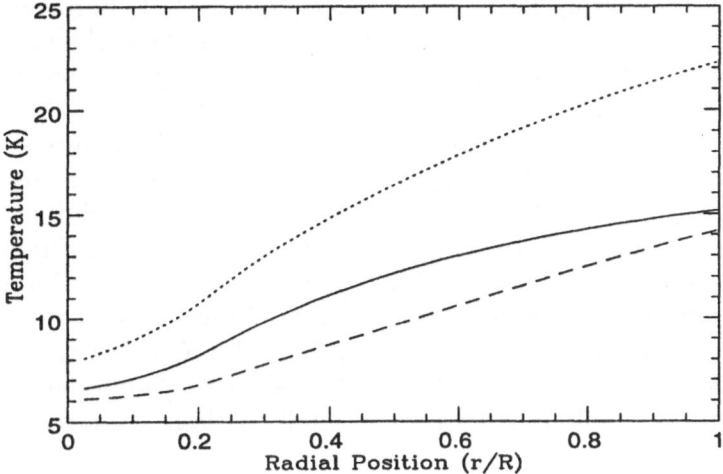

Figure 4. The dust temperature radial distribution within a molecular cloud described in the text. R is the radius of the cloud : solid line — large core-mantle particles; dotted line — graphite; dashed line — silicate.

3.2. THE TEMPERATURES OF HUMP PARTICLES AND *PAH'S*

One should keep in mind that the above "steady-state temperature" is only applicable to those particles whose sizes are large enough that their total heat content is much larger than the energy of a single absorbed photon. Very small particles, say, $a \leq 0.01\mu m$, have a heat content comparable with or smaller than the energy of a typical interstellar UV photon, so that each time an energetic photon is absorbed the temperature rises almost instantaneously to a peak value, then relatively quickly drops to something below the average. The grain spends most of the time at a temperature below the average temperature. Such so called "temperature fluctuations" were first suggested almost 30 years ago by Greenberg (1968) and, since

then, both theoretical and observational studies concerning this subject
have been carried out (Greenberg & Hong 1974b; Greenberg & Hong 1976;
Purcell 1976; Draine & Anderson 1985; Desert et al 1986; Guhathakurta &
Draine 1989).

As far as the hump particles and PAHs are concerned, the effect of
temperature fluctuations must be taken into account. If we use the enthalpy
of graphite (Chase et al 1985) for that of the hump particles and assume
the mean energy of an absorbed photon is $\sim$ 6 eV, a hump particle with
a size of 40 Å will have a peak temperature of 67.8 K while its thermal
equilibrium temperature is only 21.7 K. Note that this is still significantly
higher than the large core-mantle dust temperature.

For PAHs, following Leger & d'Hendecourt's (1987) approximation for
the specific heat of PAH's, one can estimate that an 80-carbon-atom PAH
would attain a peak temperature of $\sim$ 880 K, following the absorption of a
$\sim$ 6 eV photon.

In order to study the infrared emission spectra of such very small particles, knowing the peak temperatures is not enough. One has to take the
temperature evolution into account (Greenberg & Hong 1976; Purcell 1976;
Draine & Anderson 1985; Desert et al 1986; Guhathakurta & Draine 1989).
Following Purcell (1976) we calculated the temperature distribution function $D(T)$ which is the probability for a heated dust grain temperature to
be between T and $T + dT$ for the hump particles and PAHs. The results
are shown in Fig.5. We note that we only consider single-photon absorption
because multiple photon events for our small particles are rare.

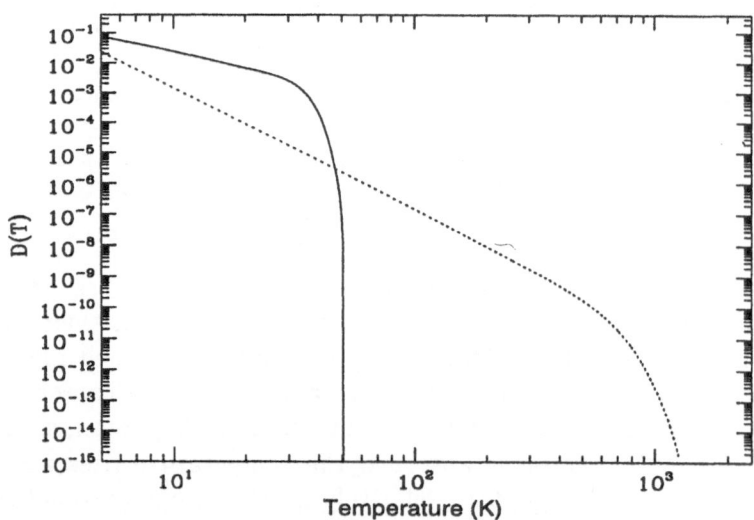

Figure 5.   The temperature distribution functions of the hump particles (solid) and
PAH's (dotted).

4. Dust emissions

Interstellar dust is a powerful source of far infrared and millimeter emission as well as extinction. Thus dust emission can also provide further tests of or constraints on interstellar dust models. It is well known that the reradiation of the heated interstellar dust grains by the ISRF photons at their thermal equilibrium temperatures $(T_d \leq 20 \text{ K})$ peaks at wavelength longer than, say $140 \mu m$ (Greenberg 1971; Mezger et al 1982; MMP; Greenberg & Li 1995). This has been confirmed by the COBE FIRAS survey (Wright et al 1991).

One of the most important results from the IRAS survey was the discovery of the presence of dust emission at wavelengths much shorter than predicted from the dust emission at its thermal equilibrium temperature (Low et al 1984; Boulanger & Perault 1988). These IRAS measurements also provided strong support for the presence of very small particles (hump particles, PAH's) in our dust grain model. As a matter of fact, this kind of phenomenon had already been observed in a wide variety of astrophysical environments such as HII regions, planetary nebulae, reflection nebulae and galaxies (Andriesse 1978; Aitken 1981; Price 1981; Sellgren 1984; Pajot et al 1986; etc).

We represent the predicted dust emission $R(\lambda)$ of our three component dust model as consisting of three parts: $R_{CM}(\lambda)$, $R_{hump}(\lambda)$ and $R_{PAH}(\lambda)$. The emission contributed by the large core-mantle particles is

$$R_{CM}(\lambda) \sim n_{CM} \pi a^2 Q_{abs}(a, \lambda) B(\lambda, T_d); \tag{7}$$

the emission contributed by the hump particles is

$$R_{hump}(\lambda) \sim n_{hump} \pi a^2 Q_{abs}(a, \lambda) \int_0^\infty B(\lambda, T) D_{hump}(T) dT; \tag{8}$$

the emission contributed by the PAH's is

$$R_{PAH}(\lambda) \sim n_{PAH} \sigma(a, \lambda) \int_0^\infty B(\lambda, T) D_{PAH}(T) dT; \tag{9}$$

where n_{CM}, n_{hump} and n_{PAH} are the number densities of the large core-mantle particles, hump particles and PAH's, respectively; $D_{hump}(T), D_{PAH}(T)$ are the dust temperature distribution functions of the hump particles and PAH's respectively; $\sigma(a, \lambda)$ is the absorption cross section of PAH's as discussed in Section.2.2. n_{CM}, n_{hump} and n_{PAH} are constrained by modeling the interstellar extinction curve (see Section 2.3.1). The best match to the extinction curve provides $n_{CM} \sim 1.3 \times 10^{-12}$ per H atom, $\frac{n_{hump}}{n_{CM}} \sim 850$

and $\frac{n_{PAH}}{n_{CM}} \sim 1.6 \times 10^5$.

Using the dust model parameters discussed in the above sections, we calculate the diffuse medium dust emission : the three separate components as well as the total dust emission spectra. The results are shown in Fig.6. Also shown are some observations, including the IRAS measurements of the high galactic latitude "cirrus" clouds which are believed to be representative of the average diffuse medium dust infrared emission of the solar vicinity (Boulanger & Perault 1988); the submillimeter observations (Lange et al 1986; Fabbri et al 1986; Pajot et al 1986); and the COBE FIRAS observations (Wright et al 1991). Our model prediction is in good agreement with the observations. It is obvious that the emission measured by IRAS is mostly contributed by PAH's and the hump particles. The large core-mantle particles are too cold to be detectable by IRAS while the far infrared ($\lambda \geq 100\mu m$) and submillimeter emission is completely contributed by the large core-mantle particles. Furthermore, we note that our predicted emission fails to exactly reproduce the COBE FIRAS observation (Wright et al 1991). This may be due to the fact that the COBE FIRAS observed spectrum includes emission from star-forming regions and mass-loss stars in which the dust is hotter than diffuse cloud dust (Draine 1994).

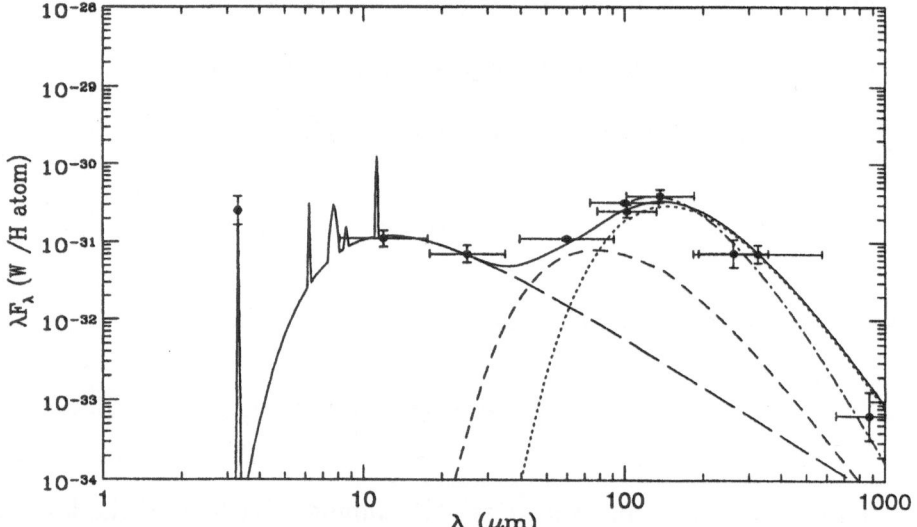

Figure 6.   The emission spectrum of interstellar dust in diffuse clouds. The points are observational data; the model predicted spectrum (solid line) is the sum of the three dust component : PAH's (long dashed); hump particles (dashed); large core-mantle particles (dotted). The COBE FIRAS spectrum ($\lambda \geq 110\mu m$) is also shown as dot-dashed line.

We have also studied the dust emission in molecular clouds, taking into account the fact that the very small particles would be depleted by being accreted onto the large core-mantle particles. Similar to the case for diffuse cloud dust emission, we calculate each component separately,

$$R_{CM}(\lambda) \sim \pi a^2 Q_{abs}(a, \lambda) \int_0^R B(\lambda, T_d(r)) n_{CM}(r) dr; \qquad (10)$$

$$R_{hump}(\lambda) \sim \pi a^2 Q_{abs}(a, \lambda) \int_0^R \int_0^\infty B(\lambda, T) D_{hump}(T) n_{hump}(r) dT dr; \qquad (11)$$

$$R_{PAH}(\lambda) \sim \sigma(a, \lambda) \int_0^R \int_0^\infty B(\lambda, T) D_{PAH}(T) n_{PAH}(r) dT dr; \qquad (12)$$

$$R(\lambda) = R_{CM}(\lambda) + R_{hump}(\lambda) + R_{PAH}(\lambda) \qquad (13)$$

Note that n_{CM}, n_{hump}, n_{PAH} and T_d are now functions of the position r within the cloud. n_{hump} and n_{PAH} are constrained by Equation 2 and 3.

Based on the dust model parameters and the dust temperature distribution functions $D(T), T_d(r)$, we derive the dust emission in the molecular cloud described in Section 3.1 and show the results in Fig.7. The most important feature in Fig.7 is that, in a molecular cloud, the dust emission is dominated by the large core-mantle (with ice mantle) particles and peaks at longer wavelength ($\lambda \geq 200\mu m$) than the diffuse cloud dust. In order to compare with the graphite-silicate model (Mathis et al 1977), we also calculated the dust emission separately of graphite and silicate and a mixture based on the Mathis et al (1977) dust model. These results are also shown in Fig.7. Obviously, the dust emission predicted by the Mathis et al dust model is higher than that of the Greenberg cyclic evolutionary model, and also peaks at a significantly shorter wavelength.

5. Concluding remarks

The trimodal evolutionary dust model consisting of large core-mantle particles, small carbonaceous particles and PAH's leads to very different infrared and submillimeter emission distributions in diffuse and molecular clouds not only in amount per unit mass but in spectral distribution. In diffuse clouds, the total flux radiated in the mid-infrared is of the order of that in the long wavelength and submillimeter region while in molecular clouds this ratio is greatly reduced. The peak emission from a molecular cloud occurs at around $200\mu m$ as compared with $\sim 120\mu m$ from diffuse clouds. In both cases the major mass fraction of this cold interarm dust ($\sim 80 - 90$ %)

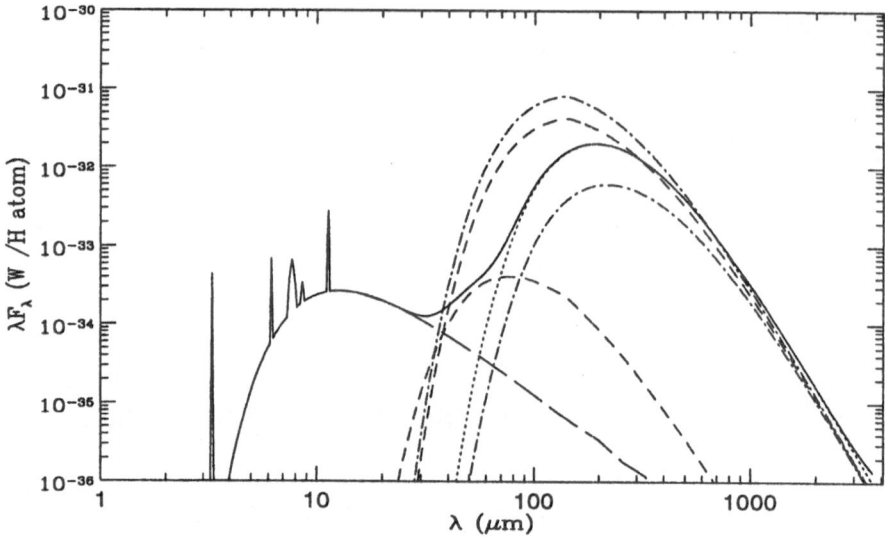

Figure 7. The dust emission spectrum in a molecular cloud described in the text. The model resulting spectrum (solid) is the sum of PAH's (long dashed), hump particles (dashed) and the large core-mantle plus ice mantle particles (dotted). For comparison, we also show the theoretical emissions of graphite (upper dot-dashed), silicate (lower dot-dashed) and a mixture (upper dashed) based on the Mathis et al (1977) dust model.

emits at wavelength not accessible to the IRAS satellite as noted by Greenberg (1971), Greenberg & Li (1995) and by Block et al (1994a, 1994b). The graphite-silicate dust model for molecular clouds predicts a peak emission wavelength $\sim 70\mu m$ shorter than that predicted by the core-mantle grains.

A natural result of the evolutionary model is that it predicts an abrupt decrease in the $60\mu m/100\mu m$ emission ratio as at the edge of a molecular cloud because of accretion of the small particles on the large ones. This is readily deduced from a comparison of Fig.6 and Fig.7 where it is seen that the $60\mu m$ emission is dominated by the hump particles so that any loss of these will immediately lead to a drop in the $60\mu m$ emission. Such an effect can not be explained within the graphite-silicate model for which a continuous size distribution leads to a fraction of "large" graphite particles contributing $\sim 50\%$ of the $60\mu m$ emission (Siebenmorgen & Krügel 1992; Laureijs et al 1991). The proposal by Laureijs et al (1991) that ice mantles on the very small "$60\mu m$ particles" can explain their temperature drop is highly unlikely because very small particles are inhibited from accreting mantles both observationally (Greenberg & Hong 1974a) and theoretically (Greenberg 1980; Aannestad & Kenyon 1979).

6. Acknowledgements

This work was supported in part by NASA grant NGR 33-018-148 and by a grant from the Netherlands Organization for Space Research (SRON). The authors acknowledge helpful advice on PAH's by Dr. Willem A. Schutte. One of us (AL) wishes to thank the World Laboratory for a fellowship.

References

Aannestad, P.A. & Kenyon, S.J., 1979, ApJ, 230, 771
Aitken, D.K., 1981, in : IAU Symposium 96, Infrared Astronomy, Wynn-Williams, C.G. & Cruikshank, D.P. (eds), Dordrecht : Reidel, p.207
Andriesse, C.D, 1978, A&A, 66, 169
Bernard, J.P., Boulanger, F., Desert, F.X. & Puget, J.L., 1992, A&A, 263, 258
Block, D.L., Bertin, G., Stiction, A., et al, 1994a, A&A, 288, 365
Block, D.L., Witt, A.N., Grosbol, P., et al, 1994b, A&A, 288, 383
Bohren, C.F. & Huffman, D.R., 1983, Absorption and Scattering of Light by Small Particles, Wiley, New York
Boulanger, F. & Perault, F., 1988, ApJ, 330, 964
Bussoletti, E., Colangelli, L. & Orofino, V., 1987, ApJ, 321, L87
Chase, M.W., Davis, C.A., Downey, J.R., Frurip, D.J., McDonald, R.A. &Syverud, A.N., 1985, J. Phys. Chem. ref. Data, 14, Suppl. No.1
Cherchneff, I., Barker, J.R. & Tielens, A.G.G.M., 1991, ApJ, 377, 541
Chlewicki, G. & Greenberg, J.M., 1990, ApJ 365, 230
de Groot, M., van der Zwet, G.P., Jenniskens, P., Bauer, R., Baas, F. & Greenberg, J.M., 1988, in : Dust in the Universe, Bailey, M.E. & Williams, D.A., Cambridge Press, p.265
Desert, F.X., Boulanger, F. & Shore, S.N., 1986, A&A, 160, 295
Desert, F.X., Boulanger, F. & Puget, J.L., 1990, A&A, 237, 215
Dorschner, J., Begemann, B., Henning, Th., Jäger, C. & Mutschke, H., 1995, A&A, 300, 503
Draine, B.T. & Lee, H.M., 1984, ApJ 285, 89
Draine, B.T. & Anderson, N., 1985, ApJ, 292, 494
Draine, B.T., 1988, ApJ 333, 848
Draine, B.T., 1994, in : The Infrared Cirrus and Diffuse Interstellar Clouds, Cutri, R. & Latters, W.B. (eds), San Francisco, ASP Conference Series 58, p.227
Eddington, A.S., 1926, Diffuse matter in interstellar space (Bakerian Lecture), Proc. Roy. Soc., 111A, p.424
Fabbri, R., Guidi, I., Natale, V. & Ventura, G., 1986, in : Proc. of Marcel Grossman Meeting, Rome
Fitzpatrick, E.L. & Massa, D., 1986, ApJ, 307, 286
Greenberg, J.M., 1968, in: Stars and Stellar Systems, vol. VII, Middlehurst, B.M. & Aller, L.H. (eds.), University of Chicago Press, p. 221
Greenberg, J.M., 1971, A&A 12, 240
Greenberg, J.M.,& Shah, G.A., 1971, A&A 12, 250
Greenberg,J.M., Yencha ,A.J., Corbett,J.W.,and Frisch,H.L., 1972, Mem. Soc. Roy. des Sciences de Liege, 6e Serie, tome IV, p. 425
Greenberg, J.M., 1973, in: Molecules in the Galactic Environment, Gordon, M.A., & Snyder, L.E.(eds.), Wiley, p. 94
Greenberg, J.M. & Hong, S.S., 1974a, in : HII regions and the Galactic center, Moorwood, A.F.M. (ed), ESRSOSP-105, p.153
Greenberg, J.M. & Hong, S.S., 1974b, in : IAU Symposium 60, Galactic and Radio Astronomy, Kerr, F. & Simonson, S.C. (eds), Dordrecht : Reidel, p.155

Greenberg, J.M. & Hong, S.S., 1976, in : Far Infrared Astronomy, (ed) Rowan-Robinson, Pergamon Press, p.299

Greenberg, J.M., 1978, in: Cosmic Dust, McDonnell, J.A.M.(ed.), Wiley, p.187

Greenberg, J.M., 1980, in : Les Spectres des Molecules Simples au Laboratoire et en Astrophysics, Universite de Liege, p.555

Greenberg, J.M. & Chlewicki, G., 1983, ApJ, 272, 563

Greenberg, J.M. & Li, A., 1995, in: The Opacity of Spiral Disks, eds. J.I. Davies and D. Burstein, Dordrecht, Kluwer, p.19

Greenberg, J.M. & Li, A., 1996, A&A, in press

Greenberg, J.M., Li, A., Mendoza-Gómez, C.X., Schutte, W.A., Gerakines, P.A., de Groot, M., 1995, ApJ, 455, L177

Guhathakurta, P. & Draine, B.T., 1989, ApJ, 345, 230

Henning, Th. & Stognienko, R., 1996, A&A, in press

Hong, S.S., Greenberg, J.M., 1980, A&A 88, 194

Jenniskens, P. 1993, A&A 274, 653

Jenniskens, P. & Greenberg, J.M., 1993, A&A, 274, 439

Lange, A.E., et al, 1989, in : Interstellar Dust (IAU Symp. 135), eds. Allamandola, L.J. & Tielens, A.G.G.M., Kluwer, Dordrecht, p.499

Laureijs, R.J., Clark, F.O. & Prusti, 1991, ApJ, 372, 185

Leger, A. & d'Hendecourt, L., 1987, in : Polycyclic Aromatic Hydrocarbon and Astrophysics, Leger et al (eds), p.223

Leger, A., d'Hendecourt, L. & Defourneau, D., 1989a, A&A, 216, 148

Leger, A., et al., 1989b, in : Interstellar Dust (IAU Symp. 135), eds. Allamandola, L.J. & Tielens, A.G.G.M., Kluwer, Dordrecht, p.173

Low, F.J., et al, 1984, ApJ, 278, L19

Mathis, J.S., Rumpl, W., Nordsieck, K.H., 1977, ApJ 217, 425

Mathis, J.S., Mezger, P.G., & Panagia, 1983, A&A 128, 212

Mathis, J.S., 1990, ARA&A, 28, 37

Maxwell-Garnett, J.C., 1904, Phil.Trans.R.Soc., London, 203A, 385

Mezger, P.G., Mathis, J.S. & Panagia, 1982, A&A 105, 372

Mendoza-Gomez, C.X., de Groot, M., Greenberg, J.M. 1995, A&A 295, 479

Pajot, F., Boisse, P., Gispert, R., Lamarre, J.M. & Puget, J.L., 1986, A&A, 157, 393

Price, J., 1981, AJ, 86, 193

Purcell, E.M., 1976, ApJ, 206, 685

Puget, L.J. & Leger, A., 1989, ARA&A, 27, 161

Sakata, A., Wada, S., Tanabe, T. & Onaka, T., 1984, ApJ, 287, L51

Savage, B.D. & Mathis, J.S., 1979, ARA&A, 17, 73

Schutte, W. & Greenberg, J.M., 1992, A&A 244,190

Schutte, W., Tielens, A.G.G.M. & Allamandola, L.J., 1993, ApJ, 415, 397

Schutte, W.A., 1996, in: The Cosmic Dust Connection, Greenberg, J.M.(ed.), Kluwer, Dordrecht, in press

Sellgren, K., 1984, ApJ, 277, 623

Siebenmorgen, R. & Krügel, E., 1992, A&A, 259, 614

Van de Hulst, H.C., 1957, Light Scattering by Small Particles, Wiley

Völk, H.J., Jones, F.C., Morfil, G.E. & Rosen, S., 1980, A&A, 85, 316

Wright, E.L., et al, 1991, ApJ, 381, 200

ORGANICS AND ICES IN GALACTIC DUST

YVONNE J. PENDLETON

NASA Ames Research Center

MS245-3,Moffett Field, CA, 94035-1000, USA

Abstract. Hydrocarbon absorption (3.4 μm) along several different sight-lines through the galaxy indicates that the dust carrier lies in the diffuse interstellar medium (DISM). Correlations of the hydrocarbon and silicate absorption features suggest an increased abundance of both in the direction of the galactic center. Dense molecular clouds do not exhibit the same type of 3.4 μm feature as seen in the DISM, which is puzzling in light of dust cycling times. Laboratory organics produced by different processes are compared to the DISM features. The 3.4 μm feature has now been detected in other galaxies, which suggests that the widespread distribution of organic material in our own galaxy may be common to other galaxies as well.

1. Introduction

Near infrared observation of interstellar sightlines in the galaxy have revealed the presence of ices in dense molecular clouds and relatively complex hydrocarbons in the diffuse interstellar medium. The evolution of dust from various birthsites (including the circumstellar environments of stars, supernovae outflows, and dense molecular clouds) is of considerable importance to the overall understanding of the physical processes which dominate the interstellar medium in our galaxy. Advances in near infrared (IR) observational capabilities over the past few years have allowed detailed studies to be made of the distribution and composition of interstellar dust. IR spectroscopy of diffuse interstellar dust has revealed the widespread distribution of absorption features (near 3.4μm) which have been attributed to saturated aliphatic (chain-like) hydrocarbons (Sandford *et al.*,1991; Pendleton *et al.*1994). The recent detection of these bands in other galaxies (Bridger *et al.*,1993; Wright *et al.*this volume) and the remarkable similarity of the

135

D. L. Block and J. M. Greenberg (eds.), New Extragalactic Perspectives in the New South Africa, 135–142.
© *1996 Kluwer Academic Publishers.*

subfeatures to those in our own galaxy suggest the organic moieties responsible for the 3.4μm absorption bands are widespread in other galaxies as well as our own. Laboratory organics produced by a variety of energetic processes exhibit absorption in the 3.4μm region which is, in general, similar to that observed in the diffuse interstellar medium (ISM). Astrophysically relevant ice mixtures, exposed to ultraviolet radiation, show a very good match to the 3.4μm observations (Greenberg *et al.*1995), although other processes such as the ion bombardment of methane (Strazzulla & Johnson 1991) and the plasma discharge production of hydrogenated amorphous carbons (HACs) (Duley 1994; Jones *et al.*1991) also provide good matches to the DISM (Pendleton et al 1994). In order to discriminate between the relevant processes responsible for the production of the DISM organics, observational studies of the 5-9 μm region are required. The laboratory organics must be able to match the observations at the longer wavelengths as well as in the 3.4 μm region.The 5-9μm spectral region cannot be observed from the ground, but can be observed by the Infrared Space Observatory (ISO). Although there are no ISO observations scheduled to observe the diffuse ISM of our own galaxy in the 5-9μm region, ISO observations are scheduled for a small number of galaxies where the 3.4μm absorption feature has been detected. This paper will briefly discuss the relationship between the observed C-H stretching band (3.4μm) and the observed Si-O stretching band (9.7μm), in which the abundance of each appears to increase in the direction of the galactic center, and problems with a current theory of the cycling of dust between the dense and diffuse clouds are discussed.

2. Hydrocarbons in the Diffuse Interstellar Medium

Observations of dust in the diffuse interstellar medium have revealed a series of absorption bands near 3.4 μm (Adamson, Whittet, & Duley 1990; Sandford *et al.*1991).The positions of the 3.38μm (2955 cm^{-1}) and 3 .42 μm (2925 cm^{-1}) subfeatures are characteristic of the symmetric C-H stretching frequencies of -CH$_3$ (methyl) and -CH$_2$- (methylene) groups in saturated aliphatic hydrocarbons and the band at 3.48 μm (2870 cm^{-1}) is characteristic of the asymmetric C-H stretching vibrations of these same functional groups when perturbed by other chemical groups. The carbonaceous material in the diffuse ISM has an average -CH$_2$-/-CH$_3$ ratio of 2.0 - 2.5 and likely contains moderate length aliphatic chains, such as -CH$_2$-CH$_2$-CH$_3$ associated with chemical groups like -OH, -CN, and aromatics. These features have been seen along a dozen different sightlines towards different types of objects in our galaxy. At least 2.5% of the cosmic carbon in the local interstellar medium and 4% toward the Galactic Center is tied up in the carrier of the 3.4 μm band.

Several materials have been suggested as candidate carriers of interstellar carbon. These include: (i) organic grain mantles formed by irradiation of ices (ii) hydrogenated amorphous carbons (HACs) having various degrees of hydrogenation, (iii) Quenched Carbonaceous composite (QCC), a material produced by quenching the plasma of methane gas and (iv) ion bombardment of gas and ices. Comparison of the diffuse interstellar C-H band profiles with the spectra of laboratory samples of candidate analog materials all show general similarities to the interstellar C-H stretching feature. Although the production of the organic residue by UV photolysis of interstellar ice mixtures is the more satisfying pathway, since it closely mimics the processing that should occur in molecular clouds, it may be the case that the material responsible for the 3.4 μm feature is so thoroughly processed that it has lost its "cosmo-chemical memory", and we will never be able to unravel the processing that such material experienced. In that case, we must not ignore the other materials (such as HAC) which are not produced in such a straightforward astrophysical method, but which nonetheless produce a material which appears to be a relatively certain component of the ISM. HAC is a continuum of materials whose properties can be altered by processing in interstellar space through chemical exposure as well as radiation. There are clearly multiple pathways by which the interstellar HAC observed in the diffuse ISM can be produced.

3. Dust Evolution in Dense Molecular Clouds

In the quiescent regions of a molecular cloud, gas phase species collide with the dust particles and stick to them due to the low temperature of the dust. These species include many atomic and molecular radicals. Reactions among these accreted species will form a mixture of simple molecules (d'Hendecourt, Allamandola, & Greenberg (1985)). Further evolution of the grain mantle occurs as it is processed by ultraviolet radiation and/or cosmic rays which penetrate the cloud. Ultraviolet radiation from embedded young stars will also contribute to the grain mantle photolysis and evolution. Such processing produces molecular subgroups and complex molecules which cannot be produced by accretion of reactive species alone (Allamandola, Sandford, & Valero, 1988). After the stellar system has been produced, the cloud remnants are disbursed into the diffuse medium where the volatile components are vaporized and the complex component probably remains on the grains. In this scenario, a new molecular cloud will form from swept up interstellar dust that will include some of this older, first generation material. The relatively short cycling time between dense and diffuse clouds, suggests that the refractory, organic component of the interstellar medium should be seen in both dense and diffuse cloud regions. However, the rel-

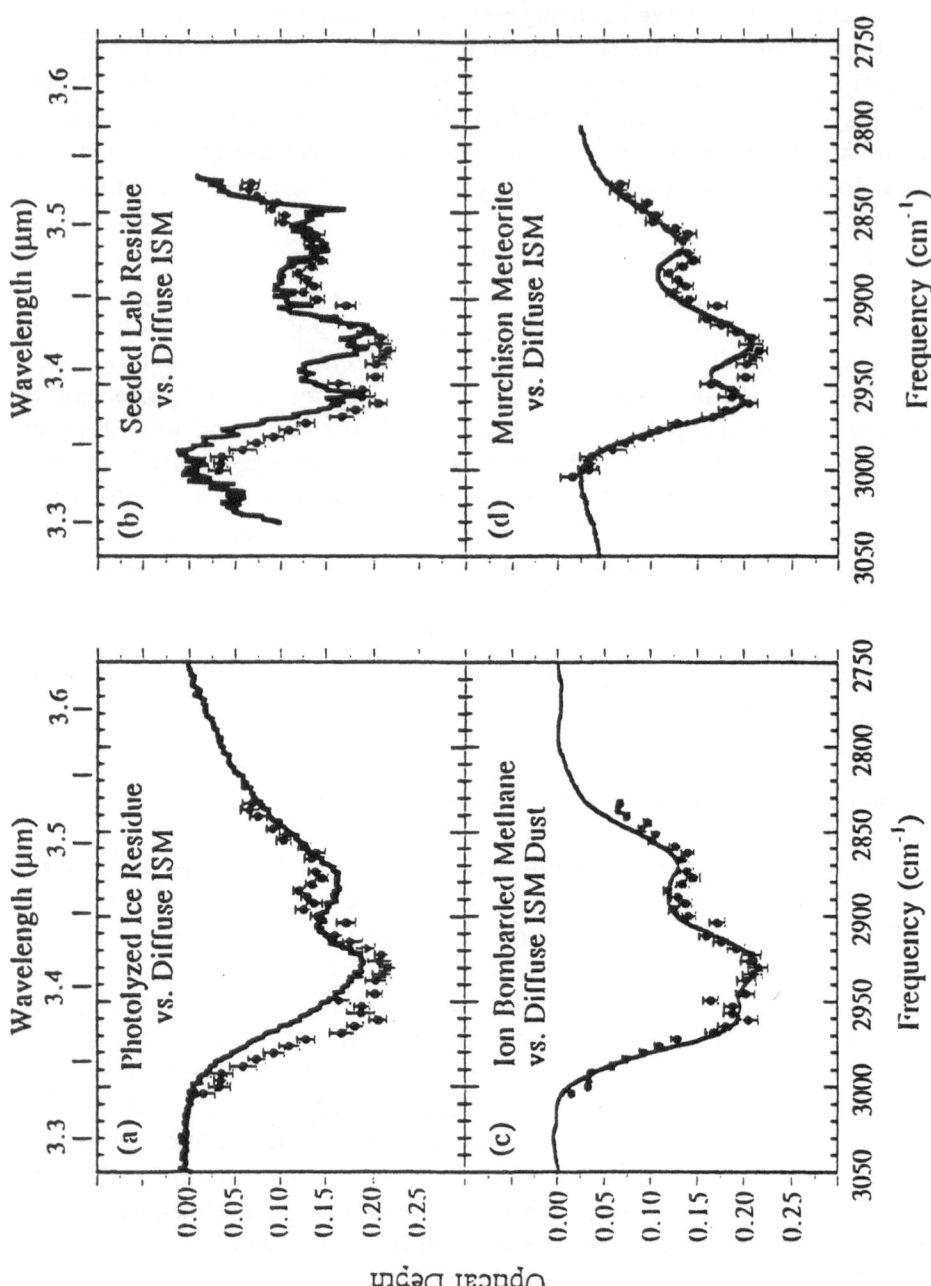

Figure 1. Lab comparisons (Pendleton *et al.* 1994)

atively complex aliphatic (CH_2 and CH_3 chains) hydrocarbons seen along several sightlines through diffuse cloud dust are notably absent in the spectra of dust embedded dense molecular cloud objects (Brooke *et al.*, 1996).

The spectral signature of $3.4\mu m$ wing of the $3.08\mu m$ water ice feature seen through dust in the Taurus dark cloud (Smith, Sellgren, and Brooke, 1993) and the $3.47\ \mu m$ absorption band observed in dense molecular cloud spectra (Allamandola *et al.*1992; Brooke *et al.*1996) are in sharp contrast with that of the diffuse interstellar medium. In comparison, the diffuse dust features straddle the $3.47\mu m$ dense cloud feature, and show substructure which is not seen in the $3.4\ \mu m$ wing of the water ice spectra in the Taurus sources. These differences indicate that there are two very different and independent solid hydrocarbon interstellar dust components. The absence of CH_2 and CH_3 bands in the spectra of embedded protostars may imply that diffuse medium C-rich materials do not get incorporated into dense clouds. It may also imply that the amount of organic refractory material produced in the dense clouds is only a minor contributor to that observed in the diffuse ISM. A possible avenue for study of the process responsible for the production of the DISM organics is to observe the predicted intermediate or resultant steps along the way of the process of interest. For instance, Bernstein *et al.*, 1995 have shown that as much as 60% of the organic residue produced through the UV photolysis of ices is attributable to the Hexamethylenetetramine (HMT), $C_6H_{12}N_4$, molecule. Upon further irradiation, the so called X-CN ($4.62\ \mu$m) feature appears. The X-CN absorption feature has been detected towards a small number of sources and is always seen through dense cloud dust. A recent study by Tegler *et al.*(1995) reports a preliminary discovery that the X-CN feature is not detected in the dust towards background stars but is apparent in the embedded star spectra when observing stars through similar amounts of dust obscurration in the Taurus cloud. This result could have a strong impact on constraints placed upon the production requirements and/or the extent to which dust grain mixing typically occurs throughout a molecular cloud. Another comparative study that would be useful is a study of the presence and detailed shape of the X-CN band in different molecular clouds. Variations from cloud to cloud might well depend upon such factors as the rate of star formation, the mass of the stars forming, etc.

3.1. HYDROCARBONS AND SILICATES IN OUR GALAXY

The A_v/tau ratio for the 3.4 μm feature is lower toward the Galactic Center than toward sources in the local solar neighborhood (Sandford, Pendleton, & Allamandola, 1995). A similar trend has been observed previously for silicates in the diffuse medium (Roche & Aitken 1985), suggesting that (i)

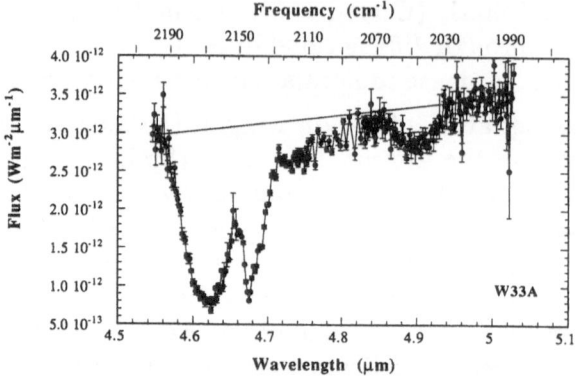

Figure 2. XCN, CO, and OCS in W33A (Pendleton, Tielens, & Tokunaga, 1996)

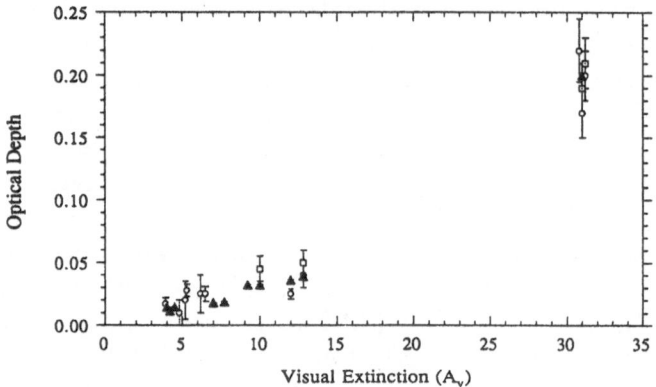

Figure 3. A_v versus optical depth of the C-H (open points) and Si- O bands (Sandford, Pendleton, & Allamandola, 1995)

the silicate and carbonaceous materials in the DISM may be physically correlated and (ii) there is either dust compositional variation in the galaxy or galactic variation in the grain population density distribution. The similar behavior of the C-H and Si-O stretching bands illustrates the possiblilty that these two components may be coupled, perhaps in the form of silicate-core, organic-mantled grains, as proposed by J. M. Greenberg years ago. Comparisons of the 3.4 μm and 9.7μm absorption bands in other galaxies will provide an additional view of this complex situation.

4. Summary

Through high resolution, high signal-to-noise observations we have learned that the interstellar medium in our galaxy contains an organic component which is widely distributed throughout the diffuse medium. This compo-

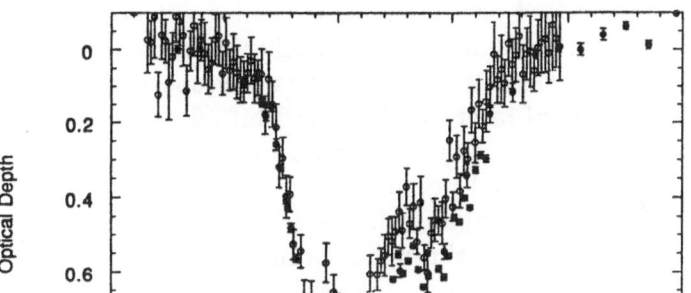

Figure 4.   Comparison of the 3.4μm feature in the embedded Seyfert Galaxy IRAS 08572+3915 and the Galactic Center source GC IRS6E

nent is not seen in dense molecular clouds, even though it is predicted to be there if dust cycles between the diffuse and dense clouds with the efficiency that has been suggested. Laboratory comparisons of organic material produced through a variety of methods have revealed that more than one end product can provide a good detailed match to the observed diffuse ISM. The residue produced through UV photolysis has the advantage of replicating an astrophysically relevant situation, but if the organic residue that we observe has lost its "cosmochemical memory", then well fitting organics produced by alternative means (such as HACs) should be given equal consideration as the possible carrier for all or some of the 3.4μm feature. In order to discriminate between the relevant processes, which in turn might reveal where the organics are produced, observations in the 5-9μm region are required. Another direction for studies of the physical processes will be to look closely at the intermediate or resultant steps which are predicted as a result of the specific process to be investigated. In the case of UV photolysis, for instance, the photolyzed organic residue composition consists of 60% HMT. The UV photolysis of HMT produces an X-CN feature. We can observe the X-CN feature in dense clouds, comparing the distribution of the carrier throughout the cloud and comparing cloud-to-cloud variations to reveal the extent of the photolytic process on the organic refractory residue. The discovery of the 3.4μm absorption feature in other galaxies is a result of the vast improvement in instrumentation that has become available just recently. The surprising discovery of a much stronger 3.4μm

feature in a distant galaxy than the strongest measurement observed in our
own galaxy has opened new vistas that show great promise. Now we may
be able to study interstellar features in other galaxies that are obscured by
telluric lines in the Earth's atmosphere (when observing galactic objects)
because the features of interest will be redshifted corresponding to the distance of the galaxy. The widespread distribution of the organic component
throughout our galaxy and possibly other galaxies underlines the possibility that the basic building blocks of life are available for incorporation into
planetary systems forming throughout the universe.

5. References

Adamson, A.J., Whittet, D., & Duley, W.W. 1990, *M.N.R.A.S.*, 243, 400

Allamandola, L. J., Sandford, S. A., Tielens, A.G.G.M., & Herbst, T. M. 1992, *ApJ*, **399**, 139

Allamandola, L. J., Sandford, S. A., & Valero, G. J. 1988,*Icarus*, **76**, 225

Bernstein, M., Sandford, S. ,Allamandola, L., & Chang, S. 1995, *Ap. J.*,**454**, 327

Brooke, T. Y., Sellgren, K., & Smith, R. G. 1996, *Ap. J.*, march issue

Bridger, A., Wright,G. & Geballe, T. 1993 In *Infrared Astronomy with Arrays: The Next Generation*, (I.S. McLean, ed.) Kluwer Academic Press, p. 537

d'Hendecourt, L. B., Allamandola, L. J. and Greenberg, J. M. 1985, *Astr. Ap.*, **152**, 130

Duley, W. W. 1994, *Ap. J. (Letters)*,**430**, L133

Greenberg, J. M. , Li, A.,Mendoza-Gomez, C.X., Schutte,W., Gerakines, P. A.& DéGroot, M. 1995, *Ap. J. (Letters)*, **455**, L177

Jones, A. P., Duley, W. W., & Williams, D. A. 1987, *M.N.R.A.S.*, **229**, 213

Pendleton, Y. 1994, in *Infrared Cirrus and Diffuse Dust*, R. Cutri and W. Latter, eds., bf 58, 255

Pendleton, Y, Sandford, S. Allamandola, L. J.,Tielens, A. G. G. M. & Sellgren, K. 1994, *Ap. J.*, **437**, 683

Pendleton, Y, Tielens, A.G.G.M, & Tokunaga, A. 1996, in preparation

Roche, P.F., & Aitken, D.K 1985, *M.N.R.A.S.*, **215**, 425

Sandford, S. A., Allamandola, L. J., Tielens, A.G.G.M., Sellgren, K., Tapia, M., & Pendleton, Y. 1991, *Ap. J.*, **331**, 607

Strazzulla, G. & Johnson, R. E. 1991, In *Comets in the Post-Halley Era*, R. L. Newburn, Jr., M. Neugebauer, and J. Rahe (eds.), (Kluwer:Dordrec 1, 243

Tegler, S.C., Weintraub, D. A. Rettig, T. W. Pendleton, Y. J. , Whittet, D., & Kulesa, C. A. 1995, *Ap. J.*, **439**, 279

STUDIES OF NIR DUST ABSORPTION FEATURES IN THE NUCLEI OF ACTIVE AND IRAS GALAXIES

G.S. WRIGHT, A. BRIDGER AND T.R. GEBALLE
Joint Astronomy Centre
660 N. A'ohōkū Place, Hilo, HI96720, USA

AND

Y. PENDLETON
NASA Ames Research Center
Space Science Division, Moffett Field, CA94035, USA

Abstract. We present the first survey of the 3.4μm hydrocarbon absorption feature in active and IRAS galaxies. Comparison of our data with galactic studies of this feature provides insight into the nature of dust grains around active nuclei, and holds the potential to improve our understanding of interstellar dust.

1. Introduction

One of the most exciting developments for understanding active galactic nuclei has been the development of unification theories in which the presence or absence of the broad permitted lines (*i.e.*type 1 or type 2 Seyfert class) is determined by the line of sight through a physically and geometrically thick obscuring torus of material which surrounds the active nucleus (*cf.* Antonucci 1993). There is a growing body of evidence that several type 2 Seyfert galaxies do indeed have hidden broad line regions. NGC1068 is an archetypal example of such a galaxy in which spectro-polarimetry has shown the broad line region in emission that is scattered into the line of sight by free electrons above and below the disk. There are now also a few direct detections of a broad line line region in Seyfert 2 galaxies (*e.g.*Goodrich *et al.* 1994), observed in IR lines which penetrate the dust. However, until now there has been little direct study of the nature of the obscuring material or its distribution. Roche *et al.* (1991) carried out a survey of the 10μm

D. L. Block and J. M. Greenberg (eds.), New Extragalactic Perspectives in the New South Africa, 143–150.
© *1996 Kluwer Academic Publishers.*

band in infrared bright galaxies and found that the spectra of AGN were either featureless or showed silicate absorption, while starforming galaxies showed strong PAH emission features.

The recent dramatic improvements in sensitivity and resolution of near-IR array spectrometers make extra-galactic observations of dust absorption features in the 3-5μm region feasible. To provide alternative diagnostics of obscuration in type 2 AGN and IRAS galaxies we have therefore begun a programme to study directly the dust and molecules. The 3.4 μm hydrocarbon absorption features, H_2O ice features (3.1 μm), CO absorption (4.6 μm and the silicate feature (9.7 μm) are all seen towards our galactic centre and have the potential to provide new information about the nature (e.g. composition, density) of the extinction in these galaxies. The 3.4 μm band is particularly important because it has been well studied in our galaxy. The shape of the feature has been shown to result from stretching modes of C-H in saturated aliphatic hydrocarbons, and Pendleton *et al.* (1994) have demonstrated that the carriers of this feature are widespread amongst the dust in our galaxy. The detection of this band in other galaxies also has the potential to increase our understanding of the grains in the ISM by extending the range of conditions in which the grains are observed. We present here the very first extra-galactic detections of the 3.4 μm band.

2. Observations

Spectra of the nuclei of a sample of about 12 galaxies covering at least 0.5μm centered at the redshifted wavelength of the 3.4μm feature were obtained in March 1994 and June 1995 at the United Kingdom Infra Red Telescope with the near-IR spectrograph CGS4 (Mountain *et al.* 1990). The galaxies were selected from a sample of all galaxies for which there was some evidence for a high degree of extinction to the nucleus, e.g. evidence for an obscured broad line region from optical polarisation measurements or near-IR spectroscopy, extremely red near- to mid- IR colours, a previous measurement of a silicate absorption feature in the 10μm band, etc., and which were estimated to be sufficiently bright to detect with CGS4 at 3-4μm from K, L or M band photometry. The sample includes nearby Seyferts and some ultra-luminous IRAS galaxies. Bright starburst galaxies have not been included as many are known from previous studies to have PAH emission features at 3.28μm (Moorwood 1986) which would make detection of the 3.4μm band more difficult.

In March 1994 the spectral resolution was $\sim$ 500, the spatial resolution was 3.1 x 3.1 arcsec, and the spectral coverage was achieved by observing several overlapping wavelength regions. By June 1995 a larger array provided more wavelength coverage with a spectral resolution of $\sim$ 800

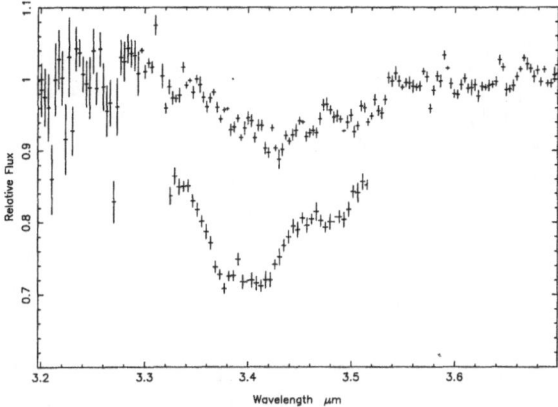

Figure 1. NGC1068 (top) compared with IRS6 (bottom)

and spatial resolution of 3 x 1.5 arcsec. Even the brightest galaxies we observed are 2-3 orders of magnitude fainter than the galactic sources in which the 3.4μm feature has been studied, and accurate sky subtraction at these wavelengths is essential. With a long slit spectrometer such as CGS4, subtraction of residual sky emission (after the usual IR technique of chopping, nodding and subtraction in pairs), may be achieved by fitting to the blank sky positions in the slit - a technique which greatly improves the reliability of weak features in extra-galactic sources. In addition, at least two independent observations of each galaxy were obtained. Wavelength calibration is accurate to better than 0.003μm across the spectra. The galaxy spectra were ratioed by F/G stellar spectra at the same mean airmass to remove atmospheric absorption. Repeated observations and spectra taken at overlapping wavelength regions were then merged to produce a final spectrum. Galaxies for which the independent spectra were not consistent and which indicate a sky subtraction problem will be re-observed. A linear continuum was fitted to data outside the absorption feature and removed.

Reduction of all the data from the survey has not yet been completed. Moreover, additional observations of some of the galaxies are required for a definitive result. The discussion in the following sections is based on the galaxies for which reduction and analysis of the spectra has been completed to date.

3. Results

The two best detections of the 3.4μm feature obtained to date are in the classic example of an obscured active nucleus, NGC1068 (Bridger *et al.* 1994), and in the ultraluminous IRAS galaxy, 08572+3915, which has

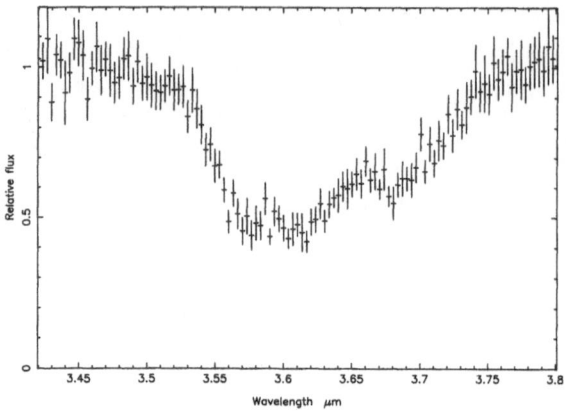

Figure 2. Spectrum of the 3.4μm absorption in IRAS 08572+3915

an unusually strong feature.

Figure 1 shows the spectrum of NGC1068 after the continuum has been ratioed out, along with a spectrum of the galactic centre source IRS6 (Pendleton *et al.* 1994) which is plotted on the same scale, but offset for clarity. The two spectra are remarkably similar. In particular the sub-peaks in the galactic spectra seen at 3.38, 3.42 and 3.48μm are clearly seen in the NGC1068 spectrum at 3.395, 3.43 and 3.50μ m consistent with the recession velocity, 1137 km/s, of NGC1068. In Figure 2 the spectrum of IRAS 08572+3915 at a redshift of 0.058 is shown after removal of the continuum. The sub-peaks of the galactic interstellar feature can again be seen, as expected at 3.57, 3.61 and 3.70μm. This feature is four times stronger than that seen towards IRS6, and is the deepest 3.4μm absorption ever measured.

Noisier detections of the absorption have also been obtained in Mkn463E and IRAS05189-25. The feature was not detected in NGC4418, while in Mkn231 a very weak feature may have been seen. Figure 3, a spectrum of IRAS05189-25, gives an indication of the more typical quality of 3.4μm spectra of galaxies, in which a weak absorption feature is seen against a noisy continuum. In IRAS05189-25, it is likely that both a 3.28μm PAH emission feature and the 3.4μm absorption have been detected. A summary of the results of our survey to date is given in Table 1.

4. Comparison of galactic and extra-galactic dust features

In NGC1068 and IRAS 08572+3915 the detailed shape of the 3.4μm absorption band is amazingly similar to that seen from the diffuse ISM in our galaxy. Indeed, as the figure in Pendleton (1996, this volume) comparing the

TABLE 1. Measurements of the 3.4μm absorption in galaxies.

Galaxy	Redshift	Optical Depth at 3.42μm band
NGC1068	0.003	0.1
IRAS08572+3915	0.058	0.5
Mkn463E	0.05	≤ 0.05
NGC4418	0.007	No features
Mkn231	0.041	≤ 0.02
IRAS23060+0505	0.173	No features
IRAS05189-25	0.043	0.05

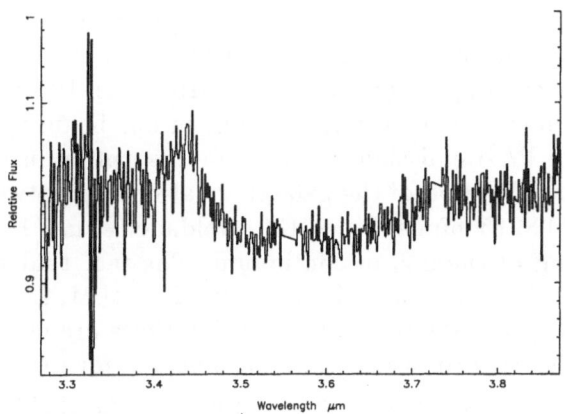

Figure 3.   Spectrum of IRAS 05189-25

IRS6 spectrum with a de-redshifted plot of IRAS08572+3915 shows, even the relative strength of the 3.38 and 3.42μm sub-features is approximately the same, indicating that the extra-galactic dust has a similar CH_2-CH_3 ratio. The signal to noise ratios of the other detections of the absorption band are too low for a detailed comparison of shape and sub-features, but the general width is the same, suggesting that the same grains are present in these galaxies also. Pendleton *et al.* (1994) have argued that the organic component of the diffuse ISM is widespread and these detections extend that result to other galaxies.

The Seyfert galaxies in which we have detected the feature are almost face on, and it is extremely unlikely in these galaxies that the absorption arises from the diffuse ISM in the disk of the galaxy. It is more plausible that the absorption arises in the torus surrounding the active core. In both NGC1068 and Mkn 463E the optical depth of the 3.4μm feature is consistent with other estimates of the extinction in the circumnuclear torus, if the galactic relationship between $\tau_{3.4}$ and A_v in the ISM (Sandford et al. 1995) is valid for the obscuring torus. For NGC1068, $\tau_{3.4} \sim 0.09$, implies $A_v \sim$ 22, consistent with $A_v \sim$ 20-25 derived by Bailey et al (1988) to account for the polarisation at 2.2μm. Since the extinction to the broad line region in NGC1068 is much higher than this, the best model is that the 3.4μm absorption is measuring the extinction through the edge of the torus to a region of warm dust emitting in the NIR (cf. Efstathiou et al 1996). For Mkn 463E $\tau_{3.4} \leq 0.05$, implies $A_v \leq 12$, consistent with $A_v \sim 10$, estimated from the direct detection of the broad line region in the near-IR in this galaxy (Goodrich et al. 1994).

In the two IRAS galaxies in which we have detected the feature the location of the dust is more uncertain. The higher redshift of these galaxies means that the slit includes a larger region of the surrounding galaxy. Moreover the galaxies have disturbed optical morphologies, with dust lanes crossing the nuclear regions (Sanders et al. 1988). In IRAS05189-25 the inclusion of near nuclear star-forming regions in our beam would explain the presence of both PAH emission and 3.4 μm absorption in the spectrum. A simple minded application of the galactic relationship between A_v and $\tau_{3.42}$ for IRAS 08572+3915 implies $A_v \sim 130$. Goldader et al. (1995) use $A_v \sim 38$ for the reddening of the 2.2μm continuum. The fact that the 3.4μm band is so strong in this galaxy may be an indication that, as may be the case in our galactic centre (Sandford et al. 1995), there are more carriers of the 3.4μm feature relative to other grains responsible for extinction.

Recent studies in our galaxy have shown a remarkable correlation between organic (optical depth at 3.4 μm) and silicate (optical depth at 9.7 μm) dust (Sandford et al. 1995). The distributions of the carriers of the two features are so similar that it has been suggested that these two components of the ISM may be coupled in the form of silicate-core, organic-mantle grains. (see the discussion in Pendleton 1996, for more details). Some of the galaxies in which we have measured the 3.4 μm band have measurements of the silicate feature obtained in similar apertures (Roche et al. 1984, Dudley, private communication). Table 2 gives the ratio of optical depths in the 3.4 and 9.7 μm bands for galaxies in our survey, which should be compared to the galactic value of 0.06 +/− 0.01.

There appears to be little correlation between the the depths of the Si-O and C-H bands for this small sample. The most striking demonstration of this is that we did not detect a 3.4 μm absorption band in NGC4418 which has one of the strongest silicate absorptions measured (Roche et al. 1986).

TABLE 2. Ratio of optical depths for
organinc and silicate dust

Description 1	$\tau_{3.42}/\tau_{9.7}$
NGC1068	0.16
IRAS08572+3915	0.02
Mkn463E	≤ 0.10
NGC4418	≤ 0.03

It may be that the lack of correlation is an indication of real abundance differences between the carriers, perhaps due to more or different grain processing around the AGN in these galaxies. However, it is more likely to be due to geometrical effects in the distribution and temperature of the dust surrounding the AGN. Observations of normal galaxies and more detailed modelling in these galaxies will be required to clarify the interpretation of this result.

Spectra of galactic centre sources towards which the 3.4μm C-H absorption band is strong, also show a strong broad absorption from 2.9 to 3.23μm due to the O-H stretching band, which is attributed H_2O ices or water hydration in silicates. Sandford *et al.* (1991) show that the O-H and C-H features probably arise from independant carriers and that the O-H is associated with dense cloud material while the C-H material arises from the diffuse interstellar medium (DISM). We have not detected the O-H band in either NGC1068 or IRAS 08572+3915, the only two galaxies in which we have searched for it. This suggests that like the DISM, the dust in the central regions of these galaxies does not contain significant amounts of OH. If the O-H band in the galactic observations is largely due to water ices, the absence of such a feature in AGN may be a result of the substantially warmer dust temperatures.

5. Conclusions

We have presented the first survey of the 3.4 μm dust absorption band in external galaxies. These observations provide new insight into the nature of the dust in obscured AGN and IRAS galaxies. They also provide tantalising hints that dust properties may vary between galaxies. However, they are only the beginning of what will be possible for extra-galactic studies of dust. Detections of this feature in the ISM of normal spiral galaxies are just becoming possible. A few detections of the 3.4μm feature in normal galaxies are the key to improved understanding of the chemistry of grains in the ISM and in the circumnuclear dust discussed here. We also intend to obtain higher resolution, higher signal-to-noise comparisions of the detailed structure in the galactic and extra-galactic features. ISO spectra of these galaxies will provide a wealth of information about related features at wavelengths that are inaccessible from the ground. New mid-infrared spectrometers will soon provide better spatial and spectral resolution in the 10 and 20 μm bands, and enable a more detailed comparison of near and mid-IR dust features in external galaxies. Obtaining such data for a wide range of galaxies may provide crucial details of the destruction and survival of dust grains in galaxies.

6. Acknowledgements

We would like to thank Chris Dudley for kindly providing an estimate of the depth of the 9.7 μm silicate absorption in IRAS08572+3915 in advance of publication. The United Kingdom Infra-Red Telescope is operated by The Royal Observatories on behalf of the UK Particle Physics and Astronomy Research Council.

References

Antonucci, R.R. (1993), *Ann. Rev. Astron. Astrophys.*, **31**, 473
Bailey, J.A., Axon, D.J., Hough, J.H., Ward, M.J., McLean, I and Heathcote, S.R. (1988)*M.N.R.A.S.*, **234**, 899
Bridger, A., Wright, G.S. and Geballe, T.R. (1994) *Infrared Astronomy with Arrays*, (I.S. McLean, ed.), Kluwer Academic Press, p. 537
Efstathiou, A., Hough, J.H., and Young, S. (1996) *M.N.R.A.S.*, preprint
Goldader, J.D., Joseph, R.D., Doyon, R. and Sanders, D.M. (1995) *Ap. J.*, **444**, 97
Goodrich, R.W., Veilleux, S. and Hill, G.J. (1994) *Ap. J.*, **422**, 521
Moorwood, A.F. (1986) *A & A*, **166**, 4
Mountain, C.M., Robertson, D.J., Lee, T.J and Wade, R. (1990) *Instrumentation in Astronomy VII, ed. D.L. Crawford, SPIE*, **1235**, 25
Pendleton, Y., Sandford, S.A., Allamandola, L., Tielens, A.G.M.M. and Sellgren, K. (1994) *Ap. J.*, **437**, 683
Pendleton, Y., (1996) *This volume*
Roche, P.F., Aitken, D.K., Phillips, M.M. and Whitmore, B (1984) *M.N.R.A.S.*, **207**, 35
Roche, P.F., Aitken, D.K., Smith, C.H. and James, S.D. (1986) *M.N.R.A.S.*, **218**, 19p
Roche, P.F., Aitken, D.K., Smith, C.H. and Ward, M.J. (1991) *M.N.R.A.S.*, **248**, 606
Sanders, D.B., Soifer, B.T., Elias, J.H., Madore, B.F., Matthews, K., Neugebauer, G., Scoville, N.Z (1988) *Ap. J.*, **325**, 74
Sandford, S.A., Allamandola, L., Tielens, A.G.M.M., Sellgren, K., Tapia, M and Pendleton Y. (1991) *Ap. J.*, **371**, 607
Sandford, S.A., Pendleton, Y.J. and Allamondola L.J. (1995) *Ap. J.*, **440**, 697

TINY GRAINS AND LARGE MOLECULES
IN THE MILKY WAY AND OTHER GALAXIES

K. SELLGREN
Astronomy Department, Ohio State University
174 W. 18th Av., Columbus, OH 43210 USA

Tiny grains and large molecules are important constituents of the interstellar medium of our own and other galaxies. These small particles are intermediate in size between molecules and classical interstellar grains, with radii of ~10 Å and ~100 atoms per particle (Sellgren 1984; Léger & Puget 1984; Allamandola, Tielens, & Barker 1985).

Emission features at 3.3, 6.2, 7.7, 8.6, and 11.3 μm and continuum emission at 1–25 μm are both attributed to small particle emission. The emission features and their associated continuum are observed together in many individual sources within the Milky Way, including proto-planetary nebulae, planetary nebulae, reflection nebulae, H II regions, and young stellar objects (see Aitken 1981; Willner 1984; Allamandola 1984, 1989; Bregman 1989; Allamandola, Tielens, & Barker 1989; Puget & Léger 1989; Sellgren 1990, 1994 for previous reviews). Studies of the diffuse emission from the Milky Way (Boulanger & Pérault 1988; Bernard et al. 1994) and of individual galaxies (Gillett et al. 1975; Russell et al. 1977; Willner et al. 1977; Phillips et al. 1984; Roche & Aitken 1985; Helou 1986; Pajot et al. 1986; Cohen & Volk 1989; Helou et al. 1991; Roche et al. 1991; Soifer & Neugebauer 1991; Helou et al. 1991; Sauvage & Thuan 1994) demonstrate that the emission from these small particles dominates the mid-infrared emission of normal spiral galaxies. Because the small particles *emit* up to 40% of the total luminosity radiated by dust in galaxies, they also must *absorb* up to 40% of the starlight re-radiated by dust.

A variety of proposals have been made to explain the near infrared continuum emission from tiny particles. One idea is thermal emission from transiently heated tiny grains, with a radius of ~ 10 Å, which are briefly heated to temperatures near ~ 1000 K by the absorption of single ultraviolet (UV) photons (Sellgren et al. 1983; Sellgren 1984). Another suggestion is a quasi-continuum of overlapping overtone and combination bands arising from vi-

D. L. Block and J. M. Greenberg (eds.), New Extragalactic Perspectives in the New South Africa, 151–154.
© 1996 Kluwer Academic Publishers.

brational fluorescence in polycyclic aromatic hydrocarbon (PAH) molecules
(Léger & Puget 1984; Allamandola, Tielens, & Barker 1985; Puget & Léger
1989). A third explanation is an electronic fluorescence or phosphorescence
in PAH molecules (Allamandola, Tielens, & Barker 1985, 1989). A fourth
proposal is luminescence in hydrogenated amorphous carbon (HAC) grains
(Duley & Williams 1988; Duley 1988).

The emission features seen at 3.3, 6.2, 7.7, 8.6, and 11.3 μm are generally believed to be due to some sort of aromatic hydrocarbon. Proposed
laboratory analogs for the material which emits these features include amorphous carbon (Duley & Williams 1981; Borghesi, Bussoletti, & Colangeli
1987), $C_{60}H_{60}$ (Webster 1991), coal (Papoular et al. 1989), HAC (Blanco,
Bussoletti, & Colangeli 1988; Ogmen & Duley 1988), nitrogenated amorphous carbon (Saperstein, Metin, & Kaufman 1989), the carbonaceous
residue from Orgueil meteorite (Wdowiak, Flickinger, & Cronin 1988), PAH
molecules (Léger & Puget 1984; Allamandola, Tielens, & Barker 1985, 1989;
Léger, d'Hendecourt, & Défourneau 1989), quenched carbonaceous composite (Sakata et al. 1987), and a spiral carbonaceous microparticle with an
internal hydrogen (Balm & Kroto 1990). The emission mechanism for the
aromatic hydrocarbon features has been proposed to be thermal emission
from transiently heated tiny aromatic grains (Sellgren et al. 1983; Sellgren
1984), or a UV-excited vibrational fluorescence in PAH molecules (Léger &
Puget 1984; Allamandola, Tielens, & Barker 1985, 1989).

Infrared spectroscopy of galaxies has shown an interesting dichotomy
between normal spiral galaxies and galaxies with an active nucleus. The
spectra of normal spirals, particularly those with giant H II regions near
the nucleus, show the aromatic emission features (Gillett et al. 1975; Russell
et al. 1977; Willner et al. 1977; Phillips et al. 1984; Roche & Aitken 1985;
Moorwood 1986; Cohen & Volk 1989; Mouri et al. 1990; Dennefeld & Désert
1990; Roche et al. 1991; Mizutani et al. 1994), as does the diffuse interstellar
medium of the Milky Way (Giard et al. 1988, 1989, 1994; Ristorcelli et al.
1994). In contrast, active galaxies, particularly quasars and Seyfert 1 galaxies, have featureless mid-infrared spectra (Cutri et al. 1981; Lee et al. 1982;
Roche et al. 1984; Aitken & Roche 1985; Roche et al. 1991). The aromatic
emission features in normal spirals is naturally explained as emission from
either star formation regions or from the diffuse interstellar medium, both
of which are associated with these emission features in the Milky Way. The
lack of aromatic emission features in active galaxies has been attributed to
destruction of the small particles by UV radiation (Aitken & Roche 1985;
Désert & Dennefeld 1988) or X-ray radiation (Voit 1991, 1992) from the active nucleus. Some active galaxies, such as the Seyfert 1 galaxy NGC 7469,
do show the aromatic features, but the emission comes from surrounding
star formation regions, not the active nucleus (Cutri et al. 1984; Mazzarella

et al. 1994; Miles et al. 1994).

The IRAS emission from spiral galaxies has been shown to follow the same trends as individual clouds within the Milky Way. The relative amount of small particle emission is traced by $R(12/25)$ and $R(12/100)$, where $R(\lambda_1/\lambda_2)$ is the ratio of flux densities at λ_1 and λ_2. Observations show $R(12/25)$ and $R(12/100)$ both decrease with increasing $R(60/100)$ in individual clouds (Ryter et al. 1987; Boulanger et al. 1988) and among galaxies (Pajot et al. 1986; Helou 1986; Soifer et al. 1989; Soifer & Neugebauer 1991; Helou et al. 1991; Sauvage & Thuan 1994). Observations show $R(60/100)$ increases with increasing UV energy density within an individual cloud (Ryter et al. 1987; Boulanger et al. 1988) and with galaxy luminosity (Soifer & Neugebauer 1991).

The observed decrease of $R(12/25)$ with increasing $R(60/100)$ within individual clouds is attributed to destruction of small particles in intense UV radiation fields (Ryter et al. 1987; Boulanger et al. 1988; Désert et al. 1990). The observed decrease of $R(12/25)$ with increasing $R(60/100)$ and increasing luminosity in galaxies has been interpreted either as UV destruction of small particles (Soifer & Neugebauer 1991), or as due to changes in the relative importance of the colder diffuse interstellar medium and warmer star formation regions amongst galaxies (Pajot et al. 1986; Helou 1986; Sauvage & Thuan 1994).

References

Aitken, D. K. 1981, in *IAU Symposium 96, Infrared Astronomy*, ed. C. G. Wynn-Williams & D. P. Cruikshank (Dordrecht: Reidel), p. 207.
Aitken, D. K., & Roche, P. F. 1985, *MNRAS*, **213**, 777
Allamandola, L. J. 1984, in *Galactic and Extragalactic Infrared Spectroscopy*, ed. M. F. Kessler and J. P. Phillips (Dordrecht: Reidel), p. 5.
Allamandola, L. J. 1989, in *IAU Symposium 135, Interstellar Dust*, ed. L. J. Allamandola and A. G. G. M. Tielens (Dordrecht: Reidel), p. 129.
Allamandola, L. J., Tielens, A. G. G. M., & Barker, J. R. 1985, *ApJL*, **290**, L25.
Allamandola, L. J., Tielens, A. G. G. M., & Barker, J. R. 1989, *ApJS*, **71**, 733.
Balm, S. P., & Kroto, H. W. 1990, *MNRAS*, **245**, 193.
Bernard, J. P., Boulanger, F., Désert, F. X., Giard, M., Helou, G., & Puget, J. L. 1994, *A&A*, **291**, L5
Blanco, A., Bussoletti, E., & Colangeli, L. 1988, *ApJ*, **334**, 875.
Borghesi, A., Bussoletti, E., & Colangeli, L. 1987, *ApJ*, **314**, 422.
Boulanger, F., & Pérault, M. 1988, *ApJ*, **330**, 964.
Boulanger, F., Beichman, C., Désert, F. X., Helou, G., Pérault, M., & Ryter, C. 1988, *ApJ*, **332**, 328.
Bregman, J. D. 1989, in *IAU Symposium 135, Interstellar Dust*, ed. L. J. Allamandola & A. G. G. M. Tielens (Dordrecht: Kluwer), p. 109.
Cohen, M., & Volk, K. 1989, *AJ*, **98**, 1563
Cutri, R. M. et al. 1981, *ApJ*, **245**, 818
Cutri, R. M., Rieke, G. H., Tokunaga, A. T., Willner, S. P., & Rudy, R. J. 1984, *ApJ*, **280**, 521
Dennefeld, M. & Désert, F. X. 1990, *A&A*, **227**, 379

Désert, F. X. & Dennefeld, M. 1988, *A&A*, **206**, 227
Désert, F. X., Boulanger, F., & Puget, J. L. 1990, *A&A*, **237**, 215.
Duley, W. W. 1988, *MNRAS*, **234**, 61P.
Duley, W. W., & Williams, D. A. 1981, *MNRAS*, **196**, 269.
Duley, W. W., & Williams, D. A. 1988, *MNRAS*, **230**, 1P
Giard, M. et al. 1988, *A&A*, **201**, L1.
Giard, M., Pajot, F., Lamarre, J. M., Serra, G., & Caux, E. 1989, *A&A*, **215**, 92.
Giard, M., Lamarre, J. M., Pajot, F., & Serra, G. 1994, *A&A*, **286**, 203
Gillett, F. C., Kleinmann, D. E., Wright, E. L., & Capps, R. W. 1975, *ApJL*, **198**, L65
Helou, G. 1986, *ApJL*, **311**, L33
Helou, G., Ryter, C., & Soifer, B. T. 1991, *ApJ*, **376**, 505
Léger, A. & Puget, J. L. 1984, *A&A*, **137**, L5.
Léger, A., d'Hendecourt, L., & Défourneau, D. 1989, *A&A*, **216**, 148.
Lee, T. J., Beattie, D. H., Gatley, I., Brand, P. W. J. L., Jones, T., & Hyland, A. R. 1982, *Nature*, **295**, 214
Mazzarella, J. M., Voit, G. M., Soifer, B. T., Matthews, K., Graham, J. R., Armus, L., & Shupe, D. 1994, *AJ*, **107**, 1274
Miles, J. W., Houck, J. R., & Hayward, T. L. 1994, *ApJL*, **425**, L37
Mizutani, K., Suto, H., & Maihara, T. 1994, *ApJ*, **421**, 475
Moorwood, A. F. M. 1986, *A&A*, **166**, 4
Mouri, H., Kawara, K., Taniguchi, Y., & Nishida, M. 1990, *ApJL*, **356**, L39
Ogmen, M. & Duley, W. W. 1988, *ApJL*, **334**, L117.
Pajot, F., Boissé, P., Gispert, R., Lamarre, J. M., Puget, J. L., & Serra, G. 1986, *A&A*, **157**, 393
Papoular, R., Conard, J., Guiliano, M., Kister, J., & Mille, G. 1989, *A&A*, **217**, 204.
Phillips, M. M., Aitken, D. K., & Roche, P. F. 1984, *MNRAS*, **207**, 25
Puget, J. L. & Léger, A. 1989, *ARA&A*, **27**, 161.
Ristorcelli, I., Giard, M. Mény, C., Serra, G., Lamarre, J. M., Le Naour, C., Léotin, J., & Pajot, F. 1994, *A&A*, **286**, L23
Roche, P. F., & Aitken, D. K. 1985, *MNRAS*, **213**, 789
Roche, P. F., Whitmore, B., Aitken, D. K., & Phillips, M. M. 1984, *MNRAS*, **207**, 35
Roche, P. F., Aitken, D. K., Smith, C. H., & Ward, M. J. 1991, *MNRAS*, **248**, 606
Russell, R. W., Soifer, B. T., & Merrill, K. M. 1977, *ApJ*, **213**, 66
Ryter, C., Puget, J. L., & Pérault, M. 1987, *A&A*, **186**, 312
Sakata, A., Wada, S., Onaka, T., & Tokunaga, A. T. 1987, *ApJL*, **320**, L63.
Saperstein, D. D., Metin, S. S., & Kaufman, J. H. 1989 *ApJL*, **342**, L47.
Sauvage, M., & Thuan, T. X. 1994, *ApJ*, **429**, 153
Sellgren, K. 1984, *ApJ*, **277**, 623.
Sellgren, K. 1990, in *Dusty Objects in the Universe,* ed. E. Bussoletti & A. A. Vittone (Kluwer: Dordrecht), p. 35.
Sellgren, K. 1994, in *The Infrared Cirrus and Diffuse Interstellar Clouds*, eds. R. M. Cutri and W. B. Latter (San Francisco: ASP), p. 243
Sellgren, K., Werner, M. W., & Dinerstein, H. L. 1983, *ApJL*, **271**, L13.
Soifer, B. T. & Neugebauer, G. 1991, *AJ*, **101**, 354
Soifer, B. T., Boehmer, L., Neugebauer, G., & Sanders, D. B. 1989, *AJ*, **98**, 766
Voit, G. M. 1991, *ApJ*, **379**, 122
Voit, G. M. 1992, *MNRAS*, **258**, 841
Wdowiak, T. J., Flickinger, G. C., & Cronin, J. R. 1988, *ApJL*, **328**, L75
Webster, A. 1991, *Nature*, **352**, 412.
Willner, S. P. 1984, in *Galactic and Extragalactic Infrared Spectroscopy*, ed. M. F. Kessler & J. P. Phillips (Dordrecht: Reidel), p. 37.
Willner, S. P., Soifer, B. T., Russell, R. W., Joyce, R. R., & Gillett, F. C. 1977, *ApJL*, **217**, L121

THE ROLE OF UV OBSERVATIONS IN UNDERSTANDING DUST AND ITS MORPHOLOGY

N. BROSCH AND B. BILENKO
School of Physics and Astronomy, Beverly and Raymond Sackler
Faculty of Exact Sciences
Tel Aviv University, Tel Aviv 69978, Israel

1. Introduction

Observations in the space-UV domain promise, for the first time, a better sampling of the distribution of dust grains in our Milky Way and in other galaxies. We developed a method to map the three-dimensional distribution of dust in the nearest kpc, which combines optical information with UV photometry. This is an update and extension of the method used only in the optical by FitzGerald (1968) and Lucke (1978). The advantages are sensitivity to relatively small amounts of extinction and applicability to large data sets in a semi-automatic manner. A treatment of a combined data set, using common objects from the TD-1 catalog and in the Hipparcos Input Catalog, yielded strong information about the "granularity" of the distribution of dust clouds.

2. Deriving extinction from databank observations

In the late 1970s a few attempts were made to define an average law for the wavelength dependence of the extinction in the Galaxy (*e.g.*, Savage & Mathis 1979 [SM79], Seaton 1979). With increased knowledge of the UV spectral energy distribution (SED) of stars, it became clear that the extinction law differs significantly from one direction to another (Fitzpatrick & Massa 1990). The deviations from an average relation appear minor in the optical and near-IR, and it is even possible that the extinction law changes with galactic latitude (Kizskurno-Koziej & Lequeux 1987).

We use TD–1 observations and data from the Hipparcos input catalog (HIC) to determine the distribution of extinction in the one kpc region

155

D. L. Block and J. M. Greenberg (eds.), New Extragalactic Perspectives in the New South Africa, 155–158.
© *1996 Kluwer Academic Publishers.*

near the Sun. Our method is based on the determination of UV properties
of a star from its spectral type and luminosity class (from HIC), coupled
with the UV measurements listed for this star in the TD–1 catalog. The
first part of the method was demonstrated by Brosch (1991). It consists of
the empirical derivation of a relation between the (B–V) color index and a
(UV–V) or (UV–B) color index, determined by synthetic photometry from
IUE spectra. The relation is derived from stars with very low to negligible
reddening [E(B–V)<0.1] as a third-degree polynomial, and is applied to a
large set of stars in the TD–1 catalog with counterparts in HIC.

In addition to extinction in the UV, we also determine the distance modulus from the predicted absolute magnitude and the measured magnitude
to locate each object in three-dimesional space. Knowing the extinction to
each object and its location, we can map the 3D distribution of dust clouds
within the nearest kpc. The limitation is imposed by the depth of the two
data sets we use.

The joint TD-1+HIC sample contains 12,348 stars of which we use
~6,700, an average of one star pèr ~6 square degrees (6□°). However, the
distribution is far from uniform on the celestial sphere. Most UV-bright
stars are close to the Galactic plane. To improve the statistics, we consider
sky areas of 10×10□° in which at least four stars with well-determined
extinction are present.

In each 100□° area we fit a linear regression for the extinction vs. distance in the four TD-1 bands, from which we determine the extinction
gradient of the line of sight (LOS). From the distribution of values among
~150 LOS we find the average and the most probable values for the extinction in the four TD-1 bands. The values SM79 extinction values are close
to the **most probable** extinction values, but the distribution of A(λ) has
wide wings. The average UV extinction is generally smaller that the values
of SM79.

3. Analysis of one line of sight

We used our method to analyze one LOS where the stars in the joint TD-
1+HIC catalogue are well-distributed with distance. This LOS is to the
center of the Galaxy, within l=0° ± 10° and |b| < 30°. In this 1200 square
degree area there are 325 stars, which we bin in 50 pc distance bins to 550
pc. Farther than this the number of stars in each distance bin is four or
less, and the errors in the average extinction per bin become large.

We derive the average extinction gradient in each of the TD-1 filters
by fitting a linear relation for the extinction vs. distance. The expectation
is that the further a star is the larger the extinction to it. However, there
are distant bins where the extinction is lower than in nearer bins. The only

physically acceptable explanation is that the ISM is patchy, as mentioned already by Lynds & Wickramasinghe (1968). While some stars in a nearby distance bin may be extinguished, other stars further away may be shining through a clear ISM patch and show less extinction than the average of the nearer distance bin. The ISM patchiness, on scales of 30 arcsec, was recently demonstrated by Stanek (1996).

The extinction-distance distribution can be fitted with a linear relation up to ~ 500 pc, where the number of stars per bin decreases sharply. We find that the extinction per kpc is 2.5±0.2 mag for the 1965Å TD-1 band, 2.7±0.2 for the 2365Å band, and 2.1±0.3 for the 2740Å band. The shortest TD-1 band yielded unreliable results and has not been considered here. All three gradient values correspond to ~ 1 mag/kpc in the V band, using the SM79 extinction law, somewhat lower than accepted as the typical gradient in the galactic disk.

There are a number of points, in the 100 and 150 pc bin, which are definitely above the linear relation. These distance bins have a relatively high number of stars, thus the discrepant extinction cannot be attributed to a statistical fluke. In our attempt to understand this we identify a high value of extinction to ω Sco, a B1V star at 183 pc [log N(HI)=21.24; Fruscione et al. 1994]. Realizing this, we analyzed separately the 33 stars near ω Sco. We find the extinction there $\sim 3 \times$ higher than for the other stars in this direction. Another star 6° away and at 140 pc, σ Sco (B1III), has a strange extinction: SM79 give $\frac{E(1250-V)}{E(B-V)} = 2.8$ while the average value is 6.55. The HI column density to σ Sco is also high: log N(HI)=21.28±0.07 cm^{-2}. The extinction values derived by our method for all four TD-1 bands, for both ω and σ Sco, differ considerably from the SM79 values and are close to the values one would obtain with R$_V$=3.1 and E(B-V)=0.56 (for σ Sco).

Our results indicate that the LOS to the Scorpio region samples the same small ISM cloud. The stars with abnormal extinction gradient are located between 50 and 150 pc. The region has an approximate angular size of 10°. With a depth of ~ 100 pc and N(HI)$\approx 10^{21.2}$ cm^{-2}, the total mass of the cloud is $\sim 8 \ 10^4$ M$_\odot$ and its average density ~ 100 atoms cm^{-2}. A dust cloud was identified in this location by FitzGerald (1968) and Lucke (1978).

While the HI density of the Scorpius cloud is orders of magnitude higher than average, the extinction is only $\sim 15 \times$ larger than in the field, indicating that the dust density is only $\sim$one order of magnitude higher than in the field. The wavelength dependence of the extinction in the Scorpio cloud is $\sim$flat, which is usually interpreted as a signature of large grains (a>0.2 μm). If this is the case, the grain temperature must be lower than usually encountered in the ISM. This could perhaps be tested with IRAS data.

4. Conclusions

We describe a method to derive extinction in the UV using a combination of TD-1 UV photometry and information from the Hipparcos Input Catalog. The method relies on the ability to predict the UV emission from a star, which is then compared with that actually observed. From the optical information we also derive the distance to the star. In a statistically significant sample this allows one to map the 3D distribution of the extinction. Instead of the accepted picture of uniformly distributed dust we find that the extinction is extremely patchy. The extincting dust clouds are smaller than our angular resolution of $\sim 10°$. We identify a cloud with abnormal extinction only $\sim$100 pc away; the dust grains in this $\sim 10^5 M_\odot$ cloud are larger than those of the typical ISM.

References

Brosch, N. 1991 MNRAS **250**, 780.
Cardelli, J.A., Clayton, G.C. & Mathis, J.S. 1989 ApJ **345**, 245.
FitzGerald, M.P. 1968 AJ **73**, 983.
Fitzpatrick, E.L. & Massa, D. 1986 ApJ **307**, 286.
Fruscione, A., Hawkins, I., Jelinsky, P. & Wiercigroch, A. 1994 ApJ Suppl **94**, 127.
Lucke, P.B. 1978 A & A **64**, 367.
Lynds, B.T. & Wickramasinghe, N.C. 1968 Ann. Rev. A & A **6**, 215.
O'Donnell, J.E. 1994 ApJ **422**, 158.
Savage, B.D. & Mathis, J.S. 1979 Ann. Rev. A & A **17**, 73.
Seaton, M.J. 1979 MNRAS **187**, 73.
Stanek, K.Z. 1996 astro-ph preprint 9512137.

JDIES OF INTERSTELLAR DUST AND GAS WITH THE
₹ ULTRAVIOLET CAMERAS AND FAR ULTRAVIOLET
₄GING SPECTROGRAPH SPACE SHUTTLE INVESTIGATIONS

GEORGE R. CARRUTHERS

E. O. Hulburt Center for Space Research
Naval Research Laboratory, Washington DC 20375-5320

₁tract.

The Naval Research Laboratory's Far Ultraviolet Cameras experiment
ɣn on space shuttle mission STS-39 in April-May, 1991) and Far Ultra-
₁t Imaging Spectrograph (FUVIS), flown as the Spartan-204 payload on
-63 in February, 1995, were primarily intended to study diffuse sources
₁r-UV radiation, including those due to the interstellar medium (dust-
tered starlight and gas-phase emissions, including molecular hydrogen
₁escence). Among results of the Far UV Cameras was imagery of a very
nsive reflection nebulosity in the upper Scorpius region, including Rho
iuchi. The FUVIS experiment objectives included an imaging spec-
raphic survey of the Orion Nebula and other emission and reflection
ɪlae in the wavelength range 97-200 nanometers. The current status of
, analysis for the Far UV Cameras and FUVIS observations of inter-
ar dust are discussed. Future plans include a new version of the Far
Cameras, using electron-bombarded CCD detectors, to be flown on a
-duration unmanned Air Force satellite, ARGOS, in 1997. This mis-
will obtain all-sky surveys of point and diffuse sources in the 92-105,
160, and 131-200 nm wavelength ranges. Also proposed is a reflight of
Spartan-204 FUVIS, and/or a new version of FUVIS as a small long-
ɪtion satellite payload, also using EBCCD detectors.

₃lock and J. M. Greenberg (eds.), *New Extragalactic Perspectives in the New South Africa*, 159–169.
6 *Kluwer Academic Publishers.*

1. INTRODUCTION

Astronomical imagery and spectroscopy in the ground-inaccessible ultra-violet wavelength range (below 300 nm) has great potential to improve our understanding of the interstellar medium, both gaseous and dust components. Several previous space experiments, including the IUE satellite, the Hubble Space Telescope, and the Astro shuttle missions, have provided valuable information on the composition and physical properties of interstellar gas and dust. In particular, absorption spectroscopy of interstellar material has provided much more comprehensive information about the composition and properties of the gaseous component, and the extinction properties of the dust component.

However, two areas which have been relatively neglected are (a) spectroscopy of light emitted by diffuse objects, including gaseous emission and dust- scattered starlight; and (b) wide-field imaging photometry of diffuse objects, including the general diffuse background, in selected UV wavelength bands. These provide information on the properties and distribution of interstellar material which are complementary to the above-mentioned observations.

Part of the reason for this neglect is that these latter studies require instrumentation with (a) very high diffuse-source sensitivity, and (b) wide two- dimensional field of view (for imaging instruments), or (c) wide field along the slit, with imaging capability in this dimension (for spectroscopic instruments). These design attributes are not compatible with the primary objectives of instruments such as IUE and HST, which are designed for high point-source sensitivity and resolution, for the study of stars and compact sources. Our group at NRL has focused on the development and flight application of instruments in the first category. Here, we discuss the space flight investigations we have completed or plan for the near future, which are applicable to studies of interstellar material.

2. THE FAR ULTRAVIOLET CAMERAS SHUTTLE EXPERIMENT

The Naval Research Laboratory's Far Ultraviolet Cameras experiment (Carruthers 1986; Carruthers et al. 1992a; Carruthers et al. 1994) was part of the Air Force Program 675 payload, operated in Earth orbit during the STS-39 mission, April 28-May 6, 1991. The experiment objectives included far-UV (105-200 nm) measurements of Earth's upper atmosphere, celestial targets (stars and diffuse background), and the shuttle environment (including "shuttle glow" and thruster firings).

The two electrographic Schmidt cameras were sensitive in the 105-160 and 123-200 nm wavelength ranges, with auxiliary filters providing narrower bands of 123-160 and 165-200 nm. They had 20° diameter fields of view and about 3 arc min angular resolution.

The instrument observed 12 star fields, including regions at high and low galactic latitudes. A major objective was is to measure and map the naturally- occurring far-UV background of point and diffuse sources, over a large portion of the sky as observed from near-Earth space. The images provided photometric measurements of stars (Schmidt and Carruthers 1995, 1996), diffuse sources (including regions of reflection and emission nebulae), and the Magellanic Clouds. The diffuse UV background radiation at high and low galactic latitudes was also measured.

One of the most interesting fields is that centered on ρ Ophiuchi, which includes the head of the constellation Scorpius, where vast interstellar dust clouds were imaged for the first time in far-UV wavelengths (Carruthers et al. 1992a; Gordon et al. 1994) (see Figs. 1 and 2). Comparisons of the images in the 123-160 and 165-200 nm ranges, and with ground- based images, are expected to provide new information regarding the scattering and absorbing properties of dust particles as a function of wavelength (which ultimately may lead to greater insight into the physical properties and chemical composition of the dust particles), as well as the spatial distribution of dust density. The images in Figures 1 and 2 give qualitative support to the findings of A. Witt and co-workers, that the dust scattering phase function becomes more forward-scattering toward shorter wavelengths. This is seen by comparing (in the two images) the "bright rim" features near 22 i Sco (largely small-angle scattering of light from τ Sco) at lower center, with the reflection nebula associated with ζ Oph (outside the images to the upper left) which exhibits large-angle scattering. Note, however, that differences in foreground absorption between and within the two regions need to be taken into account.

Other targets in the galactic plane included the region of η Carinae, the Gum Nebula, the region of the Coalsack, and an overlapping set of three fields including the galactic center region. As can be seen in Figure 3, the "dark rift" of the Milky Way is apparent in deep exposures in the 165-200 nm wavelength range. Another target, the region of the North America Nebula, was observed for comparison with imagery obtained previously by the Apollo 16 Far UV Camera (Carruthers and Page 1976) and in a sounding rocket flight (Carruthers, Heckathorn and Gull 1980).

The Magellanic Clouds were observed to very low surface brightness levels in the 165-200 nm wavelength range, showing faint features in the outer regions of these galaxies. Figure 4 is an image of the LMC which also shows the gradient of (our) galactic diffuse background with galactic latitude (the galactic plane is outside the image to the lower left).

Further analysis of the diffuse background and nebular observations re-

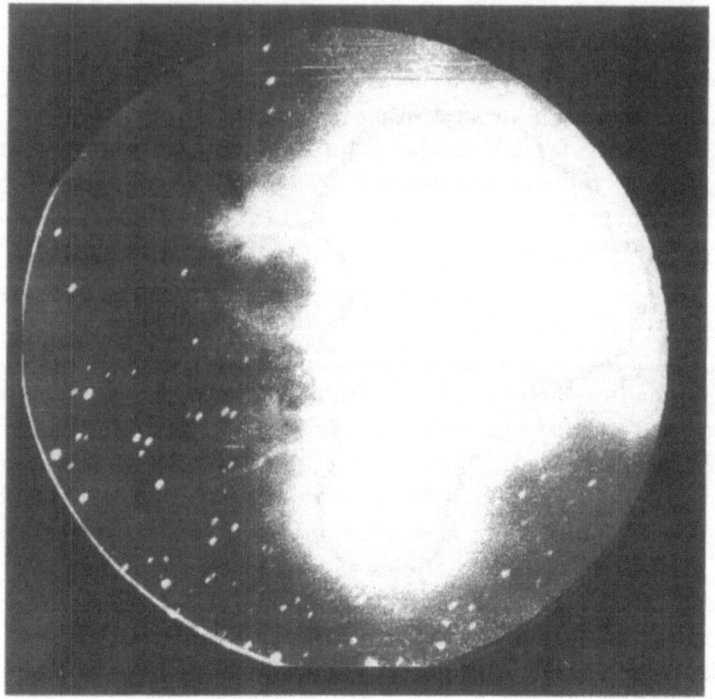

Figure 1.   Image of the region centered near ρ Ophiuchi, a 300-second exposure obtained with the Far Ultraviolet Cameras in the 123-160 nm wavelength range. The image
has been processed to enhance the contrast in the faint outer regions of the reflection
nebulosity. Note the "bright rim" feature at lower center.

quires further refinement of the camera absolute calibrations (Carruthers et
al. 1994), especially in the filtered mode of camera #2 (165-200 nm wavelength range). We are currently analyzing observations of stars previously
observed by OAO-2, IUE, and other space experiments (Carruthers et al.
1994; Schmidt and Carruthers 1995), as a basis for "in-flight" calibrations.
Also, the contribution of local upper-atmospheric airglow emissions (i.e. O
130.4 and 135.6 nm emission lines, and NO δ-band emission) must be determined and subtracted from the measured background to determine the
true celestial background component.

3. THE FAR ULTRAVIOLET IMAGING SPECTROGRAPH SHUTTLE EXPERIMENT

The Far Ultraviolet Imaging Spectrograph (FUVIS) was the primary science instrument in the NASA-Air Force Space Test Program Spartan-204

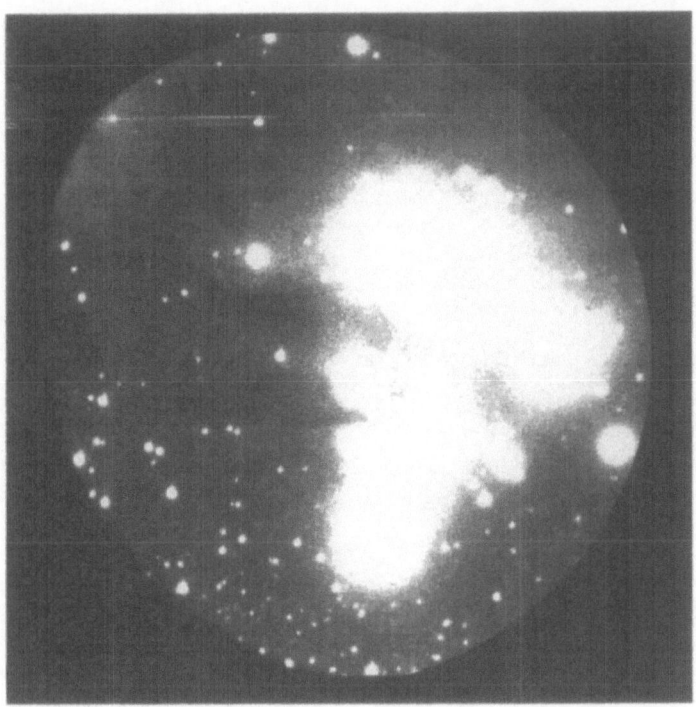

Figure 2.   As in Fig. 1, but a 100-second exposure in the 165-200 nm wavelength range. Note the reflection nebulosity in the upper left of the image, which is associated with the star ζ Ophiuchi (which is outside of the image).

shuttle payload, flown on the STS-63 mission in February, 1995. The scientific objectives were studies of diffuse nebulae (including H II regions, reflection nebulae, and supernova remnants) and the general diffuse galactic background.

The FUVIS instrument (Carruthers et al. 1988a) was designed to cover the wavelength range 97-200 nm with two interchangeable spectral resolutions. With its "narrow" 1.5'-wide x 2.7°-long slit, it has a spectral resolution of about 0.5 nm and an imaging resolution of 3 arc min along the slit. With an alternate "wide" (15') slit, it has a spectral resolution of 3 nm, but has 10 times higher sensitivity to diffuse sources filling the slit field of view. For objects (such as stars or planetary nebulae) smaller than the wide-slit field of view, FUVIS can also be used to obtain 2-dimensional, spectrally dispersed images, with 3 arc min resolution and with spectral resolution determined by the apparent angular size of the source.

The instrument also operates in two interchangeable wavelength ranges. The range 123-200 nm is covered using a calcium fluoride Schmidt corrector

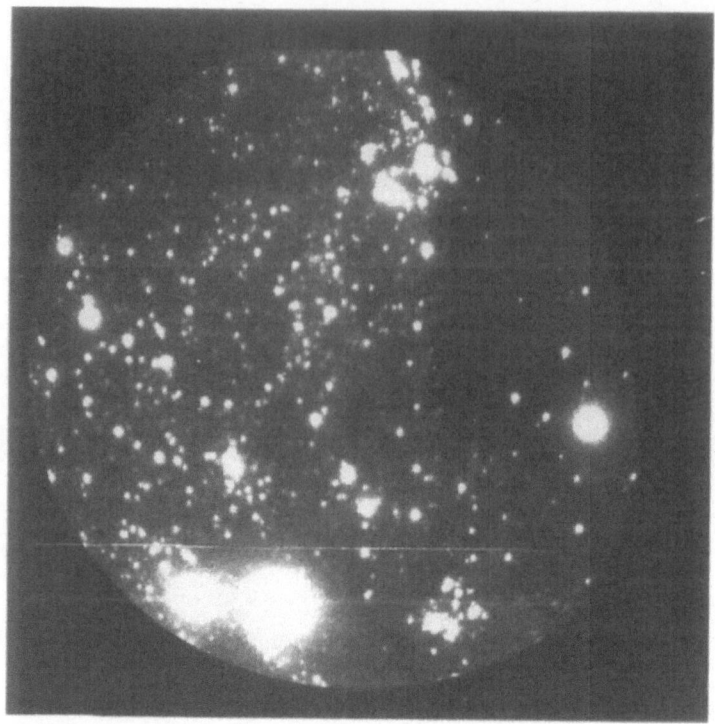

Figure 3.   Image of a region near the galactic center, a 100 second exposure obtained
with the Far UV Cameras in the 165-200 nm wavelength range. Note the low diffuse
background level in the right portion of the image, corresponding to the "Dark Rift" in
the visible Milky Way.

on the spectrograph camera, which maximizes sensitivity in this wavelength
range while blocking the intense hydrogen geocoronal Lyman-α (121.6 nm)
nightglow emission. The astrophysically important range below Lyman-α
(extending down to 97 nm) is covered, but at lower sensitivity and including
Lyman-α, using a windowless annular aperture in the camera.

Studies of the physical properties, composition, and spatial distribution
of interstellar dust by its reflection (scattering) of ultraviolet starlight is an
important objective of this investigation. Studies of dust scattering complement observations of interstellar extinction in the lines of sight to hot
stars, in that the extinction is made up of two parts, pure absorption and
scattering. Separate examination of the scattering component can provide
new information which can help determine the properties and composition
of the interstellar dust particles, and their variation in different regions of
space.

The observed spatial and spectral intensity distribution of dust-scattered

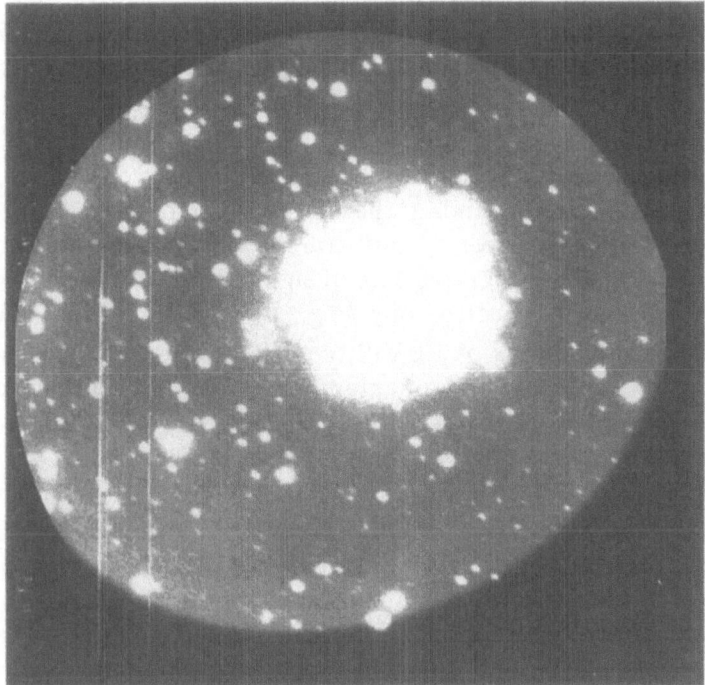

Figure 4.   A 300-second exposure of the Large Magellanic Cloud, obtained with the Far UV Cameras in the 165-200 nm wavelength range. Note the trend toward increasing diffuse background intensity in the direction of our galactic plane, toward the lower left.

starlight in a reflection nebula depends in a complicated way on the dust particle albedo and scattering phase function, the geometry of illumination (location of the particle with respect to the observer and the illuminating star), extinction in the lines of sight from the observer to the particle and from the illuminating star to the particle, and multiple scattering effects. Therefore, in practice, it is necessary to obtain *complete two-dimensional maps* of the dust-scattered light intensity, as well as its wavelength variation. Also it is necessary to compare the integrated intensity at each wavelength with that directly observed from the illuminating star. The UV wavelength range below 300 nm, and especially that below 120 nm, have been lacking in sensitive and comprehensive measurements.

Another primary objective is to detect and measure UV emission lines in supernova remnants, H II regions, planetary nebulae, and the "hot" interstellar medium. The UV spectral range offers enhanced capabilities for studying the energetics of such diffuse emission regions, since the excitation of UV lines requires more energy than is needed for most ground-accessible nebular lines. Hence, measurements of the absolute fluxes in UV lines and comparison with those of visual lines provide more sensitive measurements of electron temperature or other sources of excitation. In addition, UV lines provide more sensitive and reliable measurements of the abundances of carbon and silicon than possible in the ground-accessible wavelength range.

Observations of far-UV fluorescence of interstellar H_2 can provide alternate methods for determining the distribution of interstellar H_2, particularly its location relative to hot stars which excite the fluorescence, which complement the integrated line of sight concentration measurements obtained by absorption spectroscopy (e.g., by Copernicus and HUT).

The unique capabilities of FUVIS, compared to other instruments, include its large field of view with imaging capability, and sensitivity to wavelengths as short as 97 nm (including, for example, the important O VI resonance lines near 103 nm).

Targets planned for observation by FUVIS during the Spartan 204/STS-63 mission included the Orion Nebula, IC 410, the region of 15 Monoceros, and several other diffuse nebulae. Unfortunately, some problems were encountered with the FUVIS instrument during the flight, which compromised several of the observations. The primary difficulty was that the slit exchange mechanism (which alternates between the narrow-slit and wide-slit modes) malfunctioned such that, during much of the free-flight mission, the instrument was in an intermediate state with reduced throughput (relative even to the narrow-slit mode). Nevertheless, good data were obtained for some of the Orion Nebula and other target observations. At the time of this writing, the microdensitometry of the FUVIS flight film, including pre-flight and post-flight laboratory calibrations, has been completed and data reduction is currently in progress.

We are proposing a reflight of the Spartan-204 FUVIS (which would have been highly desirable even if the first flight had been 100% successful). In addition to correcting problems encountered in the first flight, we plan to replace the electrographic (film recording) spectrograph camera with a similar one based on an electron-bombarded charge-coupled device (EBCCD) (Carruthers et al. 1988b; Opal and Carruthers 1989). This latter sensor type is also planned for use in a far-UV camera experiment (GIMI) to be flown in a long-duration satellite mission (ARGOS), described in the following section. In the FUVIS Spartan reflight, the data from the EBCCD camera would be recorded on-board for post-mission retrieval, rather than telemetered to the ground. As a followon to this, the obvious long-term goal would be to develop and fly a version of FUVIS on a long-duration satellite. This would allow observations of objects over the entire sky (not possible on any one shuttle mission), and with longer exposure times and more varied observing geometries on each target.

4. THE GIMI EXPERIMENT ON THE ARGOS SATELLITE

The Global Imaging Monitor of the Ionosphere (GIMI) is one of several remote-sensing instruments under development for flight on the Air Force Space Test Program's P91-1 Advanced Research and Global Observation Satellite (ARGOS), planned for launch in early 1997. Carruthers and Seeley (1992b) give a description of the GIMI instrument and its scientific objectives. The primary objectives of GIMI are to map and monitor the ionospheric O^+ and electron density on a global basis, by means of wide-field imaging of ionospheric far-ultraviolet emissions; and to measure the neutral atmospheric density distributions of N_2 and O_2 by observations of stellar occultations by Earth's atmosphere. However, another major objective is to map and monitor the ultraviolet background in near-Earth space due to *extraterrestrial* sources (including diffuse nebulae and diffuse galactic background). This objective is of importance to basic astrophysics as well as to DoD applications. The astrophysically-related objectives of GIMI include all-sky celestial surveys in three far-UV passbands, 92-105 nm, 131-160 nm, and 131-200 nm.

The GIMI /ARGOS mission affords the first opportunity to perform an all-sky far-UV imaging survey in these passbands, including the first survey of any type in the 92-105 nm band, and sensitive surveys of *diffuse* sources and diffuse galactic background in the two longer-wavelength bands. Here, we will discuss only the latter GIMI objective: diffuse-source imagery and photometry in the 131-160 and 131-200 nm wavelength ranges over the entire sky.

The GIMI investigation is a follow-on to partial surveys obtained with similar cameras in sounding rocket flights and the Apollo 16 mission, and with the STS-39 Far UV Cameras. The wide fields of view of the cameras (10.5° square), and the near-polar, sun-sychronous orbit and long (1 to 3 year) duration of the ARGOS mission, will allow complete sky coverage. Although GIMI does not have the spatial resolution capabilities of an optimal sky survey instrument, its long-wavelength camera will extend to considerably fainter diffuse-source brightnesses than previous imaging and photometric surveys.

The GIMI observations will determine brightness distributions of dust-scattered starlight in reflection nebulae, H II regions, and the general interstellar medium; and of ultraviolet emission line sources such as planetary nebulae and supernova remnants. Comparison of imagery in the 131-160 and 131-200 nm ranges, in combination with pre-flight and in-flight calibration data, will provide some information on the spectral distribution of detected diffuse radiation. The objectives of the all-sky survey of the diffuse galactic background are to map the spatial distribution of UV-scattering dust, and by comparing the diffuse brightness distribution with the distribution of UV-bright stars, to determine the absorptivity and scattering properties (albedo and scattering phase function) of the dust in large regions of space and as a function of galactic latitude and longitude.

5. ACKNOWLEDGEMENTS

The space flight investigations described here are supported and sponsored by the DoD Space Test Program, the Office of Naval Research, and NASA. The author thanks his co-workers at NRL, NASA, and elsewhere for their contributions to this research program. Science co-investigators include Drs. Edward Schmidt, Adolf Witt, Reginald Dufour, John Raymond, and Mr. Karl Gordon. Mr. Brian Dohne and others at NRL have provided data reduction support.

References

Carruthers, G.R. (1986) The Far UV Cameras (NRL-803) Space Test Program Shuttle Experiment, *Ultraviolet Technology*, Vol. 687 of Proceedings of the S.P.I.E., pp. 11-27

Carruthers, G.R. and Page, T. (1976) Far-Ultraviolet Brightness of Nebulae in Cygnus, *ApJ*, **205**, pp. 397-404

Carruthers, G.R., Heckathorn, H.M. and Gull, T.R. (1980) Rocket-Ultraviolet Imagery of the North America Nebula, *ApJ*, **237**, pp. 438-449

Carruthers, G.R., Heckathorn, H.M., Dufour, R.J., Opal, C.B., Raymond, J.C. and Witt, A.N. (1988a), Spartan-281 Far-UV Imaging Spectrograph, S.P.I.E., pp. 87-111

Carruthers, G.R., Heckathorn, H.M., Opal, C.B., Jenkins, E.B. and Lowrance, J.L. (1988b) Development of EBCCD Cameras for the Far Ultraviolet, *Advances in Electronics and Electron Physics*, Vol. 74, ed. B. L. Morgan, Academic Press, New York, pp. 181-200

Carruthers, G.R., Morrill, J.S., Dohne, B.C. and Christensen, S.A. (1992a) The AFP-675 Far Ultraviolet Cameras Experiment: Observations of the Far-UV Space Environment, *Ultraviolet Technology IV*, Vol. 1764 of Proceedings of the S.P.I.E., pp. 21-35

Carruthers, G.R. and Seeley, T.D. (1992b) Global Imaging Monitor of the Ionosphere (GIMI): An Ultraviolet Ionospheric Imaging Experiment for the ARGOS Satellite, *Instrumentation for Planetary and Terrestrial Atmospheric Remote Sensing*, Vol. 1745 of Proceedings of the S.P.I.E., pp. 322-333; updated version in preparation for S.P.I.E. Conference *Ultraviolet Atmospheric and Space Remote Sensing: Methods and Instrumentation*, August 1996.

Carruthers, G.R., Dohne, B.C., Shephard, K.K., Reeb, S.A.C. and Schmidt, E.G. (1994) Photometric Calibrations of the AFP-675 Far Ultraviolet Cameras Experiment, *Ultraviolet Technology V*, Vol. 2282 of Proceedings of the S.P.I.E., pp. 184-201

Gordon, K.D., Witt, A.N., Carruthers, G.R., Christensen, S.A. and Dohne, B.C. (1994) The Far-Ultraviolet Dust Albedo in the Upper Scorpius Subgroup of the Scorpius OB2 Association, *ApJ*, bf 432, pp. 641-647

Opal, C.B. and Carruthers, G.R. (1989) Evaluation of Large Format Electron Bombarded Virtual Phase CCDs as Ultraviolet Imaging Detectors, *Ultraviolet Technology III*, Vol. 1158 of Proceedings of the S.P.I.E., pp. 96-103

Schmidt, E.G. and Carruthers, G.R. (1995) Far-Ultraviolet Stellar Photometry: Fields in Sagittarius and Scorpius, *ApJ Suppl.*, **96**, pp. 605-626

Schmidt, E.G. and Carruthers, G.R. (1996) Far Ultraviolet Stellar Photometry: Fields Centered on ρ Ophiuchi and the Galactic Center, *ApJ Suppl*, in press

HIGH SPATIAL RESOLUTION 50 AND 100 MICRON KAO OBSERVATIONS OF INFRARED-BRIGHT GALAXIES

BEVERLY J. SMITH
IPAC/Caltech
MS 100-22
Pasadena CA 91125

1. Abstract

In this review, the Kuiper Airborne Observatory 50 and 100 μm measurements of galaxies made by the University of Texas far-infrared group are summarized. These high spatial resolution observations have in some cases resolved galaxies which are pointlike in the IRAS data, and in others, shown details not visible with the IRAS data. We have found strong far-infrared emission from the bulge areas of quiescent spirals where little star formation is observed at other wavelengths. Thus dust heating by non-OB stars may be important in these bulges. In high luminosity starburst galaxies, in contrast, the far-infrared distribution tends to trace the observed morphology of young stars. In a few active galaxies, the far-infrared emission is quite extended, arguing against the AGN as the dominant power supply for the far-infrared emission. We find that L(FIR)/L(Hα) for the central regions of these galaxies is not correlated with CO $(1-0)$ flux; quiescent bulges have high far-infrared excesses relative to that expected by dust heating by OB stars following a Salpeter IMF. On the other hand, L(FIR)/L(H) correlates with I(CO), increasing faster than expected by extinction. These relationships support a scenario in which dust heating by young stars becomes increasingly important with increasing gas surface density.

2. Introduction

Early airborne observations, along with the successful IRAS mission in 1983, have shown that radiation in the far-infrared regime can be an important part of the total energy budget of galaxies. Emission in the 60 and 100 μm IRAS bands, which arises from warm dust in the interstellar

D. L. Block and J. M. Greenberg (eds.), New Extragalactic Perspectives in the New South Africa, 170–177.
© 1996 *Kluwer Academic Publishers.*

medium ($T_d \sim 20-100$K), can in extreme cases contribute up to 99% of the total luminosity of the galaxy, and in more typical cases the luminosity in the far-infrared may equal that in the visible. Determining the spatial distribution within galaxies of the warm dust traced by IRAS has been an important goal for infrared astronomers, as it is related to the issue of what is powering the far-infrared luminosity. Based on data from the IRAS mission, a two-component model for the far-infrared emission from non-active galaxies has been suggested, where both OB stars and the general interstellar radiation field contribute to dust heating in galaxies, in proportions differing from galaxy to galaxy (Helou 1986; Persson and Helou 1987). In active galaxies, dust heating by the AGN may also contribute. To test and refine this picture, it is important to determine the spatial distribution of the far-infrared emission within galaxies and compare this with other components in the galaxy. The IRAS spatial resolution, however, is relatively poor (1.5 × 5' at 100 μm), so without special processing, only the most nearby galaxies can be studied in detail. The development of the HiRes deconvolution method for IRAS data (Aumann et al. 1990) has allowed studies of more distant galaxies, but at best, the effective resolution reached is only ~80" at 100 μm (Rice 1993).

To obtain information at smaller physical scales or in more distant galaxies even higher resolution is needed. Until its recent decommission, this was possible using the 0.9m telescope of the Kuiper Airborne Observatory. At the University of Texas, we used the KAO to conduct a program of diffraction-limited observations of galaxies. These data are presented in a series of papers on individual galaxies (Joy et al. 1986, 1987, 1988, 1989; Lester et al. 1986, 1987, 1995; Brock et al. 1988; Smith et al. 1991, 1994, 1995). A recent paper (Smith and Harvey 1996) presents data for eleven additional galaxies. In this review, the KAO data for these 22 galaxies are summarized and compared with optical, near-infrared, and millimeter data.

3. The Data

The observations were made using either a single-spatial, dual-spectral system (Wilking et al. 1984), a linear 8-detector array (Smith et al. 1991), or a 2×10 array (Smith et al. 1994). The beamsizes were 18" × 25" at 50 μm and 30" × 40" at 100 μm, varying somewhat from instrument to instrument.

In this section, the results for various types of galaxies are discussed.

3.1. INTERACTING AND MERGING GALAXIES

One class of galaxies of particular interest are the interacting or colliding galaxies. With the KAO, it was possible to measure the sizes of the far-infrared emitting regions in a few of these galaxies which were unresolved

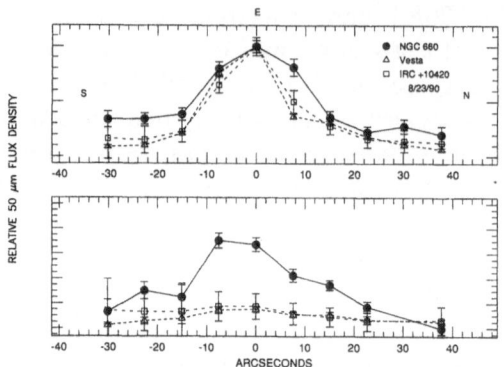

Figure 1. The 50 μm data for NGC 660, compared to the data for the point source objects. This data was obtained with the 2×10 array system. The top panel is the data for the first bank of 10 detectors, which crosses the nucleus of the galaxy; the lower panel is the data for the second bank (from Smith and Harvey 1996).

by IRAS. These include the probable merger remnants NGC 660 (Figure 1; Smith and Harvey 1996) and NGC 2146 (Smith et al. 1995), which have FWHM extents of 1 − 2 kpc. The far-infrared sizes are in agreement with the distribution of the observed OB stars. The peculiar galaxies Arp 220 and NGC 3256 are unresolved with both IRAS and the KAO; the KAO source size limits for these more distant galaxies are ≤3−4 kpc (Joy et al. 1986; Smith and Harvey 1996), significantly better than the IRAS limits. The interacting pair Arp 299 (NGC 3690/IC 694) is separated into two galaxies in the KAO data (Joy et al. 1989); NGC 3690 itself is resolved in the KAO scans, while IC 694 is compact (Joy et al. 1989).

3.2. ACTIVE GALAXIES

Another galaxy which has recently undergone a galactic collision is the nearby active galaxy Centaurus A (NGC 5128). The KAO data shows that this peculiar galaxy has an extended far-infrared disk (FWHM ∼ 5 kpc) surrounding a compact central source (Joy et al. 1988). The point source provides ∼10% of the total far-infrared luminosity, thus the AGN does not dominate dust heating. For the Seyfert galaxy NGC 1068, a single extended far-infrared source is seen; this emission may be powered in approximately equal measure by the AGN and a circumnuclear starburst (Lester et al. 1987). In contrast to these two galaxies, the Seyfert/starburst galaxies NGC 4945 and Circinus are compact in the KAO data (Brock et al. 1988; Smith and Harvey 1996), in spite of being closer. Thus, these circumnuclear starbursts are more centrally concentrated than those in Cen A or NGC 1068.

Another active galaxy of great interest is the distant giant elliptical

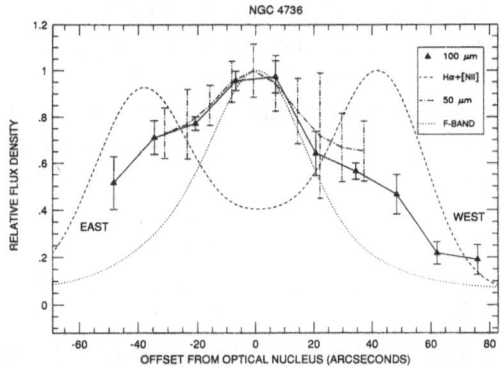

Figure 2. The KAO 50 and 100 μm data for NGC 4736, for the first bank of detectors. The 50 μm data has been smoothed to the 100 μm resolution. Hα+[N II] and F-band (red) cuts across the galaxy (from the images of Pogge [1989] and Kent [private communication]), smoothed to the 100 μm resolution, are also shown (from Smith et al. 1994).

galaxy NGC 1275. It is resolved with the KAO, with a FWHM size of 10 kpc (Lester et al. 1995). NGC 1275 is the dominant galaxy in the Perseus cluster and contains the variable radio source 3C 84. Our 100 μm flux density obtained a decade after IRAS has a flux density consistent with that of IRAS, while the millimeter continuum flux density has decreased by an order of magnitude over that time. This observation, along with the large 100 μm extent, implies that the far-infrared radiation arises from dust in the interstellar medium rather than directly from the active nucleus. An intriguing possibility is that the dust in NGC 1275 may be be heated by the hot X-ray emitting intracluster gas rather than by starlight (Lester et al. 1995).

3.3. QUIESCENT SPIRAL GALAXIES

In the bulges of some relatively quiescent spiral galaxies, few OB stars are found, however, the stellar density is high enough that significant dust heating may nevertheless occur. With the KAO, in a few cases we have been able to distinguish this bulge far-infrared emission from that powered by star formation in the disk. The best example of this is the Sab galaxy NGC 4736, where far-infrared radiation from a circumnuclear ring of H II regions (radius ~ 40″) is clearly separated from that of the bulge (Figure 2; Smith et al. 1991, 1994). These data show a bright central source with secondary 'shoulders' at the radius of the star forming ring. Optical Hα+[N II] and F-band (red) profiles across the galaxy are also shown in Figure 2. These demonstrate that the far-infrared emission in NGC 4736 does not trace the observed distribution of OB stars. The Hα+[N II] is strongest at the ring, while the far-infrared peaks at the center of the galaxy. The far-infrared distribution traces that of the older stars in the bulge, with an additional enhancement at the star forming ring. This morphology supports the two-component picture of extragalactic far-infrared radiation, with dust heating being dominated by older stars in the bulge and young stars in the ring. An analysis of the extinction and energetics of this galaxy (Section 4; Smith et al. 1991, 1994) is consistent with this hypothesis.

The Sb galaxy NGC 3627 is another galaxy for which bulge and spiral arm emission have been separated (Smith et al. 1994); the bulge contributes ∼20% of the total far-infrared luminosity. In both NGC 3627 and NGC 4736 the bulge far-infrared emission is extended with respect to the instrumental point source profile, with FWHM source sizes of ∼1 kpc, comparable to that of the older stars. This argues against dust heating being due to an obscured AGN. A third possible candidate for bulge star heating is the Sb galaxy NGC 7331, where we find a very broad flat-topped 100 μm distribution (Smith and Harvey 1996).

3.4. SPIRAL GALAXIES WITH STARBURST NUCLEI/NUCLEAR RINGS

In contrast to these relatively low luminosity quiescent galaxies, many of the brightest IRAS galaxies are spirals with powerful nuclear/circumnuclear starbursts. The KAO has provided an opportunity to measure the sizes and luminosities of these starbursts in the far-infrared. An extreme example of a nuclear starburst is that found in the SABc galaxy NGC 253, where ≥70% of the far-infrared emission is arising from the inner ≤30″ (≤500 pc) (Figure 3; Smith and Harvey 1996). Milder nuclear starbursts are found in M 83 and NGC 6946, where only ∼20% of the total far-infrared luminosity arises from the inner 400 pc and ≤750 pc, respectively (Smith and Harvey 1996).

As noted previously, a circumnuclear ring of far-infrared emission is resolved in NGC 4736 (Smith et al. 1991); M 51 also appears to have a far-infrared ring of radius ∼ 0.5 kpc (Lester et al. 1986). We have also observed three more distant galaxies with circumnuclear rings, NGC 1097, NGC 1808, and NGC 7552, which were unresolved with the KAO. In these cases, the KAO data show that the ring/nuclear region dominates the far-infrared luminosity of these galaxies (Smith and Harvey 1996).

4. Star Formation, Old Stars, and Extinction

As seen in the previous section, the far-infrared morphology of a galaxy provides clues as to what is powering the far-infrared luminosity. To make a more quantitative assessment, we have compared our KAO data with optical, near-infrared, and millimeter measurements. We have also included

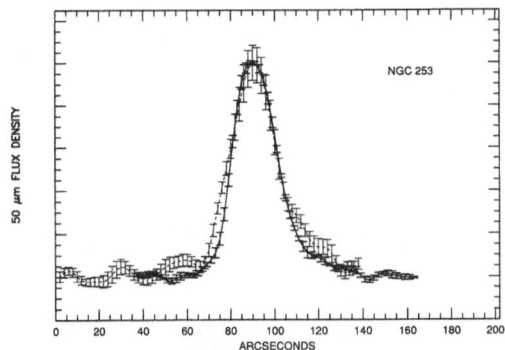

Figure 3. The 50 μm data for NGC 253 (dashed line) and the point source profile (solid line) (from Smith and Harvey 1996). These data were obtained with a scanning mode; the profile shown is for the detector which crossed the nucleus. The peak KAO flux density is 620 Jy, compared to a total 50 μm flux density of 739 Jy, from interpolation of the IRAS data.

the IRAS data for the bulges of M 31 and M 81 (Soifer et al. 1986; Smith and Harvey 1996), since these are sufficiently resolved with IRAS.

In Figure 4, we plot log L(FIR)/L(Hα) vs. log I(CO) for the central regions of the sample galaxies, while Figure 5 gives log L(FIR)/L(H) vs. I(CO). Note that L(FIR)/L(H) correlates with CO flux, and therefore with extinction, with starburst galaxies having higher L(FIR)/L(H) and CO values. In contrast, L(FIR)/L(Hα) does not increase monotonically with I(CO); the bulge areas of the quiescent galaxies NGC 4736, NGC 3627, and NGC 7331 have similar or higher L(FIR)/L(Hα) ratios but lower CO fluxes compared to the starbursts. In Figure 4, we have plotted the expected relationship if dust heating is due to OB stars following a Salpeter IMF and the standard Galactic $I(CO)/n(H_2)$ ratio and dust/gas ratio holds. The quiescent bulges lie above this curve, implying an infrared excess above that expected by massive star dust heating. This suggests important contributions to dust heating by non-OB stars (Smith et al. 1994). For these quiescent bulges, the star formation rate derived from the Hα luminosity is likely an upper limit; the LINER-like optical spectra of these galaxies may be due to ionization by an AGN, shocks, or some other mechanism rather than OB stars (see Smith et al. 1994). The starburst galaxies are to the right of this line, showing that they have sufficient OB stars to account for the far-infrared luminosity; in fact, the CO appears to over-estimate the Hα extinction. The $I(CO)/n(H_2)$ ratio may be enhanced in starburst galaxies (Maloney and Black 1988).

In Figure 5, the standard extinction curve is plotted, normalized to L(FIR)/L(H) ~ 2 at A_V = 0. This curve is the expected relationship if

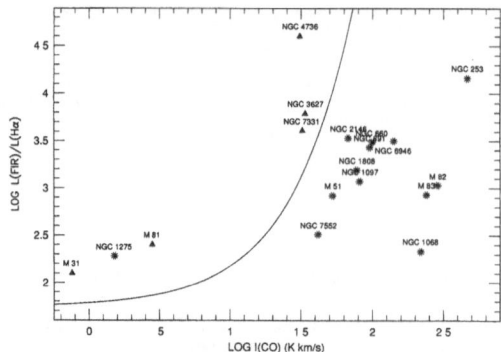

Figure 4.   L(FIR)/L(Hα) vs. I(CO) for the central regions of the 22 galaxies observed by the UT KAO group, along with the IRAS bulge data for M 31 (Soifer et al. 1986) and M 81 (Smith and Harvey 1996). The Hα and CO fluxes were obtained from the literature, and their beamsizes are similiar to our KAO beams. The curve is the expected relationship if dust heating is dominated by OB stars and I(CO) traces the extinction (see text). The asterisks are galaxies with strong observed star formation in the KAO beam, while the triangles are the data for the quiescent bulges.

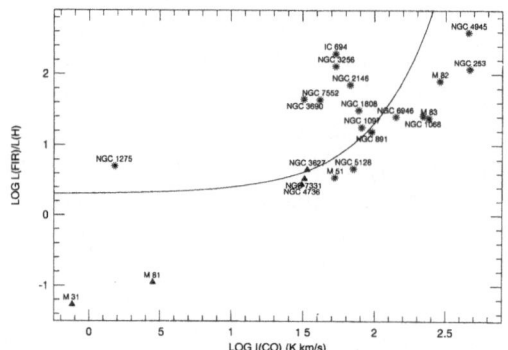

Figure 5.   L(FIR)/L(H) vs. I(CO) for the central regions of the 24 sample galaxies. The curve is the expected relationship due to extinction (see text).

I(CO) traces the extinction and ∼1/5 of the total bolometric luminosity of the old stars is absorbed by dust and re-emitted in the far-infrared. The bulges of the quiescent galaxies NGC 4736, NGC 3627, and NGC 7331 lie near this curve, while several of the starburst galaxies lie above it and M 31 and M 81 below. This trend, in which L(FIR)/L(H) increases with I(CO) at a rate faster than expected by extinction alone, suggests both a younger stellar population and a larger dust cross-section with increasing I(CO). Figure 4 and 5 together support a two-component model for dust heating in these galaxies, with larger contributions from OB stars for higher gas surface densities. The separation between the starbursts and the quiescent bulges occurs at a gas surface density equal to the critical gas density at which star formation occurs in a rotating disk with a rotational velocity of 150 km s^{-1} at a radius of 500 pc (Kennicutt 1989). Thus the lack of star formation in the quiescent bulges may be due to a threshold effect.

Many people have contributed to these studies. In particular, I would like to thank Paul Harvey for his long-standing support and encouragement. Dan Lester and Marshall Joy pioneered these studies and gave much help and advice, while a number of students and post-doctoral fellows helped with the observations over the years, including Cecilia Colomé, James DiFrancesco, David Brock, Cheng-Yue Zhang, and Chris Koresko. I am also very grateful to the KAO ground and flight staff for their help in acquiring these data. This work was supported by NASA grant NAG 2-67. Part of this work was performed while the author held a National Research Council-JPL Research Associateship. This research has made use of the NASA/IPAC Extragalactic Database (NED), which is operated by the Jet Propulsion Laboratory, Caltech, under contract with the National Aeronautics and Space Administration.

REFERENCES

Aumann, H. H., Fowler, J. W., and Melnyk, M. 1990, AJ, 99, 1674.

Brock, D., et al. 1988, ApJ, 329, 208.

Helou, G. 1986, ApJ, 311, L33.

Joy, M., Lester, D. F., and Harvey, P. M. 1987, ApJ, 319, 314.

Joy, M., Lester, D. F., Harvey, P. M., and Ellis, H. B. 1988, ApJ, 326, 662.

Joy, M., Lester, D. F., Harvey, P. M., and Frueh, M. 1986, ApJ, 307, 110.

Joy, M., et al. 1989, ApJ, 339, 100.

Kennicutt, R. C., Jr. 1989, ApJ, 344, 685.

Lester, D. F., Harvey, P. M., and Joy, M. 1986, ApJ, 302, 280.

Lester, D. F., et al. 1987, ApJ, 321, 755.

Lester, D. F., et al. 1995, ApJ, 439, 185.

Maloney, P., and Black, J. H. 1988, ApJ, 325, 389.

Persson, C. J. L., and Helou, G. 1987, ApJ, 314, 513.

Rice, W. 1993, AJ, 105, 67.

Smith, B. J., et al. 1991, ApJ, 373, 66.

Smith, B. J., et al. 1994, ApJ, 425, 91.

Smith, B. J., Harvey, P. M., and Lester, D. F. 1995, ApJ, 442, 610.

Smith, B. J. and Harvey, P. M. 1996, ApJ, submitted.

Soifer, B. T., et al. 1986, ApJ, 304, 651.

Wilking, B. A., et al. 1984, ApJ, 279, 291.

INFRARED PHOTOMETRY AND DUST ABSORPTION
IN HIGHLY INCLINED SPIRAL GALAXIES

L.E. KUCHINSKI AND D.M. TERNDRUP

Department of Astronomy, The Ohio State University
174 W. 18th Ave., Columbus, OH 43210, USA

1. Introduction

We have begun an extensive study of the structure and kinematics of bulges, especially those with boxy or peanut shaped morphology. In particular, we wish to investigate the possible connection between boxy bulges and bars (*e.g.*, Kuijken & Merrifield 1995).

We have initially studied the effects of dust on infrared photometry of highly inclined galaxies. Our goal is to provide quantitative estimates of the observed extinction and reddening in these systems. This is essential for deriving a reliable mass model from the stellar light distribution.

2. Data

We have obtained JHK images of 15 galaxies with $i > 65°$ and Hubble Types ranging from Sab through Sc. Eight of these have boxy– or peanut–shaped bulges (Jarvis 1986; de Souza and dos Anjos 1987; Shaw *et al.* 1990). The rest were selected from the $RC3$ (de Vaucouleurs *et al.* 1991) to span a similar range of absolute magnitudes, major axis dimensions, and inclination angles. The data were taken with the Ohio State Infrared Imager and Spectrograph (OSIRIS) on the 1.5 m telescope at CTIO. We have also obtained and are currently analyzing optical images of these galaxies.

3. Method

We computed color maps in $(J - K)$, $(H - K)$, and $(J - H)$ by aligning the appropriate pairs of images for each galaxy, then calculating the colors

D. L. Block and J. M. Greenberg (eds.), New Extragalactic Perspectives in the New South Africa, 178–181.
© 1996 *Kluwer Academic Publishers.*

at each pixel. Minor axis color profiles were determined by measuring the colors in one-pixel intervals along the minor axis of each galaxy.

The observed color profiles are compared to predictions from several radiative transfer models. Model calculations yield values for total extinctions and color excesses as a function of some measure of dust quantity, here the total V-band optical depth τ_V through the galaxy. We consider two simple models that neglect scattering and have analytical solutions to the transfer equation: a screen of dust between the galaxy and the observer and a uniformly mixed slab of dust and stars. We also consider three of the numerical models calculated by Witt et $al.$ (1992), which include both absorption and scattering for spherically symmetric systems: the dusty galaxy model, a uniformly mixed sphere of dust and stars; the starburst galaxy model, a centrally concentrated sphere of stars with a uniform sphere of dust embedded in it; and the dusty galactic nucleus model, a sphere of stars surrounded by a spherical shell of dust. Models that include scattering require more dust to produce a given extinction.

The model comparison is carried out for $(J-H)$ and $(H-K)$ simultaneously by looking at the distribution of minor axis colors in the two-color plane. We estimate unreddened bulge colors $(J-H)_0$ and $(H-K)_0$ for each galaxy by assuming that extinction is negligible in the outer bulge. For each dust model, the reddened colors are spread out along a path given by $(J-H)_0 + E(J-H)$ and $(H-K)_0 + E(H-K)$ as a function of τ_V. In the simple screen model, $E(J-H) \propto E(H-K)$ and we have the traditional reddening "vector" in the two-color plot. In other dust geometries or when scattering is included, the ratio $E(J-H)/E(H-K)$ is not constant and the behavior of the two colors with increasing optical depth follows a curved path, termed the model color "trajectory".

We also measure the maximum color excess in well-defined galaxy dust lanes. The maximum value of τ_V can then be inferred for each model that adequately describes the minor axis color trajectory.

4. Results

Seven galaxies in our sample have sufficient reddening in the dust lane to distinguish between the different radiative transfer models. The observed color trajectories resemble those predicted by the simple screen and uniform mixture models and the numerical starburst galaxy model of Witt et $al.$ (1992). Two of these galaxies, NGC 3390 and A 0908-08, are shown in Figure 1. Four other galaxies have dust features that are not red enough to extend into the regime on the two-color plot where different models can be distinguished. One of these, NGC 1515, is shown in Figure 1. All of the models predict similar trajectories at low τ_V. Of the remaining four galaxies,

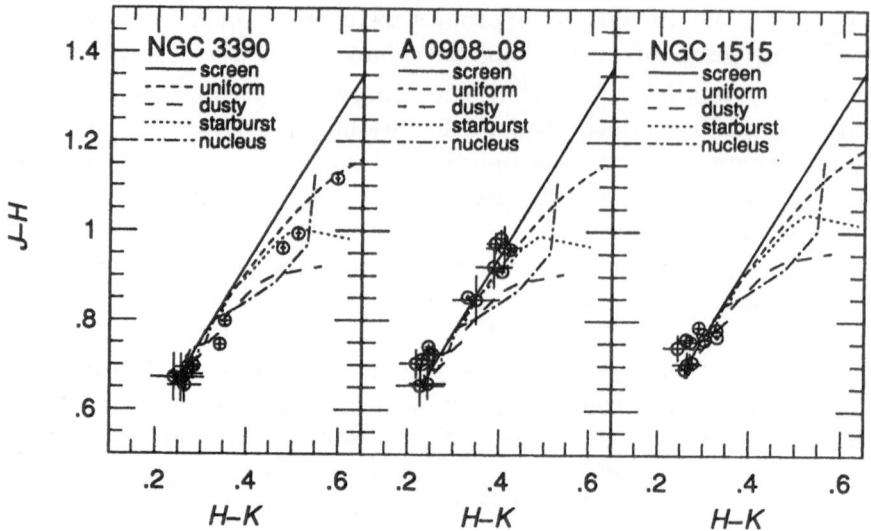

Figure 1.   Two-color plots for NGC 3390, A 0908-08, and NGC 1515, showing galaxy minor axis colors and model reddening trajectories.

three have no identifiable dust lanes or dust features located where the signal-to-noise ratio is too low to obtain reliable colors. One galaxy, NGC 1055, is heavily reddened and does not appear to match any of the model color trajectories.

For three galaxies which are almost exactly edge-on and have well-defined dust lanes, we infer a value of $\tau_{V,max} = 4.5$ in the dust lane center using the simple screen model. The value predicted using the simple uniform mixture model is $\tau_{V,max} = 15$. These values correspond to $A_K = 0.5 - 0.6$. For three galaxies which are not edge-on but have well-defined dust lanes, the inferred $\tau_{V,max}$ from the screen model ranges from $1.6 - 2.7$, while the uniform mixture and starburst galaxy models predict higher values of $\tau_{V,max} = 4 - 7.5$. These values correspond to $A_K = 0.2 - 0.3$.

5. Discussion

Trajectories in the two-color plot are influenced both by the relative distribution of dust and light sources and the effects of scattering. Dust geometries with a significant fraction of unobscured light sources, such as the Witt *et al.* (1992) dusty galaxy model, never produce enough reddening to match the observed colors in the reddest disk absorption features seen in our galaxy sample. Although the dusty galactic nucleus and dusty galaxy models differ from the simple screen and uniform models only in the inclusion of scattering and the spherical (rather than plane parallel) geometry, they predict very different JHK colors as τ_V increases. The lack of agreement between the models with scattering and the IR dust lane colors of highly inclined galaxies suggests that in the IR, scattering effects are not significant for the objects that make up our sample. However, both scattering and models with some unobscured sources would likely be required to describe face-on systems at any wavelength or galaxies of any inclination at blue wavelengths where the young disk stars are prominent.

The total optical depth τ_V needed to produce the observed reddening is highly dependent on the relative distribution of dust and stars. We cannot derive a single value of $\tau_{V,max}$ in the galaxy dust lanes because three different dust models all describe the observed color trajectories. We instead use our model comparisons to place limits on this quantity. The screen model always provides a *lower* limit on τ_V; other models require more dust to produce the same reddening because light sources near the edge of the system do not interact with dust and/or scattering effects partially compensate for absorption.

Our study of dust extinction in late-type spirals can help pinpoint where mass models based on the light distribution are likely to be inaccurate. Even in the K-band, where dust effects are often thought to be minimal, the extinction A_K is $0.5 - 0.6$ in the centers of edge-on galaxies and 0.2 in the dust lanes of other highly inclined systems. Once a plausible dust model is specified and an unreddened color is adopted, the extinction A_λ at any location can be inferred from the reddening at that point. These extinction values can be used to understand how the mass to light ratio (M/L) changes over galaxy dust features or even to crudely correct the light distribution so that a constant (M/L) can be invoked. Combined with kinematical data, this information will yield an improved picture of the stellar distribution in the inner regions of spirals, an important step in understanding dynamical influences on galaxy bulges.

A more complete discussion of this work has been accepted by the Astronomical Journal, tentatively scheduled for publication March 1996. This project was supported by NSF grants AST91-57038 (DMT), AST90-16112 and AST92-18449 (OSIRIS), and AST92-17716 (OSU galaxy survey).

References

de Souza, R. E., & dos Anjos, S. 1987, A&AS, 70, 465

de Vaucouleurs, G., de Vaucouleurs, A., Corwin, H. G., Buta, R. J., Paturel, G., & Fouque, P. 1991, *Third Reference Catalogue of Bright Galaxies*, (New York: Springer Verlag)

Jarvis, B. J. 1986, AJ, 91, 65

Kuijken, K., & Merrifield, M. R. 1995, ApJ, 443, L13

Shaw, M., Dettmnar, R.-J., & Barteldrees, A. 1990, A&A, 240, 36

Witt, A. N., Thronson, H. A., & Capuano, J. M. 1992, ApJ, 393, 611

THE NATURE OF INTERSTELLAR DUST IN LOCAL GROUP GALAXIES FROM OBSERVATIONS OF EXTINCTION AND POLARIZATION

GEOFFREY C. CLAYTON
Center for Astrophysics and Space Astronomy,
University of Colorado

Abstract. The assumption of Galactic Interstellar Dust properties in modeling the effects of Extragalactic Interstellar Dust will lead to large errors. Properties of Galactic dust have been amenable to study through observations of scattering and absorption along individual sightlines toward point sources, and through high spatial resolution observations of extended sources. Observations of Interstellar Dust in other galaxies are much more difficult since the light detected in a given aperture is the sum of a complex geometry of sources and dust which can only be separated using Radiative Transfer models. With the aid of NASA observatories such as *Hubble Space Telescope* (HST) and *Astro-2*, Interstellar Dust in Local Group galaxies can be subjected to the same kind of observations used to study Galactic dust. Individual sightlines can be observed at high spatial resolution allowing simple geometries to be studied.

1. Interstellar Dust in the Galaxy

There is an average Galactic extinction law, $A(\lambda)/A(V)$, over the wavelength range 3.5 μm to 0.125 μm, which is applicable to a wide range of Interstellar Dust environments in the UV (Cardelli, Clayton, & Mathis 1989 (CCM); Cardelli & Clayton 1991; Mathis & Cardelli 1992). This includes lines of sight through diffuse dust, dark cloud dust and dust associated with star formation. This mean extinction law depends on only one parameter which was chosen to be R_V $[=A(V)/E(B-V)]$. The existence of this law, valid over a large wavelength interval, suggests that the environmental

183

D. L. Block and J. M. Greenberg (eds.), New Extragalactic Perspectives in the New South Africa, 183–186.

processes which modify the grains are efficient and affect all grains. When changes in the grain size distribution which are reflected in the FUV extinction occur due to coagulation, the entire size distribution apparently varies in a systematic way. Dust in most Galactic environments amenable to UV observations adheres to the CCM relationship. Sightlines through dark clouds and star formation regions show significant but different deviations from CCM which may result from the presence or absence of coatings on grains. The standard Galactic extinction laws of Seaton (1979) and Savage & Mathis (1979) represent the mean extinction law for one value of $R_V \sim 3.1$.

Interstellar polarization has been described by the semi-empirical Serkowski relation (Wilking et al. 1982; Whittet et al. 1992). Clayton et al. (1995) re-examined the entire sample of 14 interstellar lines of sight for which there are now UV polarization data. We find that the previously suggested relationship between λ_{max} and the wavelength dependence of the polarization in the UV is strongly supported by the data for this larger sample (Clayton et al. 1992; Wolff, Clayton, & Meade 1993; Somerville et al. 1994). In fact, the amount of polarization in the UV is correlated with a single parameter, λ_{max}. This may indicate that there is a mean interstellar polarization law analogous to the mean CCM interstellar extinction law which is based on R_V. The data are consistent with a linear relationship between λ_{max}^{-1} and $p(UV)/p_{max}$ but more data are need to define the functional form. Both R_V and λ_{max} are proportional to the average grain size.

Over the lifetime of an average dust grain, it will cycle from the Diffuse Interstellar Medium to Dark Clouds and back again many times. While in the Dark Cloud environment, the mean grain size will typically get larger as grains coagulate. This is seen in the lower FUV extinction seen along Dark Cloud sightlines (CCM). The small grains have been removed and they are not returned to the gas phase. So it is likely that they coagulate into larger grains. If coagulation is rapid, composite grains must follow, perhaps similar to the Brownlee particles (Brownlee 1987). Therefore, to understand the absorption and scattering properties of Interstellar Dust, we must model composite grains.

Wolff et al. (1993) reviewed the applicability of several current dust grain models (all consisting of separate grain populations) with the constraints provided by the new UV polarimetry, finding that an extension of the Mathis, Rumpl, & Nordsieck (1977 (MRN)) model is the most successful at fitting both interstellar extinction and linear polarization data over the large wavelength range now available. The various wavelength dependences can be fit by varying the lower cutoff to the size distribution, n(a). The sharp cutoffs on both sides of the size distribution are likely to be unphysical. In addition, one must make assumptions regarding a particular model of the size-dependence of the grain alignment. These concerns can be circumvented through a solution of the inverse problem using the Maximum Entropy Method (Kim, Martin, & Hendry 1994; Kim & Martin 1994). Because separate populations of spherical dust grains likely do not reflect reality, it is necessary to model realistic composite grains including the effects of porosity on electromagnetic scattering, real inclusions (silicate, graphite, superparamagnetic etc.), and alignment (Mathis & Whiffen 1989; Wolff et al. 1994).

2. Extragalactic Interstellar Dust

Most galaxies even many ellipticals contain dust but almost nothing is known about how the dust differs from galaxy to galaxy. Very little is known beyond the Local Group. The CCM relation does not fit the observed extinction in other galaxies (the Magellanic Clouds and M31) (e.g. Clayton & Martin 1985; Fitzpatrick 1985, 1986; Hutchings et al. 1992; Clayton et al. 1996; Bianchi et al. 1996). The UV (1200-3200 Å) extinction in the Magellanic Clouds displays the same general wavelength dependence as in the Galaxy, but shows significant deviations from the CCM relation. The 2175 Å bump is weaker and the FUV extinction is steeper in the Magellanic cloud dust (Clayton et al. 1996 and references within). New HUT observations provide the first information on the FUV wavelength dependence of extinction for dust in the LMC (Clayton et al. 1996). In addition, two distinct wavelength dependences of UV extinction have been found in the LMC which are found in dust inside and outside the 30 Dor region. The deviations from CCM seen in the Magellanic clouds while somewhat similar to those seen for dark cloud sightlines in the Galaxy do not seem to have the same relationship to environment. Therefore, some other factor perhaps related to known metallicity (heavy element abundance) differences between these three galaxies may be involved in affecting the dust characteristics (Clayton & Martin 1985; Fitzpatrick 1986; Clayton et al. 1996).

If the Interstellar Medium properties are related to the metal content, one would not expect M31 to have similar dust properties to the Magellanic Clouds, but rather be similar to the Milky Way. (Metal abundance values for M31 range average about 0.4 dex lower than Galactic, compared with 0.7 for the LMC and 1.2 for the SMC. It is important to determine whether there is an overall correlation with metal content (and hence star-formation history), or whether other parameters or local variations in a galaxy are more important. Therefore, the nature of the dust extinction in M31 is of great interest because its metallicity is similar to that in the Galaxy (Blair & Kirshner 1985). Observations with FOS on HST of several reddened sightlines in M31 show a CCM-like FUV extinction and a weak

2175 Å bump (Bianchi et al. 1996).

Interstellar Polarization has been observed along a few lines of sight in the LMC. In the visible and UV, we find the wavelength dependence to be quite similar to that seen in the Galaxy (Clayton, Martin & Thompson 1983; Clayton et al. 1996). Martin & Shawl (1982) using globular clusters in M31 as point sources measured the wavelength dependence of the Interstellar Polarization in that direction to be similar to the Galaxy but implying slightly smaller than average dust (low λ_{max}).

For more distant galaxies, characterizing the nature and effects of Interstellar Dust becomes significantly more difficult. Observational diagnostics now depend upon the use of surface photometry, where the effects of emission, extinction and scattering in the aperture must be disentangled.

References

Bianchi, L., Clayton, G.C., Hutchings, J.B., Massey, P., & Bohlin, R.C. 1996, *ApJ*, submitted

Blair, W., & Kirshner, R. 1985, *ApJ*, **289**, 582

Brownlee, D.E. 1987, *in Interstellar Processes*, ed. D.J. Hollenbach & H.A. Thronson (Dordrecht: Reidel), p. 513

Cardelli, J.A., Clayton, G.C., & Mathis, J.S. 1989, *ApJ*, **345**, 245

Cardelli, J.A., & Clayton, G.C. 1991, *AJ*, **101**,1021

Clayton, G.C., Martin, P.G., & Thompson 1983, *ApJ*, **265**, 194

Clayton, G.C., & Martin, P.G. 1985, *ApJ*, **288**, 558

Clayton, G.C., et al. 1992, *ApJ*, **385**, L53

Clayton, G. C., Wolff, M. J., Allen, R. & Lupie, O. L. 1995, *ApJ*, **445**, 947

Clayton, G.C., Green, J., Wolff, M.J., Zellner, N.E.B., Code, A.D., & Davidsen, A.F. 1996, *ApJ*, in press (March 20)

Fitzpatrick, E.L. 1985, *ApJ*, **299**, 219

Fitzpatrick, E.L. 1986, *AJ*, **92**, 1068

Hutchings, J., Bianchi, L., Lamers, J., Massey, P., & Morris, S 1992, *ApJ*, **400**, L35

Kim, S.-H., Martin, P.G. & Hendry, P.D. 1994, *ApJ*, **422**, 167

Kim, S.-H., & Martin, P.G. 1994, *ApJ*, **431**, 783

Martin, P. G., & Shawl, S.J. 1982, *ApJ*, **253**, 86

Mathis, J. S., & Cardelli 1992, *ApJ*, **398**, 610

Mathis, J. S., Rumpl, W., & Nordsieck, K.H. 1977, *ApJ*, **217**, 425

Mathis, J. S., & Whiffen, G. 1989, *ApJ*, **341**, 808

Savage, B.D., & Mathis, J.S. 1979, *ARA&A*, **17**, 73

Seaton, M.J. 1979, *MNRAS*, **187**, 73

Somerville, W.B. et al. 1994, *ApJ*, **427**, L47

Whittet, D.C.B., Martin, P.G., Hough, J.H., Rouse, M.F., Bailey, J.A., & Axon, D.J. 1992, *ApJ*, **386**, 562

Wilking, B. A., Lebofsky, M. J., & Rieke, G. H. 1982, *AJ*,**87**, 695

Wolff, M. J., Clayton, G. C., & Meade, M. R. 1993, *ApJ*, **403**, 722

Wolff, M. J. et al. 1994, *ApJ*, **423**, 412

STRUCTURE IN THE DISTRIBUTION OF THE DUST AND ITS IMPACT ON EXTRAGALACTIC STUDIES

P. BOISSÉ AND S. THORAVAL

Radioastronomie/CNRS, École Normale Supérieure
24, rue Lhomond, 75231 PARIS Cedex 05 FRANCE

1. Introduction

The apparent morphology of galaxies (and therefore their classification scheme) appears to be much influenced by the presence of dust extinction, as indicated by near infrared (NIR) observations (see e.g. Block *et al.*, 1994). Further, in many studies of galaxies, it is critical to determine the intrinsic (i.e. unextinguished) spectral energy distribution emerging from the whole system. Indeed, the latter carries essential information about the old and young stellar populations and hence about the star formation activity. Since internal extinction may induce a very substantial attenuation, it is important to correct the observed energy distribution in an appropriate way. The so-called "uniform screen" model has long been employed to this purpose. In the latter, extinction corrections can be made as for nearby stars extinguished by interstellar clouds. Indeed, in both cases the observed flux (or brightness) writes $F_{obs} = F_0 \, e^{-\tau}$. As a consequence, a simple relation exists between reddening and extinction and estimates of the former from colors or line ratio can then be used to infer the extinction. Although it was clear that the uniform screen model is a gross oversimplification of the real situation, it was thought that the corrections thus obtained were roughly correct. Further, this procedure requires only one parameter, the slab opacity, while more realistic modelling involves many other (unknown) quantities.

However, it has been realized during the past decade that internal extinction in galaxies is a phenomenon much more complex than previously thought and that very important modifications may be induced by the effects neglected. Let us briefly mention the latter:

D. L. Block and J. M. Greenberg (eds.), New Extragalactic Perspectives in the New South Africa, 187–194.
© *1996 Kluwer Academic Publishers.*

- a) since the source is an extended one, light scattered towards the observer can be significant,

- b) stars and dust are in fact mixed together. Various stars suffer from very different extinctions depending on their location which alters very significantly the reddening - extinction relation,

- c) a noticeable fraction of the dust is enclosed within clouds, in late-type spirals at least,

- d) young stars (O and B types) are preferentially located near their parent molecular clouds which are in part confined in spiral arms.

Many studies have been devoted to the modelling of the above a) and b) effects. The radiation emerging from a mixture of stars and dust have been estimated by (among others) Bruzual et al. (1988), Witt et al. (1992), Byun et al. (1994) and Di Bartolomeo et al. (1995). Witt et al. (1992) stress that the combined effect of mixing stars and dust and of scattering is to produce little reddening even when considerable extinction is present. Puxley and Brandt (1994) use NIR line observations of two galaxies and show that the uniform screen model cannot account for the data. From a sample of 39 galaxies with both UV-visible continuum and line data, Calzetti et al. (1994) determine an effective extinction curve which displays marked differences with that measured in the Milky Way (absence of 2175Å feature in particular). From these studies, it appears that colors or line ratio cannot be used reliably to estimate the extinction.

Closely related to the correction of dust extinction is the question of the opacity of galaxy discs. Disney et al. (1989) have shown that commonly used arguments are strongly model dependent and propose that discs are optically thick. Although this question remains a much debated one (see e.g. Xu and Buat, 1995), it is clear that more realistic modelling of internal extinction is required to solve this problem. Similarly, the interpretation of data on overlapping galaxies also involves the effects listed above (a) and c) in particular: see White and Keel, 1992).

Another aspect of the problem is the determination of the optical properties of dust grains in external galaxies. As noted previously, the description of internal extinction requires the knowledge of at least the extinction curve (and, when scattering is not negligible, the albedo and asymmetry parameter as well). Besides, dust is not only a cumbersome component and is interesting to study in itself. For instance, it is of importance to know how far the extinction law is universal and how the dust properties and amount depend on environmental factors. With these motivations, many attempts have been made to measure $A(\lambda)$ in external galaxies (see e.g. Brosch et al., 1985; Goudfrooij et al., 1994; Jansen et al., 1994). To circumvent the difficulties listed above, galaxies displaying "simple" features like dust lanes are generally considered. In most of these studies, the data are interpreted within the frame of the uniform screen model which may not be always valid.

Here, we shall consider effects related to the cloudy nature of the dust distribution (effect c) above) which has not been treated at a galactic scale yet (see however Witt and Gordon, 1996a). To some extent, this problem is similar to that of the radiation transfer inside clumpy clouds. We therefore begin by summarizing previous work performed in this field and discuss effects expected in galaxies. Since the interaction of clouds with radiation strongly depend on their internal structure, we also report recent results obtained on the small scale structure in the distribution of the dust within molecular clouds. Finally, to illustrate our discussion, we present CCD observations of a galaxy with dust lanes, NGC 2685, and analyze these results to get the dust properties in a frame which allows the spatial distribution to be cloudy.

2. Summary of results on radiative transfer in clumpy media

Most of the work performed in that field and in an astrophysical context has been devoted to the interaction of radiation with clumpy interstellar clouds. Natta and Panagia (1984) considered the extinction of an extended source by a clumpy cloud. An analytical treatment of the problem including scattering and results from Monte-Carlo simulations for two-phase media have been given by Boissé (1990). Hobson and Scheuer (1993) extend the analytical formalism to media comprising more than two phases. Recently, Spaans (1995) have developed simulations in order to model the chemistry of fragmented molecular clouds. Finally, Witt and Gordon (1996b) consider the escape of photons emitted at the center of a cloudy spherical system.

The main modifications introduced by clumpiness (with respect to a uniform medium with the same geometry and amount of dust, i.e. same average opacity) is that the transparency for direct radiation is increased and the relative importance of scattering is reduced. For markovian media, it is possible to define an effective extinction coefficient and albedo such that the variation in depth of average intensities or fluxes is given by the solution of the uniform problem (cf Appendix A in Boissé, 1990). The above property probably still holds for non-markovian media as far as regions with different densities are more or less randomly mixed. Then, the increased transparency and decreased importance of scattering in a fragmented medium is reflected in a lower effective opacity and albedo. The analytical approach provides algebraic expressions for the effective opacity and albedo in terms of the dust opacity and albedo, and parameters describing the structure of the medium (Boissé, 1990; Hobson and Scheuer, 1993; see also Boissé, 1992 for media in which the dust albedo itself fluctuates). These formulas allow to assess in a simple way whether the transfer is governed by the clumps, the low density phase or both and estimate for any given structure, how much the transfer is affected by density fluctuations.

When considering application of the above computations to the internal extinction in galaxy discs, one has first to face the difficult problem of how the interstellar medium is to be described. If two phases are used, the high density phase might be identified with the molecular component and the low density phase with the diffuse atomic gas but the parameters characterizing the structure (density contrast, filling factor, ...) are not known for any galaxy. Further, the analytical model assumes that the density field is statistically homogeneous whereas in real disc galaxies, much of the absorbing material may be concentrated along spiral arms. It is therefore not easy to model the internal extinction in a realistic way by using these methods. A first step could be to replace in the sandwich model considered by Di Bartolomeo *et al.* (1995) the uniform dust layer by a two-phase medium. More detailed account of spiral arms, galactocentric variations of the star number density and interstellar medium, and of the broad range of cloud sizes certainly requires the use of Monte-Carlo simulations.

However, guided by results from previous work, it is possible to make a few general remarks about the anticipated effects of a fragmented dust distribution. A common feature of a system comprising stars mixed with dust and an extended light distribution seen through a clumpy cloud is that various parts of the source suffer from different extinctions. This results in an apparent extinction curve which is less dependent on wavelength than would be expected from the dust properties alone (in the extreme case of a source partially covered by opaque fragments, the apparent opacity is completely wavelength independent). This is in part why features in the extinction curve like the 2175Å bump appear less pronounced in both situations (Natta and Panagia, 1984; Calzetti *et al.*, 1994). The presence of clouds within a galaxy disc will tend to still increase the scatter of extinction values and therefore increase the difference between the true and apparent extinction curve. However, the "bluing" effect noted by Witt *et al.* (1992) should be relatively smaller than if the dust were distributed smoothly.

A second point is that effects of clumpiness are expected to be strong only when individual fragments are optically thick ($\tau \geq 1$; see Boissé, 1990). We can thus expect that departures from uniformity in the distribution of the dust will have gradually less importance when going from the UV to the NIR. In the near infrared, extinction is therefore not only less severe but also more simple to model, even in the presence of density fluctuations. We also note that structure strongly affects the conversion of optical light into far infrared radiation (Witt and Gordon, 1996a). Modelling of this effect in a spiral galaxy would require to take into account the relative location of molecular clouds and young stars (point d) in section 1 above).

A final point of importance is that the interaction of clouds with radiation strongly depends on their internal structure. It is therefore of interest for our purpose to know whether dense clouds globally behave regarding radiation transfer, like uniform or fragmented objects. To answer this question, we have performed several observational studies which we summarize in the next section.

3. Small scale structure in the distribution of the dust within clouds

Basically, the question is to determine how much $< e^{-\tau} >$ (which determines the effective opacity and hence, the overall transmission) is larger than $e^{-<\tau>}$ (the transmission that the cloud would have if it were uniform). If $e^{-<\tau>} \geq 0.8 < e^{-\tau} >$, the cloud is quasi-uniform ($e^{-<\tau>} = < e^{-\tau} >$ in the homogeneous case) while if $e^{-<\tau>} \leq 0.2 < e^{-\tau} >$ the cloud is markedly non-uniform. We have conducted observations aiming at the determination of the effective opacity for local Galactic clouds. The first method used involves the statistical analysis of B, V, R and i CCD data for densely populated stellar fields and relies on the accurate determination of magnitude and color distributions (Thoraval, 1995; Thoraval et $al.$, 1996a). We find that over the very broad range of spatial scales explored by this technique, the nearby (d $\approx$ 200 pc) translucent ($A_V \approx$ 2.) cloud studied displays extinction fluctuations with a relative amplitude $\sigma(A_V)/A_V$ of 25% only. With such a low value, the effective extinction is quite close to the average value since $A_{eff} \approx 0.8 < A >$. Further, most of the (weak) fluctuations appear to occur at relatively large scales ($\geq$ 40" i.e. 0.04pc) and are not correlated with ^{12}CO emission observed at a similar angular resolution.

The second method involves external galaxies that happen to lie behind molecular clouds (this is similar to the case of overlapping galaxies, one member of the pair being our own Milky Way !). The symmetrical and smooth appearance of several such objects appear to be very little distorted by the foreground cloud and imply that extinction fluctuations are quite small, in agreement with the previous study (Thoraval, 1995). Finally, motivated by the detection of secular variations of H_2CO radio absorption lines detected by Marscher et $al.$ (1993), we could get an upper limit of about 1% for the fluctuations of the extinction at a few AU scale (Thoraval et $al.$, 1996b). Overall, extinction fluctuations are found to be too small to affect the penetration of the radiation at scales less than $\approx$ 0.1pc. Below this value, molecular clouds can therefore be considered uniform. Effects of non-uniformity may however be present at larger scales.

4. Determining dust properties inside NGC 2685

In order to illustrate the difficulties encountered when attempting to determine dust properties in external systems, we present a preliminary analysis of data on the S0 galaxy NGC 2685. The latter is considered as a typical polar ring galaxy. We are interested here in the inner ring (Peletier and Christodoulou, 1993) which is molecular rich (Watson *et al.*, 1994) and induces a pronounced asymmetry in the brightness profile. Indeed, while the SW side of the NGC 2685 disc displays a smooth light distribution, the NE side is marked by several lanes which run perpendicular to the major axis.

We have obtained B, V, R and i CCD images of NGC 2685 using the 2m telescope at the Pic du Midi Observatory (France). The seeing was about 1.5 " (pixel size: 0.48"). The two SW and NE half brightness profiles in V are shown in Fig. 1a where dust lanes appear as dips in the NE curve. We then derive the variation of the apparent extinction in each band on the NE side, A_{app}, by adopting a locally uniform screen model for which $A_{app} = -2.5log(I_{NE}/I_{SW})$ where I_{NE} and I_{SW} correspond to symmetrical locations about the galaxy center. The variation of A_{app} is displayed in Fig. 1b as well as that of $A_{app}(\lambda)/A_{app}(V)$ (Fig. 1c). The latter ratio appears to be well uniform in the range 8" $< \Delta\theta <$ 25" with $A_{app}(B)/A_{app}(V) = 1.43$, $A_{app}(R)/A_{app}(V) = 0.77$ and $A_{app}(i)/A_{app}(V) = 0.57$. For comparison, these ratio take the respective value 1.32, 0.83 and 0.57 for the standard Milky Way extinction curve. A_{app} would represent a true extinction only if the following four conditions are fulfilled: C1) the SW side is unobscured and if dust were absent on the NE side, the galaxy would be symmetrical; C2) scattering is negligible; C3) no stars are present in front of the ring; C4) no marked structure is present at a scale smaller than the resolution.

Let us briefly discuss each of these assumptions. That some emission is associated with the ring is indicated by disortions appearing NW and SE of the disc and by the detection of a few HII regions (Hodge, 1974); however, these features are at a quite low level and represent only about 5% of the disc emission underlying the innest lane. C1 should be then satisfied except far away from the center. An increase of the apparent extinction in the R and i bands is seen at $\Delta\theta >$ 30 and 25" respectively with no associated excess in the B and V bands (Fig. 1c). In fact, this appears to be due to weak bumps in the unobscured SW R and i profiles, indicating that C1 is not strictly verified at $\Delta\theta \geq$ 25" in these bands. If scattering played a significant role (C2), the ratio $A_{app}(\lambda)/A_{app}(V)$ should vary across the lanes and correlate with the variation of A_{app}, especially at B where the albedo is expected to be larger than in other bands. Although $A_{app}(B)/A_{app}(V)$ is essentially constant in the range where it is accurately determined, small amplitude minima can be seen where A is maximum. This could be due to

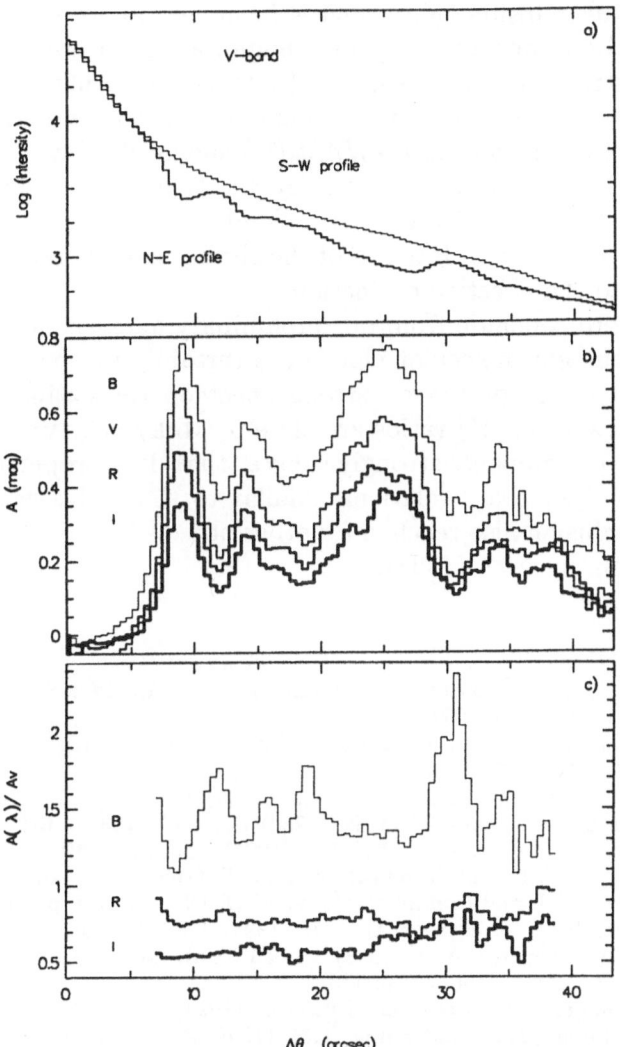

Figure 1. Variation along the major axis of the brightness (a; obscured NE side: heavy line; unobscured SW side: thin line), of the apparent extinction in B, V, R and i (b) and of the ratio $A_{app}(B)/A_{app}(V)$, $A_{app}(R)/A_{app}(V)$ and $A_{app}(i)/A_{app}(V)$.

scattering, but the effect appears to be unimportant.

If stars foreground to the dust lanes were present (condition C3) they would produce an apparent extinction which is both lower and more neutral than the true extinction. Since the inner ring has a radius which is comparable to that of the disc, it intersects the latter at a large radial distance and the contribution of foreground stars should therefore be small. Finally, from CO observations, Watson *et al.* (1994) conclude that most of the gas present inside NGC 2685 is in molecular form. Then structure effects might be important (C4) and should, as foreground stars, induce a relatively neutral apparent extinction (note that 1" is 60pc at the distance of NGC 2685). Since the structure is unlikely to be the same throughout the inner NE side, the ratios $A_{app}(\lambda)/A_{app}(V)$ should display variations (with $A_{app}(\lambda)/A_{app}(V) \approx 1$ where strong fragmentation effects are present). This is not observed which indicates that if clouds are present they are optically thin or cover only a small fraction of the surface and therefore give a small contribution to the effective extinction.

We therefore conclude from our preliminary analysis of the NGC 2685 data, that although molecular clouds are certainly present in this galaxy, fragmentation does not have a strong effect on the extinction. The fact that the $A_{app}(\lambda)/A_{app}(V)$ ratios are close to Milky Way values, associated to all the above arguments strongly suggest that dust properties in the ring encircling NGC 2685 are relatively similar to those observed in our own Galaxy, a conclusion also reached for other objects by e.g. Goudfrooij et al. (1994) and Jansen et al. (1994).

References

Block, D.L., Witt, A.N., Grosbol, P. et al. (1994) A&A, **Vol. 288**, p. 383

Boissé, P. (1990) A&A, **Vol. 228**, p. 483

Boissé, P. (1992) in Series on Advances in Mathematics for Applied Sciences, World Scientific, **Vol. 9**, p. 28

Brosch, N., Greenberg, J.M. and Grosbol, P.J. (1985) A&A, **Vol. 143**, p. 399

Bruzual, A.G., Magris, G. and Calvet, N. (1988) ApJ **Vol. 333**, p. 673

Byun, Y.I., Freeman, K.C. and Kylafis, N.D. (1994) ApJ **Vol. 432**, p. 114

Calzetti, D., Kinney, A.L. and Storchi-Bergmann, T. (1994) ApJ **Vol. 429**, p. 582

Disney, M.J., Davies, J. and Phillips S. (1989) MNRAS **Vol. 239**, p. 939

Goudfrooij, P., de Jong, T., Hansen, L. et al. (1994) MNRAS **Vol. 271**, p. 833

Hobson, M.P. and Scheuer, P.A.G. (1993) MNRAS **Vol. 264**, p. 145

Hodge, P.W. (1974) ApJS **Vol. 27**, p. 113

Jansen, R.A., Knapen, J.H., Beckman, J.E. et al. (1994) MNRAS **Vol. 270**, p. 373

Marscher, A.P., Moore, E.M. and Bania, T.M. (1993) ApJ **Vol. 419**, p. L101

Natta, A. and Panagia, N. (1984) ApJ **Vol. 287**, p. 228

Peletier, R.F. and Christodoulou D.M. (1993) AJ **Vol. 105**, p. 1378

Puxley, P.J. and Brandt, P.W.J.L. (1994) MNRAS **Vol. 266**, p. 431

Spaans, M. (1995) PhD. Thesis

Thoraval, S. (1995) PhD. Thesis

Thoraval, S., Boissé, P. and Duvert, G. (1996a) A&A, submitted

Thoraval, S., Boissé, P. and Stark, R. (1996b) A&A, in press

Watson, D.M., Guptill, M.T. and Buchholz, L.M. (1994) ApJ **Vol. 420**, p. L21

White, R.E. and Keel, W.C. (1992) Nature **Vol. 359**, p. 129

Witt, A.N., Thronson, H.A. and Capuano, J.M. (1992) ApJ **Vol. 393**, p. 611

Witt, A.N. and Gordon, K.D. (1996a) in Unveiling the Cosmic Infrared Background, ed. E. Dwek, AIP conf. Pro. 348

Witt, A.N. and Gordon, K.D. (1996b) ApJ, in press

Xu, C. and Buat, V. (1995) A&A, **Vol. 293**, p. L65

DETERMINATION OF THE 3D DUST DISTRIBUTION IN SPIRAL GALAXIES

N. D. KYLAFIS
University of Crete
Physics Department
P.O. Box 2208
714 09 Heraklion, Greece

AND

E. M. XILOURIS
University of Athens
Physics Department
Section of Astrophysics,
Astronomy & Mechanics
Panepistimiopolis
157 83 Zografos, Greece

1. Introduction

The amount and distribution of dust in spiral galaxies remains an unsolved problem. For its solution, two approaches appear to be promising: The small N approach and the large N one.

The small N approach involves detailed modeling of the nearby spiral galaxies, which are well resolved. In this way, one can derive the parameters that describe the distribution of dust in these galaxies. Thus, the opacity of a specific galaxy can be determined not in an average way, but with full spatial information. The small N approach is therefore attractive, but its main limitation is the practical upper limit on N. The larger N is, the more secure the generalization of the conclusions to all spiral galaxies is.

The large N approach involves the statistical study of a large number of galaxies and the inference of the mean opacity of these galaxies. Because N is large, this approach is extremely useful for general conclusions. Extreme care must be exercised, however, to avoid errors. We believe that the small N approach will help a lot our understanding of the dust distribution in some

D. L. Block and J. M. Greenberg (eds.), New Extragalactic Perspectives in the New South Africa, 195–202.
© *1996 Kluwer Academic Publishers.*

galaxies and therefore in the minimization of the errors in the statistical studies (see also Byun, Freeman & Kylafis 1994).

We in Crete (i.e., E. Paleologou, J. Papamastorakis, E. Xilouris and I) have initiated a study of the opacity problem in galaxies with the small N approach.

2. Observations

The observations have been made at the Skinakas Observatory in Crete, which is at 1750m above sea level. The recently established 1.3m optical telescope is of Ritchey-Cretien type with f7.7 and is equipped with a CCD camera and a Thomson chip with 1024 x 1024 pixels. Each pixel subtends 0.4 x 0.4 arcsec^2.

We have initiated observations of nearby, edge-on or nearly edge-on spiral galaxies in the B, V and I bands. Here we present images of UGC 2048 (NGC 973), which has been observed for 20 min in the I band, 45 min in the V band and 75 min in the B band. The typical seeing during the observations was 1 arcsec. Additional observations will be made in the V and B bands.

3. Model

We fit the surface photometry of spiral galaxies with the assumption that the emissivity of the stars in the disk and the bulge is axisymmetric and described by

$$L(R,z) = L_s \exp\left(-\frac{R}{h_s} - \frac{|z|}{z_s}\right) + L_b \exp(-7.67B^{1/4})\left[B^{7/8} + 0.11\right]^{-1} , \quad (1)$$

where R and z are the cylindrical coordinates, L_s is the stellar emissivity at the center of the disk and h_s and z_s are the scalelength and scalehight respectively of the stars in the disk. A 3D stellar emissivity, producing an $R^{1/4}$ law in projection, is given by the second term of the right hand side of equation (1) (Young 1976; Bottema, van der Kruit, & Valentijn 1991), with $L_b/0.11$ being the emissivity at the center of the bulge and

$$B = \frac{\sqrt{R^2 + z^2(b/a)^2}}{R_e} , \quad (2)$$

with R_e the effective radius of the bulge and a and b are the semi-major and semi-minor axis of bulge respectively.

The absorption coefficient due to dust is assumed to be of the form

$$A_\lambda(R,z) = A_\lambda \exp\left(-\frac{R}{h_d} - \frac{|z|}{z_d}\right) , \quad (3)$$

where A_λ is the absorption coefficient at wavelength λ at the center of the disk and h_d and z_d are the scalelength and the scalehight respectively of the dust.

If the above model galaxy is seen edge-on, and for the moment we ignore completely the effects of dust, the surface photometry due to the disk alone is

$$I_{\text{disk stars}}(R, z) = 2L_s R K_1(R/h_s) \exp(-|z|/z_s) , \qquad (4)$$

where $K_1(x)$ is the modified Bessel function of the second kind, first order (Abramowitz & Stegun 1965).

The optical depth through the disk, in directions parallel to the plane of the disk, is

$$\tau_\lambda(R, z) = 2A_\lambda R K_1(R/h_d) \exp(-|z|/z_d) , \qquad (5)$$

where R is the distance perpendicular to z in the plane of the sky. Thus, the central optical depth of an edge-on galaxy is $\tau_\lambda(0,0) = 2A_\lambda h_d$ and the central optical depth of the same galaxy seen face-on is $\tau_\lambda(0) = 2A_\lambda z_d$.

For specific values of the parameters in equations (1) - (3), the radiative transfer is done in the way described by Kylafis & Bahcall (1987) and a 2D image of a model galaxy is produced. The goal is to find those values of the parameters which create an image of the model galaxy as close as possible to the image of the observed galaxy. A Henyey-Greenstein phase function has been used for the scattering of the dust (Henyey & Greenstein 1941) with anisotropy parameter $g = 0.4$ and an albedo of the dust $\omega = 0.7$.

4. Model Fitting

According to equations (1) - (3), a fit to the surface photometry of a spiral galaxy should produce values of the parameters a) L_s, z_s and h_s for the stars in the disk, b) L_b, R_e and b/a for the stars in the bulge, c) A_λ or equivalently $\tau_\lambda(0,0)$, z_d and h_d for the dust and d) the inclination angle θ of the galactic disk with respect to our line of sight. This, however, is easier said than done. The search for a minimum (in a least-squares sense) in a space of ten (the number of the parameters) dimensions is not only time consuming, but also contains the danger to end up in a local minimum rather than the global one. For these reasons, it is helpful to get good estimates of as many parameters as possible before attempting a global fit to the galaxy.

We will start with the approximate determination of the parameters of the stars in the disk and the bulge of UGC 2048. We assume a distance of 63 Mpc for this galaxy (Gourgoulhon, Chamaraux & Fouque 1992). At this distance, the radius of the galaxy in the I band is 30 kpc. The pixel size corresponds to 0.12 kpc.

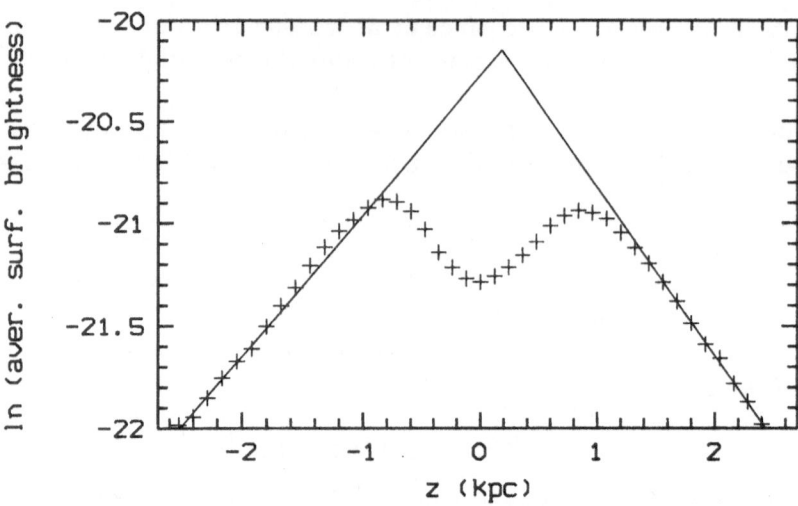

Figure 1. Natural logarithm of the average surface brightness (crosses) as a function of z for the parts of the galaxy away from the bulge. The solid line gives the slope of the average surface brightness at large z.

4.1. PARTIAL FITTING OF THE STARS

An inspection of the image of UGC 2048 reveals that this galaxy is seen approximately edge-on (the exact value of the inclination angle θ will be determined below). For an edge-on galaxy, the surface brightness away from the dust lane is proportional to $\exp(-|z|/z_s)$ at *all radial distances R* (see eq. [4]). Thus, excluding the central part of UGC 2048, which is affected by the bulge, the rest of the galaxy can be collapsed into one dimension parallel to the z axis. We do so in Figure 1, where we show the average value of the surface brightness in the I band as a function of z. It is evident that for large $|z|$, the surface brightness falls approximately exponentially with $|z|$, with a scaleheight $z_s = 1.2$ kpc (for the assumed distance) derived from $z > 0$ and $z_s = 1.4$ kpc derived from $z < 0$. None of the two values is accurate (as we will see below), but they are good initial guesses for a global fit. The inaccuracy of z_s is due to the bulge light, which is present at faint levels even at large distances from the center.

For an estimate of the scalelength h_s of the stars in the disk, we collapse the galaxy perpendicular to the z axis. By avoiding the bulge and the dust lane, the rest of the galaxy should be well described by equation (4). Integrating this equation over z, we then ask for that value of h_s that will make the ratio

$$\frac{I_{\text{obser}}}{I_{\text{model}}} = \frac{I_{\text{obser}}}{C\,R\,K_1(R/h_s)} , \tag{8}$$

approximately unity at all R. Here C is a constant that can be computed

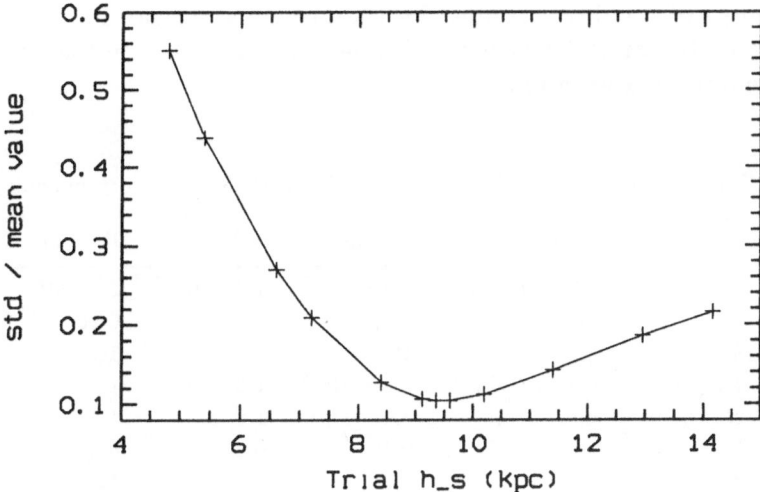

Figure 2. Standard deviation over mean value versus trial h_s. The minimum corresponds to the best estimate for h_s.

from equation (4). For each trial value of h_s we evaluate the ratio (8) for every R and compute the mean value and the standard deviation. Figure 2 shows the ratio of the standard deviation to the mean value as a function of trial h_s. At the minimum we have the best estimate of the true h_s. At the assumed distance, this is 9.4 kpc in the I band. Again, this value is not accurate enough due to the contamination by the bulge.

Having derived estimates for z_s and h_s, we have found $L_s = 1.8 \times 10^{-8}$ c/s/kpc^3 from equation (4) and the observed image in the I band.

Subtracting the derived disk from the image of the galaxy, we are left with the bulge away from the dust. From it we have found $b/a = 0.45$, $L_b = 3.1 \times 10^{-5}$ c/s/kpc^3 and $R_e = 2.8$ kpc in the I band.

4.2. PARTIAL FITTING OF THE DUST

Since galaxy models with dust require a lot of computer time, we tried to model the surface brightness at a few cuts of the galaxy parallel to the z axis. With the assumption that $h_d = h_s$, we have been able to determine the remaining parameters, i.e., $z_d = 0.36$ kpc, $\tau(0,0) = 8.5$ and $\theta = 89.5$ degrees.

4.3. GLOGAL FITTING WITH ABSORPTION ONLY

If one takes into account only the absorption by the dust and not the scattering, one can create a 2D image of a model galaxy on a workstation within a few minutes. It is thus possible to attempt a global fit to the surface photometry of a galaxy. Guided by the previous partial fits, we were able to find a global fit to UGC 2048 relatively quickly and the following values of the parameters were found.

TABLE 1. Global model fit parameters for UGC 2048 (absorption only)

Parameter	Units	I band	V band	B band
L_s	counts/s/kpc^3	1.3×10^{-8}	7.5×10^{-9}	3.8×10^{-9}
z_s	kpc	1.0	0.89	0.90
h_s	kpc	12	9.8	10
L_b	counts/s/kpc^3	5.8×10^{-5}	1.2×10^{-4}	1.1×10^{-4}
R_e	kpc	3.4	1.8	1.4
b/a	—	0.48	0.48	0.48
$\tau_\lambda(0,0)$	—	6.5	12.6	15.5
z_d	kpc	0.42	0.52	0.59
h_d	kpc	16	14.6	13
θ	degrees	89.5	89.5	89.5

4.4. GLOGAL FITTING WITH ABSORPTION AND SCATTERING

When scattering is taken into account, as it should, it requires about a cpu-day on a workstation to create a 2D image of a model galaxy. Thus, one cannot search blindly for a minimum in a ten-dimensional space. Educated initial guesses must be made! From our experience so far we have found the following: For edge-on spiral galaxies, a global fit to a galaxy with absorption only produces excellent values for the parameters z_s, h_s, z_d, h_d, b/a and θ and approximate values for L_s, L_b and R_e and $\tau_\lambda(0,0)$.

The values of the parameters derived for UGC 2048, with scattering and absorption taken properly into account, are shown in Table 2. As it can be seen, we have kept the values of the parameters z_s, h_s, z_d, h_d, b/a and θ the same as in the global fit with absorption only and varied the other four parameters.

5. Conclusions

Through three-dimensional modeling of the stars and the dust in UGC 2048, we have been able to determine the parameters that describe the distribution of dust in this galaxy. Figure 3 shows the 2D image of UGC 2048 in I, V and B band (top to bottom) with the model in the left half of the frame and the folded real galaxy in the right half of each panel. The model galaxy reproduces very well the real galaxy. We have verified this by

TABLE 2. Global model fit parameters for UGC 2048 (absorption and scattering)

Parameter	Units	I band	V band	B band
L_s	counts/s/kpc^3	1.2×10^{-8}	6.7×10^{-9}	3.4×10^{-9}
z_s	kpc	1.0	0.89	0.90
h_s	kpc	12	9.8	10
L_b	counts/s/kpc^3	9.8×10^{-5}	1.1×10^{-4}	1.1×10^{-4}
R_e	kpc	3.0	1.8	1.5
b/a	—	0.48	0.48	0.48
$\tau_\lambda(0,0)$	—	8.5	15.6	19.4
z_d	kpc	0.42	0.52	0.59
h_d	kpc	16	14.6	13
θ	degrees	89.5	89.5	89.5

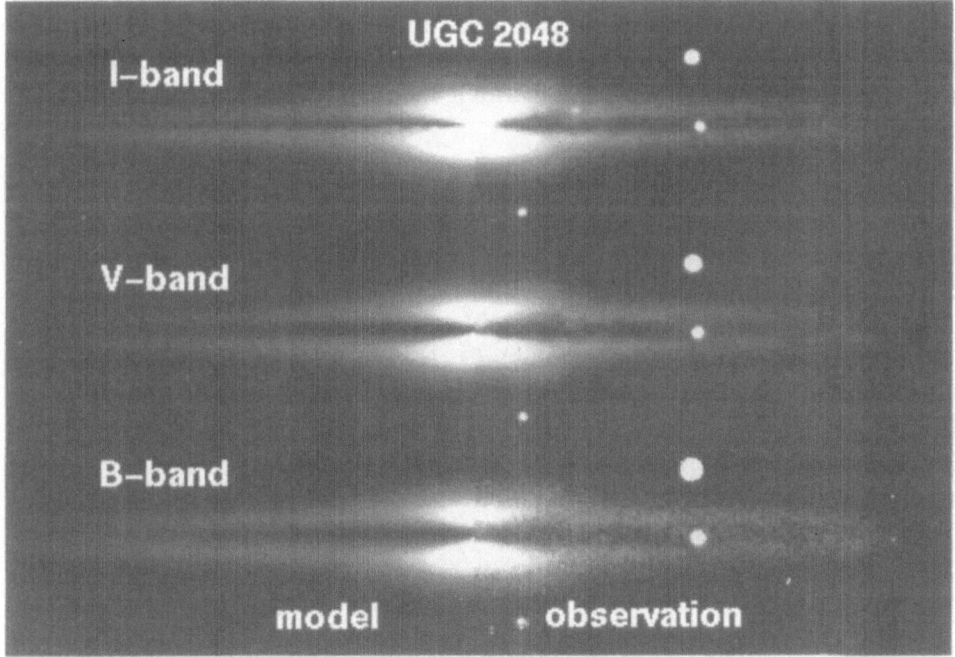

Figure 3. Image of UGC 2048 in I band (top), V band (middle) and B band (bottom). The left half in each panel is the model image and the right half is the real galaxy image.

subtracting one from the other. The difference between the model and the observations shows no systematics. In view of the large values inferred for z_s and h_s, we dare say that the assumed distance may be an overestimate of the real one.

Knowing A_λ and z_d, we find for the central, face-on optical depth $\tau_\lambda(0) = 2A_\lambda z_d$ the values: 0.22 in I band, 0.56 in V band and 0.87 in B band. Thus, if UGC 2048 were to be seen face-on, it would be essentially transparent.

The derived values of A_λ in the I, V and B bands are in the ratios 0.5:1.0:1.5, which are in good agreement with Rieke & Lebofsky (1985).

The obvious question now is: Could a different set of parameters produce a more or less equally good image? In principle, one can perturb *slightly* the set of parameters given in Table 2 and produce an equally acceptable image. However, *major departures from Table 2 produce unacceptable images.*

One may think that an acceptable image could be produced by a significantly larger scalelength h_d for the dust than that of Table 2, with a corresponding decrease in A_λ, so that $\tau_\lambda(0,0) = 2A_\lambda h_d$ is the same as that in Table 2. Fortunately this cannot be done. The image thus produced has obvious differences from the image of the galaxy and it is unacceptable.

We thank J. Papamastorakis and E. Paleologou for useful discussions. This work has been supported in part by the Greek Secretariat of Research and Technology and the British Council through a Greek-British Joint Research Program.

6. References

Abramowitz, M. & Stegun, I. A. 1965, *Handbook of Mathematical Functions* (New York: Dover).
Bottema, R., van der Kruit, P. C., & Valentijn, E. A. 1991, *A&A*, **247**, 357.
Byun, Y. I., Freeman, K. C., & Kylafis, N. D. 1994, *ApJ*, **432**, 114.
Gourgoulhon, E., Chamaraux, P., & Fouque, P. 1992, *A&A*, **255**, 69.
Henyey, L. G., & Greenstein, J. L. 1941, *ApJ*, **93**, 70.
Kylafis, N. D., & Bahcall, J. N. 1987, *ApJ*, **317**, 637.
Rieke, G. J., & Lebofsky, M. J. 1985, *ApJ*, **288**, 618.
Young, P. J. 1976, *AJ*, **81**, 807.

A 3D DUST MODEL FOR THE SOMBRERO GALAXY

Evidence for a bar-driven secular evolution

ERIC EMSELLEM
Sterrewacht Leiden
Postbus 9513
2300 RA Leiden
The Netherlands

Abstract. Using high resolution ground-based photometry and the Multi-Gaussian Expansion Technique, we built a realistic 3D photometric model of the Sombrero Galaxy (M 104) and determined the spatial distribution of its dust. It exhibits a double ring structure which peaks at $\sim 135''$, the central $100''$ being almost entirely devoid of dust. Taking into account the effect of light scattering, we derived a total dust mass of $3.2\ 10^7\ M_\odot$, which is probably a lower limit although it is already 6 times higher than the value predicted by IRAS fluxes. We conclude that the major part ($> 85\%$) of the dust in M 104 is cold (T $\sim$ 10–15 K). We also suggest that the gas and dust rings observed in the Sombrero were formed through the gravitational interaction between a stellar bar and the interstellar medium. This hypothesis could explain many of the observed features, e.g. the metallicity enhancement of the inner disc[1].

1. Introduction

The study of the properties of extragalactic dust requires realistic models of galaxies. The construction of such models is a difficult task since the luminosity distributions of spiral galaxies cannot be well represented by a simple combination of the "classical" functions (e.g. $r^{1/4}$ law, exponential discs). Furthermore, it is always difficult to disentangle the observational effects due to different dust characteristics from other phenomena (e.g. clumpy distribution).

---

[1] A detailed analysis can be found in Emsellem 1995, and Emsellem et al. 1996.

D. L. Block and J. M. Greenberg (eds.), New Extragalactic Perspectives in the New South Africa, 203–206.
© 1996 Kluwer Academic Publishers.

We examined the case of the Sombrero galaxy, an Sa with a prominent dust lane, using the Multi-Gaussian Expansion (hereafter MGE) formalism (Emsellem et al. 1994). We show that such a technique can indeed be applied to early-type spiral galaxies with complex morphologies, and derive the spatial distribution of the dust, as well as its main physical characteristics.

2. The dust model

We used the MGE technique to build a photometric model of the Sombrero galaxy from high resolution HRCAM/CFHT images (B, V, R_C and I_C bands). This formalism permits a fully analytical deprojection if the inclination angles are known. In the case of M 104, we thus obtain an axisymmetric model of the luminosity distribution which fits the data from $0\rlap{.}''05$ to $1200''$, by masking large portions of the images where dust absorption was significant. These regions were determined by computing the "absorption"[2] profiles perpendicular to the dust lane.

We then assumed that the dust in the Sombrero galaxy has uniform physical properties and its distribution a constant height. We proceeded by sampling the main dust lane in radial bins of $5''$ from $0''$ to $380''$. The value of the optical depth in each spatial bin was determined by minimizing the difference between the predicted and the observed absorption profiles.

3. Results

3.1. SCATTERING

It appeared that all available bands (B, V, R_C and I_C) could not be fitted simultaneously with the same dust model. For a dust distribution adapted to the R_C band absorption profiles, the peak of the V band absorption is overestimated in the model with respect to the observations. We have proved that this "missing extinction problem" could not be due to the effect of foreground light. Monte Carlo simulations by Witt et al. (1992) showed that the actual extinction can be significantly underestimated in disc galaxies when directly derived from the observed absorption, because of the light scattering by dust particles. Taking this effect into account, we have found that this indeed solves the "missing extinction problem" (Fig. 1): the observed absorption peaks in the B and V bands are saturated, and are therefore weakened relatively to the corresponding peaks in the R_C and I_C bands. The extinction curve derived from this dust model ($R_V \sim$ 2.86) is consistent with "Galactic dust" properties.

---

[2]The ratio of the southern flux with respect to its northern symmetric point.

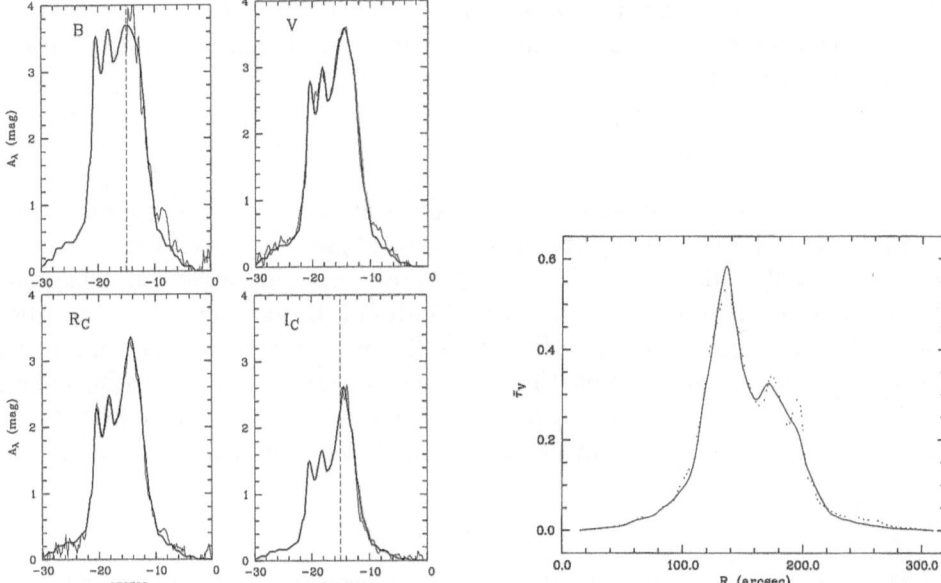

Figure 1.   *Left panel*: fit (thick line) of the central extinction profile of M 104 ($x' = 0$, thin line) with the dust model, including light scattering by dust particles. The vertical dashed lines mark the extent of the B and I_C data. *Right Panel*: radial optical depth profile $\bar{\tau}_V^0$ obtained from the observed absorption along the minor axis of the galaxy ($x' = 0$, pointed line), and averaged on all values from $x' = -15''$ to $x' = 45''$ (solid line). The values shown here are integrated over the whole vertical extent of the dust lane in the model.

3.2. THE SPATIAL DISTRIBUTION OF THE DUST

The dust is distributed in a double ring-like structure which peaks at $\sim 135''$ (the maximum) and $\sim 175''$ (Fig. 1). There is very little dust in the central $100''$, consistently with the HI distribution observed by Bajaja et al. (1984). The dust lane has an extremely small vertical extent of $0''.7$! Although our model may artificially underestimate the dust height, the thin peaks observed in the absorption profiles imply a very thin lane ($< 1''$ or ~ 90 pc). Near-infrared photometry (Wainscoat et al. 1990) revealed the presence of a cospatial stellar ring which is significantly thicker.

3.3. MASS AND TEMPERATURE

We adopted the same grain size distribution than Mathis et al. (1977), and assumed the dust in M 104 to be an equally weighted composition of graphite and silicates (Draine & Lee 1984). Our dust model of M 104 then predicts a total dust mass of $3.2\ 10^7\ M_\odot$, probably a lower limit, although it is already 6 times higher than the dust mass derived from the IRAS fluxes. This is not surprising considering that we derived an *upper limit* of 18 K for the temperature of the dust, which therefore should emit at large[3] wavelengths ($\lambda > 150\mu$m).

4. Secular evolution

The analysis of spectrographic TIGER data as well as colour gradient images (HRCAM) revealed the presence of ionized gas associated with a metal-rich inner stellar disc ($R < 15''$). We propose that the outer dust/gas and stellar rings correspond to the Outer Lindblad Resonance of a (now dissolved?) stellar bar, the Inner Lindblad Resonance thus corresponding to the transition region between the inner and outer stellar discs (Emsellem et al. 1996). We also detected a secondary plateau in the surface brightness profile which could correspond to the Ultra Harmonic Resonance (4:1).

5. Conclusions

The luminosity and dust distributions of the Sombrero galaxy can be nicely modelled using the MGE technique. This model shows the importance of light scattering by dust particles, and that the major part ($> 85\%$) of the dust in M 104 is cold (T $\sim$ 10–15 K). A more refined treatment of the scattering effect is now required, e.g. using Monte Carlo simulations and MGE, to confirm the result presented here. It would also be interesting to apply the MGE method to a limited sample of nearly edge-on spiral galaxies to determine the physical properties of their dust component. We finally have shown that a careful modelling of the spatial dust distribution of these galaxies can provide interesting hints regarding the involved evolution processes.

References

Bajaja, E., van der Burg, G., Faber, S. M., Gallagher, J. S., Knapp, G. R., Shane, W. W., 1984, A&A 141, 309
Draine, B. T., Lee, H. M., 1984, ApJ 285, 89
Emsellem, E., Monnet, G., Bacon, R., 1994, A&A 285, 723
Emsellem, E., 1995, A&A 303, 673
Emsellem, E., Bacon, R., Monnet, G., Poulain, P., 1996, submitted to A&A
Mathis, J. S., Rumpl, W., Nordsiek, K. H., 1977, ApJ 217, 425
Wainscoat, R. J., Hyland, A. R., Freeman, K. C., 1990, 348, 85
Witt, A. N., Thronson Jr., H. A., Capuano Jr., J. M., 1992, ApJ 393, 611

---

[3]IRAS principally detected the dust warmed by the star forming regions of M 104.

DOUBLING OF THE INFRARED FLUX FROM NGC 1068:
A CIRCUMNUCLEAR DUST TORUS?

I.S. GLASS
South African Astronomical Observatory
PO Box 9, Observatory 7935, South Africa

1. Introduction

The near-infrared (*JHKL*-region) fluxes of about 30 active galaxies, including Seyferts type 1 and 2, have been monitored for periods of up to 20 years at SAAO.

Infrared variability amongst Seyfert 1 galaxies is now well established, for example in Fairall 9 (Clavel, Wamsteker & Glass, 1989), NGC 3783 (Glass, 1992) and NGC 4595 (Santos-Lleó *et al.*, 1994). However, the conventional viewpoint has been that variability should not be expected in a Seyfert 2, arising from the belief, supported by mid-infrared photometry and spectrophotometry, that the source of the infrared radiation was a cool dust cloud having spatial dimensions of many light-years.

Near infrared (*HKL*) photometric monitoring stretching over two years, by Penston *et al.* (1974), appeared to confirm the position for at least one Seyfert 2 galaxy, namely NGC 1068.

Apparent variations in the nucleus of NGC 1068 in the mid-infrared were noted by Rieke & Low (1972) and Low & Rieke (1971), but different focal-plane apertures were used during their programme and the results were seriously affected by the extended nature of the source. Also, Chelli *et al.* (1987) combined isolated results at around 3.5 μm from different authors and found that a flux increase had occurred. However, the techniques of the various observers differed considerably and the possibility of an incorrect conclusion was significant.

In the SAAO programme, great care has been taken to ensure the long-term consistency of the observations. This has meant that the same filters, aperture sizes and standard stars were used throughout, and that the chopping distances (excepting for some of the earliest observations) and direc-

D. L. Block and J. M. Greenberg (eds.), New Extragalactic Perspectives in the New South Africa, 207–210.
© 1996 *Kluwer Academic Publishers.*

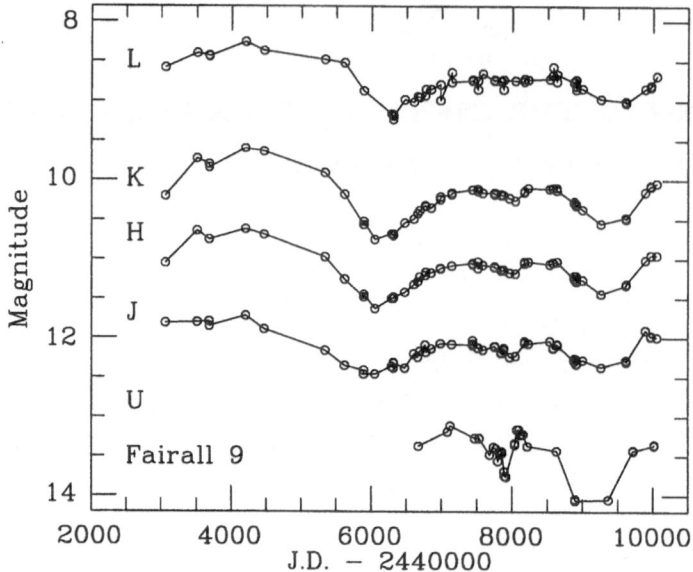

Figure 1. *JHKL-* and *U*-band photometry of Fairall 9. See text for discussion.

tions were always the same. The centering of the infrared aperture with the visual graticule was verified frequently, and the position of the telescope on the sky was checked at least 5 times during each observation. Offset guiding was used to correct for drive errors.

2. Behaviour of Fairall 9

The Seyfert 1 galaxy Fairall 9 is amongst the most luminous of the sample and has one of the largest amplitudes of variation. In addition, it has been observed extensively by the International Ultraviolet Explorer. Clavel *et al.* (1989) showed that a strong decline in its ultraviolet light was followed after 400 ± 100 days by similar behaviour in the *L*-band ($3.5\,\mu$m). An analysis of the varying infrared energy distribution suggested that the near-infrared flux originates in a shell or torus of refractory dust lying at about 400 light-days distance from the nucleus. The delay was found to be in rough quantitative agreement with the radiative properties of graphite dust and the estimated luminosity of the central XUV (x-ray and ultraviolet) source.

Since the work of Clavel *et al.*, monitoring of Fairall 9 has continued in the near infrared and *UBV* regions, the latter being undertaken by Hartmut Winkler of Vista University (Soweto). Some of the results obtained to date are given in Figure 1. There is little or no offset (as was noted by Clavel *et*

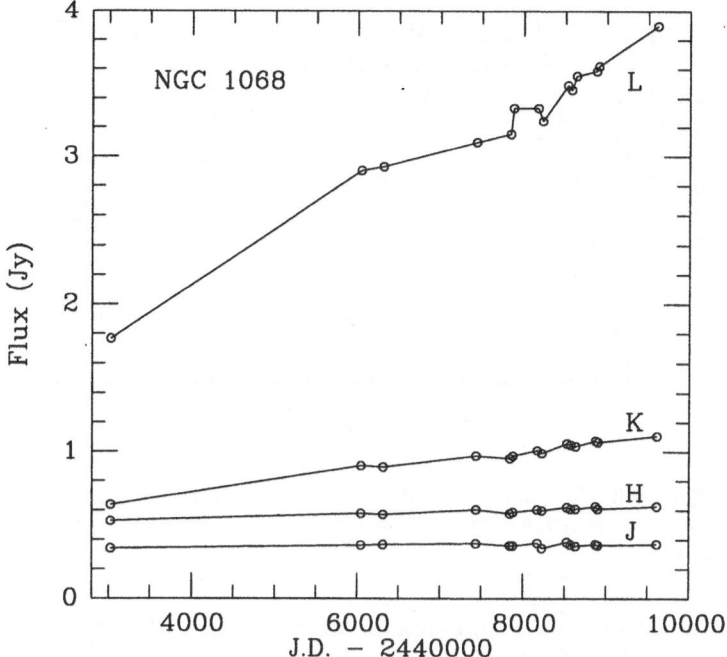

Figure 2. Fluxes from NGC 1068 in Jy, plotted against time. The photometric errors do not exceed 3% for *JHK* and 6% for *L*.

al.) between the UV light curve and that at J (1.2 μm), but the substantial delay between these and the L band is still apparent. This shows that the cross-correlation results obtained by Clavel *et al.* were not merely the result of independent fluctuations, fortuitously spaced in time.

3. Variation in NGC 1068

Infrared monitoring of NGC 1068, originally included in the programme as a probable non-variable galaxy for checking purposes, showed the surprising result that this object has increased its output steadily over 16 years. Further, the spectral shape of the variable part has remained constant. These two facts constitute further evidence that there is a hidden Seyfert 1 nucleus in this galaxy, the original spectropolarimetric evidence for which was found by Antonucci & Miller (1985).

The extremely high reddening of the nucleus of NGC 1068 makes determination of the dust temperature quite uncertain, but it may plausibly be set at about 1000K. The possibility that a hotter component, comparable to that at around 1500K in Fairall 9, cannot be excluded because of the

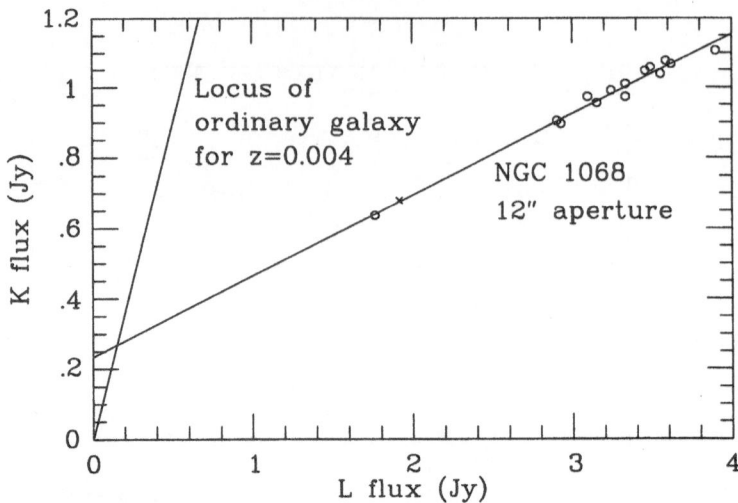

Figure 3.   *K* vs *L* fluxes from NGC 1068. The cross is from Penston *et al.* (1974). The fact that the points lie along a straight line means that the colour of the variable part remains constant.

higher extinction in the *J* and shorter wavebands. Unfortunately, the dusty torus models of Pier & Krolik (1992) do not constrain the temperature and other conditions of the infrared-emitting region by very much, owing to the limited wavelength range of the observations and the large number of parameters which have to be set.

More details of the NGC 1068 results can be found in Glass (1995).

4. ACKNOWLEDGEMENT

I wish to thank Dr H. Winkler for permitting me to reproduce his *U*-band results concerning Fairall 9 in advance of publication.

References

Antonucci, R.R.J. and Miller, J.S. (1985) *ApJ* **297**, 621.
Chelli, A., Perrier, C., Crux-Gonzalez, L., Carrasco, L. (1987) *A&A* **177**, 51
Clavel, J., Wamsteker, W. and Glass, I.S. (1989) *ApJ* **337**, 236
Glass, I.S. (1992) *MNRAS* **256**, 23p
Glass, I.S. (1995) *MNRAS* **276**, L68
Low, F.J. and Rieke, G.R. (1971) *Nature* **233**, 256
Penston, M.V., Penston, M.J., Selmes, R.A., Becklin, E.E. and Neugebauer, G. (1974)
    MNRAS *169*, 357
Pier, E.A. and Krolik, J.H. *ApJ* **401**, 99
Rieke, G.R., and Low, F.J. (1972) *ApJ* **177**, L115
Santos-Lleó, M., Clavel, J., Barr, P., Glass, I.S., Pelat, D., Peterson, B.M., and Reichert,
    G. (1994) *MNRAS* **270**, 580

RESOLVING THE FAINT GALAXY EXCESS WITH HST:

RESULTS FROM THE MEDIUM DEEP SURVEY

R. E. GRIFFITHS, K. U. RATNATUNGA, S. CASERTANO, M. IM,
N. ROCHE AND L. W. NEUSCHAEFER
JHU, Homewood Campus, Baltimore, MD 21218, USA

R. S. ELLIS, G. F. GILMORE, R. A. W. ELSON, A. ABRAHAM,
K. GLAZEBROOK AND B. SANTIAGO
Institute of Astronomy, U. Cambridge

R. F. GREEN AND V. SARAJEDINI
NOAO, Tucson, AZ

AND

J. P. HUCHRA
Harvard-Smithsonian CfA, Cambridge, MA

Abstract

WFPC2 on HST has provided the means to make rapid progress towards solving a long-standing problem in observational cosmology, viz. the nature of the objects contributing to the faint blue galaxy counts. The solution to this problem has been found in the increasing number of irregular and peculiar systems seen as HST probes to greater depths. Galaxies in the Medium Deep Survey (MDS) have been reliably classified to magnitudes $I_{814} \lesssim 22.0$ in the F814W band, at a mean redshift $\bar{z} \sim 0.5$. A high proportion ($\sim 40\%$) of these objects are irregular or anomalous and they have great diversity. They include compact galaxies, galaxies or protogalaxies with superluminous starforming regions, interacting pairs, and diffuse low surface brightness galaxies of various forms. These diverse objects contribute most of the excess counts in the I-band at our limiting magnitude, and when MDS images with BVI data are taken into account, they also explain most of the 'faint blue galaxy' excess. Of the irregular population (40% of the total at I=22), about 30% show multiple, high surface brightness components indicative of starburst activity.

But roughly half of the faint galaxies ($I < 22$) appear to be similar to regular Hubble-sequence examples observed at low redshift, with relative

D. L. Block and J. M. Greenberg (eds.), New Extragalactic Perspectives in the New South Africa, 211–226.
© *1996 Kluwer Academic Publishers.*

numbers of spheroidal and disk systems roughly consistent with nearby samples, indicating that the bulk of the local giant galaxy population was in place at half the Hubble time. Moreover, the giant population has certainly faded since z = 1, by about 1 mag. for the E/S0s and about the same for the early-type spirals. For the first time, we have observed directly the evolution of the luminosity function of E/S0 galaxies. We have found evidence for weak gravitational shear in the vicinity of E/S0 field galaxies, and have used this effect to constrain their masses and the cut-off radii of their dark matter halos.

The "major" merger rate has evolved only slowly with look-back time, as $(1 + z)^{1.2}$ out to z = 1 or 1.5, although there is already evidence from the deeper surveys that the rate may have been greater at higher redshifts. "Major" merging is probably not a major component of the explanation for the faint blue galaxy excess (to $I = 22$. "Minor merging" is a different story, and more difficult to address observationally. The deeper HST surveys, addressing the redshift range from 1 to 2, show increasing evidence of "conglomerations", i.e. groupings of what appear to be sub-galactic 'clumps'.

The clear picture which has thus emerged from the HST MDS is one in which the giant ellipticals and spirals have passively faded but have otherwise been relatively stable since z = 1, whereas there has been rapid evolution of the irregular/peculiar galaxy population – about 20% of them were brighter by about 2 mags. only 6 Gyr. ago. This latter population is not at all homogeneous, however, and seems to comprise galaxies in formation as well as fading dwarf irregulars.

1. Introduction

Before the advent of HST, ground-based imaging could not be used to separate the galaxy populations at moderate to high redshift, but now we have an instrument which can, and we can see how they have evolved individually. If we look at the total galaxy number counts, for example, then the effect of cosmological geometry, e.g. a range in the value of Ω from 0 to 1, has an effect which is smaller than the differential between blue and red galaxy counts at B = 24 (Broadhurst, Ellis & Glazebrook (1992), Metcalfe et al. (1995), Colless et al. (1994, Koo and Kron (1992), Lilly et al. (1993)). In order to make progress on these problems, we clearly needed large samples. Although the field of view of HST is small, the opportunity presented itself to take pictures of random fields with the Wide Field and Planetary Camera (WFPC2) when another HST instrument was targeted towards a known object of interest.

The Medium Deep Survey (MDS) is the HST Key Project which uses these parallel observations of random fields taken with WFPC2 in broad-band filters. Primarily, and for pointings of only a few orbits, exposures are taken in V and I; but for the longer pointings of six orbits or more, the B-band has been given priority, especially in HST observing cycles 4 and 5, when the scheduling system became more compliant. The overall goals are wide-ranging, but they are all based on statistical studies of the properties of a large sample of faint galaxies, as well as Galactic star counts and studies within nearby extragalactic objects. In particular, the Survey was designed to resolve the problem of the abundant population of faint blue galaxies. Over the past decade or more, ground-based observations have been largely frustrated in being unable to discriminate between hypotheses based on a fading population of dwarf galaxies (e.g. Broadhurst, Ellis and Shanks 1988; Cowie, Songaila and Hu 1991; Babul & Rees 1992) and those based on galaxy merging (Rocca-Volmerange and Guiderdoni 1990; Broadhurst, Ellis and Glazebrook 1992). The recent CFHT *spectroscopic* survey by Lilly and the CFRS collaboration (Lilly et al. 1995), however, showed clear evolution of the blue population and little or no evolution of the red population. More recently, and contemporaneously with the MDS, pointed HST observations (for morphology) have been combined with ground-based spectroscopic surveys to address this same problem from a different angle (Lilly et al. 1995, Cowie et al. 1995a,b).

The very first MDS observations, even with aberrated HST data (Griffiths et al. 1994a), were used to show that small galaxies were responsible for the excess, i.e. compact galaxies at or near the HST resolution limit. Morphologies have been studied with a precision adequate for classification on the normal Hubble scheme to a limiting magnitude of about $I_{814} \sim 22.0$ mag for most MDS fields (Abraham et al. 1996). Cruder information (e.g. scale lengths) has been determined to considerably fainter limits, at least $I_{814} \sim 24$ - 25 mag. (Casertano et al. 1995, Roche et al. 1996). Unlike the pre-refurbished images, morphological classifications for the WFPC2 objects are not based on parametric fits to $I_{814} \sim 22.0$ mag, nor do they rely on any deconvolution technique.

The MDS team is in the process of correlating the morphological properties of HST-selected galaxies with redshift and star formation rates derived from ground-based follow-up spectroscopy and photometry. Redshift surveys of other fields to this depth have been undertaken by Lilly et al. (1995) and indicate a median redshift $\overline{z} \simeq 0.5$ with a high fraction of objects within the range $0.4 < z < 0.7$. Consequently, we expect our sample to be representative of the field galaxy population at a lookback time of $\simeq 5$–7 Gyr ($H_0 = 50$ km s^{-1} Mpc^{-1} assumed throughout).

1.1. OBSERVATIONS

Typical MDS observations range from 600 to 2000 seconds per exposure, and may consist of 1–20 exposures of the same field. Parallel exposures may be 'registered' (exactly aligned) or not, depending on the needs of the observer using the primary HST instrument. In some cases, up to 12,000 seconds have been accumulated at a single pointing. In Cycles 1–3, 146 fields were observed with the WFC in a total exposure time of 240 hours. To December 1995, Cycle 4/5 MDS observations had been made in nearly 400 fields, with a total of 400 hours in the two predominant filters: F606W, somewhat redder and broader than Johnson V, and F814W, close to Kron-Cousins I. Of these fields, 140 were exposed to both filters, and over 30 had at least three exposures in each. Eight fields had exposures in the BVI set. The filters have been chosen to maximize the sensitivity for typical galaxies at intermediate redshifts ($z \sim 0.5$). A typical WFPC2 MDS field contains 50–400 detectable sources within the Wide Field Camera (WFC) to a limiting magnitude of $I_{814} = 24$–26 depending on the total exposure. The precision with which morphological properties can be measured is, of course, a function of both angular size and apparent magnitude.

The cycle 1–3 HST MDS object catalog (11,000 objects) is available via the STScI's Electronic Information System, STEIS, in the directory observer/catalogs/mds. Super-sky flat fields for WF/PC and WFPC2 are also available on STEIS (Ratnatunga et al. 1994a).

The results of morphological typing into three crude galaxy classes (E/S0, early-type spirals and late-type spirals/irregulars) is shown in Fig. 1 (from Abraham et al 1996), where it is shown that eyeball classification is as good as computerised classification down to I of at least 21. Fainter than $I = 21$, the classifications of late-type systems become subjective - i.e. the decision between late-type systems (Hubble T-type > 7), merging systems and peculiar systems is a function of observational bias. In the deeper HST surveys, this will become an even larger problem. Abraham et al. present objective classifications based on measurements of central concentration and asymmetry, calibrated using artificially redshifted images of local objects. Despite the problems with eyeball classification, however, the objective system confirmed the eyeball results of Glazebrook et al. (1995) who had grouped the irregulars/mergers/peculiars into one class, which was shown to be responsible for the excess number counts.

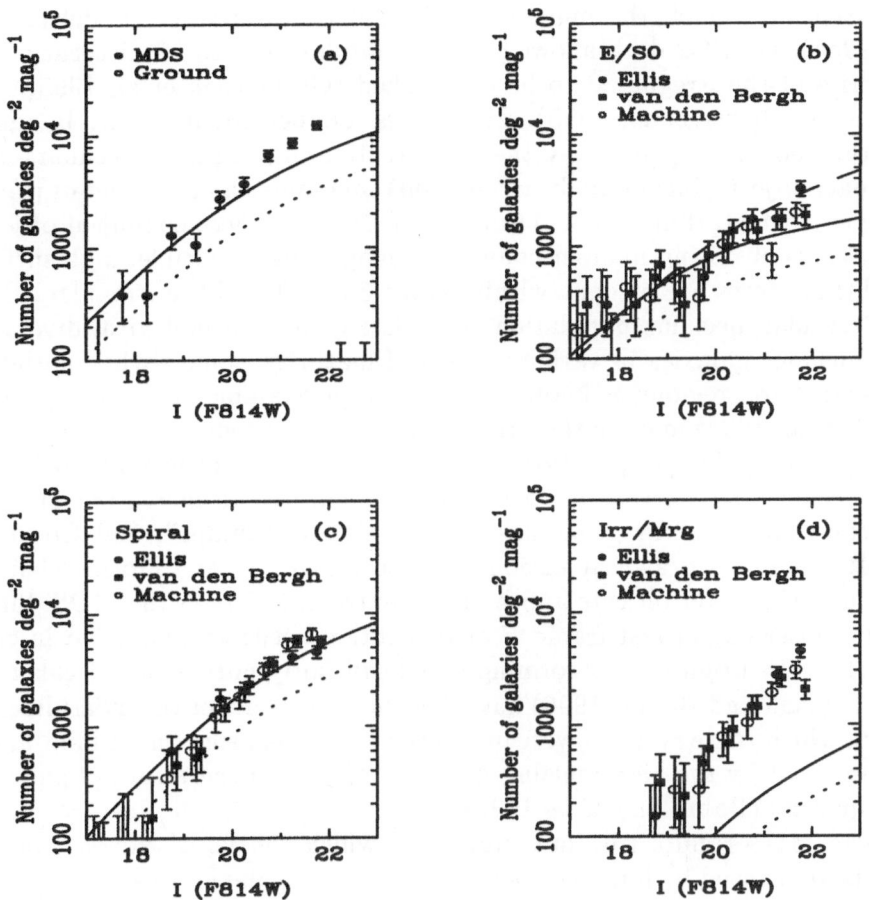

Figure 1. Number-magnitude relations for (a) all galaxies, (b) E/S0, (c) Spirals and (d) Irr./mergers/peculiars. No-evolution model curves are shown for low and high normalisation. Open circles are automated measurements; solid symbols are eyeball estimates. For details, see Abraham et al. (1996)

2. Results from Morphological Classification

2.1. THE EVOLUTION OF DWARF AND IRREGULAR GALAXIES

The number counts of dwarfs and irregulars demand a steep faint-end slope to the luminosity function (LF), like that suggested by Marzke et al. (1994) and a steepening at moderate redshift, like that recently suggested by Ellis et al (1996), i.e. the universe is dwarf-rich at $z = 0.3 - 0.5$. The initial HST result on the morphological number counts, using WF/PC, came from the marginal distribution of size vs. magnitude when compared with the predicted distributions based on different galaxy evolution models, i.e. the no-evolution model, the merger model, and the dwarf-rich model (Im et al 1995a). WFPC2 data allowed actual morphologies to be discerned and the 'rise of the irregulars' to be established (Glazebrook et al. 1995, confirmed by Driver et al. 1995). The excess number counts in B, V and I are thus largely explained by the high fraction of irregulars/peculiars and compact objects, including dwarf (dE, Im) candidates and galaxies or protogalaxies with starburst knots (Roche et al 1996b). These combined objects show a steeply rising number count with magnitude. Multiple, high surface brightness cores are evident within about 30 – 40% of the irregulars. But the irregular/peculiar population comprises great morphological diversity, and includes galaxies in various stages of maturity, some with superluminous starburst regions or knots. The prototype example at $z = 0.7$ (Glazebrook et al. 1994) may be the first observational evidence of a young spiral galaxy undergoing major starburst activity at this intermediate redshift. The deeper HST surveys seem to show evidence of an increasing number of "conglomerates" or groupings of sub-galactic "clumps", and Cowie et al. (1995a,b) has drawn attention to those which appear to form "chains" although the latter objects appear to be extremely rare in the MDS. Since we are observing in rest-frame B or U, caution needs to be applied in case we select the brighter star-forming regions of fairly normal spiral galaxies.

Ferguson and Babul (1996) have developed a model of the dwarf irregulars in which these systems were prevented from collapsing out of mini-halos before $z = 1$ by the high ionizing flux from the first generations of galaxies and quasars (Babul and Rees 1992).

The MDS results rule out models in which the faint, excess number counts are caused by large, low-surface brightness galaxies (e.g. Ferguson & McGaugh 1995) since we do not see the predicted numbers of large objects. The median half-light radius for all galaxies to $I_{814} = 24$ is 0.4 arcsec (Im et al. 1995a).

2.1.1. *Dwarf Ellipticals*

If dwarf elliptical candidates are identified by their exponential luminosity profiles, round shapes, blue colors and small sizes ('small exponential ellipticals'), the characteristics of dEs or Ims, then these constitute about 20 % of all galaxies at $I = 20 \sim 21$. Taken together with the irregulars, these appear to be responsible for the excess number counts in the bright magnitude range of the MDS (Im et al 1995b).

At this brighter end of the MDS distribution, such objects may be the evolved versions of the dwarf irregulars at higher redshift. Low surface brightness galaxies (some nucleated) are common amongst this population.

2.2. THE EVOLUTION OF THE GIANT GALAXIES

The mild-evolution models (e.g. Guiderdoni and Rocca-Volmerange 1990) predict luminous galaxies of large angular size at high redshift: such objects are certainly seen in the data, consistent with a brightening at higher redshift. Still, the excess number counts, at least to $I = 22$, are largely unexplained by 'giant' spirals or ellipticals, which are observed to have only passive (luminosity) evolution. Their number counts (Glazebrook et al. 1995, Driver et al. 1995b, Abraham et al. 1996), size vs. redshift relationship (Mutz et al. 1994), and structural parameters (Windhorst et al. 1994; Phillips et al. 1995, Forbes et al. 1995, 1996) all indicate a relatively benign population: the bulk of the local giant population was apparently in place at half the Hubble time.

Furthermore, as shown below (§3), this population has undergone relatively little 'major merging' (about 10%) since $z \sim 1$. 'Minor' mergers (with gas-rich dwarfs) may have been common but have not caused major disruptions (Driver et al. 1995a). For those galaxies which do show evidence of merger activity, photometry shows bluer colors and thus increased star formation (Forbes et al. 1995).

Spiral galaxies are, however, somewhat bluer in the past when their K-corrections are taken into account; their apparent (V-I) color does not change between $I = 18$ and $I = 22$ (Casertano et al. 1995). This population has faded since $z \sim 1$, by about 0.7 mag. (Roche et al. 1996a – see §2.2.2). The bulk of the microJansky radio population may be identified with these spiral galaxies with enhanced star formation, especially with the MDS galaxies showing evidence of interaction (Windhorst et al. 1995).

Elliptical galaxies have a higher angular correlation than spirals, but this difference in correlation amplitude is smaller than that observed in local samples (Neuschaefer et al. 1995). The two-point correlation function (all galaxies, irrespective of morphology) shows a constant slope down to arcsec scales, with no evidence for the excess galaxy pairs that might result from a high rate of 'major' mergers (see §3). The deepest of the MDS fields and the deeper HST surveys indicate, however, that mergers of galaxy components ('conglomerates') may have been common at epochs which are presumably earlier (i.e. at $z \gtrsim 1$ – see Cowie et al. 1995,a,b).

2.2.1. *Luminosity Evolution of the E/S0 population*

Perhaps the easiest galaxy type to separate out from the general galaxy population is the E/S0s. The automated galaxy classification software developed for the MDS and based on the maximum likelihood method (Ratnatunga et al. 1994a) allows us to select galaxies which have the 'de Vau-

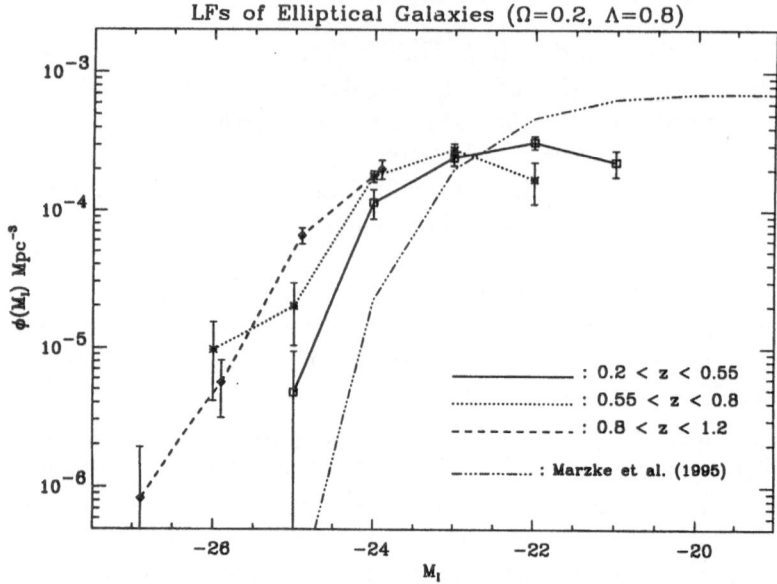

Figure 2. Evolution in the Luminosity Function of E/S0 galaxies, from Im et al. (1996c).

couleurs' $r^{1/4}$ profile. We have thus constructed the luminosity functions of elliptical galaxies using data from the MDS together with archived HST surveys (Groth et al. 1994, with spectroscopy from Lilly et al. 1995). Photometric redshifts of these E/S0s have been determined using $V - I$ colors and sizes to an accuracy of ~ 0.1 up to $z \sim 1$, and checked with MDS follow-up spectroscopy or published values (Lilly et al. 1995) for about 10% of the sample. The luminosity functions, constructed in 3 different redshift bins ($0.2 < z < 0.55$, $0.55 < z < 0.8$, $0.8 < z < 1.2$) are shown in Fig.2 (from Im et al. 1996b).

These independent luminosity functions show the brightening in the luminosity of E/S0s by about 1 ± 0.5 magnitude at $z \sim 1$, and no sign of significant number evolution. This is the first direct measurement of the luminosity evolution of E/S0 galaxies, and our results support the hypothesis of a high redshift of formation ($z \gg 1$) for elliptical galaxies, together with weak evolution of the major merger rate at $z < 1$. The high redshift of formation is consistent with that of cluster ellipticals (Bower et al. 1992, Ellis 1996) as determined from the small scatter in the (U-V) colors, and with HST studies of ellipticals in high redshift clusters (Dickinson et al. 1995). The stars in ellipticals are as old as those in globular clusters.

If the evolution of ellipticals was driven by a merging of galaxies of similar mass (the so called 'major merging' model), there would be a paucity of large and bright ellipticals at $z \gtrsim 1$. In contrast, the observed size distribution shows a considerable number of large ellipticals and is well matched if the minimal amount of merging is assumed. The majority of ellipticals must have formed at $z \gg 1$, and their number evolution was not significant at $z < 1$. Our data are consistent with a merger rate of $< 0.014 \ Gyr^{-1}$ for the $\Omega = 1$ model, and $< 0.02 \ Gyr^{-1}$ for the model with $\Omega = 0.2 \ and \ \Lambda = 0.8$ with no strong major merger rate evolution (§3).

Given the negligible number evolution, the observed color distribution of ellipticals can be matched with the predictions of stellar synthesis models. The observed $V - I$ color distribution for $I < 22$ requires evolution in the spectral energy distribution (SED), otherwise there is an excess of very red ellipticals with respect to the prediction of the no-evolution (NE) model (Im et al. 1996a). Luminosity evolution models based on a single starburst at the redshift of formation $z_{for} > 5$ (if $\Omega = 1$) or $2 < z_{for} < 30$ (if $\Omega = 0.1$) can comfortably match the observed color distribution, suggesting that most ellipticals are neither very young nor very old ($9 < $ age < 20 Gyr). We have used the number-magnitude counts in an attempt to constrain not only the evolution of ellipticals, but also the cosmological parameters Ω or Λ. Our results favor low Ω or non-zero Λ models for a minimal (local) merger rate. The $\Omega = 1$ model marginally fits the observed E/S0 number counts (Im et al. 1996a).

2.2.2. Dust Extinction in High-Redshift Ellipticals

We consider models for the number counts of E/S0 galaxies only. These are divided into equal numbers of 'hot' and 'cold' ellipticals, differing slightly in colour. The 'cold' ellipticals form stars at an exponentially declining rate, with a short 0.5 Gyr timescale, forming 95% of their stars prior to $z = 4$ (15.0 Gyr ago). The 'hot' ellipticals form 95% of their stars over the same timescale but 5% over a longer 8 Gyr exponential timescale, so they form 91% of their stars prior to $z = 4$ and are slightly bluer.

We consider these models with and without a simple approximation for dust extinction in high redshift ellipticals. The dust extinction is assumed to have a wavelength dependence of $\lambda^{-1.16}$, derived from a fit to the SMC extinction law, and to be proportional to the star formation rate. The effects of the dust would then be large at $z > 3$, during the initial starburst, but negligible at low redshifts where star-formation has almost ceased. The dust extinction is normalized to be approximately 1.0^m in the rest-frame V band at $z = 4$, approximately the lower limit from the non-detection of emission lines from these galaxies (Pahre and Djorgovski 1995). Due to the strong wavelength dependence, this corresponds to a much larger extinction of $\sim 4^m$ in the observer-frame I-band.

Figure 3 shows the I-band number counts of E/S0 galaxies as predicted by the non-evolving (dotted), evolving with dust (solid) and evolving with-

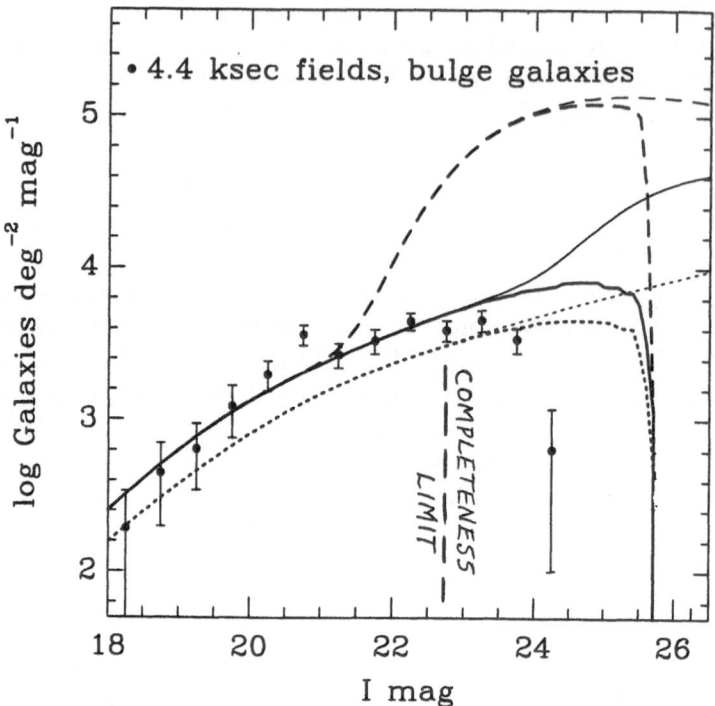

Figure 3.   Effect of Dust Extinction in Elliptical Galaxies

out dust (dashed) models, together with the observed counts of bulge-profile galaxies from the MDS and archived data. The heavy lines show the models with the inclusion of detection thresholds appropriate to the datasets, the thin lines show the models without these limits. The evolving models without dust predict that large numbers of high-redshift ellipticals will appear beyond $I \sim 22$, contrary to observations. With the inclusion of the modelled dust extinction, the evolving models give a good fit to the bulge-galaxy counts over the magnitude range where profile classification remains reasonably complete. Hence these observations seem consistent with the $\sim 1^m$ brightening to $z \simeq 1$ (Im et al. 1996) but inconsistent with the stronger brightening expected at $z > 3$ if the initial starburst is not significantly obscured by dust. Even with the modelled dust extinction, these models predict $\sim 3^m$ of evolutionary brightening of ellipticals at $z \sim 3$ and we would expect some high redshift ellipticals to appear at fainter magnitudes of $I > 24$, but of low surface brightness.

2.2.3. *Statistical Properties at the Survey Limit*

Statistical properties of galaxies are measured to I = 24 with WF/PC and 25 with WFPC2. For the pre-refurbishment WF/PC images, the structural parameters of about 13,000 objects are presented by Casertano et al. (1995), using data taken from about 112 fields. Sizes, magnitudes, colors and crude classifications are based on two-dimensional model fitting to undeconvolved images (Ratnatunga et al. 1994a). Number counts in the range $18 < I < 22$ exceed the ground-based numbers by about 50%, a range over which many ground-based objects may have been misclassified as stars. The ellipticals become redder for $I_{814} > 20$ mag, but the spirals show no trend of apparent color with magnitude, indicating that they are intrinsically slightly bluer at higher redshift when K-corrections are taken into account. For galaxies with $I \leq 22$ in WFPC2 data, the maximum-likelihood fits have been found for combined disk-plus-bulge models of all galaxies.

The angular size distribution certainly favors a steep luminosity function, but the faint data also domonstrate that the numbers of galaxies with both blue ($V - I \leq 1.2$) colours and moderately large ($0\rlap{.}''4 \leq r_{hl} \leq 1\rlap{.}''5$ at $22 \leq I \leq 24$) angular sizes exceed the non-evolving predictions by a factor of two, and are in good agreement with the simple pure luminosity evolution model (Roche et al. 1996a). This model incorporates the steep luminosity function ($\alpha = -1.65$) for later-type galaxies and moderate luminosity evolution ($\Delta(I) \simeq 0.7\text{--}1^m$ at $z \sim 1$) for spiral galaxies of all luminosities (similar to the models of Campos and Shanks 1996).

The excess of large blue galaxies in the deeper fields is positive reinforcement of our finding that large, $L \sim L^*$ spiral galaxies do undergo significant luminosity evolution to $z \sim 1$, and that evolutionary brightening is not confined to dwarf galaxies. This interpretation is supported by the spectroscopic results of Schade et al. (1995), Lilly et al. (1995) and Forbes et al. (1996), all of which suggest a similar brightening of $L \sim L^*$ spirals with redshift.

Thus, although we have no redshifts at the survey limit (and likewise for the deeper HST surveys) we can nevertheless extract useful constraints on the evolution of the different galaxy populations by using the galaxy sizes and colors (Roche et al. 1996a), and by comparison with galaxy evolutionary models which are consistent with the shallower surveys where redshifts have been obtained as part of the MDS and other spectroscopic surveys.

3. Slow Evolution of the Major Merger rate

In order to examine or constrain the evolution in the galaxy merger rate, we have examined the morphological and statistical properties of close galaxy pairs from two sets of 28 WFPC2 fields, those in the MDS and in the

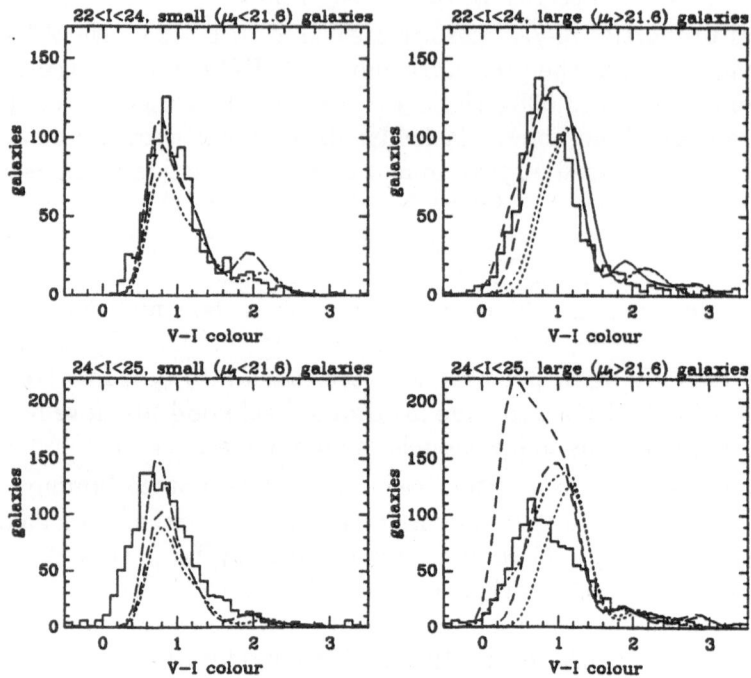

Figure 4. $V - I$ color distributions for small and large galaxies – from Roche et al.
(1996a)

Groth/Westphal Survey (GWS) in V and I passbands. Each field is 95%
complete down to $I \lesssim 24.5$ mag, $V \lesssim 25.5$ mag, within which ~ 400 galaxies
per $\simeq 5$ square arcmin field are detected. The MDS classification into disk
or bulge galaxies has been applied down to $I \leq 23$ mag ($V \lesssim 24$ mag), and
also used to differentiate galaxies from stars one magnitude fainter.

Down to $I \leq 25$ mag the number of galaxy pairs with separations $\theta \leq 3\rlap{.}''0$
is consistent with the inward extrapolation of the angular two-point correlation function $w(\theta) \propto \theta^{-0.8}$ observed from the same data; the fraction
of such pairs showing morphological evidence for physical association accounts for two-thirds of the amount indicated by $w(\theta)$. Moreover, relative
to field galaxies, the $(V\text{-}I)$ color and I-magnitude difference distributions
of $\theta \leq 3\rlap{.}''0$ pairs are similar.

We have used recent galaxy redshift surveys to estimate the rate of
galaxy merging occurring in these galaxy pair samples. We find that merging has a moderate dependence on redshift: we estimate the rate of merging
$R_{merge} \propto (1 + z)^m$, with $m = 1.2 \pm 0.4$ for galaxies with $I \leq 25$ mag. There
are two models consistent with these data: (a) a low density universe with
strong clustering evolution parameterized by a clustering exponent $\epsilon \simeq 1.0$;

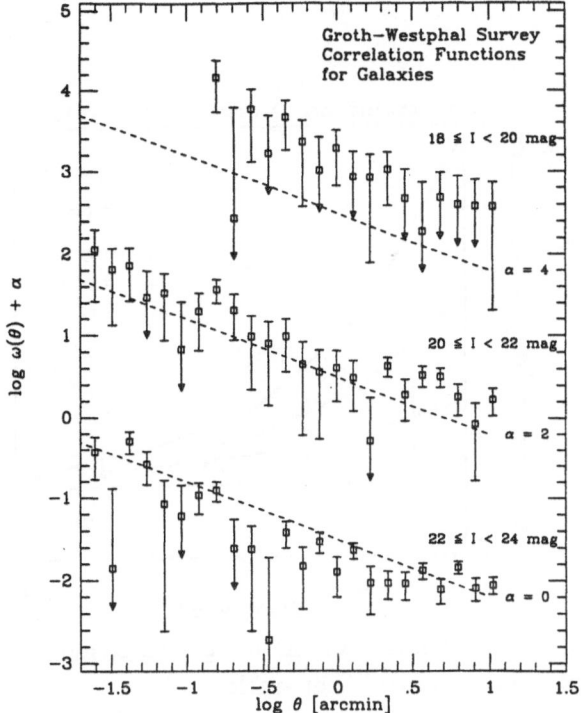

Figure 5. Angular correlation functions of galaxies in HST survey data. The lines have slope -0.7. Note the continuation to small scales – from Neuschaefer et al. (1995, 1996).

or (b) two galaxy populations identified by their clustering properties, in which the more weakly clustered population fades or dissipates but does not take part in widespread merging prior to $z \lesssim 0.3$, whereas the more strongly clustered population, which we associate with local ellipticals and disks within two magnitudes of L^*, takes part in minor mergers with dwarf galaxies at the uniform merging rate observed locally, out to $z \leq 1.0$.

4. Conclusions

Parallel HST observations have been highly successful in providing a database which is extremely useful for achieving the goals of the Medium Deep Survey: the measurement of field galaxy properties to I=22 and fainter, number counts as a function of morphology, studies of the evolution of galaxies to at least z =0.5–0.7, and study of morphology as a function of environment. The detailed morphological studies have included the search for compact and multiple nuclei, the frequency of mergers, interactions and groups, and the frequency of irregularities, starburst knots, bars, arms and rings, and gravitationally lensed background galaxies. Ongoing work includes the

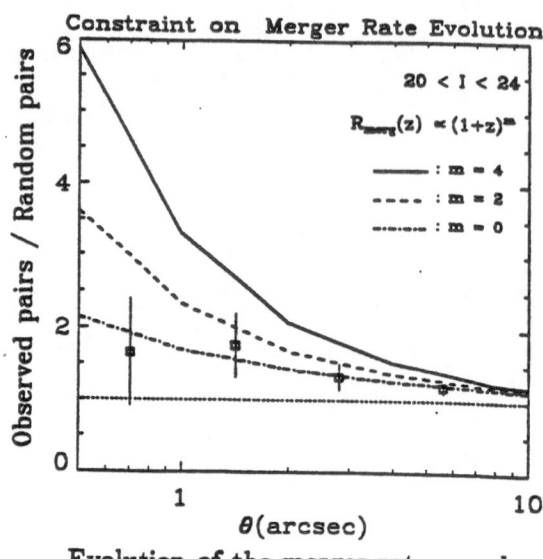

Evolution of the merger rate: number
of galaxy pairs vs. separation.

Figure 6.   Nearest-neighbour pair counts versus angular separation, θ, normalised by random sample statistics. The curves show the expected numbers of pairs for the given evolution in the merger rate, assumed to take the form $(1+z)^m$.

application of automated galaxy classification methods (e.g. Owens et al 1995).

The "excess number counts" above the predictions of no-evolution models are caused largely by the evolution of low-luminosity systems, which have a higher space density than that predicted from local ground-based surveys. These objects manifest themselves as large numbers of irregular galaxies in the data, a large fraction of which seem to have multiple starburst components. At the faint end of the survey, an increasing frequency of irregular systems and conglomerations are seen, including what appear to be high redshift galaxies or protogalaxies in various stages of formation and evolution.

Early-type galaxies are seen to undergo luminosity evolution and were brighter by about 1 mag. at $z = 1$. Their numbers and sizes are consistent with a high redshift of formation, $z \gg 1$. Nevertheless, it is difficult to use them to probe the cosmological parameters because of two problems: (i) the faint end slope of the LF for these galaxies is not well determined, and (ii) the actual epoch of formation is unknown. Although the observations support a low value for Ω, this result is tentative. Dust extinction in ellipticals at high redshift may be required to fit the faint elliptical number counts.

The rate of galaxy merging has not evolved rapidly since $z = 1$, in contrast with several claims. However, the deeper HST surveys may already indicate that the merger rate was much higher at z between 1 and 3 than it was at $z < 1$.

References

Abraham, R. G., van den Bergh, S., Glazebrook, K., Ellis, R. S., Santiago, B. X., Surma, P. & Griffiths, R. E., 1996, *Mon. Not. R. Astr. Soc. in press*

Babul, A., & Rees, M. J., 1992, *Mon. Not. R. Astr. Soc.* **255**, 346

Bower, R. G., Lucey, J. R., & Ellis, R. S., 1992, *Mon. Not. R. Astr. Soc.* **254**, 601

Broadhurst, T., Ellis, R. S., & Shanks, T., 1988, *Mon. Not. R. Astr. Soc.* **235**, 827

Broadhurst, T., Ellis, R. S., & Glazebrook, K., 1992, *Nature* **355**, 55

Bruzual, A. G. & Charlot, S., 1993, *Astrophys. J.* **405**, 538

Campos, A. and Shanks, T., 1996, *Mon. Not. R. Astr. Soc. in press*

Casertano, S., Ratnatunga, K. U., Griffiths, R. E., Neuschaefer, L. W., & Windhorst, R. A., 1995, *Astrophys. J.* **453**, 599

Colless, M., Schade, D., Broadhurst, T. J., & Ellis, R. S., 1994, *Mon. Not. R. Astr. Soc.* **267**, 1108

Cowie, L. L., Songaila, A., & Hu, E. M., 1991, *Nature* **354**, 460

Cowie, L. L., Hu, E. M. & Songaila, A., 1995a, *Nature* **377**, 603

Cowie, L. L., Hu, E. M. & Songaila, A., 1995b, *Astron. J.* **110**, 1576

Dickinson, M., 1996, in *Fresh Views on Elliptical Galaxies*, Buzzoni, A., Renzini, A. & Serrano, A., eds., *in press*

Driver, S. P., Windhorst, R. A. & Griffiths, R. E., 1995a, *Astrophys. J.* **453**, 48

Driver, S. P., Windhorst, R. A., Ostrander, E. J., Keel, W. C., Griffiths, R. E., Ratnatunga, K. U., 1995b, *Astrophys. J. Lett.* **449**, L23

Ellis, R. S., Broadhurst, T. J., Colless, M. M., Heyl, J. S. & Glazebrook, K., 1996, *Mon. Not. R. Astr. Soc. in press*

Ellis, R. S., 1996, in Proc. *Unsolved Problems in Astrophysics*, J. N. Bahcall & J. Ostriker, eds., Princeton University Press

Ferguson, H. C. & Babul, A., 1996, *Astrophys. J. in press*

Ferguson, H. C. & McGaugh, S. S. 1995, *Astrophys. J.* **440**, 470

Forbes, D. A., Elson, R. A. W., Phillips, A. C., Illingworth, G. D., & Koo, D. C., 1995, *Astrophys. J. Lett.* , **437**, L17

Forbes, D. A., Phillips, A. C., Illingworth, G. D., & Koo, D. C., 1996, *Astrophys. J. Lett.* , *in press*

Glazebrook, K., Lehar, J., Ellis, R., Aragon-Salamanca, A., & Griffiths, R., 1994, *Mon. Not. R. Astr. Soc.* **270**, L63

Glazebrook, K., Ellis, R. S., Santiago, B. and Griffiths, R. E., 1995, *Mon. Not. R. Astr. Soc.* **275**, L19

Griffiths, R. E., et al., 1994a, *Astrophys. J.* **437**, 67

Griffiths, R. E., et al., 1994b, *Astrophys. J. Lett.* **435**, L49

Griffiths, R. E., Ratnatunga, K. U., Casertano, S., Im, M., 1996 *Mon. Not. R. Astr. Soc.* , *submitted*

Groth, E. J., Kristian, J. A., Lynds, R., O'Neil, E. J., Balsano, R. & Rhodes, J., 1994, BAAS **26**, 1403

Guiderdoni, B. and Rocca-Volmerange, B., 1990, *Astron. & Astrophys.* **227**, 362

Guiderdoni, B. and Rocca-Volmerange, B., 1991, *Astron. & Astrophys.* **252**, 435

Im, M., Casertano, S., Griffiths, R. E., Ratnatunga, K. U. & Tyson, J. A., 1995a, *Astrophys. J.* **441**, 494

Im, M., Ratnatunga, K. U., Griffiths & R. E. Casertano, S., 1995b, *Astrophys. J. Lett.* **445**, L15

Im, M., Griffiths, R. E. & Ratnatunga, K. U., 1996a, *Astrophys. J. submitted*

Im, M., Griffiths, R. E., Ratnatunga, K. U., Sarajedini, V., & Green, R. F., 1996b, *Astrophys. J. Lett. , submitted*

Im, M., Griffiths, R. E. & Ratnatunga, K. U., 1996c, *Astrophys. J. Lett. , submitted*

Koo, D. C., & Kron R. G. 1992, *Ann. Rev. Astr. Astrophys.* **30**, 613

Lilly, S. J. 1993, *Astrophys. J.* **411**, 501

Lilly, S. J., Tresse, L., Hammer, F., Crampton, D., & Le Fevre, O., 1995, *Astrophys. J.* **455**, 108

Lilly, S. J., Hammer, F., Le Fevre, O. & Crampton, D., 1995, *Astrophys. J.* **455**, 75

Marzke, R. O., Geller, M. J., Huchra, J. P. & Corwin, H. G., Jr., 1994, *Astron. J.* **108**, 437

Metcalfe, N., Shanks, T., Fong, R. & Roche, N. 1995, *Mon. Not. R. Astr. Soc.* **273**, 257

Mihos, J. C., & Hernquist, L. (1994), *Astrophys. J.* **425**, L13

Mutz, S. B., Windhorst, R. A., Schmidtke, P. C., Pascarelle, S. M., Griffiths, R. E., Ratnatunga, K. U., Casertano, S., Im, M., Ellis, R. S., Glazebrook, K., Green, R. F., & Sarajedini, V. L. 1994, *Astrophys. J. Lett.* **434**, L55

Neuschaefer, L. W., Casertano, S., Griffiths, R. E. & Ratnatunga, K. U., 1995, *Astrophys. J.* **453**, 559

Neuschaefer, L. W., Im, M., Ratnatunga, K. U., Griffiths, R. E. & Casertano, S., 1996, *Astrophys. J. submitted*

Owens, E., Ratnatunga, K. U., Griffiths, R. E., 1996, *Mon. Not. R. Astr. Soc. , submitted*

Pahre, M. A. & Djorgovski, S.G., 1995, *Astrophys. J. Lett.* **449**, L1.

Phillips, A. C., Bershady, M. A., Forbes, D. A., Koo, D. C., Illingworth, G. D., Reitzel, D. B., Griffiths, R. E., & Windhorst, R. A., 1995, *Astrophys. J.* **444**, 21

Ratnatunga, K. U., Griffiths, R. E. and Casertano, S., 1994a, in *The Restoration of HST Images and Spectra II*, eds Hanisch, R. J., White, R., Proc. Space Telescope Science Institute Workshop, p. 333.

Ratnatunga, K. U., Griffiths, R. E., Casertano, S., Neuschaefer, L.W., & Wyckoff, E. W., 1994b, *Astron. J.* **108**, 2362

Ratnatunga, K. U., Ostrander, E. J., Griffiths, R. E. & Im, M., 1995, *Astrophys. J. Lett.* **453**, L5

Rocca-Volmerange, B. and Guiderdoni, B., 1990, *Mon. Not. R. Astr. Soc.* **247**, 166

Roche, N., Ratnatunga, K. U., Griffiths, R. E., Im, M., & Neuschaefer, L. W., 1996a, *Mon. Not. R. Astr. Soc. , submitted*

Roche, N., Griffiths, R. E., Ratnatunga, K. U., & Im, M., 1996b, *Mon. Not. R. Astr. Soc. , submitted*

Schade, D., Lilly, S. J., Crampton, D., Hammer, F., Le Fevre, O. & Tresse, L., 1995, *Astrophys. J. Lett.* **451**, L1

Tresse, L., Hammer, F., Le Fevre, O., & Proust, D., 1993, *Astron. & Astrophys.* **277**, 53

Windhorst, R. A., Franklin, B. E., Pascarelle, S. M. Fomalont, E. B., Kellermann, K. I., Griffiths, R. E., Partridge, R. B., & Richards, E., 1995, *Nature* **375**, 471

SPIRAL STRUCTURE IN GALAXIES:
COMPETITION AND COOPERATION OF GAS AND STARS

G. BERTIN
Scuola Normale Superiore
piazza dei Cavalieri, I-56126 Pisa, Italy

Abstract.
   The different dynamical behaviors of the collisionless stellar component and of the cold dissipative gas in galaxy disks are reviewed in the light of the new perception of spiral structure in galaxies put forward by the recent near-infrared images, with the general conclusion that most observed morphological features can be explained in terms of intrinsic mechanisms, as the result of the different types of interaction that may occur between the two main components of galaxy disks.

1. Near-infrared Observations: a New Extragalactic Perspective

Spiral structure in galaxy disks has long been recognized to owe its existence and its appearance to a subtle interplay of actions by stars and gas. The optical morphology is generally delineated by bright young stars and dust lanes and is thus dominated by the properties of the interstellar medium; in turn, dynamical theories aim at determining the properties of the gravitational field, which is largely dominated by relatively old stars that give a less prominent contribution to optical images. Thus, in the context of galactic dynamics, one major goal of the observations in the last three decades has been to find an empirical way to separate the contribution of the stars to the overall spiral gravitational field from that of the gas, i.e., to settle directly the question whether spiral arms are mostly "stellar" or "gaseous" features. As pioneered by Zwicky (1957), stressed by Oort (1962, p. 235), and later developed by Schweizer (1976), the main road to a quantitative appreciation of the properties of the underlying stellar disk has been to compare images in different wavebands and, especially, to move towards the long wavelength part of the spectrum.

D. L. Block and J. M. Greenberg (eds.), New Extragalactic Perspectives in the New South Africa, 227–242.
© *1996 Kluwer Academic Publishers.*

After the considerable progress made through the studies around 1μ (e.g., see Elmegreen 1980, Elmegreen & Elmegreen 1984, Pierce 1986), it now appears that the recently acquired imaging capability around 2μ is finally bringing us very close to the desired target of a direct diagnostics of the mass distribution in galaxy disks (Rix 1993, Rix & Rieke 1993). A quantitative assessment, for typical or for specific cases, of the precise amount of contamination to near-infrared images by relatively young objects, such as red supergiants, is still a matter of discussion. On the other hand, given the low extinction levels with respect to optical wavebands, there is little doubt that a new extragalactic perspective is growing out of the K and K' images of spiral galaxies. These are giving us the best view of the underlying mass distribution in galaxy disks and have already provided a number of impressive, and sometimes surprising, results of great dynamical interest (e.g., see Block & Wainscoat 1991, Rix & Rieke 1993, Zaritsky, Rix & Rieke 1993, Block et al. 1994, Regan & Vogel 1994, Peletier et al. 1994, Knapen et al. 1995a, Rix & Zaritsky 1995, González & Graham 1995; the list of relevant references is rapidly growing):

- One crucial point of the density wave theory, that spiral arms (at least for grand-design structure) are "stellar", i.e., associated with a significant density perturbation, is now definitely proved.
- Especially for gas-rich galaxies, spiral morphology in the near-infrared can be very different from that of optical images.
- Smooth and coherent, grand-design spiral structure is rather common in the infrared.
- Infrared spiral structure is generally dominated by low-m features, with the contribution of an $m = 1$ component showing up prominently in several cases.
- The amplitude modulation of spiral arms is found to persist in infrared images, which confirms that this is a property of the underlying gravitational potential.
- Barred structure is more frequent and more prominent in the near-infrared, although not always present (e.g., in M 99; even in the case of M 33, where the continuation of optical spiral arms down to the center has often been interpreted in terms of a weak bar, a clear evidence of a bar is not found); long spiral arms without a dominant large-scale bar are also frequent and in some cases (such as NGC 4622) they are very tight.
- In spite of its smooth appearance, large-scale spiral structure is sometimes characterized by a strong arm-interarm contrast (values between 2 and 3 have been reported for M 51 and M 99) and thus it must have reached a highly non-linear stage; however, this should not be taken as a general rule (for instance, M 100 has a lower contrast, as also

indicated by its kinematics; for the bisymmetric structure in M 33, the contrast has been found to be about 1.2).
- The exponential scalelength of the disk may change significantly as a function of the waveband of observation, being shorter in the near-infrared.

Systematic surveys of morphological properties in the near-infrared are in progress and will soon determine the statistical significance of these results, which set important empirical constraints for a proper modeling of spiral galaxies. At the present stage of this rapidly growing research area, the modal theory of spiral structure, which interprets the large-scale properties of spiral arms in terms of a *quasi-stationary* structure determined by the *intrinsic characteristics* of the individual galaxies, is definitely strengthened by this observational breakthrough.

The aim of this review is to outline some of the basic points that may help us put this new extragalactic perspective into a coherent theoretical framework. The general picture is based on a long-term research effort that is summarized in a recent monograph (Bertin & Lin 1996), which the reader is referred to for full explanations and quantitative discussions.

2. Different Physical Characteristics of Stars and Gas

For dynamical purposes, it is convenient to imagine a galaxy disk to be made of two components, which may be called Population I and Population II by analogy with the astronomical notation. One component is dominated by cold gas, in atomic or molecular form, but contains significant amounts of stars recently born in the interstellar medium. This component is in a thin layer (at least within the optical disk) and is characterized by very low velocity dispersion ($< 10 \ km/s$) with respect to the circular motions associated with the differential rotation. In the following, it will sometimes be referred to more simply as "gas". The other component, Population II, is dominated by relatively old stars, in a thicker layer, and is characterized by higher velocity dispersions, i.e., it is warmer from the dynamical point of view. It will often be called "stars". Such separation of a galaxy disk in two components with these characteristics is only a symplifying tool for dynamical investigations, while, in reality, continuous changes of dynamical properties are associated with the many components of a galaxy disk.

For the star-dominated component we note that:

- Population II objects are characterized by relatively large epicycles. Thus any large-scale structure observed in a disk should rely on this component for its dynamical support; in particular, large-scale spiral structure should require an active dominant role of the stellar disk.

- Population II objects define an essentially collisionless subsystem. Landau damping acts efficiently in the stellar disk and enforces complete absorption of waves at the Lindblad Resonances (Mark 1971, Lynden-Bell & Kalnajs 1972, Mark 1974); in turn, such efficient absorption at the Inner Lindblad Resonance severely limits the number of growing modes available in the stellar disk (so that the few existing global modes can support *coherent* large-scale spiral structure). In particular, ILR inhibits high-m modes (thus discouraging the development of large-scale multiple-armed structure). For these reasons, stars are an essential ingredient for the establishment of quasi-stationary spiral structure in grand-design galaxies, which gets its dynamical foundation in the dominance of a very small number of global spiral modes.
- Relatively heavy disks are expected to develop a large-scale bar, unless the disk is too hot or too thick to begin with (see §4 below). The conditions for the excitation of such large-scale bars are not too sensitive to the amount of random motions present in the disk, which are expected to increase on the secular timescale.
- Relatively light disks, instead, are able to support normal, non-barred spiral structure; however, this is allowed only if the disk is sufficiently cool (see §4 below). Since the stellar disk is dissipationless, if, for any reason (such as is expected on a relatively long timescale as a result of collective instabilities or of other "external" scattering agents), its random motions are increased, there is no way for it to cool back to its original state. Thus, the conditions for the existence of large-scale normal spiral structure cannot persist long enough for a purely stellar disk; they may be maintained only with a significant contribution from a dissipative component.

In contrast, for the gas-dominated component:

- Population I objects are characterized by small epicycles. Thus, from the dynamical point of view, they are naturally responsible for small-scale features in the disk.
- On the small scale most of the arguments in favor of a quasi-stationary structure do not apply; indeed, also by analogy with the physics of ordinary fluids, this component should be prone to small-scale, transient (rapidly evolving) spiral activity.
- The gaseous disk, being cold, gives an important contribution to the Jeans instability of the disk. Thus, it is an important source of excitation also for large-scale spiral modes.
- The Population I subsystem behaves mostly as a fluid. As such, it is less vulnerable to Lindblad Resonance absorption (e.g., see Bertin 1993); in particular, given the partial failure of ILR in a fluid, it is more easily subject to the presence of many modes, especially to modes with higher

m, that may outline less regular, multiple-armed spiral structure (see Haass 1983).

- The Population I subsystem is characterized by significant dissipation, which acts on a short timescale via cloud-cloud collisions. Being cold and dissipative, it is subject to shocks, which can provide a saturation mechanism at finite amplitudes for the growing global spiral modes associated with grand-design spiral structure.

- Being dissipative, the gas component can provide a dynamical "thermostat" for the two-component disk, via a process of *self-regulation*, so that the conditions for the occurrence of large-scale normal spiral structure can be maintained for a relatively long time.

- In many galaxies, the gas distribution is rather diffuse, so that its dynamical impact is concentrated mostly in the outer regions, beyond two exponential scalelengths of the stellar disk. Thus, the processes mentioned in the third and sixth items above, which are very important for relatively light disks, tend to operate mostly in the outer disk; this explains why corotation, for normal (non-barred) spiral structure, is generally expected in the outer disk.

The real disk and its morphology result from the mutual interaction between these components and from their share in the galaxy collective behavior that is controlled by the common gravitational field. Starting with the two-component study by Lin & Shu (1966), several papers have addressed interesting aspects of the dynamical interaction between stars and gas (Lynden-Bell 1967, Graham 1967, Miller et al. 1970, Quirk 1971, Kato 1972, Jog & Solomon 1984a,b, Sellwood & Carlberg 1984; Roberts 1969, Kalnajs 1972, Roberts & Shu 1972; Ostriker 1985, Lin & Bertin 1985, Shu 1985; Lubow et al. 1986). One useful and important concept is that of the "effective stability parameter" for a two-component disk (see Bertin & Romeo 1988, especially the (β, α)-diagrams of Figs. 4 and 5; Bertin et al. 1989a, especially Fig. 9); this has also led to a quantitative demonstration of the process of *self-regulation* (Bertin 1991, especially Fig. 1). We briefly recall here that the dynamical properties of a two-component, zero-thickness disk made of stars and gas can be characterized by the gas-to-star density ratio α and by the gas-to-star "temperature" ratio β. For any combination of α and β, it is possible to determine the condition of marginal stability with respect to axisymmetric disturbances, in the form $Q_* = \bar{Q}(\alpha, \beta)$ (which reduces to $Q_* = 1$ for $\alpha = 0$, i.e. in the absence of gas). In a fluid model Q_* is defined as $Q_* = c_* \kappa / \pi G \sigma_*$. Then the equivalent-$Q$ parameter (sometimes denoted by Q_{eff} or just Q) is given by

$$Q = \frac{Q_*}{\bar{Q}(\alpha, \beta)}.$$

The efficiency of a small amount of cold gas to destabilize a disk is best appreciated by referring to an approximate relation, applicable for a reasonably wide range of regimes, that is found to hold between the individual stability parameters of stars (Q_*) and gas (Q_g) and the effective stability parameter: $1/Q \approx 1/Q_* + 1/Q_g$. The analysis of Bertin & Romeo (1988) also clarifies the process of *dynamical decoupling* of the two components, which takes place when the stellar component is too hot, i.e. when the parameter β is too small (see their Figs. 3 and 4), with a demonstration that the cold component tends to respond and to dominate at short wavelengths (see their Figs. 1 and 3). The study of self-regulation in a two-component disk (Bertin 1991) clearly shows that the corresponding process *if* invoked *for the gas component alone* (Kennicutt 1989; see Bertin 1991, p. 97) is generally "incomplete" and, as such, it should be relevant only for some flocculent galaxies.

3. Modeling

The many detailed results already available from the study of two-component systems vividly prove one obvious point, that their dynamics is much more complex than that of one-component systems. Therefore, the use of reduced models or reduced equations (much like the use of *MHD* equations to describe an ionized gas) can be extremely fruitful, but should be made with great caution and should always be supplemented by a thorough physical discussion. Some concepts, such as the effective stability parameter or the "active disk mass", conveniently capture and summarize many physical properties of a disk made of stars and gas, but cannot fully represent *all* the degrees of freedom available in a two-component disk. In particular, when finite thickness effects are included, the mapping of a two-component disk $2(FT)$ into a one-component, zero-thickness "effective disk" $1(ZT)$ is justified under special conditions not only on the properties of the basic state, but also on the properties of the perturbations under investigation; such mapping, or "modeling procedure", is not yet available for time-dependent or non-linear perturbations (Bertin 1991). Thus it should be emphasized that the modeling step is even *more difficult* for scenarios of fast evolution considered as alternative to the modal theory.

These difficulties are not at all overcome by a direct use of numerical simulations of disks made of stars and gas, which in recent years have become easily accessible (see Salo 1991, Combes & Elmegreen 1993, Elmegreen & Thomasson 1993), simply because the stumbling-blocks that are found in the modeling of a $2(FT)$ by a $1(ZT)$ disk are to be faced, to an even higher level of complexity, when mapping the rich variety of available astronomical data (on HI and molecular gas and on stars of different types) into the choice of parameters to be used as an initialization of a "realistic" $2(FT)$ (to be studied by the numerical simulations). This difficult step and the inherent limitations of the modeling procedure add up to other more specific issues that are to be faced in order to judge the adequacy of N-body simulations (see Romeo 1994), especially on timescales significantly longer than the relevant dynamical time.

4. Modal Theory of Spiral Structure

The modal theory of spiral structure has grown into a wide framework able to interpret a large number of astrophysical phenomena (Bertin & Lin 1996). Here I will try to highlight only some important aspects of it.

4.1. MORPHOLOGY

The starting point is that spiral arms should be interpreted as the manifestation of *density waves* that move in the disk in a manner that is, to a large extent, decoupled from the motion of the individual particles (stars or gas clouds) that support it, much like a sound wave in an ordinary gas. Because of the differential rotation and of the resulting shocks in the interstellar medium, in this scenario different age-groups and colors are expected for an observer at different phases with respect to the peak of the spiral gravitational field. While age-groups have long been noted and identified (see Lin 1970), a convincing demonstration of the existence of significant color gradients across spiral arms has been provided only recently (for M99, see González & Graham 1995; this paper also includes an interesting determination of the pattern frequency in M 99, derived from the drift of the young stars away from their birth-place).

Since the initial formulation of the hypothesis of Quasi-Stationary Spiral Structure (Lin & Shu 1964, 1966), it has been clear that, in order for this to be a viable framework for the description of the existing morphology, the large-scale dynamics of the disk should be dominated by a very small number of global spiral *modes*. In practice, the resulting modal theory has been developed quantitatively mostly starting in the mid-seventies. The term "mode" is sometimes used interchangeably with "Fourier m-component" or with "density wave". In the present context, it should be stressed that the term "mode" has a much more specific dynamical meaning. A mode is essentially a *standing wave* that can be supported by the disk; thus it is associated with a spiral density perturbation and gravitational field that rotate rigidly at a *given frequency* around the galaxy center and do not propagate in the radial direction.

Differently from commonly known standing waves in other mechanical systems, linear spiral modes may not only be damped (i.e., they may decay in time) but sometimes be *unstable* or *self-excited* (i.e., they may grow exponentially in time as a global instability), at the expense of the free energy stored in the differential rotation of the disk; it is part of a *non-linear analysis* to demonstrate how they can "saturate", i.e., equilibrate, at finite amplitudes. Much like for other types of standing wave, the amplitude of the density perturbation associated with a global spiral mode, for an observer moving along the arms, is in general *modulated*; this, like the presence of nodes for a standing wave, is due to the interference of the elementary waves that can be imagined to *maintain* the mode. This amplitude modulation was immediately recognized (Bertin et al. 1977) to be an important feature for comparison with the observations.

The finding of self-excited global modes in a given galaxy model is telling us that, under the physical conditions that are considered, spiral structure is energetically favored and expected to be realized by intrinsic mechanisms, even without external help (such as that of tidal interactions). When moderate instability is involved, the structure of the unstable modes gives an approximate description of the properties of the non-axisymmetric state, much like the unstable $m = 2$ modes on a Maclaurin ellipsoid give an indication of the properties of the triaxial Jacobi ellipsoids (in the vicinity of the bifurcation between the two equilibrium sequences).

A given basic state is subject to a *spectrum* of global spiral modes (i.e., a set of pattern frequencies, with the related self-consistent spiral gravitational field), which can be, in principle, calculated from the relevant dynamical equations; the spectrum just reflects the intrinsic characteristics of the basic state. Thus, given a perfect knowledge of the galaxy disk one could predict the morphology and the pattern frequency (or frequencies) of its spiral structure (much like one can calculate the eigenstates of the hydrogen atom). In practice, the galaxy disk is too complex (see §3) and its basic state and general dynamical conditions are known only in an incomplete form from the data available, so that a deductive approach has no realistic chance of applicability. This point becomes even more obvious if we note that the physical system owes its current state to a variety of non-linear mechanisms and evolutionary processes that are impossible to check by direct observation. This somewhat discouraging situation is by no means specific of our problem for spiral galaxies, but rather it is well known to many other research fields (plasma physics, fluid dynamics, geophysics, meteorology, etc.) that also deal with complex collective macroscopic systems.

In this situation one productive way to proceed is to take a *semi-empirical approach*. An extensive survey of realistic models of galaxy disks (Bertin et al. 1989a) has shown that the morphology types of the global spiral modes that can be generated in a disk match the general morphological categories that are found along the Hubble sequence. Depending on the *parameter regime* that characterizes a given galaxy disk (Bertin et al. 1989b), the dominant mode may be of the *A-type* (with normal, non-barred morphology and corotation generally in the outer disk) or of the *B-type* (with a generally prominent bar and corotation predicted to occur just outside the tip of the bar). Different excitation mechanisms operate for the two classes of modes: long-short or leading-trailing wave overreflection at corotation (the latter type of overreflection is often called *swing amplification*; Toomre 1981). A-modes rely on a combined support of gas and stars, while B-modes are star dominated, with the gas often playing mostly a passive role. A-modes are rather tight. B-modes can have a plain two-lump structure, as in many SB0 galaxies like NGC 2859, or a pair of rather tight arms departing from the central bar, like in NGC 1300 (see also M 100), or even an almost normal, but rather open, spiral appearance for those systems that have a significant active contribution from the gas (see Figs. 11 and 14 in Bertin et al. 1989a). Since the three-dimensional geometry of the two-component disk has a different impact on tight or on open waves, under special circumstances, the *same* galaxy disk can support *large-scale* modes of the two classes, thus displaying a mixed or *dual* type of morphology. [An interesting but different type of dual morphology may occur when *spatially separate* zones of the disk are marked by distinct dynamical effects (e.g., see the inner structure in NGC 4622, Block et al. 1994, or the striking inner structure of M 100, on a scale of one sixth of the disk exponential scalelength, studied by Knapen et al. 1995a,b).]

By relating these findings of the modal survey and the concepts of the modal theory to a discussion of the various possible physical conditions that may characterize a galaxy disk, a general framework for the dynamical classification of spiral galaxies has been proposed (Bertin 1991, Fig. 2), with good support from the new near-infrared data (Block et al. 1994). In this picture, the transition from SA to SB morphology is controlled by the active mass of the disk (relative to the total mass, which includes that of the bulge-halo), the transition along the Hubble sequence (from a to c galaxies) is mostly controlled by the gas content, while the transition from grand-design to flocculent galaxies (in relatively light disks) is controlled by the size of the stellar epicycles (in the sense that the stellar component, when it is too hot, becomes decoupled from the gas and remains inactive (see also §6 below); in this case the disk is only subject to small-scale spiral activity generated by the gas). This framework should not be meant as a rigid scheme to reduce all the observed morphological types to a simple, small set of dynamical prototypes; rather, somewhat like the original Hubble classification scheme, the modal approach offers a useful set of broad prototypes to guide us in the understanding of the rich variety of morphologies and in the construction of more specific and complex models as required by the observations.

The new near-infrared data offer an unparalleled opportunity to test many of the arguments given above and, in general, the soundness of this dynamical framework, since they give a direct way to probe the underlying gravitational potential and to separate the contribution of the stars from that of the gas. The following are four major general points that are found to give considerable support to the modal perception:

- *Regularity of spiral structure.* Grand-design morphology is rather frequent in the near-infrared. This shows that the conditions for the support of a regular large-scale structure are rather common, and, as such, must reflect the intrinsic characteristics of galaxy disks. Even in cases like M 51, where some interaction with another galaxy is obviously taking place, the extreme coherence observed in the underlying stellar disk (Rix & Rieke 1993) suggests that the interaction is relatively gentle and may just bring out the intrinsic modal characteristics of the disk that pre-existed the encounter.

- *Morphology of large-scale structure.* The morphology of large-scale spiral structure nicely matches the morphology of the calculated spiral modes. This has been noted also for the non-trivial case of barred galaxies, like NGC 1300 (see Bertin 1993, Fig. 1). This point is confirmed and actually strengthened by near-infrared images; for instance, one might compare the near-infrared morphology of NGC 4622 (Block et al. 1994, Figs. 6 and 7) with some spiral modes like that shown by Fig. 8 in the article by Bertin et al. (1989a).

- *Amplitude modulation along the arms.* This important observed feature, which has been confirmed in the infrared for several objects (see the cases of NGC 4622 and M51), is a "signature" naturally predicted by the modal theory (see the survey of models and modes shown by Bertin et al. 1989a).

- *Setting the corotation circle.* Any positive identification of a well-defined *pattern frequency* in a galaxy disk is automatically a strong case for the modal theory. The alternative scenarios of regenerative structure and of tidally-induced transient spiral structure, for which the large-scale spiral structure is not associated with discrete modes, do not anticipate the existence of a well-defined pattern frequency. Thus all the tests devised to measure the location of the corotation circle from the data (see the set of papers introduced by Allen et al. 1993; see also Tremaine & Weinberg 1984, Elmegreen et al. 1992, Canzian 1993) implicitly support the modal theory. In this respect, some notable cases are: NGC 936 (see Kent 1987), M 81 (see Lowe et al. 1994 and references therein), M 99 (González & Graham 1995), and M 100 (Canzian & Allen 1995, Sempere et al. 1995; see also Knapen et al. 1995b). Note that for the B-modes corotation is expected to occur well inside the optical disk (at $1 - 2$ exponential scalelengths), just outside the tip of the bar, which is empirically verified for objects with a prominent large-scale bar; the case of M 99 (with corotation at $0.6 - 0.7\ R_{25}$) also appears to be within the general expectations of the modal theory. However, as stressed earlier in this article, a detailed *prediction* of corotation for an individual object would require a very laborious and difficult modeling procedure.

4.2. EVOLUTION

To a large extent, questions related to the evolution of spiral galaxies can be considered separately from the morphological issues described above. On the relatively short timescale, one might ask how the spiral structure observed in a given galaxy was generated in detail, or, in different terms, what the observed structure should have been a few dynamical timescales ago, given its current morphological characteristics. On the long timescale, one might wish to investigate how the overall characteristics of the galaxy (not only morphology, but also thickness, mass and kinetic energy distribution) would change as a result of internal mechanisms and interactions with the environment.

It is well known that for complex, dissipative systems it may be practically impossible to answer the first question, especially since a given current physical state may result from a variety of initial conditions. The modal theory is particularly suited to bypass our ignorance of the evolutionary process. As in other complex systems (such as those found in meteorology or plasma physics), modes tend to stand out and to dominate after the rapid initial transient, independently of the exact "initial conditions".

As to the long-term evolution of galaxy disks, a comprehensive predictive picture would probably require a tremendously detailed knowledge of the galaxy structure, which, even if we might imagine it to become available from the data, would most likely be impossible to handle (much like for long-term weather predictions, in spite of the very accurate and complete sets of "initial conditions" now available at any given time). One more limited, and still interesting, question that can be tackled is the determination of the timescale where long-term evolutionary effects *of a given type* are expected; in particular, the effects associated with the presence of significant torques and with the related angular momentum transport may be better estimated now that the amplitude of the spiral gravitational field is better diagnosed (Bertin 1983, Gnedin et al. 1995, Zhang 1996; these analyses still have to be extended to the regime of open spiral structure).

5. Interpreting Specific Observed Features

From the above discussion, it should be clear that the cooperation of stars and gas is essential for the support of large-scale normal, non-barred spiral structure: stars ensure large-scale coherence of the pattern, while gas provides the possibility of maintaining, via self-regulation, a relatively cool outer disk, as required for the excitation of global modes and the responsiveness of the disk to waves. Gas can also provide a saturation mechanism for the growing linear modes at finite amplitudes. Cases where this cooperation process is particularly evident are galaxies like NGC 4622 and M 81. For NGC 4622 the arms are very long, thin, and tight (see Strom & Strom 1978; Byrd et al. 1989; Buta et al. 1992) and it is extraordinary that the overall optical morphology is basically preserved in the near-infrared (Block et al. 1994). This shows empirically that the stars in NGC 4622 must be characterized by small epicycles, forming a thin light disk.

The competition between the two main components of the galaxy disk is best shown by those features where the morphology in one component does not have a counterpart in the other. Clearly such discrepant behavior should be more evident in gas-rich galaxies: (a) Small-scale irregular features are generally observed in the Population I component and tend to disappear in the near-infrared. This is very common. (b) The different behavior at ILR explains why low-m large-scale structure dominates the near-infrared images. Thus morphological changes from optical to K-band images, as the ones shown by NGC 309, have a natural interpretation. Note the beautiful example of NGC 2997, where a third arm, prominent in the Population I, is practically absent in the near-infrared. In this galaxy we may witness a combination of two different effects. On the one hand, the $m = 3$ in the Population I is not surprising; on the other hand, given the large amplitude of the $m = 1$ and $m = 2$ components in the stellar disk (as noted in the near-infrared image: see Fig. 2 of Block et al. 1994), the third arm may get considerable support from the non-linear coupling of the low-m modes present; a non-linear coupling of $m = 1$ with $m = 2$ may also be at the basis of other morphologies, such as that of M 99 (González & Graham 1995; see also Iye et al. 1982). (c) Very far in (for M 51, see Zaritsky et al. 1993; see also the case of NGC 2997) or outside the optical disk (e.g., for M 101, NGC 6946, NGC 628: Sancisi 1990, Kamphuis 1993) spiral arms are often found to continue in the gas, without a direct stellar counterpart; in both cases, this may again result from the different behavior at the Lindblad Resonances, where the fluid disk is rather "transparent" (see §2). (d) An underlying bar is often present in the Population II component (see NGC 309 and NGC 1637 in Block et al. 1994; for M 100, see Pierce 1986, Knapen et al. 1995a,b, and also the interesting related observations of molecular spiral structure by Rand, 1995), and often with a well-developed two-blob signature; these open features are very natural in relatively heavy stellar disks; the frequency of this underlying bar morphology may shed light on the general issue of whether disks tend to conform to the maximum-disk *ansatz* (van Albada et al. 1985; van Albada & Sancisi 1986). (e) It is likely that some *ring* structures originate in the vicinity of the Lindblad Resonances (see discussion in the papers introduced bu Allen et al. 1993), and should be associated with the gas (see Bertin 1993); if so, one would expect rings to be less prominent in near-infrared images. A possible example of this situation may be the inner structure of NGC 309.

A rather frequent feature in near-infrared images is the presence of a significant $m = 1$ component (see NGC 2997, NGC 1637 in the article by Block et al. 1994; see also Rix & Zaritsky 1995), often in the form of a lopsidedness of the disk. Even before such direct views of the underlying stellar disk were available, it had already been noted that the frequency of $m = 1$ asymmetries demands an intrinsic mechanism that should explain the *persistence* of the observed structure (Baldwin et al. 1980). This is basically the same argument that is used, in general, to support the modal perception for grand-design bisymmetric spiral structure. Indeed, the linear modal theory predicts that $m = 1$ modes should generally be dominated by $m = 2$ modes when available, since the latter are more efficient in transporting angular momentum outwards, but that they should be relatively frequent, especially because they do not suffer from ILR inhibition (see also Zang 1976). Within the modal theory, one may argue that one-armed structures and lopsidedeness are all the *non-linear* result of $m = 1$ modes; lopsidedness may be the $m = 1$ equivalent of the broad bar of the $m = 2$ context.

From the observational point of view, probably the most urgent task to be completed is a systematic investigation, by means of near-infrared images of large samples of galaxies of different morphological class, of the statistical significance of points of dynamical interest of the type noted earlier in this paper (see §1). From the theoretical point of view, the linear analysis of the dynamics of two-component disks should be completed soon. Non-linear processes, such as the complex mechanism of self-regulation (beyond the preliminary analysis reported by Bertin 1991) or the detailed processes of "mode saturation" at finite amplitudes, are probably more difficult to tackle, but they should be within reach in the near future; here some work carried out on the properties of orbital support to a given spiral field (see Patsis et al. 1994 and references therin) may turn out to be of indirect help for those cases where gas cannot cooperate in the saturation process.

6. A Test for Flocculent Galaxies

One of the main points in the framework for the classification of spiral galaxies mentioned earlier in §4 is that in those disks where the stellar component is not heavy enough (to generate a bar) and is dynamically too hot (to cooperate with the gas to generate normal grand-design spiral structure) the Population II component is dynamically decoupled from the Population I, which is left free to its small scale spiral activity; here we expect a flocculent structure to be present. In this interpretation, what thus makes a galaxy look like NGC 2841 (a prototype of flocculent galaxies) and not like M 81 is the size of the *stellar* epicycles.

This point of view is quite natural, but it would be very nice to check it by a quantitative observational test. Unfortunately, there are two basic difficulties with such a test: (a) A full modeling of the decoupling of the two components for a given galaxy disk, with the obvious role played by the three-dimensional geometry of the system, is going to be very laborious, in spite of the many quantitative effects already learned from several theoretical analyses (see §2 and §3). (b) Measuring the relatively low velocity dispersion in the radial direction of the older stars directly is very hard (see Bottema 1995).

Here I point to a direct measurement that appears to be feasible and that would, at least, be a test of the *trend* anticipated by the theoretical analysis. The basic principle has already been applied by Bottema et al. (1987) for the study of the edge-on galaxy NGC 5170 and in a few following articles discussed by Bottema (1995). The argument is based on the *asymmetric drift* (e.g., see Woltjer 1967; Shu 1969), which quantifies the difference between the average rotation speed u_θ measured in a medium and the rotation of a single particle on a circular orbit $V = \sqrt{r\, d\Phi/dr}$ in the same gravitational field: in the medium, the rotation speed should be lower because of the pressure gradient effect that contributes to the momentum balance equation. The velocity difference is $V - u_\theta = O\left(c_r^2/V^2\right)$, and is thus very small for a cold component with small radial velocity dispersion c_r. The test would anticipate the possibility of a positive detection of the drift from the difference between the *rotation curve* measured in the stars by spectroscopic studies in absorption (similar to the accurate kinematical measurements now amply demonstrated for elliptical galaxies and S0's) and the HI rotation curve measured in the radio; this drift should be practically absent in galaxies with a non-barred grand-design structure and should generally be detectable in flocculent galaxies. In other words the decoupling of the *observed rotation curves* in the two components would be the proof of their *dynamical* decoupling. Note that the drift gives an indirect measurement of c_r, without the problems of deconvolution of vertical motions and of other geometrical effects inherent in an attempted direct measurement of c_r. Work is in progress to make such a test (Bertin & Stiavelli, in preparation).

This work was partially supported by MURST and by CNR of Italy.

References

Allen, R.J., Canzian, B. and Lubow, S.H. (1993), *Pub. Astron. Soc. Pac.*, **105**, 638

Baldwin, J.E., Lynden-Bell, D. and Sancisi, R. (1980), *Mon. Not. R. Astr. Soc.*, **179**, 23

Bertin, G. (1983), *Astron. Astrophys.*, **127**, 145

Bertin, G. (1991), in *IAU Symp. 146*, eds. F. Combes and F. Casoli, Kluwer, Dordrecht, p. 93

Bertin, G. (1993), *Pub. Astron. Soc. Pac.*, **105**, 640

Bertin, G., Lau, Y.Y., Lin, C.C., Mark, J.W-K. and Sugiyama, L. (1977), *Proc. Natl. Acad. Sci. USA*, **74**, 4726

Bertin, G. and Lin, C.C. (1996), *Spiral Structure in Galaxies*, MIT Press, Cambridge

Bertin, G., Lin, C.C., Lowe, S.A. and Thurstans, R.P. (1989a), *Astrophys. J.*, **338**, 78

Bertin, G., Lin, C.C., Lowe, S.A. and Thurstans, R.P. (1989b), *Astrophys. J.*, **338**, 104

Bertin, G. and Romeo, A.B. (1988), *Astron. Astrophys.*, **195**, 105

Block, D.L., Bertin, G., Stockton, A., Grosbøl, P., Moorwood, A.F.M. and Peletier, R.F. (1994), *Astron. Astrophys.*, **288**, 365

Block, D.L. and Wainscoat, R.J. (1991), *Nature*, **353**, 48

Bottema, R. (1995), *Ph. D. Thesis*, Groningen University

Bottema, R., van der Kruit, P.C. and Freeman, K.C. (1987), *Astron. Astrophys.*, **178**, 77

Buta, R., Crocker, D.A. and Byrd, G.G. (1992), *Astron. J.*, **103**, 1526

Byrd, G.G., Thomasson, M., Donner, K.J., Sundelius, B., Huang, T-Y. and Valtonen, M.J. (1989), *Celest. Mechan.*, **45**, 31

Canzian, B. (1993), *Astrophys. J.*, **414**, 487

Canzian, B. and Allen, R.J. (1995), preprint

Combes, F. and Elmegreen, B.G. (1993), *Astron. Astrophys.*, **271**, 391

Elmegreen, B.G., Elmegreen, D.M. and Montenegro, L. (1992), *Astrophys. J. Suppl.*, **79**, 37

Elmegreen, B.G. and Thomasson, M. (1993), *Astron. Astrophys.*, **272**, 37

Elmegreen, D.M. (1980), *Astrophys. J. Suppl.*, **43**, 37

Elmegreen, D.M. and Elmegreen, B.G. (1984), *Astrophys. J. Suppl.*, **54**, 127

Gnedin, O.Y., Goodman, J. and Frei, Z. (1995), *Astron. J.*, **110**, 1105

González, R.A. and Graham, J.R. (1995), *Astrophys. J.*, in press

Graham, R. (1967), *Mon. Not. R. Astr. Soc.*, **137**, 25

Haass, J. (1983), in *IAU Symp. 100*, ed. E. Athanassoula, Reidel, Dordrecht, p. 121

Iye, M., Okamura, S., Hamabe, M. and Watanabe, M. (1982), *Astrophys. J.*, **256**, 103

Jog, C.J. and Solomon, P.M. (1984a), *Astrophys. J.*, **276**, 114

Jog, C.J. and Solomon, P.M. (1984b), *Astrophys. J.*, **276**, 127

Kalnajs, A.J. (1972), *Astrophys. Lett.*, **11**, 41

Kamphuis, J. (1993), *Ph. D. Thesis*, Groningen University

Kato, S. (1972), *Pub. Astron. Soc. Jap.*, **24**, 61

Kennicutt, R.C. (1989), *Astrophys. J.*, **344**, 685

Kent, S. (1987), *Astron. J.*, **93**, 1062

Knapen, J.H., Beckman, J.E., Shlosman, I., Peletier, R.F., Heller, C.H. and de Jong, R.S. (1995a), *Astrophys. J. Lett.*, **443**, L73

Knapen, J.H., Beckman, J.E., Heller, C.H., Shlosman, I., and de Jong, R.S. (1995b), *Astrophys. J.*, **454**, 623

Lin, C.C. (1970), in *IAU Symp. 38*, ed. W. Becker and G. Contopoulos, Reidel, Dordrecht, p. 377

Lin, C.C. and Bertin, G. (1985), in *IAU Symp. 106*, ed. H. van Woerden, Reidel, Dordrecht, p. 513

Lin, C.C. and Shu, F.H. (1964), *Astrophys. J.*, **140**, 646

Lin, C.C. and Shu, F.H. (1966), *Proc. Natl. Acad. Sci. USA*, **55**, 229

Lowe, S.A., Roberts, W.W., Yang, J., Bertin, G. and Lin, C.C. (1994), *Astrophys. J.*, **427**, 184

Lubow, S.A., Balbus, S.A. and Cowie, L.L. (1986), *Astrophys. J.*, **309**, 496

Lynden-Bell, D. (1967), *Lect. Appl. Math.*, **9**, 131

Lynden-Bell, D. and Kalnajs, A.J. (1972), *Mon. Not. R. Astr. Soc.*, **157**, 1

Mark, J.W-K. (1971), *Proc. Natl. Acad. Sci. USA*, **68**, 2095

Mark, J.W-K. (1974), *Astrophys. J.*, **193**, 539

Miller, R.H., Prendergast, K.H. and Quirk, W.J. (1970), *Astrophys. J.*, **161**, 903

Oort, J.H. (1962), in *Interstellar Matter in Galaxies*, ed. L. Woltjer, Benjamin, New York, p. 234

Ostriker, J.P. (1985), in *IAU Symp. 106*, ed. H. van Woerden, Reidel, Dordrecht, p. 638

Patsis, P.A., Hiotelis, N., Contopoulos, G. and Grosbøl, P. (1994), *Astron. Astrophys.*, **286**, 46

Peletier, R.F., Valentijn, E.A., Freudling, W. and Moorwood, A.F.M. (1994), *Astron. Astrophys. Suppl.*, **108**, 621

Pierce, M.J. (1986), *Astron. J.*, **92**, 285

Quirk, W.J. (1971), *Astrophys. J.*, **167**, 7

Rand, R.J. (1995), *Astron. J.*, **109**, 2444

Regan, M.W. and Vogel, S.N. (1994), *Astrophys. J.*, **434**, 536

Rix, H-W. (1993), *Pub. Astron. Soc. Pac.*, **105**, 999

Rix, H-W. and Rieke, M.J. (1993), *Astrophys. J.*, **418**, 123

Rix, H-W. and Zaritsky, D. (1995), *Astrophys. J.*, **447**, 82

Roberts, W.W. (1969), *Astrophys. J.*, **158**, 123

Roberts, W.W. and Shu, F.H. (1972), *Astrophys. Lett.*, **12**, 49

Romeo, A.B. (1994), *Astron. Astrophys.*, **286**, 799

Salo, H. (1991), *Astron. Astrophys.*, **243**, 118

Sancisi, R. (1990), in *Windows on Galaxies*, ed. G. Fabbiano et al., Kluwer, Dordrecht

Schweizer, F. (1976), *Astrophys. J. Suppl.*, **31**, 313

Sellwood, J.A. and Carlberg, R.G. (1984), *Astrophys. J.*, **282**, 61

Sempere, M.J., Garcia-Burillo, S., Combes, F. and Knapen, J.H. (1995), *Astron. Astrophys.*, **296**, 45

Shu, F.H. (1969), *Astrophys. J.*, **158**, 505

Shu, F.H. (1985), in *IAU Symp. 106*, ed. H. van Woerden, Reidel, Dordrecht, p. 530

Strom, S.E. and Strom, K.M. (1978), in *IAU Symp. 77*, eds. E.M. Berkhuijsen and R. Wielebinski, Reidel, Dordrecht, p. 69

Toomre, A. (1981), in *The Structure and Evolution of Normal Galaxies*, eds. S.M. Fall and D. Lynden-Bell, Cambridge University Press, Cambridge, p. 111

Tremaine, S. and Weinberg, M.D. (1984), *Astrophys. J. Lett.*, **282**, L5

van Albada, T.S., Bahcall, J.N., Begeman, K. and Sancisi, R. (1985), *Astrophys. J.*, **295**, 305

van Albada, T.S. and Sancisi, R. (1986), *Phil. Trans. R. Soc. London A*, **320**, 447

Woltjer, L. (1967), *Lect. Appl. Math.*, **9**, 1

Zang, T.A. (1976), *Ph. D. Thesis*, MIT

Zaritsky, D., Rix, H-W. and Rieke, M.J. (1993), *Nature*, **364**, 313

Zhang, X. (1996), *Astrophys. J.*, Jan. 20 issue

Zwicky, F. (1957), *Morphological Astronomy*, Springer, Berlin

COLOR GRADIENTS IN M99:
STELLAR POPULATIONS OR DUST?

ROSA A. GONZÁLEZ
Astronomy Department, University of California
Berkeley. CA 94720-3411

GUSTAVO BRUZUAL A. AND GLADIS MAGRIS C.
Centro de Investigaciones de Astronomía
Apartado Postal 264, Mérida, Venezuela

AND

JAMES R. GRAHAM
Astronomy Department, University of California
Berkeley, CA 94720-3411

Abstract. González & Graham (1996a, b) discovered an **azimuthal** age gradient in the stellar population across one of the arms of M99, in what amounts to the first observational confirmation of a fundamental prediction of spiral density wave theory. A more exhaustive comparison of the color gradient with stellar population synthesis models seems to rule out a wave as old as 10^{10} yr, in favor of a duration of a few times 10^8 yr. By using new models that include the effects of mixed-in dust, we confirm that dust plays a minor role in the determination of the azimuthal color changes in this region of M99. We advance some possible new explanations for the overwhelming non-detection of convincing stellar population gradients across spiral arms.

1. Introduction

Recently, González & Graham (1996a, b; GG96 hereafter) combined deep g, r_S, i, and J data of the ScI galaxy M99 into the photometric parameter

$$Q(r_S J g i) = (r_S - J) - \frac{E(r_S - J)}{E(g - i)}(g - i) = (r_S - J) - 0.82(g - i) \equiv Q. \quad (1)$$

243

D. L. Block and J. M. Greenberg (eds.), New Extragalactic Perspectives in the New South Africa, 243–250.
© *1996 Kluwer Academic Publishers.*

Q is reddening-insensitive for screen absorption, assuming the extinction curves of Schneider et al. (1983), and Rieke & Lebofsky (1985); conversely, Q is extremely sensitive to star formation, since it has a higher value for a mixture of blue and red stars than for just about any single star (GG96). The main motivation for constructing Q with the data was to search for color gradients, due to stellar population age gradients, across the arms of M99, as predicted by spiral density wave theory. González & Graham detected a gradient in Q at 6 kpc galactocentric distance across one of the northern arms of M99; by comparing the color gradient to stellar population synthesis models (Charlot & Bruzual 1991; Bruzual & Charlot 1993, BC93 hereafter) to determine age as a function of distance from the arm, they performed the first measurement of the angular frequency of the spiral pattern (Ω_p) and of the location of the corotation radius (R_{CR}), derived from the drift velocity of the young stars away from their birth site.

This paper presents the results of a more exhaustive comparison of the data to stellar population synthesis models. We consider other reddening insensitive indices, vary the parameters of the models (the initial mass function, IMF, of the stellar population; the upper mass limit, M_{upper}, of the IMF; the fraction of young stars; and the duration of the star formation burst, t_{burst}), compare models that include mixed-in dust to reddening sensitive combinations of data, and look closely at the other two arms of M99. This time, besides the E+A type model of star formation (Belloni et al. 1995) used in GG96, where a burst of star formation with constant rate (SFR) and finite length occurs against a background of close to 100 percent by mass of stars 5×10^9 yr old, we consider a "blender" model of star formation: every 2×10^8 yr, i.e., $(3 \times \Omega_p)^{-1}$, since M99 has 3 arms, the passage of the spiral density wave induces identical bursts of star formation (with constant SFR and finite length); when the galaxy is 5×10^9 yr old, for example, the most recently formed stars represent $\simeq 4$ percent by mass of the total disk population.

2. Data, Reduction, and Previous Results

Deep, broad band (g, r_S, i, J, and K_s), photometrically calibrated images of M99 were deprojected to a face-on view (conserving flux, i.e., assuming that the galaxy is optically thin everywhere at all wavelengths, but the exact same results are obtained by assuming that the galaxy is optically thick at every point at all wavelengths!); the arms were unwound by plotting the images in a θ vs. $log R$ map (cf. Iye et al. 1982; Elmegreen et al. 1992), and a Q bidimensional map was constructed with the unwrapped images. The gradient across arm N1 (following the nomenclature of Iye et al. 1982) of M99 was detected in this map, but to increase the signal-to-noise ratio, the patch from $r \sim 60''$ to $\sim 95''$ was collapsed to a unidimensional plot of Q vs. distance across the arm, increasing downstream and away from the dust lane. This plot was compared to stellar population synthesis models (Charlot & Bruzual 1991; BC93) to determine age as a function of distance from the arm; the fit of the models to the data placed R_{CR} at 0.6–0.7 R_{25} and implied a pattern speed of $\Omega_p = 17.2$–15.7 km s^{-1} kpc^{-1}.

3. The Parameters of the E+A Model

The model used to fit the data in GG96 consists of a single burst of star formation with constant SFR and finite length against a background of close to 100 percent by mass of stars 5×10^9 yr old; only models with a Salpeter IMF were tried, and the best fit to the data was achieved with $M_{upper} = 10$ $M_\odot$, $t_{burst} = 2 \times 10^7$ yr, and with 2 percent by mass of young stars. The maximum change in Q occurs when the most massive stars of the young population become red supergiants; the size of the change is directly correlated both with M_{upper} and with the percentage of young stars, and inversely correlated with t_{burst}. The width (in age of the model stellar population) of the gradient is directly correlated with t_{burst} and inversely correlated with M_{upper}. Ω_p and R_{CR} are found by mapping the age, t_{age}, of the young population in the models to distance from the arm in the data through the difference between the angular rotation speed, $\Omega(R)$, and Ω_p, i.e. $d = [\Omega(R) - \Omega_p]Rt_{age}$. Inside corotation, a model with a wider Q feature will need less "stretching" to fit the data, and will yield a larger Ω_p and a smaller R_{CR}.

Of the several reddening-free parameters that could have been constructed using g, r_S, i, J, and K_s, $Q(r_S J g i)$ was used in GG96 for two reasons: the quotient of relevant color excesses is close to unity, which is desirable in order to give similar weight to all four bands involved, and because stellar population synthesis models predict a detectable gradient. The data and the models overlap each other perfectly. There are, however, three other indices that could have been used:

$$Q(r_S K_s g J) = (r_S - K_s) - 0.78(g - J) \tag{2}$$

$$Q(g i r_S K_s) = (g - i) - 0.89(r_S - K_s) \tag{3}$$

$$Q(i K_s r_S J) = (i - K_s) - 0.94(r_S - J). \tag{4}$$

These three Q-like parameters work in the same way as $Q(r_S J g i)$ does, i.e., by detecting a mixture of red and blue stars. Except for a vertical shift (additive constant) of ~ 0.1 mag (data bluer than the model!) in the three cases, the data and the model profiles follow each other in all four Q-like parameters (Fig. 1).

For this paper –both for the E+A and the "blender" models–, we use a revised version of the BC93 population synthesis code which uses the same stellar spectra but incorporates an updated set of evolutionary tracks. Hereafter, this new set of models will be referred to as BC95 (see Charlot, Worthey. & Bressan 1996. for details). This time. the model $Q(r_S K_s g J)$ and $Q(i K_s r_S J)$ overlap the data better than $Q(r_S J g i)$, which shows a shift of ~ 0.15 between model and data; $Q(g i r_S K_s)$ again displays a shift of ~ 0.1. Contrary to what was observed with the old models. in both cases the data are redder than the model. The model profiles. however. follow the data qualitatively almost as well in all reddening-free indices as in the past (the slower second change in color is less steep in the BC95 models than in the data and in the BC93 models). The model "equilibrium" color or zero-point depends on the percentage of young stars but not on M_{upper} for the E+A model; it does depend on M_{upper} in the "blender" model (the dependency is of the order of the observational error in both cases). Given that we observe shifts of roughly the same magnitude and in opposite senses. depending on the set of evolutionary tracks used in the population synthesis models. we choose to neglect the zero-point and focus on the profiles.

4. The "Blender Model"

A model where the passage of the spiral density wave induces star formation with periodicity $(m \times \Omega_p)^{-1}$, with m being the number of arms. is more consistent with a long-lived wave. We try such "blender model": every 2×10^8 yr. the density wave induces identical bursts of star formation (with constant SFR and finite length). We compare the data with the model Q profile at 5×10^9 and 10^{10} yr, when the youngest population represents, respectively. $\simeq 4$ percent and 2 percent by mass; the model profiles are the same within observational errors. In addition to varying M_{upper}. we try a Scalo IMF. We find that only a "blender" model with a Salpeter IMF and $M_{upper} = 30\ M_\odot$ fits the **size** of the Q gradient nearly as well as the original E+A model. However, the width of the feature is not fit well for the values of Ω_p and R_{CR} derived in GG96. The effect in the width of increasing t_{burst} to 3×10^7 yr is negligible; increasing it further does not help, either, since: *a.* it starts to conflict with data from OB association studies; *b.* the size of the peak diminishes. making it necessary to increase M_{upper}. which then reduces the width of the gradient, etc. A good fit of the "blender" model with a Salpeter IMF, $M_{upper} = 30\ M_\odot$. and $t_{burst} = 2 \times 10^7$ yr to the data implies $\Omega_p \simeq 12$ km s^{-1} kpc^{-1}, $R_{CR} \simeq 0.9 R_{25}$ (Fig. 1).

If we prefer the E+A model, the implications are that the wave is a recent phenomenon. perhaps triggered by an interaction. Indeed. HI data of M99 shows signatures of gas infall (Phookun, Vogel, & Mundy 1993;

Schulman, Bregman, & Roberts 1994).

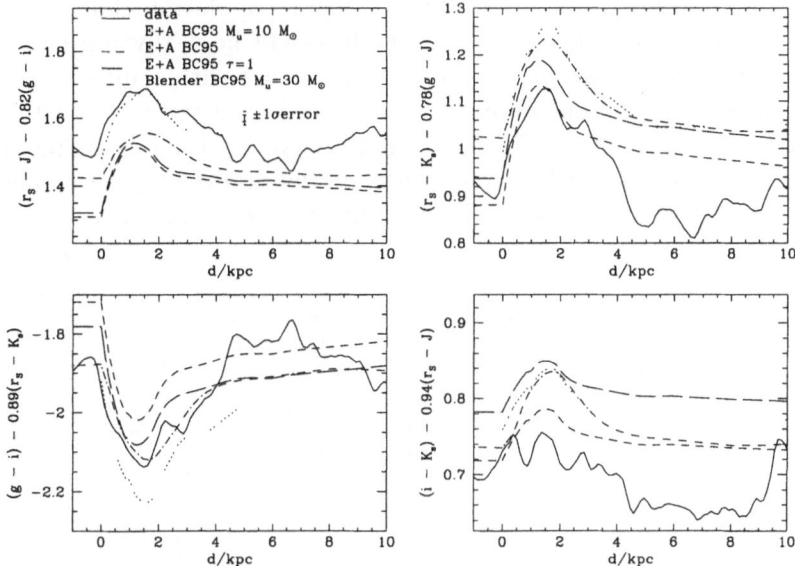

Figure 1. **Reddening-free parameters, data vs. models.** Observed profiles (*solid line*) vs. distance from the dust lane in kpc for arm N1 of M99 at R $\sim$ 5.9 kpc; all models have a Salpeter IMF and $t_{burst} = 2 \times 10^7$ yr. *Dotted line:* E+A model used in GG96 (BC'93), 2 percent by mass of young stars evolve against a background of stars 5×10^9 yr old ($M_{upper} = 10 M_{\odot}$); *short-dashed line:* E+A model with BC95 SEDs; *long-dashed line:* E+A model with BC95 SEDs and uniform $\tau_V = 1$, for a galaxy with $\mu = 0.74$; *dotted-short-dashed line:* "blender" model (BC95 SEDs) with a burst every 2×10^8 yr and $M_{upper} = 30 M_{\odot}$, after 5×10^9 yr. E+A models yield $\Omega_p \simeq 16$ km s^{-1} kpc^{-1}. "blender" model implies $\Omega_p \simeq 12$ km s^{-1} kpc^{-1}. Inner error bars show data statistical error only.

5. Dust

GG96 showed that Q is reddening insensitive and that the Q gradient in arm N1 of M99 is not due to dust. Here, to try to measure how much dust there is, we compare the arm profile in reddening **sensitive** indices to models that include reddening by dust mixed-in with plane parallel geometry and uniform extinction for all stellar types (Bruzual, Magris & Calvet 1988). Predictably, reddening sensitive model colors bluen perceptively (a uniform shift occurs) both with a bigger percentage of young stars and with a higher M_{upper}. Hence, unless the parameters of star formation are known, the optical depth and/or dust mass derived from using these models are tentative at best. Even worse, as discussed in the previous section, there could be a zero-point problem of the order of the reddening predicted by the models! What is very interesting, however, is that the E+A model used in GG96 (unreddened even, or with a uniform $\tau_V \lesssim 1$) provides an astonishing fit to the data profiles in reddening sensitive colors, confirming that the color changes across arm N1 are due to stellar population changes (Fig. 2). In particular, according to the models, it would be really hard to get the ~ 0.6 mag deep feature in $(g - J)$ and $(g - K_s)$ with reddening alone (a maximum of 0.2 mag of reddening is achieved with $\tau_V = 2$-3, after which the reddening decreases). The fit is not as good with the BC95 spectral energy distributions (SEDs), regardless of the model (E+A or "blender") used; the models exhibit a sharp dent that is not in the data.

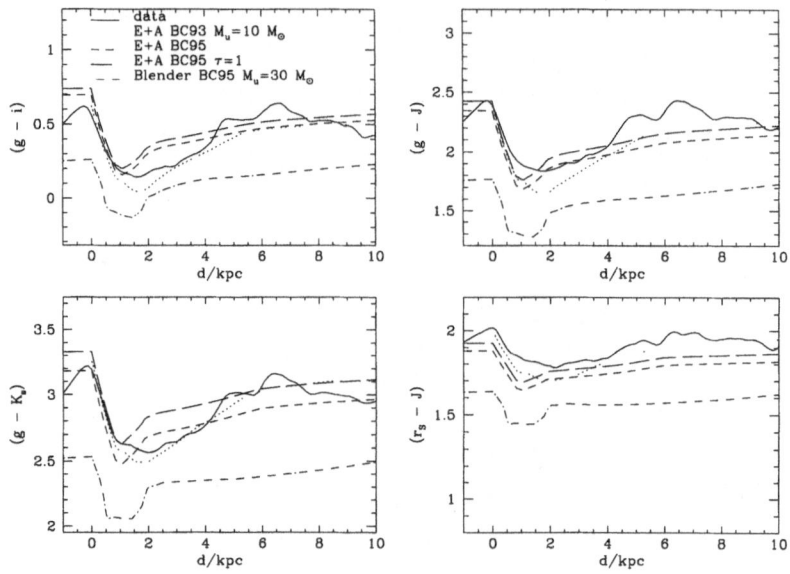

Figure 2. **Reddening-sensitive colors, data vs. models.** Parameters, models and symbols as in Figure 1; error is roughly $\sqrt{2}$ smaller than for Q-like parameters.

6. Arms N2 and S

The massive S arm displays color changes in the reddening insensitive indices that are not fully consistent with the models: there is an initial steep change, but the smooth second change of different sign is not detected. Furthermore, there is no narrow dust lane that can be associated with the shock. If we choose to believe that the initial steep change is real, and we use the feature itself to align the data and the models, the profiles of the feature in the reddening insensitive indices constructed with the data are consistent with an E+A model that has the exact same parameters as the best fit to arm N1.

The flocculent N2 arm is dusty to the extent that a structure in ($J - K_s$), a good dust tracer, completely overlaps it and is brighter than any neighboring feature that could be associated with a dust lane upstream of the arm. It is probably wrong to apply a model with the same parameters to an arm that seems so different from the other two. If we do, though, and we compare the equilibrium reddening sensitive colors of the model to the data, once again $\tau_V = 1 - 2$ seems to fit the data.

7. Conclusions

Again and again in astronomy, as the search and discovery of planets has recently reminded us, the confirmation of straightforward predictions is not easy, and objects and phenomena are found in strange places and circumstances. We have seen that only a Salpeter IMF yields a color gradient across the arm that fits the one observed in M99. We have not considered models with a Miller-Scalo IMF. We did consider Scalo IMF models, which fail to fit the data, given a reasonable combination of the remaining parameters. Also, if the passage of the density wave induces star formation of a "blender" model type (the most consistent with the assumption of a long-lived wave), the contrast is harder to detect in general, given the same IMF and M_{upper}, than in the E+A case. The region in M99 that exhibits a convincing color gradient is exceptional only because, in addition to a combination of parameters of the star formation burst that facilitated detection, a few other favorable conditions occurred to make it a textbook case. It is relatively dust free, at the same time that a very well defined dust lane has allowed, in the absence of information about the gas kinematics, a reasonable guess about the location of the shock. It is also the only patch free of H II regions in M99 (Van Dyk 1995), a fact that is consistent with our best fit using $M_{upper} = 10\ M_{\odot}$. We conclude that the color gradient predicted by theory might often be there, but undetectable due to the parameters of star formation, confused by dust, or masked by star formation that is too "energetic". These causes of non-detection ought to be added to the absence of a gradient when spiral structure is not due to a density wave and to the blurring of the gradient by radial drift of the young population (Yuan & Grosbøl 1981). We can guess that not only different mechanisms must be behind the spiral structure in different galaxies, but might be simultaneously at work in any given galaxy. In M99, just from optical and IR broad-band photometry, we see that the three arms have not been created equal. Therefore, Ω_p, R_{CR}, and the parameters of star formation induced by the wave have been derived from the only region in M99 that allowed a convincing comparison with the theory.

References

Belloni, P., Bruzual A., G., Thimm, G. J., & Röser, H.-J. 1995, A&A, 297, 61

Bruzual, G., & Charlot, S. 1993, ApJ, 405, 538

Charlot, S., & Bruzual, G. 1991, ApJ, 367, 126

Charlot, S., Worthey, G., & Bressan, A. 1996, ApJ, 457, in press

Elmegreen, B. G., Elmegreen, D. M., & Montenegro, L. 1992, ApJS, 79, 37

González, R. A., & Graham, J. R. 1996a, ApJ, April 1, in press, "Tracing the Dynamics of Disk Galaxies with Optical and Infrared Surface Photometry: Color Gradients in M99"

——————————.1996b, in Proc. of Conf. on Spiral Galaxies in the Near IR, Garching, ed. D. Minniti, Springer Verlag, in press, "Azimuthal Color Gradients in M99"

Iye, M., Okamura, S., Hamabe, M., & Watanabe, M. 1982, ApJ, 256, 103

Phookun, B., Vogel, S. N., & Mundy, L. G. 1993, ApJ, 418, 113

Schulman, E., Bregman, J. N., & Roberts, M. S. 1994, ApJ, 423, 180

Van Dyk, S. 1995, private communication.

Yuan, C., & P. Grosbøl 1981, ApJ, 243, 432

AMPLITUDE AND SHAPE OF SPIRAL ARMS IN K'

P. GROSBØL
European Southern Observatory
Karl-Schwarzschild-Str. 2, D-85748 Garching

AND

P.A. PATSIS
Max-Planck Institut für Astronomie
Königstuhl 17, D-69117 Heidelberg

1. Introduction

Full understanding of the dynamics of spiral galaxies requires accurate models of their potentials including non-axisymmetric perturbations *e.g.* spiral patterns. Although a detailed kinematic mapping of the Population II stars in their disks would provide this information, it is not currently feasible. An alternative is to use cause kinematic data to derive the general axisymmetric potential and surface photometry to deduce the density perturbations in the disk at higher spacial frequencies. Light at wavelengths blue of 1μm is strongly effected by dust extinction and population variations whereas these problems are much less serious in near-infrared bands like K' at 2.1μm (Rix and Rieke, 1993). This makes it possible to estimate surface density variations in the old stellar disk population of galaxies from K' maps and thereby study their dynamics (Block *et al.*, 1994).

Here we analyze the shape and amplitude of spiral patterns in three Sb-Sc galaxies to see whether or not they are consistent with the density wave theory (Lin and Lau, 1979).

2. Data and Analysis

Surface photometry maps of the three galaxies NGC 3223, NGC 5085 and NGC 5861 were obtained in BVI and K' at 2.2m ESO/MPI telescope, La

---

[1]Based on observations collected at European Southern Observatory, La Silla, Chile.

D. L. Block and J. M. Greenberg (eds.), New Extragalactic Perspectives in the New South Africa, 251–254.
© *1996 Kluwer Academic Publishers.*

TABLE 1. Derived parameters for the galaxies

Galaxy	P.A.	I	I_o^K	a_b	r_b	α_d	i_2^K	i_2^I	range
NGC 3223	127°1	44°4	14^m8	1″8	18″	-0.97	-11°3	-8°9	28-40″
NGC 5085	60°0	47°7	15^m3	1″7	8″	-0.32	-18°0	-13°4	26-41″
NGC 5861	152°7	55°0	16^m3	1″1	11″	-0.62	-13°7	-13°4	18-37″

Silla. The observations and basic reductions are described by Grosbøl and Patsis (1996) who also estimate parameters for the sky projections as shown in Table 1.

Deprojecting the K′ images, a central spherical bulge component shows up as an artificial bar. This component was fitted by a modified Hubble profile minimizing the sum of $a_2 \cos(2\theta_2)^2$ out to a radius of r_b where a_2 and θ_2 are relative amplitude and phase from the major axis of the azimuthal m=2 Fourier component. A bulge component with central intensity I_o^K in K′ and scale length a_b (see Table 1) was then subtracted from the images. Deprojected, face-on maps are shown in Fig. 1 for (V-K′) and relative azimuthal intensity variation in K′.

3. Discussion

All three galaxies show a grand design spiral pattern in K′ which confirms that the dust distribution is decoupled from the the old stellar disk population (Block and Wainscoat, 1991; Block et al., 1994). Although a modified Hubble profile does not fit the bulge perfectly, it is clear after the bulge was subtracted that only NGC 5085 has a weak oval distortion in the center. The radial profile of the disk component increases steeper than an exponential law in the central region. This inner disk can be fitted by a power law with the power α_d given in Table 1.

Relative amplitudes a_m and phases θ_m of the mth harmonics of azimuthal intensity variation as function of radius are shown in Fig. 2 while the radial range of the main two armed spiral pattern and its pitch angles i_2^I, i_2^K measured in I and K′ are given in Table 1. The spirals in I are tighter and has an offset compared with the ones in K′ suggesting that the perturbations are cause by a density wave. A weak spiral pattern with larger absolute pitch angle can be traced into the inner disk and may be associated with a gaseous spiral which can pass the ILR inwards. The a_4/a_2 ratio with two distinct peaks at each end of the radial range of the main pattern suggests that this corresponds to the interval between -2:1 and -4:1 resonances (Contopoulos and Grosbøl, 1988; Grosbøl, 1993).

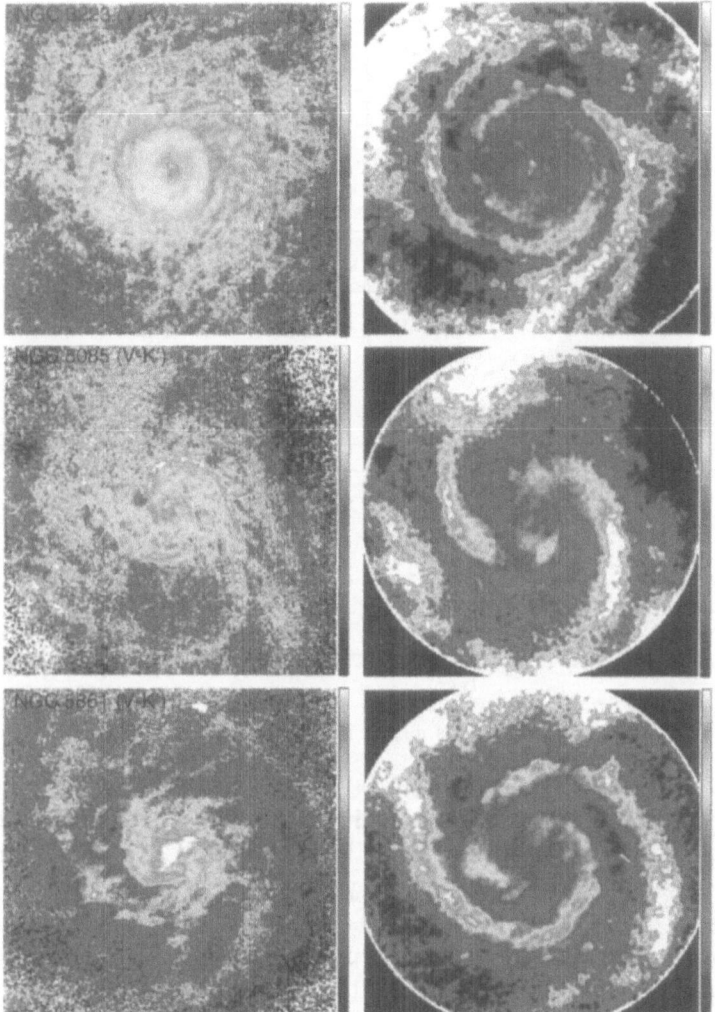

Figure 1. Maps of (V-K′) and intensity variation relative to the axisymmetric background in K′ for a) NGC 3223, b) NGC 5085 and c) NGC 5861. See also color plate 7.

4. Conclusions

The analysis of the three galaxies suggests the following conclusions:

- the inner part of the disks in K′ cannot be fitted by an exponential disk but show a significant residual possibly representing a steep inner disk component,
- the main two armed pattern follows a logarithmic spiral in the exponential part of the disk but the absolute value of its pitch angle increases

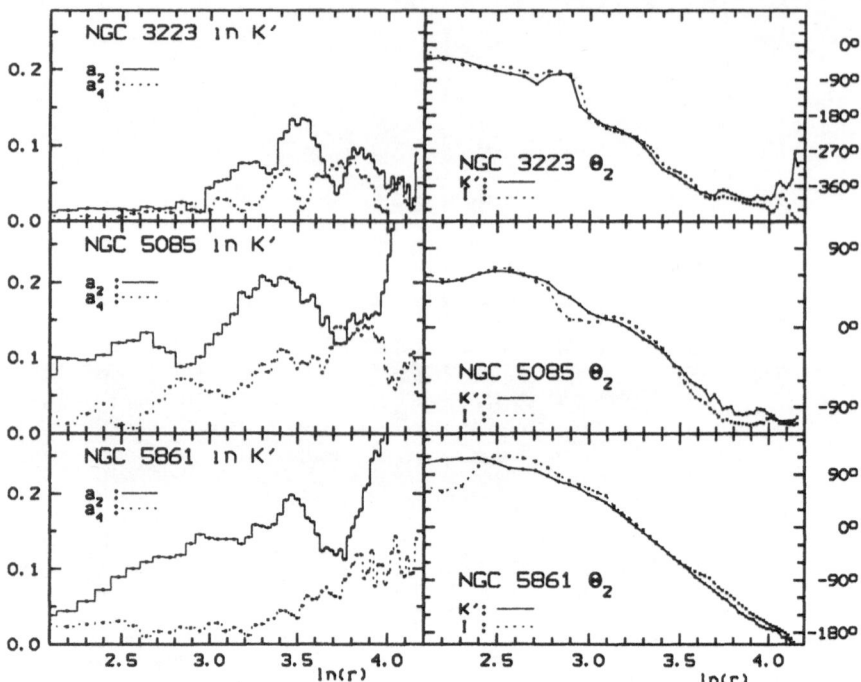

Figure 2.   Relative amplitudes a_2, a_4 in K′ and phase θ_2 in I and K′-bands

inwards at the radius where the inner disk becomes significant,
- the spiral arms measured in I is tighter wound than in K′ suggesting the pattern is caused by a density wave,
- the ratio a_4/a_2 in the range 0.2-0.5 suggests a marginal non-linear perturbations. Further it has steep peaks reaching values near 1.0 which may indicate locations of the -2:1 and -4:1 resonances, and
- the galaxies NGC 3223 and NGC 5085 show a significant gradient in their (V-K′) color along a direction perpendicular to their major axis suggesting a projection effect possibly due to dust extinction.

References

Block, D.L., Bertin, G., Stockton, A., Grosbøl, P., Moorwood, A.F.M., Peletier, R.F. (1994) *A&A* **288**, 365

Block, D.L., Wainscoat, R.J. (1991) *Nat* **353**, 48

Contopoulos, G., Grosbøl, P. (1988) *A&A* **197**, 83

Grosbøl, P. (1993) PASP **105**, 651

Grosbøl, P., Patsis, P.A. (1996) in *Spiral Galaxies in the Near-IR*, Eds. D.Minniti and H.W. Rix, ESO Astrophys. Symp., Springer, 174

Lin, C.C., Lau, Y.Y. (1979) *Stud. Appl. Math.* **60**, 97

Rix, H.-W., Rieke, M.J. (1993) *ApJ* **418**, 123

THE BULGE/DISK CONNECTION IN LATE-TYPE SPIRALS

Observational Evidence for Models of Secular Evolution

STÉPHANE COURTEAU

NOAO/KPNO

Kitt Peak National Observatory, 950 N. Cherry Ave., Tucson, AZ 85726

Abstract.
Recent ground-based photometric investigations suggest that central regions of late-type spirals are closely coupled to the inner disk and probably formed via secular evolution. Evidence presented in support of this model includes the predominance of exponential bulges, the correlation of bulge and disk scale lengths, blueness of the bulge and small differences between bulge and central disk colors, detection of spiral structure into the core, and rapid rotation. Recent HST observations show that our own bulge and that of M31, M32, and M33 probably harbor both an old and intermediate-age populations in agreement with models of early collapse of the spheroid plus gas transfer from the disk. Secular evolution provides a mechanism to build-up central regions in late-type spirals; mergers or accretion of small satellites could explain the brighter, kinematically distinct bulges of Sa's and SO's.

1. Introduction

In recent years, many studies have addressed the formation and evolution of bulges in spiral galaxies. Models proposed range from the classic picture of dissipational collapse and spheroid formation (Eggen, Lynden-Bell, & Sandage 1962, hereafter ELS; Sandage 1986) to accretion of small satellites (Pfenniger 1991, Zaritsky 1995) and mergers + starbursts (Schweizer 1990). Population studies of extragalactic bulges have been hampered by the difficulty to resolve them into stars. HST observations of our own bulge now provide deeper color-magnitude diagrams and observers have started paying greater attention to properties of bulges, in preparation perhaps for the new era of adaptive optics and high-resolution infrared photometers and

D. L. Block and J. M. Greenberg (eds.), New Extragalactic Perspectives in the New South Africa, 255–270.
© *1996 Kluwer Academic Publishers.*

spectrographs, both in space and ground-based. Here, I will limit myself to considering only one of the few possible scenarios for the formation of "bulges"[1]. The evidence in favor of secular evolution in late-type spirals is compelling which is why I have chosen to focus on it. This article is based in part on a Letter by Courteau, de Jong, and Broeils (1996). Recommended reading material on the topic of exponential disks, secular evolution, or bar dynamics include Struck-Marcell (1991), Sellwood & Wilkinson (1993), Pfenniger (1993), Martinet (1995), and Pfenniger (1996, this conference). Kormendy (1993) offers an excellent account of observational evidence that "some bulges are really disks".

2. An Overview of Secular Evolution

Secular evolution models provide a way to transfer material from the disk into the central regions of a spiral galaxy via angular momentum transfer and redistribution of the initial gas. Models of viscous evolution were first invoked to explain the nature of the exponential distribution of the stars in galactic disks (Silk & Norman 1981, Lin & Pringle 1987, Yoshii & Sommer-Larsen 1989, Saio & Yoshii 1990, Struck-Marcell 1991, hereafter SM91; Olivier et al. 1991). Given comparable timescales for star formation and viscous redistribution of the mass and angular momentum in the disk, one automatically recovers a disk with an exponential luminosity profile. In these models, other timescale combinations would lead to truncated or power-law profiles. Modern N-body realizations of angular momentum transfer which are independent of a viscosity parametrization also yield disk exponential profiles (Pfenniger & Friedli 1991, von Linden et al. 1996). It was also realized that an exponential profile in the central regions is expected from the non-axisymmetric disturbances which will induce inward radial flow of disk material (Hohl 1971, Combes et al. 1990, Saio & Yoshii 1990, Pfenniger & Friedli 1991, SM91, Kormendy 1993, hereafter K93; Pfenniger 1993, hereafter P93).

Efficiency of transport will be improved with a bar or oval distortion which can be triggered by the global dynamical instability of a rotationally supported disk or induced by interactions with a satellite (K93, P93, Martinet 1995, hereafter M95, and references therein). This in turn, will catalyze funneling of disk material into the central regions. Disk material will be heated vertically up to 1–2 kpc above the plane via resonant scattering of stellar orbits by the bar-forming instability. A "bulge-like" component

---

[1] From this point on, I shall distinguish "bulges" as kinematically and probably chemically distinct entities from the disk, from the "central regions" of late-type disks which will be understood as a central accumulation of disk material. To further clarify this distinction, "central regions" apply to 1-2 kpc radii whereas "core" would be more appropriate for 500 pc or less.

with a nearly exponential profile will emerge due to relaxation induced by the bar. The properties of the disk's central regions are directly coupled to the relative time-scales of star formation and angular momentum transfer. Such a model is expected to produce correlated scale lengths and colors between the disk and its central regions.

Gas redistribution by the bar can cause its own dissolution. Secular accumulation or satellite accretion of only 1-3% of the total stellar disk mass near the center is sufficient to induce dissolution of the bar into a lens or triaxial component and later into a spheroid (Kormendy 1982, Pfenniger & Norman 1990, P93, Friedli 1994, M95, Norman, Sellwood, & Hasan 1996). It is estimated that about two-thirds of disk galaxies currently have a bar, especially as revealed in the infrared (Block & Wainscoat 1991, Zaritzky, Rix & Rieke 1993, Sellwood & Wilkinson 1993, Martin 1995) and that most spirals have probably harbored a self-destructive bar at one time or another during their evolution (Friedli & Benz 1993, hereafter FB93; Friedli et al. 1994). Once a bar has formed, thickened and subsequently dissolved, no more thickening of the central dissipationless material is expected unless triggered by starbust activity or stellar accretion. Secular evolution is a viable mechanism for producing the small, central accumulations of material in late-type disks. Bigger bulges, however, could not be formed this way without disrupting the disk. The energy required to heat up the central material is far greater than the total bar and disk's mechanical energy supply. Accretion of a small satellite to explain the bigger bulges of SO-Sa's provides an appealing alternative (P93).

3. Luminosity Profiles

Recently, the groups of Peletier, Balcells, and Andredakis (Andredakis & Sanders 1994, Andredakis, Peletier & Balcells 1995; hereafter APB95), and Courteau, de Jong, Broeils, and Holtzman (de Jong & van der Kruit 1994, de Jong 1995, hereafter dJ95; Courteau, de Jong & Broeils 1996, hereafter CdJB96; Broeils & Courteau 1996, hereafter BC96; Courteau & Holtzman 1996, hereafter CH96) have used their high-quality surface brightness (SB) profiles to show that *central regions of disks are generally best described by an exponential luminosity profile.*

In modeling the stellar density distributions in spiral galaxies, one has often assumed an $r^{1/4}$ brightness law for the central regions (de Vaucouleurs 1948) and an exponential surface density for the outer regions of the disk (de Vaucouleurs 1959, Freeman 1970). Departures from the standard de Vaucouleurs profile in the central light distribution of early and late-type systems are however not new (de Vaucouleurs 1959, van Houten 1961, Frankston & Schild 1976, Kormendy & Bruzual 1978, Burstein 1979, Shaw

& Gilmore 1989, Andredakis & Sanders 1994). For example, Kent, Dame & Fazio (1991) used the Space Shuttle Infrared Telescope to show that the Milky Way bulge is best described by an exponential luminosity profile with a scale length of 500 pc.

A reliable approach to bulge-to-disk (B/D) decompositions is to fit for the shape of the luminosity profile as well. This method was first proposed by Sérsic (1968) with a generalized law of the form

$$\Sigma(r) = \Sigma_o exp\{-(r/r_o)^{1/n}\}$$

where Σ_o is the central surface brightness (CSB), r_o a scaling radius, and n is the shape parameter. If $n = 1$, one has a pure exponential profile with r_o as the scale length; with $n = 4$, one recovers a deVaucouleurs profile. As n becomes large, the Sérsic profile approaches a power law. Caon, Capaccioli, & D'Onofrio (1993) used their data on ellipsoids (E/S0s) and the low surface brightness (LSB) dwarf galaxies of Davies et al. (1988) to show that the parameter n correlates with absolute luminosity and half-light radius, such that bigger, brighter systems have larger values of n. This result was extended to brightest cluster galaxies by Graham et al. (1996); see Fig. 1.

Following Caon et al., APB95 and CdJB96 applied the Sérsic law to large samples of spiral galaxies for the first time. APB95 used near-infrared J and K images for 30 early-type spirals and r-band data from Kent (1986) for 21 late-type systems and performed 1 dimensional B/D decompositions using the technique of Kent (1985). The shape parameter n is fitted as a free variable and the disk is a fixed exponential. CdJB96 have combined the collection of deep r-band profiles of Sb/Sc galaxies by Courteau (1992, 1996; hereafter C96) and $BVRIHK$ photometry of dJ95 for 86 face-on Sa–Sm galaxies. Courteau's Tully-Fisher sample comprises 350 spirals but 243 were kept for final decompositions; the rest had too small a bulge to be resolved successfully. B/D decompositions were done independently by dJ95 and BC96. dJ95 used both major-axis profiles (1D) and full image B/D decompositions[2] of his thesis sample with fixed $n = 1$, 2, and 4 for the central regions and a standard disk exponential. BC96 decompose the light profile of Courteau's galaxies as "bulges" with $n = 1$ and 4 plus an exponential disk, following the method introduced by Kormendy (1977).

"Bulges" of late-type spirals are small and their luminosity profile can be severely affected by atmospheric blur. Both teams used extensive simulations with a wide range of input parameters and various values of n

---

[2]2D decompositions offer the advantage of fitting for any additional central component, like a bar, ring or lens. They also yield a more robust recovery of simulated input parameters (Byun & Freeman 1995, dJ95). Still, CdJB96 find that their results do not depend strongly on the type of techniques adopted.

From R°(1/n) profiles

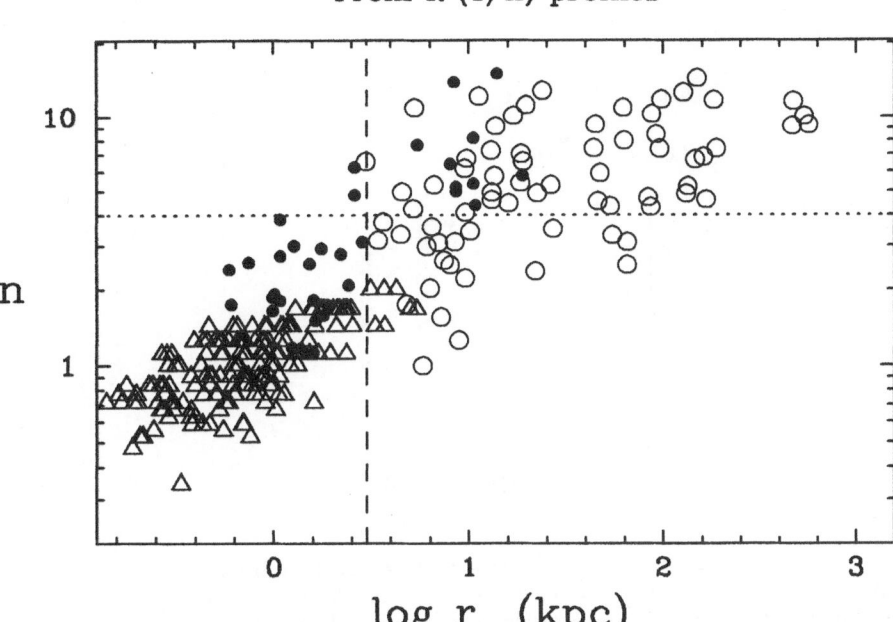

Figure 1.   Updated figure for n vs $\log(r_e)$ by Graham *et al.* (1996) based on Fig. 5 of Caon *et al.* (1993). The open circles show brightest cluster members from Lauer & Postman (1994), filled circles represent a sample of 33 E/S0 galaxies from Caon *et al.* (1993), and the triangles are for 187 LSB dwarf galaxies from the study of Davies *et al.* (1988). Note that these fits cover the entire extent of the luminosity profile and thus, unlike Figure 2, are not confined to the galaxies' central regions. This distribution follows a luminosity trend with LSBs and faint spheroids (E/S0s) at the bottom and giant Es and BGCs at the top. The dashed line at $r_e = 3$ kpc serves to distinguish the fainter and brighter galaxy families. The de Vaucouleurs profile ($n = 4$) is shown with a dotted line; it roughly delineates the hot stellar systems ($n < 4$) from objects that have grown by accretion or mergers ($n > 4$) (Caon *et al.* 1993).

to derive a space of recoverable parameters under specific seeing conditions. Sky subtraction errors, which represent another fundamental source of uncertainty in fitting the SB profile in central regions, were examined carefully. Seeing is accounted for by convolving the model profiles with a Gaussian PSF with a dispersion measured from field stars.

IR data are extremely useful for studying the central light, unobscured by the dust; profile decompositions are thus less likely to suffer from internal absorption effects (Phillipps *et al.* 1991). Furthermore, near-IR images are less influenced by star formation or starburst activity near the galaxy center.

APB95 showed that bulge profiles vary with Hubble types and are cor-

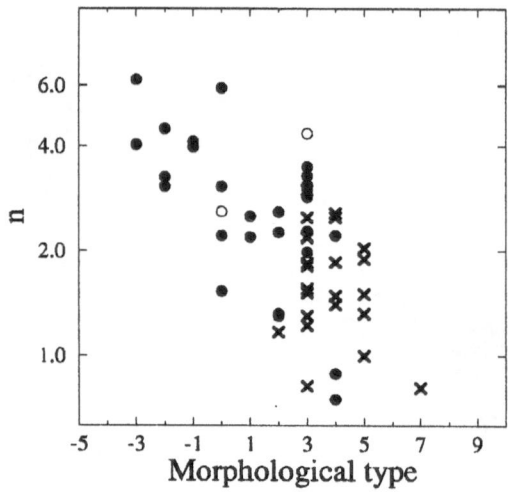

Figure 2. Best-fitted parameter *n* versus morphological type from Andredakis, Peletier, & Balcells (1995; APB). Filled circles consist of 30 S0-Sbc (T=-3 to 4) galaxies with $i > 50°$ from APB. The two open circles are barred galaxies. Crosses represent a small sample of Kent's (1986) photometry of late-type systems. Intrinsically brighter systems have greater *n*; the scatter in this figure is mostly likely correlated with luminosity.

related with B/D ratios, in good agreement with Davies *et al.* (1988) and Graham *et al.* (1996) for LSBs and spheroids (see Figure 2). CdJB96 provide supporting evidence with their larger sample. From examination of the reduced χ^2 values, they confirmed that most late-type spirals are best fitted by the double-exponential fit. Sa and Sab's are also generally best modeled with a $n = 2$ bulge. Central regions of earlier-type galaxies and of a small fraction ($\sim 15\%$) of the later types are more appropriately fit by a deVaucouleurs law. These results should serve to firmly establish the notion that central regions of late-type spirals are best described by an exponential profile.

Given that $n = 1$ for most late-type spirals, CdJB96 adopted double-exponential decompositions for all galaxies in their sample to compute a B/D scale length ratio. (Their results is unchanged if restricted to the subsample of pure $n = 1$ central profiles only.) Figure 3 show the measured scale lengths for the joint samples. Combining the two r-band data sets, CdJB96 find $r_b/r_d = 0.08 \pm 0.05$ while de Jong galaxies alone at K' yield $r_b/r_d = 0.09 \pm 0.04$ (not shown). Although dust is more conspicuous at r for central regions, r-band results statistically reproduce the same range of values than at K', though with wide error bars. Effects of dust are thus not

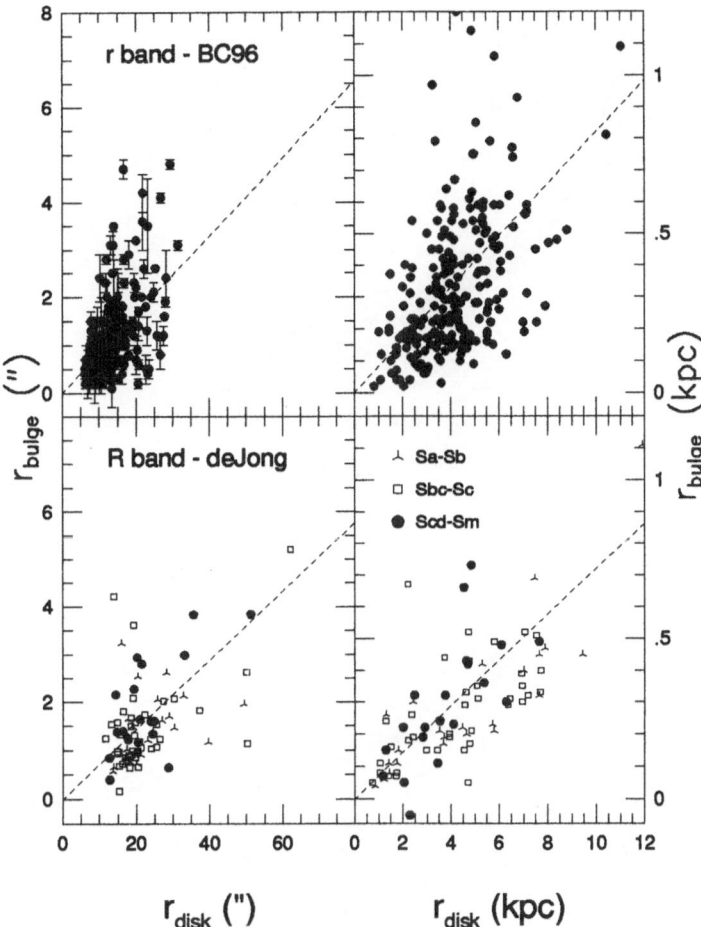

Figure 3.   Fitted scale lengths for the bulge and disk from the 1D decomposition of BC96 (top) and the 2D decomposition of de Jong (bottom). Figures on the left are plotted in arcseconds and the ones on the right use absolute coordinates. Apparent units show that the correlation is not affected by resolution effects, and the physical scale allows for a clearer comparison between the two samples. The dashed lines are fits to the data; their slope is, of course, distance independent. All SB profiles were fitted with a double-exponential.

alarming (they become severe at B or V).

Taken at face value, a correlation of B/D scale lengths is best understood in a model where the disk forms first and the "bulge" that naturally emerges is closely related to the disk. In a scenario where the bulge forms first, it would be hard to understand how a small dynamically hot component could

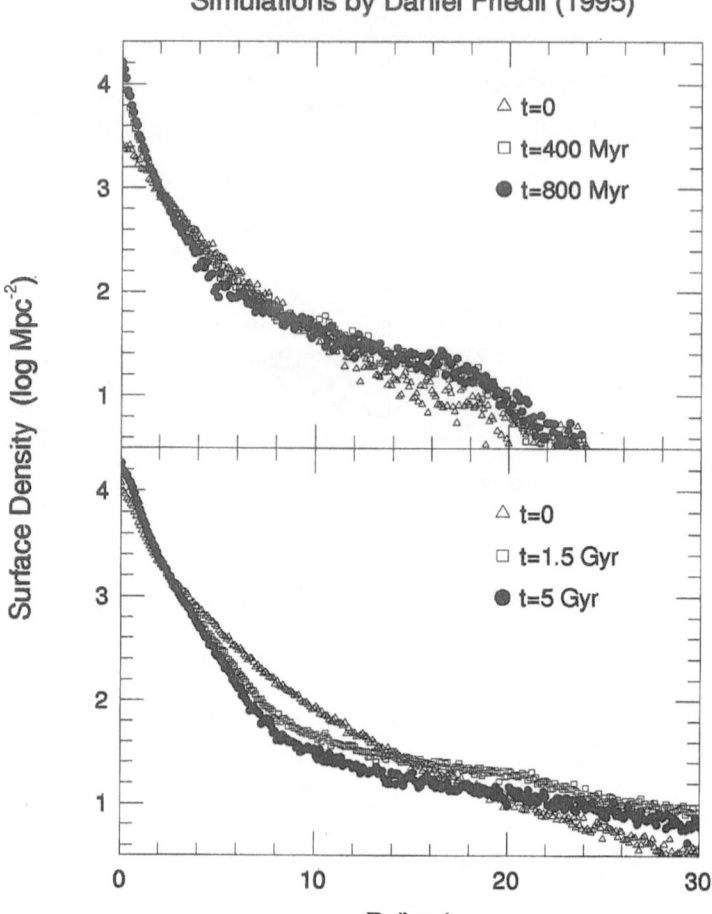

Figure 4. The secular evolution of the face-on stellar surface density profile is shown for two independent numerical simulations (top: model B_{no} by Friedli & Benz 1995. Bottom: model by Pfenniger & Friedli 1991). Both simulations originate from an axisymmetric (but bar unstable) Mihamoto-Nagai model and evolve toward an exponential bulge and disk. Though extensive testing of the simulations is still in progress, preliminary checks indicate values for the B/D scalelength ratio close to 0.09. Once the bar has formed, vertical heating of the central regions occurs on short timescales $\lesssim 1$ Gyr.

directly influence the disk global structure.

Self-consistent numerical simulations of secular evolution in disk galaxies evolve toward a double-exponential profile with a typical ratio between bulge and disk scale lengths near 0.1 (Friedli 1995, private communication). Two examples of such simulations are shown in Figure 4 (see also Fig. 7 of Norman, Sellwood, & Hasan 1996). It is interesting that some of these models naturally produce a Freeman (1970) type II disk profile. [3]

Subtraction of an elliptical profile fit from the original galaxy image also shows *residual spiral structure that extends all the way into the center of the galaxy* for the majority of Courteau's thesis sample (Courteau 1992). This provides further support for kinship between the disk and its central regions. (K93, Zaritzky, Rix & Rieke 1993). A bulge could not produce its own spiral structure.

Note that APB95 reject secular evolution on the basis of their continuous spectrum of the index n versus morphological type. They propose that the smooth sequence they observe (Fig. 2) can only result from a single mechanism of bulge formation. Given the large scatter in that diagram, such a conclusion seems ill-based. A scenario in which bright bulges (as in S0 and Sa's) form principally from a minor merger and smaller bulges (Sd's $\rightarrow$ Sab's) form mainly via secular evolution is not likely to leave any bi-modal imprint on the spectrum of "n" as both processes will operate to some degree of efficiency for all Hubble types.

4. Color Gradients

To the extent that dust and stellar population effects (age and metallicity) can be disentangled, the successful enterprise of dating disks and their central regions requires optical and IR photometry **and** line-strength gradients[4] (Searle *et al.* 1973, Frogel 1985, Evans 1992, Silva & Elston 1994, Peletier *et al.* 1994, Worthey 1994, de Jong 1995, Just, Fuchs, & Wielen 1996, Peletier & Balcells 1996a,b,c, hereafter collectively as PB96). Accurate photometry of radial profiles or wedge apertures on multicolor images of a disk galaxy indicate that optical color gradients in the disk are relatively small ($\leq 0\overset{m}{.}5$) in $B - V$ and $V - R$ but can be as large as 2 mags in $B - H$ or $B - K$ (Elmegreen & Elmegreen 1984, 1985; Courteau & Holtzman 1994, Peletier *et al.* 1994, de Jong 1995). This range is somewhat larger than that reported by Kent (1986) or Terndrup *et al.* (1994). Disk colors also vary greatly among late-spirals.

---

[3] Dust extinction will also conspire to making Type II disk profiles; see Evans (1992), Courteau and Holtzman (1994), Giovanelli *et al.* (1994).

[4] Ideally, one would like to use $H\alpha$ fluxes to measure the current star formation rate, CO $2.36 \mu m$ bandhead measurements to constrain the relative contributions of M dwarf, giant, and supergiant light (even in the absence of dust) (Silva 1996), global IRAS fluxes at $60 \mu m$ and $100 \mu m$ from warm dust heated primarily by O and B stars (Evans 1995, Devereux 1995), FAR-IR fluxes to measure the re-rediated stellar flux absorbed by cold dust, and Balmer spectroscopy without diffuse gas contamination plus the Calcium doublet and the 4000 Å break to lift the age-metallicity degeneracy, . Measurements of the $H - K$ color are necessary to constrain the contribution of intermediate-age AGB stars (Wise & Silva 1996, Silva 1996). Stellar population models which incorporate both age and metallicity effects are necessary.

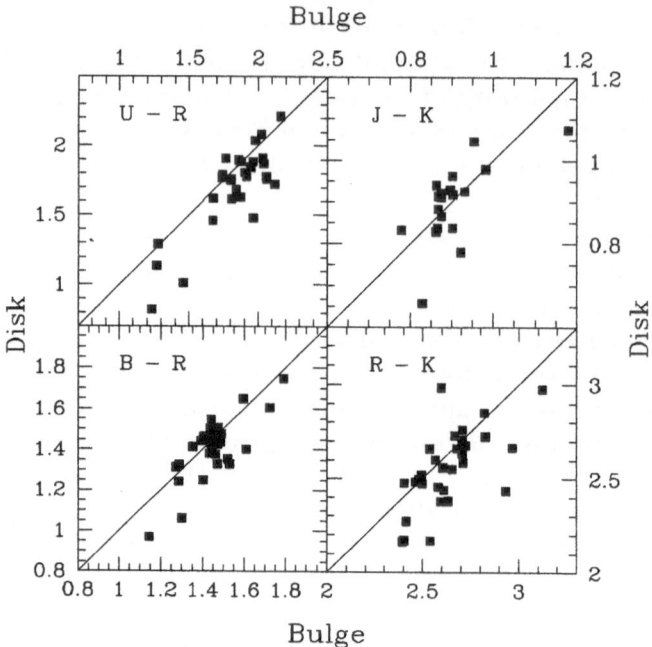

Figure 5.   Disk colors against bulge colors from Peletier & Balcells (1996). Bulge colors are measured at $r_e/2$ or $5''$, whichever is larger, and disk colors are measured at 2 major-axis scale lengths. Color differences are, on average, of order $0.^m1$ for all passband combinations. Central colors are measured in special regions where extinction is minimal.

Measurement and interpretation of disk *central* colors has essentially neglected the effects of dust reddening prior to the recent work of Balcells and Peletier (1994). Then followed the work of Terndrup *et al.* (1994), dJ95, PB96, and CH96 (see also the article by Frogel *et al.* in these proceedings). The samples of Peletier and Balcells, and dJ95 have been described above. Terndrup *et al.* (1994) observed 43 SO and later-type spirals at J and K with matching r-band photometry from Kent (1986) (see also Terndrup 1996). CH96 includes $BVRH$ color gradients for a few hundred Sb-Sc Northern galaxies.

A key question is whether the *stellar population* of the "bulge" is the same as, or related to, that of the disk. Collectively, the authors quoted above have shown, using local colors instead of integrated light, that *central regions of disks are bluer than ellipticals of the same luminosity* and reported *small, negative color gradients. The color difference between the inner disk and its central regions is very small* ($\sim 0.^m1$) *at all colors*. Figure 5 from PB96 summarizes this last statement well. Although PB96 realize that similar colors would support a model of secular evolution, they disregard that possibility on the basis that starbursts would be the only physical process able to build up the observed central space densities in disks. However, secular evolution with bars can manufacture "bulges" with a high degree of rotational support, for late-type systems at least.

PB96 examine the scale length variations in passbands sensitive or "immune" to dust (Evans 1994, Beckman *et al.* 1994) to show that dust effects appear to be small for galaxy types earlier than Sab. Thus for early-type systems, it is assumed that dust effects are negligible and effects of stellar population can be studied directly. Using the population synthesis models of Vazdekis *et al.* (1995), PB96 show that the age difference between the central regions and the inner disks of *early-type spirals* at 2 scale lengths is less than 30 %.

It is worth pointing out that current spectrophotometric population synthesis models for early-types agree poorly in the IR (Charlot, Worthey, & Bressan 1996; hereafter CWB96). These disagreements stem partly form the fact that i) temperatures and numbers of M giants are not well known, and ii) infrared atomic and especially molecular line opacities are not as accurately determined as in the optical. Author-to-author scatter can be large. Scatter in $V - K$ or $R - K$ model colors (*e.g.* see PB96, Table 3) suggests an age uncertainty of 35% even if all other parameters are determined (CWB96). The situation is even worse for younger populations (Charlot 1996). This should not be interpreted as the demise of color gradient studies but current investigations based on such models remain difficult to interpret. The presence of dust in late-type systems makes the analysis of colors + population synthesis models even harder (see e.g. Terndrup 1996). De Jong (dJ95) nonetheless attempted to combine his new 3D radiation transfer code with multiple dust geometries and stellar population models to explain his observed color gradients. His main conclusion, similar to PB96 for early-types, is that dust plays a minor role and the gradients are best explained by a combined stellar age and metallicity gradient across the disk.

In light of the large uncertainties inherent to this sort of population analysis (see CWB96 for details), the safest concluding remark is simply that the similarity between central and inner disk colors, and the fact that bulges are bluer than ellipticals of the same luminosity, are qualitatively consistent with a picture of inward gas transport and subsequent star formation from the inner disk to the center of the galaxy. Age dating of extragalactic bulges will require accurate line-strength gradients and far more robust and complete spectral evolution models (Charlot 1996).

5. Line Strengths, Kinematics and Chemical Evolution

Although bulges are known to be bluer than ellipticals of equal absolute luminosity, it has also been known for some time that some "bulges", mostly in barred galaxies, have smaller velocity dispersion than ellipticals of the same M_B (Whitmore, Kirshner, & Schechter 1979, Whitmore & Kirshner 1981, Kormendy & Illingworth 1983, K93). Most of these systems are actually rapid rotators, as shown in the classic $V_{max}/\sigma - \epsilon$ diagram (Kormendy 1985, K93 and references therein). *Unlike ellipticals, bulges' kinetic energy comes mostly in rotation which must be imparted from the disk,* in agreement with a causal link between the disk and the "bulge".

Chemical evolution is also likely to be affected by gas flows in the center of galaxies. For galaxies undergoing central star formation bursts, Friedli *et al.* 1994 show that the gas-phase abundance gradient should be characterized by two separate slopes corresponding to the inner and outer regions of the disk. A metallicity-velocity dispersion relation for the core is expected as well, though current nuclear stellar abundances are too uncertain to provide conclusive evidence (Friedli & Benz 1995). If metallicity is controlled by the depth of the potential well, low-L disks, which will only create small "bulges", will then have low dispersions and low metallicities (augmented perhaps by self-enrichment during star formation that follows inward transport of disk material). Few studies have addressed the issue of abundances and existence of a fundamental plane for the center of extragalactic disks. Boroson (1980) measured central metallicities (Mg_2 index) in the bulge of 24 spirals and found a greater correlation with bulge light than with total light though with poor statistics. A similar correlation and matching of Mg_2 with central velocity dispersion by Jablonka *et al.* (1995) also suggest that bulges and ellipticals would occupy the same locus in the fundamental plane (Bender *et al.* 1992, 1993). Such a picture, if true, would suggest a decoupling of the bulge and disk, in stark contrast with models of secular evolution.

Sil'chenko (1993) studied Fe I 5270, Mg I 5175 absorption features in the central regions of many early and late-type spirals and finds that [Mg/Fe] is mostly solar, contrary to the observation of Mg to Fe enrichment in ellipticals, SO galaxies (Fisher, Franx & Illingworth 1996) and our own Galactic Bulge (McWilliam & Rich 1994). Aided by the Balmer/Ca II age diagnostic, she inferred that disk nuclei contain a significant number of stars less than 5×10^9 years old. However, Sil'chenko's data set contains only a small number of late-types and her measurements of a mean Balmer equivalent width may be spoiled by diffuse ionized hydrogen emission at $H\beta$. In comparing trends in Mg_2-σ, Jablonka *et al.* are also limited to a small number of late-type spirals (8) and, unlike Sil'chenko, they do not distinguish between age, metallicity and luminosity effects. Note that Fisher, Franx & Illingworth (1996) find comparable bulge and disk ages for SO galaxies but reject formation scenarios in which bulges formed from heated disk or accreted material at late times. However, their analysis may also suffer, like Sil'chenko's, from the effects of emission contamination at $H\beta$.

At the moment, the status of absorption lines studies for central regions of late-type disks is inconclusive, but suggestive of the co-existence of an old and intermediate-age population. This would be expected, for example, if an old bulge was first formed by dissipational collapse of the primordial gas, and would be later enriched by inward transport of disk material.

Recent HST studies in Baade's window by Ortolani et al. (1995) and Rich (1992) and McWilliam & Rich (1994) suggest that *the Milky Way bulge is as old as 47 Tuc and possibly populated by intermediate age stars with some amount of ongoing star formation* (see also Renzini 1996). Minniti et al. (1996) show that there is a large spread of metal abundances in the Galactic bulge which could be interpreted as mixing of different components in the inner disk. Finally, Rich (1996, and this conference) interprets the existence of extended giant branches in the bulge of M31, M32, and M33 for a younger population.

6. Conclusion

Several simple observational arguments have been presented in favor of models of inward gas transfer for late-type spirals. These are:

- Central surface density best described by an exponential profile
- Restricted range of bulge and disk scalelengths
- Blueness of the central regions and small color difference with the inner disk.
- Rapid rotation and small dispersion for central regions.
- Co-existence of old and younger populations in the center of our Milky Way and other nearby galaxies.

Future tests should include accurate measurement of stellar abundances in extragalactic bulges to unravel the stellar populations and test models of star formation and chemical enrichment (Steinmetz & Müller 1994, Friedli, Benz, & Kennicutt 1994, Friedli and Benz 1995, Mollá & Ferrini 1995, Zaritzky 1995)

7. Acknowledgments

I wish to thank my collaborators, Adrick Broeils and Roelof de Jong for their comments on an early draft of this paper and permission to reproduce joint results. I am indebted to Daniel Friedli for sending me various simulations and for valuable comments. Dante Minitti was kind enough to provide many contributions in advance of publication from his ESO proceedings edited with Hans-Walter Rix, "Spiral Galaxies in the Near-Ir". This work also benefited from conversations with Sandra Faber, John Kormendy, Reynier Peletier, Daniel Pfenniger, and Guy Worthey.

References

Andredakis, Y. C., & Sanders, R. H. 1994, MNRAS, 267, 283
Andredakis, Y. C., Peletier, R. F., & Balcells, M., 1995, MNRAS, in press [APB95]
Balcells, M., & Peletier, R. F. 1994, AJ, 107, 135
Beckman. J. E., Peletier, R. F., Knapen, J. H., Mate, M. J., & Gentet, L. J. 1994, in *The Opacity in Spiral Disks*, ed. J.I. Dãvies & D. Burstein, (Nato Series; Kluwer:Dordrecht), 197
Bender, R., Burstein, D., & Faber, S. M. 1992, ApJ, 399, 462
Bender, R., Burstein, D., & Faber, S. M. 1993, ApJ, 411, 153
Block, D. L., & Wainscoat, R. J. 1991, Nature, 353, 48
Boroson, T. A. 1980, Ph.D. thesis, Univ. Arizona
Broeils, A. H., & Courteau, S. 1996, in preparation [BC96]
Burstein, D. 1979, ApJ, 234, 829
Byun, Y. I., & Freeman, K. C. 1995, ApJ, 448, 563
Caon, N., Capaccioli, M., & D'Onofrio, M. 1993, MNRAS, 265, 1013
Charlot, S. 1996, in ASP Conf. Ser., *From Stars to Galaxies*, ed. C. Leitherer & U. Fritze-von Alvensleben (San Francisco: ASP), in press
Charlot, S., Worthey, G., & Bressan, A. 1996, ApJ, in press [CWB96]
Combes, F., Debbash, F., Friedli, D., & Pfenniger, D. 1990, A&A, 233, 82
Courteau, S. 1996, ApJS, 103, March issue [C96]
Courteau, S. 1992, Ph.D. thesis, Univ. California, Santa Cruz
Courteau, S., de Jong, R. S., & Broeils, A. H. 1996, ApJ, 457, 73 [CdJB96]
Courteau, S., & Broeils, A. H. 1996, in preparation
Courteau, S., & Holtzman, J. 1994, in *The Opacity in Spiral Disks*, ed. J.I. Davies & D. Burstein, (Nato Series; Kluwer:Dordrecht), 211
Courteau, S., & Holtzman, J. 1996, in preparation [CH96]
Davies, J. I., Phillipps, S., Cawson, M. G. M., Disney, M. J., & Kibblehwite, E. J. 1988, MNRAS, 232, 239
Davies, R. L., Frogel, J. A., & Terndrup, D. M. 1992, AJ, 102, 1729
de Jong, R. S. 1996, A&A, in press
de Jong, R. S. 1995, Ph.D. thesis, Univ. Groningen [dJ95]
de Jong, R. S., & van der Kruit, P.C. 1994, A&AS, 106, 451
de Vaucouleurs, G. 1948, Ann. d'Astrophysique, 11, 247
de Vaucouleurs, G. 1959, Hand. der Physik, 53, 311
Devereux, N. 1995, in *The Opacity in Spiral Disks*, ed. J.I. Davies & D. Burstein, (Nato Series; Kluwer:Dordrecht), 269
Eggen, O. J., Lynden-Bell, D., & Sandage, A. 1962, ApJ, 136, 748 [ELS]
Elmegreen, B. G., and Elmegreen, D. M. 1985, ApJ, 288, 438

Elmegreen, D. M., and Elmegreen, B. G. 1984, ApJS, 54, 127

Evans, Rh. 1992, Ph.D. thesis, Univ. Wales, College of Cardiff

Evans, Rh. 1994 MNRAS, 266, 511

Evans, Rh. 1995, in *The Opacity in Spiral Disks*, ed. J.I. Davies & D. Burstein, (Nato Series; Kluwer:Dordrecht), 281

Faber, S. M., Worthey, G., & González, J. J. 1992, in IAU Symp. 149, *The Stellar Populations of Galaxies*, ed. B. Barbuy & A. Renzini (Dortrecht: Kluwer), 255

Fisher, D., Franx, M., & Illingworth, G. 1996, ApJ, 459, 110

Frankston, M. & Schild, R. 1976, AJ, 81, 500

Freeman, K. C. 1970, ApJ, 160, 811

Friedli, D. 1994, in *Mass-Transfer Induced Activity in Galaxies*, ed. I. Shlosman, Cambridge Univ. Press, 268

Friedli, D. & Benz, W. 1995, A&A, 301, 649

Friedli, D. & Benz, W. 1993, A&A, 268, 65 [FB93]

Friedli, D., Benz, W., & Kennicutt, R. 1994, ApJ, 430, L105

Frogel, J. A. 1985, ApJ, 298, 528

Giovanelli, R., Haynes, M. P., Salzer, J. J., Wegner, G., da Costa, L. N., & Freudling, W. 1994, AJ, 107, 2036

González, J. J., & Gorgas, J. 1995, in ASP Conf. Ser., *Fresh Views on Ellipticals*, ed. A. Buzzoni, A. Renzini, & A. Serrano (San Francisco: ASP), 225

Graham, A., Lauer, T. R., Colless, M., & Postman, M. 1996, ApJ, submitted.

Hohl, F., 1971, ApJ, 168, 343

Holtzman, J. A., *et al.* 1993, AJ, 106, 1826

Jablonka, P., Martin, P., & Arimoto, N. 1995, in ASP Conf. Ser., *Fresh Views on Ellipticals*, ed. A. Buzzoni, A. Renzini, & A. Serrano (San Francisco: ASP), 185

Just, A., Fuchs, B., & Wielen, R. 1996, A&A, submitted

Kent, S. M. 1985, ApJS, 59, 115. 301

Kent, S. M. 1986, AJ, 93, 1301

Kent, S. M., Dame, T., & Fazio, G. 1991, ApJ, 378, 131

Kormendy, J. 1993, in IAU Symp. 153, *Galactic Bulges*, ed. H. Dejonghe & H. J. Habing (Kluwer:Dordrecht), 209 [K93]

Kormendy, J. 1985, ApJ, 295, 73

Kormendy, J. 1977, ApJ, 217, 406

Kormendy, J. & Bruzual, A. 1978, ApJ, 223, L63

Kormendy, J. & Illingworth, G. 1983, ApJ, 265, 632

Lauer, T. R., & Postman, M. 1994, ApJ, 425, 418

Lin, D. N. C., & Pringle, J. E. 1987, ApJ, 320, L87

Martin, P. 1995, AJ, 109, 2428

Martinet, L. 1995, Fund. Cosmic Physics, 15, 341 [M95]

McWilliam, A. & Rich, R. M. 1994, ApJS, 91, 749

Mollá, M., & Ferrini, F. 1995, ApJ, 454, 726

Olivier, S. S., Blumenthal, G. R., & Primack, J. R. 1991, MNRAS, 252, 102

Peletier, R. F., & Balcells, M. 1996a, in *Evolutionary Phenomena in Galaxies*, ed. R. Bender & R.L. Davies ((Kluwer:Dordrecht), in press.

Peletier, R. F., & Balcells, M. 1996b, in *Spiral Galaxies in the Near-Infrared*, ed. D. Minniti & H. W. Rix (Springer-Verlag:Berlin), in press

Peletier, R. F., & Balcells, M. 1996c, submitted to A&A.

Peletier, R. F., Valentijn, E. A., Moorwood, A. F. M., & Freudling, W. 1994, A&AS, 108, 621

Pfenniger, D. 1991, in *Dynamics of Disc Galaxies*, ed. B. Sundelius, Göteborg, 191

Pfenniger, D. 1993, in *Galactic Bulges*, ed. H. Dejonghe and H. J. Habing (Kluwer:Dordrecht), 387 [P93]

Pfenniger, D., & Friedli, D. 1991, A&A, 252, 75

Pfenniger, D., & Norman, C. 1990, ApJ, 363, 391

Phillipps, S., Evans, Rh., Davies, J. I., & Disney, M. J. 1991, MNRAS, 253, 496

Rich, R. M., Mould, J. R., & Graham, J. 1993, AJ, 106, 2252
Renzini, A. 1993, in *Galactic Bulges*, ed. H. Dejonghe & H. J. Habing (Kluwer:Dordrecht), 151
Saio, H., & Yoshii, Y. 1990, ApJ, 363, 40
Sandage, A. 1986 Ann. Rev. Astron. Ap 24, 421
Schweizer, F. 1990, in *Dynamics and Interactions of Galaxies*, ed. R. Wielen (Springer-Verlag:Berlin), 60
Searle, L., Sargent, W. L. W., & Bagnuolo, W. G. 1973, ApJ, 179, 427.
Sellwood, J. A., & Wilkinson, A. 1993, Rep. Prog. Phys., 56, 173
Sércic, J.-L. 1968, Atlas de galaxias australes (Observatorio Astronomica, Cordoba)
Schombert, J. M., & Bothun, G. D. 1987, AJ, 93, 60
Shaw, M. A., & Gilmore, G. 1989, MNRAS, 237, 903
Sil'chenko, O. K. 1993a, Astron. Lett., 19, 279
Sil'chenko, O. K. 1993b, Astron. Lett., 19, 283
Silk, J.,and Norman, C. 1981, ApJ, 247, 59
Silva, D. R., & Elston, R. 1994, ApJ, 428, 511
Silva, D. R. 1996, in *Spiral Galaxies in the Near-Infrared*, ed. D. Minniti & H. W. Rix (Springer-Verlag:Berlin), in press
Steinmetz, M. & Müller, E. 1994, A&A, 281, L97
Struck-Marcell, C. 1991, ApJ, 368, 348 [SM91]
Terndrup, D. M. 1996, in *Spiral Galaxies in the Near-Infrared*, ed. D. Minniti & H. W. Rix (Springer-Verlag:Berlin), in press
Terndrup, D. M., Davies, R. L., Frogel, J. A., DePoy, D. L., & Wells, L. A. 1994, ApJ, 432, 518
van Houten, C. J. 1961, Bull. Astron. Inst. Neth., 16, 1
Vazdekis, A., Peletier, R. F., Casuso, E., & Beckman, J. 1996, in *Spiral Galaxies in the Near-Infrared*, ed. D. Minniti & H. W. Rix (Springer-Verlag:Berlin), in press
von Linden, S., Lesch, H., & Combes, F. 1996, A&A, submitted
Whitmore, B. C., Kirshner, R. P., & Schechter, P. L. 1979, ApJ, 234, 68
Whitmore, B. C., & Kirshner, R. P. 1981, ApJ, 250, 43
Wise, M. W., & Silva, D. R. 1996, ApJ, in print
Witt, A. N. & Gordon, K. D. 1996, ApJ, June issue.
Worthey, G. 1994, ApJS, 95, 107
Worthey, G. 1996, in *Evolutionary Phenomena in Galaxies*, ed. R. Bender & R.L. Davies ((Kluwer:Dordrecht), in press
Wyse, R. F. G. 1995, in IAU Symp. 164, *Stellar Populations*, ed. P. C. van der Kruit & G. Gilmore (Dortrecht: Kluwer), 133
Yoshii, Y., & Sommer-Larsen, J. 1989, MNRAS, 236, 779
Zaritsky, D., Rix, H.-W., & Rieke, M. 1993, Nature, 364, 313
Zaritsky, D. 1995, ApJ, 448, 17.

DUST IN STARBURST GALAXIES:
FROM THE ULTRAVIOLET TO THE NEAR INFRARED

D. CALZETTI

Space Telescope Science Institute
3700 San Martin Drive, Baltimore, MD 21218, USA

Abstract. A study of the dust reddening in starburst galaxies is presented. The wavelength dependence of the dust extinction is exploited by using multiwavelength data (ultraviolet, optical, and near infrared) to investigate the effects of dust on the emerging radiation. The main conclusion is that dust foreground to the starburst region (but not necessarily foreground to the galaxy) can account for most of the reddening. This simple configuration doesn't need to correspond to a physical distribution of the dust inside the starburst region; it is a convenient way to correct the radiation from galaxies for the effects of obscuration.

1. Introduction

Whenever we want to investigate the stellar and gas content of a galaxy using ultraviolet or optical data, we face the problem of how to correct the emerging radiation for the dust obscuration. The obscuration depends on the optical properties of the dust grains and on the geometrical distribution of the dust inside the galaxy. As an added complication, the galaxy intrinsic stellar population is usually unknown, therefore the zero-order term for the comparison with the observed light is missing.

Classes of galaxies whose stellar populations have similar properties offer the best chance to understand the characteristics of dust obscuration and pin down systematic effects. We concentrate here on galaxies undergoing bursts of star formation. The massive stars produced in the burst dominate the ultraviolet emission from the galaxy, and ionize the interstellar gas. Multiwavelength information on the stellar continuum and nebular emission

D. L. Block and J. M. Greenberg (eds.), New Extragalactic Perspectives in the New South Africa, 271–274.

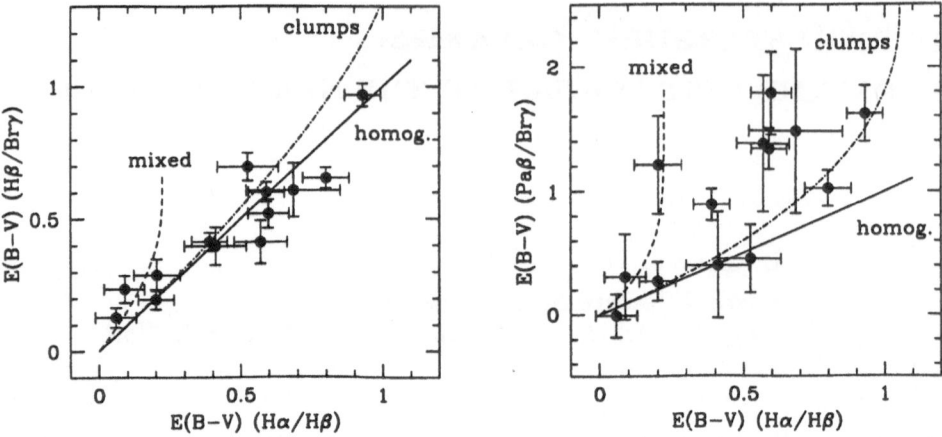

Figure 1. The color excess of the nebular hydrogen emission at optical and infrared wavelengths. Three models are superimposed to the data: a foreground homogeneous screen (continuous line), a foreground clumpy distribution (dot-dashed line), and a homogeneous mixture of dust and gas (internal dust, dashed line).

can be used to constrain the reddening which affects the massive stars and the ionized gas.

2. Data and Results

Infrared J, H, and K spectra and images, and optical and ultraviolet spectra (Calzetti et al. 1996; Calzetti 1996; McQuade et al. 1995; Storchi-Bergmann et al. 1995; Calzetti et al. 1994; Kinney et al. 1993), are used to investigate the dust extinction characteristics of 19 starburst galaxies. The data cover the wavelength range 0.12-2.25 μ.

Values of the color excess E(B−V) are derived from the observed ratios of the hydrogen recombination lines at different wavelengths, namely Hα/Hβ, Hβ/Brγ, and Paβ/Brγ, and are used to constrain the geometry of the dust obscuring the gas (Puxley 1991, Puxley & Brand 1994, Calzetti et al. 1996). Figure 1 shows that the obscuration characteristics of the ionized gas are compatible with a foreground dust distribution, either clumpy or homogeneous. A model with homogeneously mixed dust and gas is inadequate to explain most of the data.

Stars and gas are not necessarily affected by the same dust extinction, since they may have different geometrical distributions within a galaxy.

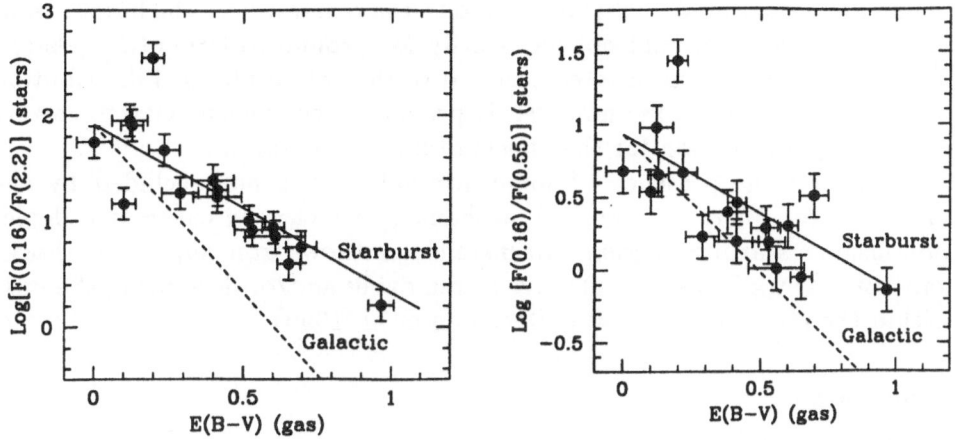

Figure 2. Ratio of the stellar continuum fluxes F(0.16 μ)/F(2.2 μ) (left panel), and F(0.16 μ)/F(0.55 μ) (right panel), as a function of the color excess E(B−V) of the ionized gas. The predicted flux ratios for a foreground dust distribution, using both the Starbust obscuration curve (Calzetti et al. 1994, continuous line), and the Galactic extinction curve (homogeneous screen, dashed line) are shown. The zero-extinction stellar continuum is given by the combination of a stellar population undergoing constant star formation for 2×10^7 yr and a 10^{10} yr old underlying stellar population.

Calzetti et al. (1994; see also Mayya & Prabhu 1996, Lançon et al. 1996, Fanelli et al. 1988) find that the UV continuum emission from starburst galaxies is less obscured, by roughly a factor 2, than the emission from the ionized gas. However, stellar and gas reddening trace each other, since the reddening of the stellar continuum correlates with the reddening of the ionized gas. The present multiwavelength data confirm those results. Figure 2 shows that the ratio of the UV to the K-band or optical stellar continuum correlates with the color excess of the ionized gas. The correlation is thus produced by the dust reddening of the UV stellar continuum and of the nebular hydrogen lines. The data are compared with two models for the foreground dust: one which uses the Starburst obscuration curve derived by Calzetti et al. (1994) and one which uses the Galactic extinction curve with an homogeneous screen (see Figure 2). The zero-extinction flux ratios are given by a stellar mixture of a 2×10^7 yr old continuous star formation event and a 10^{10} yr old population, producing 20% and 80% of the K-band flux, respectively. The mixture is derived from the observed equivalent widths of the Hα and Brγ emission lines, which have on average values around EW(Hα)~90 Åand EW(Brγ)~10 Å(Leitherer & Heckman 1995).

3. Conclusions

The reddening of the radiation from starburst galaxies appears well explained by foreground dust (foreground to the region where the burst of star
formation is located, and not necessarily foreground to the entire galaxy).
The model does not need to correspond to the actual physical distribution
of dust inside the galaxy; however, it provides a convenient way to correct
the emerging radiation for the effects of dust obscuration.

Although the presence of some internal dust is not excluded by the
present data , there are physical mechanisms which may cause dust depletion inside a starburst region. For instance, outflows from supernova shocks
and hot star winds can be effective at removing and/or destroying the dust
within the star-formation site (Heckman et al. 1990).

References

Calzetti, D., 1996, in preparation.
Calzetti, D., Kinney, A.L., & Storchi-Bergmann, T. 1994, *ApJ*, **429**, 582.
Calzetti, D., Kinney, A.L., & Storchi-Bergmann, T. 1996, *ApJ*, **458**, 132.
Carico, D.P., Soifer, B.T., Beichman, C., Elias, J.H., Matthews, K., & Neugebauer, G.,
      1986, *AJ*, **92**, 1254.
Fanelli, M.N., O'Connell, R.W., & Thuan, T.X., 1988, *ApJ*, **334**, 665.
Heckman, T.M., Armus, L., & Miley, G.K., 1990, *ApJS*, **74**, 833.
Kinney, A., Bohlin, R., Calzetti, D., Panagia, N.,& Wyse, R.F.G., 1993, *ApJS*, **86**, 5.
Lançon, A., Rocca-Volmerange, B., & Thuan, T.X., 1996, *A&A*, in press.
Leitherer, C., & Heckman, T.M., 1995, *ApJS*, **96**, 9.
Mayya, Y.D., & Prabhu, T.P., 1996, *AJ*, in press.
McQuade, K., Calzetti, D., & Kinney, A.L., 1995, *ApJS*, **97**, 331.
Puxley, P.J. 1991, *MNRAS*, **249**, 11P.
Puxley, P.J., & Brand, P.W.J.L. 1994, *MNRAS*, **266**, 431.
Storchi-Bergmann, T., Kinney, A.L., & Challis, P., 1995,*ApJS*, **98**, 103.

NEAR-INFRARED SURFACE PHOTOMETRY OF BARRED SPIRAL GALAXIES

DEBRA MELOY ELMEGREEN
Vassar College, Department of Physics & Astronomy,
Poughkeepsie, NY 12601 USA

Abstract. Photometric properties of barred galaxies in BIJHK bands are compared in early and late Hubble types. Early-type bars have a flat light distribution and are large relative to the optical disk, whereas late-type bars have an exponential light distribution and are shorter. Isophotal twists in the central regions are observed only in early-type galaxies, which may be related to the presence of an inner Lindblad resonance. Bar-interbar contrasts increase steadily in all galaxies. Arm-interarm contrasts increase out to mid-disk (near corotation) and then level off, near bar's end for flat bars but beyond the bar for exponential bars.

1. Introduction

The morphology of spiral galaxies may be influenced by the presence of central bars. Strong bars, which are associated with early Hubble types, can lead to mass inflow followed by enhanced star formation, and may also generate grand design spiral structure. The photometric properties of barred galaxies vary with Hubble type. Here we present optical and near-infrared observations in a variety of early and late barred spiral galaxies in order to examine the central parts of the bars, the light distribution along the bars, and the arm and interarm surface brightnesses in order to compare their properties.

2. Observations

Near-infrared (JHK) images of 12 galaxies were made with the 1.3-meter telescope and of 9 galaxies in B and I with the Burrell-Schmidt 0.6-meter

D. L. Block and J. M. Greenberg (eds.), New Extragalactic Perspectives in the New South Africa, 275–282.
© *1996 Kluwer Academic Publishers.*

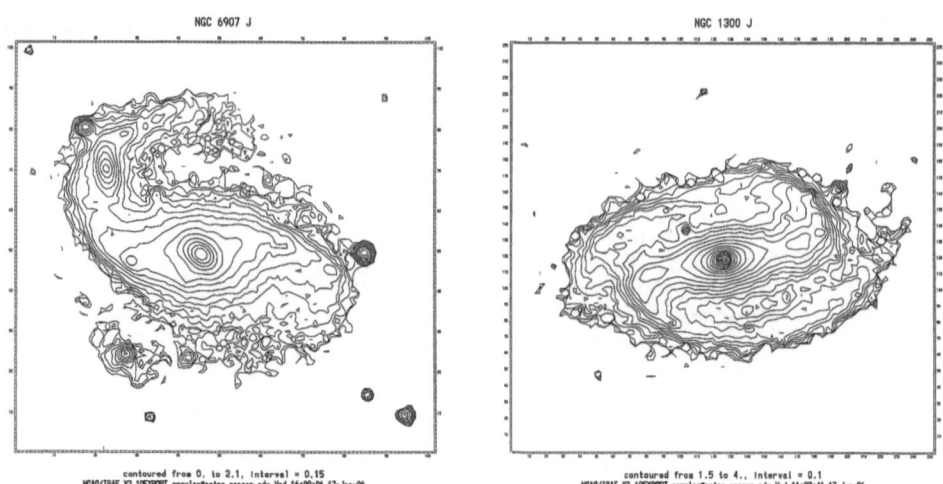

Figure 1. Contours are shown on logarithmic J band images of (a) NGC 6907 and (b) NGC 1300.

telescope at Kitt Peak National Observatory in order to compare morphology and surface photometry of early and late-type barred galaxies. In addition, we have compiled published JHK data on 80 galaxies to compare their properties as a function of Hubble type, and have examined high resolution atlas images.

3. Isophotal Twists

Isophotal twists are commonly observed in the central regions of barred galaxies, and may be signposts for gas inflow which is related to mechanisms powering high star formation activity (Shlosman et al. 1989, Pfenniger & Norman 1990, Friedli & Martinet 1993, Shaw et al. 1993). Figure 1 shows contour plots of the J band images for two early-type galaxies: NGC 6907, which has a twist, and NGC 1300, which has no twist. Twists appear to be common in early-type galaxies but absent in late-types.

The twist morphology may reveal something of the nature of the twist: position angles which change gradually with radius may represent triaxial bulges, while position angles with abrupt changes may be bars-within-bars, each with discrete pattern speeds (Wozniak et al. 1995; Pfenniger & Norman 1990, Friedli & Martinet 1993), or stellar orbits offset from bars because of gas trapping which affects their motions near the bars (Shaw et al. 1993;

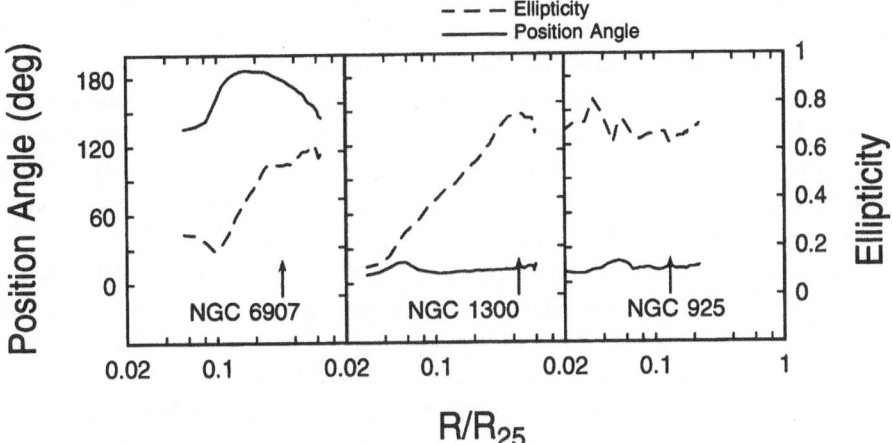

Figure 2. Ellipticities and position angles are shown as a function of radius for (a) NGC 6907, a flat-barred galaxy with a twist; (b) NGC 1300, a flat-barred galaxy with no twist; and (c) NGC925, an exponential-barred galaxy with no twist.

Combes & Gerin 1985).

Figure 2 shows ellipticities and position angles versus radius for NGC 6907, NGC 1300, and NGC 925 based on the IRAF task ELLIPSE (cf. Jedrzejewski 1987). We find that the ellipticities increase with radius throughout the bar length for early-type bars, whereas the ellipticities are nearly constant within late-type bars. This result may be due to stellar orbits and the locations of the bars with respect to resonances. The maximum change in position angle occurs at approximately the same radius as the maximum ellipticity.

There are 80 galaxies in the literature with JHK observations (Jarvis et al. 1988, Kormendy 1983, Buta 1990, Buta & Crocker 1993, Wozniak et al. 1995, Quillen et al. 1994, Pompea & Rieke 1990, Bushouse & Stanford 1992), of which 51 have isophotal twists of more than 10 degrees (see Elmegreen et al. 1996a). Another two dozen galaxies in published atlases (Sandage & Bedke 1994) appear to have twists, at least in their B images. Of these, about half of the galaxies earlier than SBbc have twists; none of the later-type galaxies have twists. This lack of twisting in late-type galaxy bars may be due to the lack of substantial bulges in these galaxies.

4. Flat and Exponential Bars

Early Hubble types have bars that are large relative to the optical size of the galaxy, whereas late Hubble types have relatively smaller bars (Elmegreen & Elmegreen 1985; hereafter EE85; Martin 1995). Early-type flat bars end at 0.3-$0.6R_{25}$, whereas late-type bars end at 0.1-$0.3R_{25}$. The relative bar length is independent of absolute blue magnitude within early or late types. The dichotomy occurs around Hubble type SBbc. Early-type bars have a light distribution which is nearly constant along their length, so they are called "flat" bars, whereas late-type bars have a light distribution which decreases exponentially with nearly the same scale length as the disk, so they are called "exponential" bars.

Figure 3 shows J and K band radial profiles for NGC 1300, a flat-barred galaxy, and NGC 2500, an exponential-barred galaxy. Profiles are shown perpendicular to the bar and along the bar. The similar flatness along the bar of the J profile compared to the K profile in NGC 1300 indicates that the color remains approximately constant along the bar. This result suggests that the old and young stars are gathered by orbit trapping. B band images show not just a flatness but sometimes an increase at bar's end, presumably because of enhanced star formation there. Azimuthally-averaged radial profiles are exponential even in flat-barred galaxies, because the arm and bar excess is balanced by an interarm deficit.

Exponential bars show a decline in their light which is very similar to that of the underlying disk; on a logarithmic scale such as in Figure 3, it is difficult even to distinguish the bar. In most galaxies, the disk scale length is the same in J and K within the measuring errors. The disk scale lengths for spiral arms are slightly longer than for the interarm regions.

Previous studies showed that the ends of flat bars are near or beyond the ends of the rising parts of the rotation curves, while the exponential bars fall short of the turnovers in the curves (EE85). The ends of bars may be related to the locations of resonances in the disks. Flat bars apparently end between the inner 4:1 resonance and corotation (EE85, Elmegreen & Elmegreen 1995), which is in accord with orbit theory (Contopoulos 1980). In contrast, exponential bars evidently end sooner, near or possibly inside the inner 4:1 resonance.

5. Arms in Barred Galaxies

Spiral arms have different morphologies in flat and exponential-barred galaxies; flat bars are associated with grand design structure, such as in NGC 1300, whereas exponential bars tend to have flocculent arms, such as in NGC 1359. Even K-band images of late-type bars still show no spiral arm symmetry. In the former case, the bars appear to be driving the structure,

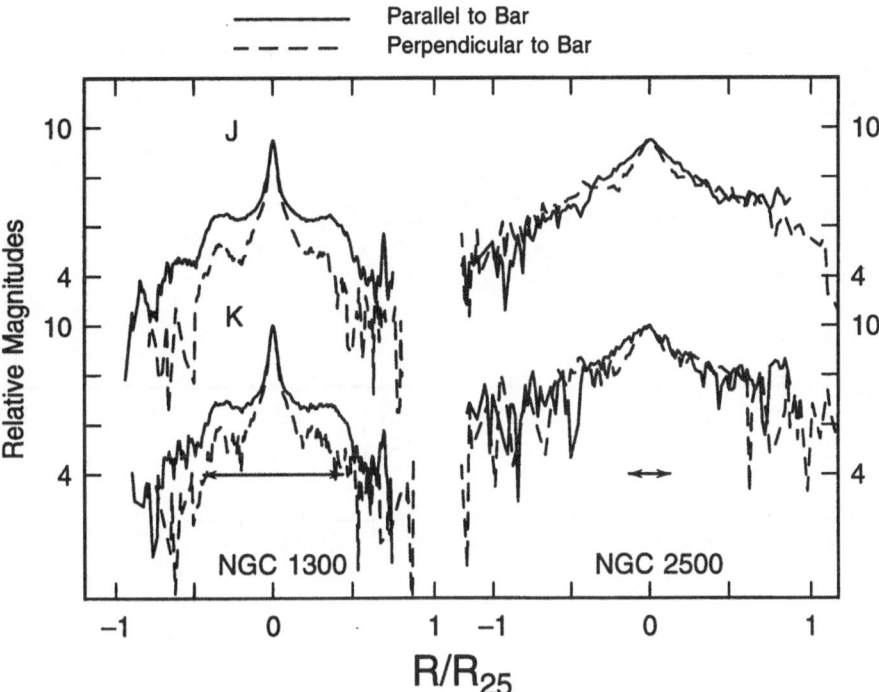

Figure 3. Radial profiles along and perpendicular to the bars are shown for (a) NGC 1300 and (b) NGC 2500.

while in the latter, the arms are not affected by the presence of a small bar (Elmegreen & Elmegreen 1985). This result is consistent with our previous survey (Elmegreen & Elmegreen 1989), in which 80% of the barred early-type galaxies have a grand design, compared with 45% of non-barred galaxies. Intermediate and late-type galaxies have 60% and 5% grand design, respectively, regardless of the presence of a bar.

Figure 4 shows the J band contrast between the bar and the interbar region and the arm and interarm regions as a function of radius, averaged over the 17 galaxies included in the survey (see Elmegreen et al. 1996b for details). The errors are 0.1 mag in the inner regions and 0.5 mag in the outer. Arm-interarm and bar-interbar contrasts are similar in all barred galaxies: the bar-interbar contrasts increase with increasing radius; the arm-interarm contrasts increase till mid-way in the disks, then remain fairly constant. This trend is present in the B, I, and K bands also. The trend

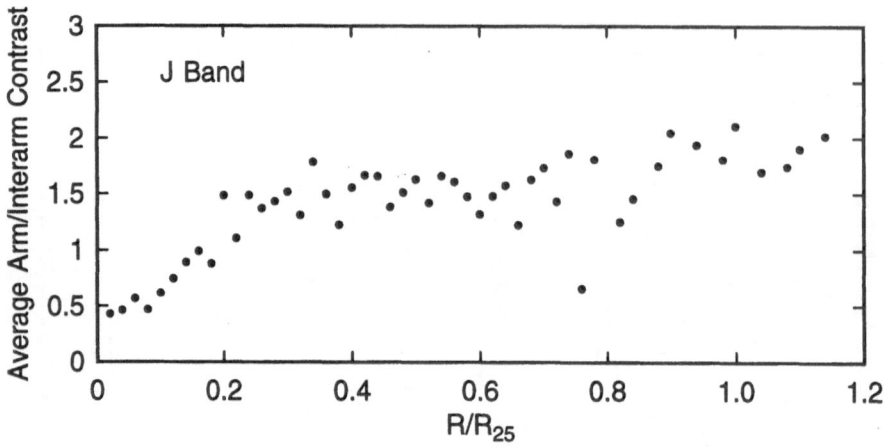

Figure 4. The average value of the bar-interbar and arm-interarm contrast is shown as a function of radius based on J band images of 17 galaxies.

of increasing amplitude in B and I bands was previously observed for non-barred and exponential-barred spirals (see Elmegreen & Elmegreen 1984, 1985), but the levelling off had only been noted for flat-barred galaxies. The results in the near-infrared are significant since they reflect the older disk population and not just the prominent star-forming regions that can dominate B band measurements.

Evidently the arms reach a maximum amplitude of 1.5-2 mag greater than the interarm; these results suggest that nonlinear density waves saturate. The peak near mid-disk is at the same radial location as the beginning of bifurcations in galaxies with inner symmetry, noted by Elmegreen & Elmegreen (1995). This location is near corotation in all cases. For flat bars, which end near corotation, the peak occurs near bar's end; for exponential bars, which are relatively smaller, the peak is well beyond bar's end. It is not clear whether the amplitude peaks are causally related to corotation or bars or bifurcations. The levelling off may have to do with density waves being reinforced out to corotation, after which their strength decreases (Lin & Lowe 1990).

6. Conclusions

We find that early and late barred Hubble types have several differences in morphology. In the central regions, early-type bars tend to show central twists, whereas late-type bars do not appear to have twists. In the bars, the radial profiles are nearly flat for early-type bars, but are exponential for late-type bars. The disk scale lengths are the same in J and K bands, and longer for the arms than for the disk alone. In the spiral arms, grand design structure is present in early-type barred galaxies, but only flocculent structure is observed for late-type barred galaxies. The bar-interbar contrast increases along the bar in all galaxies. The arm-interarm contrast increases out to mid-way in the disk, which is near corotation and near where 2-arm spirals bifurcate. This position also corresponds to the approximate end of flat bars, but is well beyond the end of exponential bars.

The author gratefully acknowledges an International Travel Grant from the American Astronomical Society, and additional travel support from Dr. David Block and the conference organizers, and Vassar College.

References

Bushouse, H. A. & Stanford, S. A. (1992), A Near-Infrared Imaging Survey of Interacting Galaxies: The Small Angular-Size Arp Systems, *ApJS*, **13**, pp. 213-253

Buta, R. (1990), Weakly-Barred Early-Type Ringed Galaxies. I. The Seyfert galaxy NGC 3081, *ApJ*, **351**, pp. 62-74

Buta, R. & Crocker, D. (1993), Metric Characteristics of Nuclear Rings and Related Features in Spiral Galaxies, *AJ*, **105**, pp. 1344-1357

Combes, F. & Gerin, M. (1985), Spiral Structure of Molecular Clouds in Response to Bar Forcing: A Particle Simulation, *A&A*, **150**, pp. 327-338

Contopoulos, G. (1980), How Far do Bars Extend?, *A&A*, **81**, pp. 198-209

Elmegreen, B. G. & Elmegreen, D. M. (1985), Blue and Near-Infrared Surface Photometry of Barred Spiral Galaxies, *ApJ*, **288**, pp. 438-455

Elmegreen, B. G., & Elmegreen, D. M. (1989), On the Relative Frequency of Flocculent and Grand Design Spiral Structure in Barred Galaxies, *ApJ*, **342**, pp. 677-679

Elmegreen, B. G., Elmegreen, D. M., Chromey, F. R., Hasselbacher, D. A., & Bissell, B. A., (1996b), Near-IR Surface Photometry of Early and Late-Type Barred Spiral Galaxies, *AJ*, submitted

Elmegreen, D. M., & Elmegreen, B. G. (1984), Blue and Near-Infrared Surface Photometry of Spiral Structure in 34 Nonbarred Grand Design and Flocculent Galaxies, *ApJS*, **54**, pp. 127-149

Elmegreen, D.M., & Elmegreen, B. G. (1995), Inner Two-Arm Symmetry in Spiral Galaxies, *ApJ*, **445**, pp. 591-598

Elmegreen, D.M., Elmegreen, B. G., Chromey, F. R., Hasselbacher, D. A., & Bissell, B. A., (1996a), Near-Infrared Isophotal Twists in Barred Spiral Galaxies, **AJ**, submitted

Friedli, D. & Martinet, L. (1993), Bars within Bars in Lenticular and Spiral Galaxies: a Step in Secular Evolution?, *A&A*, **277**, pp. 27-41

Jarvis, B. J., Dubath, P., Martinet, L., & Bacon, R. (1988), The Dynamics of SB0 Galaxies. I: The Data, *A&AS*, **74**, pp. 513-528

Jedrzejewski, R. I. (1987), CCD Surface Photometry of Elliptical Galaxies - I. Observations, Reduction and Results, *MNRAS*, **226**, pp. 747-768

Kormendy, J. (1983), The Stellar Kinematics and Dynamics of Barred Galaxies. I. NGC 936, *ApJ*, **275**, pp. 529-548

Lin, C. C., & Low, S. A. (1990), Models of Galaxies - The Modal Approach, *Annals of the NY Acad. of Sciences*, **596**, pp. 80-100

Martin, P. (1995), Quantitative Morphology of Bars in Spiral Galaxies, *AJ*, **109**, pp. 2428-2443

Pfenniger, D. & Norman, C. (1990), Dissipation in Barred Galaxies: The Growth of Bulges and Central Mass Concentrations, *ApJ*, **363**, pp. 391-410

Pompea, S. M., & Rieke, G. H. (1990), A Test of the Association of Infrared Activity with Bars, *ApJ*, **356**, pp. 416-429

Quillen, A. C., Frogel, J. A., & Gonzalez, R. A. (1994), The Gravitational Potential of the Bar in NGC 4314, *ApJ*, **437**, pp. 162-172

Sandage, A., & Bedke, J. (1994), Carnegie Atlas of Galaxies (Washington: The Carnegie Institution of Washington with the Flintridge Foundation)

Shaw, N., Axon, D., Probst, R., & Gatley, I. (1995), Nuclear Bars and Blue Nuclei within Barred Spiral Galaxies, *MNRAS*, **274**, pp. 369-387

Shaw, M. A., Combes, F., Axon, D. J., & Wright, G. S. (1993), Isophote Twists in the Nuclear Regions of Barred Spiral Galaxies, *A&A*, **273**, pp. 31-44

Shlosman, I., Frank, J., & Begelman, M. (1989), Bars within Bars: A Mechanism for Fuelling Active Galactic Nuclei, *Nature*, **338**, pp. 45-47

Wozniak, H., Friedli, D., Martinet, L., Martin, P., & Bratschi, P. (1995), Disc Galaxies with Multiple Triaxial Structures. I. BVRI and Hα Surface Photometry, *A&AS*, **111**, pp. 115-152

BARRED GALAXIES IN THE NEAR-IR: OBSERVATIONS AND DYNAMICAL IMPLICATIONS

E. ATHANASSOULA

Observatoire de Marseille
2, Place Le Verrier
13248 Marseille cedex 04
France

1. Introduction

Morphological classifications using frames in the visual wavelengths (Nilson 1973, de Vaucouleurs *et al.* 1976, Sandage and Tammann 1981) show that one third of all disc galaxies present big bars, like those seen in NGC 1365, 5383, or 1300, while another third have either small bars or ovals. Observations in the near-IR reveal yet more bars in galaxies which were initially thought non-barred (e.g. Hackwell & Schweizer 1983, Scoville *et al.* 1988, Block & Wainscoat 1991). All these bars show important morphological differences. Thus their axial ratios cover a wide range, while their shape in the disc plane varies from elliptical-like to rectangular-like. Similar comments can be made about their relative amplitude and length with respect to the size of the disc, the existence of dust lanes or ansae, how symmetrically located they are with respect to the disc center *etc.* Furthermore Elmegreen & Elmegreen (1985) find from a photometric study of 15 barred galaxies that early type galaxies tend to have bars with constant intensity along their semimajor axis and stellar spiral arms with radially decreasing amplitudes, while late type barred galaxies tend to have bars with exponential-like intensity profiles and constant or radially increasing stellar spiral arm amplitudes.

All these differences will reflect themselves on the potentials of the bars, and therefore on the corresponding orbital structure of the galaxies, though it is not yet known how. In order to examine the question further we need to know the distribution of matter in barred galaxies, and this information can

D. L. Block and J. M. Greenberg (eds.), New Extragalactic Perspectives in the New South Africa, 283–286.
© 1996 *Kluwer Academic Publishers.*

be best obtained with the help of near-IR photometry, preferably coupled with a detailed velocity field.

2. Potentials

Two-dimensional photometry does not allow us to decompose a barred galaxy into a disc, a bar and a bulge component (Athanassoula 1991), basically because the various components are not rigid. Thus the best way of proceeding is to calculate the total potential and its Fourier components directly from the projected light intensity. Even this, however, is not straightforward, for mainly two reasons:

What we observe is the *projected* light distribution and what we need in our calculations is the *volume* density. In order to calculate the one from the other it is necessary to make some assumptions about the three dimensional shape of the various components, and in particular the bar. Unfortunately the result may be heavily influenced by the assumptions we make. So far only cases of constant thickness have been considered, although N-body simulations (Combes *et al.* 1990, Raha *et al.* 1991) show that this need not necessarily be true.

A second problem is that we observe the *light* distribution, while what we need is the *mass* distribution. The one may of course be calculated from the other, provided the M/L ratio is a known function of position. Unfortunately the lack of more precise information has so far always led to the assumption of a constant M/L ratio, and this also could influence the results of the calculations.

3. Results from a small sample of barred galaxies observed in the near-IR

Notwithstanding the two main problems outlined above, the use of near-IR photometry is at present our only way to obtain some information on the potentials of real barred galaxies. For this reason Alice Quillen and I have started a thorough study of a small sample of barred galaxies (Quillen 1996a, 1996b, Athanassoula 1996, Quillen & Athanassoula in preparation). The data we use form part of the Ohio State University Survey. Only a few barred galaxies have been so far observed and fully analysed. Thus we have BVRJHK photometry of NGC 1097, 3351 and 7479, JHK photometry of NGC 4314 and J photometry of NGC 1365. Assuming that the galaxies have a constant thickness we calculated the Fourier components of the potential using the method outlined by Quillen, Frogel & González (1994)

One of our most interesting results comes from the comparison of the axisymmetric component of the densities and potentials, calculated from frames taken at different wavelengths. The axisymmetric intensity profiles show clearly that our galaxies become bluer in the outer parts. This means that the contribution of the disc and bulge "rotation curves" to the total rotation curve will be smaller and less extended than the corresponding ones calculated from B photometry. Thus the missing mass problem is aggravated and more halo material is needed to account for the observed rotation curves. Of course our sample is very small but this result is interesting enough to warrant a verification with a larger sample. This sample need not be restricted to barred galaxies, since the results of Peletier et al. (1994), De Jong & Van der Kruit (1994) and Peletier & Balcells (1996) show that disc galaxies are in general more extended in B than in the near-IR.

In calculating the potential we used three different vertical density profiles, namely exponential, $sech$ and $sech^2$. We find that the results change little, provided the profiles are normalised so that the projected mass and $< z^2 >$ are the same. Changing the scaleheight does not change the phases of the various Fourier components in any noticeable way. Its effect on their amplitude depends on how much structure the galaxy has in scales of the order of the considered wavelengths.

4. Applications

Potential calculations give interesting information on barred galaxies. Their main interest, however, resides in their various applications. I have already mentioned one of them, concerning the amount and radial distribution of dark matter in disc galaxies. Several more are underway. Thus Patsis, Athanassoula & Quillen (in preparation) analyse the orbital structure in NGC 4314 and find a number of interesting families of periodic orbits, particularly in the central parts and in the regions of the 3/1 and 4/1 resonances. Information on potentials is also necessary in order to model the gas flow in barred galaxies. Such studies have been performed for NGC 1365 (Lindblad, Lindblad & Athanassoula 1996) and NGC 1530 (Teuben et al. 1996), while a similar study for NGC 1300 is underway (P. Lindblad, private communication. For these galaxies detailed velocity fields are available. For NGC 1365 a satisfactory agreement between observations and simulation results is achieved. In the best fitting models corotation is at a distance 1.2 or 1.3 R_{bar}, where R_{bar} the length of the bar semimajor axis, and the offset dust lanes are well reproduced by the shock loci, provided an inner Lindblad resonance exists at an appropriate radius. Both these results are in good agreement with the predictions of Athanassoula (1992). In order for the form of the spiral arms across the corotation region to be well reproduced the contribution from the spiral arms themselves has to be included in the potential.

Acknowledgements. I would like to thank Albert Bosma, Panos Patsis and Alice Quillen for many interesting discussions. The Ohio State University galaxy survey is being supported in part by NSF grant AST 92-17716.

References

Athanassoula, E. (1991) *Dynamics of Disc Galaxies*, ed. B. Sundelius, Goteborg University Press, p. 149.

Athanassoula E. (1992) *Mon. Not R. Astr. Soc.*, **259**, 345.

Athanassoula, E. (1996) *Spiral Galaxies in the Near-IR* , eds. D. Minniti and H. W. Rix, Springer Verlag, p. 147.

Block, D. L. and Wainscoat, R. J. (1991) *Nature* , **353**, 48.

Combes, F., Debbash, F., Friedly, F. and Pfenniger, D. (1990): *Astr. Ap.*, **233**, 82.

De Jong, R. S. and Van der Kruit, P. C. (1994) *Infrared Astronomy with Arrays* ed. I. Mc Lean, Kluwer Pub, Dordrecht, p. 123.

De Vaucouleurs, G., de Vaucouleurs, A. and Corwin, H. G. (1976) *Second Reference Catalogue of Bright Galaxies*, University of texas Press.

Elmegreen, B. G. and Elmegreen, D. M. (1985) *Ap. J.* , **288**, 438.

Hackwell, J. A. and Schweizer, F. (1983) *Ap. J.* , **265**, 643.

Lindblad, P., Lindblad, P. O. and Athanassoula, E. (1996) *Astr. Ap.* , in press.

Nilson, P. (1973) *Uppsala General Catalogue of Galaxies.*

Peletier, F. F. and Balcells, M. (1996) *Spiral Galaxies in the Near-IR* , eds. D. Minniti and H. W. Rix, Springer Verlag, p. 48.

Peletier, F. F., Valentijn, E. A. and Moorwood, A. F. M. and Frendling, W. (1994) *Astr. Ap.*, **292**, 369.

Quillen, A. C. (1996a) *Barred Galaxies, Astron. Soc. Pac. Conf. Ser.* , eds. R. Buta, D. A. Crocker and B. G. Elmegreen, p. 390.

Quillen, A. C. (1996b) *Spiral Galaxies in the Near-IR* , eds. D. Minniti and H. W. Rix, Springer Verlag, p. 157.

Quillen, A. C., Frogel, J. A. and González, R. A. (1994) *Ap. J.* **437**, 162

Raha, N., Sellwood, J. A., James, R. A. and Kahn, F. D. (1991) *Nature*, **352**, 411.

Sandage, A. and Tammann, G. A. (1981) *A Revised Shapley-Ames Catalogue of Bright Galaxies*, Carnegie Institution of Washington.

Scoville, N. Z., Matthews, K., Carico, D. P. and Sanders, D. H. (1988) *Ap. J. Lett* , **327**, L61.

Teuben, P., Regan, M. W. and Vogel, S. N. (1996) *Barred Galaxies and Circumnuclear Activity* , eds. Aa. Sandqvist, S. Jörsäter and P. O. Lindblad, Springer Verlag, in press.

MORPHOLOGY AND DYNAMICS OF A FEW GIANT GALAXIES WITH LOW SURFACE BRIGHTNESS DISKS

A. BOSMA
Observatoire de Marseille
2 Place Le Verrier, 13248 Marseille Cedex 4, FRANCE

Abstract.

I present mass models for a few spiral galaxies with low surface brightness disks, which dynamically are giants. A comparison of the dark matter content and its relation with morphology between these galaxies and "ordinary" spiral galaxies shows that they must have had a different formation history than ordinary spirals.

1. Introduction

Recent studies of disk central surface brightnesses have questioned the universal validity of "Freeman's law", i.e. the result that for every bright spiral galaxy the extrapolated central surface brightness in the B-band is 21.65 magn arcsec^{-2}. Freeman (1970) found this result for 28 out of 36 galaxies for which photometry was available at the time. Van der Kruit (1987), from his photometry of 51 field galaxies argued that the distribution is broader towards the faint end, while galaxies with higher central surface brightnesses are not found.

De Jong (1995) has carried out an extensive study of 86 spiral galaxies of various morphological types. He has obtained extrapolated central surface densities both in the B-band and in the near infrared K-band. Even though he also concludes that Freeman's law needs modification, another reading of his result is possible. If one concentrates only on the bright Sa - Sc galaxies, Freeman's law is found back both in B and in K, although the scatter is about 0.7 magn. arcsec^{-2}. If one goes further down the sequence of late-type spirals (Sc and later), there is a progression in the sense that the more luminous Sc's have higher central surface brightnesses, while the

D. L. Block and J. M. Greenberg (eds.), New Extragalactic Perspectives in the New South Africa, 287–290.
© *1996 Kluwer Academic Publishers.*

(few) magellanic dwarfs have lower central surface brightnesses. Sd's are somewhere in between. Such a progression can be expected already from the discussions on luminosity classes, e.g. in the Revised Shapley Ames Catalog (Sandage & Tammann 1981).

2. Low surface brightness disk galaxies

One of the running arguments about Freeman's law is that it is a selection effect (cf. Disney 1976). At present most people think that an upper bound exists for the central surface brightness, while a lower bound does not seem to exist. The discovery of Malin 1 (Bothun et al. 1987) stimulated a lot of work on catalogs of galaxies with faint disks, in particular using the Second Palomar Sky Survey. Such LSB (low surface brightness) galaxies seem at first sight to fall in the sequence between Sc's and Im's (e.g. galaxies like IC 1613) but a fraction of them have a large physical extent. A suitable recent review of their properties is given by McGaugh (1996).

Bosma & Freeman (1993) use another approach : they measured sizes of galaxies on various Sky Surveys, each having a different surface brightness level. For two given surveys, one expects a fixed ratio of sizes of every disk galaxy provided there is a unique central surface brightness. They impose a minimum size of 2.2 arcmin on the deep SRC IIIaJ-survey, and by this constraint alone eliminate most of the low surface brightness dwarfs. They do find a number of galaxies which are much larger then expected on the deeper survey, and for which therefore Freeman's law does not hold. From follow-up work on these galaxies, as well as from the study of NGC 5963 by Bosma, Van der Hulst & Athanassoula (1988), they conclude that these low surface brightness galaxies are dynamically giants, which somehow do not have the same amount of baryons in their visible disks as do the galaxies obeying Freeman's law.

What is the relationship between these two lines of work ? It seems to me that the findings are more complementary than contradictory. In fact, it becomes now clear that there are two kinds of low surface brightness giants : one for which the surface brightness is low over the whole of the disk, which is very extended, like UGC 128, and one perhaps rarer kind for which the central parts are very bright and the faint outer parts are of low surface brightness, like NGC 5963. This is illustrated in Figure 1, where I present on the same radial scale mass models for the galaxies NGC 5963, NGC 2403 and UGC 128. The data for NGC 5963 comes from the work of Bosma et al. (1988), the data for NGC 2403 from Begeman, Broeils & Sanders (1991), and the data for UGC 128 from Van der Hulst et al. (1993) and De Blok et al. (1995).

Several things are immediately obvious in Fig.1 : the rotation curves

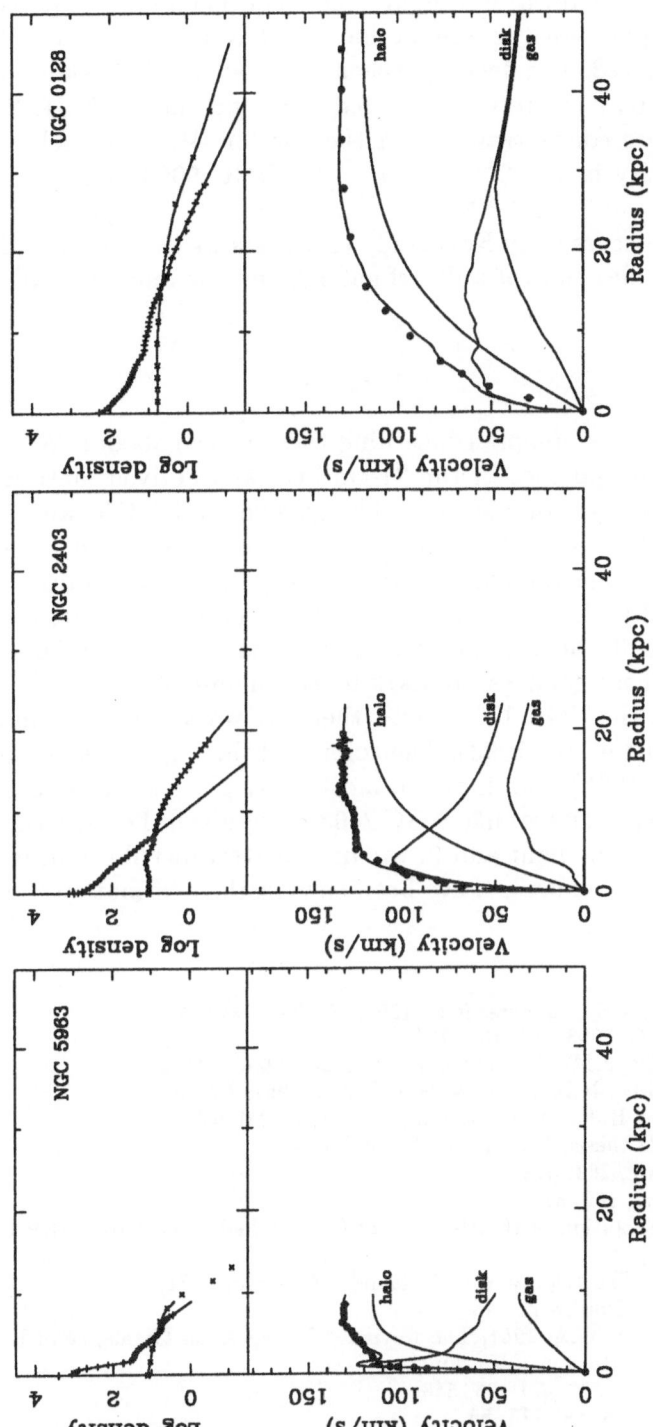

Figure 1. Composite maximum disk mass models for NGC 5963, NGC 2403 and UGC 128, all on the same radial scale. The top panels give the stellar density (the luminosity surface density times a mass-to-light ratio) (plusses) and the gas density (crosses), both in units of $M_\odot$ pc^{-2}. The bottom panels show the observed rotation curves as circles, the labelled curves are the rotation curves of each component, and the upper curves represent the total circular velocity of each of the models. Note the difference in radial extent between the galaxies, despite their similar maximum rotational velocity.

of all three galaxies rise towards roughly the same maximum amplitude, but in order of steepness of the rise we have first NGC 5963, then NGC 2403, and then UGC 128. The decomposition into disk and halo shows that the halo of NGC 5963 is more concentrated than the one for NGC 2403, which is in turn more concentrated then the one for UGC 128. The halo central volume density is 97 10^{-3} $M_\odot$ pc^{-3} for NGC 5963, 9.1 10^{-3} $M_\odot$ pc^{-3} for NGC 2403 and 3.2 10^{-3} $M_\odot$ pc^{-3} for UGC 128. Comparison with the scaling relation discussed by Kormendy (1991) shows that the dark halo of NGC 5963 ressembles that of a dwarf galaxy, and the one for UGC 128 that of a giant galaxy.

3. Discussion

This state of affairs is not unique concerning these three galaxies. We have observed several other galaxies in the 21-cm line, and derived mass models in the same manner as for Figure 1. The galaxies NGC 3657 and ESO 442-G13 are similar to NGC 5963. In fact, these galaxies pose the so-called conspiracy problem : even though the surface brightness distribution indicates the presence of an inner bright zone and an outer faint zone, the HI distribution and velocity field do not give any signature of such a "break". This suggests that these galaxies are dark matter dominated.

As for galaxies of the UGC 128 variety, there are several other examples given in Van der Hulst et al. (1993). The colours of these galaxies are quite blue (De Blok et al. 1995), and indicate most likely an mean age younger than that of "ordinary" galaxies like NGC 2403. Thus given their low central concentration, both in the light and in the dark matter distribution, it can be concluded that these galaxies have been slow in their formation.

References

Begeman, K.G., Broeils, A.H., Sanders, R.H. 1991, MNRAS, 249, 523
Bosma, A., Freeman, K.C. 1993, AJ, 106, 1394
Bosma, A., Van der Hulst, J.M., Athanassoula, E. 1988, A&A, 198, 100
Bothun, G.D., Impey, C.D., Malin, D.F.,& Mould, J.R. 1987, AJ, 94, 23
De Blok, W.J.G, Van der Hulst, J.M., Bothun, G.D. 1995, MNRAS 274, 235
De Jong, R.S. 1995, PhD thesis, University of Groningen
Disney, M.J. 1976, Nature, 263, 573
Freeman, K.C. 1970, ApJ, 160, 811
Kormendy, J. 1991, in Evolution of the Universe of Galaxies, ed. R.G. Kron, ASP Conf. Series, vol 10, p.33
McGaugh, S.S., 1996, in The Evolution of Galaxies, IAU Symp. 171, eds. R. Bender & R.L. Davies, (Kluwer, Dordrecht), in press
Sandage, A.R. & Tammann, G.A. 1981, The Revised Shapley Ames Catalogue of Bright Galaxies, Carnegie Institution of Washington
Van der Hulst, J.M., et al. 1993, AJ 106, 548
Van der Kruit, P.C., 1987, A&A, 173, 59

SECULAR EVOLUTION OF GALAXY MORPHOLOGIES

D. PFENNIGER, L. MARTINET
Geneva Observatory, University of Geneva
CH-1290 Sauverny, Switzerland

AND

F. COMBES
DEMIRM, Observatoire de Paris
61 Av. de l'Observatoire, F-75014 Paris, France

Abstract. Today we have numerous evidences that spirals evolve *dynamically* through various secular or episodic processes, such as bar formation and destruction, bulge growth and mergers, sometimes over much shorter periods than the standard galaxy age of $10 - 15$ Gyr. This, coupled to the known properties of the Hubble sequence, leads to a unique sense of evolution: from Sm to Sa. Linking this to the known mass components provides new indications on the nature of dark matter in galaxies. The existence of large amounts of yet undetected dark gas appears as the most natural option. Bounds on the amount of dark stars can be given since their formation is mostly irreversible and requires obviously a same amount of gas.

1. Introduction

Until the recent years the concepts prevailing in understanding galaxy evolution have been largely dominated by the ELS scenario (Eggen, Lynden-Bell & Sandage 1962). The views issued from it favour a *synchronous* and rapid ($\approx 100\,\mathrm{Myr}$) formation of the galaxies at a particular early time in the universal expansion. In this context, the properties of present galaxies, such as their typical scale and mass, must depend directly on the initial conditions fixed by the physical state of the Universe at the "galaxy formation epoch". The only subsequent significant evolution in galaxies to be discussed was the slow changes in their stellar populations. Dynamical processes could be assumed to be settled, and therefore ignored.

D. L. Block and J. M. Greenberg (eds.), New Extragalactic Perspectives in the New South Africa, 291–298.
© 1996 *Kluwer Academic Publishers.*

Since then major observational and theoretical progresses of direct relevance occurred, which resulted in a gradual shift in the meaning of galaxy evolution. It suffices to remind that basically all the advances in non-linear dynamics, including the recognition of the fundamental rôle of chaos, and in computer simulations of galaxies have been made in between.

Today, in all its steps the ELS scenario is no longer tenable, as explained in the following. This is still not perceived as a necessity in fields loosely connected with the recent advances in gravitational dynamics.

An instance of inadequation between the ELS scenario and more recent works appears in cosmological simulations at several 100 Mpc scale (see e.g., White 1994). In such simulations nothing like homogeneous collapses do occur, instead hierarchical clustering proceeds at all the computable scales with different speeds. This implies that galaxies and galaxy clusters, exactly as stars, form in different regions of the sky *asynchronously*. The formation process covers several dex of time-scales, so not only the free-fall time, but merging and later infall participate to it. For many galaxies the formation/evolution should be considered as not terminated even now. The galaxy age looses its original meaning because the aging, traced by various observables, may occur at widely different speeds.

A central aspect not considered in the ELS picture is the likely coupling of dynamics with stellar activity. In the recent years the large FIR emission of spirals, which even largely dominates the light in starburst galaxies, was found substantial particularly over the "late" part of the spiral sequence. The FIR emission, coming mostly from UV and visible light absorbed and recycled in the FIR by dust, is consistent with the also recent recognition of the partial opacity of the optical region of spirals (e.g. Davis et al. 1993). It turns out that the total bolometric luminosity is comparable to the power that dynamics can exchange (Pfenniger 1991b). This coincidence is best explained by a coupling of star formation and dynamics via a feed-back mechanism (Quirk 1972; Kennicutt 1989). The interesting aspect of this coupling is that the systematic global properties of galaxies are then no longer necessarily determined by the initial conditions of collapse. As for stars, the galaxy properties would then be encoded in their internal small scale physics, i.e. star formation and ISM physics. This may solve the old problem of the absence of galactic scale at the radiation-matter decoupling epoch, particularly if galaxy formation covers a sizable fraction of the Hubble time. The galactic scale would result mainly from the balancing of stellar activity effects, particularly during starburst phases, and dynamics.

The Hubble sequence is most probably an incomplete sample. For example many low surface brightness galaxies may well be missed (e.g. Bothun et al. 1990). However, whenever fast morphological changes do occur in normal galaxies (bars, mergers), they must correspond to shifts *within the sequence*, because the already existing stars certainly survive the changes.

The general properties of the Hubble sequence have been progressively determined (e.g. de Vaucouleurs 1959; Broeils 1992; Roberts & Haynes 1994; Zaritsky et al. 1994). In order to extract useful information of this "zoo", we must consider only the most general properties, keeping in mind that galaxies form a variety of objects with different ages and aging speeds. For example mergers (Schweizer 1993) certainly accelerate morphological changes at speeds which depend on the fortuitous interaction strengths.

The following list summarises the main trends and properties along the spiral sequence **from Sm to Sa**, that are useful for the discussion.

Total mass	$\nearrow$	$M_{\text{tot}} \approx 1 \to 100 \cdot 10^{10}\,M_\odot$
Kinetic energy	$\nearrow$	$\frac{1}{2}V_{\text{max}}^2 \approx 25 \to 500\,\text{eV/nucleon}$
Bulge-disk ratio	$\nearrow$	$L_{\text{sph}}/L_{\text{tot}} \approx 0 \to 0.6$
Symmetry	$\nearrow$	
Metallicity	$\nearrow$	$12 + \log(\text{O/H}) \approx 8.3 \to 9$
Detected gas	$\searrow$	$\dfrac{M_{\text{HI}+\text{H}_2}}{M_{\text{tot}}} \approx 0.10 \to 0.07,\ \dfrac{M_{\text{HI}+\text{H}_2}}{M_{\text{stars}}} \approx 1.4 \to 0.1$
Dark matter	$\searrow$	$M_{\text{dark}}/M_{\text{lum}} \approx 10 \to 1$

2. Sense of Evolution from Irreversible Processes

In fact, already a systematic sense of evolution is clear by making a list of the major irreversible processes known in spirals. Each of them gives a possible criterion and a sense of aging.

1) The energy dissipation in gravitating systems is measured by the present amount of kinetic energy, which equals the minimum energy the system had to release in order to reach the present bound state. In spirals the rotation speed is an excellent indicator of the specific energy dissipated since the rotation curves vary slowly with radius. Because disks are systems having lost the maximum of energy while conserving angular momentum well, further energy dissipation necessarily implies dissipation of angular momentum, which is best achieved by some mass transport and breaking of axisymmetry. Bars and spiral arms are just manifestations of this necessity. The energy factor already indicates clearly that the Sa side of the spiral sequence is energetically more evolved than the Sm side.

2) Building a central bulge or spheroid by heating a disk, by whatever process (see next Sect.), is also a stellar dynamical irreversible process, because there is no way to "cool" stars following fat orbits back toward circular orbits. From the stellar dynamical point of view Sa galaxies with big bulges are thus more evolved than bulgeless Sm galaxies.

3) Overall, if galaxian shapes tend toward some attractors, the degree of organisation and symmetry toward these shapes is a sign of evolution. Clearly the spiral sequence looks increasingly organised in the sense Sm to Sa, which is also consistent with the decreasing spiral pitch angle.

4) The transformation of gas into stars in cold clouds is mainly an irreversible process, star formation locking most of the mass for time-scales longer than galaxy ages. So the ratio of stellar to gas mass is a tracer of evolution, and Sa's are more evolved in this respect than Sm's. In this context, the dark matter fraction decreasing systematically along the spiral sequence to low "non-problematic" values on the Sa side is remarkable.

5) Obviously related to the previous criterion, the nucleosynthesis within stars is also irreversible, the more metal rich and dusty galaxies have a longer history through the internal activity of their stars. Sa's are more enriched than Sm's, which again indicates a sense of evolution.

Finally, the total mass along the sequence increases strongly from Sm to Sa in average, though with a considerable spread at constant type. This is not astonishing because the classification criteria are mass independent. We interpret the mass trend as indicative that massive proto-galaxies evolve in average faster than light ones. The proto-galaxies of today (Im's, Sm's) are probably lighter than the corresponding ones in the past.

In summary, the only consistent sense of aging along the spiral sequence is from Sm to S0. The important consequence is that proto-galaxies would then be mostly bulgeless gas rich disks like Sm-Sd's, or even pure gas disks, instead of initial spheroid dominated galaxies of the ELS scenario.

3. Evolution from Dynamics and Observations

Contemporary to the ELS scenario, Safronov (1960) and Toomre (1964) realised the unexpected fact that gaseous and stellar disks with too much circular motion are gravitationally unstable. So energy dissipation with angular momentum conservation brings first collapsing gravitating systems toward disk shapes with an increasing fraction of kinetic energy in rotational motion. But subsequent dissipation brings disks ineluctably toward a global instability. A first ground was found that disk galaxies may be dynamically unstable, so may evolve with dynamical time-scales.

Shortly hereafter computer simulations of stellar disks (e.g. Miller & Prendergast 1968; Hohl & Hockney 1969) allowed to simulate the non-linear phase of disk instability. They showed a systematic tendency to produce a robust bar. These results illustrated an example of major and fast morphology change (within a couple of rotational periods) of galaxy type from non-barred to barred.

The other significant proposition in the 70's came from Toomre & Toomre (1972) in which ellipticals may result from the merging of spirals or other ellipticals. Another case of major and fast change of galaxian morphology was put forward. Despite many resistances this scenario appears today as the most natural way of forming ellipticals, although it is not necessarily the only one. In fact, the more violently a galaxy is shaken, for whatever reason, the more it ends up like an elliptical.

In the 80's the bar phenomenon was investigated in more depth. From observational material, Kormendy (1982) pushed forward the idea of secular evolution in barred galaxies. The reason why bars do exist in the first place was understood by studies of their periodic orbits in the plane (Contopoulos 1980, Athanassoula et al. 1983) and in 3D (Pfenniger 1984, 1985). It was discovered that bars may evolve into boxy bulges via vertical resonances boosting bending instabilities transverse to the plane (Combes & Sanders 1981; Combes et al. 1990; Raha et al. 1990). It became also clear that chaotic orbits are playing an important role in bars. Later it was understood how the accretion of only a few percents of mass within the Inner Lindblad Resonance (ILR), either by dissipation of gas (Hasan & Norman 1990; Pfenniger & Norman 1990) or by dynamical friction on galaxy satellites (Pfenniger 1991a), may rapidly destroy a bar into a spheroidal component of similar size. This led to an increased confidence that many of the non-linear events and morphologies observed in galaxies and in N-body simulations, can be interpreted, and even predicted, via the knowledge of the underlying periodic orbits (Pfenniger & Friedli 1991). These studies of bars showed that isolated galaxies must also be seen as structures with possible fast dynamical evolution phases.

In the recent years more works have completed the above picture. Secondary bars (Friedli & Martinet 1993), gaseous and star formation effects (e.g. Friedli & Benz 1993, 1995), interactions with external galaxies and mergers (e.g. Barnes 1992) continue to be investigated. In any cases, these additional complications make even harder to freeze galaxy morphologies beyond a few Gyr. The obvious requirement is then to understand the general time-sequence of galaxy morphologies.

Independently of dynamics, several observational results strongly indicate significant secular evolution within much less than 10 Gyr:

1) From halo stellar cluster abundances Searle & Zinn (1978) arrived to an alternative scenario to ELS. From the observations they concluded that the Milky Way stars did form inside-out over several Gyr.

2) In galaxy clusters the Butcher-Oemler (1978) effect (see also Rakos & Shombert 1995) indicates that galaxies are increasingly bluer at higher redshifts. Furthermore, the morphology-radius relationship (Dressler 1980; Whitmore 1993) shows that the majority of galaxies at cluster periphery

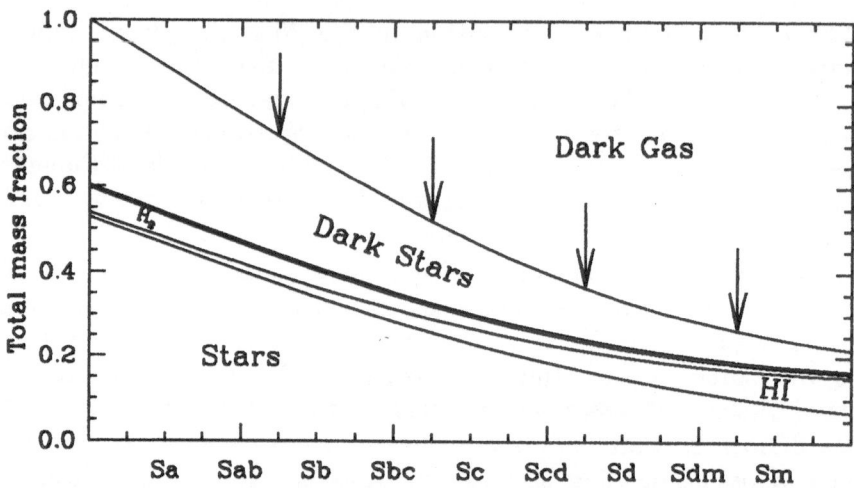

Figure 1. Composition diagram along the spiral sequence. The thick line separates the detected matter from the dark matter. The dark stellar-like objects (DS) are assumed here to be proportional to the stars. The arrows indicate that the DS mass is an upper limit and the dark gas a lower limit.

are spirals, as in the field, but these are replaced by lenticular and then ellipticals at smaller radii. To a dynamicist this relationship tells that spirals do not survive to a single center crossing or so, because the galaxies within a cluster are expected to move on rather elongated orbits. If correct the spiral morphological evolution accelerated by environmental disturbances is then directly revealed. Either spirals end in part as ellipticals, and/or they are largely dissolved and contribute to the cluster hot gas and metallicity.

4. Constraints from an Evolutive Spiral Sequence

If we take the published data about the ratios of the different known matter forms along the spiral sequence (e.g. Broeils 1992 for the stars and HI; Young & Knezek 1989 for H_2 derived from CO emission), we can solve for the mass fraction of each component: stars, HI, H_2 from CO, and the rest, dark matter (for more detail, see Pfenniger 1996). We obtain the composition diagram of the spiral sequence shown in Fig. 1.

Now if spirals do evolve along the sequence from Sm to S0, then several constraints follow (Pfenniger, Combes & Martinet 1994; Martinet 1995). We must consider that stars or jupiters are made from the gas and lock most of it for $\gg 5-10$ Gyr, and that for dynamical reasons little accretion (less than a few % in mass) can occur transversally to a stellar disk without heating it to too high values (Tóth & Ostriker 1992).

Most of the usual dark matter candidates such as CDM particles, neutrinos, jupiters, brown dwarfs, and black-holes, are inconsistent with the spiral sequence properties including dynamical evolution, mainly because evolutive phases (e.g. a merger) can be sporadic. Since the star fraction increases from Sm to Sa in a proportion exceeding the detected gas content, everything happens as if dark matter is transformed into stars, i.e. a substantial fraction of dark matter should be dark gas. We have explained elsewhere (Pfenniger & Combes 1994, and elsewhere in this volume) the many reasons why today we can consider this conservative candidate, for long totally neglected, as worth to be more investigated.

Others did arrive also to the conclusion that much more gas should exist in spirals. Toomre (1981) argued that the cold dynamics and chaotic structure of Sc's require much more gas than observed to maintain their patchy structure. Also just to sustain the present star formation rates over several Gyr Sc's require much more gas (Larson et al. 1980).

Not all the dark matter in spiral is necessarily dark gas, because the uncertainties on the stellar mass contributed by brown dwarfs and dark remnants, or by stars obscured by dust, are still substantial ($\sim 50\%$). We call these dark or obscured stars DS for short. However, the fact that S0's and ellipticals are dynamically evolved systems at the end of the gas-to-star transformation process, but with still $\sim 40\%$ of dark matter, hints toward the presence of non-gaseous components such as DS. If we assume that a population of DS pre-exists to a galaxy (the same rôle would be played by CDM particles), its maximal fraction is determined by the final S0 stage, $\sim 40\%$. In previous stages the difference must be gas to make future stars. Less speculatively, as shown in Fig. 1, if instead we assume that DS's form proportional to the stars as galaxies evolve (for example just from the increasing fraction of dust), the amount of dark gas required is correspondingly increased. For the Milky Way (an SBbc) the fraction of DS's would be $< 40\%$ in the first case, and $< 20\%$ in the second case.

5. Conclusions

Galactic dynamics forces us to see spirals as structures with possible "bursts" of dynamical evolution involving short time-scales. Taking into account to-day's observational and theoretical constraints, the only possible sense of evolution is from Sm to S0. During evolution everything happens as if dark matter in galaxies is transformed into stars, that is dark gas is required.

A possible solution to the dark matter problem in galaxies is that gas in outer HI disks clumps along a fractal structure down to solar system sizes, as explained elsewhere in this volume. The smallest clumps are then very dense and cold which makes them presently hard to detect. A sizable amount of dark (or dust obscured) stars can also be argued just from the fact that evolved galaxies (Sa-S0's) still contain about 40% of dark matter.

References

Athanassoula E., Bienaymé O., Martinet L., Pfenniger D. 1983, *A&A* **127**, 349
Barnes J.E. 1992, *ApJ* **393**, 484
Bothun G.D., Shombert J.M., Impey C.D., Schneider S.E. 1990, *ApJ* **360**, 427
Broeils A. 1992, *Dark and visible matter in spiral galaxies*, PhD Thesis, U. Groningen
Butcher H., Oemler A. 1978, *ApJ* **219**, 18
Combes F., Debbasch F., Friedli D., Pfenniger D. 1990, *A&A* **233**, 82
Combes F., Sanders R.H. 1981, *A&A* **96**, 164
Contopoulos G. 1980, *A&A* **81**, 198
Davies J.I., Phillips S., Boyce P.J., Disney M.J. 1993, *MNRAS* **260**, 491
de Vaucouleurs G. 1959, in *Handbuch der Physik LIII, Astrophysik IV: Sternsysteme*, S.
    Flügge (ed.), Springer-Verlag, Berlin, 275
Dressler A. 1980, *ApJ* **236**, 351
Eggen O.J., Lynden-Bell D., Sandage A.R. 1962, *ApJ* **136**, 748 (ELS)
Friedli D., Benz W. 1993, *A&A* **268**, 65, and 1995, *A&A*, **301**, 649
Friedli D., Martinet L. 1993, *A&A* **277**, 27
Hasan H., Norman C. 1990, *ApJ* **361**, 69.
Hohl F., Hockney R.W. 1969, *J. Comput. Phys.* **4**, 306
Kennicutt R.C. 1989, *ApJ* **344**, 685
Kormendy J. 1982, in *Morphology and Dynamics of Galaxies*, 12th Advanced Course
    Swiss Soc. Astr. Astrophys., Martinet L., Mayor M. (eds.), Geneva Observ., 115
Larson R.B. 1981, *MNRAS* **194**, 809
Larson R.B., Tinsley B.M., Caldwell C.N. 1980, *ApJ* **237**, 692
Martinet L. 1995, *Fundamental of Cosmic Physics* **15**, 341
Miller R.H., Prendergast K.H. 1968, *ApJ* **151**, 699
Pfenniger D. 1984, *A&A* **134**, 373
Pfenniger D. 1985, *A&A* **150**, 112
Pfenniger D. 1991a, in *Dynamics of Disc Galaxies*, B. Sundelius (ed.), Göteborg U., 191
Pfenniger D. 1991b, in *Dynamics of Disc Galaxies*, B. Sundelius (ed.), Göteborg U., 389
Pfenniger D. 1996, in *Third Paris Cosmology Colloquium*, H.J. de Vega, N. Sánchez (eds.),
    World Scientific, Singapore, in press
Pfenniger D., Combes F. 1994, *A&A* **285**, 94
Pfenniger D., Combes F., Martinet L. 1994, *A&A* **285**, 79
Pfenniger D., Friedli D. 1991, *A&A* **252**, 75
Pfenniger D., Norman C.A. 1990, *ApJ* **363**, 391
Quirk W.J. 1972, *ApJ* **176**, L9
Raha N., Sellwood J.A., James R.A., Kahn F.D. 1991, *Nat.* **352**, 411
Rakos K.D., Schombert J.M. 1995, *ApJ* **439**, 47
Roberts M.S., Haynes M.P. 1994, *ARAA* **32**, 115
Safronov V.S. 1960, *Annales d'Astrophysique* **23**, 979
Schweizer F. 1993, in *Physics of Nearby Galaxies, Nature or Nurture?*, T.X. Thuan, C.
    Balkowski, J.T.T. Van (eds.), Editions Frontières, Gif-sur-Yvette, 283
Searle L., Zinn R. 1978 *ApJ* **225**, 357
Toomre A. 1964, *ApJ* **139**, 1217
Toomre A. 1981, in *The Structure and Evolution of Normal Galaxies*, S.M. Fall, D.
    Lynden-Bell (eds.), Cambridge Univ. Press, 111
Toomre A., Toomre J. 1972, *ApJ* **178**, 623
Tóth G., Ostriker J.P. 1992, *ApJ* **389**, 5
White S.D.M. 1994, *Formation and Evolution of Galaxies*: Les Houches Lectures, preprint
    astro-ph/9410043
Whitmore B.C. 1993, in *Physics of Nearby Galaxies, Nature or Nurture?*, T.X. Thuan,
    C. Balkowski, J.T.T. Van (eds.), Editions Frontières, Gif-sur-Yvette, 425
Young J.S., Knezek P.M. 1989, *ApJ* **347**, L55
Zaritsky D., Kennicutt R.C., Huchra J.P. 1994, *ApJ* **420**, 87

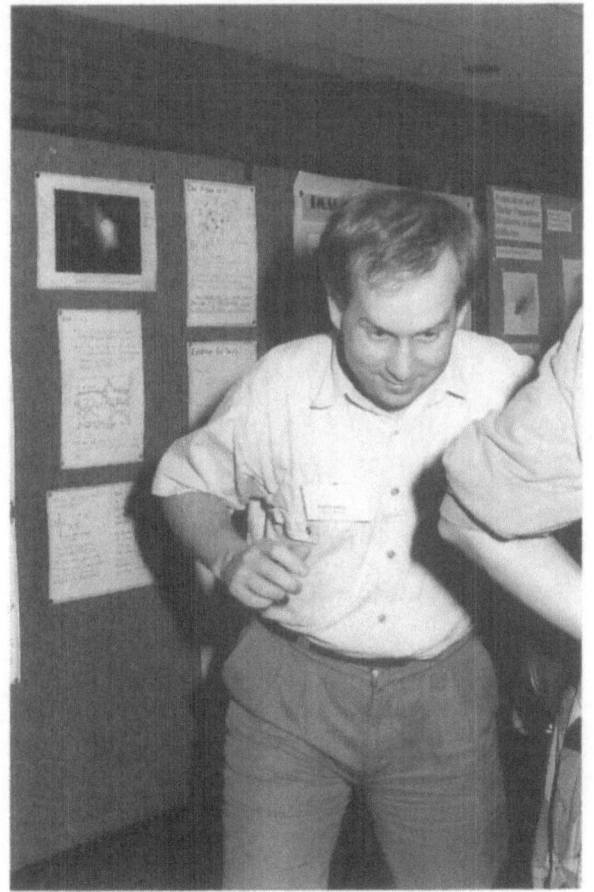

DUSTY DISKS AND THE STRUCTURE OF EARLY-TYPE GALAXIES

THOMAS Y. STEIMAN-CAMERON
NASA Ames Research Center
MS245-3, Moffett Field, CA 94035-1000

1. Introduction

Our eventual understanding of the formation and evolution of galaxies is critically dependent upon achieving knowledge of the distribution of matter within these systems. This is an excedingly difficult task, compounded by the fact that most of the matter in galaxies is not directly visible. Therefore, luminosity distributions are not sufficient to determine mass distributions. We must rely upon dynamical probes for this task. Fortunately, a potentially powerful dynamical probe of mass distributions is provided by disks which lie wholly or partially outside of a symmetry plane of the galactic light distribution. Many early-type galaxies, including dust-lane ellipticals and polar-ring galaxies, possess embedded gas disks. These disks, which are generally believed to be the result of accretion, often display warps and twists or have angular momentum vectors that are misaligned with the kinematic axes of the galaxy. The structure of these disks can be understood by examining the distribution of dust associated with them. Here we discuss the structure and use of such disks.

When an interaction involves the accretion of gas by a larger galaxy, dissipative processes and differential rotation can be expected to smear the captured material into a disk or annulus within a few orbital periods. These newly formed disks are still strongly time-dependent and are driven towards a more relaxed configuration by the interplay between gravitational and dissipational processes. The manner in which a captured disk evolves is a complex function of the gravitational potential and rotation state of the galaxy and the character of the accreted material. Physical processes which can serve to alter disk structure include precessional forces arising from the nonsphericity of the host galaxy, dissipative interactions between

301

D. L. Block and J. M. Greenberg (eds.), New Extragalactic Perspectives in the New South Africa, 301–308.

mass elements of the disk, internal torques arising from the self-gravity of the accreted disk, and effects arising from rotation of the figure or potential of the galaxy. The relative importance of each of these processes will vary from case to case. In what follows, it is assumed that steady-state mass distributions are approximately achieved by the galaxies examined.

2. Structural Evolution of Accreted Disks

Disks captured outside of a steady-state (*preferred*) orientation will precess as a result of the nonsphericity of the galactic potential. Precession is more rapid at smaller radii, thus *differential precession* causes a smooth, continuous twist to develop in the disk. All captured disks will initially experience a phase of differential precession. Subsequent disk evolution will depend upon the relative importance of the individual processes outlined in the previous section. In the absence of dissipation and self-gravity, differential precession continues unimpeded, ultimately causing the disk to twist to such an extent that it is no longer recognizable as a disk.

Dissipation within a twisted disk leads to the transport and redistribution of angular momentum, causing changes in the orientation of the disk. and inflow of material toward the center. When dissipation is important and it dominates disk self-gravity, the disk settles into a preferred orientation. Here the orbit of a disk element does not precess relative to a coordinate frame fixed with respect to the figure of the galaxy. Orbits slightly inclined to a stable preferred orientation precess about it; disks settle into the plane about which they precess. In an axisymmetric galaxy with a static surface figure, preferred orientations correspond to orbits lying in the equatorial plane. In static triaxial galaxies, disks which have their angular momentum vectors aligned with either the long axis or the short axis of the galaxy lie in a preferred orientation (*cf.* , Tohline & Durisen 1982; Steiman-Cameron & Durisen 1982, 1984). The existence of both minor and major-axis dust lane ellipticals is one of the indicators that ellipticals are globally triaxial in nature. Rotation of the isopotential surfaces causes some of the possible preferred orientations to be warped. (Tohline and Durisen 1982; de Zeeuw & Merritt 1983; Durisen *et al.* 1983; Steiman-Cameron & Durisen 1984; David *et al.* 1984, 1985; Ostriker & Binney 1989).

Torques arising from the self-gravity of a disk also serve to redistribute angular momentum within the disk once the disk has been rendered nonplanar due to differential precession. These torques can be dynamically important when they are comparable to the precessional torques arising from the galaxy and/or to the viscous torques which tend to settle the disk. In this case, the internal torques lead to changes in disk orientation as a function of radius. In axisymmetric galaxies, modes which involve the uniform precession of a warped disk are possible (*cf.* Hunter and Toomre 1969; Sparke 1984, 1986, 1991; Sparke and Casertano 1988; Sackett and Sparke 1990; Hoffner and Sparke 1994; see also Arnaboldi & Sparke 1994).

When multiple physical processes are important or when the dominant physical process governing disk evolution is unclear, the interpretation of specific galaxies can become difficult. For example, the relative importance of dissipation and disk self-gravity depends critically upon the amount of mass within the disk, its distribution, composition and physical characteristics, and the strength of the precessional forces. Self-gravity and dissipation can lead to effects of opposite sign under some conditions. If they are not both properly accounted for, the geometry of galaxies with warped and twisted disks could be misinterpreted. This in turn could lead to incorrect models for the galactic mass distributions and to grossly inaccurate estimates of disk ages and of the frequency of disk-forming capture events.

The resolution of this issue is a matter for further discussion and is beyond the scope of this article. The interested reader is referred to Steiman-Cameron & Durisen (1982, 1988, 1992), Steiman-Cameron (1995), Colley & Sparke (1996), and references therein for a more detailed discussion of models and their predictions for individual galaxies and classes of galaxies. Here we will focus on the simplest case, where the precessional forces are the primary determinant of the time-dependent geometry of the disk. In this situation, one can fairly unambiguously interpret the shape of the underlying galaxy from the structure of the accretion disk. Within this context, we will examine the case of NGC 4753, a galaxy where neither dissipation nor self-gravity appears to have played a role in disk evolution.

3. Gravitational Potentials and Precession

Guided by theoretical studies of fluid figures of equilibrium, and by numerical collapse calculations, it is commonly assumed that isodensity surfaces of an *unperturbed* galaxy are well represented by ellipsoids. Aligning the coordinate axes along the principal axes of the surface figure, and retaining terms to the quadrupole, the galactic gravitational potential can be written as

$$\Phi(r, \theta, \phi) = \Phi_0(r) + \Phi_{20} P_2^0(\cos \theta) + \Phi_{22} P_2^2(\cos \theta) \cos 2\phi, \qquad (1)$$

where P_ℓ^m are the Associated Legendere functions. The rotation curve is principally determined by the monopole term $\Phi_0(r)$. Except in highly nonspherical galaxies, quadrupole terms are expected to be typically less than a few percent of the monopole term. The quadrupole terms, arising from the nonsphericity of the galaxy, are responsible for precessional motion.

The orientation of a disk can be described by the inclination $i(r)$ and longitude of ascending node $\Omega(r)$ of annular mass elements comprising the disk. For a potential of the form of eq. (1), precessional torques lead to

$$
\begin{aligned}
(di/dt) &= (3\Phi_{22}/rv_c)\sin i \sin 2\Omega, \\
(d\Omega/dt) &= (3/rv_c)\left[\Phi_{22}\cos i \cos 2\Omega - (\Phi_{20}/2)\cos i\right],
\end{aligned}
\tag{2}
$$

where v_c is the circular velocity. Note that for axisymmetric galaxies $\Phi_{22} = 0$ and precession is purely nodal in character.

By assumption, an accreted disk is initially planar and any deviation from planarity can be ascribed to precession (a reasonable assumption since the accreted material presumably all shared a common orbit plane; this assumption can be relaxed, however). Understanding the three-dimensional mass distribution of a galaxy with a disk twisted by differential precession can now be expressed as: (1) determining the rotation curve, (2) determining $i(r)$ and $\Omega(r)$, and (3) finding the form of the potential which would produce this structure starting from a planar disk. The scale-free logarithmic potential has been frequently used to model galaxies (cf. Richstone 1980; Levison 1987; Steiman-Cameron & Durisen 1990). This potential has the properties that the rotation curve is flat, isopotential surfaces are similar and concentric ellipsoids, and isodensity surfaces of the galactic mass distribution are also similar and concentric. The potential is given by

$$
\Phi(s) = v_c^2 \ln(ks),
\tag{3}
$$

where v_c is the constant circular velocity, k is a scaling factor, $s^2 = x^2 + y^2/p^2 + z^2/q^2$ represents isopotential surfaces, and p and q are the axis ratios of these surfaces. If the galactic mass distribution is not highly flattened, then this potential can be written in the form of equation (1), with the resultant precession rate for circular orbits given by

$$
\begin{aligned}
(di/dt) &= (3v_c/r)\,\beta \sin i \sin 2\Omega, \\
(d\Omega/dt) &= (3v_c/r)\left[\beta \cos i \cos 2\Omega - (\eta/2)\cos i\right].
\end{aligned}
\tag{4}
$$

Here η and β are functions of the axis ratios (b/a) and (c/a) of the galactic mass distribution. If a, b, and c are semiaxes of an isodensity surface along the x, y, and z axes, respectively, and b is the intermediate length axis, then

$$
\eta = \frac{[1 + (b/a)^2 - 2(c/a)^2]}{6[1 + (b/a)^2 + (c/a)^2]}; \qquad
\beta = \frac{[(b/a)^2 - 1]}{12[1 + (b/a)^2 + (c/a)^2]}.
\tag{5}
$$

Note from eq. (4) that the precession rate is proportional to r^{-1}.

To understand the observational implications of equation (4), we examine the visual appearance of twisted dusty disks. Models generated for a range of twisting and viewing orientation are seen in Figure 1. The models

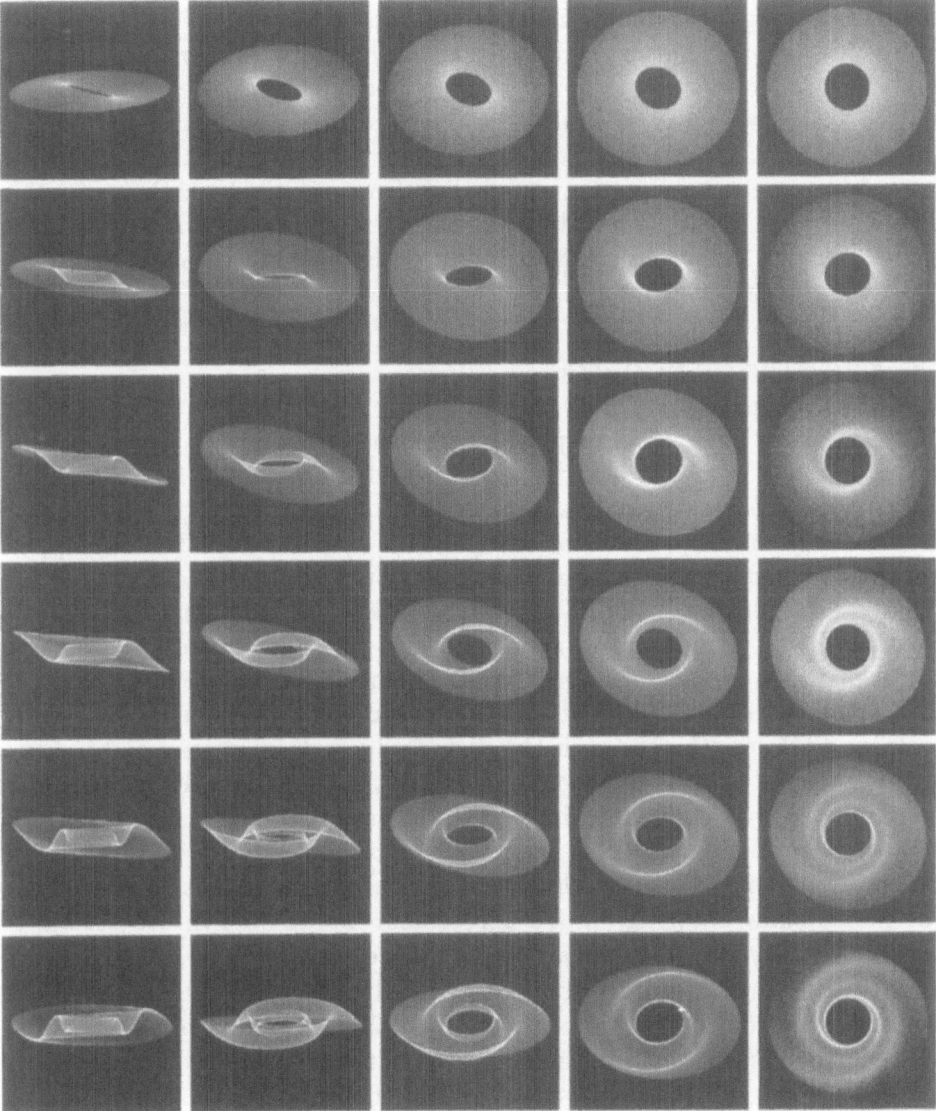

Figure 1. The appearance of a disk twisted by differential precession (see text) for a variety of twisting angles and viewing orientaions. Rows show disks differentially twisted by $n\pi/2$ from inner to outer edge; n is the row number from the top. The leftmost column has the line of sight inclined by 0° to the galaxy's equatorial plane. Viewing angles increase by 20° in successive columns.

represent r^{-1} precession of a disk inclined by 15° to the equatorial plane of an axisymmetric galaxy. At low viewing angles, the twist manifests itself in pronounced curved caustics whose appearance is a strong function of the precession law. The sharpness of the caustics and their detail, particularly in the more strongly twisted disks, thus provide very strong constraints on the precession law and hence the derived mass model. Unfortunately, although systems similar to these may not be rare (see NGC 4753 below and discussion of Steiman-Cameron *et al.* 1992), only certain viewing orientations are conducive to easy identification of a highly twisted disk. More commonly, we might expect to view a twisted disk from relatively high inclinations where, as seen in the figure, dust extinction more strongly resembles spiral waves. Indeed, it is possible that many disks which have been ascribed as possessing such waves may instead harbor twisted disks.

3.1. THE CASE OF NGC 4753

The S0 galaxy NGC 4753 displays prominent and complex appearing dust lanes. A CCD image of the inner portion of the galaxy, processed to emphasize the dust features, is shown in the top panel of Figure 2. Steiman-Cameron, Kormendy, and Durisen (1992), demonstrated that the appearance of the dust lanes can be explained as optical depth effects in a strongly twisted disk. The model implies that the disk is the product of an accretion event and was formed at an inclination of $\sim$ 15° with respect to the equatorial plane of the accreting galaxy. Differential precession has caused the disk to be twisted by 3.8π radians over a factor of seven in radius, leading to large path-lengths through the disk along certain lines-of-sight. Excellent agreement is found between the increased path-lengths in the model and the observed dust lanes in NGC 4753 (see Figure 2). A nodal precession rate proportional to r^{-1}, suggesting a scale-free potential, matches the observations with considerable accuracy. The observed dust structure and the model show amazingly good agreeement out to 90″. In projection, the model disk beyond 90″ appears as an ellipse whose major axis is roughly aligned with the outermost portions of the dust lanes fit by the model. Luminous extensions of the galaxy beyond $\sim$ 120″ appear on both sides of the galaxy at the same position angle as the extended disk model, suggesting either that star formation has occurred in the outer regions of this disk or that a stellar component was accreted along with the gas.

The properties of the twisted disk constrain the shape of the total mass distribution of NGC 4753 over a large range in radius. The model requires that most of the galaxy's mass is unseen, is nearly spherically distributed, and has a nearly scale-free spatial distribution. The shape of the galaxy's total mass distribution, including the dark matter, is oblate. The flattening

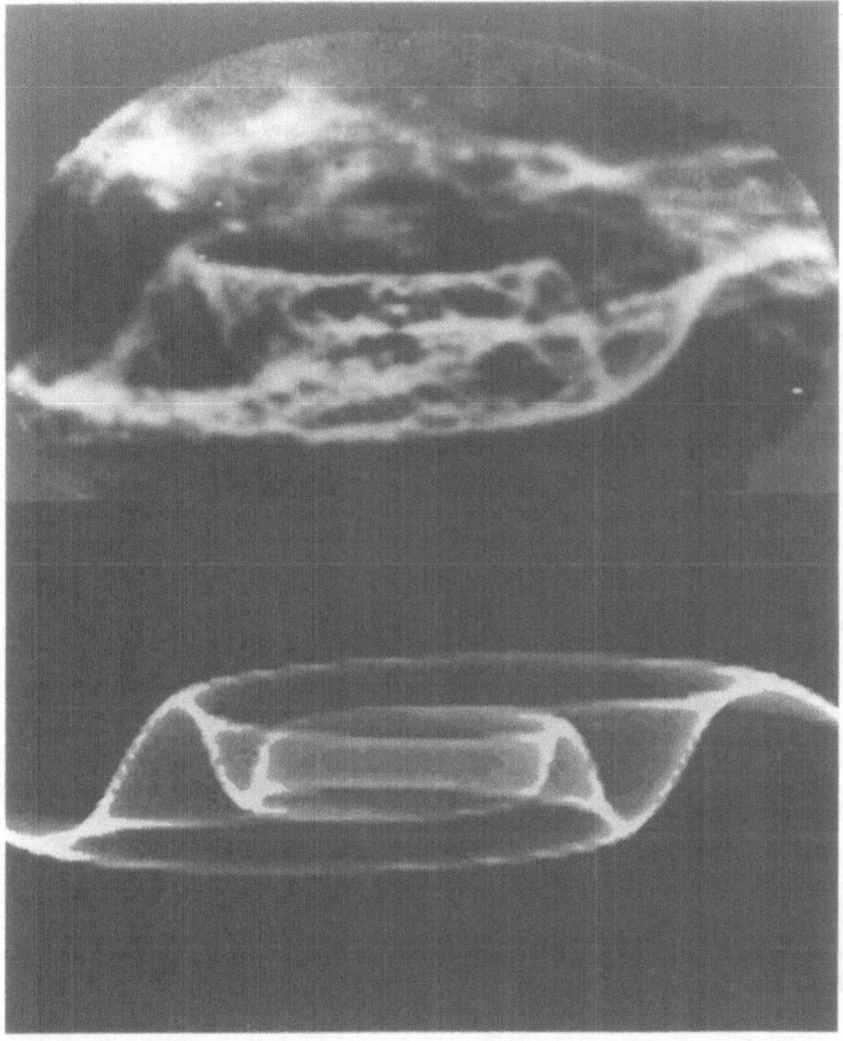

Figure 2. Top: V-band CCD image of the inner regions of NGC 4753 processed to enhance the complex dust features (Steiman-Cameron, Kormendy, & Durisen 1992). North is towards the top and west to the left. The dimensions of the oval area are $84'' \times 58''$. Bottom: Model fit to NGC4753.

of the total mass distribution and the age of the accretion disk cannot be determined independently. However, physical arguments imply that the shape of this distribution is between $\sim$ E0.1 and E1.6. In any case, the flattening of the mass distribution is much less than the flattening of the light distribution.

4. References

Arnaboldi, M., & Sparke, L. S. (1994), *AJ*, **Vol. 107**, 958.

Colley, W. N., & Sparke, L. S. (1996), preprint.

David, L.P., Durisen, R.H., & Steiman-Cameron, T.Y. (1984) *ApJ*, **Vol. 286**, 53.

David, L.P., Steiman-Cameron, T.Y., & Durisen, R.H. (1985) *ApJ*, **Vol. 295**, 65.

de Zeeuw, T., & Merritt, D. (1983) *ApJ*, **Vol. 267**, 571.

Durisen, R.H., Tohline, J.E., Burns, J.A. & Dobrovolskis, A.R. (1983) *ApJ*, **Vol. 264**, 392.

Habe, A., & Ikeuchi, S.. (1985) *ApJ*, **Vol. 289**, 540.

Habe, A., & Ikeuchi, S.. (1988) *ApJ*, **Vol. 326**, 84.

Hoffner, P., & Sparke, L. S. (1994), *ApJ*, **Vol. 428**, 466.

Hunter, C., & Toomre, A. 1969, *ApJ*, **Vol. 155**, 747.

Levison, H. (1987) *ApJL*, **Vol. 320**, L93.

Ostriker, E. C., & Binney, J. (1989) *MNRAS*, **Vol. 237**, 785.

Peletier, R. F., & Christodoulou, D.M. (1992) *AJ*, **Vol. 105**, 1378.

Richstone, D. 1980, *ApJ*, **Vol. 238**, 103.

Sackett, P. D., & Sparke, L.S. (1990) *ApJ*, **Vol. 361**, 408.

Sparke, L. S., & Casertano, S. (1988) *MNRAS*, **Vol. 234**, 873.

Sparke, L.S. (1984) *MNRAS*, **Vol. 211**, 911.

Sparke, L.S. (1986) *MNRAS*, **Vol. 219**, 657.

Sparke, L.S. (1990) in *Dynamics an, Interactions of Galaxies*, ed. R. Wielen, (Berlin, Springer), 338.

Sparke, L.S. (1991) in *Warped Disks an, Inclined Rings Around Galaxies*, eds. S. Casertano, P. Sackett, & F. Briggs (Cambridge, Cambridge Univ. Press), 85.

Steiman-Cameron, T.Y. (1995) in *Three-Dimensional Systems*, Annals of the New York Academy of Science, **Vol. 751**, ed. H. E. Kandrup, S. T. Gottesman, & J. R. Ipser (New York, New York Academy of Sciences), p. 65.

Steiman-Cameron, T.Y., Kormendy, J., & Durisen, R.H. (1992) *AJ*, **Vol. 104**, 1339.

Steiman-Cameron, T.Y., & Durisen, R.H.. (1982) *ApJ*, **Vol. 263**, L51.

Steiman-Cameron, T.Y., & Durisen, R.H.. (1984) *ApJ*, **Vol. 276**, 101.

Steiman-Cameron, T.Y., & Durisen, R.H.. (1988) *ApJ*, **Vol. 325**, 26.

Steiman-Cameron, T.Y., & Durisen, R.H.. (1990) *ApJ*, **Vol. 357**, 62.

Tohline, J.E., Simonson, G.F. & Caldwell, N.. (1982) *ApJ*, **Vol. 252**, 92.

Tohline, J.E., & Durisen, R.H. (1982) *ApJ*, **Vol. 257**, 94.

DUST AND GAS IN LOCAL GROUP GALAXIES

PAUL HODGE

Astronomy Department, University of Washington
Seattle, WA 98195

Abstract. The high spatial resolution possible in studies of Local Group galaxies makes them especially valuable sources of information about the details of the dust and gas content and distribution for galaxies of different Hubble types and absolute magnitudes. Even many of the Local Group elliptical galaxies have detectable gas, measured at 21 cm, and dust, measured both optically and at mm wavelengths. The spirals are now being mapped at high radio resolution with the VLA, at shorter resolution with mm arrays, and at high optical resolution with HST. Irregular galaxies, from the LMC down to extreme dwarfs such as LGS 3, have high-resolution maps of interstellar matter. There is a correlation between the gas to dust ratio and a galaxy's elemental abundances.

1. Introduction

The Local Group of galaxies includes more than 30 objects, the characteristics of which include most of the more common types. It is rich in dwarf galaxies, which are coming to be recognized as probably the cosmically most common kind of galaxy. All of its members are within 1.5 Mpc, so that all can be subjected to detailed scrutiny.

This review briefly discusses our present knowledge about the interstellar matter – the dust and gas – in Local Group (hereafter LG) members. It is shown that there remain gaps in our knowledge of some of the most important components of the interstellar material for some members and that there are large discrepancies for some others. Nevertheless, the collected information is rich enough to be useful for examining certain correlations that can be important for understanding galaxy formation and evolution.

After briefly reviewing the sources of information available for the in-

D. L. Block and J. M. Greenberg (eds.), New Extragalactic Perspectives in the New South Africa, 309–323.
© 1996 *Kluwer Academic Publishers.*

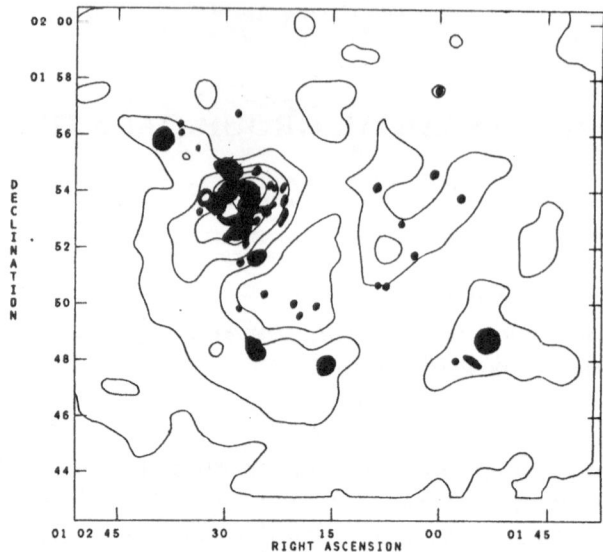

Figure 1. The spatial relationship between the HI and the HII regions in the LG dwarf irregular galaxy IC 1613 (from Hodge et al. 1990).

terstellar matter in LG galaxies, the review turns to an accounting of our knowledge for 31 galaxies that are certain or probable members. **References are given only as examples**, because complete references for LG galaxies are available elsewhere (Layden, Smith and Storm 1994; Hodge and Patterson 1996). Then the review turns to the question of the dust-to-gas ratio for the members for which this important quantity is known.

2. The Data

2.1. GAS

Gas in Local Group galaxies has been detected in the forms of HI, H_2, HII, CO and other molecules (Wilson 1994), although for most galaxies the only data available are for HI and HII. For most galaxies, we have HI maps, which are being obtained at increasingly better resolution with arrays (e.g., Brinks 1984). For all LG members, we have total HII region fluxes, allowing us to estimate the current star formation rates in each (Kennicutt 1983; Hodge 1993). Furthermore, maps of excited hydrogen can be compared with the HI maps to find relationships between gas density differences and star formation activity (Elmegreen and Lada 1977; Hodge, Lee and Gurwell 1990).

2.2. DUST

Interstellar dust in LG galaxies has been detected at optical, infrared and millimeter wavelengths. Potential information can come optically from reddening (e.g., Massey, Armandroff and Conti 1986), although to date this method has been used only for certain limited areas in most LG galaxies. Optical measures of extinction are more difficult to determine, except in special cases, such as for Cepheids (van Genderen 1975) or discrete dust clouds (Hodge and Kennicutt 1982). Polarization measurements exist for a very limited sample of LG galaxies (e.g., Martin and Shawl 1979).

Near infrared data give minimal information about the dust in most LG galaxies because of the usually overwhelming stellar contribution. However, at IRAS wavelengths it is possible to detect dust emission from the warm dust component, and several LG galaxies have useful IRAS-derived data on dust (Jura 1986; Rice et al. 1988).

Millimeter and sub-millimeter data for LG galaxies are still rather scarce, but they hold the possibility of defining the amount and distribution of the cold dust component, as is demonstrated by several papers at this Cold Dust Conference.

3. Dust and Gas in Individual LG Members

3.1. THE SPIRALS

All three of the LG spirals are well-mapped in HI. In some cases, however, CO observations are somewhat spotty, but they are reaching an increasingly satisfactory degree of completeness (M31: Dame et al. 1993; M33: Wilson and Scoville 1989; the Milky Way Galaxy, hereafter MWG: Scoville and Sanders 1987).

Generally the HI is distributed in a disk that has a larger scale length than the optical stellar disk and there is a hole in HI at the galaxy center. At least to some extent this hole is partially filled with molecular gas. For M31, for example, the CO distribution, while concentrated to the 10 kpc ring of high HI density, is displaced inward by about a kpc and has a comparatively higher concentration inside the ring than the HI (Koper et al. 1991). Measurements of two optically-selected dust clouds in the inner disk of M31 (D268 and D478) have detected ^{12}CO and ^{13}CO emission. The ^{12}CO emission comes from low-density, very cold gas, while the ^{13}CO emission is probably from higher-density clumps inside the giant molecular clouds (Allen, Le Bourlot, Lequeux, Pineau des Forets and Roueff 1995). Compared to the MWG, the data indicate a low UV radiation field and a low cosmic ray intensity.

With regard to dust content, the integrated properties of M31 and M33 are significantly different from each other, with the MWG falling at an intermediate position. M31 is relatively deficient in far ir radiation. It has approximately 1500 solar masses of cool dust in its bulge (Soifer et al. 1986), while its disk shows a widespread cold component, with a spotty warm component that is concentrated near star-forming regions (Walterbos and Schwering 1987). The infrared structure of M33 and the distribution and heating of its dust has been explored by Rice et al. (1990).

Optical data on dust are available for these galaxies, but only for limited regions and special objects. For example, the Balmer decrement in HII region spectra for M31 has been used to show that the global dust distribution shows a minimum near the center and maximum just inside the 10 kpc ring (Kumar 1979). Reddening of early-type stars suggests that the maximum is more nearly at the 10 kpc ring (Hodge and Lee 1988).

3.2. THE MAGELLANIC CLOUDS

Because of their proximity, the Magellanic Clouds dominate our information on interstellar matter in LG irregular galaxies. Marvelously-detailed maps of the HI in them are now being generated by the Australia Telescope. Both Clouds are relatively rich in HI, with ratios of gas mass to total mass on the order of 10% or more. The HI mass is better known than the total mass, because the latter depends on rather poorly-defined rotation parameters. CO maps of the Clouds are available (LMC: Cohen et al. 1988; SMC: Rubio et al. 1991). For the LMC, it is found that the densest molecular features lie in regions of intense star-forming activity, such as the 30 Doradus area and the region just south of it. The SMC has several molecular clouds, which are similar in many respects to the smaller examples in the MWG. In mass they average about 5×10^4 solar masses. As in other low-heavy-element-abundance galaxies, the clouds are underluminous in CO compared to their dynamical masses.

Infrared observations of the Magellanic Clouds by IRAS have shown that these galaxies have readily-detected cold and warm dust, but that the dust mass is relatively small (Schwering 1988). The total mass of dust detected by IRAS in the LMC, for example, is 6×10^5 solar masses, which is 10^{-4} of the total dynamical mass of the galaxy.

Optical data on dust content are available for the Magellanic Clouds, most completely for the SMC. For the LMC, in spite of hundreds of measures of optical reddening, extinction and the recognition of many visible dust clouds, the total dust content and its optical effects are still rather poorly known.

The history of this topic is relatively sparse. The first attempt to be quantitative about it was made by Shapley and Nail (1951), who counted galaxies seen behind the SMC and concluded that the Cloud is optically thin, showing no detectable extinction (as a footnote, it is probably this observation that lead Baade to the conclusion that the SMC must be a Population II galaxy). The Shapely-Nail result was disputed by the results of Wesselink (1961), who used larger scale plates and found a substantial decrease in galaxy density near the core of the SMC. This disagreement remained unresolved until 1974, when a third study showed that Shapley and Nail's assistants had included both star clusters and gaseous nebulae in their counts of non-stellar objects in the SMC fields, thereby cancelling out the effects of extinction (a result reported briefly in Hodge 1974). New observations, based on a much larger-scale survey, corroborated Wesselink's data and indicated that the extinction by dust varies from 1.3 mag (in V) in the core to about 0.2 mag at a distance of 2 degrees from the core (Hodge 1974). A similar result was obtained by MacGillivray (1975).

For the Large Cloud, only one attempt has been made so far (Gurwell and Hodge 1990). On ground-based photographic plates, its more complex structure and content makes it particularly difficult to distinguish background galaxies from faint clusters, nebulae and asterisms. Therefore, although the paper cited did seem to detect the effects of the total extinction due to dust (about 1.5 mag in V at the center), it was published as a highly-tentative result.

All of these studies used crude methods on non-linear data: searches by eye on photographic plates. They are subject to large systematic errors, both because of the largely non-reproducable and uncontrollable nature of "by eye" searches, but also because of the vagaries of the photographic emulsions. And, of course, they suffer from the crudity of ground-based seeing. Obviously, what is needed to do the job right is HST images analyzed by impartial, automated techniques, and this is now in progress.

There are three different methods for measuring the total extinction by dust in the Magellanic Clouds using background galaxies:

1. For the brighter galaxies the integrated colors can be determined and compared to those of similar Hubble types in the field, which follow a fairly narrow color-type relationship (de Vaucouleurs and Longo 1983). This technique was first tried for the LMC by Hodge (1969) and lists of bright galaxies have been assembled for this purpose for the SMC (Hodge and Snow 1975) and the LMC (Gurwell and Hodge 1990). With HST images, we are extending these lists to much fainter levels (from the limit of $V = 16.5$ to at least $V = 18.5$), as Hubble types are possible for much smaller, fainter images. This increase in the number of classifiable galaxies is important, because the method must have a reasonable number of galaxies to make up for the intrinsic spread in the color-type relationship; for example, de

Vaucouleurs (1961) quantitatively analyzed this feature of the relationship and derived a mean systematic color residual of 0.02 mag for his sample of 600 galaxies.

2. For a selection of galaxies that goes to somewhat fainter limits, it is possible to measure magnitudes in three (and in some cases, four) colors, allowing a determination of the reddening by comparison with the field galaxy color-color relations (de Vaucouleurs and Longo 1983).

3. Galaxies of all brightnesses can be counted, using automated techniques and the counts can be compared to galaxy densities in the field. The decrement can then be interpreted in terms of extinction using the apparent magnitude function for galaxies in the comparison fields. (In the earlier studies, this transformation was made assuming a uniform spatial distribution for galaxies, but one needs to be more precise to account for the effects of the red-shifts and evolution in the deeper sample that has been provided by HST.)

3.3. THE OTHER IRREGULAR GALAXIES

Besides the Magellanic Clouds, there are seven other traditional irregular galaxies in the LG (non-traditional examples are the "hybrid" galaxies discussed in Section 3.5). These are IC10, NGC 6822, IC 1613, Sextans A, Sextans B, SagDIG, and WLM. The first three have been mapped at 21 cm at high resolution, IC 10 by Hodge, Wilcots and Miller (1995), NGC 6822 by Skillman (1987) and IC 1613 by Wilcots (1996). All have large HI envelopes that extend considerably beyond the optical images of the galaxies. The map of NGC 6822 has a quasi-barred structure that lies on top of the optical bar, but that has extensions, almost arm-like in shape, that extend out to more than twice the distance from the center of the outermost detectable stars (Hodge et al. 1991).

HII regions have been studied in all seven of these irregular galaxies. Those in NGC 6822 have been analyzed the most extensively (references are in Collier et al. 1995). Their structural properties have been subjected to multivariate statistical analyses, allowing comparisons with evolutionary models. A remarkable example is the HII region known as Hubble III, which is found to be a wind-blown bubble, expanding at 50 km/sec, with two sub-bubbles within the ring (Clayton 1987). The galaxy poorest in excited hydrogen is SagDIG, which has three emission objects detected, two of which are, however, thought to be chance superpositions of Galactic emission-line stars (Skillman et al. 1989; Strobel et al. 1991).

Klein and Grave (1986) have mapped radio continuum emission from NGC 6822, IC 1613 and IC 10. Radio emission is concentrated, as expected, to the star-forming areas of these galaxies. Four objects, IC 10, NGC 6822,

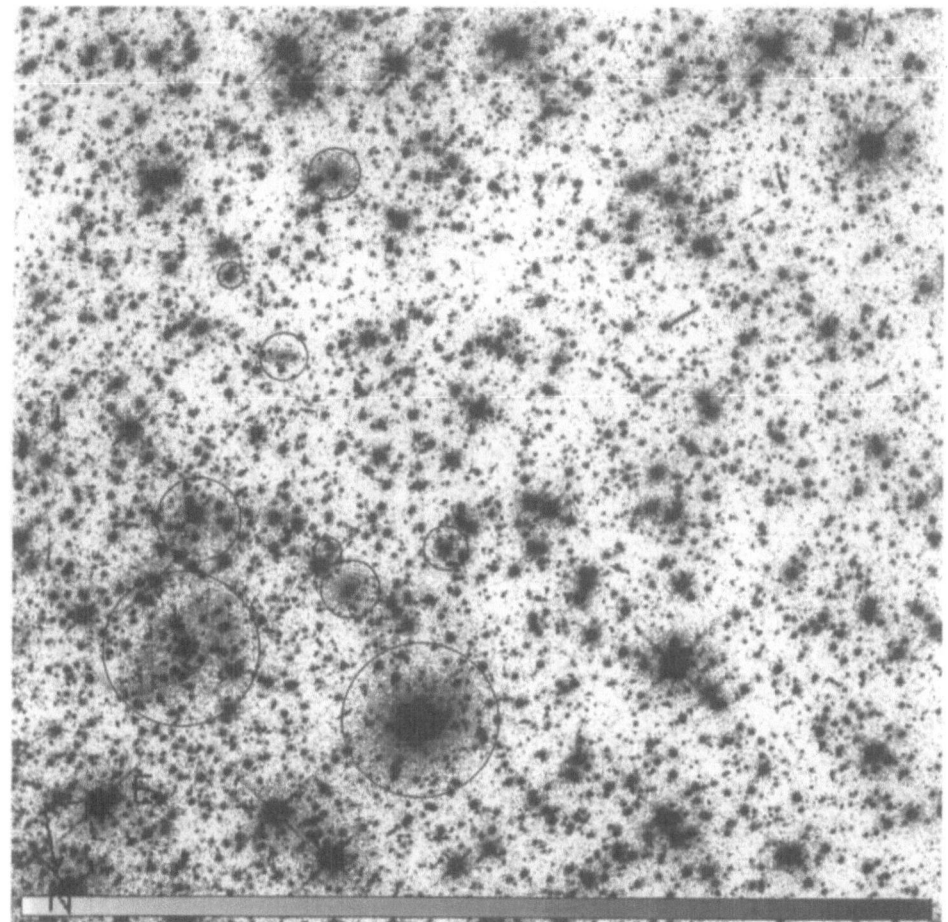

Figure 2. An HST image of a cluster of galaxies seen through a central area of the LMC.

IC 1613 and WLM, have been detected at IRAS wavelengths and have estimated cool dust masses (Schmidt and Boller 1992). Optical dust clouds are very weak and difficult to detect (see Hodge 1977 for NGC 6822 and Hodge 1978 for IC 1613), apparently because of the low average dust content of these low-metal abundance objects (Section 4).

3.4. M32 AND NGC 205

The two brightest elliptical galaxies in the LG are the two nearest companions to M31, M32 and NGC 205. The former's image is seen superimposed on that of M31 and this circumstance makes the detection of any subtle optical evidence for interstellar material very difficult. There is at present

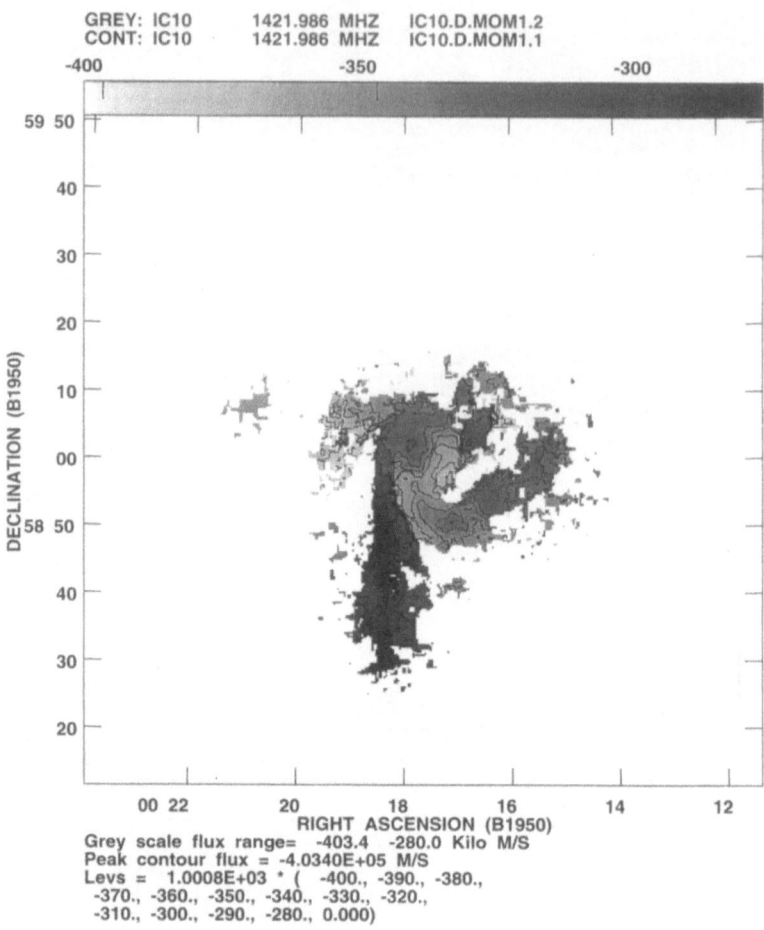

GREY: IC10   1421.986 MHZ   IC10.D.MOM1.2
CONT: IC10   1421.986 MHZ   IC10.D.MOM1.1

Grey scale flux range=   -403.4   -280.0 Kilo M/S
Peak contour flux = -4.0340E+05 M/S
Levs =   1.0008E+03 * (-400., -390., -380.,
   -370., -360., -350., -340., -330., -320.,
   -310., -300., -290., -280., 0.000)

Figure 3.   A VLA velocity map of the HI in IC 10 (courtesy E. Wilcots).

no evidence for any appreciable dust or gas in this galaxy. Its orbit may carry it through the plane of M31, in which case it may be periodically stripped of any gas generated by stellar evolution or accumulated from the intergalactic reservoir.

NGC 205, on the other hand, contains both gas and dust. There is an HI cloud of mass 3.7×10^5 solar masses and slightly displaced from the center of the galaxy (Johnson and Gottesmann 1983). Weak CO emission has also been detected (Sage and Wrobel 1989). Twelve optical dust clouds have been mapped and they represent a total mass of approximately 800 solar masses (Hodge 1973; Price and Grasdalen 1983). NGC 205's dust has also been detected at infrared and mm wavelengths (Fich and Hodge 1991). A

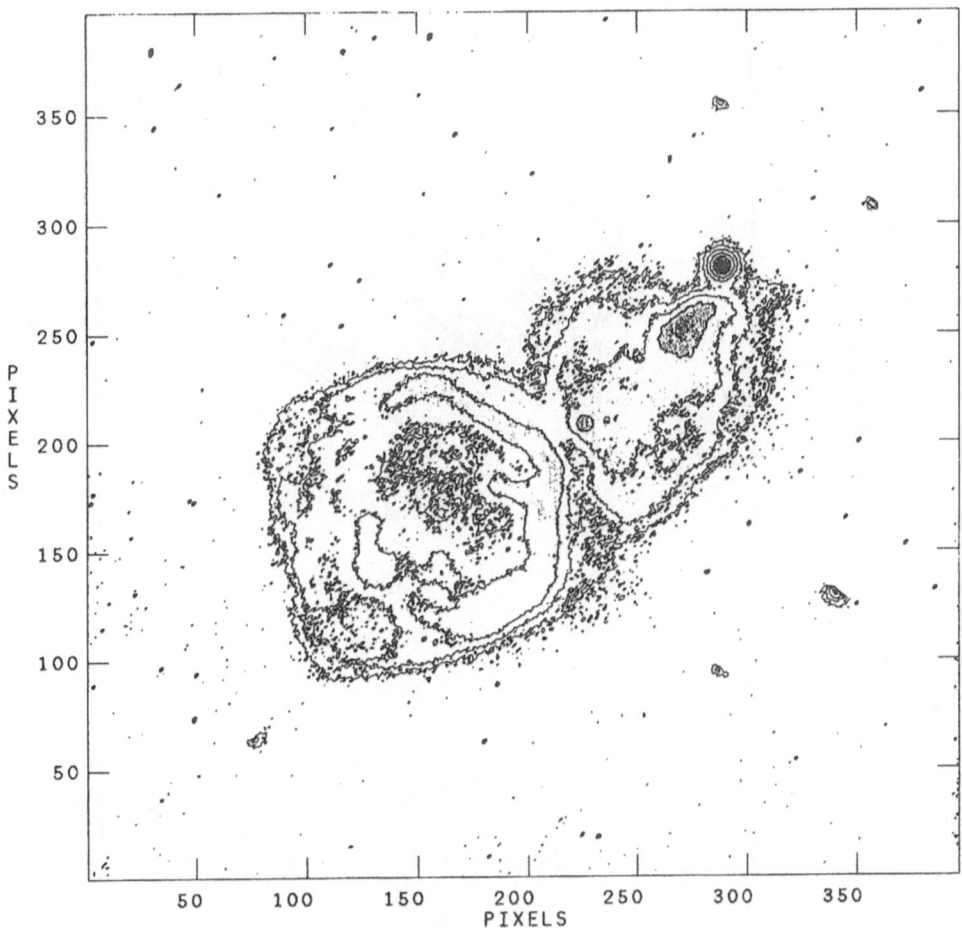

Figure 4. Isophotal maps of the contrasting HII regions, Hubble I (right) and Hubble III (left) in NGC 6822. Hubble I is excited by a single hot star that apparently does not have a strong wind, as there is no detectable expansion, while Hubble III is a wind-blown bubble.

fit to the spectrum determined from measurements with the IRAS satellite and the JCMT mm telescope indicate a mass of $\sim 2 \times 10^{-3}$ solar masses of cold dust at a temperature of approximately 23 K.

3.5. OTHER ELLIPTICAL AND SPHEROIDAL GALAXIES

Among the 15 other elliptical (and/or dwarf spheroidal; it is not now clear how these two terms should be applied) galaxies of the local group, only three have detected HI gas. One of these is NGC 185, which has very sim-

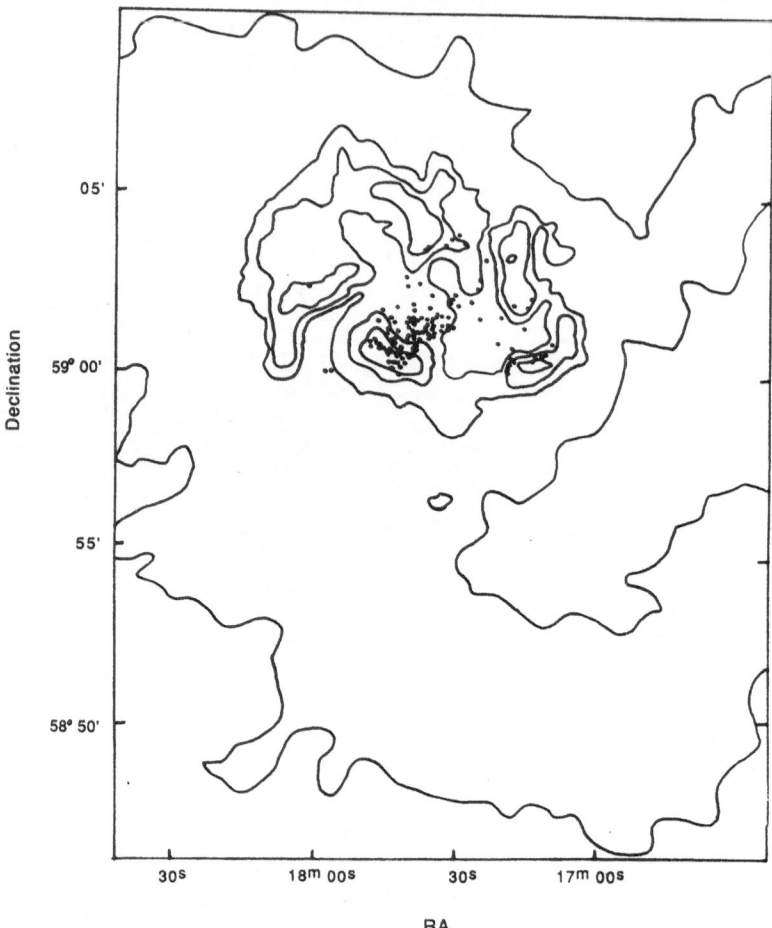

Figure 5.   The HII regions in the central area of IC 10 (from Hodge and Lee 1990; see also Hunter and Gallagher 1985). This highly-obscured galaxy, with a galactic latitude of only 3 degrees, is the only starburst galaxy in the LG. In addition to these luminous HII regions, it is rich in WR stars (Massey et al. 1992), it has two water vapor masers (Henkel et al. 1996), molecular gas (Ohta et al. 1992), and multiple supernova remnants (Yang and Skillman 1993).

ilar interstellar components to those of NGC 205, including HI (Johnson and Gottesmann 1983), CO (Wiklind and Rydbeck 1986) and dust clouds (Hodge 1963). The other two are Sculptor and Tucana (Carignan et al. 1996), two dwarf spheroidals that otherwise are made up primarily (but not entirely) of very old stars. Dust is also rare in these low mass objects; none has significant detected ir or mm radiation and none has optical dust clouds or measurable extinction. Background galaxies are seen through

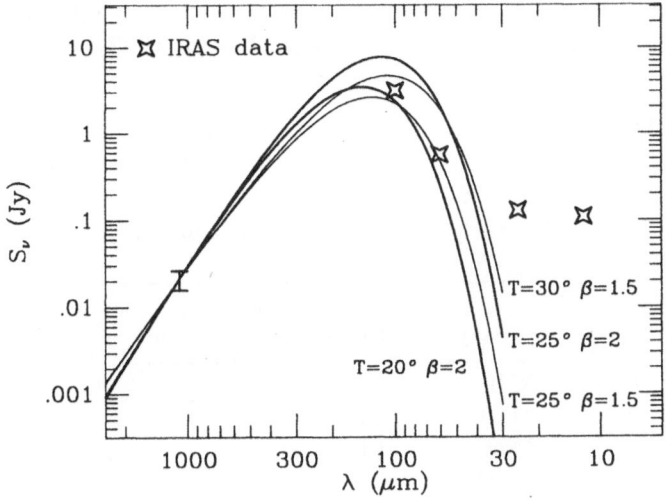

Figure 6.   The flux density from NGC 205's inner region at 1.1 mm, shown with 1σ
uncertainties. IRAS points for its four wavelengths are also shown. The lines are fits for
different temperatures and different emissivity power law indices (from Fich and Hodge
1991).

them apparently undiminished in brightness.

3.6. HYBRID GALAXIES

As the borderline between elliptical and irregular galaxies becomes less and
less well-defined, some galaxies, even in the LG, are being found to have
intermediate (or composite) characteristics. One of these is the Phoenix
dwarf galaxy, which looks optically like a dwarf elliptical galaxy with a
small superimposed population of young stars. Its HI gas is distributed
irregularly and shows primarily turbulent motions (Carignan et al. 1991).
There are about 10^5 solar masses of HI. Although there are some fairly
bright main sequence stars present (van de Rydt et al. 1991), no HII regions
have been reported.

The other "hybrid" case is the object known as LGS 3 (Thuan and
Martin 1979). Optically it looks like a typical dwarf spheroidal and among
its stars there are no apparent young members, just a red giant branch
(Cook and Olszewski 1989). However, it contains HI, which is spread out
over a 1 kpc area, with a central concentration and two local maxima (Lo et
al. 1993). The HI mass is $\sim 2 \times 10^5$ solar masses. There are no HII regions.

It is likely that these hybrid objects are not really anomalous, but rather that the LG's low-mass galaxies frequently generate or acquire enough gas (and dust) to have episodes of star formation at various times during their existence. Based on LG statistics, it appears that so far they have been quiescent for about 75% of their lifetimes (Hodge 1993).

4. Gas to Dust Ratios

TABLE 1.   Table I. Dust and Gas Masses for LG Galaxies

| Galaxy | log M_{HI}/M_{dust} | | |
	IRAS	Optical	mm
M31	2.0	–	–
MWG	2.0	–	–
M33	1.8	–	–
LMC	3.0	–	–
IC 10	3.5	–	–
NGC 6822	3.9	–	–
NGC 205	–	2.7	2.3
SMC	4.3	2.5	–
NGC 185	–	2.9	–
IC 1613	5.9	–	–
Sextans B	>5.0	–	–
WLM	5.1	–	–
LGS 3	4.1	–	–

The gas to dust ratio for a galaxy is a useful parameter with which to examine certain predictions and correlations (Israel 1984). Because dust must be manufactured in galaxies through star formation, it might be expected that this ratio would change with time in a galaxy, more or less in parallel with the change in metal abundances. Various attempts have been made to trace this change indirectly by looking for a correlation between this ratio and the mean metal abundances for different galaxies (Issa et al. 1990; Schmidt and Boller 1993). Table 2 lists data on the inferred dust masses for galaxies in the LG for which data are available. The IRAS dust mass estimates are clearly lower limits, as they do not include cold dust, which may dominate in galaxies with low ambient ultraviolet flux. That this is true is suggested by the large amount of cold dust in NGC 205 and NGC 185 that is not detected by the IRAS measurements.

Figure 9 shows a plot of gas to dust ratio versus heavy element abundance (plotted in terms of [Fe/H], but based in most cases on the measured

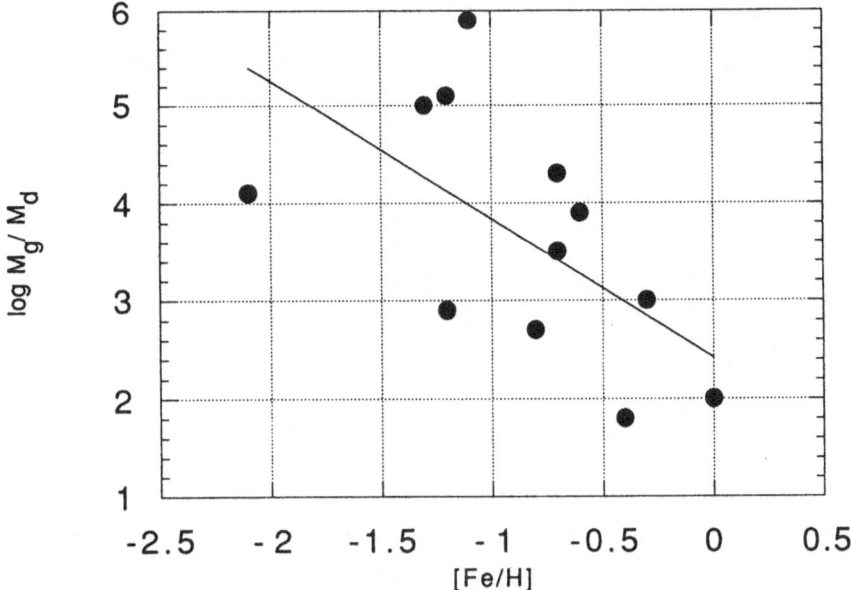

Figure 7. The relation between the calculated gas to dust ratios and the heavy element abundances for LG galaxies.

O/H abundance) from the data for the LG galaxies of Table 2. There does seem to be a correlation, but not a very strong one (for comparison, Fig. 10 shows the much better correlation between absolute magnitude and heavy element abundance for the same sample of LG galaxies). The line plotted in Figure 9 is a linear least squares fit that implies a slope of −1.4. This suggests that, relative to the gas content, the dust mass of a galaxy varies more steeply than the heavy element abundance. If we make the assumption that galaxy (chemical) evolution can be traced by looking along the galaxy luminosity sequence (e.g., along Fig. 10's points), then it can be concluded that the dust is created at a higher rate, in the mean, than the heavy elements.

References

Allen, R. J., LeBourlot, J., Lequeux, J., Pineau des Forets, G. and Roueff, E (1995) *Ap. J.* **444** 157

Brinks, E. (1984) Ph.D. Thesis Univ. Leiden

Carignan, C., Demers, S. and Cote, S. (1991) *Ap. J.* **381** L13

Carignan, C., Cote, S., Demers, S. and Mateo, M. (1996) *preprint*

Clayton, C. A. (1987) *MNRAS* **226** 493

Collier, J. and Hodge, P. W. (1994) *Ap. J. Suppl.* **92** 119

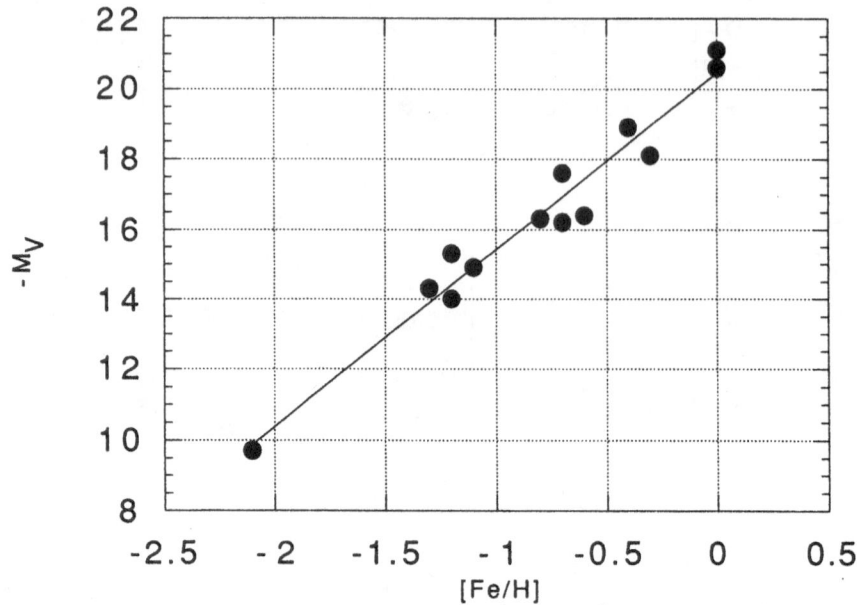

Figure 8.   The relation between the absolute magnitude of the galaxies and their heavy element abundances for the sample of LG galaxies in Table 1.

Collier, J., Hodge, P. W., and Kennicutt, R. C. (1995) *PASP* **107** 361

Cohen, R. S., Dame, T., Garay, G., Montani, J., Rubio, M. and Thaddeus, P. (1988) *Ap. J.* **331** 95

Cook, K. H. and Olszewski, E. (1989) *BAAS* **21** 775

Dame, T. M., Koper, E., Israel, F. P. and Thaddeus, P. (1993) *Ap. J.* **418** 730

de Vaucouleurs, G. (1961) *Ap. J. Suppl.* **5** 233

de Vaucouleurs, A. and Longo, G. (1983) *U. Tex Monog. in Ast.* No. 3

Elmegreen, B. and Lada, C. (1977) *Ap. J.* **214** 725

Fich, M. and Hodge, P. W. (1991) *Ap. J.* **374** L17

Gurwell, M. and Hodge, P. (1990) *PASP* **102** 849

Henkel, C., Wouterloot, J. and Bally, J. (1986) *A. and Ap.* **155** 193

Hodge, P. (1963) *A. J.* **69** 438

Hodge, P. (1969) *SAO Sp. Reports* 306

Hodge, P. (1973) *Ap. J.* **182** 671

Hodge, P. (1974) *Ap.J.* **192** 21

Hodge, P. W. (1977) *Ap. J. Suppl.* **33** 69

Hodge, P. W. (1978) *Ap. J. Suppl.* **37** 145

Hodge, P. W. (1993) in *ESO Conference and Workshop Proceedings* **49**, G. Meylan and
    P. Prugniel ed(s)., *501*

Hodge, P. W. and Kennicutt, R. C. (1982) *A. J.* **87** 264

Hodge, P. W. and Lee, M. G. (1988) *Ap. J.* **329** 651

Hodge, P. W. and Lee, M. G. (1990) *PASP* **102** 26

Hodge, P., Lee, M. G. and Gurwell, M. (1990) *PASP* **102** 1245

Hodge, P. W. and Patterson, B. (1996) in *Atlas of Local Group Galaxies*, Kluwer: Dordrecht, in press

Hodge, P. W., Smith, T., Eskridge, P., MacGillivray, H. and Beard, S. (1991) *Ap. J.* **379** 621

Hodge, P. and Snow, T. (1975) *A. J.* **80** 9

Hodge, P., Wilcots, E. and Miller, B. (1994) *BAAS* **26** 1436

Hunter, D. and Gallagher, J. (1985) *Ap. J. Suppl.* **58** 533

Israel, F. (1984) in *Structure and Evolution of the Magellanic Clouds*, S. van den Bergh and K. de Boer, eds., Reidel: Dordrecht, 319

Issa, M., Mac Laren, I. and Wolfendale, A. (1990) *A. and Ap.* **236** 237

Johnson, D. and Gottesmann, S. (1983) *Ap. J.* **275** 549

Jura, M. (1986) *Ap. J.* **306** 483

Kennicutt, R. C. (1983) *Ap. J.* **272** 54

Klein, U. and Grave, R. (1986) *A. and Ap.* **161** 155

Koper, E., Dame, T., Israel, F. and Thaddeus, P. (1991) *Ap. J.* **383** L11

Kumar, C. (1979) *Ap. J.* **230** 386

Layden, A., Smith, R. C. and Storm, J., ed(s). (1994) in *The Local Group: Comparative and Global Properties*, ESO Conference and Workshop Proceedings, 51

Lo, K. Y., Sargent, W. L. W. and Young, K. (1993) *A. J.* **16** 507

MacGillivray, H. (1975) *MNRAS* **170** 241

Martin, P. and Shawl, S. (1979) *Ap. J.* **231** L57

Massey, P., Armandroff, T. and Conti, P. (1986) *Astron. J.* **92** 1303

Massey, P., Armandroff, T. and Conti, P. (1992) *A. J.* **103** 1159

Ohta, K., Sasaki, M. and Saito, M. (1988) *P.A.S. Japan* **40** 653

Price, J. and Grasdalen, G. (1983) *Ap. J.* **275** 559

Rice, W., Boulanger, F., Vialleford, F., Soifer, B. T. and Fredman, W. (1990) *Ap. J.* **358** 418

Rice, W., Lonsdale, C., Soifer, B., Neugebauer, G., Kopen, E., Lloyd, L., de Jong, T. and Habing, H. (1988) *Ap. J. Suppl.* **68** 91

Rubio, M., Garay, G. Montani, J. and Thaddeus, P. (1991) *Ap. J.* **368** 173

Sage, L. and Wrobel, J. (1989) *Ap. J.* **344** 204

Schmidt, K.-H. and Boller, T. (1992) *Astro. Nach.* **313** 189

Schmidt, K.-H. and Boller, T. (1993) *Astr. Nach.* **314** 361

Schwering, P. (1988) *Ph.D. Thesis* Univ. Leiden

Scoville, N. and Sanders, D. B. (1987) in *Interstellar Processes*, D. Hollenbach and H. Thronson, eds., Kluwer: Dordrecht, 21

Shapley, H. and Nail, V. (1951) *PNAS* **31** 133

Skillman, E. (1987) in *Star Formation in Galaxies*, C. Lonsdale Persson, ed., *NASA*, CP-2466, 263

Skillman, E., Kennicutt, R. C. and Hodge, P. W. (1989) *Ap. J.* **347** 875

Skillman, E., Terlevich, R. and Melnick, J. (1989) *MNRAS* **240** 563

Soifer, T. B., Rice, W., Mould, J., Gillett, F., Robinson, M. and Habin, H. (1986) *Ap. J.* **304** 651

Strobel, N. V., Hodge, P. W. and Kennicutt, R. C. (1991) *Ap. J.* **383** 148

Thuan, T. and Martin, G. (1979) *Ap. J.* **232** L11

van de Rydt, F., Demers, S. and Kunkel, W. (1991) *A. J.* **102** 30

van Genderen, A. (1973) *A. and Ap.* **24** 47

Walterbos, R. and Schwering, P. (1987) *A. and Ap.* **180** 27

Wesselink, A. (1961) *MNRAS* **122** 503

Wiklind, T. and Rydbeck, G. (1986) *A. and Ap.* **164** L22

Wilcots, E. (1996) in preparation

Wilson, C. D. (1994) in *ESO Conference and Workshop Proceedings*, G. Meylan and P. Prugniel, eds., **51**, 15

Wilson, C. D. and Scoville, N. (1989) *Ap. J.* **347** 743

Yang, H. and Skillman, E. (1993) *A. J.* **106** 1448

THE NUCLEAR REGIONS OF M31, M32, AND M33 IMAGED BY HST

R.M. RICH, K.J. MIGHELL AND J.D. NEILL
Columbia University
538 W. 120th St. NY NY 10027

AND

W.L. FREEDMAN
Observatories of the Carnegie Instiution of Washington
813 Santa Barbara St., Pasadena, CA 91101

The nearest massive members of the Local Group– M31, M32, and M33– give us a window on the stellar populations and dust in the central 100 pc of galaxies. We report new imaging of the stellar populations and dust in the central 100 pc of these galaxies using HST and WFPC2.

Our main interest is in the stellar populations. Early photometry of the M31 bulge found a population of centrally concentrated luminous giants (Mould 1986) which were shown by Rich & Mould (1991) to be unusually bright in the infrared $M_{bol} < -4$. This result was attributed to disk contamination (Davies et al. 1991) and to crowding (Depoy et al. 1993) which could in principle add spurious bright star populations. Larger infrared surveys (Rich et al. 1993), pre-refurbishment HST imaging (Rich & Mighell 1995), and discovery of an IR bright extended giant branch in M32 (Freedman 1992) supported these early findings. However the M31 bulge is as red and strong-lined as luminous giant elliptical galaxies: no young population can be hidden. Bica et al. (1991) show that the most metal rich old stars in M31 should be no brighter than $M_I = -2$, or $I = 22.5$. The brightest metal poor globular cluster giants should be at $I = 20.5$, not $I = 19$ as observed. Our HST/WFPC2 images in V, I, and F1042M($= 1\mu$m) are intended to measure the brightest stars in the M31, M32, and M33 bulges.

Photometry in the less calibrated 1μm F1042M filter confirms that stars in the M31 bulge, M32, and M33 extend > 1 mag brighter than those in the old cluster G1. Secure results can come only from imaging in the the standard I=F814W bandpass (Holtzman et al. 1995) but we are still at the

D. L. Block and J. M. Greenberg (eds.), New Extragalactic Perspectives in the New South Africa, 325–327.
© *1996 Kluwer Academic Publishers.*

confusion limit, even with full HST resolution. Every detected star in the M31 bulge fields is brighter than the canonical $M_I = -4$ red giant branch tip of old metal poor Galactic Globular clusters. We use the same methods to photometer the M31 globular G1 and show that its prinicipal sequences match 47 Tuc, which gives us confidence in our methods. We will report our final results after extensive artificial star addition tests.

Dividing the V by the I images, we also reveal the dust environment of these nuclei. Color plates 8a and 8b (these *Proceedings*) and Figure 1 show that M31 and M33 have considerable complex dust absorption in the inner 100 pc, but M32 has no obvious dust. The giants in the M31 and M32 nuclei should be losing mass at the same rate, yet M31 is more dusty. For the first time, the dust of M31 is seen to resolve into many relatively small knots, while that of M33 is more uniform. The spiral structure of the M31 dust is reminiscent of the Hα emission mapped by Ciardullo et al. 1988 but was actually first clearly seen in the enhanced photograph of Johnson & Hanna (1972) and is documented in Hodge's (1980) Fig. 11. However, our HST image shows for the first time that dense condensations are present and that there is no extension of the dust toward the nucleus itself.

We mapped the nuclei of these galaxies (color plate 9) to search for dust on small scales. The peculiar double morphology of the M31 nucleus is not due to normal dust, confirming Lauer et al. 1993, but our 1μm image clearly shows the double structure, strong confirmation that dust is not responsible for this morphology. In color plate 10, color-color divisions again show that dust is not present, but P2 (dynamical center) is brighter than P1 in the ultraviolet.

This is an early report on work in progress. The stellar crowding is so severe that it remains a challenge to the restored HST. We expect to report improved photometry and dust imaging within the year. We are grateful to Paul Hodge for helpful comments on this work.

References

Bica, E., Barbuy, B., & Ortolani, S. 1991, ApJ, 382, L15
Ciardullo, R., Rubin, V.C., & Jacoby, G.M. 1988, AJ, 95, 438
Davies, R.L., Frogel, J.A., & Terndrup, D.M. 1991, AJ, 102, 1729
DePoy,D.L., Terndrup, D.M., Frogel, J.A., Atwood, B. & Blum, R. 1993, AJ, 105, 2121
Freedman, W.L. 1992, AJ, 104, 1349
Hodge, P.W. 1980, AJ, 85, 376
Holtzman, J.A. 1995, PASP, 107, 156
Johnson, H.M., & Hanna, M.M. 1972, ApJ, 174, L71
Lauer, T.R. et al. 1993, AJ, 106, 1436
Mould, J.R. 1986, in *Stellar Populations*, ed. A. Renzini & M. Tosi (Cambridge: Cambridge Univ. Press) p.9
Rich, R.M., & Mighell, K.J. 1995, ApJ, 439, 145
Rich,R.M., & Mould, J.R. 1991, AJ, 101, 1286
Rich, R.M., Mould, J.R., & Graham, J. 1993, AJ, 106, 2252

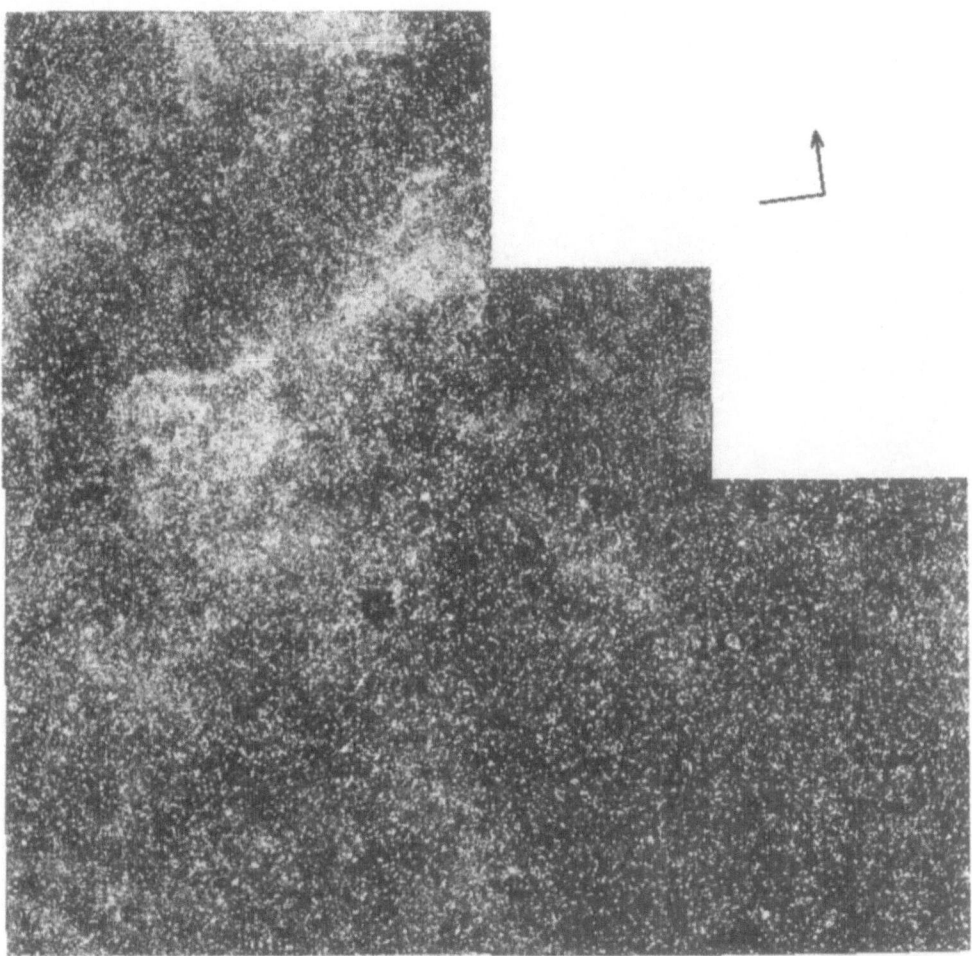

Figure 1.   M33 WFPC2 image formed by dividing V by I, as in plates 8a and 8b in the color plate section. The nucleus lies at the center of the incomplete PC quadrant. Notice that the dust is far less clumped than in M31 (color plate 8a). M33, which is a spiral with substantial star formation extending into the nucleus, also has the most dust of the three galaxies considered.

GAS AND DUST IN NORMAL AND ACTIVE GALAXIES

R. CHINI
Max–Planck–Institut für Radioastronomie
Auf dem Hügel 69, D–53121 Bonn, Germany

AND

E. KRÜGEL
Max–Planck–Institut für Radioastronomie
Auf dem Hügel 69, D–53121 Bonn, Germany

Abstract. We review mm/submm observations of dust in galaxies and show that the optically thin dust emission is a reliable tracer for the total gas mass. We discuss the various problems related to this method, in particular the determination of the dust temperature and the gas–to–dust ratio. We compare the dust emission from three complete FIR flux limited samples of normal spirals, active Mkn galaxies and radio–quiet QSOs. The temperature of the coldest dust component turns out to be a function of both, the luminosity and the activity of a galaxy; it varies from 15 K in non–active galaxies to 80 K in highly redshifted radio galaxies. We propose that the ratio of IR luminosity over gas mass is a unique signature for the global activity of a galaxy.

1. Introduction

The total amount of cold dust M_d in galaxies and its temperature T_d are important quantities because they allow to estimate the gas content M_{gas}, one of the fundamental parameters for a galaxy. While optical and IR methods often suffer from heavy extinction, observations in the mm/submm regime are certainly a potential tool to detect directly the dust and gas content of galaxies, both in continuum and spectroscopic mode.

Continuum observations offer the unique opportunity to trace simultaneously the atomic and molecular gas content of external galaxies via the radiation of interstellar dust. For far IR wavelengths this emission is opti-

D. L. Block and J. M. Greenberg (eds.), New Extragalactic Perspectives in the New South Africa, 329–344.

cally thin and there is a direct proportionality between observed flux and dust mass

$$M_{\mathrm{d}} = \frac{S_\lambda}{\kappa_\lambda \cdot B_\lambda(T_{\mathrm{d}})} \cdot D^2 \ , \tag{1}$$

κ_λ is the mass absorption coefficient of dust, D the distance to the source and S_λ the received flux. For various practical reasons, $\lambda = 1300\,\mu\mathrm{m}$ has turned out to be the ideal wavelength. The uncertainties of this method stem mainly from inadequate knowledge of T_{d} and κ. Because dust grains are small compared to the observational wavelength of $1300\,\mu\mathrm{m}$, κ is independent of grain size and only a function of the optical constants of the dust material. As discussed in detail by Krügel & Chini (1994), a value of $0.4\,\mathrm{cm^2}$ per gram of dust is appropriate for the diffuse interstellar medium. Adopting a gas–to–dust ratio R, which is a further source of uncertainty, one may obtain the gas mass. The few existing mm/submm extragalactic continuum data on M_{d} and T_{d} are heavily debated in the literature. Apart from the above mentioned uncertainties, there are several reasons for the confusion in the field:

1. Measurements are scarce. Therefore, lacking data, galaxy samples of different properties are mixed leading to confusing results, particularly in T_{d}.
2. The large spatial extent of dust emission has sometimes been neglected in previous observing strategies and yielded an underestimate of the total emission due to the differential measuring mode.
3. The combination of large beam *IRAS* data with mm/submm data of high spatial resolution may lead to erroneously high dust temperatures.
4. The decomposition of the observed FIR energy distribution is not unique and results on T_{d} may strongly vary.

For the inner parts of galaxies, where most of the gas is in molecular form, an independent check on the validity of the dust method can be obtained from the lower rotational transitions of CO which also occur in the mm regime. It has been verified empirically that there is a rough proportionality between the mass of molecular hydrogen and the CO luminosity

$$M_{\mathrm{gas}}(\mathrm{H_2}) = \beta \cdot L_{\mathrm{CO}} \ . \tag{2}$$

Values for the conversion factor β from 3 to 11 have been obtained in the Milky Way.

In the following, we review the determination of the gas and dust content in galaxies and present new results on this subject. We will demonstrate that – apart from avoiding the systematic errors mentioned above – it is compulsory to separate normal from active galaxies when discussing the coldest dust component of a galaxy. Furthermore, we will show that comparison between dust and CO data gives insight into the behavior of R, β and κ in extragalactic systems.

2. Normal Spirals

Observations of normal galaxies in the mm/submm regime are difficult. Within the first 10 years of observational effort in this field only 4 active galaxies (NGC 253, NGC 1068, M 82 and Arp 220) could be detected (Hildebrand et al. 1977, Elias et al. 1978, Jaffe et al. 1984 and Thronson et al. 1987). Smith (1982) was the first to observe successfully a non–active spiral, i.e. M 51, at FIR/submm wavelengths. The general results derived from these data are that the spectra turn over around 100 μm and that the emission is due to dust of 20 – 40 K. The sample was enlarged by the observation of 26 galaxies from the IRAS Point Source Catalog at 1300 and partly at 350 μm (Chini et al. 1986, CKKM); 18 of them were classified as Sb or Sc spirals and considered to be non–active. The spectra from 25 to 1300 μm of these 18 spirals were interpreted in terms of two dust components with about 16 and 53 K, adopting a frequency dependence of dust opacity proportional to ν^2.

The measurements of CKKM and the existence of such a cold component were subsequently doubted by Eales et al. (1989, EWD) and Stark et al. (1989). EWD observed 11 spiral galaxies between 350 and 100 μm. The authors interpreted the energy distributions between 60 and 100 μm by thermal emission from dust at a single temperature ($T_d \approx 30 - 50$ K) and found no evidence for cold ($T_d \leq 20$ K) dust. Stark et al. (1989) mapped 4 galaxies at 160 and 360 μm and came to a similar conclusion. Although there is not a single galaxy in these samples which has been observed with a comparable observational setup like the one used by CKKM, i.e. a beam size of 90″, a beam separation of 300″ and a wavelength of 1300 μm, there was the general understanding that the 1300 μm flux densities from CKKM are systematically overestimated by a factor of, at least, ten, possibly due to some observational artifacts.

Carico et al. (1992) readdressed this issue and presented 1250 μm data of 17 galaxies, characterized by unusually high FIR luminosities, with only Arp 220 common to the previous samples. They compared their sample with those of CKKM and EWD and derived that the data from EWD agree well with their measurements while the results of CKKM do not fit with those of their current sample; in particular, they suggested that the 1300 μm data of CKKM are too high, on average by a factor of 20 – 30. Carico et al. claimed that all of their galaxies can be successfully modeled without invoking any dust colder than the dust responsible for the 60 and 100 μm emission that was detected by *IRAS*. Andreani and Francheschini (1992) observed a complete sample of 32 normal galaxies and gave 1250 μm results for 9 of them. Again there was no overlap with the previous samples so that no individual comparison was possible. The major conclusion of their work, however, was that estimates of beam–aperture corrections are essential. Clements *et al.* (1993) gave submm data for 1 interacting, 2 irregular and 2 normal galaxies and derived dust temperatures between 28 and 35 K.

The new bolometer arrays attached to the IRAM 30 m telescope allow an observational technique that avoids some of the uncertainties mentioned above. Several normal spiral galaxies were subsequently mapped with high spatial resolution, among them NGC 891 (Guélin *et al.* 1993), NGC 3627 (Sievers *et al.* 1994), NGC 4631 (Braine *et al.* 1995), M 51 (Guélin *et al.* 1995) and NGC 4565 (Neininger *et al.* 1996); these observations are compatible with a coldest dust component of typically 20 K.

We have embarked on a systematic investigation of the dust and gas content of normal spirals by observing the mm/submm continuum and the mm lines of CO. The data comprise multi–aperture continuum and CO line observations as well as mm/ submm grid maps for a complete sample of 138 spirals. The sample is taken from the compilation of *Cataloged Galaxies and Quasars Observed in the IRAS Survey* (Fullmer & Lonsdale, 1989) and consist of *Sa...d* galaxies; Hubble types SB, $S0$ and obviously active systems were omitted. Due to the limited beam separation only objects whose optical diameter of the major axis D_{25} was $\leq 180''$ were taken into consideration. An additional selection criterion was the availability of at least 3 high quality *IRAS* flux densities and $S_{100} > 10$ Jy. The observational data are summarized in Chini *et al.* (1993, 1995, 1996).

2.1. THE DECOMPOSITION OF DUST SPECTRA

The IR spectrum of a galaxy from 12 to 1300 μm originates from the emission of dust at different temperatures T_d. One usually tries to deconvolve the observed spectrum into a few components and discusses them in terms of their luminosity, associated mass and heating sources. Each spectral component has the form $\kappa_\nu B_\nu(T_d)$, where the dependence of the dust absorption coefficient on frequency is written as $\kappa_\nu \propto \nu^m$. Such a decomposition is not unique and the arbitrariness is exacerbated by the large observational gaps, especially between 100 and 1300 μm, and the uncertainty of the exponent m. Only the coldest component is relevant for the mass determination because it contains the bulk of matter. For the Milky Way, there is a fairly complete spectral coverage which includes the maximum of emission and it may serve as a template for the spectral appearance of normal spirals in general.

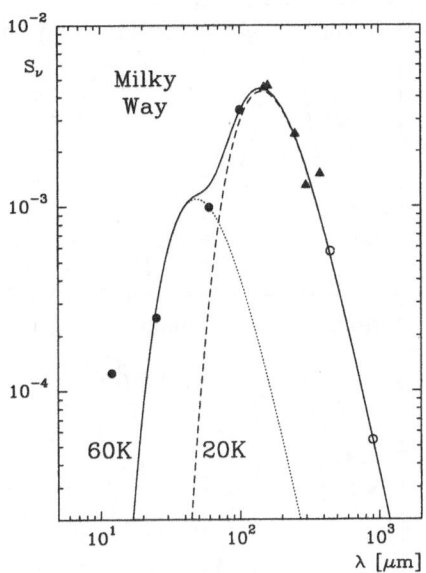

Figure 1. Infrared energy distribution of the inner part of the Milky Way showing the fluxes in the four *IRAS* bands (adapted from Fig.2 of Cox and Mezger, 1989), at 150, 250 and 300 μm (Hauser *et al.* 1984), at 160 μm (Gispert *et al.* 1982), a 380 μm (Caux *et al.* 1986), at 450 μm (Owens *et al.* 1979) and at 900 μm (Pajot *et al.* 1989). The ordinate is in arbitrary scale and refers to energy per Hz. The spectrum has been decomposed into a 20 K (dashed) and 60 K (dotted) component adopting for the dust absorptivity $m = 2$; the solid curve gives the sum of both.

2.1.1. *The Milky Way*

One important aspect of the spectrum of the Milky Way is that it still rises beyond 100 μm. Averaging its inner part ($R \leq 8$ kpc) over galactic longitude $3° - 35°$ and latitude $|b| < 1°$, the spectrum peaks around 160 μm. Adopting $m = 2$, it can be decomposed into a cold component of 20 K and a warm one of 60 K (Fig. 1); fits with $m < 2$ are not possible. There is no evidence for very cold dust; the 20 K component produces almost all emission between 100 and 1300 μm. From this analysis and a similar one for M 51, which also has the peak of its SED at around 170 μm, we conclude that the spectra of normal spirals attain their maximum beyond 100 μm and are dominated by dust not warmer than 20 K.

2.1.2. *NGC 660 and UGC 3490*

For NGC 660 and UGC 3490 there exist grid maps at 450, 800 and 1300 μm (Chini *et al.* 1995) which partly fill the gap at submm wavelengths, but detailed FIR data are still missing. Analyzing the submm data one finds different slopes for the energy distribution (Fig. 2). NGC 660 has a spectral

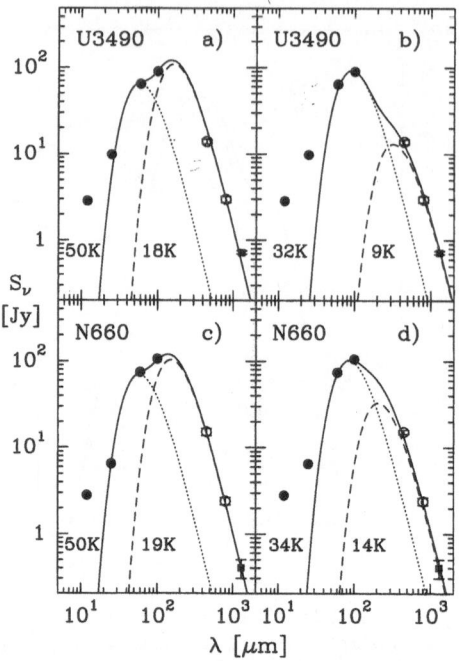

Figure 2.   Observed energy distributions of UGC 3490 and NGC 660. *IRAS* data are
from Fullmer & Lonsdale (1989), measurements at 450 and 800μm are from Chini &
Krügel (1993) and 1300 μm data come from Chini *et al.* (1995). Possible decompositions
into a cold (dashed) and a warm (dotted) component are shown assuming $m = 2$. The
temperatures of the individual components are **a)** 18 K and 50 K, **b)**, 9 K and 32 K, **c)**
19 K and 50 K, **d)** 14 K and 34 K.

index $\beta(400/1300) = 3.5$, similar to the Milky Way, whereas UGC 3490 has
a much flatter submm spectrum characterized by $\beta = 2.8$. In both objects,
the S_{450}/S_{1300} ratio is the same for the central position and for the outer
regions, indicating that there is no large temperature gradient across the
galaxy.

Due to the absence of FIR data beyond 100 μm the decomposition of the
spectra is not unique. As shown in Fig. 2a, UGC 3490 can be interpreted by
a cold component of $T_d \simeq 18$ K and $m = 2$ and a second component of 50 K
which accounts for the observation at 60 and 100 μm. Alternatively (see
Fig. 2b), there could be a very cold component of $T_d \simeq 9$ K and $m = 2$ with
a warmer component of 32 K. For the steep submm spectrum of NGC 660,
a perfect fit down to 25 μm is obtained with $m = 2$ and two components of
$T_d = 19$ K and $T_d = 50$ K (Fig. 2c). Another possible combination is $T_d =$
14 K and 33 K (Fig. 2d), all adopting $m = 2$.

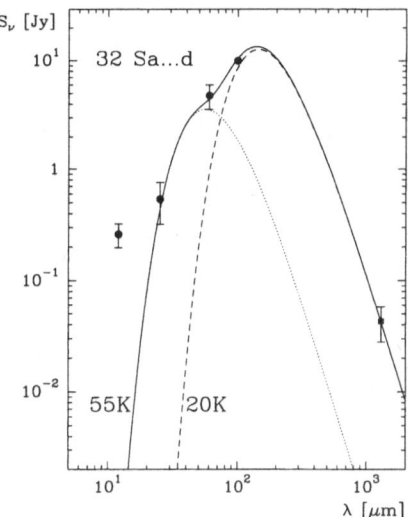

Figure 3. Normalized mean energy distribution of normal spirals; the error bars denote the scatter of the individual objects around the mean value. *IRAS* data are from Fullmer & Lonsdale (1989), 1300 μm data from Chini *et al.* (1995). A possible decomposition into a 20 K (dashed) and a 55 K (dotted) component is shown assuming $m = 2$.

2.1.3. *The Average Spectrum of Normal Spirals*

In non–active spirals the spatial extent of the dust is comparable to the optical size of the galaxies, as was found from mapping 32 of these objects at 1300 μm (Chini *et al.* 1995). About 70% of the mass is contained within half the optical radius with surface density decreasing inversely proportional to the galactocentric radius. Fig. 3 displays the averaged spectrum of these galaxies normalized at 100 μm; the vertical bars denote the scatter around the mean flux density at a given wavelength. Decomposing this average spectrum from 25 to 1300 μm one finds, similar to M 51 and the Milky Way, a temperature between 15 and 20 K for the cold dust and 55 K for the warm component.

3. Active Galaxies

Activity is a usually a phenomenon in the nuclei of galaxies, reaching from an enhanced star formation rate to *Seyferts* and quasars. The underlying physical process in starburst galaxies is generally thermal in nature; *Seyferts* represent some sort of transition stage whereas in QSOs non–thermal processes dominate. So far there is no global parameter that describes quantitatively the various stages of activity. Between 1967 and 1981 Markarian and coworkers have compiled a sample of 1500 galaxies which are characterized by a compact nucleus with UV excess and optical emission lines (Markar-

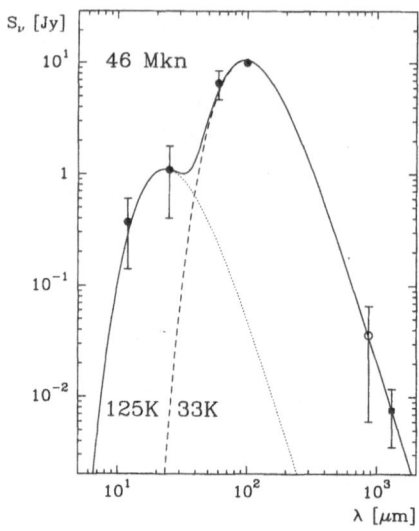

Figure 4. Normalized mean energy distribution of 46 active Mkn galaxies; the error bars denote the scatter of the individual objects around the mean value. *IRAS* data are from Fullmer & Lonsdale (1989), measurements at 870 and 1300 μm come from Krügel *et al.* (1988a,b) and Chini *et al.* (1989c, 1992a). A decomposition into a 33 K (dashed) and a 125 K (dotted) component is shown.

ian *et al.* 1989). The dust and gas content of a complete FIR flux limited subsample with good detections at all four *IRAS* bands and a 100 μm flux density > 9 Jy was studied in the continuum at 1300 μm (partly at 870 μm) as well as in the CO (1–0) and (2–1) lines Krügel *et al.* (1988a,b, 1990) and Chini *et al.* (1989b, 1992a,b). One of the major results of this project is that – in contrast to normal spirals – the interstellar matter is highly concentrated towards an inner region of ∼ 2 kpc in radius. Therefore, the minimum dust temperature should be warmer than the 20 K found in normal galaxies.

Fig. 4 shows the averaged normalized spectrum of these Mkn galaxies; the vertical bars denote the scatter of the individual galaxies around the mean value. Comparing this spectral energy distribution (SED) with Fig. 3 there are two major differences: First, the spectrum between 60 and 1300 μm is rather narrow, allowing a fit with just one dust component. Second, the ratio $S_{100/1300}$ is ∼ 1200, i.e. a factor of 5 higher than in normal galaxies, implying a higher temperature for the coldest dust component. The combination of the observations at 870 and 1300 μm made it possible to derive the frequency dependence of κ; its average value is $m = 2.0 \pm 0.2$. In these active nuclear regions a very cold component can be excluded; the average temperature of the coldest dust is $T_d = 33 \pm 3$ K.

4. The Determination of Dust and Gas Masses

It is crucial for the derivation of dust and gas masses to know the temperature of the coldest dust component in a galaxy. The influence of T_d on the dust mass after Eq. (1) is depicted in Fig. 6 and will be discussed in Section 5. Other sources of uncertainty are κ and, with respect to the gas mass, the value of R. In order to evaluate the accuracy of the conversion of S_λ into M_{gas}, Fig. 5 shows a comparison of the CO (2–1) luminosity and the 1300 μm luminosity, defined as $S_{1300} \cdot D^2$, for the samples of normal spirals and active galaxies as discussed above. L_{CO} and L_{1300} are purely observational quantities and no assumptions enter their derivation. In our observations, both quantities refer to identical regions in the center of the galaxies, where molecular hydrogen is expected to be dominant, and L_{CO} and L_{1300} correlate fairly well

$$\frac{L_{CO}}{L_{1300}} = \frac{R}{\beta \cdot \kappa_\lambda \cdot B_\lambda(T_d)} = (6.3 \pm 2.2) \cdot 10^6 \qquad (3)$$

The scatter of individual galaxies around th mean is only 50% and to a considerable degree due to observational errors, such as pointing and calibration ($\sim$ 20% for both, L_{CO} and L_{1300}), or systematical errors like contamination of L_{1300} by CO (2–1) emission, or an admixture of HI.

We briefly discuss the mutual relations of the four quantities R, β, κ and $B_\lambda(T_d)$. The mass absorption coefficient κ per g of dust is, by and large, universally uniform as we expect the same processes of dust formation (wind of red giants, SN explosions, PN) everywhere. Modifications of interstellar grains by the environment are probably unimportant for the *bulk* of the dust mass. The dust–to–gas ratio R should be proportional to the heavy element abundance and thus to the amount of star formation integrated over the history of a galaxy. The temperature of the interstellar medium is contained directly in $B_\lambda(T_d)$ and implicitly in β. However, the product $B_\lambda(T_d)\beta$ should not be temperature sensitive because, at the observing wavelength, the Planck function is close to the Rayleigh–Jeans limit, where $B_\lambda(T_d) \propto T_d$, and β, referring to an optically thick line, is expected to vary like $1/T$. As far as one theoretically understands β, its dependence on the CO abundance or gas density is weak. The fair correlation of L_{CO} /L_{1300} in Fig. 5 therefore seems to imply, first of all, a metal abundance or time–integrated star formation rate that is constant from one galaxy to another on a level better than 50%.

5. Is There Very Cold Dust in Galaxies?

Above we derived an average temperature of the coldest dust component of 15 ± 5 K for normal spirals and 33 ± 3 K for active Mkn galaxies. However,

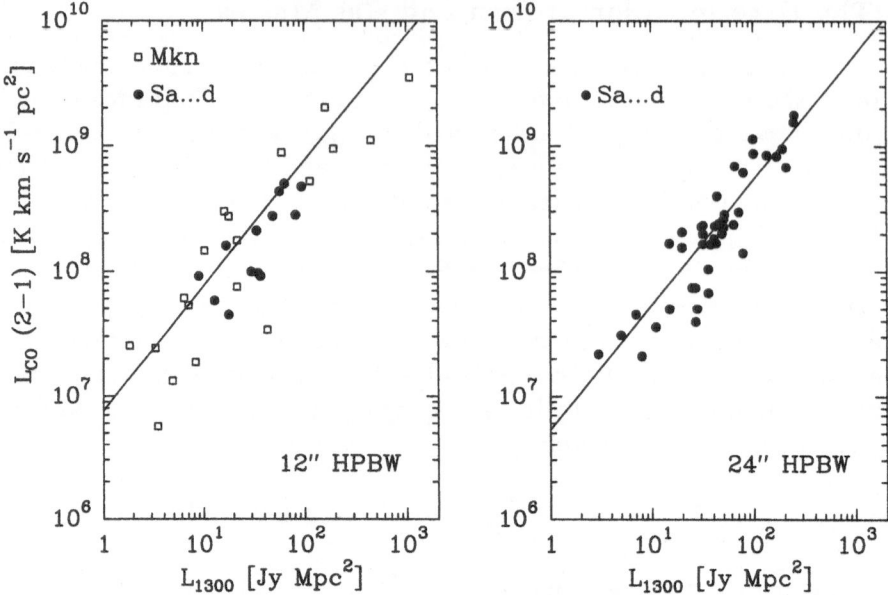

Figure 5. L_{CO} vs. L_{1300} for normal spirals (•) and active Mkn galaxies (□) referring to the inner 12 and 24″ regions, respectively. The data for normal spirals are taken from Chini *et al.* (1996), those for the Mkn galaxies are from Krügel *et al.* (1988a,b , 1990) and Chini *et al.* (1992a,b).

because of observational uncertainties and ambiguities in decomposing the SEDs, the fundamental question arises whether there exists still colder dust in substantial quantities that may have escaped our analysis and which would be associated with large amounts of undetected gas. We address this question by estimating the minimum possible temperature T_{min} of dust that is located inside the optical radius of a galaxy; we do not touch the problem of interstellar matter outside this region. Likewise, we neglect the existence of very small transiently heated particles, which are most of the time at very low temperatures, but do not contribute significantly to the total dust mass.

Very cold dust must be well protected from stellar light and shielded by an optical depth $\gg 10$ mag. It can therefore only be found deep in the interior of clouds. We assume that these clouds are nevertheless penetrated and thus heated by the FIR emission of the galaxy itself, at least for wavelengths above 25μm. There the optical depth is, according to most dust models, a factor of over 50 smaller than in the visual. The heating by FIR radiation is not very efficient because of reduced cross sections for absorption, but it is much more important than the heating by the microwave background. To further minimize the heating, we assume that the source of

FIR radiation resides in the galactic nucleus. A spatially spread–out energy source, like stars in the disk or heating by radiation of shorter wavelengths, would only increase T_{min}.

The result for T_{min} is shown in Fig. 6a. It is based on two heating components of equal luminosity with a $\nu^2 B_\nu(T_d)$ spectral shape and temperatures of 15 K and 50 K. Obviously, in a galaxy with an FIR luminosity of $5 \cdot 10^{10} L_\odot$ and an optical size of 15 kpc, like e.g. the Milky Way, the dust in the disk cannot become colder than ~ 6 K; two additional curves in Fig. 6 show how T_{min} depends on the luminosity of the galaxy.

To estimate how much dust mass can escape detection, we make the extreme assumption that the observed 1300 μm flux originates entirely from very cold dust without any contribution from a 15 K component. From the solid curve in Fig. 6b, which holds for normal dust in the diffuse interstellar medium, we can infer by how much we would underrate the mass of dust at a certain temperature assuming that T_d were 15 K; for example, dust of 6 K would yield a factor of 5. The optical properties of very cold dust, however, are inevitably modified by the deposit of ice mantles and, most likely, coagulation into fluffy agglomerates. This modification leads to an increase in the absorption coefficient κ (see Eq. (1)) by a factor of 5 to 10 (Krügel & Siebenmorgen 1994; Ossenkopf & Henning 1994). Consequently, the mass of very cold dust may even be slightly smaller or comparable to the dust at 15 K (see Fig. 6b dashed curve); dust of 6 K, e.g., would leave the mass unchanged. As the two effects, lower temperature and enhanced grain emissivity, neutralize each other we conclude that 1300 μm continuum observations yield fair results for the total dust mass.

6. Stages of Activity

Having satisfied ourselves that estimates of the gas mass from 1300 μm dust emission after to Eq. (1) are acceptable, we compare the ratio L_{IR} over M_{gas} in our samples of normal spirals and active Mkn galaxies (L_{IR} is defined as the luminosity from 12 to 1300 μm). We find a remarkable difference: the average value in solar units is L_{IR}/M_{gas} of 5 ± 2 for normal spirals and 92 ± 53 for Mkn galaxies. Although the scatter is large, both samples are clearly separated (Fig. 7), hence the ratio L_{IR}/M_{gas}, which gives the efficiency of converting mass into luminosity, turns out to be a discriminator against activity (Chini et al. 1995). The location of radio–quiet QSOs, whose source of luminosity is of entirely different origin, corroborate this result: their average L_{IR}/M_{gas} ratio is about 550.

Fig. 7 demonstrates that L_{IR} alone is not a unique signature for activity; only 5 Mkn galaxies exceed the luminosity range covered by normal spirals, while 3 active galaxies even fall below this range. Within the interval from

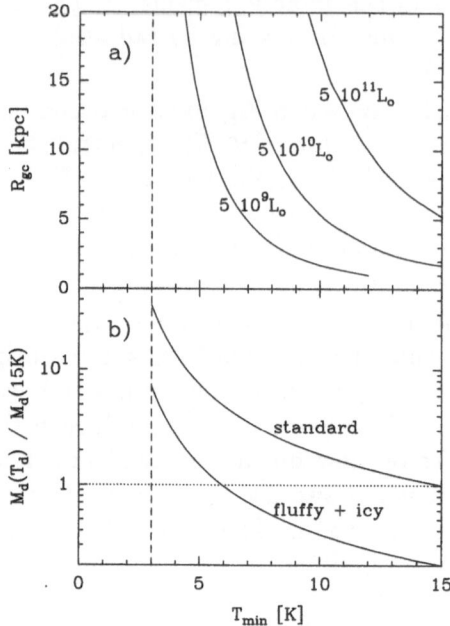

Figure 6. a)Minimum dust temperature T_{min} as a function of galactocentric radius R_{gc} for $L_{IR} = 5 \cdot 10^9$, $5 \cdot 10^{10}$ and $5 \cdot 10^{11} \, L_\odot$, respectively. b) The relative change of dust mass compared to a 15 K component for normal (solid) and fluffy + icy grains (dashed) where κ_{1300} is enhanced by a factor of 5.

10^{10} to $10^{12} \, L_\odot$, a galaxy may be normal or active. A galaxy of given gas mass M_{gas}, on the other hand, may produce different amounts of luminosity. The range $100 \geq L_{IR}/M_{gas} \geq 3$ seems indicative of star formation in general and the different efficiencies reflect the violence of the process ranging from a quiescent steady state in galactic disks to starbursts in nuclei.

In our samples of active Mkn galaxies and non–active spirals the ratio L_{IR}/M_{gas} does not seem to change with the gas mass, although the scatter is still large. Therefore a true proportionality is indicated between M_{gas} and L_{IR} and the straight lines in Fig. 7 have been drawn with a slope of one. The proportionality in both types of galaxies suggests that they derive their luminosity from star formation, although for active objects some contribution may come from a black hole.

In active galaxies, star formation and gas consumption are so high that the activity can last only for a period of $\simeq 10^8$ yr; details depend on the initial mass function (IMF). Therefore transitions must occur between the active and non–active stage. This raises the fundamental problem how normal galaxies are pushed into activity. There is consensus that transfer of gas into the nucleus is the likely stimulus for violent star formation. Dy-

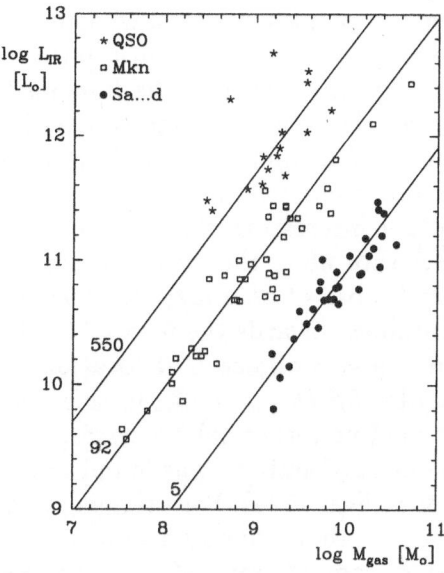

Figure 7. L_{IR} vs. M_{gas} for a sample of 32 normal spirals ● and for a sample of active galaxies □ (adapted from Chini *et al.* 1995). The solid lines through both samples are the loci where L_{IR}/M_{gas} is 5 (normal) and 92 (active), respectively. For comparison we have also included some radio–quiet QSOs ★ whose data have been taken from Chini *et al.* (1989a); the solid line there corresponds to $L_{IR}/M_{gas} = 550$.

namical calculations of the evolution of the gas in a galactic nucleus yield repetitive bursts of star formation as one possible class of solutions (Loose *et al.* 1982). A systematic variation of the basic parameters of the nucleus indicates that such cyclic bursts are even likely provided the nucleus contains enough gas and the IMF is top–heavy (Krügel & Tutukov 1993). Once triggered, the the phenomenon of repetitive bursts persists in the models for $\sim 5 \cdot 10^8$ yr. This makes them relatively easy to observe, although the duration of a burst ($\sim 2 \; 10^7$ yr) is several times smaller than the quiescent interval.

Another important question concerns the dichotomy of activity displayed in Fig. 7: Why are there so few objects with intermediate values of L_{IR}/M_{gas}? Obviously, the transition time between the two stages of activity must be short, as predicted by models for star formation in galactic nuclei (Loose *et al.* 1982, Krügel & Tutukov 1993, Tutukov & Krügel 1995). To obtain such a clear separation between active and non–active galaxies, it is also required that the nuclear starburst increases the luminosity of the whole galaxy by some fixed amount, according to Fig. 7 on the average by a factor of 20. Otherwise the active galaxies would be distributed more evenly above the line $L_{IR}/M_{gas} = 5$. This is very puzzling if we recall that the IR luminosities are dominated by the low resolution FIR fluxes of *IRAS* and thus refer to the disk *and* the nucleus, whereas only the latter becomes active. We must therefore conclude that in a burst the nucleus achieves a luminosity some 20 times greater than the luminosity of the disk. Consequently, the activity process must in some way be linked to the properties of the (non–active) disk. This link is probably provided by the mass influx of gas from the disk into the nucleus.

We mention another interesting class of solutions which occurs in the models when the SN rate after the star burst is too low to remove the gas from the nucleus (Loose *et al.* 1982; Tutukov & Krügel 1995). In this case the collapse continues towards the formation of a massive star cluster and finally to a supermassive object. The location of radio–quiet QSOs in Fig. 7 is characterized by $L_{IR}/M_{gas} \sim 550$; in this case the high efficiency is obtained from additional non–thermal processes.

Another extreme form of activity was found in highly red–shifted radio galaxies (e.g. Chini & Krügel 1994). For example, in 4C41.17 at $z = 3.8$ the coldest dust component is about 80 K. This object is not contained in Fig. 7 because the conversion from M_d into M_{gas} is unknown. Using the galactic R value, which is certainly a lower limit, implies $L_{IR}/M_{gas} = 10^3$; a lower metalicity would yield L_{IR}/M_{gas} values closer to those of QSOs.

The discussion above leads us to a view, which is summarized in Table 1, where the coldest possible dust component is strongly related to both, the luminosity and the activity of a galaxy. The indicator for activity is L_{IR}/M_{gas} as it measures the energy output per unit mass of interstellar matter. Furthermore, the temperature of the coldest dust component T_d increases with the stage of activity.

TABLE 1. Activity and dust temperature in various types of galaxies

Object	L_{IR}/M_{gas}	T_d
normal spirals	5 ± 2	15 ± 5 K
active Mkn's	92 ± 53	33 ± 3 K
radio–quiet QSOs	~ 550	~ 40 K
primeval galaxies	≤ 1000	~ 80 K

References

Andreani, P., Franceschini, A. (1992) Galaxy photometry at millimetric wavelengths, A&A, **260**, pp. 89–96

Braine, J., Krügel, E., Sievers, A.W., Wielebinski, R. (1995) 1.3mm continuum emission in the late–type spiral NGC 4631 A&A, **302**, pp. 849–860

Carico D.P., Keene J., Soifer B.T., Neugebauer G., 1992, 1.25mm observations of luminous infrared galaxies, PASP, **104**, pp. 1086–1090

Caux, E., Serra, G. (1986) Observations of the galactic disk from $l = 15°$ to $l = 82°$ in the submillimeter range, A&A, **165**, pp. L5–L8

Chini, R., Kreysa, E., Biermann, P.L. (1989a) On the nature of radio–quiet QSOs, A&A, **219**, pp. 87–97

Chini, R., Kreysa, E., Krügel, E., Mezger, P.G. (1986) Submm observations of IRAS galaxies, A&A, **166**, pp. L8–L10

Chini, R., Krügel, E. (1993) Cold dust in spiral galaxies. I, A&A, **279**, pp. 385–392

Chini, R., Krügel, E. (1994) Dust at high z, A&A, **288**, pp. L33–L36

Chini, R., Krügel, E., Kreysa E. (1992a) Star formation efficiency in active galaxies, A&A, **266**, pp. 177–182

Chini, R., Krügel, E., Kreysa, E., Gemünd, H.-P. (1989b) The submm continuum of active galaxies, A&A, **216**, pp. L5–L7

Chini, R., Krügel, E., Lemke, R. (1996) Dust and CO emission in normal spirals, A&A Suppl., , accepted

Chini, R., Krügel, E., Lemke, R., Ward–Thompson, D. (1995) Dust in spiral galaxies. II, A&A, **295**, pp. 317–329

Chini, R., Krügel, E., Steppe, H. (1992b) Carbon monoxide in active galaxies, A&A, **255**, pp. 87–99

Clements, D.L., Andreani, P., Chase, S.T. (1993) Submillimetre observations of galaxies – I. First results, MNRAS, **261**, pp. 299–305

Cox, P., Mezger, P.G. (1989) The galactic infrared/submillimeter dust radiation, A&A Rev., **1**, pp. 49–83

Eales, S.A., Wynn–Williams, G., Duncan, W.D. (1989) Cold dust in galaxies, ApJ, **339**, pp. 859–871

Elias, B. et al. (1978) 1 millimeter continuum observations of extragalactic objects, ApJ, **220**, pp. 25–41

Fullmer, L., Lonsdale, C.J. (1989) Catalogued Galaxies and Quasars Observed in the IRAS Survey, Version 2, Jet Propulsion Laboratory

Gispert, R., Puget, J.-L., Serra, G. (1982) Far–infrared survey of extended molecular clouds complexes along the galactic plane, A&A, **106**, pp. 293–306

Guélin, M., Zylka, R., Mezger, P.G., H.-P., Haslam, C.G.T., Kreysa (1995) Cold dust emission from the spiral arms of M 51, A&A, **298**, pp. L29–L32

Guélin, M., Zylka, R., Mezger, P.G., H.-P., Haslam, C.G.T., Kreysa, E., Lemke, R., Sievers, A.W. (1993) 1.3mm emission in the disk of NGC 891: Evidence of cold dust, A&A, **279**, pp. L37–L39

Hauser, M., Silverberg, M.T., Stier, T., Kelsall, T., Gezary, D.Y., Dwek, E., Walser, D., Mather, J.C., Cheung, L.H. (1984) Submillimeter wavelength survey of the galactic plane from $l = -5°$ to $l = +62°$: structure and energetics of the inner disk, ApJ, **285**, pp. 74–88

Hildebrand, R.H., Whitcomb, S.E., Winston, R., Stiening, R.F., Harper, D.A., Moseley S.H. (1977) Submillimeter photometry of extragalactic objects, ApJ, **216**, pp. 698–705

Jaffe, D., Becklin, E., Hildebrand, R. (1984) Submillimetre continuum observations of M 82, ApJ, **285**, pp. L31-L33

Krügel, E., Chini, R. (1994) Analysis of a cold cloud fragment, A&A, **287**, pp. 947–958

Krügel, E., Chini, R., Kreysa, E., Sherwood, W.A. (1988a) Millimeter observations of Markarian galaxies, A&A, **190**, pp. 47–51

Krügel, E., Chini, R., Kreysa, E., Sherwood, W.A. (1988b) Dust emission from Markarian galaxies, *A&A*, **193**, pp. L16–L18

Krügel, E., Siebenmorgen, R. (1994) Dust in protostellar cores and stellar disks, *A&A*, **288**, pp. 929–941

Krügel, E., Steppe, H., Chini, R. (1990) CO in Markarian galaxies, *A&A*, **229**, pp. 17–27

Krügel, E., Tutukov, A.V. (1993) Star formation in galactic nuclei, *A&A*, **275**, pp. 416–426

Loose, H., Krügel, E., Tutukov, A.V. (1982) Bursts of star formation in the galactic centre, *A&A*, **105**, pp. 342–350

Markarian, B.E., Lipovetskii, V.A., Stepanian, J.A., Erastova, L.K., Shapovalova, A.I. (1989) *Soobsch. Spets. Astrof. Obs*, **62**, pp. 5

Neininger, N., Guélin M., Garcia–Burillo, S., Zylka, R., Wielebinski, R. (1996) *A&A*, (in press)

Ossenkopf, V., Henning, Th. (1994) Dust opacities for protostellar cores, *A&A*, **291**, pp. 943–953

Owens, D.K., Mühlner, D.K., Weiss, R. (1979) A large–beam sky survey at millimeter and submillimeter wavelengths made from balloon altitudes, *ApJ*, **231**, pp. 702–710

Pajot, F., Gispert, R., Lamarre, J.M., Pomeranz, M.A., Puget, J.-L., Serra G. (1989) Observations of the submillimetre integrated galactic emission from the South Pole, *A&A*, **223**, pp. 107–111

Rice, W., Lonsdale, C.J., Soifer, B.T., Neugebauer, G., Kopan, E.L., Lloyd, L.A., de Jong, T., Habing, H.J. (1988) A catalog of *IRAS* observations of large optical galaxies, *ApJS* **68**, pp. 91–127

Sievers, A.W., Reuter, H.-P., Haslam, C.G.T., Kreysa, E., Lemke, R. (1994) Cold dust emission from the spiral galaxy NGC 3627, *A&A*, **281**, pp. 681–684

Smith, J. (1982) The far–infrared disk of M 51, *ApJ*, **261**, pp. 463–472

Stark, A.A., Davidson, J.A., Harper, D.A., Pernic, R., Loewenstein, R., Platt, S., Engargiola, G., Casey, S. (1989) Far–infrared and submillimeter photometric mapping of spiral galaxies in the Virgo cluster, *ApJ*, **337**, pp. 650–657

Thronson, H., Walker, C.K., Walker, C.E., Maloney, P. (1987) Observations of 1.3 millimeter continuum emission from the centers of galaxies, *ApJ*, **318**, pp. 645–653

Tutukov, A.V., Krügel, E. (1995) The main types of star formation in galactic nuclei, *A&A*, **299**, pp. 25–33

MM OBSERVATIONS OF IRAS GALAXIES: DUST PROPERTIES, LUMINOSITY FUNCTIONS AND CONTRIBUTIONS TO THE SUB-MM BACKGROUND

P.ANDREANI & A.FRANCESCHINI

Dipartimento di Astronomia di Padova
vicolo dell'Osservatorio 5, I-35142 Padova, Italy.

Abstract. We have studied the FIR/mm spectrum of IR galaxies by combining IRAS photometry with new mm data on a complete southern IRAS galaxy sample. The observed spectra and a dust model emphasize a dicothomy in the galaxy population: half of the objects with a lot of warm dust are characterized by higher values of the bolometric (UV-FIR) luminosity, of the dust-to-gas mass ratio, of the dust optical depths and extinction, while those dominated by cold (*cirrus*) dust show opposite trends. From these data we derive the mm luminosity function of galaxies and estimate their contribution to the sub-mm background (BKG).

1. Introduction

Observations of diffuse dust in galaxies impact on some basic questions about their present structure and past history. In particular, are we missing significant amounts of luminous matter because of the effects of dust extinction? How much severe are the corresponding selection effects?

Traditional approaches to investigate dust in galaxies rely on either *(a)* indirect estimates based on dust extinction effects in the optical and near-IR, or *(b)* direct measurements of dust emission from FIR/mm observations. Neither have provided conclusive results at the moment: the former method is based on model-dependent and controversial assumptions about the optical-IR spectrum of various galactic components, the latter still lacks a large enough statistical basis and suffers from the observational uncertainties of the sub-mm/mm data.

345

D. L. Block and J. M. Greenberg (eds.), New Extragalactic Perspectives in the New South Africa, 345–348.
© 1996 *Kluwer Academic Publishers.*

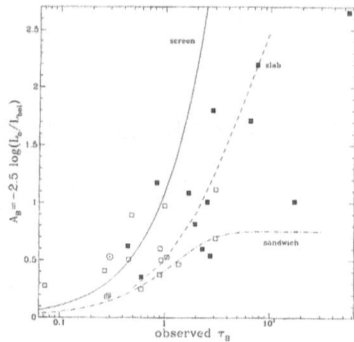

Figure 1.   Observed dust optical-depth τ_B versus overall extinction A_B. Open and filled squares refer to the inactive *cirrus*-dominated and to the starbursting objects, respectively. Position marked (⊙) corresponds to our Galaxy. Predicted dependences for a screen, a slab, and a sandwich (zero dust scale-height) model are shown for comparison (reprint from Franceschini & Andreani 1995).

We have tackled the question of the dust content of spirals by observing the 1.25 *mm* continuum emission from a complete sample of IRAS galaxies with the SEST Telescope. The SEST telescope was chosen as it provided the best compromise between detector sensitivity and spatial resolution (see for more details Franceschini & Andreani 1995).

2. The FIR/mm spectrum of galaxies: cold dust and extinction

Our observations allowed us to directly estimate, for the first time, the long-λ spectrum of galaxies (see Figure 3 below) in a poorly explored spectral domain. We interpret the FIR/mm spectra exploiting a standard dust model. Two major dust components are assumed to be present in the ISM: a cold "cirrus" component and warm dust in star-forming regions.

Sixteen galaxies, hereafter called *"cirrus"-dominated*, have a negligible contribution of warm dust at 100 μm, while the other 14 objects require high values of the warm dust fraction and their spectrum at $\lambda < 100$ μm is dominated by the starburst component. FIR/*mm* data and a reliable dust model provide a straightforward way to classify galaxies according to their star formation activity. The starburst component contributes only a couple of percent on average of the total dust mass, while the cold component bears 97 % of the total mass.

Figure 1 shows the extinction estimated from our data versus the observed B-band optical depth. This was found from the total dust mass divided by the galaxy projected area and measures the amount of dust *available* to absorb the optical light. A_B measures the overall *actual* effect of extinction and was estimated from the logarithmic ratio of the bolometric

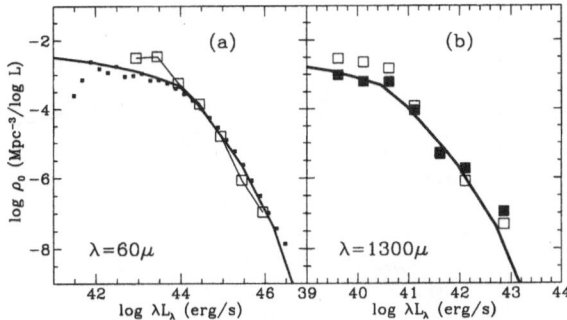

Figure 2. The local luminosity functions at 60 μm, compared with published data (panel [a]), and at 1330 μm (panel [b]), estimated from our IRAS galaxy sample. Open boxes in panel (b) are based on our new 60 μm LLF, whereas the filled ones are based of the Saunders's et al. LLF. The thick line is the 1300 μm function transformed from the 60 μm one through a constant L_{1300}/L_{60} ratio.

optical-UV luminosity to the bolometric UV/FIR light.

Two galaxy classes are significantly segregated over this plane: the inactive *"cirrus"-dominated* objects being confined to lower values of dust optical depth ($\tau_B < 2$) and low extinction ($A_B \leq 1$). The active star-forming galaxies, on the contrary, are spread over much larger values in both axes. Appreciable amounts of dust and extinction seem to characterize only the population of starbursting galaxies. This may reflect a more exhaustive processing of the ISM through star-formation in the latter objects than for the average *inactive* spiral.

3. The *mm* luminosity function and contributions to the BKG

We find the mm emission for our sample galaxies to be very well correlated with the 60 and 100 μm ones. The estimate of the mm local luminosity functions (LLF) is then straightforward, in the presence of a complete flux-limited sample.

We report in Figure 2 the LLFs at 60 and 1300 μm derived from our sample. The good match of our 60 μm LLF with published data (Saunders et al. 1990) emphasizes the completeness and reliability of the sample. The 1300 μm LLFs were derived from those at 60 μm by means of the bivariate IR/mm luminosity distribution (open and filled squares), and from a simple scaling of the 60 μm one to 1300 μm through a constant L_{1300}/L_{60} luminosity ratio (thick line). The three estimates are consistent within the errors.

A precise knowledge of the LLF allows a model-independent estimate of the minimal contribution of galaxies to the IR-mm extragalactic BKG. This

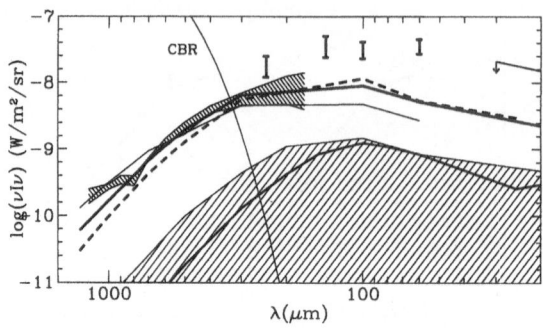

Figure 3.   Contributions of galaxies to the extragalactic BKG (see text).

minimal estimate, which is proportional to the local volume emissivity $j_\nu = \int dL\ \rho(L)L$, was computed assuming no evolution and that galaxies exist only up to z=1. The minimal galaxy BKG is shown in Fig. 3 as the lower shaded region: given such conservative assumptions, the real extragalactic BKG is quite likely in excess of this. Note that, because of the strong K-correction effect for even such low-z objects, this spectrum is much broader than the average local long-λ spectral emissivity of galaxies j_ν.

The upper shaded region in Fig. 3 is a recent estimate (Puget et al. 1996) of the extragalactic BKG in the sub-mm cosmological window, where the foreground emissions from the Galaxy and the interplanetary dust and the Cosmic BKG Radiation are at the minimum. There is a wide margin between this and our minimal BKG, which may indicate strong evolution.

In principle, the comparison between the local emissivity j_ν and the BKG spectrum allows inferences about the average redshift of the emitting sources. In practice, significant uncertainty is introduced by the spectral evolution with cosmic time. Fig. 3 reports two predictions corresponding to quite different models of galaxy formation and evolution: a merging scheme (with most of the flux contributed by $z < 1$, dashed line) and a pure luminosity evolution model (with contributions at $z > 1$, continuous line). The two predict quite similar BKG's, the former assuming a constant source spectrum, the latter one getting warmer in the past following the higher star-formation rate and the increased average radiation field.

References

Franceschini A. & Andreani P. 1995, ApJ 440, L5
Puget J.L., et al., 1996, AA Letters, submitted
Saunders et al., 1990, MNRAS 342, 318

MAPPING THE COLD DUST IN EDGE-ON GALAXIES AT 1.2 MM WAVELENGTH

NIKOLAUS NEININGER
IRAM, F-38406 St.Martin d'Hères, France and
MPIfR, D-53121 Bonn, Germany

AND

MICHEL GUELIN
IRAM, F-38406 St.Martin d'Hères, France

Abstract. Using the IRAM 30-m telescope, we have mapped the λ 1.2 mm continuum emission in the edge-on spiral galaxies NGC 891, NGC 5907 and NGC 4565. Generally, the λ 1.2 mm continuum correlates remarkably well with the CO emission; the correlation with HI is however different for the observed galaxies: in NGC 891, there is no obvious correlation; in NGC 5907 the continuum emission is extending a bit further out than the CO and seems to be correlated with HI peaks.

In NGC 4565, however, the dust emission not only shows a central peak and an inner ring like the CO, but also, like HI, a weaker, extended plateau. Comparable to the HI, the 1.2 mm contours are warped near the NW edge of the galaxy.

The average dust temperature in this galaxy is 18 K near the center and 15 K in the HI plateau. From the 1.2 mm continuum intensity and the HI line integrated intensity, we derive a dust absorption cross section per H atom $\sigma_{1.2mm}^{H} = 5 \times 10^{-27}$ cm^2 in the plateau. This value is very close to that predicted for the local diffuse clouds.

1. Motivation of the study

Cold dust represents most of the interstellar dust in normal galaxies and may be used as a tracer of both molecular and atomic gas (see Cox & Mezger 1989 and references therein). Guélin et al. (1993, 1995) mapped

D. L. Block and J. M. Greenberg (eds.), New Extragalactic Perspectives in the New South Africa, 349–352.
© *1996 Kluwer Academic Publishers.*

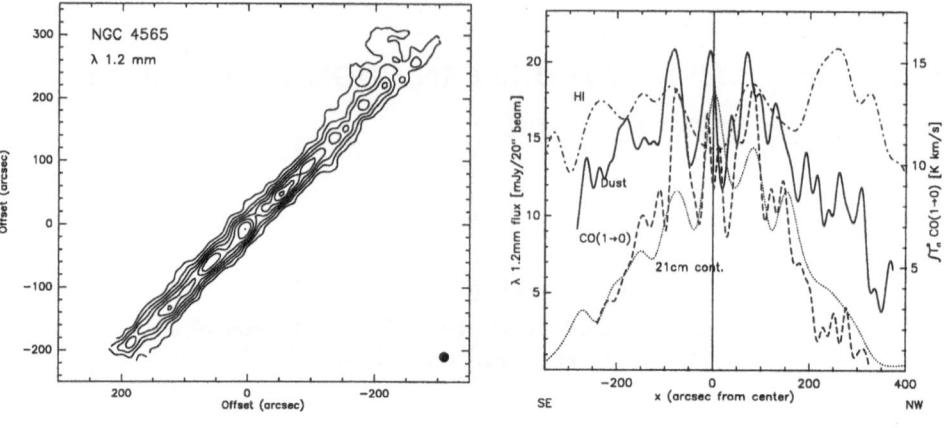

Figure 1. a) The λ 1.2 mm continuum emission of NGC 4565 smoothed to a resolution of 20″; contour levels are 3, 6, ... 21 mJy/beam. b) The brightness distributions of the integrated line intensities of atomic and molecular gas along the major axis of NGC 4565 together with the λ 1.2 mm intensity and the 21-cm continuum.

the λ 1.2 mm continuum emission of two nearby spirals, NGC 891 and M 51. This emission there was found to correlate tightly with CO and poorly with Hɪ. It was not even clearly detected beyond NGC 891's molecular 'ring', in a region where Hɪ emission is still strong. The mean dust temperatures derived from the λ 1.2 mm and FIR flux densities were found to be $\leq$ 20 K.

In order to further study the properties of the ISM, we observed two more edge-on galaxies of similar type, NGC 4565 (Neininger et al. 1996) and NGC 5907 (Dumke et al. 1996). In particular, NGC 4565 was chosen because of its weak CO emission: the dust emissivity per H atom is on the average 2 − 4 times larger for the molecular clouds than for the Hɪ clouds. Thus, the dust associated with the atomic gas becomes predominant only when the H_2 column density becomes very small − which is the case in the outer parts of NGC 4565. The edge-on geometry ($i \geq 85°$) ensures long lines of sight in the disk which helps to detect weak components of the ISM.

2. Observations

The observations of NGC 891 were carried out in February 1993 with a 7-channel bolometer array (Kreysa 1992), those of NGC 5907 and NGC 4565 with an instrument upgraded to 19 channels in March 1995. The beamwidths are 11″ (HPBW), and the spacing between two adjacent channels 20″.

The equivalent bandwidth Δ_ν and central frequency ν_0 should be close to 70 GHz and 245 GHz (1.2 mm), respectively (see Guélin et al. 1995).

The maps are mosaics of up to 26 overlapping submaps, each of the size of a few arcmin^2. During the observations, the subreflector was wobbled at 2 Hz in azimuth with a beam throw between 30″ and 56″. For each channel, a second order baseline was fitted to every azimuth scan, the scans were combined and restored into single beam maps and regridded in equatorial coordinates; after correcting the intensities for atmospheric absorption, they were calibrated with respect to the planets. Finally, the different channel maps were combined to yield a single map which reached a maximum sensitivity of 1 ... 1.5 mJy per 12″ beam for the observations of NGC 4565.

3. Findings

In all three objects, the bulk of the molecular gas lies within a radius of 4–5 kpc from the center. The central component is relatively bright in NGC 891 and strongly dominating in NGC 5907. The central 3 kpc region of NGC 4565 hosts little interstellar matter, except for a compact 'ring' of molecular gas (of diameter $\simeq$ 1 kpc). In NGC 891 and NGC 4565 a strong molecular ring is visible which contains most of the molecular gas. The bulk of the atomic gas is situated further out, forming a broad 'plateau' peaking at $R \simeq 9 - 15$ kpc and extending up to $R \simeq 15 - 20$ kpc. The outer plateau and molecular ring of NGC 4565 show narrow density structures, which are probably spiral arms.

The λ 1.2 mm emission, which is the most reliable tracer of interstellar dust, follows closely the CO brightness distribution in the central region and in the molecular ring; this holds for all three galaxies. In NGC 891 there is no sign of λ 1.2 mm emission further out (Guélin et al. 1993); in NGC 5907, this emission is a bit more extended than the CO disk and two bumps at a radius of about 10 kpc coincide with the HI at a place where no CO is left (Dumke et al. 1996). In NGC 4565, however, the λ 1.2 mm emission follows HI in the outer regions of the disk as soon as the CO emission becomes weak. This way, it extends significantly further out than the emission of the molecular gas and it is possible to derive the properties of the cold dust that is associated with the atomic gas.

The outer HI disks of all three galaxies are warped. Because of the lack of detected emission, no trace of it can be seen in the λ 1.2 mm maps of NGC 891 and NGC 5907. However, the onset of NGC 4565's warp is clearly visible in the 1.2 mm cold dust emission at the NW side (see also color plate 11a in these *Proceedings*). In the SE, the HI warp is by far less prominent, but a hint at it is seen just at the edge of our map.

4. Results

The most interesting result is the observation of λ 1.2 mm cold dust emission
at galactocentric distances > 10 kpc in the HI ring and in the warp of
NGC 4565. In these outer regions, where the gas is mostly in the atomic
form (this is indicated by the weakness of the CO emission), it is possible
to measure the dust emissivity per H-atom. For the average dust temperature of 15 K (derived from a fit to 160 μm, 200 μm and 1.2 mm data) the
comparison of the integrated HI line intensity with the observed λ 1.2 mm
flux intensity yields an absorption cross section per H-atom $\sigma_\lambda^H = 5 \times 10^{-27}$
cm^2 (H-atom)$^{-1}$ and a mean dust absorption coefficient $\kappa = 0.002$ cm^{2}g^{-1}
(Neininger et al. 1996). These cross sections and temperatures are similar
to those predicted in local diffuse clouds (see Draine & Lee 1984).

The comparison of the three galaxies does not hint at a correspondence
between the Hubble type of the galaxy and the relative extent of the cold
dust emission; a trend is however visible for the distribution of the molecular
gas: there is a stronger ring in the galaxy of earlier type and a stronger
central component in the later type. NGC 4565 is possibly a barred galaxy
(it has a peanut-shaped bulge and also the non-circular motions observed
in CO hint at this – see Neininger et al. 1996); this could influence the
distribution of its ISM. Observations at mm and FIR wavelengths are in
progress to further study these aspects.

The mass of the molecular gas in NGC 4565 is low compared to the mass
of atomic gas. The molecular gas mass, inside the strip along the major axis
covered by our CO observations, is found to be 1.0×10^9 M$_\odot$, when using
the Galactic CO-to-H$_2$ conversion factor ($X = 2.3 \times 10^{20}$cm^{-2}K^{-1}km^{-1}s
– Strong et al. 1988), and $\simeq 0.4 \times 10^9$ M$_\odot$ when using the values derived
from the λ 1.2 mm emission. This corresponds respectively to $\simeq 1/2$ and
1/5 of the HI mass in the same area (2×10^9 M$_\odot$).

References

Cox, P. Mezger, P.G.M., 1989, A&A Review 1, 49
Draine, B.T., Lee, H.M., 1984, ApJ 285, 89
Dumke, M., Braine, J., Krause, M., Zylka, R., Wielebinski, R., Guélin, M., 1996, A&A,
    in preparation
Guélin, M., Zylka, R., Mezger, P.G., Haslam, C.G.T., Kreysa, E., 1995, A&A 298, L29
Guélin, M., Zylka, R., Mezger, P.G., Haslam, C.G.T., Kreysa, E., Lemke, R., Sievers, A.,
    1993, A&A 279, L37
Kreysa, E, 1992, in *ESA-Symposium on Photon Detectors for Space Instrumentation*,
    Noordwijk, p. 207 (ESA SP-356)
Neininger, N., Guélin, M., García-Burillo, S., Zylka, R., Wielebinski, R., 1996, A&A (in
    press)
Strong, A.W., et al., 1998, A&A 207, 1

INTERNAL ABSORPTION IN SPIRAL GALAXIES USING FOUR COLOURS

BARBARA CUNOW AND WALTER F. WARGAU
Department of Mathematics, Applied Mathematics and Astronomy, University of South Africa, PO Box 392, 0001 Pretoria, South Africa

Abstract. The internal absorption in spiral galaxies is investigated statistically using data in the UV, optical and near-infrared wavelength regions. It is shown that a three-component model of the galaxy structure leads to absorption values which are consistent with values found from studies of individual galaxies using surface-brightness profiles and rotation curves.

1. Introduction

Although the internal absorption in spiral galaxies has been studied for several decades, the value of the mean absorption is, however, still a matter of controversy. The reasons for that may be found in the following: (i) statistical observations are subject to selection effects and observational biases; (ii) only a small number of galaxies are studied in detail, and (iii) the absorption values obtained largely depend on the models constructed.

For individual galaxies using surface brightness profiles and rotation curves, mostly optically thin cases are found for face-on view. With statistical studies, however, larger absorption values are obtained. Some authors find that galaxy discs are optically thick, others claim that they are not optically thick or that the galaxies show strong absorption in the central parts but are optically thin in the outer disc regions.

2. Statistical determination of absorption

The present study investigates internal absorption statistically by applying the method proposed by Holmberg (1958) to a sample of spiral galaxies with

D. L. Block and J. M. Greenberg (eds.), New Extragalactic Perspectives in the New South Africa, 353–356.
© *1996 Kluwer Academic Publishers.*

Table 1. Results for central face-on optical depth and for face-on absorption.

τ_0^U	τ_0^J	τ_0^F	τ_0^I	$A_U(0)$	$A_J(0)$	$A_F(0)$	$A_I(0)$
6.1	5.0	3.1	2.3	$0\overset{m}{.}64$	$0\overset{m}{.}53$	$0\overset{m}{.}34$	$0\overset{m}{.}28$
±2.4	±2.0	±1.3	±0.9	$\pm0\overset{m}{.}22$	$\pm0\overset{m}{.}19$	$\pm0\overset{m}{.}11$	$\pm0\overset{m}{.}09$

data in four colours. The catalogue contains of 556 Sb-galaxies with $b_J \leq 18$. For each galaxy, photographic data in the colour filters U, b_J, r_F and I are available. Total magnitudes are determined for each filter. Semimajor and semiminor axes, a and b, are measured from intensity-weighted moments of the J-images because the average signal-to-noise ratio of these images is larger than those of the other bands. The apparent ellipticity ϵ, which is used as a measure of the inclination angle i, is obtained from a and b, $\epsilon = 1 - \frac{b}{a}$. Details are found in Cunow (1992, 1993a,b) and Sommer (1994).

For each galaxy, the apparent ellipticity ϵ, the projected surface brightness S (as defined by Holmberg 1958) for the J-band and the colours $b_J - r_F$ and $U - I$ are measured. The random errors are determined from overlap regions of neighbouring fields.

The model (three-component model) is taken from Christensen (1990). The model galaxy is specified by the distributions of volume emissivities in the bulge and the luminous disc and by the spatial distribution of the dust density of the absorbing disc. This allows the calculation of the model images in each filter for various inclination angles. This model has one free parameter, τ_0^J, which is the central face-on optical depth in the J-band. The parameter values for the J-images are taken from Cunow (1992). It is assumed that the values of the geometrical parameters do not change with wavelength. The extinction law of the Galaxy is adopted for the variation of the extinction coefficient with wavelength. The bulge-to-disc ratio B/D for U, r_F and I are calculated from B/D for b_J given in Cunow (1992) by assuming mean colours for the disc as well as for the bulge (van der Kruit & Searle 1981a,b, 1982a,b). The following values are adopted: $(B/D)_U = 0.28$, $(B/D)_J = 0.29$, $(B/D)_F = 0.38$ and $(B/D)_I = 0.40$.

A set of realisations of the three-component model is calculated for different optical depths τ_0^J: 0, 0.5, 1, 3, 5, 7, 10 and 15 at various inclination angles. For all images, ϵ, S, $b_J - r_F$ and $U - I$ are determined using the same algorithms as for the real data. It is not necessary to consider corrections for the apparent semimajor axes because the diameters are affected in the same way as the diameters of the actual galaxies.

The model with $\tau_0^J = 5$ gives the best fit between model and observations (reduced $\chi^2 = 0.66$). Table 1 shows the results.

Table 2. Results for individual galaxies taken from the literature and transferred to face-on view and to the J-band. Values for the central face-on optical depth τ_0^J and the mean face-on optical depth $< \tau_J >$ are given. They are obtained by applying the three-component model.

Source	Object	τ_0^J	$< \tau_J >$
Kylafis & Bahcall (1987)	NGC 891	0.74	0.084
Disney et al. (1989)	Galaxy	5	0.57
Phillipps et al. (1991)	M101	3.8	0.43
White & Keel (1992)	AM1316-241	1.8	0.21
Byun (1993)	5 galaxies [1]	≤ 5.5	≤ 0.63

3. Discussion

The result $\tau_0^J = 5$ for face-on view shows that the galaxy center is optically thick. However, the mean optical depth for face-on view is only $< \tau_J >= 0.57$. This is due to the strong decrease of the dust density with increasing distance from the galaxy center. For face-on view, $\tau_J = 1$ is found at a distance of $r = 1.6\,r_0$, where r_0 is the scale length. We conclude that galaxies are optically thick in the center but are optically thin in the outer regions.

Table 2 gives the internal absorption of individual galaxies for which data are taken from the literature. These are transferred to face-on view and the J-band using the three-component model. The statistical results, $\tau_0^J = 5$ and $< \tau_J >= 0.57$, are within the ranges found for the individual galaxies.

The apparent diameters in the present analysis are not corrected for inclination effects. If a change of the apparent semimajor axis with inclination angle exists, the variation of the projected surface brightness with inclination is also changed. The colours are not changed.

Here the semimajor and semiminor axes are determined using intensity-weighted moments. The systematic variation with inclination of the observed semimajor axis in the J-band is determined from the model for various optical depths. Fig. 1 shows the variation. The observed semimajor axis increases with increasing τ_0^J. This is caused by the fact that the surface brightness profiles are flatter for high than for low optical depth. For small and for large values of τ_0^J, the apparent semimajor axis does not change with inclination. For the intermediate cases, however, a systematic change is present. This behaviour is similar to the variation expected for the effective diameters (e.g., Disney et al. 1989, Giovanelli et al. 1994). The largest difference is found for $\tau_0^J = 3$, where the observed semimajor axis increases by 16%.

[1]ESO 48-G2, MCG 4-35-3, NGC 1055, NGC 4594, NGC 5746

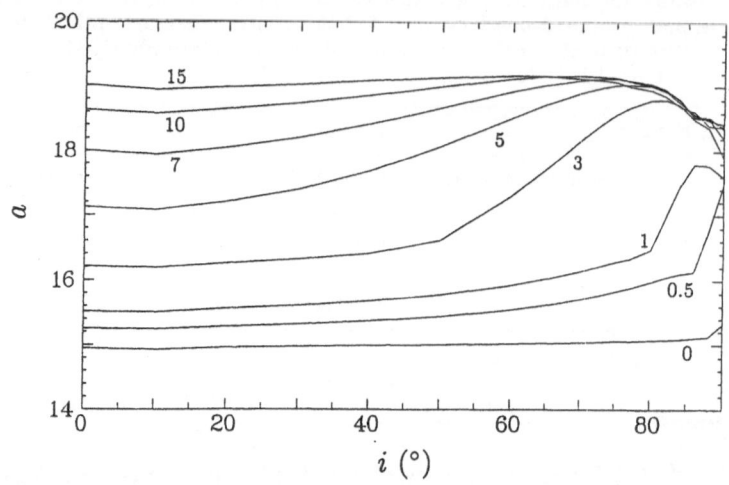

Figure 1. Variation of the apparent semimajor axis a in the J-band with inclination angle i for the three-component model. a is scaled arbitrarily. Eight different values for the central optical depth for face-on view τ_0^J are used: 0, 0.5, 1, 3, 5, 7, 10 and 15.

Acknowledgements

We thank Prof. W. Seitter, Dr. H. Horstmann and Mr. G. Spiekermann for many useful discussions. Special thanks go to N. Sommer for digitising and reducing the U- and I-data. This work is part of the Muenster Redshift Project MRSP. Financial support of the MRSP under numbers Se 345/14–1,2,3, Se 345/20–1 and Se 345/21–1,2 by the Deutsche Forschungsgemeinschaft (DFG) is gratefully acknowledged.

References

Byun Y.I., 1993, PASP 105, 993
Christensen J.H., 1990, MNRAS 246, 535
Cunow B., 1992, MNRAS 258, 251
Cunow B., 1993a, A&A 268, 491
Cunow B., 1993b, PhD thesis, Münster University
Disney M., Davies J., Phillipps S., 1989, MNRAS 239, 939
Giovanelli R., Haynes M.P., Salzer J.J., Wegner G., Da Costa L.N., Freudling W., 1994, AJ 107, 2036
Holmberg E., 1958, Medd. Lunds Obs., Ser. 2., No. 136
Kylafis N.D., Bahcall J.N., 1987, ApJ 317, 637
Phillipps S., Evans R., Davies J.I., Disney M.J., 1991, MNRAS 253, 496
Sommer N., 1994, Diploma thesis, Münster University
van der Kruit P.C., Searle L., 1981a, A&A 95, 105
van der Kruit P.C., Searle L., 1981b, A&A 95, 116
van der Kruit P.C., Searle L., 1982a, A&A 110, 61
van der Kruit P.C., Searle L., 1982b, A&A 110, 79
White R.E., Keel W.C., 1992, Nature 359, 129

DUST IN GALAXIES

A Far Infrared Perspective

NICK DEVEREUX
New Mexico State University
Dept. of Astronomy, Box 30001/Dept 4500, Las Cruces, NM
88003, USA

1. Introduction

The past decade has heralded a revolution in our understanding of dust in galaxies due largely to the extraordinary success of the Infrared Astronomical Satellite (IRAS). The satellite, launched in 1983, surveyed 96% of the sky yielding continuum flux measurements at 12, 25, 60 and $100\mu m$ for literally thousands of galaxies (Cataloged Galaxies & Quasars in the IRAS Survey 1985). Complimentary 60, 100 and $160\mu m$ observations have also been obtained for a limited number of galaxies with the Kuiper Airborne Observatory (KAO). The past decade has also witnessed a dramatic development in the field of submillimeter astronomy with the commissioning of several new submillimeter telescopes including the JCMT, CSO, IRAM and SEST, all of which have yielded new results on the 300 to $1300\mu m$ continuum emission from galaxies. The goal of this review is to discuss the new observations within the context of the origin of the far infrared luminosity, the temperature of dust in galaxies and the implications of the new data for the opacity of spiral galaxy disks.

2. The Origin of the Far Infrared Luminosity

2.1. THE CORRELATIONS

The origin of the far infrared luminosity in galaxies is a subject of continuing controversy. The controversy centers on whether the far infrared $(40\text{-}120\mu m)$ luminosity is powered predominantly by high mass stars or lower mass, non-ionizing stars. At issue, of course, is whether or not the considerable IRAS database can be exploited to measure high mass star formation rates in galaxies.

D. L. Block and J. M. Greenberg (eds.), New Extragalactic Perspectives in the New South Africa, 357–372.
© *1996 Kluwer Academic Publishers.*

On the one hand, the correlations between the IRAS 40-120μm luminosity and the molecular gas mass (Young et al. 1984, 1986; Sanders & Mirabel 1985; Stark et al. 1986; Solomon & Sage 1988) and the correlation between the IRAS 40-120μm luminosity and the Hα luminosity (Leech et al. 1988; Lonsdale-Persson & Helou 1987; Devereux & Young 1990a) reinforce the view that the far infrared luminosity is thermal emission from dust heated by high mass stars. A similar interpretation has been invoked to explain the curiously tight correlation between the IRAS 40-120μm luminosity and the non-thermal radio continuum emission (Dickey & Salpeter 1984; Helou, Soifer & Rowan-Robinson 1985; de Jong et al. 1985; Gavazzi, Cocito & Vettolani 1986; Devereux & Eales 1989; Price & Duric 1992). On the other hand, the propensity of galaxies with low far infrared luminosities that are energetically comparable to the optical luminosity led de Jong et al. (1984), de Jong & Brink (1987), Lonsdale-Persson & Helou (1987), Bothun et al. (1989), and Thronson et al. (1990) to suggest lower mass, non-ionizing stars as a significant dust heating source, particularly in those galaxies with cool 100μm/60μm color temperatures. The ambiguity concerning the origin of the far infrared luminosity that arises from the correlations described above indicates that a more sophisticated analysis is required to distinguish between the possibilities.

2.2. THE FAR INFRARED MORPHOLOGY

The versatility of the IRAS database has increased significantly following the release of the IRAS high resolution images. Application of a high resolution reconstruction algorithm (Aumann et al. 1990) increased the resolution by about a factor of 3 over that of the original survey data, yielding useful far infrared images for the half a dozen or so galaxies that are nearby and large enough to be resolved by the $\sim$ 105 arc sec FWHM reconstructed beam.

One particularly promising approach towards understanding the origin of the far infrared luminosity is to compare the high resolution far infrared images with complimentary Hα images that have been convolved to the same angular resolution. The Hα images trace the location of massive stars which are undoubtedly among the principle dust heating sources responsible for the far infrared luminosity in galaxies.

The images on the facing page show the striking similarity between the far infrared and Hα morphology for the two nearby galaxies M31 and M33.

M 31

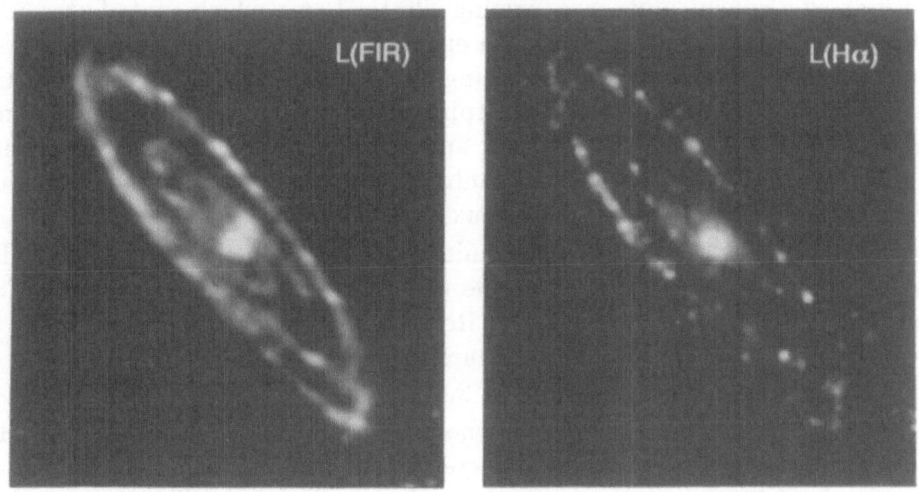

M 33

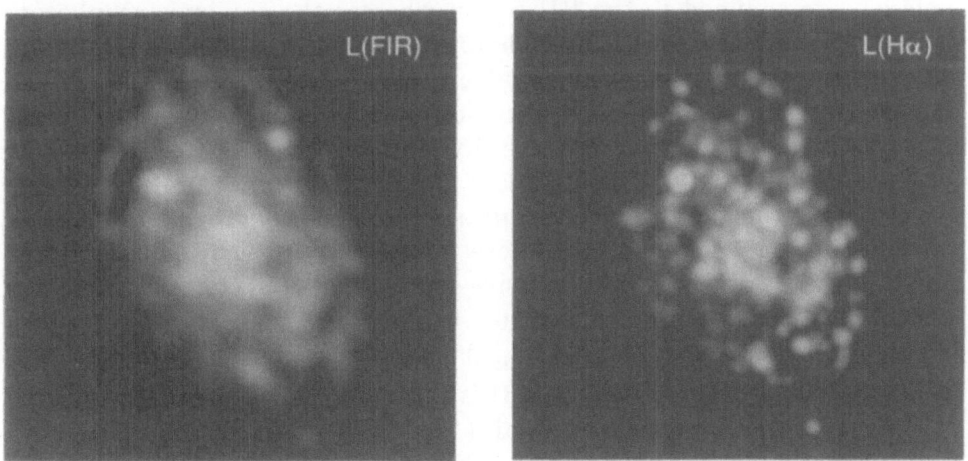

Figure 1. Upper Panel: The Andromeda galaxy (M31). Far infrared (left) and Hα (right). Lower Panel: Two views of M33. Far infrared (left) and Hα (right). The images share a common resolution of 105 arc seconds.

M31, more commonly known as the Andromeda galaxy, is the nearest (0.7 Mpc) early-type (Sb) spiral galaxy. M31 has remained at the focus of the debate concerning the origin of the far infrared luminosity because of its proximity (Habing et al. 1984; Walterbos & Schwering 1987; Devereux et al. 1994).

The far infrared and Hα images of M31 are very similar in that they both show a conspicuous star forming elliptical ring which contributes 70% of *both* the total far infrared *and* Hα emission. The bright emission centered on the bulge contributes the remaining 30% of the total far infrared and Hα emission. M33, is a late-type (Sc) spiral galaxy that is at about the same distance as M31. The far infrared and Hα images of M33 are also very similar in that they both show multiple luminosity peaks that coincide with cataloged HII regions (Devereux et al. 1996).

The fact that the far infrared luminosity peaks coincide with known HII regions qualifies massive stars as the principle dust heating source in M33 and the star forming ring of M31. However, massive stars are unlikely to be responsible for the far infrared luminosity associated with the bulge of M31 as discussed further in section 2.4.

The fact that the far infrared morphology is not identical to the Hα morphology is not entirely unexpected even though both emissions may be sustained by a common source, namely massive stars. The Hα emission suffers from extinction that can cause the relative brightnesses of features in the Hα image to be different from the corresponding features in the far infrared image. Additionally, the ionizing photons that give rise to the Hα emission are confined to the HII region whereas the non-ionizing photons that are produced by the hot star photosphere and the non-ionizing photons that result from recombination are able to escape the HII region, heat dust, and hence generate far infrared emission over an area that is 2 to 3 times larger than that associated with the Hα emission. The latter effect explains why the far infrared emission appears more diffuse than the Hα emission. Nevertheless, the fact that the far infrared emission is more diffuse than the Hα emission has led some to propose a diffuse component of far infrared emission that is not heated by massive stars (Rand et al. 1992; Xu & Helou 1996; Walterbos & Greenawalt 1996).

Xu & Helou (1996) recently argued that only ∼ 30% of the total far infrared, 8-1000μm, luminosity of M31 is produced by HII regions compared to 70% measured by Devereux et al. (1994). The main reason for the difference can be traced to the fact that Xu & Helou adopt an Hα luminosity for the star forming ring that is a factor of 1.7 lower than that measured by Devereux et al. They also adopt a total (8-1000μm) far infrared luminosity that is a factor of 1.6 higher than that measured by Devereux et al. These two factors cause Xu & Helou to underestimate the contribution of the star forming ring to the total far infrared luminosity by a factor of 2.7 with respect to the results reported in Devereux et al. Regardless of the

numerical differences, one just has to look at the striking correspondence
between the far infrared and Hα morphology (see Figure 1) to realize that
Xu & Helou's claim can not possibly be correct.

2.3. THE L(FIR)/L(Hα) RATIO

Further evidence that the O and B stars, which are required to generate the
Hα emission, are also able to sustain the far infrared emission, is provided by
a quantitative analysis of the far infrared and Hα luminosities. A histogram
illustrating the distribution of L(FIR)/L(Hα) luminosity ratios measured
at 114 independent locations within M31 is presented in Figure 2.

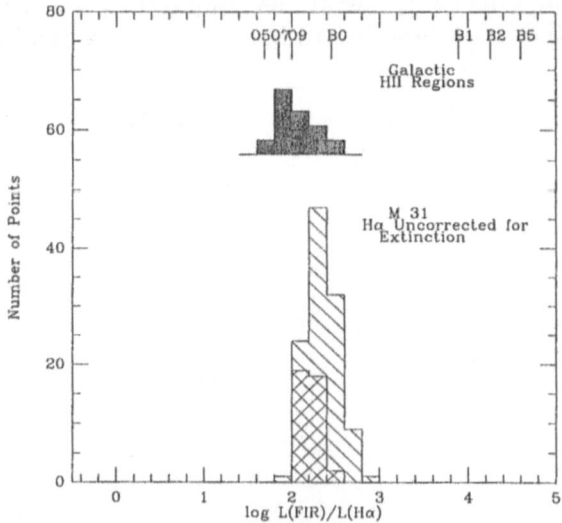

Figure 2.   Histogram illustrating the similarity between the distribution of
L(Hα)/L(FIR) ratios measured within M31 and Galactic HII regions. The single hatched
histogram pertains to the star forming ring, the double hatched histogram to the nucleus
and associated filamentary emission interior to the star forming ring

The histogram exhibits several notable features. First, the range of
L(FIR)/L(Hα) ratios measured within M31 is similar to that determined
for Galactic star forming regions. Correcting the Hα luminosities for extinction only strengthens the result. For example, correcting for a plausible
1 magnitude of extinction would shift the histogram to the left by 0.3 dex
causing it to more closely match the Galactic HII regions histogram.

The range of L(FIR)/L(Hα) luminosity ratios is also consistent with theoretical expectations based on stellar atmosphere models. The tick marks
at the top of the figure identify the theoretical L(FIR)/L(Hα) ratios expected for HII regions powered by massive stars of a variety of spectral
types. The values are based on Panagia (1973) stellar atmospheres and a

simple model in which the Lyman continuum photons ionize the hydrogen
gas and the bolometric luminosity of the exciting star is radiated in the far
infrared. Although simple, such a model appears capable of explaining the
energetics of Galactic HII regions as noted previously by Wynn-Williams
& Becklin (1974).

The L(FIR)/L(Hα) ratio is essentially independent of radius within M31
(see Figure 2) which is somewhat surprising given that the origin of the far
infrared luminosity in the bulge of M31 is probably different from that in the
star forming ring (see section 2.4). In contrast to M31, the L(FIR)/L(Hα)
ratio does change significantly with radius in M33 (see Figure 3). Devereux
et al. (1996) attribute the radial change in the L(FIR)/L(Hα) ratio to a
corresponding metallicity gradient that apparently causes the upper mass
limit of the initial mass function to increase with radius in M33.

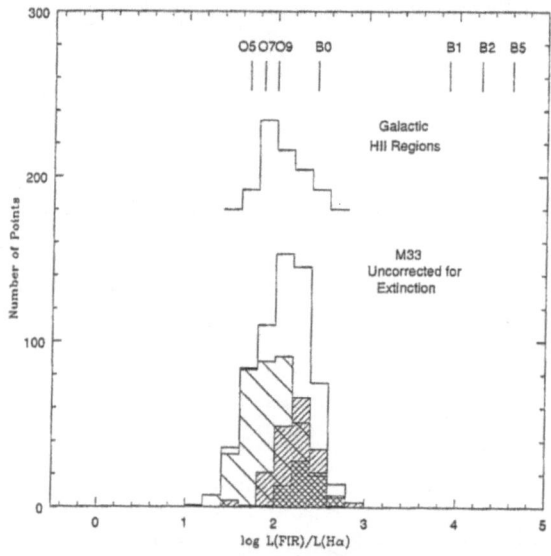

Figure 3.   Histogram illustrating the radial gradient in the L(Hα)/L(FIR) ratios measured within M33. The close double hatched histogram represents the central r < 500 arc
sec region. The wide double hatched histogram represents intermediate radii (500 ≤ r <
1000 arc sec. The single hatched histogram represents the outer region r ≥ 1000 arc sec.

Given that the L(FIR)/L(Hα) ratio is expected, indeed observed, to
depend on spectral type varying by a factor of 1000 between O5 and B5
type stars (see Figure 3), it is inappropriate to identify HII regions with
a single value of the L(FIR)/L(Hα) ratio. Nevertheless, this approach has
led some authors to attribute up to 70% of the far infrared luminosity with
a component that is not powered by massive stars (Rice et al. 1990; Rand
et al. 1992). The additional component is frequently referred to as cirrus.
The term cirrus was originally coined by Low et al. (1984) to describe the

morphology of the Galactic far infrared emission. However, the only galaxy that is near enough to identify cirrus morphologically is our own. Use of the term cirrus in other galaxies usually connotes a component of far infrared luminosity that is produced by a diffuse component of dust heated by the general interstellar radiation field. It is important to appreciate at this juncture that it is presently very difficult to determine if O and B stars contribute significantly to the general interstellar radiation field in other galaxies, but if they do, the distinction between cirrus and star formation as the origin of the far infrared luminosity becomes a semantic issue. Such a situation is most likely to arise the magellanic irregular galaxies and the disks of spiral galaxies.

Sauvage & Thuan (1992) attribute a weak Hubble type dependent trend in the $L(FIR)/L(H\alpha)$ ratio as evidence for a systematic change in the contribution of cirrus to the far infrared luminosity that increases from 3% in Sdm galaxies to 86% in Sa galaxies. Such a Hubble type dependent trend in the $L(FIR)/L(H\alpha)$ ratio was not apparent in an earlier analysis of essentially the same dataset by Devereux & Young (1991). The relationship between the far infrared and $H\alpha$ luminosity is illustrated in Figure 4 for 124 galaxies included in the emission line survey of Kennicutt & Kent (1983). The points are coded by spiral type. The figure shows that the $L(FIR)/L(H\alpha)$ ratios are indicative of HII regions powered by massive stars regardless of spiral type. In this regard, the low luminosity nearby galaxies M31 and M33 are no different than their higher luminosity counterparts.

2.4. SPIRAL BULGES

The bright patch of $H\alpha$ emission that is co-extensive with the bulge of M31 is almost certainly not powered by massive stars. The $L(FIR)/L(H\alpha)$ ratio measured for the bulge of M31 is very similar to that expected for HII regions powered by O9 stars (see Figure 2) but such stars are unlikely to be responsible for either the far infrared or the $H\alpha$ luminosity for several reasons. First, while the bulge far infrared and $H\alpha$ luminosity could, in principle, be powered by $\sim 10,000$ O9 stars, the original higher resolution $H\alpha$ image (Jacoby et al. 1985, Devereux et al. 1994) shows the bulge $H\alpha$ emission to be filamentary and unlike that of the star forming ring. Furthermore, the stars identified in an ultraviolet image of the central 44 arc second region of M31 taken with the Hubble Space Telescope (HST) (King et al. 1992) are much too faint to be O9 stars although they could be B5 stars. B5 stars would not be expected to generate large HII regions but the surface density of B5 stars that would be required to sustain the observed $H\alpha$ luminosity exceeds the surface density of stars in the HST image by at least 2 orders of magnitude. Consequently, high mass stars; the most com-

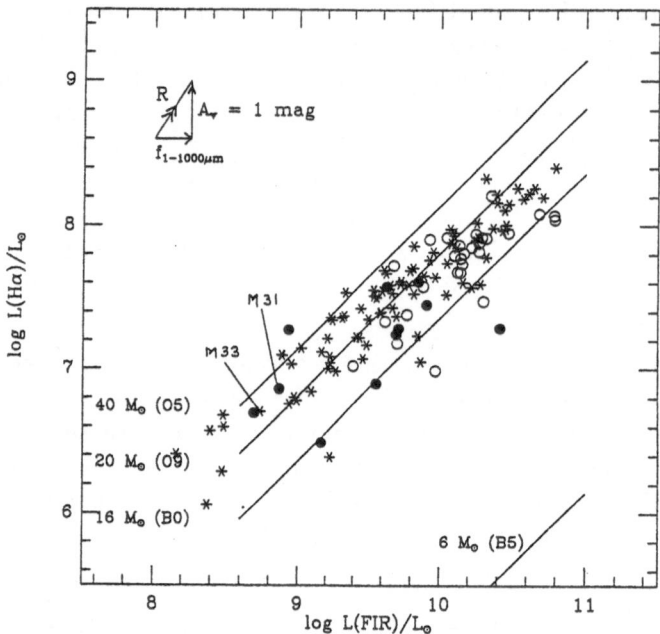

Figure 4. The correlation between the Hα and IRAS $40 - 120\mu m$ luminosity for 126 spiral galaxies. The points are coded by spiral type; filled circles identify early-types (Sa-Sab), open circles; intermediate types (Sb-Sbc) and stars; late-types (Sc and later).

mon source of ionizing radiation in galaxies, appear to be ruled out as the source of excitation for the Hα emission observed in the bulge of M31. The Hα emission is most likely excited by post asymptotic giant branch (PAGB) stars (Buson et al. 1993; Binette et al. 1994). PAGB stars are not expected to generate significant far infrared emission, thus the origin of the bulge far infrared emission may be completely unrelated to the origin of the Hα emission. Smith et al. (1991,1994) used the Kuiper Airborne Observatory to study the far infrared emission from the bulges of the early-type spiral galaxies M94 (NGC 4736) and M66 (NGC 3627) and concluded that the interstellar radiation field of the bulge stars is sufficient to heat the dust up to the observed temperatures.

The bulge of M31 radiates a similar percentage, $\sim 30\%$, of the total far infrared and Hα luminosity. Thus, adopting either the total far infrared or the total Hα luminosity as a measure of the current massive star formation rate would lead to an overestimate of 30% for this early-type spiral galaxy. Consequently, the far infrared luminosity appears to be as good a measure of the massive star formation rate as the Hα luminosity, because both measures are similarly compromised, at the 30% level, by emission from the bulge that can not be attributed to massive stars. The same may be true for other early-type spiral galaxies.

2.5. THE IRAS $S_{100\mu}/S_{60\mu}$ COLOR TEMPERATURES

The IRAS $S_{100\mu m}/S_{60\mu m}$ dust color temperatures may provide an additional constraint on the origin of the far infrared luminosity in spiral galaxies as illustrated in the following example.

2.5.1. *The Late-Type Spiral Galaxy M101*

The dust temperature morphology within the late type spiral galaxy M101 is illustrated in Figure 5 along with complimentary far infrared and Hα images, all convolved to a common resolution of 105 arc seconds (Devereux & Scowen 1994). The figure shows that the dust temperature peaks coincide with Hα luminosity peaks implicating HII regions as a major contributor to the dust heating within M101. The dust temperatures decrease radially from peak values of $\sim$ 35 - 45 K to a minimum $\sim$ 25 K in between the peaks (assuming a λ^{-1} emissivity law).

Like M31 and M33, the far infrared emission in M101 is more diffuse than the Hα emission. Since the dust temperature peaks coincide with peaks in the Hα image, the temperature of the diffuse emission, which distinguishes the far infrared morphology from the Hα morphology, must be cooler than that associated with the peaks and perhaps cool enough to be identified with diffuse dust heated by the general interstellar radiation field. The temperature of such diffuse dust can be estimated using the results of Desert et al. (1990) who calculate the $S_{100\mu m}/S_{60\mu m}$ dust color temperatures expected for a cirrus grain size distribution that is irradiated by stellar photons ranging in intensity from 0.3 to 1000 times that in the solar neighborhood. It is important to appreciate that the models of Desert et al. (1990) specifically include the small grains that are required to explain the unusually high color temperatures measured for Galactic cirrus by IRAS (Draine & Anderson 1985).

Adopting the V band surface brightness as a tracer of the interstellar radiation field intensity one finds that the maximum $S_{100\mu m}/S_{60\mu m}$ dust color temperature predicted for cirrus in M101 corresponds to a temperature of $\sim$ 27 K for a λ^{-1} emissivity law. Cirrus is unlikely to be a major contributor to the far infrared luminosity, however, because Figure 6 shows that 80% of the 40-120μm far infrared luminosity of M101 is radiated by dust with temperatures greater than that.

M101

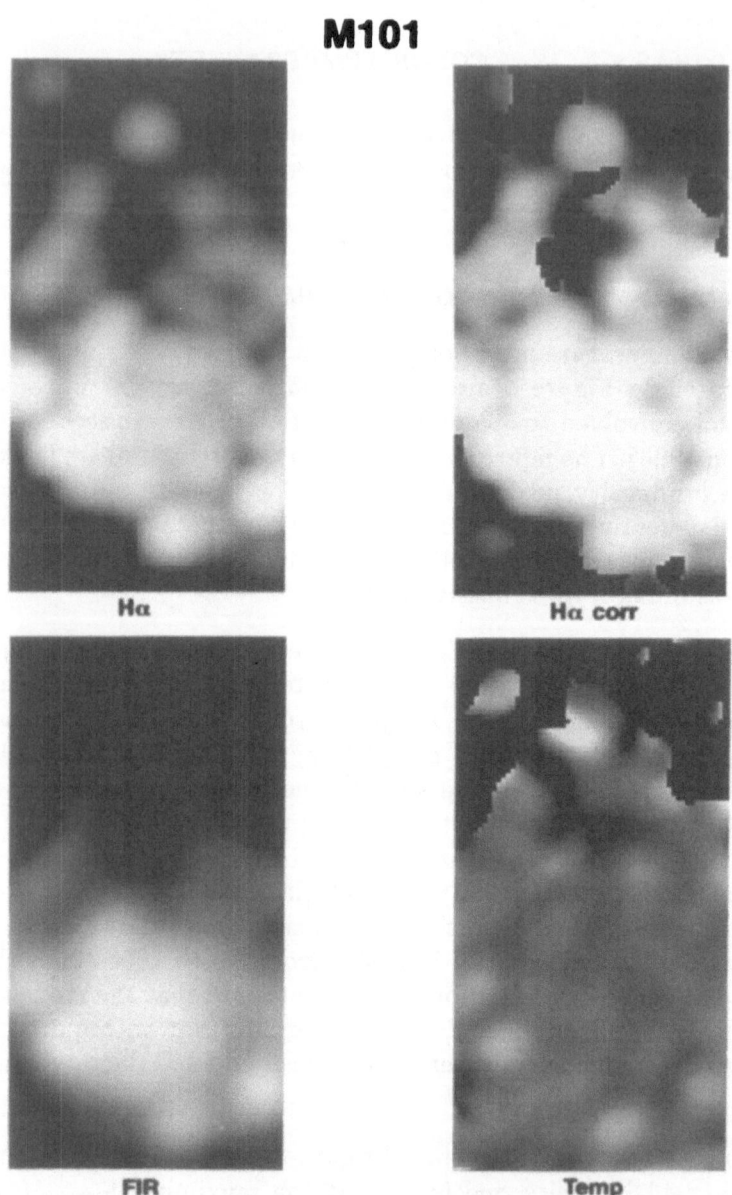

Figure 5. The distribution of the observed Hα luminosity (top left), extinction corrected Hα luminosity (top right), far infrared luminosity (lower left) and the $S_{100\mu m}/S_{60\mu m}$ dust color temperature (lower right) within the late-type spiral galaxy M101

3. The Temperature of Dust in Galaxies and the Dust Opacity

The temperature of dust in galaxies depends on a number of parameters including the intensity of the radiation field heating the dust, the dust grain

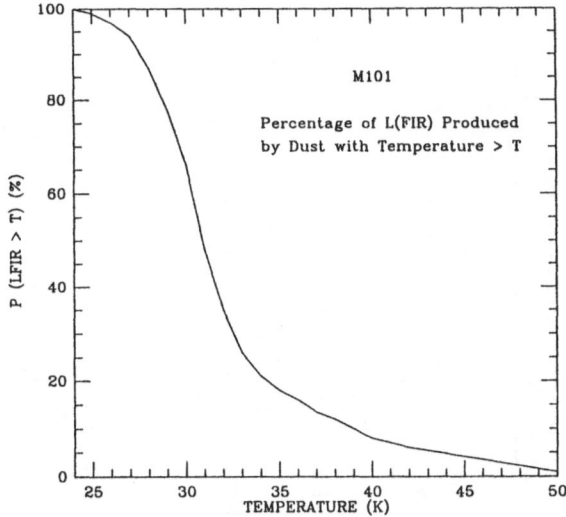

Figure 6. The percentage of the total 40-120μm far infrared luminosity of M101 that is produced by dust with temperatures greater than T. The temperature is based on the $S_{100\mu m}/S_{60\mu m}$ flux ratio and a λ^{-1} emissivity law

size and its composition. By way of illustration, the temperature expected for dust of a variety öf grain sizes and compositions that is heated by the general interstellar radiation field within the early-type spiral galaxy M81 is shown in Figure 7.

The temperatures expected for large, 0.1 and $1\mu m$, Silicate and Carbon grains are based on the models of Jura (1982). The line labelled cirrus reflects the dust temperature expected for a grain size distribution that is required to explain the high temperatures observed for Galactic cirrus (Desert et al. 1990).

The temperature of large, 0.1 and $1\mu m$, dust grains is expected to exhibit a strong radial gradient reflecting the exponential decline in the intensity of the stellar radiation field that is characteristic of spiral galaxy disks. The temperature gradient is expected to be much smaller for a grain size distribution that includes small grains, however, because of the weaker dependence of the temperature of small grains on the *intensity* of the interstellar radiation field (Draine & Anderson 1985).

Several attempts have been made over the years to use the IRAS data to investigate the opacity of spiral galaxy disks (Phillips et al. 1991; Bothun & Rogers 1992; Xu & Buat 1995). Unfortunately the approach may be flawed as there are two important results, in addition to those presented in the previous section, which suggest that the warm dust measured by IRAS has very little to do with the dust that is responsible for the optical extinction

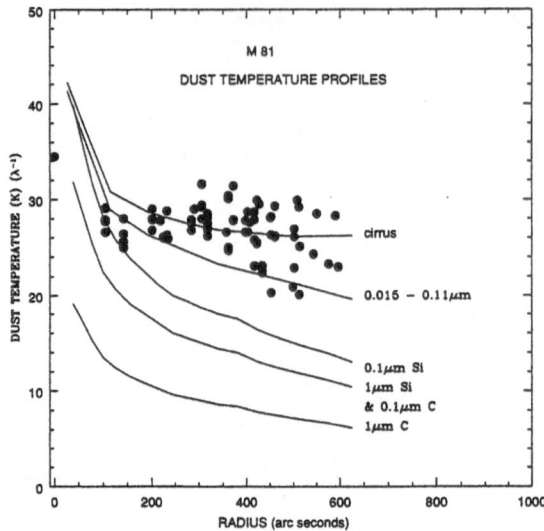

Figure 7. Temperatures expected for dust of a variety of grain sizes and compositions that is heated by the general interstellar radiation field within M81. The solid dots represent the observed temperatures derived from the $S_{100\mu m}/S_{60\mu m}$ flux ratio and a λ^{-1} emissivity law.

in galaxies.

First, the observed IRAS $S_{100\mu m}/S_{60\mu m}$ dust color temperatures are much higher than the temperatures expected for the large, $\geq 0.1\mu m$, dust grains responsible for the optical extinction in galaxies (see Figure 7). Second, the dust mass that is required to radiate the far infrared emission measured by IRAS typically represents only $\sim 10\%$ of the total dust mass inferred from the atomic and molecular gas masses and a Galactic gas mass/dust mass ratio (Devereux & Young 1990b). The implication is that whatever stars are responsible for heating the dust in galaxies up to the 27 - 45 K temperatures recorded by IRAS, they are only able to heat $\sim 10\%$ of the total dust mass to those temperatures, the bulk of the dust mass must therefore be much cooler and radiating at wavelengths longward of $100\mu m$. Continuum measurements longward of $100\mu m$ show direct evidence for such cold dust (Chini et al. 1986; Devereux & Young 1992; 1993; Chini & Krugel 1993; Chini et al. 1995) and the derived dust temperatures are consistent with those expected for the large dust grains responsible for the optical extinction in galaxies.

The evidence for cold dust in the late-type spiral galaxy NGC 6946 is presented in Figure 8 which shows the far infrared spectral energy distribution as defined by the IRAS 12, 25, 60 and $100\mu m$ fluxes in addition to 160 and $170\mu m$ measurements obtained with the KAO. A single temperature blackbody *fitted* to the IRAS 60 and $100\mu m$ data is unable to simultaneously fit the independently determined $160\mu m$ (Engargiola 1991) and $170\mu m$ (Smith et al. 1984) measurements. The discrepancy is unlikely to be attributed to a difference in calibration between the KAO and IRAS as the calibration has been checked by Engargiola (1991). The most likely explanation is that the $160\mu m$ and $170\mu m$ measurements are registering emission from dust which is significantly colder than the ~ 30K dust that is responsible for the $60\mu m$ and $100\mu m$ emission measured by IRAS.

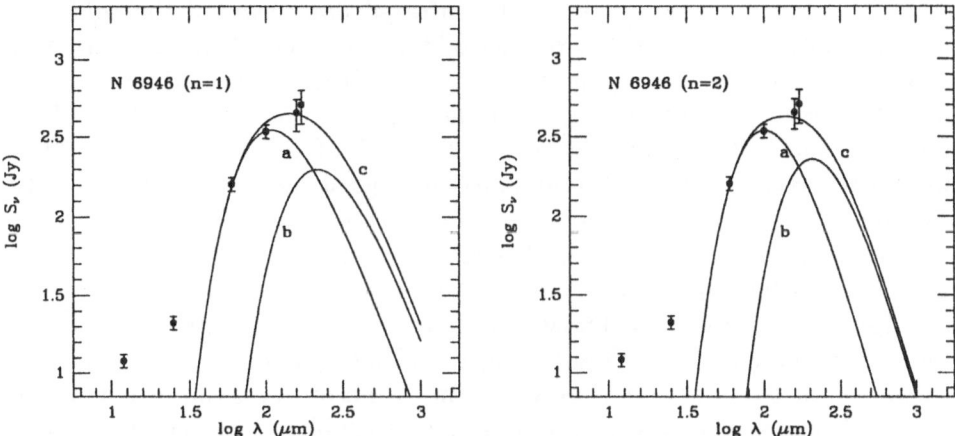

Figure 8. The global far infrared spectral energy distribution of NGC 6946. The independently measured $160\mu m$ and $170\mu m$ fluxes both exceed those expected if all the dust were radiating at the same temperature as that required to produce the IRAS 60 and $100\mu m$ emission (curve a) regardless of whether the dust emissivity index n=1 (left panel) or n=2 (right panel).

The 60, 100, 160 and $170\mu m$ photometry can be used to uniquely constrain the mass and temperature of the dust in the following way. The observations can be fit by the sum of two blackbodies, one *fitted* to the IRAS 60 and $100\mu m$ measurements (curve a) representing the *warm* dust mass and a second blackbody (curve b) representing the remainder of the dust mass that is cold. The temperature and amplitude of the second blackbody are varied to satisfy the *dual* requirements that $\sim 90\%$ of the dust *mass* be in the cold component and the *sum* of the two blackbodies (curve c) provide a satisfactory fit to the observations. It is important to appreciate that the temperature and the amplitude of the two blackbodies are completely constrained by the observational data.

For an emissivity index n=1, curve c in Figure 8 represents the sum of the emission expected if 10% of the dust mass is at a temperature of 34 K (curve a) and 90% of the dust mass is at a temperature of 17 K (curve b).

For an emissivity index n=2, curve c represents the sum of the emission expected if 4% of the dust mass is at a temperature of 29 K (curve a) and 96% of the dust mass is at 14 K (curve b). In both cases, the far infrared spectral energy distribution of NGC 6946 indicates that the bulk, 90 - 96%, of the *dust mass* is at a temperature of 14 - 17 K, regardless of the dust emissivity law, be it proportional to λ^{-1} or λ^{-2}.

The 14 - 17 K temperature range derived for the bulk of the dust is consistent with the temperature expected for large, $\geq 0.1\mu m$, grains heated by the general interstellar radiation field within NGC 6946 (Devereux & Young 1993). Such large dust grains are expected to dominate the optical extinction. Consequently, it is the far infrared *luminosity* of the *cold dust*, not the warm dust measured by IRAS, that is to be identified with that responsible for the opacity in spiral galaxy disks. The distinction is an important one because even though the cold dust represents 90 - 96% of the dust *mass*, it produces only $\sim$ 22 - 24% of the total 40 - 1000μm *luminosity*. NGC 6946 is not unique in this regard, similar results have also been obtained for the late-type spiral M51 (Devereux & Young 1992).

The method of analysis described above occasionally causes some of my colleagues to become a little agitated. One of the most frequently asked questions is "How do you know that both the 60 and 100μm emission is radiated by dust of a single temperature ?". The answer is that I don't ! But, I also hasten to point out that this is not a necessary assumption, I only adopt it because it makes the method easier to explain. It is equally valid to relax the assumption, so long as the method is followed in a self consistent way, it will always work. For example, one could posit that all of the 60μm flux but only half the 100μm flux is produced by the warm component. One then calculates a "new" temperature for the warm dust, use that to calculate a "new" warm dust mass. Next one would determine the cold dust mass by subtracting the warm dust mass from the total dust mass. The total dust mass is derived from the sum of the atomic and molecular gas masses and the assumption of a gas/dust ratio. The temperature and amplitude of the cool dust blackbody are then varied as before to satisfy the *dual* requirements that the correct fraction of the dust *mass* be in the cold component and the *sum* of the two blackbodies (curve c) provide a satisfactory fit to the photometry.

Another frequently asked question is "How can you use the $S_{100\mu m}/S_{60\mu m}$ flux ratio to calculate a dust temperature ?", the implication here being that emission from small grains compromises the determination of a physical dust temperature. A key point is that empirically, the far infrared luminosity increases as if the $S_{100\mu m}/S_{60\mu m}$ flux ratio were a physical dust temperature as discussed previously in Devereux & Young (1990b).

There has been considerable debate over the past decade as to whether or not the far infrared (40 - 1000μm) spectral energy distributions of spiral galaxies can be fit by a single temperature component (Eales et al. 1989; Stark et al. 1989) or whether additional components are required as suggested by Chini et al. (1986), Chini & Krugel (1993) and Chini et al. (1995). The situation has undoubtedly been confused by the fact that the submillimeter observations rarely measure the globally integrated flux like the IRAS measurements with which they are frequently compared. However, restricting the analysis to include only the few galaxies that have been mapped, one discovers that the cold component can be seen in some galaxies like M51 (Devereux & Young 1992), NGC 6946 (Devereux & Young 1993), NGC 660 and UGC 3490 (Chini & Krugel 1993), but not in others like the 4 Virgo spirals mapped by Stark et al. (Devereux & Young 1990b) and the Markarian galaxies (Chini et al. 1989). That the cold dust component can not be seen in some galaxies means one of two things; that the cold dust is not there or that the cold dust emission is drowned out by the much brighter emission from the warmer dust. In either case, the fact that the cold component can not always be measured places a fundamental limitation on the use of submillimeter continuum observations to determine the cold dust luminosity, and the associated dust mass in galaxies. Even if it were possible to reliably measure the cold dust luminosity, deriving the dust opacity from that is, in my opinion, a completely intractable problem.

The greatest source of uncertainty in calculating the dust opacity from a measurement of the cold dust luminosity is the 3 dimensional distribution assumed for the dust and stars. Witt et al. (1992) have shown the opacity to be *highly dependent* on the star and dust geometry. Thus, while it may be possible to construct various models, their usefulness in predicting the disk opacity will always be compromised by the uncertainty surrounding the *actual* distribution of stars and dust within the galaxies with which the models are being compared. Consequently, measuring the cold dust luminosity may not be the best way to determine the disk opacity.

References

Aumann, H.H., Fowler, J. W., & Melnick, M., 1990, AJ **99**, 1674.
Binette, L., Magris, C.G., Stasinska, G., & Bruzual, A.G., 1994, A&A **292**, 13.
Bothun, G.D., Lonsdale, C.J., & Rice, W., 1989, ApJ **341**, 129.
Bothun, G.D., & Rogers, C., 1992, AJ **103**, 1484.
Buson, L.M., et al., 1993, A&A **280**, 409.
Cataloged Galaxies & Quasars in the IRAS Survey, 1985, prepared by C.J. Lonsdale, G. Helou, & W. Rice (Washington D.C.: Jet Propulsion Laboratory).
Chini, R., Kreysa, E., Krugel, E., & Mezger P.G., 1986, A&AL **166**, 8.
Chini, R., & Krugel, E., 1993, A&A **279**, 385.
Chini, R., Krugel, E., Kreysa, E., Gemund, H.-P., 1989, A&AL 216, 5.
Chini, R., Krugel, E., Lemke, R., & Ward-Thompson, D., 1995, A&A **295**, 317.
Desert, F,-X., Boulanger, F., & Puget, J.L., 1990, A&A 237, 215.

de Jong et al., 1984, ApJL **278**, 67.
de Jong, T., & Brink, E., 1987, In *Star Formation in Galaxies*, p. 323. ed. C. Lonsdale (Washington D.C.: NASA).
de Jong, T., Klein, U., Wielebinski, R., & Wunderlich, E., 1985, A&AL **147**, 6.
Devereux, N.A., Duric, N., & Scowen, P.A., 1996, AJ (submitted)
Devereux, N.A., & Eales, S.A., 1989, ApJ **340**, 708.
Devereux, N.A., Price, R., Wells, L.A., & Duric, N., 1994, AJ **108**, 1667.
Devereux, N.A., & Scowen, P.A., 1994, AJ **108**, 1244.
Devereux, N.A., & Young, J.S., 1990a, ApJL **350**, 25.
Devereux, N.A., & Young, J.S., 1990b, ApJ **359**, 42.
Devereux, N.A., & Young, J.S., 1991, ApJ **371**, 515.
Devereux, N.A., & Young, J.S., 1992, AJ **103**, 1536.
Devereux, N.A., & Young, J.S., 1993, AJ **106**, 948.
Dickey, J.M., & Salpeter, E.E., 1984, ApJ **284**, 461.
Draine, B.T., & Anderson, N., 1985, ApJ **292**, 494.
Eales, S.A., Wynn-Williams, C.G., & Duncan, W.D., 1989, ApJ **339**, 859.
Engargiola, G., 1991, ApJ Supp. **76**, 875.
Gavazzi, G., Cocito, A., & Vettolani, G., 1986, ApJL **305**, 15.
Habing, H.J., *et al.*, 1984, ApJ Letters **278**, 59.
Helou, G., Soifer, B.T., & Rowan-Robinson, M., 1985, ApJL **298**, 7.
Jacoby, G.H., Ford, H., & Ciardullo, R., 1985, ApJ **290**, 136.
Jura, M., 1982, ApJ **254**, 70.
Kennicutt, R.C., Jr., & Kent, S. M.,1983, AJ **88**, 1094.
King, I.R., et al. 1992, ApJ L **397**, 35.
Leech, K.J., Lawrence, A., Rowan-Robinson, M., Walker, D., & Penston, M.V., 1988, MNRAS **231**, 977.
Lonsdale-Persson, C.J., & Helou, G., 1987, ApJ **314**, 513.
Low, F.J., et al., 1984, ApJL **278**, 19.
Panagia, N., 1973, AJ **78**,, 929.
Phillipps, S., Evans, Rh., Davies, J.I., & Disney, M.J., 1991, MNRAS **253**, 496.
Price, R., & Duric, N., 1992, ApJ **401**, 81.
Rice, W., Boulanger, F., Viallefond, F., Soifer, B.T., & Freedman, W.L., 1990, **358**, 418.
Rand, R.J., Kulkarni, S.R., & Rice, W., 1992, **390**, 66.
Sanders, D.B., & Mirabel, I.F., 1985, ApJL **298**, 31.
Sauvage, M., & Thuan, T.X., 1992, ApJL **396**, 69.
Smith, B.J., Harvey, P.M., Colome, C., Zhang, C.Y., DiFrancesco, J., & Pogge, R.W., 1994, ApJ **425**, 91.
Smith, B.J., Lester, D.F., Harvey, P.M., & Pogge, R.W., 1991, ApJ **373**, 66.
Smith, J., Harper, D.A., & Loewenstein, R.F., 1984, Airborne Astronomy Symposium Proceedings. ed. H. A. Thronson, Jr., & E.F. Erickson (NASA Conf. Pub. 2353) p. 277.
Solomon, P.M., & Sage, L.J., 198, ApJ **334**, 613.
Stark, A.A., Davidson, J.A., Harper, D.A., Pernic, R., Loewenstein, R., Platt, S., Engargiola, G., & Casey, S., 1989, ApJ **337**, 650.
Stark, A.A., Knapp, G.R., Bally, J., Wilson, R.W., Penzias, A.A., & Rowe, H.E., 1986, ApJ **310**, 660.
Thronson, H.A., Jr., Majewski, S., Descartes, L., & Herald, M., 1990, ApJ **364**, 456.
Walterbos, R.A.M., & Schwering, P.B.W., 1987, A&A **180**, 27.
Walterbos, R.A.M., & Greenawalt, B., 1996, ApJ (in press).
Witt, A.N., Thronson, H.A., & Capuano, J.M., Jr., 1992, ApJ **393**, 611
Wynn-Williams, C.G., & Becklin, E.E., 1974, PASP **86**, 5.
Xu, C., & Buat, V., 1995, A&AL **293**, 65.
Xu, C., & Helou, G., 1996, ApJ **456**, 152.
Young, J.S., Kenney, J.D., Lord, S.D., & Schloerb, F.P., 1984, ApJL **287**, 65.
Young, J.S., Schloerb, F.P., Kenney, J.D., & Lord, S.D., 1986, ApJ **304**, 443.

SEARCH FOR COLD MOLECULAR GAS

E.J. DE GEUS
OVRO, Caltech, MS 105-24, Pasadena, CA 91125
and Kluwer Academic Publishers, Dordrecht, the Netherlands

1. Abstract

The past several years have seen an increase in the interest for the potential presence of significant amounts of cold molecular gas in galactic disks. It has been hailed as the solution to the (Galactic) dark matter problem, a statement which merits a thorough discussion and investigation of the validity of the claims posed.

In order to present this talk, there needs to be clarity about what exactly it is that we are looking for, and what exactly it is that the different observing techniques reveal to us about the properties of the gas. As I hope to show you, it is not entirely straightforward to interpret the results from the different observations, and the basic assumptions that are made need to be understood and questioned.

2. Introduction

Knowledge of the total mass of all components of matter is important for a proper understanding of the dynamics of astronomical systems. It has been rather a sore point for a long time that it appears to be impossible even in galactic systems to account for all sources of mass, and the missing material has subsequently been dubbed dark matter, an indication intended of course to resolve astronomers of all blame for not having found it until now.

In a recent paper Pfenniger et al. (1994) argued that the "missing mass" in the Galaxy may be in the form of cold, predominantly molecular gas in a flat distribution. To have escaped previous detection in the outer parts of our Galaxy, this gas would have to be very cold and have a low emissivity. Because most substantial heating sources of gas are associated with the presence of massive stars (UV radiation and cosmic rays) the requirement

373

D. L. Block and J. M. Greenberg (eds.), New Extragalactic Perspectives in the New South Africa, 373–386.
© *1996 Kluwer Academic Publishers.*

of low available heating might be satisfied in various locations in a galaxy, such as the outer parts of the galactic disk, the Halo, or even inside the star forming ring in the inner parts of a galaxy such as M31. Cold molecular gas has been searched for in various locations in our Galaxy and external galaxies using a variety of observing techniques. The possibility that cold molecular gas might be the dark matter component required to flatten the rotation curve of galaxies certainly warrants a search for it. Nonetheless, even if little mass would be contained in it, knowledge of the distribution and properties of this component of the ISM, and its relation to the warmer, star forming component would be of interest in itself.

Using single dish emission line observations of molecular gas in the outer parts of our Galaxy Mead & Kutner (1988) have reported low kinetic temperatures, based on a direct interpretation of the antenna temperature measured, corrected for various telescope efficiencies, in terms of a kinetic temperature of the gas.

Also using single dis emission h line studies, Allen & Lequeux (1993) claim observation of very low temperature molecular gas in the region inside the star forming ring of M31. Their temperature estimates are based on the ratio of the observed CO(2-1) and CO(1-0) lines. Comparison with radiative transfer models of various levels of sophistication enable the interpretation of certain level populations in terms of a kinetic temperature of the gas. Certain other physical parameters of the gas (density and column density) need to be restricted using transitions of other molecules or other isotopes.

Lequeux, Allen & Guilloteau (1992, LAG) observed molecular absorption against two quasars, using the IRAM 30-m single-dish. From the detection of four absorption lines of CO (one double line) along these two lines of sight (all within 12 kpc from the Galactic center), and a comparison with the amount of HI absorption, they conclude that enough mass could be contained in cold molecular gas to flatten the rotation curve of the Galaxy.

De Geus & Phillips (1996a & b) used the OVRO mm-interferometer also to search the outer parts of our Galaxy for a low-emissivity molecular gas component by looking for molecular mm-absorption lines toward 8 background quasars, including the two lines of sight studied by Lequeux et al. This resulted in a total of only three absorption lines (one double system), significantly fewer. Based on these observations de Geus & Phillips concluded that half of the absorption lines seen in the Lequeux et al paper are artefacts of the observing technique, that the gas seen in absorption is not necessarily cold, and that there is no evidence for the presence of molecular gas without significantly more atomic gas.

Kobulnicky, Dickey & Akeson (1995) combined interferometer absorption observations and single-dish emission data to obtain the excitation temperature of gas in a nearby cloud. They found an excitation temperature of 2.7 K, but they concluded that differences in columns traced by the emission and the absorption are so different that the technique is invalid. Furthermore, the excitation temperature was argued to be significantly lower than the actual kinetic temperature of the gas.

It is obvious from the research mentioned above that a number of techniques can be used to attempt to find and study cold molecular gas. The claims made, however, should be subjected to the utmost scrutiny, because of the potential importance of the results. This paper will attempt to be critical of all observations resulting in claims of low kinetic temperatures of molecular gas, it will provide a thorough investigation of all the assumptions that enter into the interpretation, and discuss their validity.

3. Back to Basics

First the situation of one transition, interpreted in terms of Tkin, will be discussed. The most basic molecular line emission observation is that of the CO(1–0) line at 2.6 mm. The measured antenna power can be directly related to an emission temperature of molecular gas when the antenna pattern of the telescope is known. Kraus (1982) expresses it simply as $T_A = T_K \frac{\Omega_M}{\Omega_A}$, where T_A is the measured antenna temperature, T_K is the kinetic temperature of the emitting material,Ω_M is the main-beam solid angle, and Ω_A is the antenna solid angle. This primary measure of the temperature of the gas is highly uncertain. The most important problem is the beam filling factor. The observed temperature is only a measure of the gas temperature if the source couple perfectly to the telescope beam. In other words the emitting material needs to extend beyond the beam size of the telescope. If the gas does not fill the beam (because it is patchy or filamentary), then this method always underestimates the true gas temperature. If it does fill the beam, but the gas has a lot of structure, then the measured temperature will only be an average over the cloud. Furthermore, this method only measures the escitation temperature of the gas, and problems of subthermal excitation (see below) are ignored.

More commonly temperatures of molecular clouds are measured by comparing the line ratios of different transitions of the CO molecule. These line ratios are compared to the results of radiative transfer calculations of varying levels of complexity, and the temperature of the best fitting model is inferred to be the kinetic temperature of the gas. The model calculations involve a number of additional physical properties of the gas and its environment, such as optical depth in the visual, gas density, gas column density, cosmic ray flux density and strength of the interstellar UV radiation field (I_{UV}). The CO line ratios are not entirely independent of these other parameters, and as such the line ratios are not perfect thermometers of the gas, especially when only two transitions are used (usually 2–1 and 1–0). Therefore, in order to do a proper analysis of the kinetic temperature of the gas, further observations of isotopes of CO and more than two transitions need to be observed in order to perform a proper level analysis and determine the kinetic temperature more accurately. This method too is subject to uncertainties if the cloud is not accurately mapped. Under typical observing conditions, a molecular cloud can be followed out to larger size in CO(1–0) than it can in CO(2–1). The low column-density edges of clouds are still observable in 1–0 and no longer in 2–1, which is due to the fact that in the low-density and low column-density parts of the cloud non-LTE effects start playing a role, a problem which is known as subthermal excitation. In the low-density low-column-density regime, excitation of the levels is no longer entirely governed by collisions, but spontaneous emission (i.e. de-excitation) becomes important as well. In the low-column-density (i.e. low optical depth) case, the spontaneously emitted photons escape from the cloud and do not result in a new excitation back to the higher level. This causes a relative underpopulation of the higher levels, and the level populations then mimick a lower temperature. Cloud temperatures based on line ratios can therefore usually only be trusted in the higher density and column-density parts of clouds. If one integrates the antenna temperatures over larger areas including the low-column-density parts of the cloud, simulating a larger beam size, the drop in 1–0 is smaller than that of the 2–1 antenna temperature, resulting in lower line ratios, and underestimates of the gas temperatures. It has been defined that line ratios of different transitions give an excitation temperature (T_{ex}, and the interpretation of the excitation temperature in terms of a true gas kinetic temperature remains dependent on the density and column density of the gas. The more sophisticated radiative transfer models, i.e. the ones that do not assume LTE (which assumes $T_{ex} = T_k$), will deal properly with subthermal excitation when sufficient constraints on density and column density are available to the model.

Often, the predicate "cold" is easily used when observations reveal absorption. In order for absorption to occur a foreground parcel of gas needs to be colder than a background source of emission. Self-absorption occurs in single-dish spectra, and is the situation where the background source is a cloud of molecular gas, and colder gas exists at almost the same LSR velocity (thus interpreted as the same distance). Part of the emission line profile is then absorbed, sometimes completely. Since some molecular clouds are actually quite warm, and in star-forming clouds temperatures increase inwards, self absorption is not uncommon. Absorption can also be found due to gas in front of a continuum source, which for molecular line absorption usually is a quasar (point source). The situation in that case is best illustrated using the basic equation of radiative transfer in the Rayleigh-Jeans limit is:

$$T_b = T_o e^{-\tau_\nu} + T_{ex}(1 - e^{-\tau_\nu})   \qquad 1$$

, where T_b is the observed antenna temperature, the excitation temperature of the foreground gas is T_e, which has optical depth τ_ν, and the background source temperature is T_0. Normally this equation describes the emission, above zero intensity, of a line. However, when a background continuum source is present, part of the continuum flux is absorbed, but filled in again with emission. Absorption only results when the excitation temperature of the gas is lower than the temperature of the background source. A special observing technique employing single dish telescopes to find absorption was used already in the 1960's to detect HI absorption against quasars. In order to cancel the emission filling in of the absorption line, the average of four surrounding spectra was taken to be representative of the emission toward the source and subtracted. This then leaves an absorption spectrum. However, Radhakrishnan (1972) already pointed out that this method is fraught with error, because any structure in the emission in the "off"-positions results in an artificial absorption line. A similar situation occurs when absorption is searched with an interferometer. Since an interferometer acts as a spatial filter, removing the lowest spatial frequencies from the signal, any fairly smoothly distributed component of emission would be naturally missing from the data. As a result, merely the existence of foreground gas will result in an absorption profile. These two observing methods result in a reduced radiative transfer equation, from which the emission part is absent:

$$T_b = T_o e^{-\tau_\nu}.   \qquad 2$$

The foreground gas temperature no longer appears in this equation, so, when using either of these techniques, one cannot conclude that gas is cold simply because it is seen in absorption.

It is immediately obvious from the reduced radiative transfer equation (eq. 2), that neither of the latter two methods gives us any indication of the temperature of the foreground gas. The basic absorption method (which is still described by eq. 1) usually also does not restrict the temperature very much, for the following reason. For a background continuum source (a point source) the flux density (S_ν) is constant, while its (antenna) temperature (T_0) depends on the area of the antenna beam (A): $T_0/S_\nu \propto A^{-1}$. This implies that the smaller the single-dish beam size (the more sophisticated the telescope) the larger the background temperature, and the less restriction this implies for the temperature of the absorbing gas.

It is obvious from this discussion that great care needs to be exercised when interpreting reports of low temperatures.

4. Emission Observations

Reports of low kinetic temperatures of molecular gas appear regularly and we have to make sure that we understand the assumptions made to come to the conclusion that the gas is cold.

Searches for cold molecular gas have been carried out in various locations. All of them have been guided by the supposition that cold gas can only exist in locations where few sources of heating are available. The places searched are, therefore, the outer Galaxy, locations in external galaxies away from the presence of massive stars, and the Galactic Halo.

4.1. THE OUTER GALAXY IN SINGLE DISH

The outer Galaxy has received a lot less attention from researchers than the inner Galaxy, for obvious reasons. The inner Galaxy contains far more observable molecular material as well as a greater abundance of star forming regions. From a viewpoint of galactic dynamics, however, the outer Galaxy (or equivalently the outer parts of external galaxies) are of interest, because it is there that the problem of missing matter necessary to understand the generally flat rotation curves is most obvious.

Kutner & Mead (1981) and Mead & Kutner (1988) have observed a number of molecular clouds in the outer Galaxy in ^{12}CO (1–0) and (2–1) and in ^{13}CO (1–0), from which they derive rather low kinetic temperatures of 7 K for the envelopes of the observed clouds, as opposed to the interiors which are at higher temperatures (10 K). These data were not analyzed with a full radiative transefr code, but $CO(2-1)$ and ^{13}CO were used first to restrict the optical depth and density of the gas. The conclusion from that was that "at and near cloud peaks the isotopomer ratio indicates CO being optically thick". The excitation temperature was then directly calculated from the observed $CO(1-0)$ temperature and $CO(2-1)$ temperature independently, assuming a source coupling efficiency (of 0.8). This leaves the question whether on the cloud edges the optical depth is sufficiently high to prevent subthermal excitation from occuring. Furthermore, any presence of structure in the cloud (particularly on the edges) would result in large deviations of the source coupling efficiency from the assumed number. Presence of structure results in a decrease in the source coupling efficiency, leading to an increase in the inferred excitation temperature. Recently also Brand & Wouterloot (1995) have studied molecular clouds in the outer Galaxy. Their sample was based on IRAS point sources with colours indicating massive star formation. For two clouds they observed ^{12}CO (1–0) and (2–1) and ^{13}CO, and they found a range of kinetic temperatures from 6 to 15 K. The higher temperatures were found in the vicinity of the star forming cores, while the lowest temperatures were found in the quiescent parts of the cloud. The radiative transfer code they used was the escape probability code by Stutzki & Winnewisser (1985). The low temperatures were found only in one cloud (WB 1152), but not in a low column density region, in fact the lowest temperatures were found toward a peak in ^{13}CO (with a ratio of CO (2–1)/(1–0) of 0.57). This implies that it is unlikely that the density and column density are so low that subthermal excitation will play a role in this situation. In the same paper, Brand & Wouterloot (1995) investigated the relation of observed antenna temperatures of clouds as a function of Galactic radius. For this they used clouds that were not related to IRAS sources (but were serendipitous detections in their survey), so are not biased towards the warmer clouds. The interesting trend is that within each sample, the measured antenna temperatures appear to be constant with Galactic radius. This implies that in the outer Galaxy there is no indication for a drop of molecular gas temperatures with distance to the Galactic center.

4.2. EXTERNAL GALAXIES IN SINGLE DISH

In a number of recent papers Allen & Lequeux (1993), Loinard, Allen & Lequeux (1995) and Allen et al. (1995) have studied molecular gas associated with dark clouds in the inner parts of M31. The dark clouds were catalogued by Hodge (1980), and are situated in the relatively empty region inside the molecular/star forming ring. The observations were made with the IRAM 30-m dish, in the lines of ^{12}CO (2–1) and (1–0) and ^{13}CO (2–1) and (1–0). These data were analyzed to determine the physical properties of the gas in Allen et al. (1995), using models described by LeBourlot et al. (1993), assuming LTE. The CO (2–1) and (1–0) ratios observed are very low (between 0.2 and 0.4). The dark clouds found by Hodge (1980) extend over very large areas. Furthermore, these clouds are very filamentary and patchy, and in fact the impression one gets from these plates is that these objects are not the kind of GMCs that we find in the inner parts of our Galaxy, but rather much more like the cirrus that we see all around us at an average z-height of 100 pc. The beamsize of the IRAM 30-m translates to an linear diameter of 75 pc at 115 Ghz and 43 pc at 230 Ghz (smoothed to 75 pc for the analysis). On those scales, judging by the charts from Hodge's (1980) atlas, the 30-m beam is much larger than the size scales of these filaments and patches. This implies that in one beam one measures not only the cloud peaks, but also significant amounts of low-density and low-column-density material. The CO measurements will therefore, within the beam, also include significant areas where subthermal emission will play an important role. Drawing conclusions using the emission from components of gas with such a huge variety of densities, column densities and presumably temperatures, all gathered into one spectrum is a very risky. In fact, Allen et a. (1995) cannot reproduce the different line ratios observed by using only a single density model, and advocate a combined model of low and high density gas.This, however, makes the models poorly constrained by the data: strengths of 4 lines are are matched to two models with three free parameters each. Furthermore, two parameters (cosmic ray flux density and interstellar UV field) had to be guessed and taken to be very low. The result was that the best-fit models had very low kinetic temperatures. Because of the large number of possible parameters, these models are unlikely to be unique, and higher temperature models might fit the results just as well.

The low ratio of CO(2–1) and (1–0) that was found in M31 is also not unusual. In fact, similarly low ratios were found in our own Galaxy by Sakamoto *et al.* (1995) who compared very large-scale (8') Galactic CO(2–1) observations with the CO(1–0) observations of Dame *et al.* 1985. This supports the argument that the low ratios are not due to low temperatures but rather due to t large areas over which the data is gathered.

5. Absorption Searches

5.1. THE OUTER GALAXY

In the past few years, two attempts have been made to find evidence for the existence of very cold molecular gas in the outer parts of our Galaxy, by searching for molecular absorption towards mm-bright quasars. The first one is by Lequeux, Allen and Guilloteau (1993, LAG), who used the IRAM 30-m interferometer in CO for an absorption experiment looking toward 2 quasars. They averaged four spectra (OFF) immediately surrounding the position of the quasar (ON) and subtracted that from the ON spectrum in order to cancel the emission part of the signal, to be left with the absorption. The second one is by de Geus & Phillips (1996) , who used the OVRO mm interferometer looking for absorption in CO and HCO$^+$ against a total of 8 objects, including the two by LAG.

LAG found a total of five absorption lines in their two lines of sight, all within Galactic radius of 12 kpc. De Geus & Phillips found a total of four absorption lines in eight lines of sight, with three absorption lines along the two lines of sight in common between both studies. Based on a comparison of the column densities inferred from the molecular line data and from HI observations, LAG concluded that cold molecular gas contained more mass than atomic gas by a factor of 4 in the outer Galaxy. Interpreting this as due to the lack of sources of cloud heating in the outer Galaxy, they inferred that this mass ratio would increase even further out in the Galaxy, leading to the conclusion that possibly an order of magnitude more mass is contained in cold molecular gas than in HI, sufficient to explain the flat rotation curve.

From IRAM 30-m observations, LAG argued that gas found in emission toward 0727-115 was cold, with $T_{ex} \approx 3.5$ K. To reach this conclusion they ignored the possibility of subthermal excitation and assumed that the filling factor of CO emission in the 23" IRAM beam was unity, so it is unlikely that the actual kinetic temperature is as low as this (see also Wilson & Mauersberger 1994). The optically thick absorption line toward 0727-115,which was reported by LAG was not detected by de Geus & Phillips (1996). However, LAG also noted that there were significant emission gradients in the immediate vicinity of 0727-115, which implies that a single dish absorption experiment like that of LAG is very susceptible to creation of false absorption. In fact, follow-up absorption observations with the IRAM interferometer support the conclusion that no optically thick CO absorption is present (Allen & Lequeux priv. comm.) De Geus & Phillips did detect a faint absorption line of HCO^+, however (see Figure 1).

The other object observed by LAG, 2013+370 is a compact radio source (Duin et al. 1975) whose line of sight traverses two supernova remnants: Cygnus X and G 74.9+1.2, and three spiral arms: the Local Arm, the Perseus Arm, and the Cygnus Arm. The lines at -56 km s^{-1} are due to gas located at a Galactic radius of 11.5 kpc, which at the longitude of 74.87 deg is most likely on the near side of the Cygnus (or outer) Arm, while the gas at $+6$ km s^{-1} is associated with the gas in the Local Arm around the Cygnus X region. The absorption lines we observed toward 2013+370 are associated with molecular gas previously detected in large scale surveys of CO(1-0) emission (Leung & Thaddeus 1992, Sanders et al. 1986), and are therefore unlikely to be as cold as 3.5 K (LAG).

In the absorption observations of de Geus & Phillips (1996) both the HCO^+ lines at -56 km s^{-1} and at $+6$ km s^{-1} show a low-level, broad component. This was not found for the CO lines (only observed at -56 km s^{-1}), although owing to the noise we cannot rule out a broad, low-level absorption if it has similar optical depth to the HCO^+. The presence of Gaussian line components with low-level, broad line wings have been observed before in emission spectra, and their interpretation has been either in terms of clump and inter-clump gas (e.g. Blitz & Stark 1986; Blitz, Magnani & Wendel 1988), or in terms of intermittency in turbulent gas (Falgarone & Phillips 1990; Falgarone, Phillips & Walker 1991).

The presence of turbulence implies that kinetic energy is transferred from the largest scales in a molecular cloud to the smallest scales where it is dissipates in *heat*. Spaans *et al.* (1995) calculated the heating of molecular gas due to various sources including turbulence, cosmic rays and the photoelectric effect. At densities and turbulent velocity widths typical for diffuse clouds, turbulent heating is approximately 3 – 4 times greater than cosmic ray heating, but two orders of magnitude less than heating by UV photons. LAG, however, argue that in the outer Galaxy the UV-radiation field and the cosmic-ray flux density may be significantly lower than typical values in the Solar neighborhood. Although there is no evidence that this is true for the clouds we see in absorption, we can calculate the temperature of the gas if turbulence were the only source of heating. We assume that the two dominant sources of cooling are neutral hydrogen impact excitation of atomic carbon and line-emission by CO. We use table 4 from Dalgarno & McCray (1972) for the H – C cooling rate and the curves given by Goldsmith & Langer (1978), and Gilden (1984) for the CO cooling rate. We find that the equilibrium kinetic temperature would be approximately 15 – 20 K. Although this number is uncertain, we conclude that the gas is probably warmer than the 3.5 K derived by LAG.

All molecular absorption lines reported here were previously seen in HI absorption (Dickey et al. 1983). This is in contradiction with LAG, who claim to find an optically thick CO absorption line toward 0727–115, which does not correspond to any HI absorption. As pointed out before, we find a much lower molecular gas column density, CO is certainly not optically thick, in fact our non-detection of CO gives an upper limit of $\tau_{CO} < 0.65$. From the HCO$^+$ column density toward 0727–115, and using the HCO$^+$/CO column density ratio observed toward 2013+370, we estimate a CO column density of $1 - 8 \times 10^{15} \text{cm}^{-2}$, and thus, for a normal CO to H$_2$ abundance ratio of 10^{-4}, an H$_2$ column density of $1 - 8 \times 10^{19} \text{cm}^{-2}$. Close inspection of the data of Dickey et al. (1983) reveals an HI absorption line with a column density of approximately $3 \times 10^{20} \text{cm}^{-2}$, which implies that in this absorbing cloud 6 to 35 % of the mass is molecular. For each of the lines around -56 km s^{-1} toward 2013+370 we estimate an HI column density of approximately $1.3 \times 10^{21} \text{cm}^{-2}$. Based on the CO column density we derive H$_2$ column densities of $1 - 5 \times 10^{19} \text{cm}^{-2}$. This implies that along this line of sight 2 to 10 % of the mass of the gas is molecular. Furthermore, the lines of sight which did not reveal molecular absorption also contain significant amounts of HI absorption. We therefore conclude that, purely based on our sample, the ratio of total molecular to total atomic gas mass is no more than 5%, a measurement valid out to Galactic radius of 11.6 kpc. Unless the ratio of molecular to atomic gas mass increases dramatically beyond 11.6 kpc, molecular gas would be insufficient to be the dark matter component that flattens the rotation curve in contradiction to the statement by LAG that there is 4 – 5 times more molecular gas than atomic gas in the outer Galaxy, and that molecular gas could be the dark matter component.

We therefore conclude that there is no longer an observational basis for the two conclusions by LAG that 1.) there is *cold* molecular gas in the outer Galaxy and 2.) that this component is so massive that it can explain the flat rotation curve of the Galaxy. However, our observations do not constrain the amount of cold molecular gas in the far outer Galaxy beyond 12 kpc. If such a component does exist beyond 12 kpc, the best way to detect it is in absorption against a background source, but the small number of background sources in the 3-mm wave band limits the detection probability (additional methods for finding such a component are discussed by Combes in this proceedings). The probability (P_n) of detecting the molecular component using a limited number (n) of (pencil-beam) lines of sight depends directly on the volume filling factor (f_V) of the gas: $P_n = 1 - (1 - (f_V)^{2/3})^n$. If the cold molecular gas density is ~ 300 cm^{-3}, then $f_V \sim 10^{-3}$. The probability of detecting a cold cloud with eight lines of sight is only $P_8 \approx 8\%$. Thus, the results so far do not rule out the possibility of the existance of (abundant) cold gas further out in the Galaxy.

The HCO$^+$ absorption line detected in 3C418 is due to gas located at a distance of 11.5 kpc from the Galactic center at a Galactic longitude of 88.81 deg, and a heliocentric distance of 7.9 kpc. This implies that the absorbing gas is coincident with the Perseus spiral arm. This line of sight was included in the emission line survey by Liszt & Wilson (1993), but no CO (1–0) emission brighter than 0.3 K (T_R^*) was detected. The absorbing gas toward 3C418, lies at a z-height of 840 pc above $b = 0$ deg in the Perseus Arm. Kulkarni, Blitz & Heiles (1982) show that at this location in the Galaxy, the mid-plane of atomic hydrogen deviates from $b = 0$ deg by approximately 100 – 150 pc. The absorbing cloud then lies at a distance of $\gtrsim 700$ pc from the mid-plane of galactic HI. Furthermore, contrary to the HCO$^+$ absorption lines in 2013+370, this absorption line is fit perfectly well by a single narrow Gaussian component, and no evidence is found for a broad low-level component. The large distance of the absorbing cloud above the midplane of the HI in the Galaxy could mean that it is far enough removed from a constant supply of large-scale energy input required to sustain a turbulent energy cascade (like shock waves from various sources) or that it makes other sources inefficient (cloud-cloud collisions, shear due to differential Galactic rotation). In our present search it is therefore the best candidate for "cold" molecular gas, but we have no knowledge of the temperature whatsoever.

5.2. HIGH-VELOCITY CLOUDS

De Geus, Phillips & Akeson (in preparation) have searched for absorption by molecular gas toward a sample of 35 quasars in the general direction of High-Velocity Clouds (HVCs). This in an attempt to find a component of molecular gas associated with the Galactic Halo. No detections were made in CO or HCO$^+$. Unfortunately, because no HI absorption data are available for these lines of sight, the limits on the molecular gas column densities cannot be used to restrict the ratio of molecular gas to atomic gas in HVCs.

6. Conclusions

The search for cold molecular gas is an important one, because of the possibly large amounts of mass that could be hidden in such a component.

Methods used for determining molecular gas temperatures were discussed, and their restrictions illustrated. Specific applications of these methods were reviewed and their conclusions about low molecular-gas temperatures were examined critically. A number of claims of detections of low-temperature gas were found to be uncertain. The lowest trustworthy temperature reported is that by Brand & Wouterloot (1995), which is 6 K for a cloud in the outer Galaxy. Low temperatures of molecular gas in the inner regions of M31 (Allen et al. 1995) have been questioned. The observations used to infer the presence of large amounts of cold molecular gas in the outer parts of the Galaxy (LAG) have been shown to be flawed. New observations have shown that the amount of molecular gas is no more than 5% of the atomic component in the outer Galaxy out to distances of 12 kpc from the center.

In eight lines of sight investigated we found four absorption components, of which three are in the outer Galaxy. Toward one of the lines of sight in common with LAG (2013+370) only two of the four absorption components claimed by LAG are confirmed, while in the second one (0727–115) the absorption line reported by LAG was found in HCO$^+$ but not in CO. Two of the three absorption components that were detected with high signal-to-noise (both toward 2013+370) show evidence for broad line wings, possibly caused by intermittency in a Kolmogorov-type turbulence. From the thermal balance between the heating caused by this turbulence (a lower limit because it ignores all other sources of heating) and cooling due to CO, equilibrium temperatures of the gas in these two clouds of between 15 and 20 K were derived, suggesting that the gas is warm. All absorption lines reported in this paper have counterparts in HI absorption, and the ratio of molecular to atomic gas is estimated to be less than 5%. Unless this ratio increases dramatically beyond 11.6 kpc (the absorption line furthest from the Galactic center), molecular gas can not provide enough mass to flatten the rotation curve. All the molecular absorption data so far are insufficient to draw conclusions about whether cold molecular gas is the dark matter component.

7. Acknowledgements

I thank John Carlstrom for stimulating discussions, and for the suggestion to use HCO$^+$ as a good absorption tracer. The OVRO mm-array is supported by the National Science Foundation under grant number AST 93-14079. I acknowledge the hospitality of the Sterrewacht Leiden, and the support of David Larner, while preparing this paper.

8. References

Allen, R.J. & Lequeux, J., 1993, Astrophys. J. **410**, L15.

Allen, R.J. *et al.*, 1995, ApJ. **444**, 157.

Blitz, L., Magnani, L., & Wendel, A., 1988, Astrophys. J. **331**, L127.

Blitz, L. & Stark, A.A., 1986, Astrophys. J. **300**, L89.

Brand, J. & Wouterloot, J.G.A., 1995, Astron. Astrophys. **303**, 851.

Dalgarno, A. & McCray, R.A., 1972, Ann. Rev. Astron. Astrophys. **10**, 375.

de Geus, E.J. & Phillips, J.A., 1995. In IAU Symposium 169 *Unsolved Problems of the Milky Way*, 1994. Ed. L. Blitz.

Dickey, J.M., Kulkarni, S.R., van Gorkom, J.H. & Heiles, C.H., 1983, Astrophys. J. Suppl. **53**, 591.

Duin, R.M., Israel, F.P., Dickel, J.R. & Seaquist, E.R., 1975, Astron. Astrophys. **38**, 461.

Falgarone, E. & Phillips, T.G., 1990, Astrophys. J. **359**, 344.

Falgarone, E., Phillips, T.G., & Walker, C.K., 1991, Astrophys. J. **378**, 186.

Gilden, D.L., 1984, Astrophys. J. **283**, 679.

Goldsmith, P.F. & Langer, W.D., 1978, Astrophys. J. **222**, 881.

Hodge, P., 1980. *Atlas of the Andromeda Galaxy*, U. of Washington Press, Seattle.

Kobulnicky, H.A., Dickey, J.M. & Akeson, R.L., 1995, Astrophys. J. **443**, L45.

Kraus, Radio Astronomy, 2nd Edition.

Kulkarni, S.R., Blitz, L. & Heiles, C., 1982, Astrophys. J. **259**, L63.

Kutner, M.L.& Mead, K.N., 1981, Astrophys. J. **249**, L15.

LeBourlot, J. *et al.*, 1993, Astron. Astrophys. **267**, 233.

Leung, H.O. & Thaddeus, P., 1992, Astrophys. J. **81**, 267.

Lequeux, J., Allen, R.J., and Guilloteau, S., 1993, A&A, 280, L23.

Liszt, H.S. & Wilson, R.W., 1993, Astrophys. J. **403**, 663.

Loinard, L., Allen, R.J. & Lequeux, J., 1995, Astron. Astrophys. **301**, 68.

Mead, K.N. & Kutner, M.L., 1988, Astrophys. J. **330**, 399.

Pfenniger, D., Combes, F.& Martinet, L., 1994, Astron. Astrophys. **285**, 79.

Radhakrishnan, V., Murray, J.D., Lockhart, P. & Whittle, R.P.J., 1972, Astrophys. J. Suppl. Ser. **24**, 15.

Sanders, Clemens, Scoville & Solomon, 1986, Astrophys. J. Suppl. **160**, 1.

Spaans, M., Black, J.H. & van Dishoeck, E.F., 1994, Astron. Astrophys. submitted.

Stutzki, J. & Winnewisser, G., 1985, Astron. Astrophys. **144**, 13.

Wilson, T. & Mauersberger, R., 1994, Astron. Astrophys. **282**, L 41.

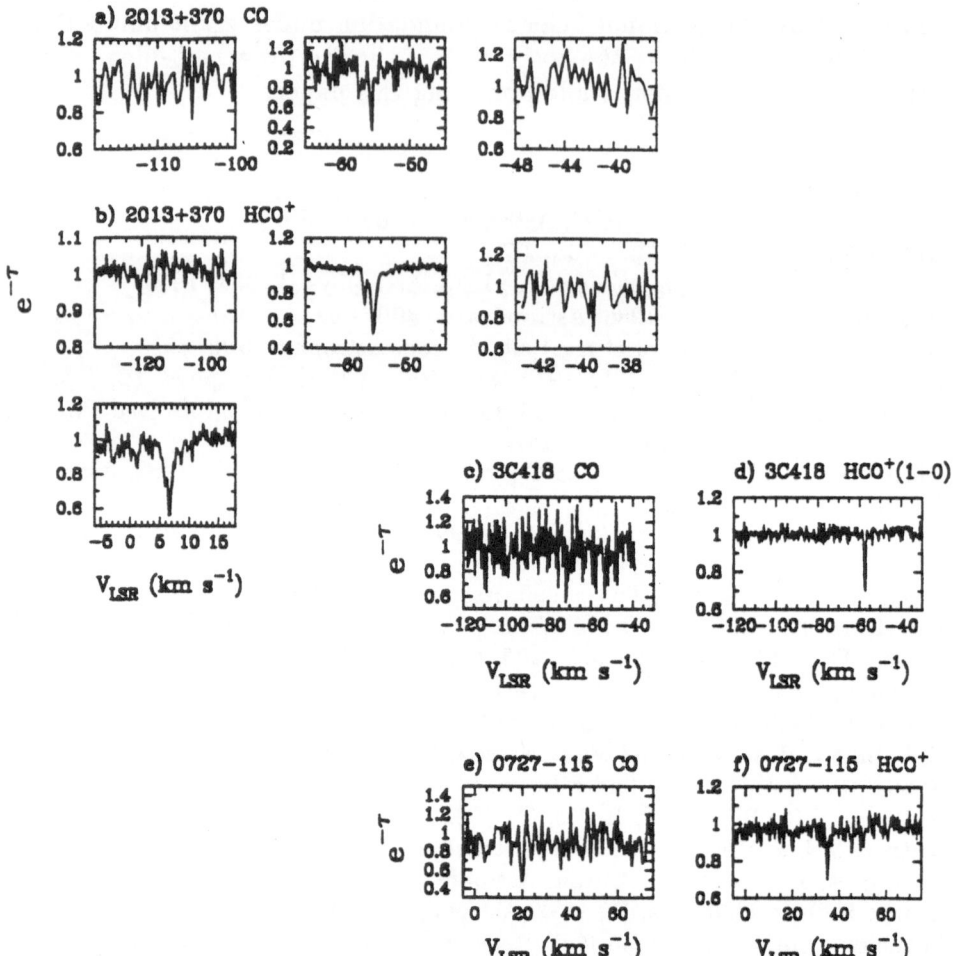

Figure 1. a.) OVRO spectra of the quasar 2013+370. The CO spectra cover the three lines at negative velocities reported by LAG. In CO we only confirm the composite line at -56 km s^{-1}. b.) The four panels show the HCO$^+$ spectra for the 4 different velocities at which LAG reported the detection of absorption. We only confirm two of those lines: the second and fourth panel show our detections of the lines at $+6$ km s^{-1} and the composite line at -56 km s^{-1}. The lines at -104 km s^{-1} and -40 km s^{-1} are not confirmed. c.) Data for the quasar 3C418 from an 8 hour track on the CO(1–0) line, and d.) the same integration time on the HCO$^+$(1–0) line. A line at a Galactic radius of 11.5 kpc is clearly detected in HCO$^+$ but it is hidden in the noise in the CO spectrum. OVRO spectra toward the low-galactic latitude quasar 0727-115 in the lines of CO (e.) and HCO$^+$ (f.). We confirm the presence of a line at $+35$ km s^{-1} in HCO$^+$, but find no CO with an optical depth greater than 0.7.

GALACTIC STRUCTURE AND MORPHOLOGY OF THE MILKY WAY FROM THE TMGS

F. GARZÓN, P. L. HAMMERSLEY, X. CALBET
T. J. MAHONEY AND M. LÓPEZ–CORREDOIRA
Instituto de Astrofísica de Canarias,
E-38200 La Laguna, Tenerife, Spain

Abstract.
We discuss the Two Micron Galactic Survey, a project which has been running since 1989 at the IAC mapping extended areas of the Galactic plane and bulge with a dedicated IR camera. With the analysis made so far we can suggest the existence of a central Galactic bar, whose receding end is found at $l = 27°$ and which is associated with a huge star formation region holding the one of most luminous IR star clusters in the Galaxy. Also the internal distribution of the stars on the plane shows that the vast majority of these belong to the young and old disc population and to a well defined internal bulge, with almost no stars found in the classical spherical distribution of the bulge, or few with high luminosity. The latter is compatible with a barred gravitational potential, which could scatter the stars from the central plane. Finally, two observational follow-up programmes dedicated to enlarging the spectral coverage and scope of the TMGS are outlined.

1. Introduction

Since 1988 a project aimed at surveying large areas of the sky in the near IR, mainly located in the Galactic plane and bulge visible from the Observatorio del Teide, started at the Instituto de Astrofísica de Canarias (IAC) using the new 7 channel InSb camera, developed and built by the group of M. J. Selby at the Blackett Laboratory Imperial College, London. Observational work began in 1989, after several comissioning periods at the Carlos Sánchez Telescope (TCS, 1.5 m) during 1988 and 1989, and have been carried out continously since then, making use of about 25% of the total available

D. L. Block and J. M. Greenberg (eds.), New Extragalactic Perspectives in the New South Africa, 388–395.
© 1996 *Kluwer Academic Publishers.*

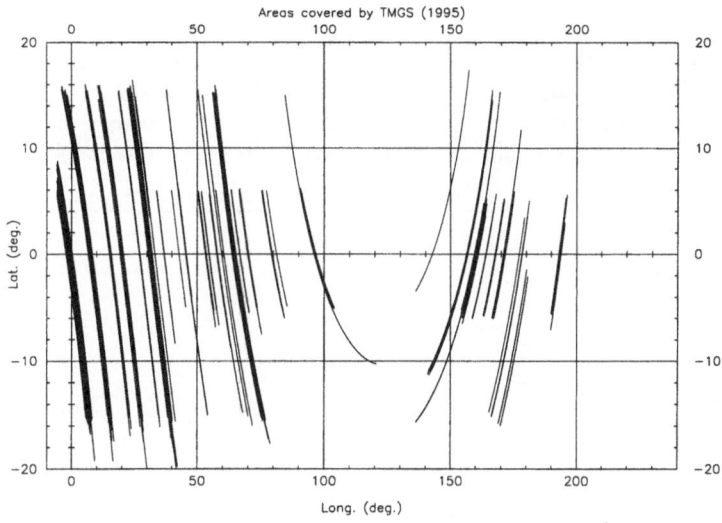

Figure 1. TMGS map in the (l, b) plane of the sky

time at the telescope. With the formal start of DENIS (*Deep European Near Infrared Survey of the Southern Sky*), to which consortium we belong to, we have stopped the observations. More than 400 square degrees of sky have been mapped, leading to a detection of more than 600.000 IR sources, a great percentage without visible counterparts in the GSC. Near infrared photometry at 2.2 micron on the scale of the TMGS ($| b | < 15°$) is a powerful means of seeking out highly reddened sources in the Galactic plane and bulge, where visible photometry alone is restricted to a few kpc, and provides information on the spectral types and luminosity classes of sources buried deep in the plane and bulge. This is particularly useful for the bulge, where most studies to date have of necessity been restricted to windows of exceptionally low extinction.

A detailed description of the project itself can be found in Garzón *et al.* (1993). For the sake of simplicity, we summarize in Table 1 the main features of the survey. In figure 1 we plot a diagram in the (l, b) plane of the areas covered by TMGS. It is seen that the bulk of the map is concentrated along the central bulge, with the scans covering the region of $| b | < 15°$, and also in two regions around $l \sim 65°$ and towards the anticentre.

It is worth to note that the drift–scanning method at sidereal rate used in the TMGS imposes several restrictions on the way the histograms presented in the next sections are built. Even when we have made the choice of the (l, b) Galactic coordinate system for analysing the star count distributions, the scans were effectively taken following trajectories of increasing α, at constant δ, as can be inferred from figure 1. That means that in every

TABLE 1. Main features of TMGS

Filter: K band $(2.2\mu m)$	7–channel InSb linear array
$15''(\alpha) \times 17''(\delta)$ pixels	dynamic range(3σ) : $3.8\mathrm{mag} < m_K < 10.8\mathrm{mag}$
completeness: $m_K \leq 9.8\mathrm{mag}$	confussion limit: $m_K = 8 - 9\mathrm{mag}$
photometric accuracy $\approx 0.2\mathrm{mag}$	positional accuracy: $\alpha \approx 2''; \delta \approx 7''$

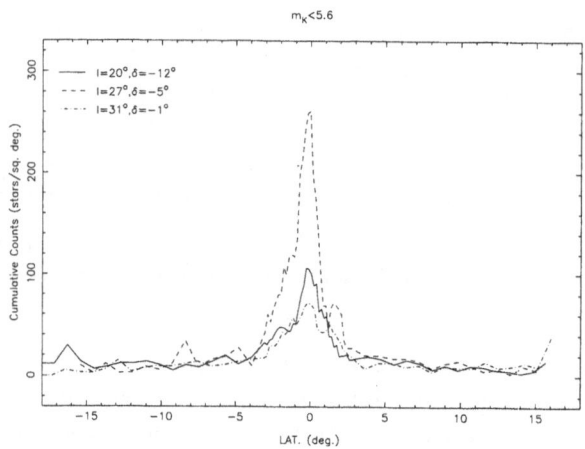

Figure 2.   Histograms of the TMGS areas centered at $l = 20°$, $l = 27°$ and $l = 31°$ in cumulative star counts per square degree to $m_K = 5.6\mathrm{mag}$.

histogram the two ends differ from each other in several degrees over l. This effect might be ruled out away from the centre, where a smooth gradient along l is expected, but has to be borne in mind when looking at the inner Galaxy.

2. Evidence for the Galactic bar

In the following subsections we will introduce briefly three lines of argument which point towards the existence of the central Galactic bar. These direct observational evidence, being considered individually, might be attributed to other causes, but, all together, strongly indicate at the bar.

2.1. THE STAR FORMATION REGION AT $l = 27°$

In figure 2 we plot the histograms of star distribution in TMGS regions centred at $l = 20°$, $27°$ and $31°$. It can be seen (Hammersley *et al.*, 1994) that the graphs for $l = 20°$ and $l = 31°$ basically coincide except for the prominent central spike at $l = 20°$, which, in its turn, is much less prominent than

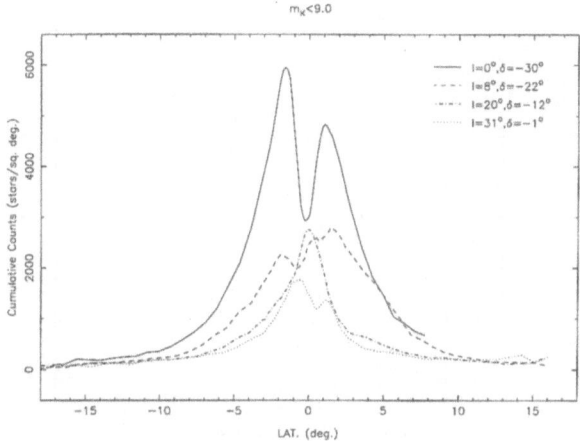

Figure 3. Histograms of the TMGS areas centred at $l = 0°$, $l = 8°$, $l = 20°$ and $l = 31°$ in cumulative star counts per square degree to $m_K = 5.6$.

the one at $l = 27°$. In Hammersley *et al.* (1994) we discuss several scenarios and conclude that this feature is due to a huge star formation region (SFR) containing some of the most luminous IR stars in the Galaxy. such large SFR's are also present in most external barred galaxies. In addition, this SFR permits us to estimate the position angle of the bar which is $75° \pm 5°$ for a circular symmetric Galaxy (see Hammersley *et al.* (1994) for details), compatible with that determined by Blitz & Spergel (1991), who place the near end of the bar in the quadrant $0° < l < 90°$.

2.2. THE CENTRAL BULGE

The structure of the inner Galaxy can be investigate by the TMGS due to its high sensitivity and large spatial coverage. In figure 3 we show the distribution of star counts for four TMGS areas located in the central part of the Galaxy. By comparing the density of star in the plane at the $l = 0°$ with that at $l = 8°$, which is smaller by a factor larger than 2, two main conclusions can be extracted:

1. There are too many stars in both distributions to be explained by disc sources alone. That is, the TMGS is penetrating into the central star distribution.
2. By only assuming axial symmetry in the distribution of the stars on the plane of the Galaxy, at least 50% of the sources towards $l = 0°$ are clustered in the central 6°, forming a separated morphological component which can be called the inner bulge.

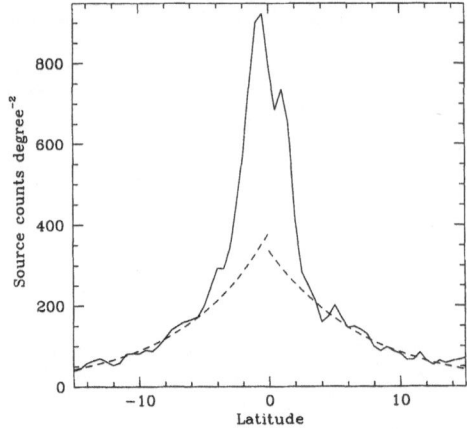

Figure 4.   TMGS counts at $l = 31°$ with exponentials fitted to both sides for $| b | > 5°$

2.3. THE DISTRIBUTION OF STARS IN THE DISC TOWARDS THE CENTRE

Another important consequence can be extracted from the star count distributions shown in figure 3. If one accepts that a minimum of 50% of the stars towards the centre of the Galaxy are located within a few hundred pc of the centre, then the remaining stars are enough to account only for the well accepted standard exponential old disc and young (mainly spiral arms, see Hammersley *et al.* 1994) disc up to roughly 5 kpc from the Sun. Hence there is a gap in the star distribution in the central 3 kpc or, at least, a severe decrease in the luminous IR star density.

3. Is the local disc tilted?

The TMGS star counts of the inner Galaxy ($l < 40°$) can be separated into the old exponential thin disc, which dominates the counts for $| b | > 5°$, and the young disc and spiral arms, which can be distinguished at $| b | < 5°$, as seen in figure 4 (see Hammersley *et al.* 1995 for details).

Steadily in the TMGS counts the fittings for $b < 0°$ are higher than the ones for $b > 0°$ in the same region. When the TMGS data are examined in conjunction with the DIRBE 2.3 and 3.5μm maps we find that the plane of the local stellar disc is flat within about 4 kpc, but tilted around 0.4° with respect to the $b = 0°$ plane. Also the position of the Sun is found to be at $z = 15.5 \pm 3$pc above the plane. These two effects can produce errors of up to 20% in the star counts predicted by models if they are not allowed for.

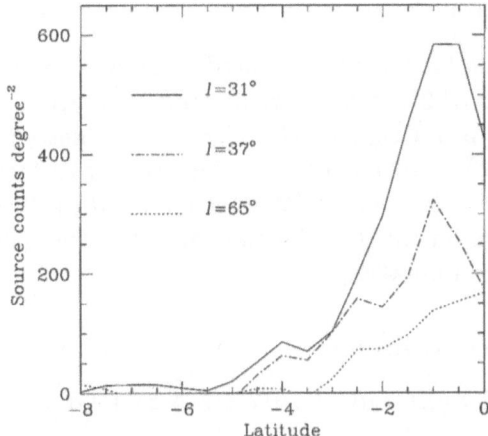

Figure 5. Cumulative star counts at different longitudes after substractig the exponential function

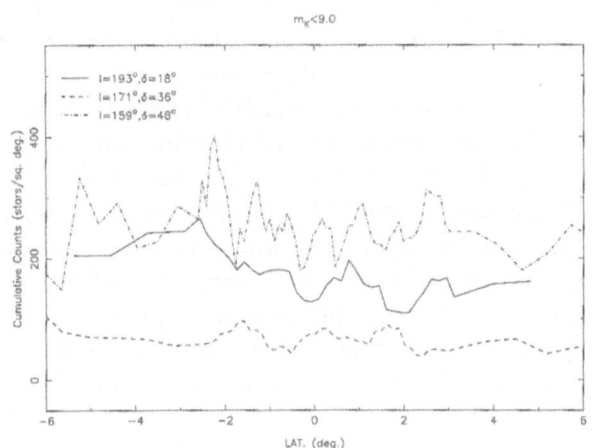

Figure 6. Cumulative star counts at different longitudes in the anticentre region

4. The distribution of stars in the disc

4.1. THE INNER DISC

As outlined in the previous section, in the TMGS histograms in regions of the inner Galaxy the contributions of the different structural components of the Galaxy can be nicely extracted by just fitting standard exponentials to the wings for the old disc and analysing the remaining distribution. This is shown in figure 5, where the plots show the counts mainly due to the Scutum, Sagittarius and Perseus spiral arms. From this, a general scale height of 325 pc and length of 2650 pc can be deduced for the thin disc.

4.2. THE ANTICENTRE

The TMGS has also looked at the distribution of stars in directions outside the solar circle ($l > 150°$), which are plotted in figure 6, with the surprising results that there is no sign of the disc in the histograms (i.e., they are basically flat over b), and that the star densities are a factor of two, roughly, below standard models predict (Wainscoat *et al.*, (1992)). This might be explained by a flare on the disc, by the cut–off in the disc being closer than accepted or by a combination of both.

5. Follow–up programmes

5.1. INT OPTICAL SPECTROSCOPY OF A SET OF SUPERGIANT CANDIDATES

As outlined in the preceding section, the TMGS has found a giant SFR which is embedding a cluster of ultraluminous IR stars, the majority of which are likely to be classified as supergiants, according to their K magnitude and estimated distances. An observational programme devoted to identifying the luminosity classes and, whenever possible, the spectral types of a TMGS source subset selected according to its K magnitude, began in 1995 at the Isaac Newton Telescope (INT, 2.5m), in the Observatorio del Roque de los Muchachos (La Palma), using the Intermediate Dispersion Spectrograph. The primary goal is to measure the equivalent width of the IR CaII triplet and thus deduce the luminosity class (Díaz *et al.*, 1989).

After two runs at the INT we have taken around 50 spectra, with the promising result that 15 sources appear to be supergiants. This programme is still under way. In figure 7 we have shown four of the most common types that we have found, together with two calibration stars of known types, for comparison.

5.2. VRI PHOTOMETRY OF SELECTED AREAS OF THE GALACTIC PLANE AND BULGE

The VRI CCD photometry is part of a programme aimed at calibrating photographic plates taken on the UK Schmidt Telescope and measured on the COSMOS facility at the Royal Observatory, Edinburgh. We have made extensive use of the CCD camera of the IAC80 telescope at the Observatorio del Teide (Tenerife) and that of the JKT (1m) at the Observatorio del Roque de los Muchachos (La Palma)to map several zones of the TMGS for which can have the optical counterpart on photographic plates. Apart from its intrinsic capabilities for studying of the distribution of the different populations along the plane, this programme will permit the cross–correlation of both the TMGS catalogue and COSMOS data base.

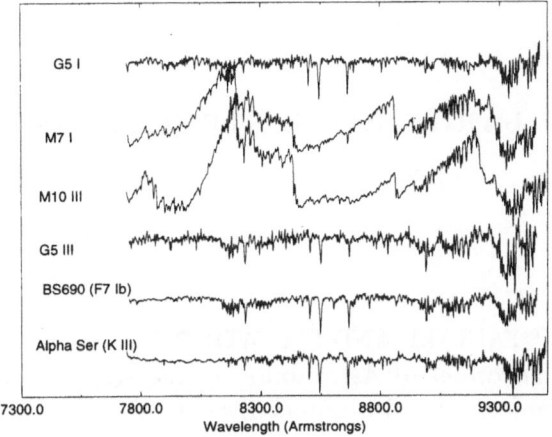

Figure 7. Some spectra taken at INT, with its identified types. The two at the bottom are calibration stars shown for comparison

6. Acknowledgments

The TCS and IAC80 are operated on the island of Tenerife by the Instituto de Astrofísica de Canarias in the Spanish Observatorio del Teide of the Instituto de Astrofísica de Canarias. The INT and JKT are operated on the island of La Palma by the Royal Greenwich Observatory in the Spanish Observatorio del Roque de los Muchachos of the Instituto de Astrofísica de Canarias. We particularly wish to thank the late M. J. Selby without whom the TMGS would not have been possible.

This project has been supported by IAC Grants 28/86 and 5/94 and by the Science and Engineering Research Council. PLH and TJM have been supported by the Dirección General de Investigación Científica y Técnica, Spain.

References

Blitz L., Spergel D. N. (1991) *ApJ*, **379**, 631
Calbet X., Mahoney T., Garzón F., & Hammersley P. L. (1996), *MNRAS*, **276**, 301
Calbet X., Mahoney T., Hammersley P. L., Garzón F. & M. López–Corredoria (1996), *ApJ*, **457**, L27
Díaz A. I., Terlevich E., Terlevich R., 1989, *MNRAS*, **239**, 325
Garzón F., Hammersley P. L., Mahoney T., Calbet X., Selby M. J. & Hepburn, I. (1993), *MNRAS*, **264**, 773
Hammersley P. L., Garzón F., Mahoney T. & Calbet X. (1994), *MNRAS*, **269**, 753
Hammersley P. L., Garzón F., Mahoney T. & Calbet X. (1995), *MNRAS*, **273**, 206
Wainscoat R. J., Cohen M., Volk K., Walker H. J. & Schwartz D. E. (1992), *ApJS*, **83**, 111

UNVEILING LARGE-SCALE STRUCTURES BEHIND THE MILKY WAY

A.P. FAIRALL AND P.A. WOUDT
Department of Astronomy, University of Cape Town
Rondebosch, 7700 South Africa

AND

R.C. KRAAN-KORTEWEG
Departement d'Astrophysique Extragalactique et de Cosmologie,
Observatoire de Paris-Meudon
F-92195 Meudon Cedex, France

The band of the Milky Way obscures about 25% of the extragalactic sky. It hides some of the most crucial nearby structures in the large-scale distribution of galaxies. This is especially true in the southern sky where a local overdensity - the "Great Attractor" is believed to exist and is probably mainly responsible for the large-scale steaming motion of our galaxy and neighbouring regions relative to the Cosmic Microwave Background. A recent determination of the centre of the Great Attractor (Dekel 1994) puts it at a redshift of cz=4500 km s^{-1} in the direction l=320°, b=0°, right behind the galactic equator, where it is totally obscured, at optical wavelengths, by the dust of the southern Milky Way.

The general distribution of galaxies in the southern sky is shown in Figure 1. Except for the Puppis region, where a few galaxies show through, the Milky Way appears as a conspicuous "Zone of Avoidance" more than ten degrees wide. The figure shows galaxies from the Lauberts catalogue which is based on a lower diameter limit of 1.0 arcmin. As the dust of the Milky Way reduces the apparent luminosities and diameters of galaxies, so they fail to be included in the catalogue. An obvious solution is therefore to survey the vicinity of the Milky Way ($-10° < b < 10°$) with a reduced diameter limit. Consequently two of us have been engaged in an optical search of the SRC IIIa-J Sky Survey for galaxies with diameters > 0.2 arcmin (Kraan-Korteweg 1989, Kraan-Korteweg and Woudt 1994). The search, undertaken with 50-times magnification, is tedious in that the fields

D. L. Block and J. M. Greenberg (eds.), New Extragalactic Perspectives in the New South Africa, 396–399.
© 1996 *Kluwer Academic Publishers.*

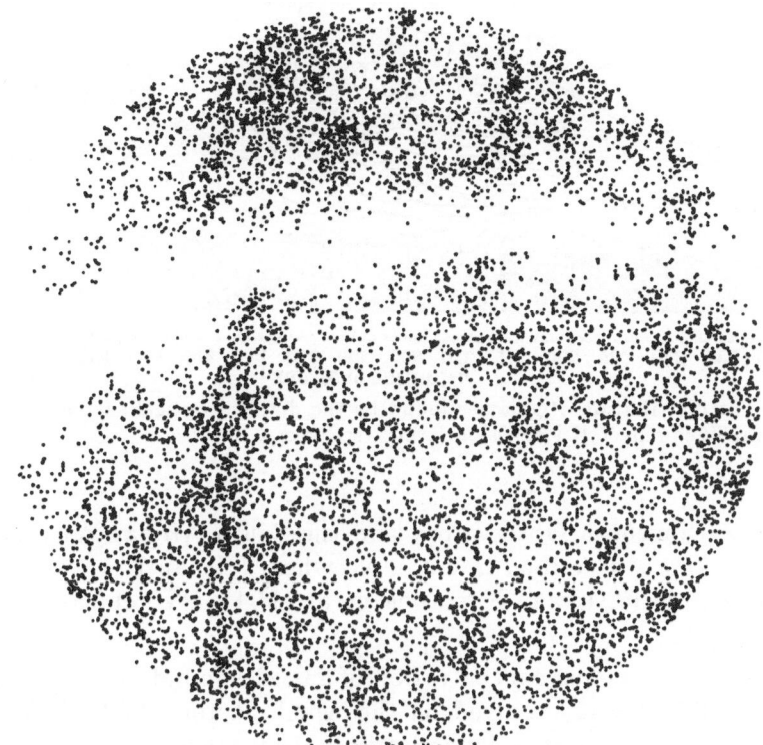

Figure 1.   The distribution of galaxies south of Declination -17.5° (Lauberts 1982)

are heavily contaminated by foreground stars (such that it is still beyond the capabilities of an automated survey). So far, the survey, conducted over $266° < l < 340°$, has revealed some 10500 galaxies.

Figure 2 shows the distribution of these galaxies. They fill in much of the "Zone of Avoidance" in the Lauberts catalogue, but there still remains a band 5° wide where the Milky Way appears to be completely opaque. It is also centred on the dust equator, which here is generally displaced about 1° south of the galactic equator. Figure 2 also includes H I emission contours (Lahav et al 1989). The Milky Way appears almost completely opaque at the contour corresponding to a column density $N(H\ I) = 6.4\ 10^{21}$ atoms cm^{-2} which translates (via Burstein and Heiles 1982) to an extinction $A_v = 4.8m$.

The H I contours parallel the dust equator, leading us to believe that the effects of the obscuration are generally independent of galactic longitude. Consequently, variations in galaxy density with longitude are more likely indicative of large-scale overdensities beyond the Milky Way. This has been explored by the follow-up redshift observations we have carried out from

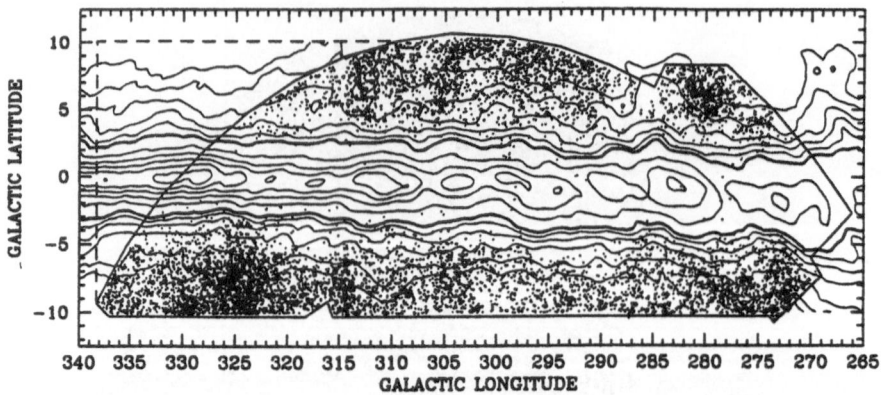

Figure 2. The distribution of galaxies found by the detailed optical search. The super-posed contour lines reflect H I emission levels - see text

the European Southern Observatory and the South African Astronomical Observatory, and is described in detail elsewhere (Kraan-Korteweg, Fairall and Balkowski 1995, Kraan-Korteweg et al 1994). We report here on a recent highlight.

The most conspicuous concentration of galaxies found in our survey lies towards l = 325° b= -7° centred on the Abell cluster A3627. This is the only Abell cluster to lie at so low a latitude. Nevertheless, when examined on the sky survey, the cluster seems unimpressive at first sight. However, at this latitude, the extinction in the centre of the cluster would be A_v = 1.4 m, while many smaller cluster members would be lost amongst the stellar images. Correcting the galaxy diameters for extinction would in fact provide 139 galaxies with D>1.0 arcmin (the Lauberts catalogue criterion) within 1.75° of the cluster centre. These galaxies are incorporated into a revised version of Figure 1 - presented as Figure 3 overleaf. It shows that lifting the veil of the Milky Way's dust obscuration would reveal this cluster as the most prominent overdensity in the southern sky.

Our redshift observations (Kraan-Korteweg et al 1996) show that this cluster has a mean velocity of cz=4882 km s^{-1} and a velocity dispersion of 903 km s^{-1}. The general properties of the cluster make it comparable to the well-known Coma cluster - often termed the nearest rich cluster - while its virial mass is an order higher than the neighbouring Hydra and Centaurus clusters. Although recognised as an Abell cluster, its significance has not been appreciated until now. Its redshift matches the redshift of the Great Attractor, and its position in the sky is only a few degrees away from the predicted centre. Nevertheless, massive though the cluster is ($\sim 3 \cdot 10^{15} \mathcal{M}_\odot$), it is an order of magnitude too small to be the Great Attractor itself. It may prove however to mark the bottom of the potential well of the Great

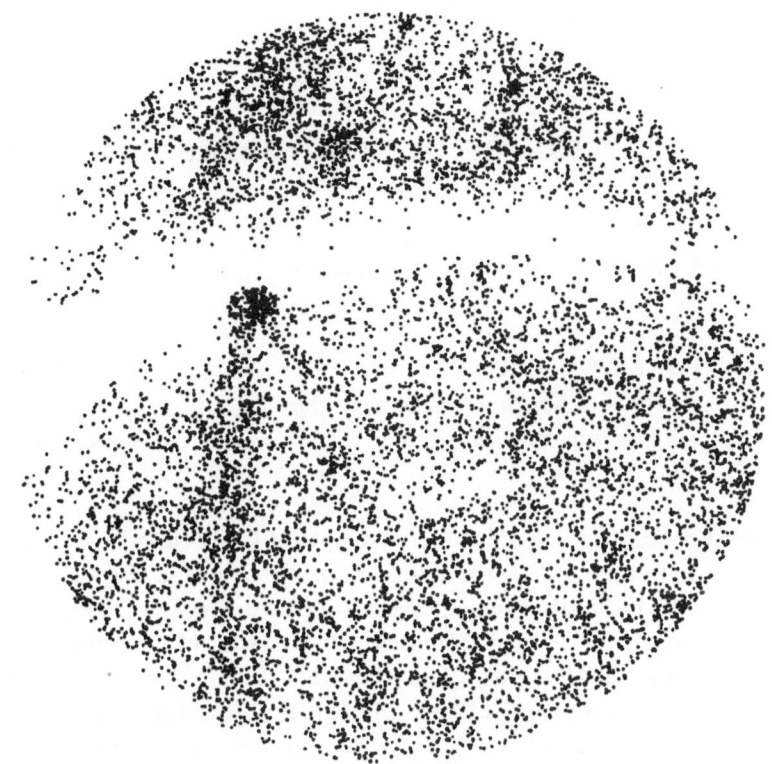

Figure 3. The same Lauberts galaxies as Figure 1, but with the addition of galaxies within 5° of the centre of A3627, which when corrected for extinction would meet the Lauberts crterion. The cluster is now revealed as a dominant feature.

Attractor, and its recognition adds considerably to our understanding of the large-scale distribution of galaxies in the southern sky.

References

Burstein, D. & Heiles, C., *Astron. J.*, **87**, 1165.

Dekel, A., 1994. *Ann. Rev. Astron. Astroph.*, **32**, 371.

Kraan-Korteweg, R.C., 1989. In *"Rev. in Modern Astr. 2"*, ed G. Klare, Springer (Berlin), p119.

Kraan-Korteweg, R.C. & Woudt, P.A., 1994. In *"Unveiling Large-Scale Structures behind the Milky Way, Proceedings of the 4th DAEC Meeting"* eds. C. Balkowski & R.C. Kraan-Korteweg, Conf. A.S.P., p89.

Kraan-Korteweg, R.C., Cayatte, V., Fairall, A.P., Balkowski, C. & Henning, P.A., 1994. In *"Unveiling Large-Scale Structures behind the Milky Way, Proceedings of the 4th DAEC Meeting"* eds. C. Balkowski & R.C. Kraan-Korteweg, Conf. A.S.P., p99.

Kraan-Korteweg, R.C., Woudt, P.A., Cayatte, V., Fairall, A.P., Balkowski, C. & Henning, P.A., 1996. *Nature, in press.*

Lahav, O., Edge, A.C., Fabian, A.C., & Putney, A., 1989. *Mon. Not. R. astr. Soc.*, **238**, 881.

Lauberts, A., 1982. *The ESO/Uppsala Survey of the ESO(B) Atlas*, ESO (Garching).

THE DISTRIBUTION OF DUST AND GAS IN ELLIPTICAL GALAXIES

PAUL GOUDFROOIJ

European Southern Observatory
Karl-Schwarzschild-Strasse 2, D-85748 Garching, Germany

Abstract. Results from *IRAS* and recent optical CCD surveys are examined to discuss the distribution and origin of dust and ionized gas in elliptical galaxies. In strong contrast with the situation among spiral galaxies, masses of dust in elliptical galaxies as derived from optical extinction are an order of magnitude *lower* than those derived from *IRAS* data. I find that this dilemma can be resolved by a *diffusely distributed component* of dust which is not detectable in optical data.

The morphology of dust lanes and their association with ionized gas in elliptical galaxies argues for an external origin of both components of the ISM.

1. Introduction: Dust and Gas in Elliptical Galaxies

Recent advances in instrumental sensitivity have challenged the very definition of elliptical galaxies in Hubble's galaxy classification scheme. It is now clear that ellipticals contain a complex, diverse ISM, primarily in the form of hot ($T \sim 10^7$ K), X-ray-emitting gas, with masses $\lesssim 10^{11}$ M$_\odot$. Small amounts of H I, H$_2$, ionized gas, and dust have been detected as well in many ellipticals (*e.g.*, Bregman *et al.* 1992).

Unlike the situation in spiral galaxies, physical and evolutionary relationships between the various components of the ISM in ellipticals are not yet understood. A number of theoretical concepts have been developed for the secular evolution of the different components of the ISM of ellipticals. The two currently most popular concepts are *(i)* the "cooling flow" picture in which mass loss from stars within the galaxy, heated to $10^6 - 10^7$ K by supernova explosions and collisions between expanding stellar envelopes during the violent galaxy formation stage, quiescently cools and condenses (cf. the review of Fabian *et al.* 1991) and *(ii)* the "evaporation flow" pic-

D. L. Block and J. M. Greenberg (eds.), New Extragalactic Perspectives in the New South Africa, 400–407.

ture in which clouds of dust and gas have been accreted during post-collapse galaxy interactions. Subsequent heating (and evaporation) of the accreted gas is provided by thermal conduction in the hot, X-ray-emitting gas and/or star formation (cf. de Jong *et al.* 1990; Sparks *et al.* 1989).

The first direct evidence for the common presence of cool ISM in ellipticals was presented by Jura *et al.* (1987) who used *IRAS* ADDSCANs and found that $\gtrsim 50\%$ of nearby, bright ellipticals were detected at 60 and 100 μm. Implied dust masses were of order $\sim 10^4 - 10^6$ $M_\odot$ (using $H_0 = 50$ km s^{-1} Mpc^{-1}). Interestingly, there are several X-ray-emitting ellipticals with suspected cooling flows (cf. Forman *et al.* 1985) among the *IRAS* detections. The presence of dust in such objects is surprising, since the lifetime of a dust grain against collisions with hot ions ("sputtering") in hot gas with typical pressures $nT \sim 10^5$ cm^{-3} K is only $10^6 - 10^7$ yr (Draine & Salpeter 1979). What is the origin of this dust, and how is it distributed?

In order to systematically study the origin and fate of the ISM of elliptical galaxies, we have recently conducted a deep, systematic optical survey of a complete, blue magnitude-limited sample of 56 elliptical galaxies drawn exclusively from the RSA catalog (Sandage & Tammann 1981). Deep CCD imaging has been performed through both broad-band filters and narrowband filters isolating the nebular Hα+[N II] emission lines.

In this paper I combine results from this survey with the *IRAS* data to discuss the distribution and origin of dust and gas in ellipticals. Part of this paper is based on Goudfrooij & de Jong (1995, hereafter Paper IV).

2. Detection Rates of Dust from Optical Surveys

Optical observations are essential for establishing the presence and distribution of dust and gas in ellipticals, thanks to their high spatial resolution. A commonly used optical technique to detect dust is by inspecting color-index (*e.g.*, $B - I$) images in which dust shows up as distinct, reddened structures with a morphology different from the smooth underlying distribution of stellar light (*e.g.*, Goudfrooij *et al.* 1994b, hereafter Paper II). However, a strong limitation of optical detection methods (compared to the use of *IRAS* data) is that only dust distributions that are sufficiently different from that of the stellar light (*i.e.*, dust lanes, rings, or patches) can be detected. Moreover, detections are limited to nearly edge-on dust distributions (*e.g.*, no dust lanes with inclinations $\gtrsim 35°$ have been detected, cf. Sadler & Gerhard 1985; Paper II). Thus, the optical detection rate of dust (currently 41%, cf. Paper II) represents a firm lower limit. Since an inclination of 35° is equivalent to about half the total solid angle on the sky, one can expect the *true* detection rate to be about twice the measured one (at a given detection limit for dust absorption), which means that *the vast majority of ellipticals could harbor dust lanes and/or patches.*

3. The Dust/Ionized Gas Association: Clues to their Origin

Optical emission-line surveys of luminous, X-ray-emitting ellipticals have revealed that these galaxies often contain extended regions of ionized gas (*e.g.*, Trinchieri & di Serego Alighieri 1991), which have been argued to arise as thermally instable regions in a "cooling flow". As mentioned before, the emission-line regions in these galaxies are suspected to be dust-free in view of the very short lifetime of dust grains. However, an important result of our optical survey of ellipticals is the finding that emission-line regions are essentially *always* associated with substantial dust absorption (see Color Plate 12 (this volume); Paper II; see also Macchetto & Sparks 1991), which is difficult to account for in the "cooling flow" scenario. This dilemma can however be resolved in the "evaporation flow" scenario (de Jong *et al.* 1990) in which the ISM has been accreted from a companion galaxy.

Closely related to the origin of the dust and gas is their dynamical state, *i.e.*, whether or not their motions are already settled in the galaxy potential. This question is, in turn, linked to the intrinsic shape of ellipticals, since in case of a settled dust lane, its morphology indicates a plane in the galaxy in which stable closed orbits are allowed (*e.g.*, Merritt & de Zeeuw 1983). These issues can be studied best in the inner regions of ellipticals, in view of the short relaxation time scales involved, allowing a direct relation to the intrinsic shape of the parent galaxy. A recent analysis of properties of *nuclear* dust in 64 ellipticals imaged with HST has shown that dust lanes are randomly oriented with respect to the apparent major axis of the galaxy (van Dokkum & Franx 1995). Moreover, the dust lane is significantly misaligned with the *kinematic* axis of the stars for almost all galaxies in their sample for which stellar kinematics are available. This means that *even at these small scales*, the dust and stars are generally dynamically decoupled, which argues for an external origin of the dust. This conclusion is strengthened by the decoupled kinematics of stars and gas in ellipticals with *large-scale* dust lanes (*e.g.*, Bertola *et al.* 1988).

4. The "Dust Mass Discrepancy" among Elliptical Galaxies

As mentioned in the introduction, dust in ellipticals has been detected by optical as well as far-IR surveys. Since the optical and far-IR surveys yielded quite similar detection rates, one is tempted to conclude that both methods trace the same component of dust. In this Section, this point will be addressed by discussing the distribution of dust in ellipticals.

The methods used for deriving dust masses from optical extinction values and from the IRAS flux densities at 60 and 100 μm, and the limitations involved in these methods, are detailed upon in Goudfrooij *et al.* (1994c, hereafter Paper III) and Paper IV. It is found that the dust masses estimated from the optical extinction are significantly *lower* than those estimated from the far-IR emission (see Paper IV). Quantitatively, the average ratio $<M_{d,IRAS}/M_{d,opt}> = 8.4 \pm 1.3$ for the galaxies in our "RSA sample" for which the presence of dust is revealed by both far-IR emission and optical dust lanes or patches.

I should like to emphasize that this "dust mass discrepancy" among ellipticals is quite remarkable, since the situation is *significantly different* in the case of spiral galaxies: Careful analyses of deep multi-color imagery of dust extinction in spiral galaxies (*e.g.*, Block *et al.* 1994; Emsellem 1995) also reveal a discrepancy between dust masses derived from optical and *IRAS* data, *but in the other sense, i.e.,* $M_{d,IRAS} \ll M_{d,opt}$! This can be understood since the *IRAS* measurements were sensitive to "cool" dust with temperatures $T_d \gtrsim 25\,K$, but much less to "cold" dust at lower temperatures which radiates predominantly at wavelengths beyond 100 μm (*e.g.*, Young *et al.* 1986). Since dust temperatures of order 20 K and lower are appropriate to spiral galaxies (Greenberg & Li 1995 and references therein), dust masses derived from the *IRAS* data are strict *lower limits* by nature. Evidently, the bulk of the dust in spiral disks is too cold to emit significantly at 60 and 100 μm, but still causes significant extinction of optical light. Interestingly, $T_d \lesssim 20\,K$ is *also* appropriate to the outer parts of ellipticals (cf. Paper IV), underlining the significance of the apparent "dust mass discrepancy" among ellipticals. What could be the cause?

Orientation effects? If the discrepancy would be due to an orientation effect, the ratio $M_{d,IRAS}/M_{d,opt}$ would be inversely proportional to $\cos i$, where i is the inclination of the dust lane with respect to the line of sight. However, we have measured inclinations of regular, uniform dust lanes in ellipticals from images shown in homogeneous optical CCD surveys, and found that the relation between $M_{d,IRAS}/M_{d,opt}$ and $\cos i$ is a scatter plot (cf. Fig. 1 of Paper IV). Thus, the effect of orientation on the dust mass discrepancy must be weak if present at all. This suggests that the dust in the lanes is concentrated in dense clumps with a low volume filling factor.

Diffusely distributed dust? Having eliminated the effect of orientation, the most plausible way out of the dilemma of the dust mass discrepancy is to postulate an additional, diffusely distributed component of dust, which is therefore virtually undetectable by optical methods. We note that this diffuse component of dust is not unexpected: the late-type stellar population of typical giant ellipticals ($L_B = 10^{10} - 10^{11}\,L_\odot$) has a substantial present-day mass loss rate ($\sim 0.1 - 1\,M_\odot\,yr^{-1}$ of gas and dust; cf. Faber & Gallagher 1976) which can be expected to be diffusely distributed. An interesting potential way to trace this diffuse component of dust is provided

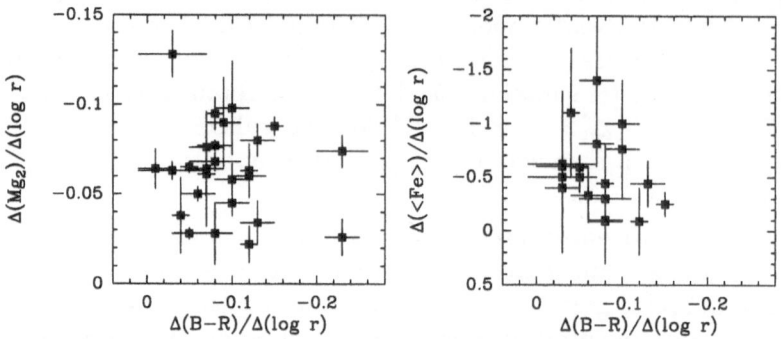

Figure 1. The relation of radial $B - R$ color gradients with radial gradients of the stellar line-strength indices Mg_2 *(left panel)* and $<Fe>$ *(right panel)* for all ellipticals for which both pairs of quantities have been measured to date. Data taken from Peletier (1989), Carollo *et al.* (1993), Davies *et al.* (1993), Carollo & Danziger (1994), and Paper I.

by radial color gradients in ellipticals. With very few significant exceptions, "normal" ellipticals show a global reddening toward their centers, in a sense approximately linear with log (radius) (Goudfrooij *et al.* 1994a (hereafter Paper I) and references therein). This is usually interpreted as gradients in stellar metallicity, as metallic line-strength indices show a similar radial gradient (*e.g.*, Davies *et al.* 1993). However, compiling all measurements published to date on color- and line-strength gradients within ellipticals shows no obvious correlation (cf. Fig. 1), suggesting that an additional process is (partly) responsible for the color gradients. Although the presence of dust in ellipticals is now beyond dispute, the implications of dust extinction have been generally discarded in the interpretation of color gradients. However, recent Monte Carlo simulations of radiation transfer within ellipticals by Witt *et al.* (1992, hereafter WTC) and Wise & Silva (1996) have demonstrated that a diffuse distribution of dust throughout ellipticals can cause significant color gradients even with modest dust optical depths.

We have used WTC's "Elliptical" model to predict dust-induced color gradients appropriate to the far-IR properties of ellipticals in the "RSA sample", as derived from the *IRAS* data. The model features King (1962) profiles for the radial density distributions of both stars and dust:

$$\rho(r) = \rho_0 \left[1 + \left(\frac{r}{r_0}\right)^2\right]^{-\alpha/2}$$

where ρ_0 represents the central density and r_0 is the core radius. The "steepness" parameter α was set to 3 for the stars, and to 1 for the dust [The reason for the low value of α for the dust distribution is that it generates color gradients that are linear with $\log(r)$, as observed (Paper I; Wise &

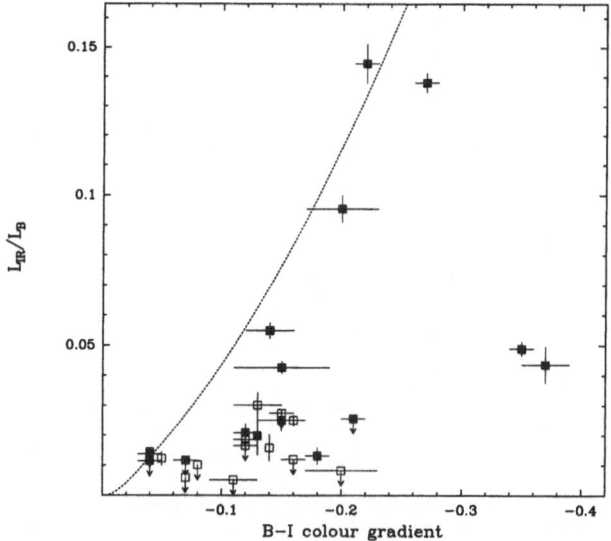

Figure 2. The relation of L_{IR}/L_B with radial $B - I$ color gradients (defined as $\Delta (B - I)/\Delta (\log r)$) for elliptical galaxies in the "RSA sample" (cf. Paper I). Filled squares represent galaxies detected by IRAS showing optical evidence for dust, and open squares represent galaxies detected by IRAS without optical evidence for dust. Arrows pointing downwards indicate upper limits to L_{IR}/L_B. The dotted line represents the color gradient expected from differential extinction by a diffuse distribution of dust (see text). Figure taken from Paper IV.

Silva 1996)]. Using this model, far-infrared-to-blue luminosity ratios and color gradients have been derived as a function of the total optical depth τ of the dust (*i.e.*, the total dust mass). The result is plotted in Fig. 2. It is obvious that color gradients in elliptical galaxies are generally larger than can be generated by a diffuse distribution of dust throughout the galaxies according to the model of WTC. This is as expected, since color gradients should be partly due to stellar population gradients as well. However, *none* of the galaxies in this sample has a color gradient significantly *smaller* than that indicated by the model of WTC. I argue that this is caused by a "bottom-layer" color gradient due to differential extinction, which should be taken seriously in the interpretation of color gradients in ellipticals.

We have checked whether the assumption of the presence of a diffusely distributed component is also *energetically* consistent with the available *IRAS* data. To this end, we computed heating rates for dust grains as a function of galactocentric radius. We assumed heating by *(i)* stellar photons, using the radial surface brightness profiles from Paper I, and *(ii)* hot electrons in X-ray-emitting gas, if appropriate. Radial dust temperature profiles are derived by equating the heating rates to the cooling rate of a

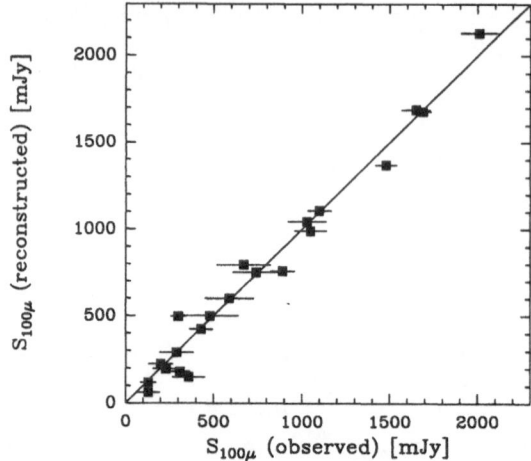

Figure 3. The 100 μm flux densities reconstructed from our calculations of the dust temperatures in elliptical galaxies (see text) versus the observed 100 μm flux densities (and their 1 σ error bars). The solid line connects loci with "reconstructed = observed".

dust grain by far-IR emission.

Using the derived radial dust temperature profiles, we reconstructed *IRAS* flux densities for both the optically visible component and the (postulated) diffusely distributed component. Apparently regular dust lanes were assumed to by circular disks of uniform density, reaching down to the galaxy nucleus. After subtracting the contribution of the optically visible component of dust to the *IRAS* flux densities, the resulting flux densities were assigned to the diffuse component. Using the WTC model calculations (cf. Fig. 2), L_{IR}/L_B ratios were translated into total optical depths of the dust (and hence dust mass column densities). Dividing the dust masses of the diffusely distributed component by the dust mass column densities, outer galactocentric radii for the diffusely distributed dust component were derived (typical values were ~ 2 kpc). Finally, the *IRAS* flux densities were constructed from the masses of the diffusely distributed component by integrating over spheres (we refer the reader to Paper IV for details).

A comparison of the observed and reconstructed *IRAS* flux densities (cf. Fig. 3) reveals that the *observed IRAS* flux densities can *in virtually all elliptical galaxies in the RSA sample* be reproduced *within the 1 σ uncertainties* by assuming two components of dust in elliptical galaxies: an optically visible component in the form of dust lanes and/or patches, and a newly postulated dust component which is diffusely distributed within the inner few kpc from the center of the galaxies.

We remind the reader that we have only considered dust which was detected by *IRAS*, *i.e.*, with $T_d \gtrsim 25$ K. In reality, the postulated diffuse component of dust in elliptical galaxies may generally be expected to extend out to where the dust temperature is lower. Observations with the *Infrared Space Observatory (ISO)* of the RSA sample of elliptical galaxies are foreseen, and may reveal this cooler dust component in ellipticals.

Acknowledgments. I am very grateful to the SOC for allowing me to participate in this great conference. It is also a pleasure to thank Drs. Teije de Jong, Leif Hansen, Henning Jørgensen, and Hans-Ulrik Nørgaard-Nielsen for their various contributions to this project.

References

Bertola F., Buson L. M., Zeilinger W. W., 1988, Nat 335, 705

Block D. L., Witt A. N., Grosbøl P., Stockton A., Moneti A., 1994, A&A 288, 383

Bregman J. N., Hogg D. E., Roberts M. S., 1992, ApJ 387, 484

Carollo C. M., Danziger I. J., Buson L. M., 1993, MNRAS 265, 553

Carollo C. M., Danziger I. J., 1994, MNRAS 270, 523

De Jong T., Nørgaard-Nielsen H. U., Hansen L., Jørgensen H. E., 1990, A&A 232, 317

Davies R. L., Sadler E. M., Peletier R. F., 1993, MNRAS 262, 650

Draine B. T., Salpeter E., 1979, ApJ 231, 77

Emsellem E., 1995, A&A 303, 673

Faber S. M., Gallagher J. S., 1976, ApJ 204,365

Fabian A. C., Nulsen P. E. J:, Canizares C. R., 1991, A&AR 2, 191

Forman W., Jones C., Tucker W., 1985, ApJ 293, 102

Goudfrooij P., *et al.*, 1994b, A&AS 104, 179 (Paper I)

Goudfrooij P., Hansen L., Jørgensen H. E., Nørgaard-Nielsen H. U., 1994b, A&AS 105, 341 (Paper II)

Goudfrooij P., de Jong T., Nørgaard-Nielsen H. U., Hansen L., 1994c, MNRAS 271, 833 (Paper III)

Goudfrooij P., de Jong T., 1995, A&A 298, 784 (Paper IV)

Greenberg J. M., Li A., 1995, in: *The Opacity of Spiral Disks*, eds. J. I. Davies & D. Burstein, Kluwer, Dordrecht, p. 19

Jura M., Kim D.-W., Knapp G. R., Guhathakurta P., 1987, ApJL 312, L11

Macchetto F., Sparks W. B., 1991, in: *Morphological and Physical Classification of Galaxies*, eds. G. Longo, M. Cappaccioli & G. Bussarello, Kluwer, Dordrecht, p. 191

Merritt D., de Zeeuw P. T., 1983, ApJL 267, L19

King I. R., 1962, AJ 67, 471

Peletier R. F., 1989, Ph. D. Thesis, University of Groningen

Phillips M. M., Jenkins C. R., Dopita M. A., Sadler E. M., Binette L. 1986, AJ 91, 1062

Sadler E. M., Gerhard O. E., 1985, MNRAS 214, 177

Sandage A. R., Tammann G. A., 1981, *A Revised Shapley-Ames Catalog of Bright Galaxies*, Carnegie Institution of Washington

Sparks W. B., Macchetto F., Golombek D., 1989, ApJ 345, 153

Trinchieri G., di Serego Alighieri S., 1991, AJ 101, 1647

Van Dokkum P. G., Franx M., 1995, AJ 110, 2027

Véron-Cetty M. P., Véron P., 1988, A&A 204, 28

Wise M. W., Silva D. R., 1996, ApJ, in press (NOAO Preprint No. 677)

Witt A. N., Thronson H. A. jr., Capuano J. M., 1992, ApJ 393, 611 (WTC)

Young J. S., Schloerb F. P., Kenney D., Lord S. D., 1986, ApJ 304, 443

THE IONIZED GAS IN EARLY-TYPE GALAXIES

F. MACCHETTO[1,2], N. CAON[1,2], W.B. SPARKS[1] & M. PASTORIZA

[1] *Space Telescope Science Institute - Baltimore MD, USA*
[2] *Affiliated to the Astrophysics Div., Space Science Dept., ESA*
[3] *Instituto de Fisica, UFRGS, Porto Alegre, Brazil*

1. Introduction

Optical surveys have shown that a large percentage of early-type galaxies contain extended dust lanes, patches or a central disk-like dust structure (Sadler & Gerhardt 1985; Sparks et al. 1985; Macchetto & Sparks 1992; Goudfrooij et al. 1994). In addition to the dust, warm (10^4K) ionized gas is present in a high fraction of ellipticals, of the order of 50%, as shown by spectroscopic and/or imaging surveys. Although relatively small in mass, 10^3–10^4 solar masses, the contribution to the energy budget of the galaxy can be very large. The optical line-emission can dominate the energy loss from X-rays and therefore it has to be understood in order to derive a correct description for the energetics of the interstellar medium in these galaxies.

In order to study the properties of the interstellar medium in early-type galaxies and elucidate its origin, we have carried out an extensive program of imaging and spectroscopic observations of a large sample of bright elliptical and lenticular galaxie ($-19.0 \leq MB \geq 23.5$). The galaxies in our sample have been selected so as to span a wide range of optical, X-ray, radio, infrared and kinematical properties. The observations have been aimed at reaching fainter limiting-flux levels than in previous studies, in order to provide a more complete and extensive mapping of the ionized-gas distribution, especially of those faint filamentary structures extending far out from the galaxy nucleus.

D. L. Block and J. M. Greenberg (eds.), New Extragalactic Perspectives in the New South Africa, 408–415.

2. Observations and Data Reduction

We observed 39 ellipticals and 34 lenticular galaxies with the 3.6m and the NTT ESO telescopes. CCD images were acquired both through a broad-band R filter matching the Cousins standard bands and through a narrow-band centered on the $H\alpha$+[NII] emission lines. For each galaxy, we gener-ated emission-line maps which were then flux-calibrated and derived the total luminosity and extension of the ionized gas. A great deal of care was taken in each case to reduce the sources of errors so as to derive reliable values for the $H\alpha$+[NII] (see Macchetto et al. 1996 for details). Using these values, we estimated the total mass of ionized gas under the assumption of case B recombination (Osterbrock 1974).

3. Results

A large fraction of E (72%) and S0 (85%) galaxies in our sample contain ionized gas; our rate of detection is larger than the results of previous imaging surveys (cf. $\sim$ 57%, Goudfrooij et al. 1994), and spectroscopic surveys ($\sim$ 50%, Phillips et al. 1986; $\sim$ 40%, Caldwell 1984). The $H\alpha$+[NII] luminosities range between $10^{38} - 10^{41}$ erg s^{-1}.

The ionized gas morphology appears to be rather smooth for most galax-ies; however a significant fraction ($\sim$ 12% of the galaxies), show a very ex-tended filamentary structure. According to the morphology and size of the line-emitting region, the galaxies were classified into three broad classes:

1) small disk (SD): in some cases with quite faint and short filaments; the mean diameter is $\lesssim$ 4 kpc;
2) regular extended (RE): similar to the previous class, but larger in size (4 to 18 kpc), with short filaments sometimes present;
3) filaments (F): if a conspicuous filamentary structure dominates the morphology, often departing from a more regular disk-like inner region. It can extend as far as 10 kpc from the galaxy center.

Of all the S0 galaxies that show gas emission, 55% belong to the RE class, 34% to the SD class, and 11% to the filamentary class. The distri-bution for ellipticals is 25% for the RE, 54% for the SD, and 21% for the filamentary class. Therefore, elliptical galaxies that show SD gas structure are more frequent than those with RE structure, while it is the other way around for S0 galaxies. This is in line with the properties of stellar disks, which are both more luminous and larger in size in lenticular than in ellipti-cal galaxies (see Scorza & Bender 1995). No apparent correlation is however found between the boxyness parameter a_4 and either the luminosity or the size of the $H\alpha$+[NII] emission.

There is no clear correlation between the gas morphology and the galaxy luminosity. While galaxies with filamentary structure tend to be more luminous, both in the optical bands and in Hα+[NII], and those with small disks are generally fainter, at any given optical luminosity there is a range of about 1 dex in the mean diameter of the emitting region. On the other hand, a tighter correlation between the Hα+[NII] luminosity and the size of the emitting region suggests that the Hα+[NII] luminosity is mainly related to the extension of the gas distribution rather than to its projected brightness.

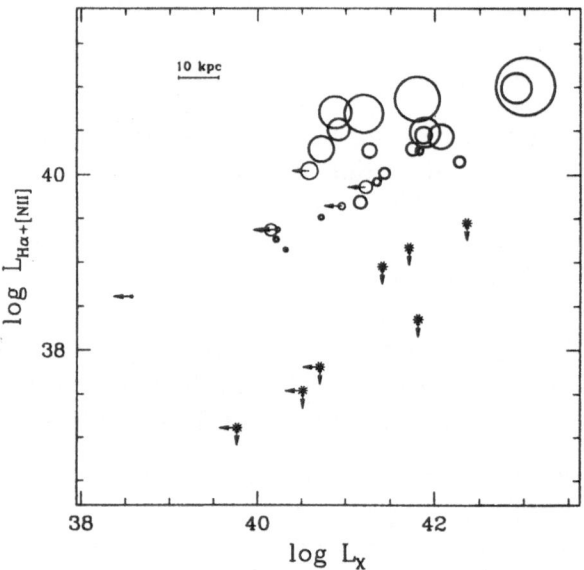

Figure 1.   Hα+[NII] luminosities plotted against X-ray luminosities. The size of each point is proportional to the size of the line-emitting region.

We have analyzed the possible relationships between the warm and the hot gaseous components of the ISM of the galaxies in our sample through the statistical correlation between the total $L_{H\alpha}$ and L_X luminosities. A significant correlation, between Hα+[NII] and X-ray luminosities is found for those galaxies (27% of the sample) for which we have detected ionized gas and are listed as X-ray sources (see Figure 1). The strongest $L_{H\alpha+[NII]}$ galaxies also have strong L_X and present a more extended ionized gas distribution. This suggests that the presence of hot gas is a condition necessary but not sufficient for the presence of warm gas.

The relationship between the warm gas and the far-infrared radiation produced by cold dust has been examined by comparing the $L_{H\alpha+[NII]}$ lumi-

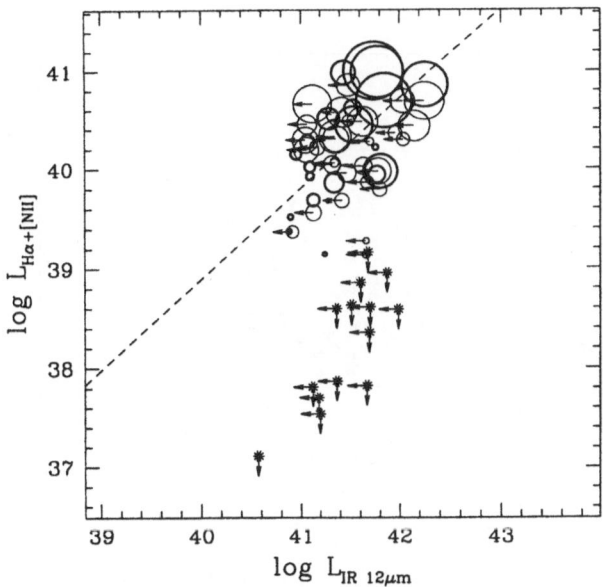

Figure 2. Hα+[NII] luminosities plotted against the 12 μm luminosities.

nosities with L_{FIR} computed from the 60 and 100 μm bands; no correlation has been found. Figure 2 shows a correlation between the infrared luminosity in the 12μm band and $L_{Hα+[NII]}$. As in the correlation between $L_{Hα+[NII]}$ and L_X there seems to be galaxies with a significant Hα+[NII] luminosity but with only upper limits on the 12μm emission, and viceversa galaxies with significant infrared emission at this wavelength but little or no warm gas.

No significant correlations were found with radio integrated and core luminosities. A weak correlation was found between $L_{Hα+[NII]}$ and the velocity dispersions σ_0, taken from the compilation by McElroy (1995). These correlations are in the sense that those systems with the deepest potential wells have the highest Hα+[NII] luminosities. A similar result was found between L_X and σ_0. No correlation was found between $L_{Hα+[NII]}$ and Mg$_2$ for the present sample. We have also looked for evidence of correlation between Hα+[NII] luminosities and the merger history as parameterized by the "fine structure" index defined by Schweizer & Seitzer (1992), finding no correlation for the nine galaxies in common with Schweizer and Seitzer's sample.

Finally we have investigated the correlation between the Hα+[NII] luminosities and the total blue luminosities of the galaxies. The weak trend present with total B luminosities becomes much more evident when $L_{Hα+[NII]}$

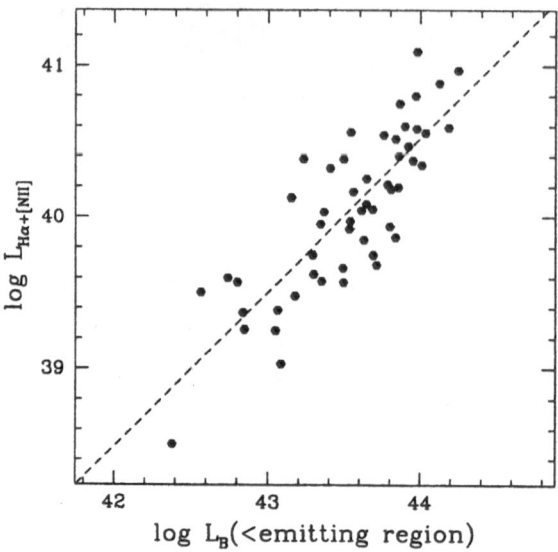

Figure 3.   Hα+[NII] luminosities plotted against the B-fluxes within the line-emitting region.

is plotted versus the blue luminosity inside the region occupied by the line-emitting gas, $L_B(<$ emitt.$)$ (Figure 3). We have defined $L_B(<$ emitt.$)$ as the total luminosity inside a centered circular region with diameter equal to the mean diameter of the line-emitting region. The same result is obtained using fluxes instead of luminosities, which supports the fact that this is a real correlation, and not an artifact of selection effects (faint emission in nearby galaxies would be easier to detect than in more luminous and distant galaxies).

4. Ionization Mechanisms

There are several potential mechanisms that can account in part or fully for the observed gas ionization. We have analyzed and rejected a number of them. There remain only two viable mechanisms which will be discussed here.

4.1. PHOTOIONIZATION BY POST-AGB STARS

The tight correlation between Hα+[NII] and optical luminosity inside the emitting region strongly suggests a stellar origin for the ionizing photons. Post-AGB stars provide sufficient ionizing radiation to account for the observed Hα luminosities and equivalent widths in very early-type galaxies (Binette et al 1994). We have calculated for our galaxies the predicted Hα luminosities from stars in the post-AGB phase and compared them with the observed Hα luminosities. We have found that for a large fraction of galaxies the ratios are of order unity, while several objects display values larger than 2, suggesting that the post-AGB stars provide a sufficient number of ionizing photons within the Hα emitting region. The presence of predicted/computed Hα luminosity ratios significantly larger than one is not an embarrassment, as there are several alternatives to explain an excess of available and unused photons. First, the covering factor of the gas can be below unity, as it is likely for ellipticals, which typically are not gas rich systems. In these cases, a significant fraction of extreme ultraviolet photons will escape to the intergalactic space, instead of being reprocessed by the interstellar medium. Another possible explanation is that some of the emitting clouds are not visible due to the large opacity of the dust along the line of sight. The galaxies in our sample have extensive dust systems, which are closely associated, at least morphologically, with the emission line gas.

4.2. THERMAL CONDUCTION

Conductive heating of the warm gas by hot electrons associated with an X-ray emitting coronal halo typically found within elliptical galaxies offers an alternative excitation mechanism, as proposed by Sparks & Collier-Cameron (1988) and Sparks et al. (1989). In this scenario, each galaxy has an associated pre-existing coronal halo, that is a diffuse hot gaseous plasma which pervades the galaxy. When the galaxy accretes cold gas and dust either in a minor merger or as a result of a tidal encounter with a gas-rich companion, thermal electron conduction will cause energy to flow from the hot coronal component into the cold gas which simultaneously excites the cold gas into emission while locally cooling the coronal gas. The energy radiated by the cold gas (the optical emission filament system is simply the heat flux into the filaments. Quantitatively, this idea has worked well with previously studied objects and explains the dustiness of the gas as well as the X-ray spectral parameters (Sparks 1992).

Under the assumption that the available energy flux is in the saturated regime , the amount of energy available to excite the optical emission filaments is (Sparks et al. 1989), $Q_{\text{sat}} = 5.2 \, 10^{40} \, T_7^{3/2} \, n_{0.01} \, D_{\text{maj}} \, D_{\text{min}}$, where $T_7 = T/10^7$ is the temperature of the coronal halo and $n_{0.01} = n_e/0.01$ is its density. In terms of the observed emission line flux, that is

$$L_{\text{Hα+[NII]}} \text{ (predicted)} = 1.5 \, 10^{39} \, T_7^{3/2} \, n_{0.01} \, D_{\text{maj}} \, D_{\text{min}} \tag{1}$$

assuming that $\sim 1\%$ of the available energy is radiated in the Hα+[NII] lines

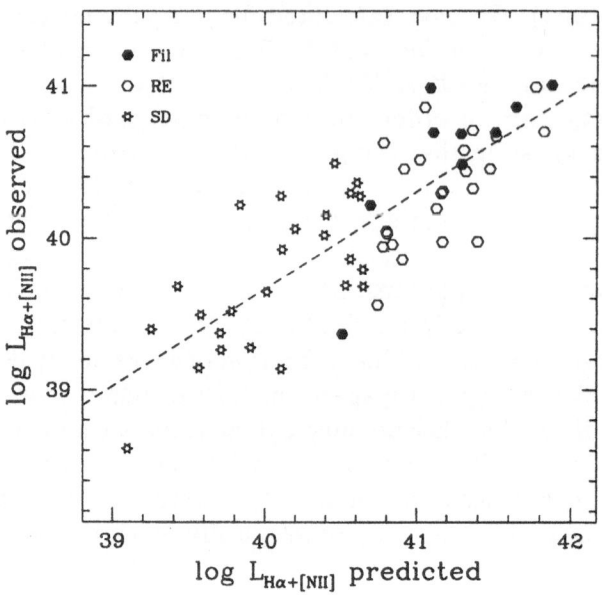

Figure 4. Hα+[NII] luminosities versus the predicted luminosity from conductive heating of the warm gas.

and that the total surface area of the system is three times the apparent projected surface area. With $T \simeq 1.5$ KeV and ne $\simeq 0.01 cm-^{3}$, we predict a filament luminosity and show the comparison to observations in Fig.4.

No account has yet been taken of the variation of coronal gas parameters such as temperature and density with galaxy size or mass, nor has any adjustment made to the *absolute* normalization of Sparks et al. (1989). The degree of correlation and the *absolute level* of the correlation are excellent, and demonstrate that an interpretation based on mergers and thermal interaction of hot and cold plasma components at this stage is perfectly consistent with the data. We must note, however, that this correlation holds even for those galaxies where there is no measured X-ray emission, and therefore it cannot be construed as a proof that X-ray conduction by hot electrons is necessarily the only mechanism responsible for the observed line-emission.

This has an important bearing on our understanding of the mechanisms responsible for the heating and ionization of the hot and warm gas, and whether we should invoke a single mechanism, and therefore expect a strong correlation between the two gas phases, or seek separate but somehow related mechanisms, and therefore expect weaker correlations. This issue is still very much open and will require further observations, as well as additional theoretical work.

References

Binette, L., Magris, C.G., Stasinŝka, G., Bruzual, A.G. 1994, A&A, 292, 13
Caldwell, N. 1984, PASP, 96, 287
Goudfrooij, P., Hansen L., Jørgensen, H.E., Nørgaard-Nielsen, H.U. 1994, A&AS, 105, 341
Macchetto, F., Sparks, W.B. 1992, in: Busarello G., Capaccioli M., Longo G. (eds), Morphological and Physical Classification of Galaxies. Kluwer, Dordrecht, p. 191
Macchetto, F., Pastoriza, M., Caon, N., Sparks, W.B., Giavalisco, M., Bender, R., Capaccioli, M., 1996, A&A, in press
McElroy, D. 1995, ApJS, 100, 105
Osterbrock, D.E. 1974, Astrophysics of Gaseous Nebulae (W.H. Freeman & Co., San Francisco)
Phillips, M.M., Jenkins, C.R., Dopita, M.A., Sadler, E.M., Binette, L. 1986, AJ, 91, 1062
Sadler, E.M., Gerhard, O.E. 1985, MNRAS, 214, 177
Schweizer, F., Seitzer, P. 1992, AJ, 104, 1039
Scorza, C., Bender, R. 1995, A&A, 293, 20
Sparks, W.B. 1992, ApJ, 399, 66
Sparks, W.B., Collier-Cameron, A. 1988, MNRAS, 232, 215
Sparks, W.B., Macchetto, F., Golombek, D. 1989, ApJ, 345, 153
Sparks, W.B., Wall, J.V., Thorne, D.J., Jorden, P.R., van Breda, I.G., Rudd, P.J., Jørgensen, H.E. 1985, MNRAS, 217, 87

HST IMAGING OF THE DUST IN KINEMATICALLY DISTINC CORE ELLIPTICALS

DUNCAN A. FORBES
Lick Observatory
University of California
Santa Cruz, CA 95064, USA

1. Introduction

About one third of nearby elliptical galaxies have central regions that are decoupled kinematically from the outer parts of the galaxy (e.g. Illingworth & Franx 1989). In some cases the core rotates in an opposite direction to the stars in the outer regions. Such galaxies with peculiar kinematics are called kinematically distinct core (KDC) ellipticals. Their origin remains somewhat of a mystery.

Originally, it was thought that these cores could have formed from the accretion of a small, compact companion (Kormendy 1984; Balcells & Quinn 1990). This would predict a blue and metal–poor core which is generally not observed (Bender & Surma 1992). Another possibility, associated with mergers, is the collision of two gas–rich galaxies and subsequent star formation in kinematically distinct gas (Hernquist & Barnes 1991). Projection effects (e.g. Statler 1991) have difficulty explaining the rapidly, counter–rotating cores such as IC 1459 and global correlations with core properties. Finally, KDCs may be a normal aspect of hierarchical galaxy formation.

The relevant spatial scales for KDCs in nearby galaxies is a few arcseconds. Thus to investigate these regions we require the spatial resolution of the *Hubble Space Telescope*. We have studied a sample of KDC ellipticals using both the WFPC1 and WFPC2 cameras. The WFPC1 data consisted of F555W (V) images of eight KDC ellipticals. Details can be found in Forbes (1994), Forbes et al. (1994) and Forbes et al. (1995). The WFPC2 data were obtained in two filters F555W (V) and F814W (I) of fifteen KDC ellipticals. Results will be published as Carollo et al. (1996) and Forbes et al. (1996). Our sample covers a range in environment and luminosity. As

D. L. Block and J. M. Greenberg (eds.), New Extragalactic Perspectives in the New South Africa, 416–421.
© 1996 *Kluwer Academic Publishers.*

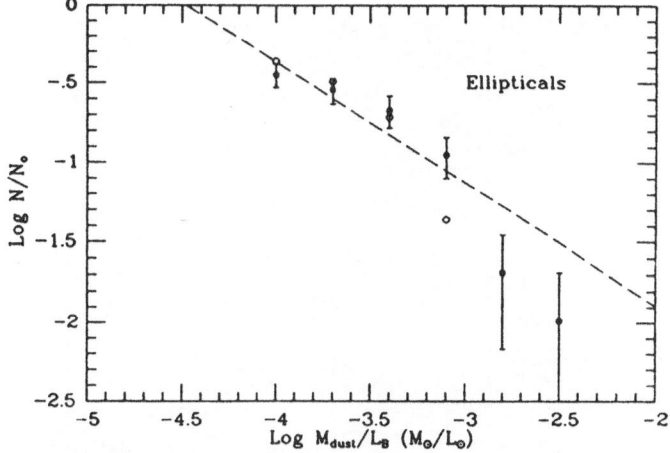

Figure 1. The distribution of dust in a large sample of ellipticals. The distribution indicates a large range in the relative dust content, unlike spirals which reveal a small scatter about a well defined mean value (see Forbes 1991).

well as examining the stellar light distribution, such data allow us to study the incidence and properties of the dust on small spatial scales in these galaxies. Dust in ellipticals is generally thought to have an external origin (e.g. Forbes 1991) rather than purely stellar mass loss (see Figure 1). Here we focus on dust in KDC ellipticals.

2. Results from WFPC1 Imaging

2.1. DUST

Of the eight galaxies, six show clear evidence for dust with a variety of morphologies from dust rings (e.g. NGC 4494), dust lanes and patches (e.g. NGC 7626) to a chaotic distribution (e.g. NGC 4589). Those with chaotic dust suggest a relatively recent encounter. Of the six dusty galaxies, only three were known to have dust from ground–based images. The HST has provided the spatial resolution to identify these small scale and filamentary features. Given the short dynamical times ($\sim 10^6$ yrs) in the central regions it is likely that this dust (and associated gas) provides a continuing source of fuel for the radio nucleus. Of the four galaxies with radio emission, all reveal small scale dust features.

2.2. STELLAR LIGHT DISTRIBUTION

Spectroscopic line–analysis of several KDC galaxies suggests that the distinct component is cold and rapidly rotating, i.e. in a disk (Franx & Illingworth 1988; Surma 1992). One aim of the HST imaging is to search for a

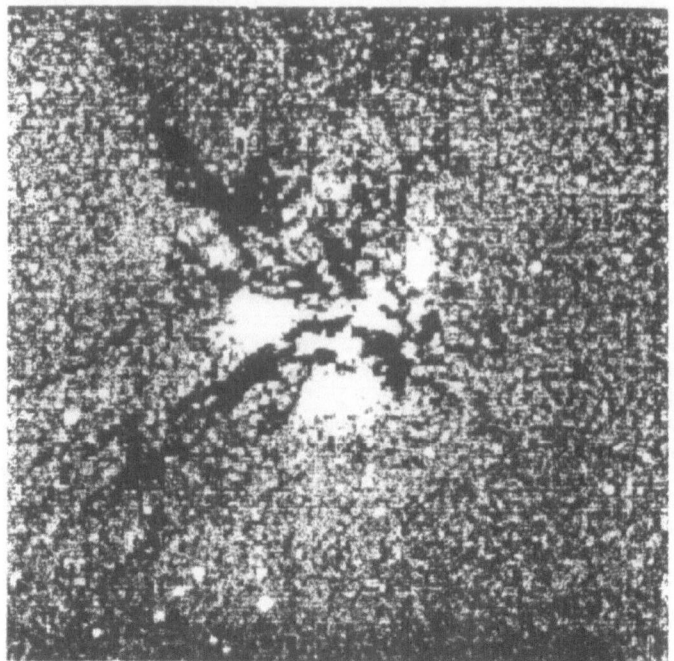

Figure 2. Model subtracted image of NGC 5813 from Carollo et al. (1996). The image is 30 × 30 arcsec in size.

photomteric counterpart to the kinematic disk. We have found clear evidence for a small scale disk in NGC 4365 that corresponds to the kinematically distinct component, but for other galaxies detection was inconclusive due to ubiquitous dust.

The surface brightness profiles revealed a variety of shapes from steeply rising to almost flat; none are isothermal. The variety of profiles and the high central stellar densities inferred argue against the accretion of a small galaxy for KDC formation. The profiles were fit with a dual–power law and compared to other 'kinematically normal' ellipticals observed by HST. The normal galaxies reveal a trend for the more luminous galaxies to have shallower inner profiles. The KDC galaxies are within the scatter of the normal galaxy trend. Thus although kinematically peculiar, the central stellar properties of KDC ellipticals are similar to other ellipticals, which are governed by fundamental–plane like relations.

3. Results from WFPC2 Imaging

3.1. DUST

Our WFPC2 data provided not only more concentrated PSFs but two filters, on a larger sample, to better investigate the dust properties. For the sample of 15 KDCs, we find dust in 13 of them again with a variety of morphologies. A striking example of dust in NGC 5813 is shown in Figure 2.

The extinction has been derived in various small apertures assuming $A_\lambda = -2.5 log \frac{F_{obs}}{F_{intr}}$, where F_{obs} = observed flux in counts and F_{intr} is the intrinsic flux assumed to be given by a smooth model fit to the galaxy isophotes. Dust masses in the central $1''$ (~ 10 pc) are calculated to be ~ 100 $M_\odot$ and $A_V \sim 0.1^m$. Most galaxies show slight reddening trends at small galactocentric radii. Could these galaxies harbour a smoothly distributed, diffuse dust component that is not detected in our modeling procedure ? Could such dust be responsible for the colour gradients observed ?

The presence of diffuse dust would flatten the surface brightness profiles in galaxies with the steepest colour gradients. We find the opposite trend in the very central regions, i.e. galaxies with redder cores tend to have steeper slopes (see Figure 3). Comparison with the spherically–symmetric mixed star–dust model of Silva & Wise (1995) gives an average $A_V < 0.1^m$ and $\tau < 0.2$. Our sample also shows a correlation of V–I colour in a $5''$ aperture with Mg_2 metallicity from the literature indicating that V–I is strongly coupled to stellar populations. So although dust is very common in KDCs, the extinction is not so high or dust so widespread, as to totally mask the effects of stellar populations.

3.2. STELLAR LIGHT DISTRIBUTION

The WFPC2 colour imaging has made detection of a photometric disk much easier. We have identified a photometric disk, on similar spatial scales to the kinematically distinct component, in seven galaxies. However in no case is there a strong colour difference between the colour of the disk and that of the surrounding galaxy. This suggests an age difference of less than 30% and argues against the dissipationless infall of a small galaxy.

Surface brightness profiles range from almost flat (high luminosity galaxies) to very steep (low luminosity galaxies) as found in the WFPC1 data. There are also correlations of inner slope with the anisotropy parameter (v/σ) and Mg_2, in the same sense as kinematically normal galaxies. Thus pressure supported high metallicity, massive galaxies have shallow cores, while fast–rotating, low metallicity small galaxies have steep cuspy density profiles. It is not yet clear whether this represents a distinct dichotomy or

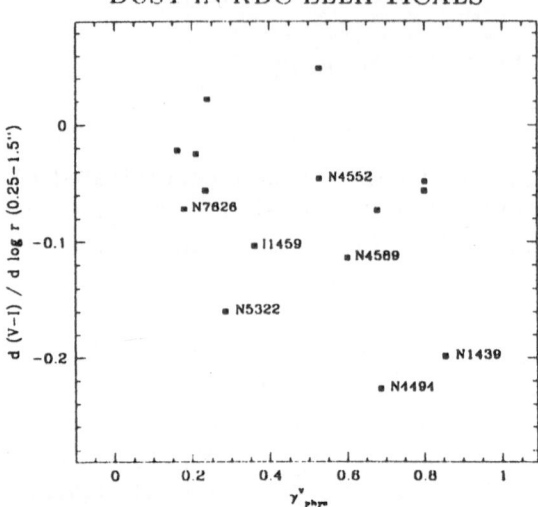

Figure 3. Central colour gradient vs inner slope of surface brightness profile from Carollo et al. (1996). Galaxies with redder cores do *not* have flatter inner slopes.

a smooth range in galactic properties. The presence, or absence, of a KDC does *not* affect the inner stellar density, velocity or metallicity of the galaxy.

3.3. GLOBULAR CLUSTERS

As well as the galaxy itself, the WFPC2 images reveal ~200 globular clusters to V ~ 25 (see Forbes et al. 1996 for details). Very little is known about globular clusters in the central regions of ellipticals due to seeing/blending effects in ground–based images. Unlike currently merging galaxies, we do *not* find a population of centrally located blue globulars in our sample ellipticals. There is no obvious difference in the globular cluster population of galaxies with dust and those without. The surface density distribution of globulars flattens off towards the galaxy centre, which does *not* appear to be due to dust or blending effects. This distribution is well fit by a core model, for which the core radius correlates with galaxy luminosity. We find that the globulars are slightly bluer (more metal poor) than the underlying galaxy starlight at any given galactocentric distance. Furthermore, the mean globular metallicity is well correlated with galaxy luminosity from dwarfs to massive ellipticals. These trends suggest that the formation of globulars is intrinsically related to the formation mechanism of the parent galaxy and that they have a similar enrichment history.

4. Concluding Remarks

We find localised dust to be a common feature of KDC galaxies. This dust probably has an external origin and indicates a source of fuel for the radio nucleus in these galaxies. The dust extinction is not so high, or dust so widespread, as to totally mask the effects of stellar populations. We have found photometric counterparts to the kinematic disks in at least seven galaxies. The lack of a colour difference between these disks and the bulges, in addition to the diversity of surface brightness profiles, argues against the accretion model for the formation of KDCs. The surface brightness profiles are correlated with galaxy luminosity in the same sense as kinematically normal galaxies. Globular clusters have a similar enrichment history as their parent galaxy and further study may yield clues to galaxy formation.

I thank M. Franx, G. Illingworth, M. Carollo for considerable help and useful discussions. This work was funded by HST grant No. GO-3551.01-91A and a travel grant from the AAS.

References

Balcells, M. and Quinn, P.J. (1990) *Ap. J.*, **361**, 381.

Bender, R. and Surma, P. (1992) *A. & A.*, **258**, 250.

Carollo, C.M., Franx, M.J., Illingworth, G.D. and Forbes, D.A. (1996), *Ap. J.*, submitted.

Forbes, D.A. (1991) *MNRAS*, **249**, 779.

Forbes, D.A., Franx, M.J. and Illingworth, G.D. (1995), *A. J.*, **109**, 1988.

Forbes, D.A. (1994) *A. J.*, **107**, 2017.

Forbes, D.A., Franx, M.J., Illingworth, G.D. and Carollo, C.M. (1996) *Ap. J.*, submitted.

Forbes, D.A., Franx, M.J. and Illingworth, G.D. (1994) *Ap. J.*, **428**, L49.

Franx, M.J. and Illingworth, G.D. (1988) *Ap. J.*, **327**, L55.

Hernquist, L. and Barnes, J. (1991) *Nature*, **354**, 210.

Illingworth, G.D. and Franx M.J. (1989) Dynamics of Dense Stellar Systems. Cambridge University Press, Cambridge.

Kormendy, J. (1984) *Ap. J.*, **287**, 577.

Silva, D. and Wise, M. (1995) *Ap. J.*, in press.

Statler, T.S. (1991) *Ap. J.*, **382**, L11.

Surma, P. (1992) Structure, Dynamics and Chemical Evolution of Elliptical Galaxies. ESO, Garching.

DUST AND GAS IN THE OUTER PARTS OF GALAXIES

J. LEQUEUX
DEMIRM, Observatoire de Paris,
61 Av. de l'Observatoire, F-75014 Paris, France

AND

M. GUÉLIN
IRAM, 300 Rue de la Piscine, Domaine Universitaire,
F-38406 St Martin d'Hères CEDEX, France

1. Abstract

We review the observations of dust in the outer disks and the halos of spiral and irregular galaxies. These observations not only yield the abundance of dust and indirectly of heavy elements, but also give indirect information on molecular hydrogen which has been proposed as a constituent of dark matter in galactic disks. Dust can be detected either through its thermal emission or through the reddening and extinction it produces on the light of background objects. Detection of the thermal emission of dust at millimeter wavelengths using bolometers on large telescopes is making considerable progress and we are close to the point where dust will be seen in this way out to the 25-26th mag. arcsec^2 blue isophote of the disk. Dust has now been detected in these outer regions of the disks from a statistics of the colors of disks as a function of their inclination, and by direct measurements of the reddening of background galaxies. New results concerning M 31 are presented. There is a surprisingly high dust/gas ratio, and also massive star formation, at these very large galactic radii. The evidence for dust in galactic halos is uncertain.

2. Introduction

Due to relatively small column densities and low temperatures it is difficult to detect molecular gas and dust in the outer parts of galaxies. However there are strong incentives to this detection. There have been suggestions for

D. L. Block and J. M. Greenberg (eds.), New Extragalactic Perspectives in the New South Africa, 422–434.
© 1996 *Kluwer Academic Publishers.*

large amounts of cold molecular hydrogen in the external parts of galactic disks and even in galactic halos (Lequeux et al. 1993; Pfenniger et al. 1994; de Paolis et al. 1995). It is almost impossible to detect such molecular gas directly except through the UV absorption lines produced by galactic disks in front of quasars, which might be detectable with the HST at redshifts larger than 0.07. In spite of much effort, only HI has been detected to radii as large as 70 kpc in big spirals, but no H_2 (e.g. Lanzetta et al. 1995). The only case where H_2 has been detected remains that of an absorption system at $z=2.77$ in front of the quasar PKS 0528-250 (Foltz et al. 1988). If this molecular gas is cold, the usual tracer, the CO molecule, may not be sufficiently excited to produce millimeter line emission. CO can be detected in absorption, but only if the few available lines of sight to sufficiently strong millimeter continuum sources intercept molecular clouds. The CO absorptions detected by Lequeux et al. (1993) and Lucas & Liszt (1996 and references herein) are no more attributed to a cold, dense molecular gas but rather to a low-density, warm interstellar gas (see e.g. Falgarone et al. 1995 and de Geus, this conference). The idea of large amounts of diffuse molecular gas in the outer parts and in the halos of galaxies has received another serious blow from the Galactic gamma-ray measurements with EGRET on board the Gamma Ray Observatory (Digel et al. 1996; Salati et al. 1996). The possibility that the molecular gas is in dense clumps partially opaque to cosmic rays and gamma rays remains open, however (F. Combes, this conference).

It is however easier to detect dust through its thermal emission and through the extinction and reddening it produces on the light of background objects. As dust seems to be always associated with gas, the detection of dust without HI 21-cm emission might indicate indirectly the presence of relatively diffuse molecular gas. When dust is observed to correlate with HI, it is possible to determine the dust/gas ratio and indirectly the metallicity of the gas as the dust/gas ratio is proportional to metallicity (Bouchet et al. 1985); this may offer the best way to determine the metallicity in the outer regions of galactic disks. In this review, we will concentrate on the detection of dust in the outer disks and halos of spiral and irregular galaxies. Section 3 deals with observations of the thermal emission of this dust and Section 4 with observations of extinction and reddening of background objects. Section 5 contains a discussion with reference to dust in objects at cosmological distances, and conclusions.

3. Thermal emission of dust in the outer parts of galaxies

IRAS has opened the possibility of observing systematically the thermal emission of interstellar dust. However its spectral range was limited to 100

μm on the long-wavelength side, making it impossible to detect cold dust. ISO, with its limitation at 200 μm, will do better but might still be relatively insensitive to the cold dust which could be present in the outskirts of galaxies where the radiation field is weak. It is necessary to observe at longer wavelengths for obtaining a good determination of the column density of dust. COBE has made such observations for our Galaxy; unfortunately our location in the galactic plane is such that the emission by the cold dust in the outer disk or in the halo cannot be easily distinguished from that of local, warmer dust. There are a few observations of submillimeter extragalactic dust emission done with the Kuiper Airborne Observatory, but this facility is now closed. SOFIA and/or FIRST will allow observations of cold dust in the outer parts of external galaxies. But the advances in millimeter bolometer arrays associated with large radio telescopes, such as the MPIfR 1.2-mm 19-channel bolometer array on the IRAM 30-m telescope, make it already possible to detect dust emission at large galactic radii with a good angular resolution.

The best example is that of NGC 4565 (Neininger et al. 1996; see also Neininger, this conference), an edge-on spiral in which dust emission at 1.2 mm has been seen as far as 19 kpc from the center (assuming a distance of 10 Mpc), i.e. at much larger radii than the CO emission. The dust emission from the outer regions correlates well with the 21-cm line emission and obviously arises from HI clouds; it shows the same out-of-plane warp as HI. Its detection offers an unique opportunity to derive the dust absorption cross section per H atom, $\tau_\lambda/N(H)=5 \ 10^{-27}$ cm^2 for $\lambda=1.2$ mm at 10-12 kpc from the center; this parameter is difficult to measure in the Milky way due to distance ambiguities. In NGC 4565, it is very close to that predicted by Draine & Lee (1984) for local diffuse clouds. Observations in progress for NGC 891 and NGC 3079 (Zykla et al., Braine et al. in preparation) seem to point to similar values.

How much farther can we hope to detect thermal emission from dust? Taking the example of NGC 4565 (data from van der Kruit & Searle 1981) and assuming that the interstellar radiation field scales with the surface brightness and that the grain properties are constant, we estimate a dust temperature of 9 K at the optical and HI cut-off radius of 24-25 kpc, corresponding to a surface brightness $\mu_J = 26.5$ mag. arcsec^{-2} in the J band of van der Kruit & Searle. The dust emissivity per H atom becomes there 2.3 times weaker than the value given above for R=12 kpc. With the 30-m telescope the present sensitivity is about 1 mJy per 12" beam at 1.2 mm; for dust at 9 K, this corresponds to N(HI) $(Z/Z_\odot)^{-1}$=4 10^{21} cm^{-2}, assuming that the abundance of dust is proportional to metallicity Z. This would make it possible to detect the dust emission at 24 kpc in NGC 4565 if the abundance of dust and heavy elements was like near the Sun. The abundance is likely to be smaller by a factor of a few (see later the discussion of M 31), but we are not far from the goal. The extension of the MPIfR bolometer to 37 channels, this winter, may make a detection within reach. The situation is similar for NGC 891.

Going to shorter wavelengths increases the intensity of the dust emission, which scales as $\lambda^{-(3.5-4)}$ in the submillimeter domain. The 99-channel SCUBA bolometer, which will be installed soon on the JCMT, will operate down to 350 micrometers, near the peak of the dust emission, and may allow to detect very low dust column densities; however the dust emission varies as T_d^5 at this wavelength and is hard to interpret. Moreover the atmosphere becomes nearly opaque making sensitive observations difficult.

The ISO satellite has a poor angular resolution at long wavelengths and, as noted above, may miss the bulk of cold dust. However the dust emission is not uniform and due to the existence of hot stars in the outskirts of galaxies (see later for M 31) there might be warm dust there. This will probably make possible a detection at 200 μm, to a limiting column density perhaps an order of magnitude better than at 1.2 mm. Such observations are planned. The observational limit will probably arise from confusion with Galactic cirruses.

4. Extinction and reddening by dust in the outer parts of galaxies

For the time being, it seems somewhat easier to detect dust in the outskirts of galaxies through extinction and/or reddening of the light of background objects than through its thermal emission. We will discuss the results obtained up to now: first by a statistics of surface brightnesses and colors of galaxies as a function of their inclination, then by observation of individual foreground objects.

4.1. STATISTICAL EXTINCTION IN GALAXIES OF VARIOUS INCLINATIONS

The analysis of surface photometry of galaxies as a function of inclination has shown that spiral galaxies, in particular Sb and Sc ones, suffer from considerable extinction at least in their central regions. But there is a controversy as to the quantitative amount of extinction, which is apparently due in part to selection effects in the galaxy samples (see e.g. Valentijn 1994). The basic idea in such studies is that inclined galaxies should have a lower surface brightness and be redder than face-on galaxies due to extinction by their internal dust. A simple, useful parameter to characterize extinction is the quantity C such that: μ(obs) = μ(face-on) - 2.5 C log(a/b) where μ is the surface brightness in a magnitude scale and a/b is the observed axis ratio. For a simple slab model, C is equal to unity for a transparent galaxy and to zero for an opaque one. The physical interpretation of C depends obviously of the respective distribution of the stars and dust. After a thorough discussion, Valentijn (1994) concludes that Sb galaxies and high-surface brightness Sc and Sd galaxies have low C values (high extinction) over the whole of their disks, at least on the average as dust can be distributed very irregularly. C is still found of the order of 0.2-0.5 in the blue at their μ_B=26 mag. arcsec^{-2} isophote while the fainter surface brightness Sc and Sd galaxies are nearly transparent (C = 0.5-0.7 in the blue) at their μ_B=26 mag. arcsec^{-2} isophote. Peletier et al. (1995) have recently given a discussion of the B-K color of galaxies as a function of inclination. This color is a good indicator of the amount of extinction (plus scattering!) since light in the K band is much less affected by dust than light in the B band. They find that while extinction is generally large (but variable from galaxy to galaxy) in the central regions, it is usually small in the outer parts, i.e. at μ_K=21 mag. arcsec^{-2} corresponding roughly to μ_B=25 mag. arcsec^{-2}. The actual figures depend on the respective distribution of dust and stars, and there is a large range of acceptable solutions for the relative scale lengths (in the disk) and scale heights (perpendicular to the disk) of dust and stars. For our purpose, we will extract from this study a maximum amount of extinction normal to the disk at μ_B=25 mag. arcsec^{-2} of A_B = 0.1 to 0.4. This conclusion is in agreement with that of Giovanelli et al. (1994) who find that extinction is small at μ_I=23.5 mag. arcsec^{-2}, about the same level. To conclude, it appears that extinction is small, but not necessarily negligible, in the outer parts of the disks of galaxies.

4.2. DIRECT DETERMINATION OF REDDENING OR EXTINCTION IN THE OUTER DISKS

4.2.1. *NGC 7814*

In a few cases, extinction in the outer disks can be seen directly. This is the case in the warped edge-on Sa galaxy NGC 7814, for which Lequeux et al. (1995) have obtained very deep images in V and I. These images reveal clearly the torsion of the disk and dust absorption marks out to 238" radius (17.5 kpc for an adopted distance of 15 Mpc), almost as far as the outermost extent of the visible disk (264" or 19.4 kpc). There are a few background galaxies seen behind the absorption marks at an average radius of 14 kpc, with a 2.4σ average color excess E(V-I)=0.10 mag. with respect to the galaxies in blank surrounding fields. Assuming a Galactic extinction law this corresponds to A_B=0.7 mag., then given the inclination of the warped disk at this location to a face-on extinction of A_B=0.2 mag., well within the range indicated by the statistical studies mentioned above. As the column density of HI is known at this location, one can determine the N(H)/E(B-V) ratio which is of the order of 5 10^{21} cm^{-2} mag.$^{-1}$, of the same order as the ratio near the Sun in our Galaxy. This suggests that the metallicity is still high at this radius. Unfortunately the measurement of extinction is only at 2.4σ, and it is not easy to determine the surface brightness of the disk at this location because of the difficulty in separating it from the enormous bulge of the galaxy; examination of fig. 6 of van der Kruit & Searle (1982) suggests $\mu_B = 25.5$ mag. arcsec^{-2}.

4.2.2. Overlapping galaxies

When two galaxies are on the same line of sight, it is possible to obtain information on the extinction in the foreground galaxy after some interpolation of the surface brightness of the background galaxy is made. Some examples have been presented by J. Frogel and R.E. White at this conference, with the result that extinction is only a few tens of magnitudes in the external regions. To this study one should add that by Andreakis & van der Kruit (1992). They have studied the extinction by the spiral galaxy NGC 450 on the background spiral UGC 807, whose center projects behind the $\mu_B = 25$ mag. cm^{-2} isophote of NGC 450. The maximum extinction allowed for by the data is $A_B = 0.3$ mag., also compatible with the studies just cited.

4.2.3. The Andromeda galaxy M 31

Very recently J.-C. Cuillandre, Y. Mellier, B. Fort, R.J. Allen and J. Lequeux (in preparation) have succeeded in detecting the extinction in the outermost regions of the Andromeda galaxy M 31 by studying statistically the color of background galaxies. They observed with a 8192×8192 pixel CCD mosaic at the prime focus of the Canada-France-Hawaii 3.6-m telescope, kindly loaned by G. Luppino (University of Hawaii). The mosaic covered a field 28'×28' whose center is located at 116'=23.2 kpc from the center of M31 on the SW side of the major axis. This region is at the extreme limit of the visible disk of the galaxy (Innanen et al. 1982, Walterbos & Kennicutt 1988) and of the distribution of neutral hydrogen (Newton & Emerson 1977). The B surface brightness there is approximately $\mu_B = 26$ mag. arcsec^{-2} (Walterbos & Kennicutt 1988). The cumulative exposure times were 4 hours in the V and in the I filters, with photometric quality, and the image quality on the resulting, combined frames is about 0.8" in both bands. An algorithm was used to separate stellar images from the images of galaxies and was quite successful in spite of the crowding of the frames, except for regions around bright objects or groups of objects. A map of the mean V-I color of the galaxies with I magnitudes between 22 and 28 was then constructed. The difference with the average V-I in a reference region believed to be free of extinction by M 31 gives the color excess due to the dust of M 31. This map is shown on fig. 1. It correlates well with the HI map of Newton & Emerson (1977) corrected for the primary beam response.

From the data of fig. 1, and assuming that the extinction law in the visible is similar for this region of M 31 and for the Solar neighborhood (a reasonable assumption according to the study of extinction in the Magellanic Clouds by Bouchet et al. 1985), the dust-to-gas ratio is found to be 0.4 times that in the Solar Neighborhood. This is a provisional value which will be refined by a more detailed study. It shows that at this large radius the gas of M 31 still contains a large amount of dust indicative of an unexpectedly large metallicity, about 0.4 solar. An alternative would be to imagine that the studied region contains not only atomic, but also molecular hydrogen with a distribution similar to that of HI, and that the dust corresponds to both components; the dust/gas ratio and the metallicity would be reduced accordingly. This molecular gas should however have a large surface filling factor in order to produce a sufficient average extinction on background galaxies, and is not the gas distributed in dense, high-density small clumps advocated by Pfenniger et al. (1994). The discussion in the introduction of this paper does not favor the existence of diffuse molecular hydrogen and we assume that the column density of H_2 is negligible. The average color excess in the most reddened region, inside the $N(H)=1.28\ 10^{21}$ atom cm^{-2} contour of fig. 1, is E(V-I)=0.18 mag., corresponding to A_B=0.4 mag.

During the reduction of the M 31 frames, stars were separated from galaxies. We noticed that the mean V-I of stars is smaller (bluer) in some regions corresponding roughly to the peaks of HI. We thus obtained color-magnitude (CM) diagrams of these stars. They are displayed on fig. 2. In a reference region with low gas column densities (the CCD at the SW of the image; right panel) one sees a red giant branch, an horizontal branch at V=25.5, and a loosely populated AGB branch. There are not many stars outside these features, showing that contamination by Galactic stars is rather small. We have checked that the contamination by galaxies is also not large. The left panel displays the CM diagram for stars in the "blue" regions. The contamination and photometric errors are somewhat larger here due to to a higher density of objects, but the giant branch and the AGB branch are still well visible. The differences between the two CM diagrams are due to different populations (mainly halo stars in the reference field and a mixture of halo and disk stars in the other field), but this has to be studied in details. These CM diagrams resemble closely some diagrams obtained on the halo of M 31 with the HST and presented by M. Rich at this conference. The most interesting feature for our present purpose is the presence in the left diagram of blue stars with V-I around 0.0, as a vertical sequence well separated from the red giant branch. These are main-sequence B1-B9 stars. Thus there is still massive-star formation in this remote part of M 31! Figure 3 is a map of the ratio of blue to red

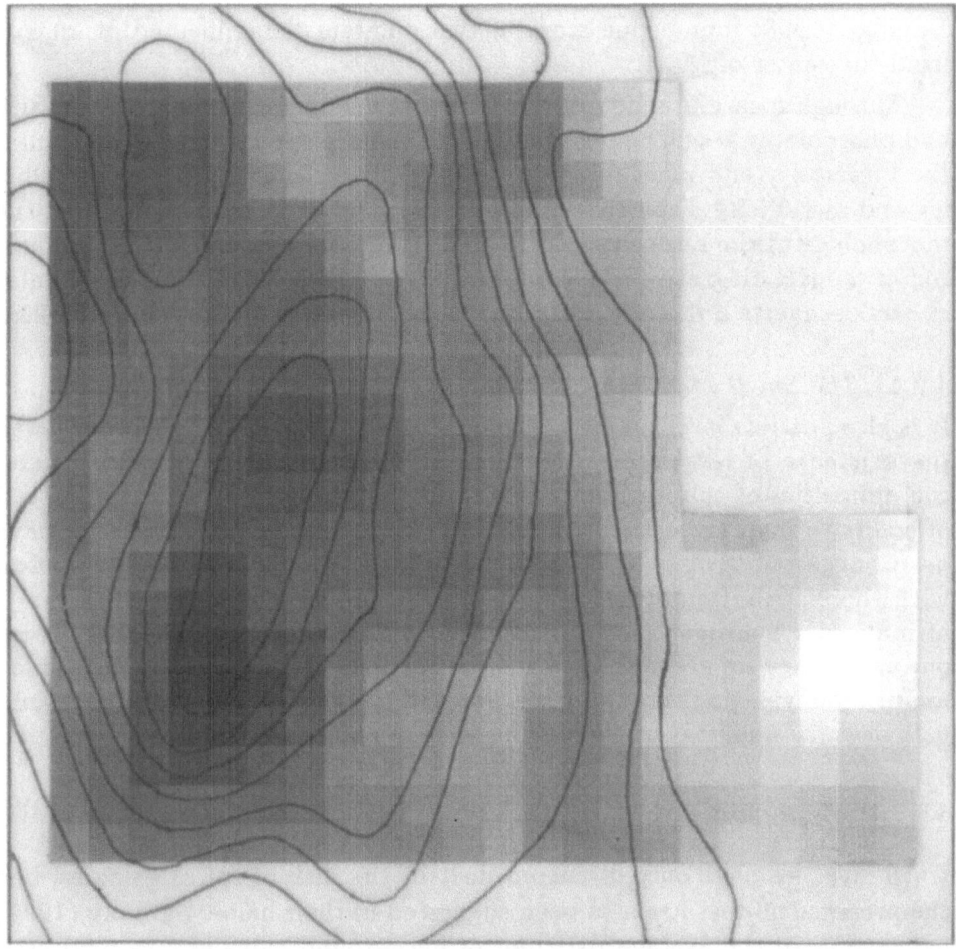

Figure 1. Map of the mean V-I color of background galaxies in a field at the extreme SW of M 31 (grey scale), compared to the column density of atomic hydrogen (contours). The field center is at $\alpha(1950)$=00h 33m 21.4s, $\delta(1950)$=+39° 32' 21". The color data are smoothed to roughly the same resolution as the HI data (3' EW × 5' NS), resulting in the loss of the outer part of the image. The size of the outer square is 28'×28' ($5.6{\times}5.6$ kpc^2). North to the top, east to the left. The part of the image at the upper right was lost due to the guiding probe. The grey scale goes from V-I=0.74 (white) to 1.06 (darkest), on a linear scale. The HI contours correspond to column densities of $(1...9) \times 1.6 \ 10^{20}$ atom cm^{-2}. Note the correlation between the reddening of background galaxies and the column density of atomic hydrogen.

stars over the whole studied field, at the same angular resolution as fig. 1. It gives a first approximation to the relative rate of massive star formation. The correlation with HI is obvious, although less good as for the reddening. An earlier study of the CM diagram of stars in a region located near the middle of the left side of our field (Richer et al. 1990) missed the blue stars which are almost absent in this area. However Hodge et al. (1988) and Davidge (1993) found blue stars in two different fields located at 20 kpc from the center of M 31.

Through a careful study of the effects of contamination on star density and photometry, it will be possible to obtain more information, in particular the luminosity and mass functions of the B stars, and the mean metallicity and metallicity dispersion of the background of older stars through the morphology of the horizontal, giant and AGB branchs. Already a comparison of the HR diagram of the comparison (halo) field with those of globular clusters suggests a metallicity comparable to that of 47 Tuc, [Fe/H]=-0.7.

4.2.4. *The Small Magellanic Cloud*

It is also possible in principle, although considerably more dangerous given the existence of large-scale structures in the distribution of galaxies and the difficulties of calibration, to use counts of background galaxies instead of colors to map extinction. In this way, large amounts of extinction have been suggested around the Small Magellanic Cloud with no correlation with the distribution of HI thus suggesting the presence of large amounts of molecular hydrogen (see a discussion in Lequeux 1994). However given our experience we are now somewhat suspicious about these results, which require confirmation through a statistical study of the colors of background galaxies now undertaken by P. Hodge using archival HST data.

4.3. DUST IN THE HALOS OF GALAXIES

Until now, we have only discussed dust in the disk of spiral galaxies, but the presence of dust has also been suggested in their halos. Zaritsky (1994) has compared the average B-I color of background galaxies at a projected radius of 60 kpc around two spiral galaxies (NGC 2835 and NGC 3521) with that at a radius of 220 kpc, and found that the former are redder by 0.067 mag. with respect to the latter, a 2σ effect. If real, this detection has profound consequences as it would correspond to a total mass of dust of $3 \; 10^8 \; M_\odot$ within 60 kpc and a corresponding mass of gas of more than $3 \; 10^{10}(Z/Z_\odot)^{-1} \; M_\odot$, Z being the metallicity, assuming that gas and dust are well mixed. This would open the possibility that the dark halo is made of such material. Unfortunately these measurements had to be performed with different exposures at different times, with considerable difficulties

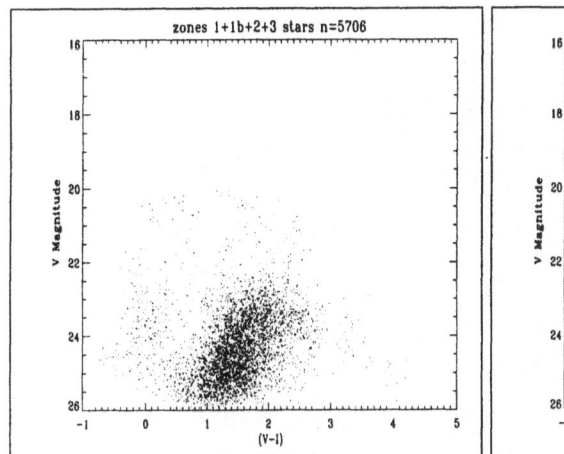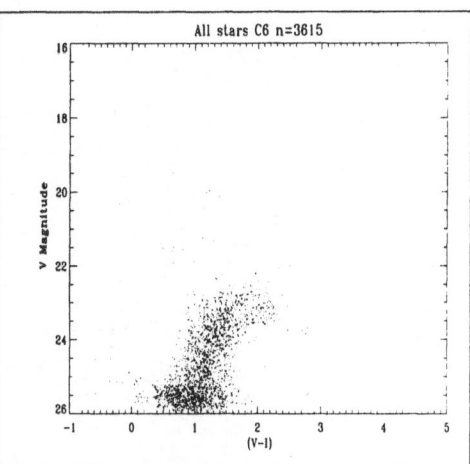

Figure 2.   V vs V-I color-magnitude diagrams for stars in a region of the outskirts of M 31 without detected atomic hydrogen (right) and with a large column density of atomic hydrogen (left). The diagram on the right shows a giant branch, the beginning of an AGB and an horizontal branch at V=25.5 mag. (M_V=1.0 mag.). Contamination by galactic stars and background galaxies is small, and the vast majority of the stars belong to M 31, probably to its halo. The number of stars is indicated above. The diagram on the left shows the same features but with conspicuous differences, probably due to a disk population. There is also a vertical sequence of blue stars around V-I=0.0, which is absent on the other diagram.

in cross-calibrating the different frames. Recently, Lequeux et al. (1995) have done a similar work in the range of distances 15 kpc to 35 kpc from the center of NGC 7814 (outside the disk) and found no gradient, and no variations in the mean V-I colors of background galaxies within 4.4 x 4.4 kpc^2 squares covering the field, within 2.4 times a rms deviation of 0.04 mag. This corresponds approximately to the same accuracy as quoted by Zaritsky (1994), but the result is considerably more secure as it was obtained with a single CCD covering all the field. Lequeux, Fort, Dantel-Fort, Cuillandre and Mellier (in preparation) have done a similar study around NGC 5907, an edge-on galaxy, and similarly found no color gradient for the background galaxies. A provisional value for the difference in the mean V-I of galaxies

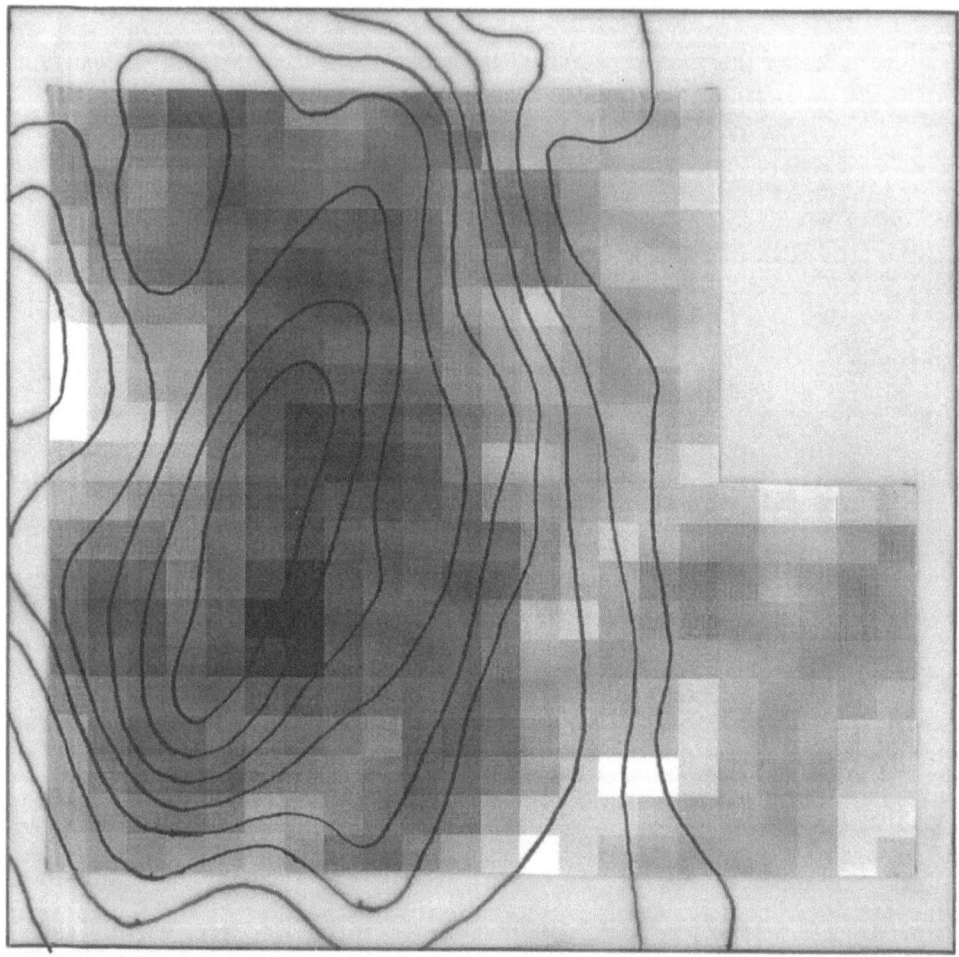

Figure 3.   Grey-scale map of the ratio of blue to red stars in the observed region of the outskirts of M 31. The blue stars are defined as having V-I≤0.4. The darker areas have more blue stars. The contours represent the column density of atomic hydrogen, as for fig. 1. Note the general correlation between the two patterns, but also the shift of the maxima with respect to each other.

between a long strip located between 2.7 and 5.5 kpc parallel to the major axis and a more distant reference region is Δ(V-I)=-0.03±0.04 mag.; this can certainly be refined by further analysis. This observation was made with MOCAM, an array of 4 2048×2048 pixel CCDs at the prime focus of the Canada-France-Hawaii Telescope, also covering the whole field at once. In conclusion, the evidence for dust in the halos of spiral galaxies is weak, but the limits are not yet sensitive enough to give stringent constraints on gas as a constituent of the dark halos.

5. Discussion and conclusions

We have shown that there are good evidences for a substantial amount of dust in the outermost regions of the disks of spiral galaxies. The typical extinction through a face-on disk is of the order of 0.2 mag. in the blue at a surface brightness μ_B=26 mag. arcsec^{-1}. For M 31, the only galaxy for which this information is available, the extinction correlates well with the column density of atomic hydrogen. This does not leave much room for molecular hydrogen unless it has a distribution similar to that of HI or is distributed in dense clumps covering a small fraction of the surface. If we neglect molecular hydrogen, the dust/gas ratio is given by the ratio of extinction to HI column density and can be calibrated on the Galactic ratio near the Sun if the dust has optical properties not too different from the Galactic dust. For NGC 7814 the dust/gas ratio (uncertain) is approximately like near the Sun at a radius of 14 kpc, and for M 31 where it is better determined it is approximately 0.4 of that near the Sun at 22 kpc radius. This implies in turn surprisingly large values of the metallicity at such large radii. These high values were not completely unexpected, however. Jacoby & Ford (1986) have measured an oxygen abundance O/H = $3.2\ 10^{-4}$ in an HII region 17 kpc from the center of M 31, and of $1.2\ 10^{-4}$ in a planetary nebula at 33 kpc projected radius (however an object with halo kinematics), in rough agreement with the present determination. In our Galaxy, de Geus et al. (1993) and Digel et al. (1994) have found a cloud with CO emission, hence heavy elements, associated with an HII region at a kinematic distance of 28 kpc, almost at the edge of the HI disk whose kinematic radius is of the order of 30 kpc.

Recent abundance measurements in the outer Galaxy indicate a surprisingly small abundance gradient (Kaufer et al. 1994), in qualitative agreement with the above findings. It is not yet clear if the formation of massive stars found at large radii at least in M 31 and in our Galaxy is sufficient to account for the observed heavy elements, or if one has to invoke radial mixing or preferential infall in the inner disk of spiral galaxies.

In general, formation of heavy elements and of dust appears to have occured early in the evolution of the Universe. The metallicity is of the order of 1/10 $Z_\odot$ in damped Lyman α absorption systems at z between 1.7 and 3, with a stronger depletion of chromium which is believed to indicate the formation of dust (Pettini et al. 1994). The dust abundance predicted in this way (also 1/10 of near the Sun) is found directly through a reddening of background quasars (Pei et al. 1991). This has been discussed by M. Fall at this conference. Although it is yet not possible to clearly relate the damped Lyman α systems to the present galaxies, this shows that relatively high abundances in the gas and large amounts of dust existed in the material from which galaxies are formed. But we still have to understand how heavy-element abundances and dust/gas ratios not much smaller than near the Sun have been built in the outskirts of spiral galaxies.

One of us (J.L.) thanks Gerry Luppino for the generous loan of the large CCD mosaic of the University of Hawaii, which made the observations of M 31 possible.

References

Andreakis Y.C., van der Kruit P.C. 1992, A&A 265, 396

Bouchet P., Lequeux J., Maurice E., Prévot L., Prévot-Burnichon M.-L. 1985, A&A 149, 330

Davidge T.J. 1993, ApJ 409, 190

de Paolis F., Ingrosso G., Jetzer Ph., Roncadelli M. 1995, A&A 295, 567

de Geus E.J., Vogel S.N., Digel S.W., Gruendl R.A. 1993, ApJ 413, L97

Digel S., de Geus E.J, Thaddeus P. 1994, ApJ 422, 92

Digel S.W., Grenier I., Heithausen A., Hunter S.D., Thaddeus P. 1996, ApJ in press

Draine B.T., Lee H.M. 1984, ApJ 441, 270

Falgarone E., Pineau des Forets G., Roueff E. 1995, A&A 300, 870

Foltz C.B., Chaffee F.H., Black J.H. 1988, ApJ 324, 267

Giovanelli R., Haynes M.P., Salzer J.J., Wegner G., Da Costa L.N., Freudling W. 1994, AJ 107, 2036

Hodge P., Lee M.G., Mateo M. 1988, ApJ 324, 172

Innanen K.A., Kamper K.W., Papp K.A., van den Bergh S. 1982, ApJ 254, 515

Jacoby G.H, Ford H.C. 1986, ApJ 304, 490

Kaufer A., Szeifert Th., Krenzin R., Baschek B., Wolf B. 1994, A&A 289, 740

Lanzetta K.M., Bowen D.V., Tytler D., Webb J.K. 1995, ApJ 442, 538

Lequeux J. 1994 A&A 287, 370

Lequeux J., Allen R.J., Guilloteau S. 1993, A&A 280, L23

Lequeux J., Dantel-Fort M., Fort B. 1995, A&A 296, L13

Lucas R., Liszt H. 1996, A&A in press

Neininger N., Guélin M., Garcia-Burillo S., Zylka R., Wielebinski R. 1996, A&A in press

Newton K., Emerson D.T. 1977, MNRAS 181, 573

Pei Y.C., Fall S.M., Bechtold J. 1991, ApJ 378, 6

Peletier R.F., Valentijn E.A., Moorwood A.F.M., Freudling W., Knapen J.H., Beckman J.E. 1995 A&A 300, L1

Pettini M., Smith L.J., Hunstead R.W., King D.L. 1994, ApJ 426, 79

Pfenniger D., Combes F., Martinet L. 1994, A&A 285, 79

Richer H.B., Crabtree D.R., Pritchet C.J. 1990, ApJ 355, 448

Salati P., Chardonnet P., Xiaochun Luo, Silk J., Taillet R., 1996, preprint

Valentijn E.A. 1994, MNRAS 266, 614

van der Kruit P.C., Searle L. 1981, A&A 95, 105

van der Kruit P.C., Searle L. 1982, A&A 110, 79

Walterbos R.A.M., Kennicutt R.C. Jr. 1988, A&A 198, 61

Zaritsky D. 1994, AJ 108, 1619

DUST SPECTRA OF THE MILKY WAY: PRIMORDIAL MOLECULAR HYDROGEN AND HELIUM IN THE OUTER GALAXY

J. SCHAEFER

Max-Planck-Institut für Astrophysik,
Karl-Schwarzschild-Str.1, 85740 Garching, Germany

1. Introduction

The far-infrared continuum spectra of the FIRAS-COBE mission [1] used in this paper show mainly dust radiation observed in the Galaxy at frequencies below 100 cm^{-1}. For a complete explanation of the spectra a second component of radiation has to be added necessarily to the true dust spectrum if the Kramers-Kronig theorem is assumed valid for the dust radiation which requires a power law emissivity $\epsilon_\nu \propto \nu^\alpha$ in the low frequency limit[2], where α is an even positive integer (=2). While studying a recent paper by Reach *et al.*[3], I decided about the fitting of the dust spectra to take advantage of the frequency range of the measurements by doing *as good fits as possible* in order to avoid relative large uncontrollable remnants of intensity contributing to the second component. The meaning of "as good as possible" is accurate agreement at the higher frequencies, where the second component contributes negligibly, and accepting even losses of intensity of the second component - with the hope of not loosing essential information about the model of this source. The following shows results obtained by applying this kind of careful partitioning successfully to the FIRAS-COBE spectra after subtracting the CMBR intensity[3].

2. The Dust Model

An empirical temperature dependent emissivity of the dust has been found by extending a correction of the simple emissivity $\epsilon_\nu = (\nu/\nu_0)^2$, $\nu_0 = 30$ cm^{-1}, used in the paper of Reach *et al.*[3] to account for the high-frequency rolloff of emissivity close to the galactic plane. The resulting dust model is

D. L. Block and J. M. Greenberg (eds.), New Extragalactic Perspectives in the New South Africa, 435–438.
© 1996 *Kluwer Academic Publishers.*

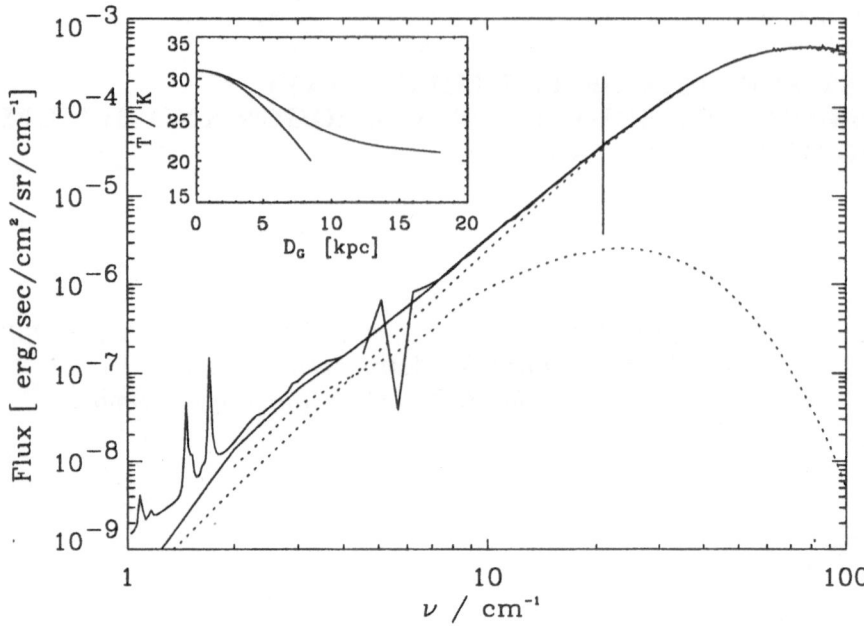

Figure 1.   The FIRAS-COBE continuum spectra from the "llss" scan mode at $(l, b)=$ (358.785°, 0.246°) and from the "rhss" scan mode at $(l, b)=$ (358.785°, 0.246°), linked at 21 cm^{-1} (vertical line), fitted by a dust spectrum (upper dotted curve) with the temperature function shown in the insert, and by the collision induced emission spectrum of the primordial mixture of normal-H_2 and helium gas of 11 K (lower dotted curve). Theoretically predicted dimer features of H_2 are shown in the lower left corner.

determined by a temperature dependent mass absorption coefficient

$$\kappa_m(\nu, T) = \kappa_0 \cdot \frac{(\nu/30)^2}{[1 + \{\nu(1 + 0.02(T - 19))/50\}^6]^{1/6}}$$

with an unknown constant κ_0, to be multiplied along the line of sight with the dust density obtained from a reasonable density function

$$\rho = \rho_0 \cdot exp\{-a_R R_G\} \cdot exp\{-a_z z^2\},$$

where ρ_0 is an effective density contained in a freely chosen factor $(\kappa_0 \rho_0)$ for each fitted spectrum, and a_R, a_z are fixed parameters: $a_R = 0.14$ kpc^{-1} and $a_z = 28.4089$ kpc^{-2} provide a density decrease from 1. at zero R_G to 0.08 at $R_G = 18$ kpc and a falling-off perpendicular to the Galactic plane with 156.2 pc HWHM. A temperature function along the line of sight is varied together with $(\kappa_0 \rho_0)$ until sufficient agreement is reached at $30 < \nu < 90$ cm^{-1}. An example of a center of Galaxy spectrum is shown in Fig.1.

3. The Gas Model

The remaining intensity is accurately fitted by the collision induced emission spectrum of the primordial mixture of normal hydrogen (86%) and helium (14%) gas of 11 K [4] times a factor which is related to the column density - *and this holds for all* FIRAS-COBE *spectra*. I may say this after about 15% of the spectra released by NASA in August '95 have been handled. However, the FIRAS-COBE spectra do not provide full proof for the hydrogen source as one can see in Fig.1: the instrument noise on the low frequency side is about $3 \cdot 10^{-7}$ in the units of this figure, i.e., it is much too large compared to the faint features of the hydrogen gas, and the frequency resolution is insufficient as well below 7 cm^{-1}.

Necessary conditions for the hydrogen source model are as follows: The hydrogen exists as gas and the ortho-para ratio has to be normal (3:1) or close to normal. The density of the observed gas should be as large as the vapor pressure of condensing normal hydrogen at 11 K: 685 Pa or 0.168 amagat or $4.5 \cdot 10^{18}$ cm^{-3}. Any significantly lower density would require an unreasonable amount of hydrogen mass. How can these high densities of condensing sources in not condensing gas clouds occur in the ISM?

Pfenniger and Combes[5, 6] provided help by favouring cold molecular hydrogen of 3K in fractal-structured low volume-filling clouds as a candidate of dark matter. They both give lectures on this subject here at this conference. I use parts of their model, i.e., fractal structured clouds and gravitationally bound elementary fragments with relative high averaged densities (they assumed a quasi-isothermal model with $\bar{\rho} = 10^9$ cm^{-3}). But I have to go beyond this model to reach the required density by assuming less cooling in the fragments and the development of enormous density gradients in gravitational turbulence due to the shortage of condensation sites. Only the top densities, occuring as rather small inhomogeneities of the gas, contribute to the observed spectra. It is plausible that a hydrogen - helium mixture with too little condensation sites cannot dissipate the gravitational energy gained in the collapse together with the sublimation heat per H_2 molecule (77 cm^{-1}), the rotational energy of the ortho H_2 species (118 cm^{-1}), and the energy of the photons absorbed from the stellar radiation field.

Nevertheless, this gas model describes clouds which are pretty close to star formation, and a realistic rate of star formation out of these clouds is a wellknown observed phenomenon because the gas sources have been found to correlate with HI in the outer Galaxy rather than with the dust. This can be seen by comparing the fitted column densities of the gas with plotted column densities of HI as e.g. published by Burton[7]. In order to demonstrate this in a short lecture I have asked a colleague, V. Spirko[8], to

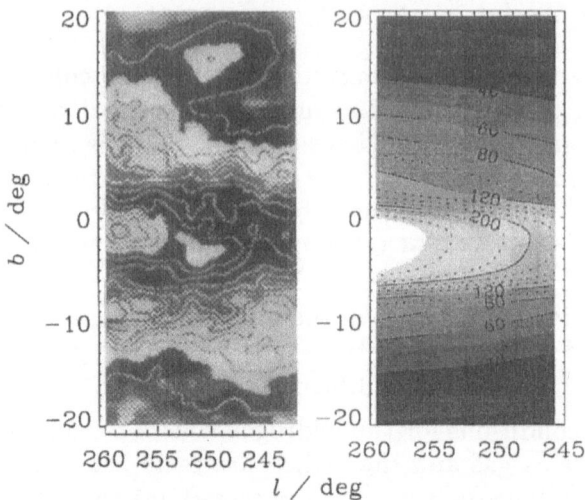

Figure 2. Comparison of the enlarged contour plot of HI with contours of the fitted H_2 - He radiation sources. Reading this figure needs still some imagination: the HI contours must be convolved with the instrument function of a 7° FWHM beam to become comparable in the larger scales and, unfortunately, the grey-scaling is in opposite directions.

produce global least square fits from the fitted data obtained in three chosen regions of Galactic longitude and for latitudes $|b| < 50°$ to be compared with the contour plots of HI. One example is shown in Fig.2.

Thus far the dust and gas model and the partitioning of the FIRAS-COBE spectra turned out to be successful. These spectra contain radiation from the visible part of the so-called dark matter.

References

1. *COBE Far Infrared Absolute Spectrophotometer (FIRAS) Explanatory Supplement,* ed. Mather J.C., Shafer R.A., Eplee R.E., Fixsen D.J., Isaacmen R.B., and Trenholme A.R., *COBE* Ref.Pub. No. 95-C (Greenbelt, MD: NASA/GSFC).
2. Wright E.L., 1993, in *Back to the Galaxy*, eds. S.S. Holt & F. Verter (AIP: New York), p.193.
3. Reach W.T., Dwek E., Fixsen D.J., Hewagama T., Mather J.C., Shafer R.A., Banday A.J., Bennett C.L., Cheng E.S., Eplee R.E., Leisawitz D., Lubin P.M., Read S.M., Rosen L.P., Shuman F.G.D., Smoot G.F., Sodroski T.J., and Wright E.L, submitted to ApJ.
4. For details of this radiation see e.g. Schaefer J., Astron.&Astrophys. **284**, (1994), 1015 and references therein.
5. Pfenniger D., Combes F., and Martinet L., Astron.&Astrophys. **285** (1994), 79
6. Pfenniger D. and Combes F., Astron.&Astrophys. **285** (1994), 94
7. Burton W.B., 1974, in *Galactic and Extragalactic Astronomy*, A&A Library, G.L. Verschuur and K.T. Kellermann edts, Springer Berlin, 2nd edition.
8. I gratefully acknowledge this help. The program is normally used for providing global fits of potential energy surfaces in quantum chemistry.

FUNDAMENTAL ASPECTS OF THE ISM FRACTALITY

D. PFENNIGER

Geneva Observatory, University of Geneva
CH-1290 Sauverny, Switzerland

Abstract. The ubiquitous clumpy state of the ISM raises a fundamental and open problem of physics, which is the correct statistical treatment of systems dominated by long range interactions. A simple solvable hierarchical model is presented which explains why systems dominated by gravity prefer to adopt a fractal dimension around 2 or less, like the cold ISM and large scale structures. This has direct relation with the general transparency, or blackness, of the Universe.

1. Introduction: Clumpiness in Astrophysical Gases

The extreme inhomogeneity, the supersonic turbulence, and multi-phase nature of the interstellar medium (ISM), meaning an extreme departure from thermodynamical equilibrium, was well recognised many decades ago (e.g. von Weizsäcker 1951). Over the years, new wavelengths and higher angular resolution have revealed several times that the interstellar gas was more clumpy than accepted before. Often, it was argued that the smallest clouds had been resolved, to be disproved later on (typically they span "three times the beamwidth", Verschuur 1993).

Clearly, the radically opposite behaviours of cosmic and terrestrial gases must be understood. Since the ISM covers many decades of density and temperature, very different physical processes, such as radiation transfer, chemistry, magnetic fields, and gravity are simultaneously involved, confusing the issue about the relative importance of each one. Star formation can even be viewed as an extreme prolongation of the ISM inhomogeneity. The concurrent interplay of so many factors makes the study of the ISM formidable. Moreover, different kinds of devices and skills are needed to unveil its fullness, increasing the communication problem across specialties.

D. L. Block and J. M. Greenberg (eds.), New Extragalactic Perspectives in the New South Africa, 439–446.
© *1996 Kluwer Academic Publishers.*

With respect to terrestrial gas, the new factor in the ISM and larger scale structures is, of course, self-gravity. Although there are discussions whether the molecular clouds are bounds by gravity or just marginally so (e.g. Blitz 1993), in any case gravity pervades cosmic structures from the largest galaxy superclusters down to stars and planets. By continuity its rôle must be expected to be generally dominant throughout the scales.

2. Homogeneity in the Classical Gas

Before discussing the inhomogeneity of cosmic gas, one should clearly understand why in terrestrial conditions gas tends toward homogeneity. In terrestrial situations a classical gas is *confined* by exterior factors (walls, etc.), which means that in the Lagrangian of the N gas particles,

$$L(q_i, \dot{q}_i) = T(\dot{q}_i) - U(q_i), \quad \text{and} \quad U(q_i)/T(\dot{q}_i) \to 0, \quad (i = 1, \ldots N), \quad (1)$$

the kinetic energy term $T(\dot{q}_i)$ dominates the interaction term $U(q_i)$. The motion of individual particles being highly chaotic, sensitive to perturbations, no integrals beside the classical integrals does exist, and the system can be assumed ergodic in good approximation. Typically, perturbations ("the walls") reshuffle the particles in phase space while preserving in average the scalar energy integral. This allows the average time invariance of the system. But the global *vector* integrals (angular momentum, centers of velocity and mass) must vanish in the reference frame of the box. Since the system (1) is locally dominated by the kinetic term, it must be translationally invariant in space. As well known to this symmetry corresponds spatial homogeneity. At scales much larger than the interparticle distances the gas must tend to be smooth, i.e. differentiable. The use of hydrodynamics and other differential equations is then justified.

In such a situation the total energy of the system is proportional to the volume, i.e. extensive. In order to be able to take the large N limit, the extensivity of energy is an additional requisite in usual statistical mechanics.

Thus, the a priori complex behaviour of a gas can be statistically simple when a global symmetry, such as translation invariance, exists. However, when the interaction potential becomes non-negligible at lower temperature this symmetry may be broken. Often a phase transition reveals then another symmetry (e.g. when crystallisation occurs). But if the interactions are short range, then energy is still extensive in good approximation.

In contrast, when the interactions are long range and attractive, as for gravity, they can never be neglected in the large N limit, and energy is no longer extensive: the ground to expect a statistical homogeneous state, even locally, is lost. Curiously, the inapplicability of thermodynamics to gravitating systems is well known by thermodynamicists (e.g. Jaynes 1957; Prigogine 1962), but often ignored in astrophysics.

The association of thermodynamics with gravity leads immediately to paradoxes, such as the occurrence of negative specific heats (Lynden-Bell & Lynden-Bell 1977), implying that a thermal equilibrium is impossible!

The ideal model of the gravitating perfect gas enclosed in a spherical box at a fixed temperature shows itself the limit of the association (see Binney & Tremaine 1987, Fig. [8-1]). The isothermal sphere equilibrium does exist only above a critical temperature, which corresponds to the Jeans' critical temperature. At sufficiently high temperature the system energy is positive and a box is required to confine the system, as for usual terrestrial gas. At some lower temperature the system energy is negative, so is bound by gravity, but a thermal equilibrium still exists: this is the regime of the stars. However, below the Jeans' critical temperature, no equilibrium solution does exist which fulfills the local homogeneity assumption. The self-gravity then dominates. The open problem is then to characterise the asymptotic statistical state of the perfect gas *below the Jeans' critical temperature*. Clearly, this concerns many astrophysical systems, from most of the ISM to the large-scale cosmological structures.

3. Statistical Equilibrium with Long-Range Interactions

Before introducing any additional physics relevant to the ISM, the statistical behaviour resulting just from the interplay between gravity and dynamics should be characterised. This partial problem is a highly idealised simplification of the full complexity of the ISM, but no progress can be expected if this fundamental aspect remains obscure.

The classical gas is based on a very rough model of particles without interaction. Yet, more important it includes the symmetry of translational invariance. In the ISM and large scale structures, the dominant symmetry is *scale-invariance*. The Lagrangian of the gas particles is

$$L(q_i, \dot{q}_i) = T(\dot{q}_i) - U(q_i), \quad \text{with} \quad U(\alpha q_i) = \alpha^p U(q_i), \quad p = -1. \quad (2)$$

The idea is to represent this empirical fact by a hierarchical system *statistically scale-invariant*. In many cases of solid state physics, systems in phase transition build also long-range correlations with universal scaling laws; interestingly, even surprisingly crude models, such as the Ising model, reproduce remarkably well the measured scaling laws (e.g. Stanley 1995). Important is therefore to include the right symmetry in the model.

Here we suppose that the system is made of hierarchical mass *clumps*, each clump being made, in average, of N sub-clumps, recursively over a number L of levels above the ground level 0 (see Pfenniger & Combes 1994, PC94). A clump is characterised by a finite mass M and a finite length scale r. The mass distribution, to be scale-invariant, scales as a fixed power

D, which defines the mass fractal dimension. Thus,

$$M_L = NM_{L-1} = N^L M_0, \qquad \text{and} \qquad M_L = M_0 \left(r_L/r_0\right)^D. \qquad (3)$$

In real physical systems a lower and upper levels must exist, which define, analogously to walls in the classical gas, the boundary conditions. The boundary conditions are convenient to abstract the complications coming from the external world. For example: 1) the lowest level of fragmentation is reached in the ISM when the heat transfer time exceeds the dynamical time (see PC94); 2) the upper level in cosmological structures is given by the time-dependence introduced at the largest scale by the universal expansion, which proceeds at a slower pace than the small-scale clustering.

The second hypothesis concerns the particle interactions. Generalising the gravitating case, the potential Φ is supposed to be a power law, i.e. $\Phi = GMr^p/p$ ($p = -1$ for gravity, and $G > 0$ for attracting interactions). Suggested by the system hierarchical organisation, we approximate the potential energy U_L at level L by

$$U_L = NU_{L-1} + \frac{G}{p}M_L^2 r_L^p \left(1 + \alpha\frac{r_{L-1}}{r_L}\right). \qquad (4)$$

The term with α represents tidal interactions. Since the scale ratio is constant, this term is constant and can be absorbed in a new coupling constant G'. We will see that the main results do not depend on the *value* of G (indeed scale invariance effects depend on the power law exponent p).

For parameters suited to the ISM ($D \approx 1 - 2$, $N = 5 - 8$, Scalo 1990) the approximation (4) has been checked to be accurate at the percent level.

The third hypothesis is of statistical character. Thermodynamics can not be used since it excludes long-range interactions. The only remaining statistical tool is the virial theorem. If a scale invariant system finds a statistical equilibrium, at each level the virial theorem must hold,

$$2T_L - pU_L = 3P_{L+1}V_L, \qquad (5)$$

where T_L is the total kinetic energy cumulated up level L, P_{L+1} is the outer pressure, and $V_L \sim r_L^3$ the clump volume. Here the outer pressure is purely kinetic: it is given by 2/3 of the kinetic energy density outside the sub-clumps. Approximately,

$$P_L = \frac{2}{3}\frac{T_L - NT_{L-1}}{V_L - NV_{L-1}}, \qquad (6)$$

which neglects the superpositions of clumps during collisions. For the sake of simplicity, in PC94 the pressure term was neglected, yet the clump "collisions" were determined to be frequent for the typical parameters of the

ISM. Clump collisions may well disrupt or merge them, but the supposed Jeans unstable medium (due e.g. to fast cooling) also fragments clouds fast, reforming the clumps before a crossing time. Therefore, in a statistical equilibrium the average N should be constant, and at any moment the fraction of colliding clumps should be small.

The above recurrences for M_L, V_L, U_L and T_L in Eqs. [3-6] can be solved exactly in finite terms. With $x \equiv r_L/r_{L-1} = N^{1/D} > 1$, we find,

$$\frac{M_L}{M_0} = x^{DL}, \qquad \frac{V_L}{V_0} = x^{3L}, \qquad \frac{U_L/M_L}{U_0/M_0} = \frac{x^{(D+p)(L+1)}-1}{x^{(D+p)}-1},$$

$$\frac{T_L/V_L}{T_0/V_0} = 1 - \beta\left[\frac{1}{1-x^{2D+p-3}} + \frac{x^{L(D-3)}}{x^{D+p}-1}\left(1 - x^{(L+1)(D+p)}\frac{1-x^{D-3}}{1-x^{2D+p-3}}\right)\right] \qquad (7)$$

where $\beta \equiv pU_o/2T_0$ is the ground virial ratio ($0 \leq \beta \leq 1$). The free parameters are D, N, p, and β. Although not very illuminating at first sight, the functional properties of the solution (7) are very interesting. We just summarise here the most important features. More detail will be presented elsewhere (Pfenniger 1996).

First, not any combinations of parameters lead to a physical solution. Not only the kinetic energy T_L must positive, but also the pressure, which is proportional to the velocity dispersion v_L squared, $\frac{1}{2}M_L v_L^2 = T_L - NT_{L-1}$.

In the "thermodynamical limit" $L \to \infty$ a striking phenomenon occurs. The range of physical solutions shrinks on a subspace of the parameter space, leading to a new constraint,

$$D = \frac{3-p}{2 - \ln(1-\beta)/\ln N}. \qquad (8)$$

So, for $p = -1$, $D < 2$ in any case. For systems not too confined by the outer pressure ($1/2 < \beta < 1$) and $N > 5$, we have $D < 1.7$, while systems with more outer pressure ("pressure confined clouds") increase D up to 2. *Therefore, we conclude that hierarchical gravitating systems in statistical equilibrium can indeed exist, but with $D < 2$.*

The scale-velocity dispersion relation takes the exact scale-free form $v \propto r^\kappa$, where $\kappa \equiv (D+p)/2$. The result for $p = -1$ in PC94 remains therefore unchanged in spite of the inclusion of the outer pressure. For fractal ISM cold clouds with substantial ambient pressure (Blitz 1993), $\beta \approx 1/2$, and with $N = 5-10$ (Scalo 1990), we expect $D \approx 1.7$ and $\kappa \approx 0.35$, which is comparable to Larson's (1981) size-linewidth relationship.

Finally, the short to long-range interaction transition occurs at $p = -3$. For $p < -3$ the typical feasible state is non-longer hierarchical, but with a single level, large outer pressure ($\beta \ll 1/2$), large N and a dimension D close to 3: the classical homogeneous gas is recovered.

4. Final Remarks: Universality of the Fractal Structure

With a simple model we have motivated here the proposition that strongly gravitating systems in statistical average tend to lower their dimensionality below 2, so to adopt a very inhomogeneous structure, as manifest in many cases. The property $D \lesssim 2$ is well documented for the cold ISM (Scalo 1990), and for the large scale structures (Coleman & Pietronero 1992). Actually, the general transparency, or blackness, of the Universe at most of the wavelengths, extending the De Chéseaux-Olbers paradox to non-stellar objects, reflects for a good part the widespread $D < 2$ fractality.

A deceptive effect occurs when observing fractals with $D < 2$, such as cold ISM clouds: their orthogonal projections have the same $D < 2$ (Falconer 1990), so less than a surface. Practically it means that over the more scales the gas does indeed behave as a fractal, the smaller is the sky fraction over which 50% of the mass projects. The bias is then obvious: when sampling the sky, most of it appears at low column density so mass *looks* well sampled. In fact the mass is poorly sampled because most of it resides in small regions of the sky at high column density, therefore likely to be optically thick. Not only increasingly higher angular resolution is required to resolve the smallest structures, but also more wavelength types for piercing all the decades of column densities. In addition, the velocity dispersion (or temperature) decreases at smaller scales, so the smallest structures are both the coldest and the densest ones, so often hard to detect.

If the fractal state in fast cooling cosmic gas is indeed universal, we have little reason not to expect that the outer HI galactic disks are just the warmer "atmosphere" of a fractal which extends down to very small scale. The lowest temperature in the Universe being 2.73 K, it is then natural to assume that much more gas mass can be hidden in very cold molecular gas, making a sizable fraction of the galactic dark matter (PC94).

References

Binney J., Tremaine S. 1987, *Galactic Dynamics*, Princeton Univ. Press
Blitz L. 1994, in *The Cold Universe*, XIIIth Astrophysics Meetings, Montmerle Th., et al. (eds.), Editions Frontières, Gif-sur-Yvette, 99
Coleman P.H., Pietronero L. 1992, *Physics Reports* **213**, No 6
Falconer K. 1990, *Fractal Geometry*, John Wiley & Sons, Chichester
Jaynes E.T. 1957, *Phys. Rev.* **106**, 620
Larson R.B. 1981, *MNRAS* **194**, 809
Lynden-Bell D., Lynden-Bell R.M. 1977, *MNRAS* **181**, 405
Pfenniger D. 1996, in preparation
Pfenniger D., Combes F., 1994, *Astron. Astrophys.* **285**, 94 (PC94)
Prigogine I., 1962, *Non-Equilibrium Statistical Mechanics*, John Wiley, New York
Scalo J., 1990, in *Physical Processes in Fragmentation and Star Formation*, R. Capuzzo-Dolcetta et al. (eds.), Kluwer, Dordrecht, 151
Stanley H.E. 1995, Nature **378**, 554
Verschuur G.L. 1993, *Astron. J.* **106**, 2580
von Weizsäcker C.F., 1951, *Astrophys. J.* **114**, 165

Figure 1. Classically distributed "clouds" in a truncated isothermal sphere $\rho \sim r^{-2}$, $r < 1$. The mass distribution follows a $dN/dM \sim M^{-1.5}$ law over 2 dex. In the top frame, the cloud radius is determined to be space filling in average, and in the bottom frame, it is ten times smaller. There are about 42 000 clouds, each of them is made of an opaque core and a diffuse envelope. Two distant spot-lights illuminate the scene perpendicularly to the checker planes. The shadows and diffuse light are calculated by the ray-tracing technique.

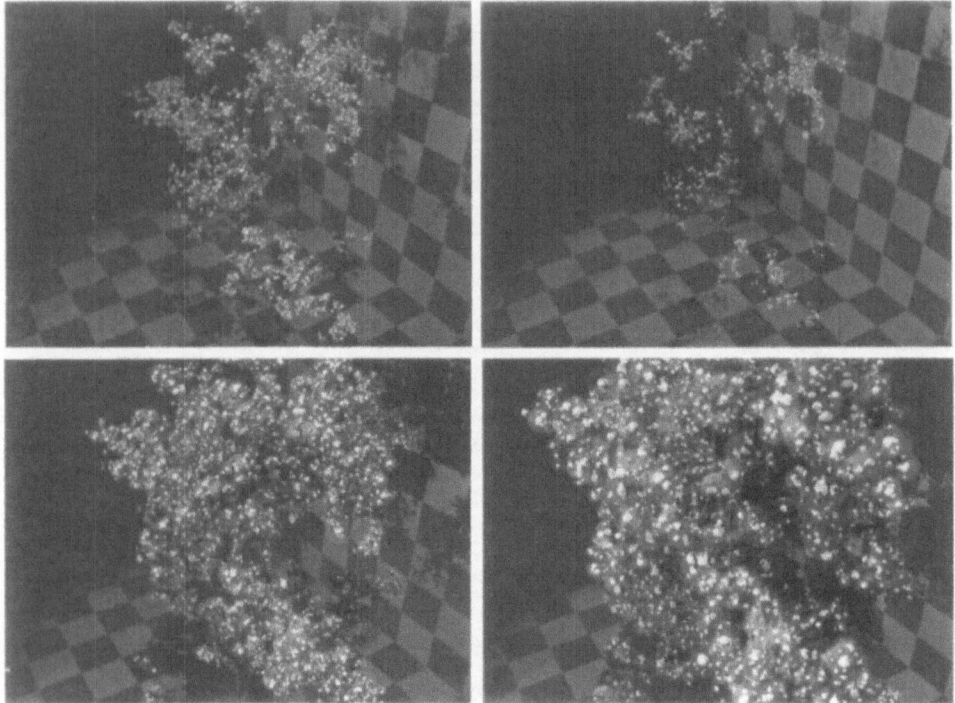

Figure 2. Fractal cloud models of dimension $D = 1.7$, 2.0, 2.5 and 3.0, from top left clockwise. At each level of the fractal the number N of sublevels is calculated by a Poisson distribution of mean 8. The length ratio x between each adjacent level is given by $x = N^{1/D}$. The position of each sub-clump is distributed according to a truncated isothermal law. Five levels are calculated; the total number of clouds at the smallest scale is about 42 000. The mass distribution is calculated as for Fig.1, and the cloud radius is fixed by the fractal scaling. Due the spatial correlations of the clouds, absent in the classical clouds of Fig. 1, it is apparent that the projection properties of the lower dimension fractals are much different from the higher dimensional fractals or the classical distributions in Fig. 1. While below dimension 3 the space filling decreases, below dimension 2 the sky coverage decreases rapidly as the number of levels increases. Then, most of the projected mass is in high column density regions which covers a small fraction of the sky. See also the Color Plate Section for a color rendition of Figure 2.

THE DM–HI CONNECTION

CLAUDE CARIGNAN
Département de physique and
Observatoire astronomique du Mont Mégantic,
Université de Montréal, C.P. 6128, Succ "centre–ville",
Montréal, Québec, Canada. H3C 3J7

1. INTRODUCTION

Since the development of radio aperture synthesis techniques, the HI gas has proved ideal to probe the gravitational potential in the outer parts of spiral and dwarf irregular galaxies. The great advantage of the 21 cm observations over optical techniques, such as long–slit spectroscopy and Fabry–Pérot interferometry, is that the HI gas extends to much larger galactocentric distances; in some cases, as much as 5 times the optical diameter (Bothun *et al.* 1987, Carignan & Freeman 1988, Hoffman *et al.* 1993). While most optical rotation curves can well be explained only by the luminous matter component, one cannot escape the necessity to postulate the existence of an unseen dark halo component to explain the fact that the HI rotation curves stay flat even at several optical radii.

Careful modeling of the mass distribution (Carignan & Freeman 1985, van Albada & Sancisi 1986, Lake & Feinswog 1989, Carignan & Beaulieu 1989) has established, without any doubt, that the DM distribution, within spiral and dwarf irregular galaxies (as determined from their HI kinematics) is *not at all* correlated with the luminous stellar disk distribution. Hence the question addressed by this paper . *Is there any relation between the way DM and the HI gas are distributed ?*

The possible existence of such a relation was first pointed out by Bosma (1981). Using a sample of 16 galaxies, he looked at the ratio between the total mass and the HI gas surface densities (Σ). For galaxies having rotation curves derived to sufficiently large radii ($r_{max} \geq 20$ kpc), this is essentially equivalent to looking at the DM–to–HI surface density ratio. In that study it was shown that, outside the disk dominated region ($r > 4$–$5\ \alpha^{-1}$; where

447

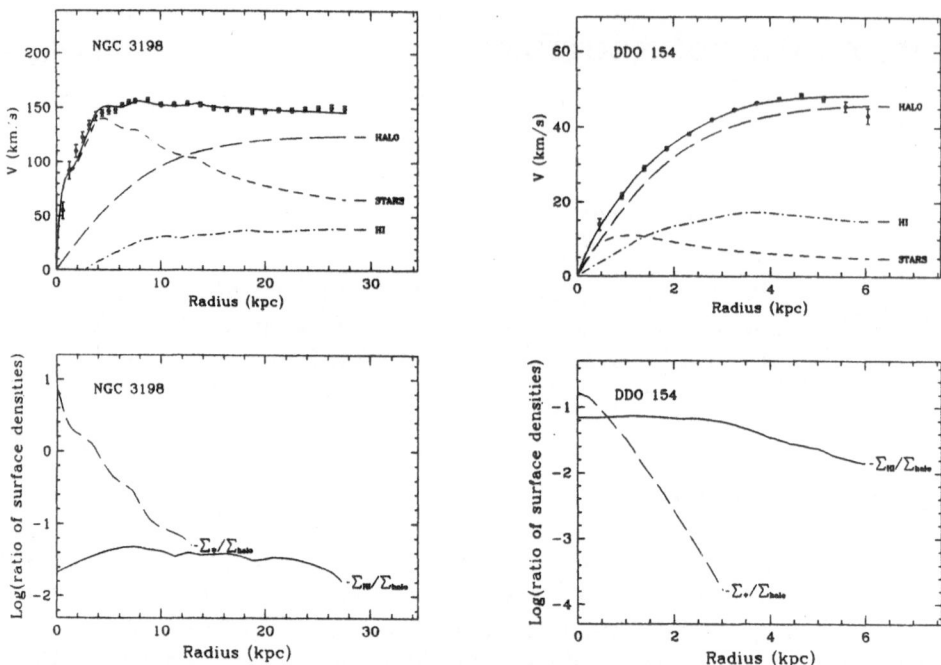

Figure 1. (top) Mass models for NGC 3198 and DDO 154. (bottom) Ratios of the integrated surface densities of the stellar and HI disks to the integrated surface densities of their dark halos.

α^{-1} is the optical scale length), $\Sigma_{total}/\Sigma_{HI}$ is very nearly constant. This was interpreted by Sancisi (1983) as suggesting that "... in the outer parts of galaxies, the HI and the dark mass may decrease similarly and may be coextensive."

However, in Bosma's study, the mass distribution was analysed as disk-only systems. Because of this, definite conclusions could not really be reached and the analysis needed to be redone with more realistic models. This was done in many studies since then (Carignan & Beaulieu 1989, Carignan *et al.* 1990, Carignan & Puche 1990, Jobin & Carignan 1990, Puche *et al.* 1990) where the relation between the dark matter component and the HI disk was looked at as a function of radius.

2. MASS MODELS

To illustrate this relation, the data on the well-studied spiral NGC 3198 (Begeman 1989) and on the dwarf irregular DDO 154 (Carignan & Beaulieu

1989) are used. These galaxies have probably the best data sets available in the literature. The HI gas extends to $\sim 11\,\alpha^{-1}$ in one case and to $\sim 5\,R_{HO}$ (Holmberg radius: isophotal radius at a level of $\mu_B = 26.6$ mag arcsec^{-2}) in the other, and the velocity fields are very regular.

In order to derive the ratio of the HI–to–DM surface densities, the mass distribution was analysed in the usual way. The stellar disk contribution was obtained by a straight inversion of the luminosity profile assuming a constant mass–to–light ratio. The HI disk mass distribution was obtained directly from the 21 cm observations (the HI surface densities are multiplied by 4/3 to correct for He). For the dark halo, an isothermal sphere potential (Carignan 1985) was used. This component has two free parameters: a radial scaling (r_c = core radius) and a velocity scaling (σ = one dimentional velocity dispersion). The central density is then given by $\rho_0 = 9\sigma^2/4\pi G r_c^2$.

Using the result from the best–fit models, Fig. 1 show the ratio of the two luminous components (stars & gas) surface densities to the integrated surface densities of the dark isothermal halo ($\Sigma_*/\Sigma_{\mathrm{halo}}$ and $\Sigma_{\mathrm{HI}}/\Sigma_{\mathrm{halo}}$). It can be seen that the DM distribution has no relation at all with the stellar disk distribution while, on the other hand, the ratio $\Sigma_{\mathrm{HI}}/\Sigma_{\mathrm{halo}}$ is fairly flat with $\langle \log(\Sigma_{\mathrm{HI}}/\Sigma_{\mathrm{halo}}) \rangle \simeq -1.5$. So, not only does the earlier finding of Bosma that *DM and HI are distributed similarly* in the outer parts is confirmed but, using more realistic models, it appears that this may also be true *on almost the whole radius range*.

3. THE DM–HI CONNECTION

Figure 2 shows DM–to–HI ratios for a sample of 20 galaxies covering $-19.88 \le M_B \le -12.71$. So, despite a range of nearly 10^3 in luminosity, this ratio varies by less than a factor of 8 ($11 \le \frac{\Sigma_{\mathrm{halo}}}{\Sigma_{HI}} \le 87$). It can be seen that it is nearly constant at $\langle \frac{\Sigma_{\mathrm{halo}}}{\Sigma_{HI}} \rangle \simeq 35 \pm 20$, except maybe for the very low luminosity systems where it seems to increase in the outer parts. It thus appears that the HI gas could be used not only to probe the gravitational potential in the outer parts of galaxies but also to trace the exact distribution of the DM component. However, one should not conclude that this relation implies that dark matter is distributed in a thin disk similar to the HI gas. As shown in Carignan & Beaulieu 1989 for the case of DDO 154, and the same is true for most galaxies in Fig. 2, just scaling up the HI distribution fails to represent the velocities in the outer parts.

Is such a relation only a coincidence ? Is it only showing us that it is the natural fate of the HI gas to settle into the dominant potential well of the dark halo ? Or, is it pointing toward possible candidates for the dark matter, maybe in the form of cold gas, as suggested recently by Pfenniger *et al.* 1994 ?

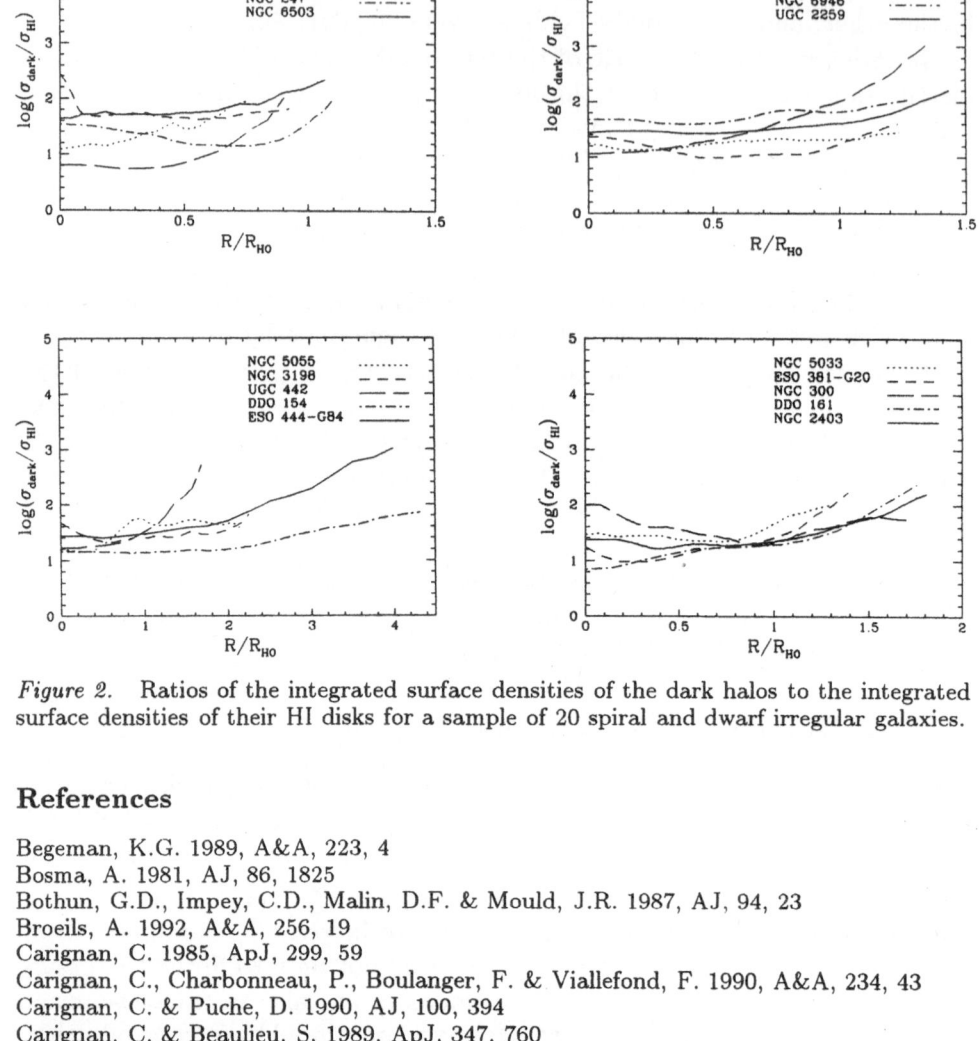

Figure 2. Ratios of the integrated surface densities of the dark halos to the integrated surface densities of their HI disks for a sample of 20 spiral and dwarf irregular galaxies.

References

Begeman, K.G. 1989, A&A, 223, 4

Bosma, A. 1981, AJ, 86, 1825

Bothun, G.D., Impey, C.D., Malin, D.F. & Mould, J.R. 1987, AJ, 94, 23

Broeils, A. 1992, A&A, 256, 19

Carignan, C. 1985, ApJ, 299, 59

Carignan, C., Charbonneau, P., Boulanger, F. & Viallefond, F. 1990, A&A, 234, 43

Carignan, C. & Puche, D. 1990, AJ, 100, 394

Carignan, C. & Beaulieu, S. 1989, ApJ, 347, 760

Carignan, C. & Freeman, K. C. 1985, ApJ, 294, 494

Carignan, C. & Freeman, K.C. 1988, ApJ, 332, L33

Jobin, M. & Carignan, C. 1990, AJ, 100, 648

Hoffman, G.L., Lu, N.Y., Salpeter, E.E., Farhat, B., Lamphier, C. & Roos, T. 1993, AJ, 106, 39

Lake, G. & Feinswog, L. 1989, AJ, 98, 166

Pfenniger, D., Combes, F. & Martinet, L. 1994, A&A, 285, 79

Puche, D., Carignan, C. & Bosma, A. 1990, AJ, 100, 1468

Sancisi, R. 1983, in Internal Kinematics and Dynamics of Galaxies: IAU Symposium 100, ed. E. Athanassoula (Dordrecht: Reidel)

van Albada, T.S., and Sancisi, R. 1986, Phil. Trans. R. Soc. Lond., 320, 15

VERY COLD GAS AND DARK MATTER

F. COMBES
DEMIRM, Observatoire de Paris
61 Av. de l'Observatoire, F-75014 Paris, France

AND

D. PFENNIGER
Geneva Observatory, University of Geneva
CH-1290 Sauverny, Switzerland

Abstract. We have recently proposed a new candidate for baryonic dark matter: very cold molecular gas, in near-isothermal equilibrium with the cosmic background radiation at 2.73 K. The cold gas, of quasi-primordial abundances, is condensed in a fractal structure, resembling the hierarchical structure of the detected interstellar medium.

We present some perspectives of detecting this very cold gas, either directly or indirectly. The H_2 molecule has an "ultrafine" structure, due to the interaction between the rotation-induced magnetic moment and the nuclear spins. But the lines fall in the km domain, and are very weak. The best opportunity might be the UV absorption of H_2 in front of quasars. The unexpected cold dust component, revealed by the COBE/FIRAS submillimetric results, could also be due to this very cold H_2 gas, through collision-induced radiation, or solid H_2 grains or snowflakes. The γ-ray distribution, much more radially extended than the supernovæ at the origin of cosmic rays acceleration, also points towards and extended gas distribution.

1. Introduction

The possibility that most of the mass of the Universe could be under the form of gas around or in between galaxies has been widely discussed in the 1960's (e.g. the review by Peebles, 1971). At that time gas was assumed to be distributed in a smooth homogeneous fashion. This was not justified by observations of the interstellar gas already then. The intergalactic material

451

D. L. Block and J. M. Greenberg (eds.), New Extragalactic Perspectives in the New South Africa, 451-466.
© *1996 Kluwer Academic Publishers.*

was supposed to be hydrogen exclusively (with some helium), either atomic, molecular, ionized, or even in a condensed snow. Self-gravity was ignored. Several tests were proposed, such as emission or absorption of HI at the 21 cm line, the Gunn-Peterson test in the Lyα line of HI or the Lyman band in molecular hydrogen; the existence of an intergalactic plasma would have been detected through free-free emission or absorption, recombination lines, chromatic phase lag, or Thomson scattering.

All the tests were negative, constraining the density of any intergalactic gas several orders of magnitude below the closure density. However, the homogeneity hypothesis is drastic, especially when comparing to the now much better known ISM, and including gravity as a major force. The Gunn-Peterson test is of course invalidated in case of a clumpy medium.

Even until recently, the hypothesis of cold gas as dark matter was eliminated quickly, without critical reflection, through stability arguments (e.g. Hegyi & Olive 1986): for an homogeneous galactic gaseous halo, the virial temperature is of the order of 10^6 K; the gas cannot remain hot, it would have been seen in X-ray; in any case, at this temperature, cooling processes are violent, the gas collapses and forms stars. Hydrogen snowballs, massive enough not to collide with each other ($m > 1$ g) quickly evaporate, unless gravitationally bound (but then they join the problem of brown dwarfs).

All the above objections disappear when a realistic hierarchical structure of cold and dense clouds is considered, closely resembling the familiar fractal structure of the detected ISM (Pfenniger et al. 1994). A model was then built to account for the baryonic dark matter around galaxies, composed of basic "clumpuscules" of molecular hydrogen, of a Jupiter mass (10^{-3} M$_\odot$), but with a much smaller density than brown dwarfs, with 10^{9-10} cm^{-3} and radius of ≈ 20 AU (Pfenniger & Combes 1994). For these basic units clumpuscule are self-gravitating, statistically in equilibrium with pressure forces, at the limit of the adiabatic regime, since then the radiation transfer time equals the dynamical time. They compose the smallest scale of a hierarchical structure, that ranges over six orders of magnitude, up to giant molecular clouds of 100 pc; above this scale, bigger gaseous complexes are torn apart by galactic shear. The ensemble of clumpuscules is in quasi isothermal equilibrium with the bath of photons of the cosmic background radiation at $T = 2.73$ K$(1 + z)$. Due to the fractal structure (of dimension $D < 2$), the clumpuscules collide together frequently, with, for such fractal dimension, a rate of the order of the cooling-heating and dynamical times.

But all the fractal structure is far from local thermodynamical equilibrium as explained in another paper in this volume, the equilibrium including gravity is only statistical. The large fluctuations prevent most of the clumpuscules to cool down and collapse further (were they distributed homogeneously they would not). Through these collision induced fluctuations,

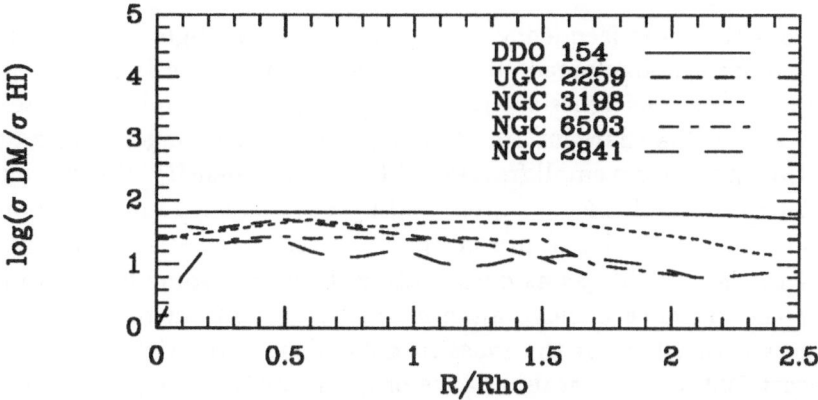

Figure 1. Ratio of HI to Dark Matter (DM) surface densities in spiral galaxies, adapted from Bosma (1981), Puche et al (1990) and Freeman (1992).

the gas maintains exchanges with the background, coalescing in giving back energy, or fragmenting and evaporating in absorbing energy. The condensed structure resembles closely the well-studied ISM gas, already known as a $D < 2$ fractal-like structure over several orders of magnitude in scale (Larson 1981; Scalo 1985; Falgarone et al. 1992). The main difference is that star-formation in the visible disk has metal-enriched the medium that can then cool down much faster, and the heating sources have partly destroyed the condensed fractal and formed a diffuse intercloud medium.

In galaxy outskirts, the condensed H_2 phase is almost only in contact with the intergalactic radiation field, which photodissociates a small fraction of it into HI gas, because at the envisaged column densities ($\sim 10^{25}\,\mathrm{cm}^{-2}$) H_2 can self-shield easily. An even smaller fraction could be ionized. Since the average surface density of the gas is falling as R^{-1}, HI must disappear into ionised gas at a critical radius; this corresponds to the sudden fall off of HI measurements, resembling an ionisation front (e.g. the case of NGC 3198, van Gorkom et al. 1987, unpublished). From this point of view, the atomic gas in the outer parts serves as a tracer of dark matter.

There is indeed some evidence that the gas and dark matter are intimately related. From the flat rotation curves, the surface density of dark matter σ_{DM} varies asymptotically as R^{-1}, and as well the HI surface density σ_{HI} (cf. Fig. 1). Bosma (1981) was the first to notice a constant ratio of σ_{DM}/σ_{HI} as a function of radius in spirals, which has been confirmed by many authors (Sancisi & van Albada 1987; Puche et al. 1990), and varies between 10 and 20 according to the morphological type (Broeils 1992; Carignan, this volume).

The presence of large gaseous extensions around galaxies can explain the widespread detection of absorption lines in front of quasars. The $Ly\alpha$

forest, and the large frequency of absorptions on a single line of sight (up to 100) remain unexplained. If these absorptions come from gas around galaxies, the derived cross-section of a galaxy corresponds to a radius of 480 kpc at $z = 2.5$ (Sargent 1988). The gas corresponding to these atomic absorptions remains a small fraction of the critical density. But already the total contribution of the gas in damped Lyα systems amounts to about the luminous matter density in present galaxies (e.g. Lanzetta et al. 1991).

The model of cold gas as dark matter sheds also some light on puzzles such as the presence of huge amounts of hot gas in clusters (the hot gas represents 8 to 10 times the mass in galaxies in some rich clusters, Edge & Stewart 1991), the overabundance of gas in interacting galaxies (Braine & Combes 1993), the necessity of gas infall for maintaining star-formation and the spiral structure, and the evolution of galaxies along the Hubble sequence (e.g. Pfenniger et al. 1994).

Recent models have also been proposed, taking up the hypothesis that H_2 gas could contribute significantly to the dark matter, based mainly in massive proto globular cluster clouds possibly mixed with brown dwarfs (de Paolis et al. 1995; Gerhard & Silk 1995). In this case, the cold H_2 is not *very* cold, but has a temperature between 5 and 20 K, which makes it easier to detect in emission. To us is unclear, however, how cold gas coexisting with a brown dwarf cluster could be maintained in the mutual gravity at this low temperature, since the brown dwarfs must be subject to a significant dynamical friction, and the cluster must undergo core collapses that inevitably relax and virialize also the gas to a higher temperature.

2. Local dark matter and gravitational stability

2.1. FIT OF THE ROTATION CURVE

The rotation curve of our own Galaxy is well-known from several tracers, including the HI, H_2 gas or ionised gas, and is typical for a SBbc galaxy: massive bulge traced by a central peak of rotational velocity, flat rotation in the outskirts, although uncertainties become large outside of the solar circle, due to the lack of precise distance indicators (e.g. Fich & Tremaine 1991). Given all the observed parameters of the bulge, stellar and gaseous disks, is it possible to constrain the model of dark matter confined in the disk, in particular in solar neighbourhood?

We have retained a simple axisymmetric model of the Galaxy, composed with a bulge (Plummer component), an exponential stellar disk, with or without truncation, and two gaseous disks. The first represents the observable molecular ring, and is modelled by a difference between two Toomre disks of index $n = 2$ (Toomre 1963). The second represents the observed HI component, and is modelled by an empirical ring like distribution, de-

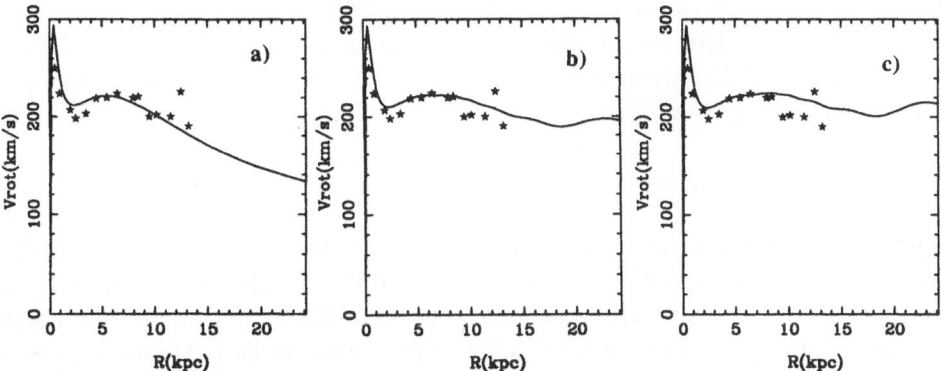

Figure 2. Fit of the Milky Way rotation curve, with the parameters of Table 1. Data points are from the compilation of Fich & Tremaine (1991). The dark matter is assumed to follow the HI distribution: a) without dark matter; b) with dark matter, with a distribution following that of HI, and $\sigma_{\rm DM}/\sigma_{\rm HI} = 15$; c) same with $\sigma_{\rm DM}/\sigma_{\rm HI}=20$

ficient in the center, and falling as R^{-1} at large radii. The mass of this component has been multiplied by an adjustable ratio $\sigma_{\rm DM}/\sigma_{\rm HI}$ to model the gaseous dark matter. The disks are modelled with a finite thickness, and the rotation curves are calculated following the integral formulation from Casertano (1983). The final fit is displayed in Fig. 2, together with the observed points. All the adopted parameters are listed in Table 1.

First, we see that it is possible to account for the rotation curve without any dark matter until $R = 10$ kpc. Afterwards dark matter is actually necessary to avoid a fall off of the circular speed (Fig. 2a). But adding the dark matter in the disk, with the bulk of it outside the solar radius reduces the rotation speed inside $R = 10$ kpc. The modelisation is therefore not unique, it depends on several free parameters of total mass and concentrations, much more than in the hypothesis of a spherical DM halo. For instance the rotational velocity at the extreme galactic radii depends on how the HI-DM disk is truncated. Fig. 2b and c display reasonable good fits, with $\sigma_{\rm DM}/\sigma_{\rm HI}=15$ and 20 respectively. We can estimate the dark matter frac-

TABLE 1. Parameters of the rotation curve fit

Component	Mass [$10^{10}\,M_\odot$]	Scale-lengths [kpc]		Type
Bulge	$M_b = 1$	$a_b = 0.25$		Plummer
Disk	$M_d = 6$	$a_d = 2.7$		Exponential
H$_2$	$M_{\rm H_2} = 0.2$	$R_1 = 6.0$	$R_2 = 6.8$	diff $n = 2$ Toomre disks
HI	$M_{\rm HI} = 0.4$	$R_0 = 8.$ [1]		$\mu \propto R^{-1}$

[1] Start of the R^{-1} behaviour

tion in the disk at the solar radius at about 50% of the total mass. This is compatible with the recent claim from Bahcall et al (1992), but not from the Kuijken & Gilmore studies (1989), but the question of the existence of local dark matter is still open.

2.2. HI SCALE-HEIGHT

The gaseous plane of the Milky Way is flaring linearly with radius, outside the solar circle, i.e. the HI scale-height $h_z \approx 0.045R$ (Merrifield 1992). If the gas is considered in gravitational equilibrium in the direction perpendicular to the plane, with a constant velocity dispersion $\sigma \approx 10$ km/s as in face-on galaxies (e.g. Dickey et al 1990), then different scale-heights are predicted by the different models for the dark matter. In the case of a spherical halo, the equilibrium requires $GM(R)R^{-3}h_z^2 \simeq \sigma^2$, and since $M(R) \propto R$, we predict $h_z = 0.09R$, which is larger than observed. A flattened dark halo is needed. In the case of a self-gravitating disk, where the dark matter follows the HI flaring, the equilibrium requires $2\pi G\mu h_z \simeq \sigma^2$, with the surface density $\mu \propto R^{-1}$; a height $h_z = 0.03R$ is predicted with the rotation curve model of the previous section. This is quite compatible with the observations, given the uncertainty in σ, and the probable overestimation of the observed height by the Milky Way warp.

2.3. OUTER DISK STABILITY

For Gerhard & Silk (1995), the main problem raised by the existence of a self-gravitating gaseous disk in the outskirts is its global stability. For a flat rotation curve, the surface density Σ falls off as R^{-1} as well as the epicyclic frequency κ, so that the critical velocity for axisymmetric stability $c_r \propto \Sigma/\kappa$ is constant. This critical velocity dispersion is of order $60 - 70$ km s^{-1} in the Galaxy, while the observed velocity dispersion *perpendicular* to the plane is of the order of $7 - 10$ km s^{-1} (e.g. Dickey et al. 1990).

We have already discussed how this can be approached (e.g. Pfenniger et al. 1994). First, gaseous disks are not razor thin: they flare with radius (Merrifield 1992), which reduces self-gravity and the critical velocity. Second, real gaseous disks *are manifestly unstable*, as witnessed by the spiral waves, asymmetries and large scale inhomogeneities of outer HI disks in *every* spiral. Third, the dispersion, averaged over the gravitational scales up to ~ 1 kpc and over a few rotation periods, is indeed close to the critical velocity. Finally, the clumpy gaseous dark matter component may still retain, as stars, a substantial dispersion anisotropy, contrary to a classical smooth gas; so the velocity dispersion, well measured perpendicular to the plane, is not necessarily a good indicator of the effective horizontal dispersion (see also Elmegreen, this volume).

3. Perspectives of detection

The above model is not only a plausible and conservative hypothesis on the nature of dark matter, but it is a falsifiable one, since a series of observations can be carried out to confirm or refute our propositions. We review below possible observational tests and try to select the most promising ones.

3.1. THE HYPERFINE STRUCTURE OF ORTHO-H_2

The hydrogen molecule can be found in two species, para-H_2, in which the nuclear spins of the two protons are anti-aligned, and the resulting spin $I = 0$, and ortho-H_2 for which the total nuclear spin is $I = 1$, with the spins of the two protons parallel. The rotation quantum number J is even for para-H_2 and odd for ortho-H_2. In the para ground state $J = 0$, there is no hyperfine splitting, but for the ortho $J = 1$, three levels can be identified, corresponding to $F = 0, 1$ and 2. This splitting comes from the interaction of the nuclear spin magnetic dipole, with the magnetic field created from the motion of charges due to rotation. To this interaction, must be added the spin-spin magnetic interaction for the two nuclei, and the interaction of any nuclear electrical quadrupole moment with the variation of the molecular electric field in the vicinity of the nucleus (Kellog et al. 1939, 1940; Ramsey 1952). Magnetic dipole transitions are possible for $\Delta F = 1$, i.e. there are two transitions, $F = 2 - 1$, and $F = 1 - 0$. The wavelength of these two transitions have been measured in the laboratory at 0.55 and 5.5 km, or more precisely at frequencies of 546.390 kHz and 54.850 KHz respectively for $F = 1 - 0$ and $F = 2 - 1$.

In fact this structure could be called ultrafine structure, since it is several orders of magnitude below hyperfine structure (cf. Field et al. 1966). Since the interaction involves two nuclear momenta, the splitting is proportional to μ_n^2, the nuclear magneton, while the hyperfine structure involves the product of μ_n, with the Bohr magneton $\mu_o = eh/(4\pi mc)$, where m is the electron mass. The ultrafine to hyperfine structure ratio is therefore $\mu_n/\mu_o = m/m_p$, where m_p is the proton mass.

3.2. THE ORTHO-PARA RATIO

Only the ortho-H_2 is concerned by the ultrafine structure. Normal molecular hydrogen gas contains a mixture of the two varieties, with an ortho-to-para ratio of 3, when the temperature is high with respect to the energy difference of the two fundamental states (171 K). At lower temperatures, the ortho-to-para ratio must be lower, if the thermodynamical equilibrium can be reached, until all the hydrogen is in para state at $T = 0$. However, due to the rarefied density of the ISM, the ortho-to-para ratio is frozen to the H_2-

formation value. Considerable densities are required for the ortho-to-para conversion, which occurs in solid H_2 for instance.

But the fractal gas must be seen as a dynamical structure far out of thermodynamical equilibrium, not only with large density contrasts, but also large temperature contrasts. The HI is then the warm interface, and H_2 the coldest component, in a mass ratio of about 1 to 10. By continuity, H_2 forms from HI and vice versa with a rate given by the clump collision time-scale at the scale corresponding to the virial temperature of the transition. This is in any case relatively short: for 3000 K, $D \sim 1.7$, we estimate a duty-cycle of transformation HI to H_2 of the order of 10^{6-7} yr.

Now the key role in ortho to para conversion in interstellar clouds is the proton exchange reaction $(H^+ + H_2(j = 1) \rightarrow H^+ + H_2(j = 0)$, cf. Dalgarno et al. 1973; Gerlich 1990). This reaction can transform the ortho in a time-scale $5 \cdot 10^{13} n(H_2)^{-1/2}$ s, if the H^+ ions in dense clouds are essentially due to cosmic ray impacts, with the ionising flux $\xi = 10^{-17}$ s^{-1} characteristic of the solar neighbourhood. The corresponding time-scale for a clumpuscule near the sun is 10 yr, and the ortho fraction is negligible, but at large distances in the Galaxy outskirts, where the cosmic-ray flux falls to zero, we can expect a significant part of ortho-H_2 in the cold gas.

3.3. DETECTABILITY OF THE H_2 ULTRAFINE LINES

On Earth, the ionosphere is reflecting the long radio wavelengths, which is useful for long distance communications. The ionospheric plasma is filtering all frequencies below the plasma frequency $\omega = e(4\pi n/m)^{1/2} \approx 100$ MHz. It is therefore necessary to observe from space. Even from space, the long wavelength radiations are somewhat hindered by interplanetary or interstellar scintillations (e.g. Cordes et al. 1986).

3.3.1. *Interstellar plasma*

In the ISM the plasma frequency can be estimated by $\nu_p = 9n_e^{1/2}$ kHz, where n_e is the electron density. Since the latter is in average of the order of 10^{-3} cm^{-3}, the plasma frequency $\nu_p \approx 250$ Hz. Radiation of frequencies below that value does not propagate in the medium. More exactly, since the ISM is far from homogeneous, low-frequency radiation propagates in rarefied regions, and is reflected and absorbed by denser condensations. For kilometric wavelengths, there is no problem of propagation, but the waves are scattered due to fluctuations in electron density. The electric vector undergoes phase fluctuations, since the index of refraction is $(1 - \nu_p^2/\nu^2)^{1/2}$, where ν is the radiation frequency. If the ISM is modeled by a gaussian spatial distribution of turbulent clumps of size a, the scattering angle can be expressed by $\theta_{scat} \approx 10^8 (L/a)^{1/2} \langle \Delta n_e^2 \rangle^{1/2}/\nu^2$ radian, where L is the

total path crossed by the radiation (e.g. Lang 1980). At a typical distance
of $L = 3\,\mathrm{kpc}$, and for the frequencies considered ($\approx 200\,\mathrm{kHz}$), θ_{scat} is of the
order of $1°$. This means that higher angular resolution should be inaccessible
below $0.1\,\mathrm{MHz}$. The scintillation problem is therefore severe, and hinders
the resolution of point sources, but still the galactic disk can be mapped.

The interplanetary medium produces somewhat less scattering, and the
total order of magnitude remains unchanged.

3.3.2. *Intensity of the H_2 ultrafine lines*

The radiation has a dipole matrix element proportional to μ_n^2; the line intensity is therefore much weaker than for usual hyperfine transitions (magnetic
dipole in μ_o^2). Since the spontaneous emission coefficient A is proportional
to ν^3, the life-time of a hydrogen molecule in the upper ultrafine states is
much larger than a Hubble time: $A \approx 10^{-32}\,\mathrm{sec}^{-1}$. It is then likely that
the desexcitation is mostly collisional. Even at the $3\,\mathrm{K}$ temperature, the
upper levels are populated in the statistical weights ratio. A weak radiation is therefore expected, but the velocity-integrated emission ($\int T_a\,dv$)
is ten orders of magnitude less than for the HI line, for the same column
density of hydrogen. The prospects to detect the lines are scarce in the
near future, since it would need an instrument of about 6 orders of magnitude increase in surface with respect to nowadays ground-base telescopes!
A solution could be to dispose a grid of cables spaced by $\lambda/4 \approx 125\,\mathrm{m}$ on
a significant surface of the Moon, e.g. an area of $(300\,\mathrm{km})^2$. This requirement could be released, however, if there exists strong coherent continuum
sources at km wavelengths. The H_2 ultrafine line could then be detected
much more easily in absorption, with presently planned km instruments.

3.4. THE HD AND LIH TRANSITIONS AND DETECTABILITY

HD has a weak electric dipole moment; it has been measured in the ground
vibrational state from the intensity of the pure rotational spectrum to be
$5.85 \pm 0.17 \cdot 10^{-4}$ Debye (Trefler & Gush 1968). The first rotational level is
at $\approx 130\,\mathrm{K}$ above the ground level, the corresponding wavelength is $112\,\mu$.
This line could be only observed in emission from heated regions, and given
the very low abundance ratio $HD/H_2 \approx 10^{-5}$ and weak dipole, does not
appear as a good tracer of the cold gas.

The LiH molecule has a much larger dipole moment, $\mu = 5.9$ Debye
(Lawrence et al. 1963), and the first rotational level is only at $\approx 21\,\mathrm{K}$
above the ground level. The corresponding wavelength is $0.67\,\mathrm{mm}$ (Pearson
& Gordy 1969; Rothstein 1969). The line frequencies in the submillimeter
and far-infrared domain have been recently determined with high precision
in the laboratory (Plummer et al. 1984; Bellini et al. 1994), and the great

astrophysical interest of the LiH molecule has been emphasized (e.g. Puy et al. 1993). A tentative has even been carried out to detect LiH at very high redshifts (de Bernardis et al. 1993). This line is unfortunately not accessible from the ground at $z = 0$ due to H_2O atmospheric absorption. This has to wait the launching of a submillimeter satellite, in which case it is a good candidate. The abundance of LiH/H_2 is at most $\approx 10^{-10}$, in the absence of photodissociation, and the optical depth should reach 1 for a column density of $10^{12}\,cm^{-2}$, or $N(H_2) = 10^{22}\,cm^{-2}$, in channels of $1\,km\,s^{-1}$.

3.5. THE H_2^+ HYPERFINE TRANSITIONS

The abundance of the H_2^+ ion is predicted to be less than 10^{-11} to 10^{-10} in chemical models (e.g. Viala 1986). But the H_2^+ ion possesses an hyperfine structure in its ground state, unfortunately in the first rotational level $N = 1$. The electron spin is $\frac{1}{2}$, and the nuclear spin $I = 1$, which couple in $F_2 = I + S = \frac{1}{2}$ and $\frac{3}{2}$; then $F = F_2 + N = \frac{1}{2}, \frac{3}{2}$ and $\frac{5}{2}$. Five transitions are therefore expected, of which the strongest is $F, F_2 = \frac{5}{2}, \frac{3}{2} \rightarrow \frac{3}{2}, \frac{1}{2}$, at 1343 MHz (Sommerville 1965; Field et al. 1966). At the interface between the cold molecular gas and the interstellar/intergalactic radiation field, one can hope to encounter a sufficient column density of H_2^+. The excitation to the $E_u = 110$ K level is problematic however.

3.6. C AND O POLLUTION OF THE QUASI-PRIMORDIAL COLD GAS

As soon as there exist some metal enrichment from stellar nucleosynthesis, cold gas could be traced by CO molecules, provided a sufficient column density can be shielded from photodissociation. The abundance [O/H] decreases exponentially with radius in spiral galaxies, with a gradient between 0.05 and 0.1 dex/kpc (e.g. Pagel & Edmunds 1981), and the $N(H_2)/I(CO)$ conversion ratio is consequently increasing exponentially with radius (Sakamoto 1996). There could be even more dramatic effects such as a sharp threshold in extinction (at 0.25 mag) before CO is detectable (Blitz et al. 1990), due to photo-dissociation. We can then estimate until which radius the dense clouds are likely to contain CO molecules, if we assume that the opacity gradient follows the metallicity gradient. Assuming the proportionality relation $N(H) \approx 2 \cdot 10^{21} A_V$ atoms $cm^{-2}\,mag^{-1}$ between the gas column density and opacity in the solar neighbourhood (Savage et al. 1977), and a column density of $10^{25}\,cm^{-2}$ for the densest fragments, their opacity A_V falls to 0.25 at $R \approx 60$ kpc, but of course the CO disappears at larger scales before.

The lack of heating sources is another effect hindering the detection of molecular tracers in emission, far from star-formation regions. It is impossible to detect emission from a cloud at a temperature close to the background temperature. Only absorption is possible, although improbable for a surface filling factor of less than 1%. Absorption is biased towards diffuse clouds (intercloud medium) with a large filling factor and a low density (and therefore a low excitation temperature). This is beautifully demonstrated in the molecular absorption survey of Lucas & Liszt (1994) in our Galaxy. But this diffuse medium is preferentially depleted in CO at low metallicity.

Galaxy clusters is a privileged environment where the cold gas might be metal-enriched. Galaxy-galaxy interactions progressively heat the cold gas coming from individual galaxies; the virialised hot medium (10^{6-7} K) experiences a strong mixing, and is enriched by the galaxy ejecta, to the observed intra-cluster abundance of $\approx 0.3\,Z_\odot$. In the cluster center, where the density of hot gas is high enough, a cooling flow is started, and the gas temperature runs away down to the background temperature again, in a fragmented structure (Pfenniger & Combes 1994). The cluster medium is therefore multiphase, with a dense phase completely screened from the X-ray flux (Ferland et al. 1994). This accounts for the apparent complete disappearance of gas in cooling flows, and may explain the high concentration of dark matter in clusters deduced from X-ray data and gravitational arcs (Durret et al. 1994; Wu & Hammer 1993). Many authors have tried to detect this gas in emission or absorption, either in HI (e.g. Dwarakanath et al. 1995) or in the CO molecule (e.g. Braine & Dupraz 1994). Maybe the best evidence of the presence of the cooling gas is the extended soft X-ray absorption (White et al. 1991). The gas has a high surface filling factor (≈ 1), and a column density of the order of $N_H \approx 10^{21}\,\mathrm{cm}^{-2}$. The total mass derived is of the order of $10^{11}\,M_\odot$ over a 100 kpc region. This diffuse phase corresponds to the interface between the very cold molecular gas and the hot medium. Part of the interface is atomic, part is ionised (as observed Hα filaments suggest). Although the HI is not detected in emission with upper limits of the order of $10^9 - 10^{10}\,M_\odot$, it is sometimes detected in absorption, when there is a strong continuum source in the central galaxy. The corresponding column densities are $> 10^{20}\,\mathrm{cm}^{-2}$.

If the molecular clouds are cold ($T \approx 3$ K) and condensed (filling factor $< 1\%$), it is extremely difficult to detect them, either in emission or in absorption, even at solar metallicity. The best upper limits reported in the literature ($N(\mathrm{H}_2) < 10^{20}\,\mathrm{cm}^{-2}$, average over 10 kpc wide regions, but assuming $T \approx 20$ K and solar metallicity, i.e. the standard $N(\mathrm{H}_2)/I(\mathrm{CO})$ conversion ratio, are perfectly compatible with the existence of a huge cold H$_2$ mass (the conversion factor tends to infinity when the temperature tends to the background temperature).

3.7. UV H_2 ABSORPTION IN FRONT OF QUASARS

For the bulk of the gas at $T = T_{bg}$, only absorption could be detected. Absorption in the vibration-rotation part of the spectrum (in infrared) is not the best method, since the transitions are quadrupolar and very weak. An H_2 absorption in Orion has been detected only recently (Lacy et al. 1994) and the apparent optical depth is only about 1%. This needs exceptionally strong continuum sources, which are rare. Electronic lines in the UV should be more easy to see in absorption.

Molecular hydrogen has been found in absorption in front of the PKS-0528-250 quasar by Foltz et al. (1988), at a redshift $z = 2.8$ where there is already a damped Lyα system; the column density is not very high, $10^{18}\,cm^{-2}$, with an estimated width of $5\,km\,s^{-1}$ and a temperature of $100\,K$. Recognizing H_2 absorption is not easy in the Lyα forest, and many tentatives have remained inconclusive. A careful cross-correlation analysis of the spectrum is needed in order to extract the H_2 lines from the confusion. Already Levshakov & Varshalovich (1985) had made a tentative detection, towards PKS0528-250, with some 13 coincident lines among the Lyman and Werner H_2 bands. High velocity resolution to reduce confusion would be helpful in the future to detect more systems. Again absorption is biased towards diffuse gas in galaxies, but in less than $f = 1\%$ of cases, a damped H_2 system should be detected. A molecular clumpuscule on the line of sight of a quasar should produce a very wide saturated absorption, since the line would be in the square-root section of the curve of growth, and nearby lines should overlap (most of the UV continuum could be absorbed). If the clumpuscule is in our own galaxy, temporal variations are expected over a few months interval. It might be the most promising way to detect the cold H_2 gas in the outer parts of galaxies.

3.8. SUBMILLIMETER CONTINUUM

The far-infrared and submillimeter continuum spectrum from $100\,\mu$ to $2\,mm$ has been derived from COBE/FIRAS observations by Reach et al. (1995). They show that in addition to the predominant warm dust emission, fitted by a temperature of $T \approx 20\,K$, there is evidence for a very cold component ($T = 4 - 7\,K$), ubiquitous in the Galaxy, and somewhat spatially correlated with the warm component (see also Mather, this volume). The opacity of the cold component, if interpreted by the same dust model, is about 7 times that of the warm component. It could correspond to those dense clumps of gas, shielded from the interstellar radiation field, that have been polluted by dust and heavy elements.

Schaefer (1996, and this volume) proposes that the cold dust component detected by COBE/FIRAS might be due in fact to molecular hydrogen emission, as collision-induced dipole transitions: in small aggregates at very high density (a fraction of Amagat= $4 \cdot 10^{18} \, \text{cm}^{-3}$), the H_2 gas can emit a continuum radiation, corresponding to free-bound or free-free transitions of weakly bound H_2 dimers, containing a large fraction of ortho-H_2. The para-para complexes do not produce the radiation, by symmetry. Schaefer finds a good fit for the COBE spectra if the dense H_2 clouds follow the HI distribution in the outer parts of the Galaxy.

The weak-dipole radiation due to H_2 collisional complexes is an interesting possibility to detect the presence of cold molecular hydrogen. Only exceptionally dense regions could explain the signal detected by COBE, since the emission is proportional to the square of the density (Schaefer 1994). The required density then imposes the temperature ($T > 11 \, \text{K}$), to avoid the transition to solid molecular hydrogen. Already we had remarked that at the present cosmic background temperature of $T_{bg0} = 2.726 \pm 0.01 \, \text{K}$, the average pressure in the H_2 clumpuscules was about 100 times the pressure of saturated vapour, and that probably a fraction of the molecular mass might be in solid form (Pfenniger & Combes 1994). Already traces of H_2 snow flakes can improve the coupling with the CBR, but the large latent heat of $110 \, \text{K}$ per H_2 molecule and the lack of nucleation sites prevent a large mass fraction to freeze out. The condition of dimerization is then largely satisfied in the conditions of the clumpuscules ($T \approx 3 \, \text{K}$, $n \approx 10^{10} \, \text{cm}^{-3}$). We expect continuum radiation to be emitted and absorbed by the H_2 collisional complexes, through collision-induced dipole moment. The absorption coefficient peaks around $\lambda = 0.5 \, \text{mm}$. The optical depth of each clumpuscule is however quite low $\tau \approx 10^{-9}$ which makes such signature hardly detectable.

4. Gamma-ray distribution

Gamma rays (γ) of high energy come mainly from the interaction of cosmic rays (CR) with the nucleons of the ISM (e.g. Bloemen 1989). Many attempts have been made in the recent years to derive the radial distribution of CR's from observation of γ's, assuming that the gas distribution is well known, derived from HI and CO measurements. Two main problems arose from these derivations: the CR distribution obtained has a very much smoother and extended radial dependence (scale-length 16 kpc) than the assumed CR sources (supernovae and stellar distribution, scale-length 4 kpc); and the H_2 mass derived from CO emission and a constant H_2/CO conversion ratio in the Galactic Center, appears too high by a factor at least 3 with respect to the γ rays detected there (Osborne et al. 1987).

The fact that the γ-ray distribution is much more extended radially than the CR sources has been interpreted in terms of CR diffusion (Bloemen 1989). However the amount of diffusion is not well known. CR particles

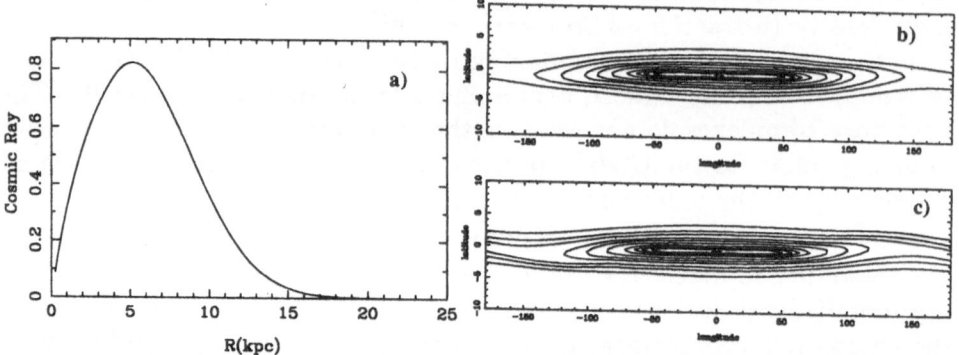

Figure 3. Fit of the γ-ray distribution; a) radial distribution of cosmic rays adopted, including diffusion; b) γ-ray l-b map obtained with the observed HI and H_2; c) same but the HI density multiplied by 20

are closely following the magnetic field lines, due to their small gyration radius, and the field intensity is a function of gas volumic density, so that it is intrinsically hard to disentangle the CR and gas distributions. The possibility of convection in the halo, that can redistribute the CR's in a flatter radial profiles, has been debated. Large halos are excluded however, from the study of primaries, secondaries, and radioactive secondaries as a function of energy (e.g. Weber et al. 1992). Also the thickness of the radio synchrotron halo has been derived around 3.6 kpc (e.g. Beuermann et al. 1985), giving the scale-height of CR electrons.

We have tried to fit the observed EGRET γ-ray distribution assuming different axisymmetric models for CR and total gas distribution. Some of the models are displayed in Fig. 3. Due to the low CR density in the outer parts of the Galaxy, the existence of large amounts of cold gas there is compatible with the data. A more detailed model, releasing the axisymmetry hypothesis, and taking into account the actual $l - b - v$ diagrams of the HI and CO emissions will be reported in a future work.

5. Conclusions

The main argument for the existence of baryonic dark matter comes from the constraints of the Big Bang nucleosynthesis, compared with the observed abundances of primordial elements. The most recent estimates find that the baryon density relative to the critical closure density must lie in the range $\Omega_B = 0.01 - 0.09$ (Smith et al. 1993). The visible baryons account for a much lower density, around $\Omega = 0.002$ (Persic & Salucci 1992). But the galactic dark matter could still be entirely baryonic (e.g. Carr 1995). The recent micro-lensing experiments conducted towards the Magellanic

Clouds have revealed that Machos can account for about 20% of these dark baryons (Aubourg et al. 1993; Alcock et al. 1995), although the constraint is loose.

Molecular hydrogen is one of the least exotic candidate (Pfenniger et al. 1994). Present observations are not incompatible with this hypothesis. The main difficulty to detect very cold gas in emission is its temperature close to the one of the cosmic background. However, we must search for observational tests to falsify the proposition. The detection of the "ultrafine" structure of the ortho-H_2 molecules at km wavelengths raises considerable difficulties for the near future, but could be a means to fix the $N(H_2)/I(CO)$ conversion ratio in our Galaxy. The LiH rotational lines will be easily detectable by submillimeter satellites.

Absorption lines detection might be the best way if the gas is indeed very cold. Since the surface filling factor of the molecular clumps is low ($f < 1\%$), large statistics are required, but the perspectives are far from hopeless. H_2 absorption in the Lyman and Werner bands has already been identified in a damped Lyα system, at $z = 2.8$. For a clumpuscule in our own galaxy falling just on the line of sight of a quasar, we expect a strongly damped and transient absorption over a few months.

Finally, it is not excluded that the cold dust component detected by COBE/FIRAS is tracing the cold H_2 component, limited to galactic radii where the cold gas is still mixed with some dust. Gamma ray data could also be interpreted with the help of radially extended gas distributions.

References

Alcock C.A. et al: 1995, Phys. Rev. Lett. 74, 2867
Aubourg E. et al: 1993, Nature 365, 623
Bahcall J.N., Flynn C., Gould A.: 1992, ApJ 389, 234
Bellini M., de Natale P., Inguscio M., et al: 1994, ApJ 424, 507
Beuermann K., Kanbach G., Berkhuijsen E.M.: 1985, A&A 153, 17
Blitz L., Bazell D., Désert F.X.: 1990, ApJ 352, L13
Bloemen J.B.G.M.: 1989, ARAA 27, 469
Bosma A.: 1981, AJ 86, 1825
Braine J., Combes F.: 1993, A&A 269, 7
Braine J., Dupraz C.: 1994, A&A 283, 407
Broeils A.: 1992, "Dark and visible matter in spiral galaxies", PhD Thesis, Rijksuniversiteit Groningen
Carr B.: 1995, in "Advances in Astrofundamental Physics", ed. N. Sanchez & A. Zichichi, World Scientific, p. 3
Casertano S.: 1983, MNRAS 203, 735
Cordes J.M., Pidwerbetsky A., Lovelace R.V.E.:1986, ApJ 310, 737
Dalgarno A., Black J.H., Weisheit J.C.: 1973, Ap. Letters, 14, 77
de Bernardis P., Dubrovich V., Encrenaz P. et al.: 1993, A&A 269, 1
de Paolis F., Ingrosso G., Jetzer Ph., Roncadelli M.: 1995, A&A 295, 567
Dickey J.M., Hanson M.M., Helou G. 1990, ApJ 352, 522
Durret F., Gerbal D., Lachièze-Rey M., Lima-Neto G., Sadat R.: 1994, A&A 287, 733

Dwarakanath K.S., Owen F.N., van Gorkom J.H.: 1995, ApJ 442, L1

Edge A.C., Stewart G.C: 1991, MNRAS 252, 428

Falgarone E., Puget J-L., Pérault M.: 1992, A&A 257, 715

Ferland G.J., Fabian A.C., Johnstone R.M.: 1994, MNRAS 266, 399

Fich M., Tremaine S.: 1991, ARAA 29, 409

Field G.B., Somerville W.B., Dressler K.: 1966, ARAA 4, 207

Foltz C.B., Chaffee F.H., Black J.H.: 1988, ApJ 324, 267

Freeman F.C.: 1992, in "Physics of Nearby Galaxies, Nature or Nurture?", ed. T.X. Thuan, C. Balkowski, Van J.T.T.; ed. Frontières, Gif-sur-Yvette, p. 201

Gerhard O., Silk J.: 1995, preprint astro-ph/9509149, submitted to ApJ

Gerlich D.: 1990, J. Chem. Phys. 92, 2377

Hegyi D.J., Olive K.A.: 1986, ApJ 303, 56

Kellogg J.M.B., Rabi I.I., Ramsey N.F., Zacharias J.R.: 1939, Phys. Rev. 56, 728

Kellogg J.M.B., Rabi I.I., Ramsey N.F., Zacharias J.R.: 1940, Phys. Rev. 57, 677

Kuijken K., Gilmore G.: 1989, MNRAS 239, 571 & 605

Lacy J.H., Knacke R., Geballe T.R., Tokunaga A.T.: 1994, ApJ 428, L69

Lang K.R.: 1980, "Astrophysical Formulae", Springer-Verlag

Lanzetta K.M., Wolfe A.M., Turnshek D.A. et al: 1991, ApJS 77, 1

Larson R.B., 1981, MNRAS 194, 809

Lawrence T.R., Anderson C.H., Ramsey N.F.: 1963, Phys. Rev. 130, 1865

Levshakov S.A., Varshalovich D.A.: 1985, MNRAS 212, 517

Lucas R., Liszt H.S.: 1994, A&A 282, L5

Merrifield M.R.: 1992, AJ 103, 1552

Osborne J.L., Parkinson M., Richardson K.M., Wolfendale A.W.: 1987, in "Physical Processes in Interstellar Clouds", ed. G.E. Morfill & M. Scholer, Reidel Pub., p. 81

Pagel B.E.J., Edmunds M.G.: 1981 ARAA 19, 77

Pearson E.F., Gordy W.: 1969, Phys. Rev. 177, 59

Peebles P.J.E.: 1971, in "Physical Cosmology", Princeton Univ. Press

Persic M., Salucci P.: 1992, MNRAS 258, 14P

Pfenniger D., Combes F., 1994, A&A, 285, 94

Pfenniger D., Combes F., Martinet L., 1994, A&A, 285, 79

Plummer G.M., Herbst E., de Lucia F.C.: 1984, J. Chem. Phys. 81, 4893

Puche D., Carignan C., Bosma A.: 1990, AJ 100, 1468

Puy D., Alecian G., Le Bourlot J., Léorat J., Pineau des Forêts G., 1993, A&A 267, 337

Ramsey N.F.: 1952, Phys. Rev. 85, 60

Reach W.T., Dwek E., Fixsen D.J. et al.: 1995, ApJ 451, 188

Rothstein E.: 1969, J. Chem. Phys. 50, 1899 (Err. 52, 2804)

Sakamoto S. 1996 ApJ in press

Sancisi R,, van Albada T.S.: 1987, in "Dark Matter in the Universe" IAU Symp. 117, J. Kormendy, G.R. Knapp (eds.), Reidel, Dordrecht, p. 67

Sargent W.L.W.: 1988, in "QSO absorption lines", ed. J. Blades, D. Turnshek & C. Norman, Cambridge University Press, p.1

Savage B.D., Bohlin R.C., Drake J.F., Budich W.: 1977, ApJ 216, 291

Scalo J.M.: 1985, in "Protostars and Planets II", ed. D.C. Black, M.S. Matthews, Univ. of Arizona Press, Tucson, p. 201

Schaefer J : 1994, A&A 284, 1015

Schaefer J.: 1996, Europhysics Letters, submitted

Smith M.S., Kawano L.H., Malaney R.A.: 1993, ApJS 85, 219

Sommerville W.B.: 1965, J. Chem. Phys. 43, 3398

Toomre A., 1963, ApJ 138, 385

Trefler M., Gush H.P.: 1968, Phys. Rev. Let. 20, 703

Viala Y.P.: 1986, A&AS 64, 391

Webber W.R., Lee M.A., Gupta M.: 1992, ApJ 390, 96

White D.A., Fabian A.C., Johnstone R.M. et al.: 1991, MNRAS 252, 72

Wu X.P., Hammer F.: 1993, MNRAS 262, 187

CONSTRAINTS ON THE SURFACE DENSITY OF GAS IN OUTER GALAXY DISKS FROM STABILITY ANALYSES

B.G. ELMEGREEN

IBM Research Division, T.J. Watson Research Center
P.O. Box 218, Yorktown Heights, NY 10598 USA

Abstract. This paper discusses conditions under which the surface density of gas in the outer disk of a galaxy can be high enough to give the rotation curve without causing severe observational consequences. Constraints related to the presence or lack of disk instabilities include the smoothness of extended HI gas, the level of outer disk star formation, the regularity and evolution time of outer resonance rings, and the regularity of self-gravitational clumping in tidal arms. These constraints are discussed here in reference to the compelling proposals by Lequeux (1993) and Pfenniger, Combes & Martinet (1994) that significant dark matter can be in a cold gaseous form in the outer parts of galaxy disks.

1. Outer Disk Instabilities

If the stability parameter $Q = c\kappa/(\pi G\sigma)$ is less than 1 for a thin disk, the interstellar medium will be unstable to form rings. According to Kennicutt (1989), stars form when $Q < 1.4$ and according to Toomre (1981), spiral arms form when Q is approximately less than 2. Here c is the 1D radial velocity dispersion of the gas (actually, c is the dispersion multiplied by $\gamma_{eff}^{1/2}$ for effective ratio of specific heats, γ, which can be significantly less than 1 – Elmegreen 1991); κ is the epicyclic frequency, and σ is the mass column density. For a given c, these stability criteria place limits on the disk surface density if the disk is relatively stable.

In a more realistic disk with finite thickness $2H$, the dispersion relation

D. L. Block and J. M. Greenberg (eds.), New Extragalactic Perspectives in the New South Africa, 467–479.
© 1996 *Kluwer Academic Publishers.*

for radial instabilities can be written approximately as

$$\omega^2 = k^2 c^2 - 2\pi G \sigma k \frac{1 - e^{-kH}}{kH} + \kappa^2. \tag{1}$$

Here ω is the growth rate and k is the wavenumber ($= 2\pi/\lambda$ for wavelength λ). This equation differs from the thin-disk relation only by the presence of the self-gravity dilution term, $(1 - e^{-kH})/kH$, which assumes a uniform density over the layer. For a pure gas disk, the scale height is $H = c_z^2/(\pi G \sigma)$ for z dispersion c_z. We let $C = (c/c_z)$. For stars, the z dispersion is about half the radial dispersion; if the same is true for gas, then $C \sim 2$. We see from this dispersion relation that the fastest growing wavenumber in the thick-disk case is $k = 0.57/H$ for $C = 1$ and $k = 0.20/H$ for $C = 2$. At these wavenumbers, $\omega < 0$ for instability requires $Q < 0.74$ or < 0.90 for $C = 1$ or 2.

Now we can ask what is the value of Q if the mass column density σ is large enough to give the rotation curve in the outer disk. For a flat rotation curve of speed V, this requires a mass $M \approx V^2 R/G$ for radius R. If all of this mass is in the disk, then locally, $\sigma = V^2/(2\pi GR)$. Setting $\kappa = 2^{0.5} V/R$ for this flat rotation curve case, we get

$$Q = 8^{0.5} c/V = 2.8 c/V. \tag{2}$$

Thus $c > 0.26V \sim 60$ km s^{-1} stabilizes a disk dominated by dark matter if $C = 1$ (with $Q > 0.74$), and $c > 0.32V \sim 70$ km s^{-1} stabilizes a such a disk if $C = 2$ (with $Q > 0.90$).

The in-plane velocity dispersion of HI in the outer disk of the Milky Way can be determined from longitude-velocity diagrams. From Figures 1 and 4 in Burton (1992), the HI dispersion at longitude $\sim 180°$ has to be less than about 25 km s^{-1}. This implies that either a dark matter disk in our galaxy is unstable ($Q < 0.3$) or fast-moving dark matter is dynamically decoupled from the HI.

Another constraint comes from the scale height. Setting $H = c_z^2/(\pi G \sigma)$ for a pure gas outer disk with a mass that gives the rotation curve, we get

$$\frac{H}{R} = \frac{2}{C^2} \left(\frac{c}{V}\right)^2 = \frac{Q^2}{4C^2}. \tag{3}$$

Thus disk stability requires the dark matter to extend to heights $H > 0.14R$ for $C = 1$ and $H > 0.05R$ for $C = 2$. This is a relatively mild constraint for an outer disk (see also Section 2).

What is the wavelength for instabilities if dark matter is in the disk? At peak growth, $kH = 0.57$ for $C = 1$, so

$$\frac{\lambda_{\text{peak}}}{R} = 2.8 Q^2; \tag{4}$$

(for $C = 2$, this is $2.0Q^2$). Thus the unstable scale is the whole outer disk unless Q is very small. With such a large scale, the equations that are used to derive Q should be modified, i.e. curvature terms and the exponential nature of the disk should be included.

From the thick-disk dispersion relation at $kH = 0.57$ for $C = 1$, the growth rate is given by

$$\frac{\omega^2}{\omega_0^2} = Q^2 - 0.54 \tag{5}$$

for $\omega_0 = \pi G\sigma/c$. Thus the fastest possible growth rate is $0.54^{1/2}\omega_0$, which gives the minimum growth time

$$T_{\text{growth}} = \frac{c}{0.74\pi G\sigma} = 2.7\frac{cR}{V^2} = 0.43\frac{c}{V}T_{\text{orbit}} = 0.15QT_{\text{orbit}}. \tag{6}$$

Thus unstable growth can occur in a time significantly less than the orbit time if Q is low (the constant on the right is 0.13 for $C = 2$).

The unstable region presumably shears into a spiral, which, after time t and N growth times, has a pitch angle i given by

$$\tan i = \frac{R}{Vt} = \frac{0.37V}{cN} \tag{7}$$

for $C = 1$ (the constant changes to 0.45 for $C = 2$). After several orbit times, which is all that is usually possible in the age of the Universe for an outer disk, the unstable region may appear as a large open spiral or asymmetry.

We are led to the following conclusions: If all of the dark matter in the outer region of a galaxy is in an extended, uniform gas disk, and the gas motions there are nearly isotropic ($C = 1$) or star-like ($C = 2$) with typically low dispersions, $c \sim 10$ km s^{-1}, then there should be large-scale asymmetries in this disk. These asymmetries may have important observational consequences for the inner disk, such as spiral arm or bar generation. If they do, then galaxies without density waves or bars in the inner disk, such as NGC 2841, might be expected to have small outer disk asymmetries and therefore either little disk dark matter or high-dispersion disk dark matter.

Conversely, we can say that a smooth, symmetric, outer HI disk with a flat rotation curve cannot contain much dark matter at normal interstellar velocity dispersions. Otherwise there would be instabilities of the type discussed above, and the HI would be forced to follow these instabilities and become distorted. NGC 2841 may again be an example of a galaxy with a flat rotation curve and a clumpy, but spiral-less outer HI disk (Bosma 1981).

There are many examples of outer HI disks that are not symmetric, as discussed by Pfenniger, Combes & Martinet (1994), but it is unclear whether these asymmetries come from self-instabilities in a dark-matter disk or from long-past interactions with other galaxies. The observation of a *spiral-less* outer HI disk implies that the dark matter has either a high in-plane dispersion or is mostly in the halo. If all galaxies have structure in their outer HI disks, then they may all have instabilities, in which case significant disk dark matter could be present; it will be difficult to know from the outer disk structure alone.

These conclusions should be valid regardless of the form of the dark matter, i.e., whether it is invisible gas as proposed by Pfenniger, Combes & Martinet (1994), or black holes, non-interacting particles, small stars, and so on. The primary assumptions are that the random motions in c are homogeneous and the gas layer is uniform. Larger velocity anisotropies can be considered by increasing $\mathcal{C}$.

2. Gaseous Scale Heights

The scale height of the outer disk may provide a constraint on the presence of dark matter. If both the z dispersion and the scale height can be observed simultaneously in a galaxy, then the mass column density in the disk can be obtained directly, from $\sigma = c_z^2/(\pi G H)$.

For example, observations of the face-on galaxy NGC 1058 show a relatively small HI velocity dispersion at large radii: ~ 7 km s^{-1} (Dickey, Hanson & Helou 1990). The same is true for NGC 6946 (Boulanger & Viallefond 1992). There are no scale height measures for these galaxies, however.

Conversely, the scale height is known for the outer part of our Galaxy, but the z dispersion is not. If the z dispersion in our Galaxy at 20 kpc is 7 km s^{-1}, then the scale height there, 600 pc (Burton 1992), gives a mass column density of 1.3×10^{-3} gm cm^{-2}. This corresponds to $\sim 6 \times 10^{20}$ cm^{-2} of HI with He mixed in. At this radius, the observed perpendicular column density of HI is only $\sim 1 \times 10^{20}$ cm^{-2} (Burton 1992), so the total column density must be $\sim 6\times$ larger than the HI column density at 20 kpc to give the scale height. Some of this extra column density could be from unseen molecules or other types of dark matter, and some could be from stars. The equivalent dark matter column density to give the rotation curve is $V^2/(2\pi G R) \sim 0.019$ gm cm^{-2}, which corresponds to $\sim 8 \times 10^{21}$. This is 15$\times$ that necessary to give the scale height at $c_z = 7$ km s^{-1} and $\sim 80\times$ larger than the observed $N(HI)$.

The observed scale height at 20 kpc in the Milky Way, combined with a reasonable z velocity dispersion for HI obtained from other galaxies, suggests that the outer disk contains ~ 6 times more matter than the HI alone, but ~ 15 times less matter than what would be necessary to give the rotation curve. This result is independent of the radial velocity dispersion of either the HI or the dark matter, and therefore independent of the degree of self-gravitational stability discussed in the previous section. It is also independent of the z velocity dispersion of the dark matter, since it is the HI with its small dispersion that has the observed large scale height. Thus it appears that 90% of the column density of dark matter in our Galaxy at 20 kpc is not within the 1200 pc full thickness of the disk.

3. Radial Instabilities in Fractal/Turbulent Disks

The assumptions used in Section 1 to assess the state of self-gravitational stability of outer galaxy disks were that the interstellar medium is homogeneous and uniform, and that it has a nearly isotropic velocity dispersion. Here we relax the first assumption and consider fractal and turbulent gas.

Vazquez-Semandeni and Gazol (1995) rederived the 3D dispersion relation for a gas with scale-dependent velocity dispersion and density: $c = c_0(L/L_0)^\alpha$, $\rho = \rho_0(L/L_0)^\beta$. Observations suggest that $\alpha \sim 1/2$ and $\beta \sim -1$.

Applying their analysis to a disk, and assuming $c = c_0(k_0/k)^{1/2}$ and $\sigma = \sigma_0(k/k_0)$, we obtain a dispersion relation

$$\omega^2 = k_0 k c_0^2 - 2\pi G \sigma_0 \frac{k^2}{k_0} \frac{1 - e^{-kH}}{kH} + \kappa^2. \tag{8}$$

We cannot convert this to the usual dimensionless form, using $\pi G \sigma_0 / c_0$ for frequency, $k_0 = \pi G \sigma_0 / c_0^2$ for wavenumber and $Q_0 = \kappa c_0 / (\pi G \sigma_0)$ for epicyclic frequency, because these scaling variables depend on the size of the region over which they are measured; they are not constant. Instead we must make this equation dimensionless using different variables that are independent of scale. These variables come from the turbulent scaling relations: $c = c_0(k_0/k)^{1/2}$ and $\sigma = \sigma_0(k/k_0)$. The combinations $c_0^2 k_0$ and σ_0/k_0 are true constants, independent of the length scale over which they are measured. The first has the units of acceleration, the second, when multiplied by πG, has the units of a velocity squared. Thus we write the turbulence-normalizing constants:

$$A_t = c_0^2 k_0 \; ; \quad V_t^2 = \frac{\pi G \sigma_0}{k_0}. \tag{9}$$

Now the unit of length becomes $L_t = V_t^2/A_t$ and the unit of time becomes $T_t = V_t/A_t$. We will evaluate these quantities numerically below.

The dimensionless form of the dispersion relation is now

$$\omega^2 T_t^2 = kL_t - 2(kL_t)^3 C^2 \left(1 - e^{-1/[CkL_t]}\right) + \kappa^2 T_t^2. \tag{10}$$

Note that $Q_t = \kappa T_t$ is the turbulence-equivalence of the Q parameter from section 1. For the exponent, we have written the true scale height $H = c^2/(C^2 \pi G \sigma) = c_0^2 k_0^2/(C^2 \pi G \sigma_0 k^2)$, which varies as the length squared.

The stability properties of such a turbulent disk are easily determined from this equation. At small kL_t, $(\omega T_t)^2 \sim kL_t + Q_t^2 > 0$, so the disk is always stable on large scales. At large kL_t, $(\omega T_t)^2 \sim -2(kL_t)^2 + Q_t^2 < 0$, so the disk is always unstable on small scales.

Now we evaluate the constants and determine the size threshold for stability. Observations suggest that $c \sim 6$ km s^{-1} ~ 6 pc My^{-1} (Stark & Brand 1989) on scales comparable to the Galactic thickness of $k_0^{-1} \sim 400$ pc, so $A_t = c_0^2 k_0 = 6^2/400 = 0.09$ pc My^{-2}. This constant also gives the size-linewidth relation for molecular clouds, $c \sim 0.3(L/\text{pc})^{1/2}$ (Blitz 1993).

Observations do not give σ_0/k_0 directly, and indeed, this type of scaling for interstellar gas is not even observed yet, but we may infer a value for σ_0/k_0 from the commonly used value of Q. We write

$$V_t^2 = \frac{\pi G \sigma_0}{k_0} = \frac{\pi G \sigma}{k} = \frac{\kappa c}{Qk} = \frac{\kappa c_0}{Q_0 k_0} \tag{11}$$

where $Q_0 = \kappa c_0/(\pi G \sigma_0)$ is the value of Q measured on the same scale at which c_0 is measured, i.e., $k_0^{-1} \sim 400$ pc. This is the value of Q that is usually quoted from observations, and is ~ 1 locally, or ~ 0.2 in an outer disk that contains enough slow-moving dark matter to give the rotation curve (cf. Eq. 2). Thus

$$V_t^2 = \frac{0.033 \text{ My}^{-1} \times 6 \text{ km s}^{-1} \times 400 \text{ pc}}{Q_0} \sim \frac{(9 \text{ pc My}^{-1})^2}{Q_0} \tag{12}$$

using $\kappa \sim 2^{1/2}V/R$ for typical $V = 200$ km s^{-1} and $R = 8.5$ kpc, where these quantities are measured.

For these locally evaluated turbulence constants, it follows that $Q_t^2 = (\kappa V_t/A_t)^2 \sim 10/Q_0$. We choose $Q_0 \sim 1$ for the conventional ISM, and $Q_0 \sim 0.2$ for a disk with dark matter. Then $\kappa^2 T_t^2 = Q_t^2 \sim 10$ or 50 in equation 10. The threshold values of kL_t that separate stability from instability can now be evaluated numerically for the cases $C = (1,2)$. They are $kL_t = (2.76, 2.57)$ for $Q_t^2 = 10$ and $kL_t = (5.51, 5.32)$ for $Q_t^2 = 50$. The unit of length is $L_t = V_t^2/A_t = 900$ pc/Q_0. Thus the threshold lengths are $k^{-1} = (320, 350)$ pc for $Q_0 = 1$ and $k^{-1} = (800, 840)$ pc for $Q_0 = 0.2$. Density concentrations larger than these scales are stabilized by high velocity dispersions, low average densities, rotation, and the gravity-dilution factor for a large scale height. Density concentrations smaller than these scales are unstable because of their low velocity dispersions and high densities.

Large-scale stability in a turbulent medium is not surprising, considering that the original stability parameter $Q = Q_0(Lk_0)^{3/2}$. If $Q = Q_0 = 0.2$ on scale $k_0 \sim 400$ pc, then $Q > 1$ when $L > 2.9k_0^{-1}$. Evidently, the large-scale stability in a turbulent medium is really the result of a large Q on that scale. Stability in a turbulent medium still requires large c.

Note that in a turbulent ISM, $H = c_z^2/(\pi G\sigma) \propto L^2$ so, according to these assumptions, larger regions have lower σ, larger c, and larger H, i.e., there is a nested hierarchy of structures both in the plane and perpendicular to the plane. Large L implies large H if c is nearly isotropic.

4. Outer Disk Pressure and Star Formation

If the outer disks of galaxies have an amount of mass given by the rotation curve, and if this mass is in the form of dynamically coupled gas, then we can say something additional about the interstellar pressure and star formation rate there. This follows from the fact that the interstellar pressure scales with σ because of the weight of the gas layer.

In a gas+stellar disk, the midplane pressure is given by (Elmegreen 1989)

$$P \approx \frac{\pi}{2}G\sigma_{\text{gas}}\left(\sigma_{\text{gas}} + \frac{c_{z,\text{gas}}}{c_{z,\text{stars}}}\sigma_{\text{stars}}\right). \tag{13}$$

In a pure gas disk, as assumed here for the outer, dark parts of galaxies, $P = (\pi/2)G\sigma^2$. With all of the dark matter for the rotation curve in the disk, this gives

$$P = \frac{V^4}{8\pi GR^2} = 1.8 \times 10^5 k_B \left(\frac{V}{200 \text{ km s}^{-1}}\right)^4 \left(\frac{R}{20 \text{ kpc}}\right)^{-2}. \tag{14}$$

This result suggests that dark matter outer disks should have pressures $10\times$ larger than the local interstellar medium. This is an enormous pressure for a gas without bright star formation. Considering the thermal balance of the diffuse clouds (Elmegreen & Parravano 1994), this pressure gives the gas both warm and cool thermal HI temperatures or, if the radiation field is typically low, only the cool phase for HI.

Without the dark matter in the disk, the pressures in the outer regions of galaxies should become so low that only the warm HI phase is possible. This is a reasonable result, as it nicely explains why active star formation stops at about 4 scale lengths, and also why star formation has always been low in low-surface brightness galaxies (Elmegreen & Parravano 1994). With dark matter in the outer disk, however, the pressure there should be comparable to that in the inner disk and the star formation rate should be just as high. Outer disks with significant amounts of normal cold gas should have high star formation rates, in analogy with the situation for inner disks. Because outer extended HI disks with flat rotation curves are optically dark, the dark matter either has to be in the halo or it has to be in a peculiar form that cannot make stars (e.g., Pfenniger, Combes, & Martinet 1994).

5. Star Formation Rates in Outer Resonance Rings

A faint outer ring has been observed around NGC 1300 (Elmegreen et al. 1996) that is similar to the outer resonance rings observed in many other galaxies (Buta 1995). The ring in NGC 1300 has an HI counterpart observed by England (1989). Elmegreen et al. (1996) used the ring colors to infer an age of $\sim 10^9$ years, which is reasonable considering that such rings form by a gradual outward motion of gas induced by bar and spiral torques. Star formation regions are in the NGC 1300 ring at the locations of HI peaks. The star formation ages are also obtained from the colors, and are in the range of $1 - 6 \times 10^7$ years.

England's HI map gives a Q value of 1.6 inside the ring if $c = 7$ km s^{-1}, which is typical for outer disks (see Sect. 2). Outside the ring, $Q \sim 3$ because of the lower $N(HI)$. Thus the ring is mildly unstable to form giant star formation regions by a segmented, parallel collapse. Such a collapse has been proposed for inner Lindblad resonance rings by Elmegreen (1993). The time scale for the instability should be $\kappa^{-1} \sim 6 \times 10^7$ yrs for $Q \sim 1$ (smaller for smaller Q), which is also the age of the star formation.

These results gives a self-consistent picture in which the outer ring in NGC 1300 forms by the gradual outward motion of gas. Star formation occurs in the ring after it reaches a high enough density to become unstable (which requires $\rho > \kappa^2/G$). The time scale for star formation is comparable to the time scale for the ring instability.

This picture has to change if the dark matter in NGC 1300, which has a flat rotation curve, resides in the outer disk. Then the Q value throughout the whole outer disk would be so small (see above) that a nearly-uniform ring would not be able to form in the first place; it would be too easily broken into large asymmetric pieces by self-gravitational instabilities. Also the time scale for instabilities inside the ring would be much smaller than the observed star formation time, by a factor of about 10.

The observation of outer rings in many galaxies (Buta 1995), and the presence of many of these rings in HI, implies that outer disks are relatively stable. The thinness of the outer rings, ΔR, also suggests that the in-plane velocity dispersion is only a small fraction of the rotation speed, $c/V \sim 2^{1/2}\Delta R/R \sim 0.1$, which is not enough for outer disk stability according to Section 1. These two constraints of ring regularity and thinness seem to exclude the possibility that a significant amount of dark matter is in the outer disk unless this dark matter has a high velocity dispersion and does not get concentrated into the ring along with the HI gas.

6. Dwarf Galaxy Formation in Tidal Arms

Tidal arms should be gravitationally unstable to form clumps the size of dwarf galaxies, 10^8 $M_\odot$ (Kaufman, et al. 1992; Elmegreen, et al. 1993; Barnes & Hernquist 1992). The ratio of the clump separation to the arm width is a measure of the self-gravitational binding. This ratio equals ~ 3 for condensations in most spiral arms (Elmegreen & Elmegreen 1983), in filamentary clouds (Schneider & Elmegreen 1979), and in galactic tidal arms (Hunsberger & Charlton 1995). For such cylindrical regions, this ratio follows from the stability conditions if the gas is marginally stable (Elmegreen 1994), at which point

$$\frac{2G\mu}{c^2} \sim 1 \tag{15}$$

for mass/length μ.

We can use this condition of marginal stability to place an upper limit on μ for reasonable c. If μ is larger than this upper limit, then the separation between condensations would be smaller than observed. For tidal arms, an upper limit on μ corresponds to an upper limit on the mass of the outer disk.

During a galaxy interaction, most of the outer disk mass is put into the two long tidal arms. We assume here that each tidal arm contains half of the disk mass between R_{min} and R_{max}. If the outer disk contains enough dark matter to give the rotation curve, then

$$\mu \sim \frac{V^2}{2G}\frac{R_{max} - R_{min}}{L_{arm}} \sim \frac{V^2}{4G}, \tag{16}$$

if L_{arm} is the length of the arm, assumed equal to $2(R_{max} - R_{min})$ for the last step in the equation.

It follows that if $2G\mu/c^2 \sim 1$ for a 3:1 spacing of the condensations in the tidal arms studied by Hunsberger & Charlton (1995), then

$$\frac{2G\mu}{c^2} \sim \frac{V^2}{2c^2} \sim 1, \tag{17}$$

or $c \sim 0.7V \sim 150$ km s^{-1}. This implied velocity dispersion is too high. In IC 2163, $c \sim 20$ km s^{-1} in the tidal arms (Elmegreen et al. 1995). Also, there is no strong clumping in the IC 2163 arms, so $2G\mu/c^2 < 1$. Thus $\mu < c^2/(2G) = V^2/(4G)(2c^2/V^2) = \mu_{max}(2c^2/V^2)$, so $\mu < 1\%\mu_{max}$, where μ_{max} is the value given by putting all the dark matter in the disk before the interaction.

Once again we see that if the velocity dispersion in the outer disk of a galaxy is much less than the rotation speed, then stability constraints imply that the disk mass must be much less than the total dark matter needed to bind the galaxy.

7. DISCUSSION

Outer disks would be significantly unstable to form stars, spirals and sheared asymmetries if all of the dark matter (DM) that gives the rotation curve is in the disk and if the DM velocity dispersions are reasonably isotropic and as low as the observed HI dispersions. Symmetric, extended HI disks with little star formation cannot contain much low-dispersion dark matter.

Dark matter in the outer HI disk of our Galaxy would also make it thinner than it is. The HI disk at 20 kpc is too thick to contain much dark matter if the perpendicular HI velocity dispersion is as low as it is in other galaxies. In fact, the disk can contain only about 10% of the outer Galaxy dark matter. This result depends only on the velocity dispersion of the HI, not the velocity dispersion of the dark matter. There is still a need for disk matter other than the observed HI, but not so much to include all of the dark matter that gives the rotation curve.

Disks containing significant dark matter also cause problems for the usual interpretation of outer rings and tidal arms, which would be more unstable than they appear to be, based on structural regularity, star formation time scales, and the geometry of the condensations, if the dark matter that gives the rotation curve is associated.

These constraints have several implications. The first is the requirement that disk dark matter have a velocity dispersion of at least $0.2\times$ the rotation speed, or ~ 60 km s^{-1} for a large galaxy. If the dispersion is much less than this, then the outer disk would be unstable and the resulting motions should bring the dispersion up to this limit. This is one of the feedback cycles usually discussed for the inner disk, and it should apply to the outer disk as well, regardless of the form of the dark matter. Note that this DM stabilization would have had to occur long ago, or else the outer disk HI would be fast-moving now too.

A second constraint is that the fast-moving dark matter should not couple dynamically to the observed HI, which is slowly moving. For example, the outer disk dark matter cannot be low-density diffuse clouds in pressure equilibrium with the HI; such clouds would stir the HI as they move.

This coupling constraint also implies that outer disk dark matter cannot be in the form of *dense* clouds if there is a magnetic field at large radii and the magnetic diffusion time is long. For a pressure-equilibrium cold gas, the density is $n \sim P/kT \sim 10^4$ cm^{-3}. For a pressure equilibrium magnetic field, $B \sim (8\pi P)^{1/2} \sim 2 \times 10^{-5}$ Gauss. The ion-neutral diffusion time of magnetic flux from a cloud with a magnetic field curvature or gradient of order B/L is approximately $4\pi\rho n_e < \sigma_{in}c > L^2/B^2$ for ion-neutral collision rate $< \sigma_{in}c >\sim 2 \times 10^{-9}$ cm^{3}s^{-1} and electron density $n_e \sim 10^{-5}n^{1/2}\zeta_{17}^{1/2}$ with cosmic ray ionization rate ζ_{17} measured in units of the local rate of $\sim 10^{-17}$ s^{-1}. The result is a diffusion time

$$\tau_{diff} \sim \frac{4 \times 10^8 \text{ yr}}{f^{1/2}} \left(\frac{L/\text{ pc}}{B/20\mu \text{ Gauss}}\right)^2 \left(\frac{n}{10^4 \text{ cm}^{-3}}\right)^{3/2} \left(\frac{\zeta}{10^{17} \text{ s}^{-1}}\right)^{1/2}.$$

(18)

If the dense DM gas has substructure at higher density, with n representing only the average density, then the diffusion time increases by $f^{-1/2}$, as indicated, for substructure filling factor f (Elmegreen & Combes 1992).

To prevent acceleration of the HI gas, which is easily dragged along by the magnetic field lines, the dense DM gas has to slip through the magnetic field very quickly, on time scales less than $\sim 0.1\times$ the momentum transfer time between the two fluids. This fraction is the ratio of the velocity dispersion of the HI (6 km s^{-1}) to the velocity dispersion of the dark matter (60 km s^{-1}). The momentum transfer time is the time it takes an Alfven wave outside the cloud to cover a mass of HI equal to the cloud's mass (Elmegreen 1981), or

$$\tau_{mom} \sim \frac{nL}{n_0 v_A} \sim 2.5 \times 10^8 \text{ yr} \left(\frac{L/\text{ pc}}{B/20\mu \text{ Gauss}}\right) \left(\frac{n}{10^4 \text{ cm}^{-3}}\right) \left(\frac{n_0}{\text{cm}^{-3}}\right)^{-1/2}$$

(19)

for cloud density n, external density n_0, and external Alfven speed $v_A = B/(4\pi n_0 m_H)^{1/2}$. For a pressure equilibrium field in a dark matter disk, $v_A \sim 40$ km s^{-1}. The dark matter disk alone has $n_0 \sim 1$ cm^{-3} and we take $n \sim 10^4$ cm^{-3} as above.

The ratio $\tau_{diff}/\tau_{mom} < 0.1$ implies

$$\left(\frac{1.6}{f^{1/2}}\right) \left(\frac{L/\text{ pc}}{B/20\mu \text{ Gauss}}\right) \left(\frac{n}{10^4 \text{ cm}^{-3}}\right)^{1/2} \left(\frac{n_0}{\text{cm}^{-3}}\right)^{1/2} \left(\frac{\zeta}{10^{17} \text{ s}^{-1}}\right)^{1/2} < 0.1.$$

(20)

This result places constraints on the form of the dark matter gas, i.e., it can be only in tiny, but not very dense clumps. The clumpuscules discussed by Pfenniger & Combes (1994) have $L \sim 10^{-4}$ pc and $n \sim 10^9$ cm^{-3} at 3K temperature. This combination makes $(L/\text{pc})(n/10^4 \text{ cm}^{-3})^{1/2} \sim 0.03$, which is barely small enough unless $B < 20\mu$ Gauss and ζ is not too high. Larger, lower density clouds might hide dark matter as well: for example, if $n \sim 10^4$ cm^{-3} in pressure equilibrium, then $L << 0.1$ pc satisfies the diffusion constraint. Such small sizes are also less than the thermal Jeans length: $L_{\text{JEANS}} \sim (kT/m_H)/(GP)^{1/2} \sim 0.1$ pc for pressure P from Section 4; this implies that sufficiently small pressure-equilibrium clumps might not form stars. Other constraints from energy dissipation of the high-dispersion gas and general connectivity to larger scale structures would have some bearing on dynamical coupling with the HI as well.

References

Barnes, J.E. and Hernquist, L. (1992) Formation of Dwarf Galaxies in Tidal Tails, *Nature*, **360**, pp. 715-7

Blitz, L. (1993) Giant Molecular Clouds, *Protostars and Planets III*, ed. E.H. Levy and J.I. Lunine, University of Arizona, Tucson, pp. 125-162

Bosma, A. (1981) 21-cm Line Studies of Spiral Galaxies I: Observations of the Galaxies NGC 5033, 3198, 5055, 2841, and 7331, *Astr. J.*, **86**, pp. 1791-1824

Boulanger, F. and Viallefond, F. (1992) Observational Study of the Spiral Galaxy NGC 6946: I. HI and Radio Continuum Observations, *AA*, **266**, pp. 37-56

Burton, W. B. (1992) Distribution and Observational Properties of the ISM, in *The Galactic Interstellar Medium*, ed. D. Pfenniger and P. Bartholdi, Springer-Verlag, Berlin, pp. 1-155

Buta, R. (1995) The Catalog of Southern Ring Galaxies, *ApJSupp*, **96**, pp. 39-116

Dickey, J.H., Hanson, M.M. and Helou, G. (1990) Gas Motions in an Extended, Quiescent Spiral Disk, *ApJ*, **352**, pp. 522-31

Elmegreen, B.G. (1981) The Role of Magnetic Fields in Constraining the Translational Motions of Giant Cloud Complexes, *ApJ*, **243**, pp. 512-525

Elmegreen, B.G. (1989) A Pressure and Metallicity Dependence for Molecular Cloud Correlations and the Calibration of Mass, *apJ*, **338**, pp. 178-196

Elmegreen, B.G. (1991) Cloud Formation by Combined Instabilities in Galactic Gas Layers: Evidence for a Q Threshold in the Fragmentation of Shearing Wavelets, *ApJ*, **378**, pp. 139-156

Elmegreen, B.G. (1993) Starbursts by Gravitational Collapse in the Inner Lindblad Resonance Rings of Galaxies, *ApJ*, **425**, pp. L73-6

Elmegreen, B.G. (1994) Supercloud Formation by Gravitational Collapse of Magnetic Gas in the Crest of a Spiral Density Wave, *ApJ*, **433**, pp. 39-47

Elmegreen, B.G. and Elmegreen, D.M. (1983) Regularly Spaced H II Regions and Superclouds in Spiral Galaxies: Clues to the Origins of Cloudy Structure, *MNRAS*, **203**, pp. 31-45

Elmegreen, B.G. and Combes, F. (1992) Magnetic Diffusion in Clumpy Molecular Clouds, *AA*, **259**, pp. 232-240

Elmegreen, B.G., Kaufman, M. and Thomasson, M. (1993) An Interaction Model for the Formation of Dwarf Galaxies and 10^8 M$_\odot$ clouds in Spiral Disks, *ApJ*, **412**, pp. 90-98

Elmegreen, B.G. and Parravano, A. (1994) When Star Formation Stops: Galaxy Edges and Low Surface Brightness Disks, *ApJ*, **435**, pp. L121-4

Elmegreen, D.M., Kaufman, M., Brinks, E., Elmegreen, B.G. and Sundin, M. (1995) The Interaction Between Spiral Galaxies IC 2163 and NGC 2207, I. Observations, *ApJ*, **453**, pp. 100-138

Elmegreen, D.M., Chromey, F., Elmegreen, B.G. and Hasselbacher, D. (1996) Surface Photometry of the Outer Ring in NGC 1300, *ApJ*, submitted.

England, M.N. (1989) High Resolution Neutral Hydrogen Observations: The Barred Spiral Galaxy NGC 1300, *ApJ*, **337**, pp. 191-208

Hunsberger, S. and Charlton, J. (1995) The Formation of Dwarf Galaxies in Tidal Debris: A Study in the Compact Group Environment, *ApJ*, in press.

Kaufman, M., Elmegreen, B.G. and Thomasson, M. (1992) Formation of Massive Clouds and Dwarf Galaxies during Tidal Encounters, in The Evolution of Galaxies and their Environment, ed. D. Hollenbach, H. Thronson, and J.M. Shull, NASA Conference Publication 3190, pp. 238-239.

Kennicutt, R.C. (1989) The Star Formation Law in Galactic Disks, *ApJ*, **344**, pp. 685-703

Lequeux (1993) Large amounts of cold molecular hydrogen in the Small Magellanic Cloud,

AA, **287**, pp. 368-70

Pfenniger, D., Combes, F. and Martinet, L. (1994) Is dark matter in spiral galaxies cold gas? I. Observational constraints and dynamical clues about galaxy evolution, *AA*, **285**, pp. 79-93

Pfenniger, D. and Combes, F. (1994) Is dark matter in spiral galaxies cold gas? II. Fractal Models and Star Non-Formation, *AA*, **285**, pp. 94-118

Schneider, S. and Elmegreen, B.G. (1979) A Catalogue of Globular Filaments, *ApJSupp*, **41**, pp. 87-95

Stark, A.A. and Brand, J. (1989) Kinematics of Molecular Clouds: II. New Data on Nearby Giant Molecular Clouds, *ApJ*, **339**, pp. 763-771

Toomre, A. (1981) What amplifies the spirals?, in The Structure and Evolution of Normal Galaxies, ed. S.M. Fall and D. Lynden-Bell, Cambridge University Press, Cambridge, pp. 111-136

Vazquez-Semandini, E. and Gazol, A. (1995), preprint

THE CHANGING CIRCUMSTELLAR ENVIRONMENT OF YOUNG STARS

J.A. GRAHAM

Carnegie Institution of Washington
Dept. of Terrestrial Magnetism
5241 Broad Branch Road N.W.
Washington, DC 20015 USA

Abstract. New ice band absorption profiles from dust around embedded young stars are discussed.

1. Introduction

Once a star forms, we expect rapid processing of the remnant circumstellar material by radiation and outflowing winds prior to planet formation. While it is easy to detect circumstellar gas, the study of dust is more difficult. One problem is to locate and isolate the different dust zones around the young star before the dust largely disperses. There is thus an incentive to observe active young stars still heavily obscured, as one expects their occasional intense outbursts to have a significant effect on their surroundings.

About 5 years ago, W.P. Chen and I began an investigation of the 3μm water-ice band which is seen in absorption by dust around young stars. We used the Infrared Spectrometer on the 4m telescope at Cerro Tololo Inter-American Observatory to obtain low resolution spectra of some young, barely visible stars as well as of some deeply embedded, still invisible protostars in the CrA association (Graham and Chen 1991, Chen and Graham 1993). With the small detecting array available at that time, we were restricted to a resolution $\lambda/\Delta\lambda$ (two pixels) of 150. In some cases, we found peculiar absorption profiles which suggested that the 2.97μm ammonia ice feature might be of comparable strength to the water ice feature itself.

The old observations were unsatisfactory in several ways and our interpretation could not be reconciled with the expected chemical makeup of the dust. In June 1995, I had the opportunity to reobserve the most interest-

480

D. L. Block and J. M. Greenberg (eds.), New Extragalactic Perspectives in the New South Africa, 480–483.
© *1996 Kluwer Academic Publishers.*

ing objects with the newly renovated spectrometer at Cerro Tololo Inter-American Observatory which incorporated a 256×256 array with $30\mu m$ pixels. The resolution was 1300, 9 times the old value.

2. Observations

Observing procedures followed those of Smith, Sellgren and Tokunaga 1989. The observing list was much smaller for the 1995 observing run because of the need to settle the controversy surrounding the earlier data. Thus, apart from standard stars, only those listed in Table 1 were observed.

TABLE 1. Objects Observed

Object	Other ID	L(mag)	T_{bb}	$\tau_{3.08}$
IR12496-7650	HH54star	3.5	700	0.6
V346 Nor	HH57star	5.1	830	0.7
HH100-IR	TS 2.6	4.7	700	1.39
TS 13.1	IRS 2	5.4	820	1.23
TS 2.4	IRS 5	7.6	600	1.59

The objects are described in Graham and Chen 1991 and Chen and Graham 1993. Following reductions, black body temperatures for the program objects were estimated from the observed fluxes at 2.8 and $3.8\mu m$ and appropriate flux distributions calculated. These temperatures are included in Table 1. After ratioing the program spectra with the black body spectra, optical depths were computed. Some results are displayed in Figures 1 and 2. The optical depth for each object at $3.08\mu m$ is listed in Table 1. The new optical depths are larger than those given in Chen and Graham 1993. There, it is now clear, the continuum levels were underestimated.

3. Discussion

The first feature noted in all 5 spectra is the smoothness of the absorption profiles between 2.8 and $3.0\mu m$. Clearly the $2.97\mu m$ feature reported by Chen and Graham 1993 in 4 of these objects is an artefact, probably caused by incomplete subtraction of the terrestrial atmospheric bands. If, as now seems likely, flexure in the spectrograph was a principal factor, it is easy to see how the error could have been a systematic one for a particular star.

The spectrum of the HH 54 star has again a lot of scatter. This is due to the high southern declination (-77°) and consequent high air mass which aggrevates the difficulty of removing the atmospheric bands. The

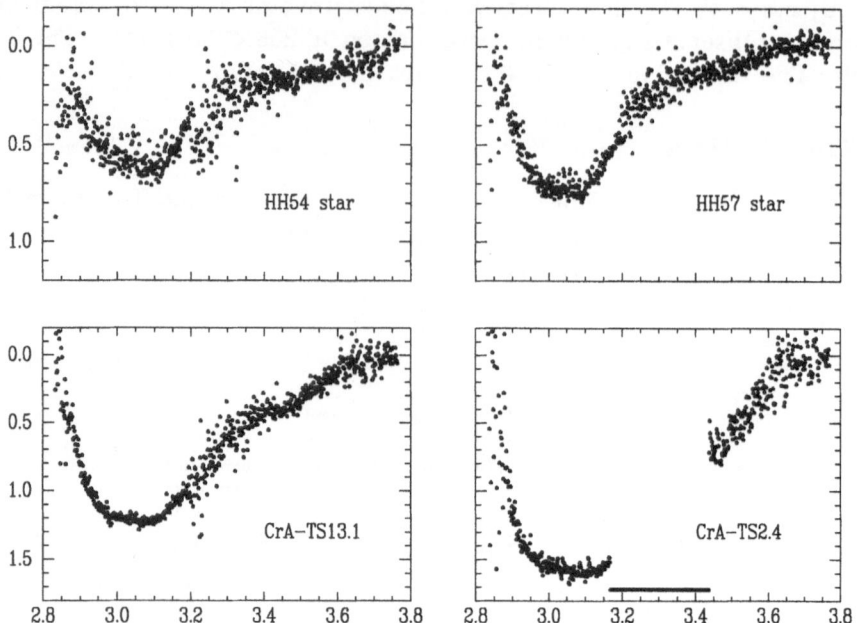

Figure 1. Observed ice band profiles for the HH54 star, the HH57 star and two sources in the CrA association. Optical depth τ is plotted against wavelength in microns.

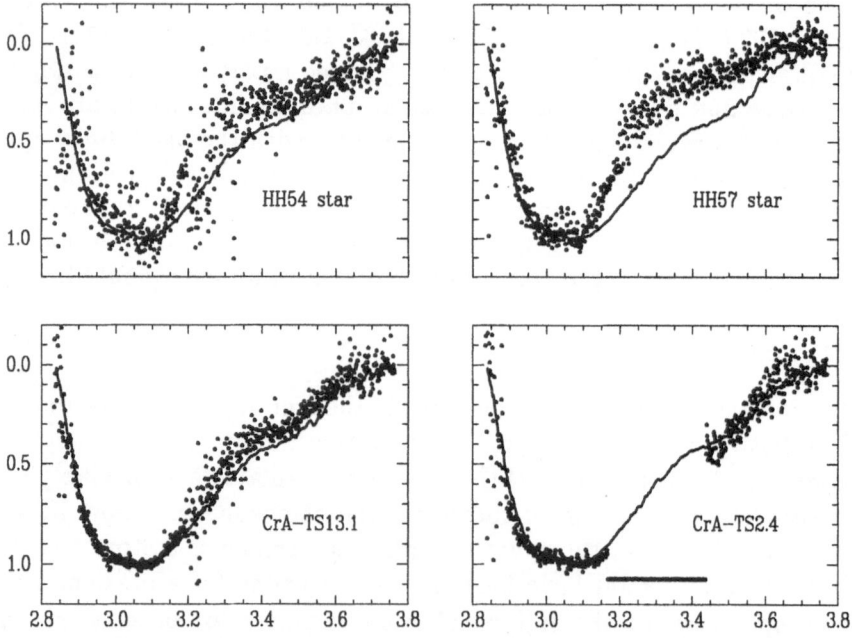

Figure 2. Optical depths normalized to $\tau=1$ compared with normalized ice band profile of HH100-IR.

HH 57 star (V 346 Nor) and HH100-IR were observed under near optimal conditions as they transited close to the zenith. The TS 2.4 object could not observed on the second night hence coverage is incomplete.

The absorption in HH100-IR is very well defined. Comparison with the profile published by Whittet *et al* based on observations made in 1988 shows remarkable agreement.

The long-wavelength wing of the $3\mu m$ ice feature has been discussed extensively in the literature (e.g. Smith, Sellgren and Tokunaga 1989, Smith, Sellgren and Brooke 1993, Brooke, Sellgren and Smith 1996) and can be seen in the spectra presented here. To compare, I have normalized each of the spectra in Figure 1 to optical depth $\tau=1$. Then I have superposed in Figure 2 a smoothed (boxcar smoothing, 9 pixel box) similarly normalized spectrum of HH100-IR. The agreement between the HH100-IR profile and those of TS 2.4 and TS 13.1 is remarkably good. However, the long-wavelength wing is very much weaker in the two spectra associated with visible stars near HH 54 and HH 57. Although both objects are still heavily obscured ($A_v=14$, 7 mag respectively), I suggest that this indicates that some processing is taking place for the dust nearest the star. Until the cause of the long-wavelength wing is more clearly understood, it is not possible to say much more. Further work is being planned to compare in detail the observed profiles and the spectra predicted by theoretical grain models. In the interim, I encourage similar observations of comparable resolution for other objects which are in the process of clearing their surroundings.

Brooke, Sellgren and Smith 1996 have described a broad absorption feature at 3.47 μm which they interpret as a C$-$H stretch absorption. This feature can clearly be seen in 3 of the objects described here. For HH100-IR, another deeper dip is seen at 3.55μm which can plausibly be linked to the solid CH_3OH feature at 3.54μm observed by Brooke *et al* in NGC 7538/IRS 9. The 3.47μm dip can be seen weakly in the HH 57 star spectrum but is too noisy to quantify. The present results agree qualitatively with Brooke *et al*'s remark that the abundance of C$-$H averaged along the lines of sight is closely related to that of water ice.

References

Brooke, T.Y., Sellgren, K. and Smith, R.G. (1996) *ApJ* (in press)

Chen, W.P. and Graham, J.A. (1993) *ApJ* 409, 319

Graham, J.A. and Chen, W.P. (1991) *AJ* 102, 91

Smith, R.G., Sellgren, K. and Tokunaga, A.T. (1989) *ApJ* 344, 413

Smith, R.G., Sellgren, K. and Brooke, T.Y. (1993) *MNRAS* 263, 749

Whittet, D.C.B., Smith, R.G., Adamson, A.J., Aitken, D.K., Chiar, J.E., Kerr, T.H., Roche, P.F., Smith, C.H. and Wright, C.M. (1996) *ApJ* (in press)

INTERSTELLAR DUST IN THE SOLAR SYSTEM

DALE P. CRUIKSHANK
NASA Ames Research Center
MS 245-6,Moffett Field, CA, 94035-1000, USA

Abstract. Interstellar (IS) dust has been detected passing through the Solar System in the current epoch, and it is found preserved from the pre-solar epoch in meteorites and comet fragments.

1. Introduction

Dust in the Solar System is derived from the comminution of asteroids in the Mars-Jupiter zone and planetesimals beyond Neptune by their mutual collisions, from the evaporation of comets when they approach the Sun, and from the passage of the Solar System through the local interstellar medium. Dust generated within the Solar System moves in heliocentric orbits spiralling toward the Sun by radiation forces, but some is trapped in a circumsolar ring that is dynamically resonant with Earth (1). Small grains of the dust from some asteroids and (perhaps) all of the comets are chemically and thermally unaltered from their time of condensation to the time of their collection on Earth or detection by spacecraft.

2. Interstellar Dust in the Solar System

The Ulysses spacecraft detected IS dust particles that are distinguished from dust in the zodiacal cloud by their velocities and direction of motion (2). At 5 AU from the Sun, Ulysses observed a flux of $1.5 \ 10^{-4} m^{-2} s^{-1}$ IS dust particles of mean mass $3 \ 10^{-13}$ g, yielding a mass flux of $5 \ 10^{-17} g \ m^{-2} s^{-1}$. The dust particles appear to be coupled to the neutral IS gas flowing through the Solar System at velocity 26 km s^{-1}, and although some velocity and directional dispersion is seen, the flux is peaked near the position of the apex of the Sun's motion. For an average particle density of $1 \ g \ cm^{-3}$, the average radius is $0.4 \mu m$.

D. L. Block and J. M. Greenberg (eds.), New Extragalactic Perspectives in the New South Africa, 485-488.
© 1996 *Kluwer Academic Publishers.*

The dust in the diffuse ISM produces an increase in the extinction toward the far UV with a maximum (bump) at 0.2175μm. It also produces partially polarized extinction in the visible and near-IR, plus infrared emission bands and non-equilibrium thermal emission. Small icy and organic-rich (radiation processed) grains produce IR spectral features of aliphatic hydrocarbons and related compounds (3,4). Very small particles ($a \leq 0.1\mu$m) are responsible for the extinction features, while grains of molecular size (0.0005-0.0002μm) produce the emission bands and non-thermal emission. Thus, the Ulysses particles are substantially larger than the average outside the heliosphere. Small grains with electrostatic charges may be prevented from entering the heliosphere by Lorenz forces. At 5 AU, the contribution of IS dust particles to the brightness of the Zodiacal Light is small (2). Even larger particles (10μm) of IS origin may have been detected by the Pioneer 10 and 11 spacecraft (5) between 3 and 18 AU. A constant flux of $3 \, 10^{-6}$ and $10^{-6}m^{-2}s^{-1}$ for particles of 10^{-9} and 10^{-8} g particles, respectively, was found. These particles are not readily explained as interplanetary dust particles (2).

3. Meteorites, Interplanetary Dust Particles, and Comets

IS dust grains from many sources were trapped in the condensing material of the solar nebula at the time of Sun and planet formation. The grains are preserved in comets, certain meteorites, and dust particles derived from the dissipation of comets. In primitive meteorites, the presolar material consists of SiC, graphite, microdiamonds, alumina, and carbon-rich organic solids. Meteoritic microdiamonds (0.001μm), contain dissolved D-enriched H, N, and O, and they are enriched in the heavy and light isotopes of Xe, suggestive of processing near supernovae. SiC grains of 0.03-10μm size found in meteorites carry the five middle isotopes of Xe in an abundance pattern indicative of the s-process that is characteristic of the atmospheres of AGB stars (6,7).

Meteorites carry several varieties of IS graphite grains, each showing its own pattern of density, crystallinity, H,N,O-content, and $^{12}C/^{13}C$ (6,7,8). Ne abundances in some graphite indicate formation in novae, while other grains show Ne abundances more typical of origin by the s-process in an AGB star. The study of meteorites has lent "...reality to some purely theoretical concepts of nucleosynthesis. For example, the distinct stellar origins of the s- and r-processes have been clearly demonstrated by the identification of the s-process signature in the abundances of Xe isotopes from SiC and the r-process signature for Ba and Nd [in] the Allende meteorite. Nearly pure s-process Ba, Nd and Sm have subsequently been found in IS SiC" (6).

Finally, a fraction of the abundant hydrocarbons in primitive meteorites appears to be retained from the organic solids formed by photolysis on the surfaces of IS grains. Polycyclic aromatic hydrocarbons (PAHs) constitute some of this material, while the aliphatic component of the organic matter in the aqueously altered Murchison carbonaceous meteorite consists of diverse branched, alkyl-substituted cycloalkanes with carbon numbers from about 15-30. The remaining organic material is a macromolecular kerogen, consisting of small aromatic structural units linked by short- chain aliphatic bridges. This material is D-rich, indicating an IS origin. Amino and carboxylic acids in some aqueously altered meteorites were derived later from simpler hydrocarbons also of IS origin, as determined by their D-enrichments (9,10,11).

Interplanetary dust particles originate from collisions among asteroids and from the dissipation of comets. Comet IDPs are typically $5-30\mu m$ in diameter and consist of porous aggregates of pyroxene, olivine, and iron-rich sulfides; their bulk elemental composition is chondritic (solar). Within cometary IDPs Bradley (12) has identified tiny ($0.1-0.5\mu m$) glassy grains showing depletions of Mg and Si and inclusions of Fe-Ni metal and Fe-rich sulfides. These building blocks of the cometary IDPs are termed GEMS (glass with embedded metal and sulfides). Solids in space are irradiated by H and He ions, producing physical erosion (mass loss), the formation of glassy rims on grains (as crystalline material is rendered amorphous or glassy), elemental imbalances, and implantation of nuclear tracks from solar flares. Cometary IDPs have irradiation- induced morphologies and all the other characteristics just noted, indicating $\sim10^4$ years of exposure to the present solar wind. Individual GEMS also have characteristics of irradiation, but for much longer time scales (12) on the order of 10^8 years, typical of IS grains. Thus GEMS were irradiated before they accreted into the planetesimals that later became active comets (12,13).

PAHs of extraterrestrial origin are also found in some IDPs (14). Phenanthrene (178 amu), pyrene (202 amu), chrysene (228 amu), etc. are found, plus evidence for methyl- and methylene-substituted components. Large D and heavy N enrichments are found in these particles.

Most of the meteoritic material accreted by Earth in the current epoch arives as some $1.6 \ 10^7$ kg yr^{-1} of micrometeorites of $50-400\mu m$ size. This material, which is excavated from the Antarctic ice, is chemically distinct from all other meteoritic material, but is most like the CM chondrites (e.g., Murchison) (15).

Grains analyzed in situ in the coma of Comet Halley by the Giotto spacecraft ranged in mass from 10^{-11} to 10^{-16} g. The denser particles are silicate-dominated and less dense particles consist of refractory organics called CHON (carbon, hydrogen, oxygen, nitrogen). The CHON, which consists of highly unsaturated polycondensates rich in C=C and C-O compounds, is probably a coating on silicate cores, in general accord with the model of Greenberg (16). The CHON elements are more abundant in comets than in CI meteorites, with C and O abundances approaching the cosmic abundance values (17).

4. References

1. Dermott, S.F., Jayaraman, S., Zu, Y.L., Gustafson, B., and Liou, J.C. 1994 *Nature* **369**, 719-723.
2. Grün, E., Gustafson, B., Mann, I., Baguhl, M. Morfill, G.E., Staubach, P., Taylor, A., and Zook, H.A. 1994 *Astron. & Astrophys.* **286**, 915-924.
3. Pendleton, Y.J., Sandford, S.A., Allamandola, L.J., Tielens, A.G.G.M., and Selgren, K. 1994 *Ap. J.* **437**, 683-696.
4. Greenberg, J.M., Li, A., Mendoza-Gomez, C.X., Schutte, W.A., Gerakines, P.A., and de Groot, M. 1995 *Ap. J.* **455**, L177-L180.
5. Humes, D.H. 1980 *J. Geophys. Res.* **85**, 5841.
6. Anders, E., and Zinner, E. 1993 *Meteoritics* **28**, 490-514.
7. Ott, U. 1993 *Nature* **364**, 25-33.
8. Zinner, E. 1988 *In* **Meteorites and the Early Solar System**, eds J.F. Kerridge and M.S. Matthews, University of Arizona Press, Tucson, USA, 956-983.
9. Cronin, J.R., Pizzarello, S., and Cruikshank, D.P. 1988 *In* **Meteorites and the Early Solar System**, eds J.F. Kerridge and M.S. Matthew Univ. of Arizona Press, Tucson, USA, 819-857.
10. Pizzarello, S., Krishnamurthy, R.V., Epstein, S., and Cronin, J.R. 1991 *Geochim. Cosmochim. Acta* **55**, 905-910.
11. Zinner, E. 1996 *Science* **271**, 41-42.
12. Bradley, J.P. 1994 *Science* **265**, 925- 929.
13. Martin, P.G. 1995 *Ap. J.* **445**, L63-L66.
14. Clemett, S.J., Maechling, C.R., Zare, R.N., Swan, P.D., and Walker, R.M. 1993 *Science* **262**, 721-725.
15. Walter, J., Kurat, G., Brandstätter, F., Koeberl, C., and Maurette, M. 1995 *Meteoritics* **30**, 592 (abstract).
16. Greenberg, J.M. 1982 *In* **Comets**, ed L. Wilkening, Univ. of Arizona Press, Tucson, USA, 131-163.
17. Jessberger, E.K., and Kissel, J. 1991 *In* **Comets in the Post-Halley Era-II**, eds R. L. Newburn, Jr., M. Neugebauer, and J. Rahe, Kluwer, Dordrecht, Holland, 1075-1092.

PRE-STELLAR CORES AND THE INITIAL CONDITIONS FOR STAR FORMATION

D. WARD-THOMPSON & N. E. JESSOP

Royal Observatory,
Blackford Hill, Edinburgh, UK

Abstract. Results are outlined of a submillimetre study of a pre-stellar core. The pre-stellar phase is one in which a dense core in a molecular cloud is gravitationally bound, but contains no embedded luminosity source. This takes place prior to the protostellar Class 0 phase. Observations of a pre-stellar core show that the radial temperature profile is consistent with external heating, while the density profile is inconsistent with a singular isothermal sphere but consistent with an ambipolar diffusion model.

1. Introduction

An empirical picture is gradually emerging of the stages through which a low mass young stellar object must pass in its evolution from molecular cloud core to main sequence star. The initial Class 0 stage represents the main protostellar accretion phase (André, Ward-Thompson & Barsony 1993). The subsequent phases of the infrared Classes I to III (Lada 1987) represent the late accretion phase and T Tauri phases respectively (André & Montmerle 1994). In this paper we explore the phase of evolution, which occurs earlier than any of the above, and which we originally termed pre-protostellar (Ward-Thompson et al. 1994), but now term pre-stellar for brevity. This may be defined as the phase in which a gravitationally bound core has formed in a molecular cloud, and evolves towards progressively higher degrees of central condensation, eventually leading to protostellar collapse, but no central protostar exists yet in the dense cloud core. We present JCMT and IRAM mm/submm data of a pre-stellar core, and discuss its nature. The significance of these observations is that pre-stellar cores represent the initial conditions for protostellar collapse.

D. L. Block and J. M. Greenberg (eds.), New Extragalactic Perspectives in the New South Africa, 489-492.
© 1996 *Kluwer Academic Publishers.*

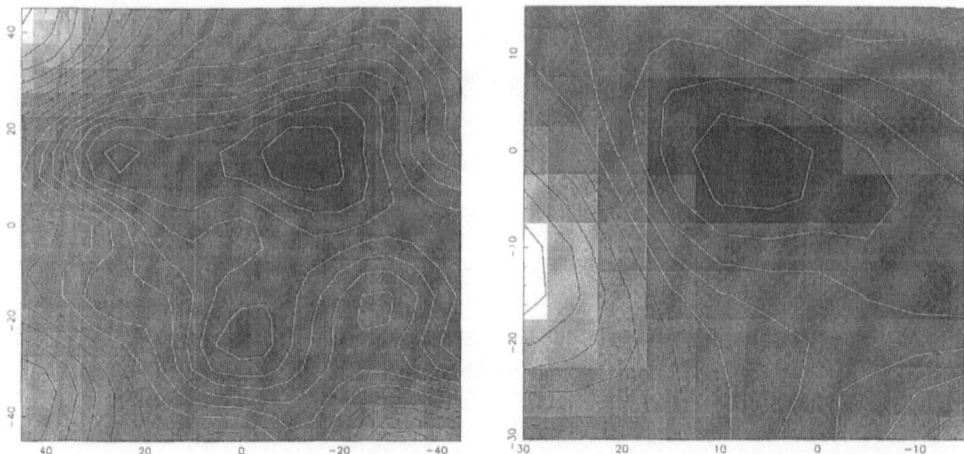

Figure 1. L1689B in the transitions of (a) $C^{18}O$ J=2→1 and (b) $C^{18}O$ J=3→2.

2. L1689B

Dense cores were catalogued by Myers & Benson (1983), and separated into 'starless cores' and cores with stars (Beichman et al. 1986). One such starless core was designated L1689B, which we observed in the submillimetre. Figure 1 (a) & (b) shows maps of L1689B made at the JCMT in July 1995 in the transitions of $C^{18}O$ J=2→1 and J=3→2 respectively. While the J=3→2 map shows a peak at the centre of the L1689B core, the J=2→1 map does not peak at the centre, but instead peaks in an apparent partial ring, or 'horse-shoe', around the core of L1689B. The simplest explanation of this is that the $C^{18}O$ J=2→1 line is optically thick towards the centre of the core, and that the core is externally heated, creating a 'limb-brightening' effect, while the J=3→2 line is optically thin and simply traces density. The departures from circular symmetry seen in Figure 1 can be explained by non-uniformities in the external radiation field.

The first submillimetre continuum maps of pre-stellar cores were made by Ward-Thompson et al. (1994). Figure 2(a) shows an 800-μm map of the pre-stellar core L1689B taken from Ward-Thompson et al. They found that pre-stellar cores have larger FWHM sizes than, but comparable masses to, the protostellar envelopes surrounding Class 0 protostars. This is consistent with pre-stellar cores being the precursors of Class 0 protostars.

Figure 2(b) shows the 1.3-mm isophotal contour map of L1689B taken from André, Ward-Thompson & Motte (1996). In this map, the L1689B core appears as an east-west elongated source (aspect ratio ∼ 0.7). In addition, a secondary component is clearly visible north-west of the main core. Its

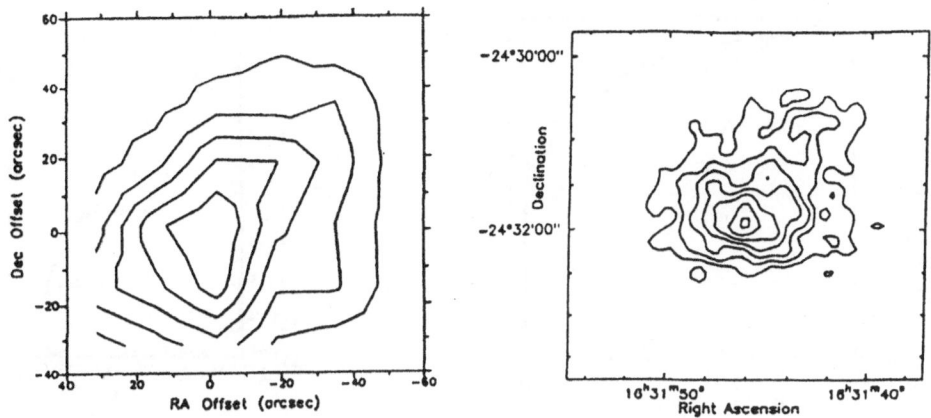

Figure 2. (a) 800-μm JCMT continuum map of pre-stellar core L1689B. (b) 1.3-mm IRAM continuum map of the same core at higher resolution.

presence probably explains the apparent SE-NW elliptical shape seen in the 800-μm map taken at lower angular resolution.

3. Discussion

The radial density profile of the core may be derived from the observed intensity profile, assuming there is no central heating source. If $\rho(r) \propto r^{-p}$ and $T(r) \propto r^{-q}$ are the density and temperature distributions as a function of radius r, and $I(\theta) \propto \theta^{-m}$ is the observed intensity profile as a function of projected radius θ, then p, q and m are related by the simple equation m=p+q−1. With no central heating source there is no temperature gradient, and hence q=0. Therefore the equation simplifies to m=p−1.

Figure 3(a) shows the azimuthally averaged flux density profile of the core L1689B, measured from Fig. 2(b) as a solid line. Note that we have measured the difference between the profiles in the major and minor axes of the core and found them not to be significantly different in form. For comparison a simulation was made of a dual-beam observation of a scale-free $\rho(r) \propto r^{-2}$ isothermal sphere profile. The dotted line in Fig. 3(a) shows its radial flux density profile and the IRAM beam is shown as a dashed line. The observed radial flux density profile shows a flat inner region up to 25 arcsec from the center and a steeper region beyond this. This is very similar to the profile observed for this core in the 800-μm data (Ward-Thompson et al. 1994). Hence we have here shown that the observed radial density profile does indeed flatten towards the centre, and this is not an effect of

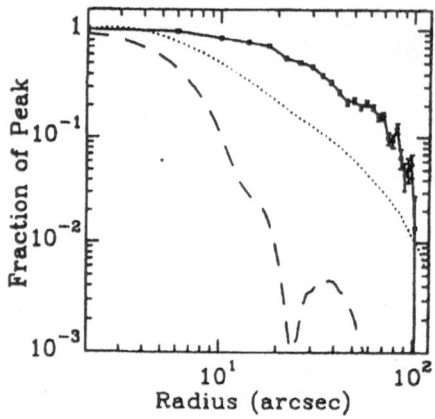 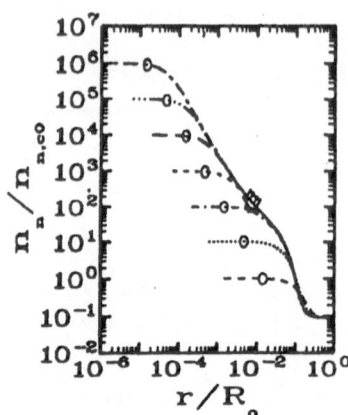

Figure 3. (a) Azimuthally averaged radial flux density profile of L1689B (solid line) compared to the IRAM beam (dashed line) and a simulated isothermal $\rho \propto r^{-2}$ profile (dotted line). (b) Predicted theoretical radial density profiles at various times for a magnetically supported pre-stellar core undergoing ambipolar diffusion.

telescope, instrumentation or observing technique. The profiles we observe are inconsistent with a single, scale-free power-law.

Figure 3(b) shows theoretical density profiles (Basu & Mouschovias 1995) of a pre-stellar core which is undergoing ambipolar diffusion. The profiles show a time sequence of evolution of the core from bottom to top, with each successive step having an order of magnitude greater central density than the previous one. The core spends the majority of its time in the lowest state. In this state the central part of the profile is flat, and it steepens to r^{-2} in the outer part. The supercritical core generally forms at a relatively early stage, when the centre still has a flat density distribution. Hence the data appear to agree reasonably well with theoretical predictions for radial core density profiles during the ambipolar diffusion phase.

References

André, P., Ward-Thompson, D., Barsony, M. (1993) *ApJ* **406**, 122
André, P., Ward-Thompson, D., Motte, F. (1996) *A&A* in press
André, P., Montmerle, T. (1994) *ApJ* **420**, 837
Basu, S., Mouschovias, T.Ch. (1995) ApJ **452**, 386
Beichman, C.A. et al. (1986) *ApJ* **307**, 337
Lada, C.J. (1987) *IAU Symp.* **115**, 1
Myers, P.C., Benson, P.J. (1983) *ApJ* **266**, 309
Ward-Thompson, D., Scott, P., Hills, R.E., André, P. (1994) *MNRAS* **268**, 276

DUST AND MORPHOLOGY OF HIGH REDSHIFT RADIO GALAXIES : CLUES FROM SCATTERING

ANDREA CIMATTI

Osservatorio Astrofisico di Arcetri
Largo E. Fermi 5, I-50125, Firenze, Italy

AND

IGPP, Lawrence Livermore National Laboratory, USA

1. Introduction

High redshift radio galaxies (HzRGs) are observable to cosmological distances rivalling those of the most distant quasars ($z_{max}=$ 4.4, Lacy et al. 1996 in preparation), and provide the opportunity to investigate the properties of galaxies at very early cosmological epochs. Their main advantage over quasars is that the continuum emission of HzRGs is spatially resolved in ground-based observations, allowing the study of their host galaxies at optical and near-IR wavelengths (McCarthy 1993 for a recent review). The two main problems present in the study of HzRGs are that they are faint in the optical ($R \sim$20-25 for 1<z<4), which limits the feasibility of detailed observations, and they are active galaxies. The active nucleus adds a component of non-stellar radiation and complicates the extraction of information on the stellar content and evolutionary state of the host galaxy.

In the radio, HzRGs have Fanaroff–Riley II (FRII) morphologies (classical double-lobed sources with $P_{radio} > 10^{25}$ WHz^{-1} at 178 MHz). In the optical, HzRGs are characterized by extended (up to $\sim$100 kpc) emission line regions (EELRs) and rest-frame UV continua both aligned with the radio source axis. In particular, when the observations are made in a fixed optical band, the rest-frame UV continua become systematically aligned with the radio axis for $z > 0.7$ (alignment effect; McCarthy et al. 1987; Chambers et al. 1987). On the other hand, the rest-frame optical continua (observed in the near-IR) are generally 'rounder', resembling that of nearby elliptical galaxies (McCarthy 1993 and references therein) (Fig. 1).

D. L. Block and J. M. Greenberg (eds.), New Extragalactic Perspectives in the New South Africa, 493–500.
© *1996 Kluwer Academic Publishers.*

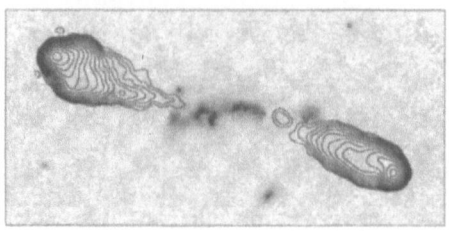

$1'' \sim \overline{10}$ kpc

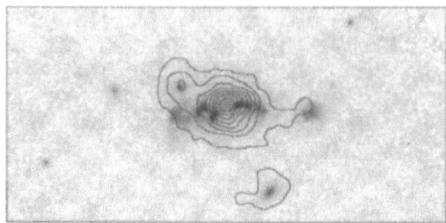

Figure 1. *(a) Top:* HST Planetary Camera image of 3C 324 through the F702W filter. The contours show the 5 GHz VLA map (Fernini et al. 1993). *(b) Bottom:* KPNO 4m K-band image superimposed on the HST image. All the figures are courtesy of Mark Dickinson, and taken from Dickinson, Dey & Spinrad (1995).

The origin of the alignment effect is still debated and many theories have been proposed to explain the nature of the UV continuum : starburst induced by the propagation of the radio source in the ISM of the host galaxy (McCarthy 1993 and references therein), scattering of anisotropic radiation (di Serego Alighieri et al. 1989), inverse Compton scattering (Daly 1992), propagation of plasma in a high density environment (Eales 1992), nebular continuum (Dickson et al. 1995). Furthermore, since the UV properties of nearby radio galaxies are poorly known, it is not established whether the alignment effect is an evolutionary phenomenon occurring only at high z (Cimatti & di Serego Alighieri 1995). Some results seem to suggest a link with the radio power, the alignment effect being weaker in lower power radio galaxies (Thompson et al. 1994).

2. The role of polarimetry

Following the idea of Tadhunter et al. (1988), the first optical polarimetric observations of two HzRGs showed the presence of high linear polarization with $\vec{E}$ oriented perpendicular to the radio-UV axis (di Serego Alighieri et al. 1989). This result suggested the presence of a non-stellar component in the UV continuum.

In order to investigate the generality of that result and to study the origin of the alignment effect, we started a systematic program of optical polarimetry in 1991. Polarimetry is one of the most powerful techniques to disentangle different radiation components, and provides important byproducts like deep total flux images and spectra in case of imaging-polarimetry and spectropolarimetry respectively.

Polarimetry is also an important tool for testing the basic hypothesis of the unified model of radio-loud AGN. In fact, if FRII radio galaxies host a quasar nucleus which is invisible because of orientation effects (e.g. obscured by a torus of optically thick gas/dust), then the quasar radiation should be emitted anisotropically and scattered by the dust and the electrons present in the ISM of the host galaxy (Antonucci 1984; Urry & Padovani 1995 for the most recent review). If this scenario is correct, and the torus axis of symmetry is oriented parallel to the radio axis, strong linear polarization with $\vec{E}$ orthogonal to the radio axis is expected. In this regard, HzRGs provide a unique laboratory because their high redshifts allow the ground-based observation of their rest-frame UV spectral region, where the effects of the diluting unpolarized light from the evolved stellar population of the host galaxy is small.

However, their faintness limits the observations to the brightest objects. In order to obtain an error on the degree of polarization P of the order of $\sigma_P \sim$1-2%, the observational limits for a 4m class telescope equipped with a CCD detector are typically $R \sim 22$ and $R \sim 20$ with imaging-polarimetry and spectropolarimetry respectively. We observed 21 galaxies with $0.05 < z < 2.6$ with different telescopes depending on the redshifts (WHT, ESO 3.6m, NOT, ESO 2.2m, Loiano 1.5m). The low redshift objects were included in order to have a comparison with the HzRGs. Other polarimetric observations of HzRGs have been made also by Tadhunter et al. (1992) and Jannuzi et al. (1991,1995). Finally, in 1995 I had the privilege to participate in a spectropolarimetry program at the W.M. Keck 10m telescope aimed to the detailed study of HzRGs.

3. Results and Implications

3.1. IMAGING–POLARIMETRY

The most recent review of the imaging-polarimetry results has been presented by Cimatti et al. (1993). Other results can be found in the papers by Tadhunter et al. (1992), Draper et al. (1993), Cimatti & di Serego Alighieri (1995) and Jannuzi et al. (1991,1995). The main findings for the HzRGs can be summarized as follows : (1) strong linear polarization is observed in radio galaxies with $z > 0.7$ at level of $P \sim$5-20%, (2) $\vec{E}$ is always perpendicular to the major axis of the UV continuum, (3) the polarized flux is spatially extended in all the cases where the S/N is sufficient. Recent Keck imaging-polarimetry has provided detailed polarization maps (M. Cohen & H. Tran, personal communication).

The non-correspondence between the UV and radio continua, and the absence of strong reddening, exclude respectively synchrotron and dichroic transmission as dominant polarization mechanisms, leaving scattering as the most likely polarization process. This conclusion is strengthened by the perpendicularity of $\vec{E}$ to the UV continuum axis.

At lower redshifts, the results are being analysed, but some premiliminary implications can already be drawn (Cimatti & di Serego Alighieri 1996, in preparation). In the *rest-frame optical* (observed typically in V or R bands), the polarization is generally low, shows a variety of $\vec{E}$ orientations , and is spatially extended (Draper et al. 1993). In the *rest-frame UV* (typically in U and B bands depending on z), we find that only a few radio galaxies with $0.1 \le z < 0.5$ show aligned and perpendicularly polarized UV continua. Therefore, at low z the alignment effect is not as ubiquitous as it is at high z, and may be considered an 'evolutionary' phenomenon, probably related to the radio power and/or to the environment properties.

In summary, imaging-polarimetry has shown the generality of the UV polarization in HzRGs, its consistency with the basic requirement of the unified model, and has suggested that the alignment effect is an environmental and/or luminosity-dependent phenomenon.

3.2. SPECTROPOLARIMETRY

Spectropolarimetry is a uniquely powerful tool because it allows the spectral analysis of the polarization and to search for signatures of the putative quasar incident spectrum, providing the most stringent test for the unified model. The early observations by Antonucci (1984) showed for the first time a few cases of low z radio galaxies with broad lines in their polarized flux spectra, and suggested the unification scenario.

The first spectropolarimetric results for HzRGs were obtained with the ESO 3.6m telescope by di Serego Alighieri, Cimatti & Fosbury (1994). Despite the faintness of the objects, the observations of 3C 226 and 3C 277.2 ($z \sim 0.8$) showed perpendicularly polarized MgII2800 line, unpolarized [OII]3727 line, and constant position angle of $\vec{E}$. A similar result has been obtained on 3C 265 ($z \sim 0.8$) with the WHT 4.2m telescope (di Serego Alighieri et al. 1996). These observations also show that spectropolarimetry with 4m class telescopes is limited to the few brightest HzRGs.

The advent of the W.M. Keck 10m telescope provided the possibility to perform high S/N spectropolarimetry (see Fig.2) and to extend the observations to the majority of the HzRGs. During 1995 we started a spectropolarimetry program, observed 6 HzRGs with $0.7 < z < 1.8$, and analysed the

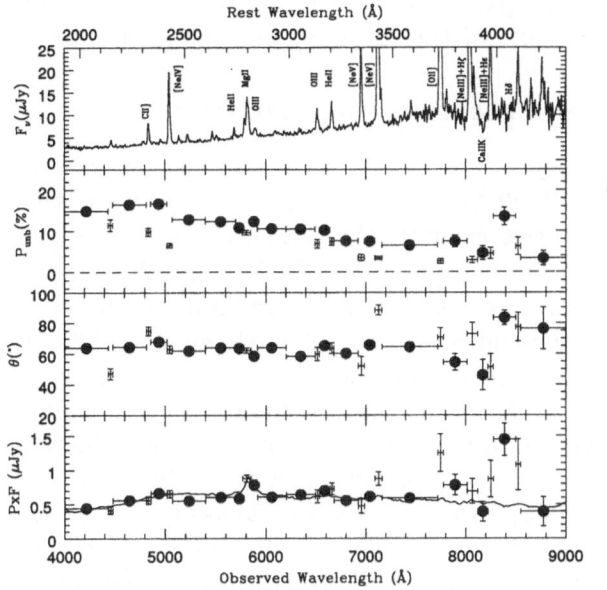

Figure 2. Keck spectropolarimetry of 3C 356a. From top to bottom : the total flux spectrum, the position angle of $\vec{E}$, the degree of polarization and the polarized spectrum. Filled circles indicate continuum bins. Crosses indicate bins including emission lines + continuum. The continuum line in the $P \times F$ plot is a radio–loud quasar average spectrum scattered by Galactic type dust (E_{B-V}=0.05) (Cimatti et al. 1996 in preparation).

data for 5 of them. The results for 3C 256 ($z \sim 1.8$) and 3C 324 ($z \sim 1.2$) have been recently pulished by Dey et al. (1996) and Cimatti et al. (1996) respectively. The general results obtained are the following : (1) high linear polarization of the UV continuum is detected in all the 5 galaxies; (2) the observed $P(\lambda)$ are either flat or blue; (3) the perpendicularity of $\vec{E}$ to the UV continuum axis and the constancy of $\theta(\lambda)$ suggest that scattering is the dominant polarization mechanism; (4) the detection of the MgIIλ2800 emission line in polarized flux in at least 3 galaxies suggests that the incident radiation comes from an obscured quasar nucleus and is emitted anisotropically along the radio axis. In particular, the broad and polarized MgIIλ2800 in 3C 324 has velocity and equivalent widths consistent with those observed in radio-loud quasars (Cimatti et al. 1996); (5) the polarized flux continua can be reproduced by a dust and/or electron scattered radio-loud quasar spectrum; (6) the lower or null polarization of the [OII]λ3727 line implies that this line is emitted isotropically outside the obscuring region; (7) in the two galaxies analysed in detail so far (Dey et al. 1996; Cimatti et al. 1996), we observe *spatially extended* polarization along the UV continuum axis, implying that scattered light is a necessary ingredient to explain the alignment effect. In particular, in 3C 324 we suggest that scattered light contributes up to 50% of the total UV continuum (Cimatti et al. 1996).

4. The nature of the scattering particles

The nature of the scattering particles is still an open question. In the cases where the MgIIλ2800 line is detected significantly in the polarized flux, its FWHM is consistent with that observed in radio-loud quasars. This excludes the scenario proposed by Fabian (1989) of hot ($T \gg 10^5$ K) electrons scattering, because the scattered lines would be strongly or even completely Doppler–smeared out (Cimatti et al. 1996). The exclusion of hot electrons leaves dust grains and cooler ($T \leq 10^5$ K) electrons as the possible scatterers. However, the present data do not allow to distinguish between these possibilities because $P(\lambda)$ (shape and degree of polarization) and $P(\lambda) \times F_\nu(\lambda)$ can be reproduced both by dust and electron scattering (Cimatti et al. 1996). An unambiguous answer may be impossible simply because *both dust and electrons* are likely to be present in the multi-phase ISM of radio galaxies. In fact, there is growing evidence for a significant amount of dust in the ISM of HzRGs. Submillimetric continuum observations have revealed the presence of dust in a few HzRGs (Ivison 1995 and references therein). Optical observations have led to the discovery of some HzRGs whose spectral properties are affected by a large amount of dust in their ISM (van Ojik et al. 1994; Dey, Spinrad & Dickinson 1995). In addition, the observed Lyα/Hα ratios (McCarthy 1993) and the recent near-IR observations of broad Hα lines in some HzRGs (Rawlings et al. 1995 and references therein) also suggest the presence of dust in HzRGs. Even though these results suggest that dust is certainly present in HzRGs, they contain no information about its spatial distribution, which would be crucial for the intepretation of the polarimetric results and the alignment effect. However, although warm electron scattering cannot be excluded, the larger scattering cross section per unit mass of the dust grains suggests a dominant contribution by dust scattering. In fact, scattering models predict realistic total dust masses of 10^{7-8} $M_\odot$, while in the case of pure electron scattering the required amount of ionized gas is usually exceedingly large (Manzini & di Serego Alighieri 1996 and references therein; Cimatti et al. 1996). Finally, we note that the dust masses inferred with sub-mm observations are in agreement with those required by the dust scattering models.

5. Stars and unpolarized light

The results discussed in the previous sections imply a scenario where the HzRGs are misdirected quasars. A significant fraction of the UV alignment effect is due to anisotropic radiation emitted by an obscured quasar and scattered toward the observer by the dust and the electrons present in the ISM of the host galaxy and in the companion galaxies intercepted by the radiation beam. Also the emission lines properties can be explained in terms of photoionization by an anisotropic nuclear source (McCarthy 1993 and references therein). The main open questions are : (1) what fraction of the UV continuum is due to unpolarized radiation ?, (2) what is the nature of the UV unpolarized radiation ?, (3) what is the stellar populations content and evolutionary state of HzRGs ?

Inverse-Compton scattering (Daly 1992) is unlikely to be dominant because the radio and UV continuum features do not show a one-to-one spatial correspondence. Also the plasma propagation scenario proposed by Eales (1992) seems unlikely because its prediction of a strong alignment effect in lower radio power galaxies has been not confirmed by recent observations (Thompson et al. 1994). Starburst and nebular continuum represent two possible contributors to the unpolarized fraction of the UV radiation. Dickson et al. (1995) have demostrated that a large fraction (up to $\sim$50%) of the total UV continuum may be due to nebular continuum. Therefore, the sum of scattered and nebular continuum does not leave much room to the starburst component predicted by the jet–induced star formation scenario. In fact, the evidence of young massive stars is marginal (McCarthy 1993). However, one interesting case is provided by 3C 324 (z=1.2), where an unpolarized companion galaxy lies along the radio axis and shows a continuum spectrum characteristic of a star forming galaxy (Cimatti et al. 1996). At low redshifts, a number of similar examples are known (van Breugel & Dey 1993 and references therein). In this regard, we notice that the rest-frame range $\sim$2000–4000 Å, (where the deepest spectra are presently available) is not the most appropriate region to search for spectral signatures of a starburst, but better clues might come from deep spectroscopy with 10m class telescopes in the rest-frame region $\sim$1000–2000 Å, where most of the UV absorbtion lines of O and B stars are expected.

A secure age estimate of the oldest stellar populations present in HzRGs might provide clues on their formation epochs and could provide stringent limits on the possible values of H_0 and q_0. The possibility that the host galaxies of HzRGs are evolved (age $\geq$1 Gyr) already at $z \sim$1–2 is supported by their K-band morphologies which appear to be similar to those of nearby ellipticals, their colors, the tight $K-z$ relation, and the spectral signatures (4000 Å break and CaII K absorption) observed in a few objects at $z \sim$1 (McCarthy 1993 and references therein; Stockton et al. 1996). Furthermore, the non-detection of CO emission in HzRGs (van Ojik 1995) has provided limits on the gas mass slightly less than 10^{11} $M_\odot$. One interpretation is that HzRGs are not forming stars at the rates required by the jet-induced star formation, and suggests that they may have already converted most of their molecular gas into the bulk of their stellar population. Even though the previous arguments favour quite strongly the presence of old stars in HzRGs, further observations are required before HzRGs may be used as probes of galaxy evolution.

6. Acknowledgements

I am grateful to Ski Antonucci, Sperello di Serego Alighieri, Bob Fosbury, Marco Salvati, and Wil van Breugel for their useful suggestions, and to Mark Dickinson for providing the figures of 3C 324. The W.M. Keck telescope observing program is in collaboration with Wil van Breugel, Arjun Dey, Ski Antonucci, Hyron Spinrad, and Todd Hurt.

References

Antonucci R. 1984, *ApJ*, 278, 499
Chambers K.C., Miley G.K., van Breugel W.J.M. 1987, *Nature*, 329, 604
Cimatti A., di Serego Alighieri S., Fosbury R. A. E., Salvati M., & Taylor D. 1993, *MNRAS*, 264, 421
Cimatti A., di Serego Alighieri S. 1995, *MNRAS*, 273, L7
Cimatti, A., Dey, A., van Breugel, W., Antonucci, R., Spinrad, H., 1996, *ApJ*, in press
di Serego Alighieri S., Fosbury R., Quinn P., Tadhunter C. 1989, *Nature*, 341, 307
di Serego Alighieri S., Cimatti A., & Fosbury R. 1994, *ApJ*, 431, 123
di Serego Alighieri S., Cimatti A., Fosbury R., Perez-Fournon I. 1996, *MNRAS* submitted
Daly R.A. 1992, *ApJ*, 399, 426
Dey A., Spinrad H., Dickinson M. 1995, *ApJ*, 440, 515
Dey, A., Cimatti, A., van Breugel, W., Antonucci, R., Spinrad, H., 1996, *ApJ*, in press
Dickinson M., Dey A., Spinrad H. 1995, in Proc. Conf. on "Galaxies in the Young Universe", ed. H. Hippelein & K. Meisenheimer, Springer-Verlag, in press
Dickson R., Tadhunter C.N., Shaw M., Clark N., Morganti R. 1995, *MNRAS*, 273, L29
Draper P.W., Scarrott S.M., Tadhunter C.N. 1993, *MNRAS*, 262, 1029
Eales S.A. 1992, *ApJ*, 397, 49
Fabian A.C. 1989, *MNRAS*, 238, 41p
Ivison R.J. 1995, MNRAS, 275, L33
Jannuzi B.T., Elston R. 1991, *ApJ*, 366, L69
Jannuzi B.T., Elston R., Schmidt G., Smith P., Stockman H. 1995, *ApJ*, in press
Manzini A., di Serego Alighieri S. 1996, *A&A*, in press
McCarthy, P.J., van Breugel, W.J.M., Spinrad, H., Djorgovski, S., 1987, *ApJ*, 321, L29
McCarthy P.J. 1993, *ARAA*, 31, 693
Rawlings S., Lacy M., Sivia D.S., Eales S.A. 1995, *MNRAS*, 274, 428
Stockton A., Kellogg M., Ridgway S. 1996, *ApJ*, in press
Tadhunter C.N., Fosbury R.A.E., di Serego Alighieri S. 1988, Proceedings "BL Lac Objects", ed. Maraschi L., Maccacaro T., Ulrich M.H., Springer-Verlag, p. 79
Tadhunter C.N., Scarrott S.M., Draper P., Rolph C. 1992, *MNRAS*, 256, 53p
Thompson D., Djorgovski S., Vigotti M., Grueff G. 1994, *AJ*, 108, 828
Urry C.M., Padovani P. 1995, *PASP*, 107, 803
van Breugel W., Dey A. 1993, *ApJ*, 414, 563
van Ojik R., Röttgering H.J.A., Miley G.K., Bremer M.N., Macchetto F., Chambers K.C. 1994, *A&A*, 289, 54
van Ojik R. 1995, *PhD Thesis*, Leiden University, The Netherland

DUST OBSCURED QUASARS

PAUL J. FRANCIS AND RACHEL J. WEBSTER
University of Melbourne, School of Physics
Parkville, Victoria 3052, Australia

BRUCE A. PETERSON
Mt Stromlo and Siding Spring Observatory
The Australian National University, Private Bag, Weston Creek
Post Office, ACT 2611, Australia

MICHAEL J. DRINKWATER
Anglo-Australian Observatory
Coonabarabran, NSW 2357, Australia

AND

FRANK J. MASCI
University of Melbourne

Abstract.

We find that nearly all radio-loud, flat spectrum quasars are obscured by a magnitude or more of dust. The dust is probably located in the quasar or its host galaxy. If radio-quiet QSOs have comparable amounts of dust, $\sim 80\%$ have been missed by all existing surveys. These missing QSOs may explain the X-ray background.

1. Introduction

It has long been folk wisdom in the field of quasar research that quasars are blue. There has been some discussion of dust in quasars, but the consensus has been that its effects are subtle.

We claim that the apparent lack of dust is a selection effect. Nearly all detailed work on quasars has concentrated on the optically brightest, which are the least dust obscured. If an unbiassed sample of quasars is studied, the effects of dust are dramatic.

D. L. Block and J. M. Greenberg (eds.), New Extragalactic Perspectives in the New South Africa, 501–504.
© *1996 Kluwer Academic Publishers.*

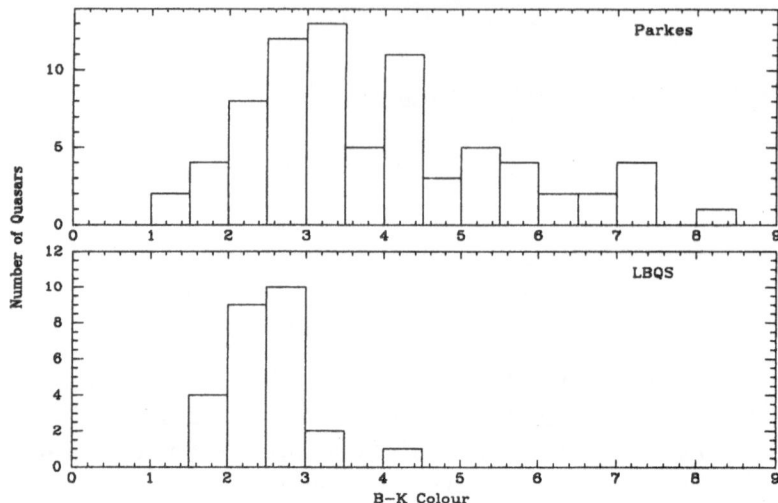

Figure 1. The distribution of $B - K$ colours for a complete sub-sample of the Parkes flat-spectrum survey, and for an optically selected QSO survey, the Large Bright QSO Survey (LBQS, Morris et al. 1991).

2. The Parkes Flat-Spectrum Survey

We are studying a complete sample of 323 flat spectrum radio sources, selected with the Parkes radio telescope to be brighter than 0.5 Jky at 2.7-GHz and covering most of the southern sky. We have accurate radio positions, and are nearing the end of a campaign to identify all the sources, solely on the basis of positional correspondence. We are obtaining spectroscopy, optical and near-IR images of all our sources. Nearly all the sources are quasars. Because the sample is radio selected, it is unbiassed by dust.

In Figure 1 we show the distribution of $B - K$ colours of complete subset of our quasars, compared to the colours of an optically selected QSO sample. It is immediately obvious that the Parkes sample contains a tail of extremely red objects not seen in the optically selected, radio-quiet sample. The colours of the optically selected QSOs lie right on the blue edge of the colour distribution of the Parkes quasars.

3. Why are our Quasars so Red?

Are our sources red because of dust obscuration, or because some red continuum emission component is abnormally strong? If the latter, equivalent widths of the emission-lines should strongly anticorrelate with the redness, as the red continuum component dilutes them. Nothing of the kind is seen;

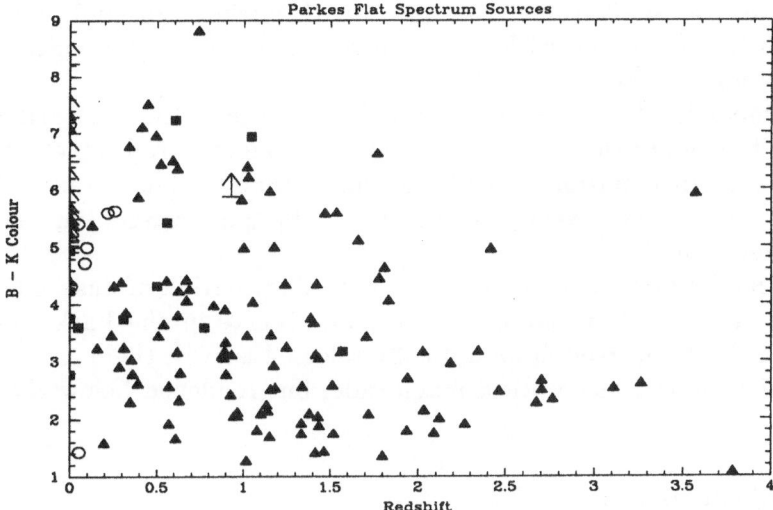

Figure 2. The distribution of $B - K$ colours for the Parkes sample, as a function of redshift.

the emission-line equivalent widths of our reddest quasars are identical to those of the bluest. *We conclude therefore that our quasars are red because of dust.*

This is also consistent with the emission-line ratios and the shape of the continuum.

4. Where is the Dust?

Is the dust in the quasars themselves, in their host galaxies, or somewhere along the line of sight in an intervening galaxy? The lack of any correlation between redness and redshift (Figure 2) suggests that the dust is not intervening; at the moment we cannot tell whether it is in the QSO or the host galaxy.

5. Why aren't there any Optically Selected Dusty QSOs?

It is possible that there are simply no dusty radio-quiet QSOs; for some reason dust may only be associated with radio-loud quasars. However, even if radio-quiet QSOs have just as much dust as the Parkes quasars, we would not expect to detect many red quasars optically. Consider an optical QSO sample with a limiting B-band magnitude of m. For a QSO with five magnitudes of B-band dust extinction to make it into the sample, it would have to have an intrinsic brightness of $m - 5$. However, due to the steepness of the QSO luminosity function, QSOs with intrinsic B-band magnitudes of $(m - 5)$ are $\sim 10^7$ rarer than QSOs with intrinsic B-band magnitudes m. Thus in any survey with a blue magnitude limit, dusty QSOs will be seriously under-represented.

We modelled this bias, and showed that even if the true distribution of $B - K$ colours of radio-quiet QSOs was like the top panel of Fig 1, the observed distribution in a B-band magnitude limited sample would be like the lower panel. Nearly all the red radio-quiet QSOs fall below the magnitude limit.

The model further predicted that *if* the distribution of dust extinctions in radio-quiet QSOs is similar to that we observe in the Parkes quasars, for every QSO observed in an optically selected survey, there are ~ 5 more with the same intrinsic optical magnitude, but reddened below the survey limit.

6. Implications

If radio-quiet QSOs are as dusty as our radio-loud ones (as suggested by the data of Hines et al. 1995), $\sim 80\%$ *of all QSOs have been missed by all existing surveys.*

These QSOs, with their dust-obscured X-ray spectra, are exactly what are needed to explain the X-ray background (Madau, Ghisellini & Fabian 1994).

If this picture is true, all QSOs, both radio-loud and radio-quiet, have very similar intrinsic continuum slopes in the rest-frame UV, optical and near-IR; roughly power-laws of the form $F_\nu \propto \nu^{-0.3}$. All the observed range of slopes may be caused by dust.

7. References

An earlier version of this work is reported in Webster et al. (1995)

References

Hines, D. C., Schmidt, G. D., Smith, P. S., Cutri, R. M. & Low, F. J. (1995) *ApJ*, **450**, L1
Madau, P., Ghisellini, G., & Fabian, A. C. (1994), *MNRAS*, **270**, L17
Morris, S. L. et al. (1991) *AJ*, **102**, 1627
Webster, R. L., Francis, P. J., Peterson, B. A., Drinkwater, M. J. & Masci, F. J. (1995), *Nature*, **375**, 469

THE OPACITY OF LOW SURFACE BRIGHTNESS GALAXIES AND THEIR CONTRIBUTION TO QUASAR ABSORPTION

C.D. IMPEY
University of Arizona
Steward Observatory, University of Arizona, Tucson
Arizona 85721, USA

1. Introduction

The sum of the mass in the luminous parts of the $\sim 10^{10}$ galaxies in the universe amounts to only 0.3% of the closure density. Since this is well below the lower bound predicted by the successful model of big bang nucleosynthesis (Walker et al. 1991), there are presumably many baryons waiting to be found. Dust can dim and redden the light from distant galaxies, removing them from optical catalogs. It can also extinguish the light from background quasars, and so obscure our view of the high redshift universe (Heisler & Ostriker 1988). In addition, there is a strong observational selection against diffuse or low surface brightness (LSB) galaxies (Disney & Phillips 1983), or galaxies that have not converted much of their gas into stars. Galaxies with intermediate surface brightness are visible out larger distances, so such galaxies must be overrepresented in any catalog. Both issues, the presence of dust and the neglect of LSB galaxies, hamper a complete census of baryons.

2. Low Surface Brightness Galaxies and Baryons

The existence of low surface brightness galaxies was first demonstrated in the Virgo cluster (Bingelli et al. 1985). In clusters, large numbers of LSB galaxies act to steepen the tail of the luminosity function. Outside of clusters, LSB galaxies cover a remarkable range from tiny, gas-poor dwarf spheroidals in the Local Group to gas-rich giant disks like Malin 1 (Impey & Bothun 1989). An R-band image of Malin 3 appears in these *Proceedings*

D. L. Block and J. M. Greenberg (eds.), New Extragalactic Perspectives in the New South Africa, 505–508.
© 1996 *Kluwer Academic Publishers.*

(see color plate 11b). Recently, we have completed the first survey of field LSB galaxies with extensive redshift information (Impey et al. 1996). Nearly 700 previously uncataloged LSB galaxies were discovered in an equatorial strip of the sky using the APM machine in Cambridge. Half of these diffuse galaxies have redshifts from either optical or 21 cm observations.

The results show that existing luminosity functions are significantly incomplete. When surface brightness selection effects are accounted for, there are equal numbers of galaxies per unit surface brightness interval (Sprayberry et al. 1996). Integrating the APM sample over the distribution of luminosity and surface brightness yields $L_{\rm LSB} = \int \int \phi(L, \mu) L \, dL d\mu = 5.2 \times 10^7 h_{100} \, L_\odot \, {\rm Mpc}^{-3}$, which is 50% of that calculated by Marzke et al. (1994) for field galaxies. This is sufficient to steepen the exponential tail of the luminosity function from $\alpha = -1$ to $\alpha = -1.25$, and perhaps to $\alpha = -1.4$, if the surface brightness corrections are extrapolated to fainter isophotes. Baryonic mass-to-light may increase slightly with decreasing luminosity, $(M/L)_{\rm baryon} \propto L^\eta$, and the predicted baryon contribution rises rapidly if $1 + \alpha + \eta < -1$ (Bristow & Phillips 1994). Based on existing evidence, it is likely that LSB galaxies can make up the shortfall between existing galaxy catalogs and the predictions of big bang nucleosynthesis.

3. Low Surface Brightness Galaxies and Dust

Little is known about the stellar populations and dust content of LSB galaxies. Typical colors are too blue ($B - V \sim 0.6$, $V - I \sim 1.0$) for them to be faded remnants of an earlier population of gas-rich galaxies. The blue colors are consistent with a young mean age, but at the moment the effects of age and metallicity cannot be disentangled. Many of the field LSB galaxies from the recent survey are not dwarfs, and most are gas-rich, with $M_{\rm HI} > 3 \times 10^8$ $M_\odot$. Along with gas and young stars, dust might be expected. The radial profiles of most LSB galaxies are well fit by exponentials, although this is not a reliable indicator of a disk system. The surface brightness vs. inclination test for opacity in spiral disks has a long history. Yet it can be a blunt tool for measuring opacity if the galaxy sample suffers from selection effects, or if the spatial distribution of the scatterers is not known (see Burstein et al. 1995, and many contributions in the same volume).

Following Valentijn (1994), we plot axial ratio as a function of local surface brightness at the effective radius, thereby avoiding distance-dependent selection effects. The top panel of Figure 1 shows that the APM sample has no correlation between effective surface brightness and axial ratio. The superimposed tracks show the coefficient in the relation $\mu^{\rm obs} = \mu^{\rm face-on} + 2.5C \log(b/a)$. In a slab model of dust well mixed with stars, $C = 1$ is transparent, and $C = 0$ is opaque. However, in realistic mod-

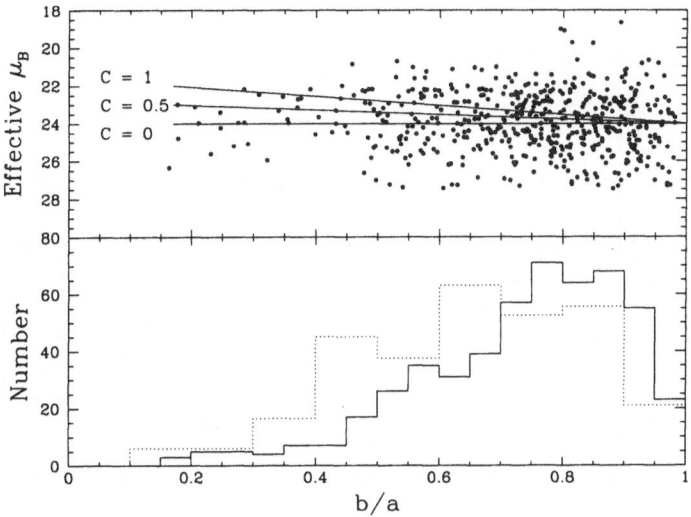

Figure 1. The distribution of effective surface brightness and axial ratio for LSB galaxies, with models of disk opacity superimposed (top panel), and the axial ratio distribution for LSB galaxies (solid) and exponential profile dwarfs (dashed) from the HST Medium Deep survey (lower panel).

els the value of C calculated from the slab model can underestimate (dust sandwiches, multiple scattering) or overestimate (bulges) the extinction. The lower panel shows that the LSB galaxies are quite round on average, the distribution of axial ratios is disimilar to magnitude limited samples of either optically thin or optically thick disks. It is, however, similar to the axis ratio distribution of the small "exponential" galaxies that make up most of the difference between the faint galaxy counts and non-evolving galaxy models (Im et al. 1995). LSB disks follow Tully-Fisher relations with the same slope as high surface brightness spirals; the much higher scatter of the LSB galaxies does not appear to be caused by dust.

4. Low Surface Brightness Galaxies and Quasar Absorption

The large space density of diffuse and often gas-rich galaxies has important implications for quasar absorption. Neutral hydrogen is difficult to detect in emission for $n_{HI} < 10^{18}$ atoms cm^{-2} or $z > 0.1$, whereas in absorption backlit by a quasar it can be detected down to $n_{HI} = 10^{12}$ atoms cm^{-2} and out to $z = 5$. Figure 2 shows how the integral of the path length to absorption depends on galaxy luminosity, the faint end luminosity function slope, and the scaling between luminosity and absorber cross section. Traditionally, people have adopted $R \propto L^{\beta}$ with $\beta = 0.4$, but $\beta = 0.2$ is indicated when surface brightness selection is minimized. The integral diverges and

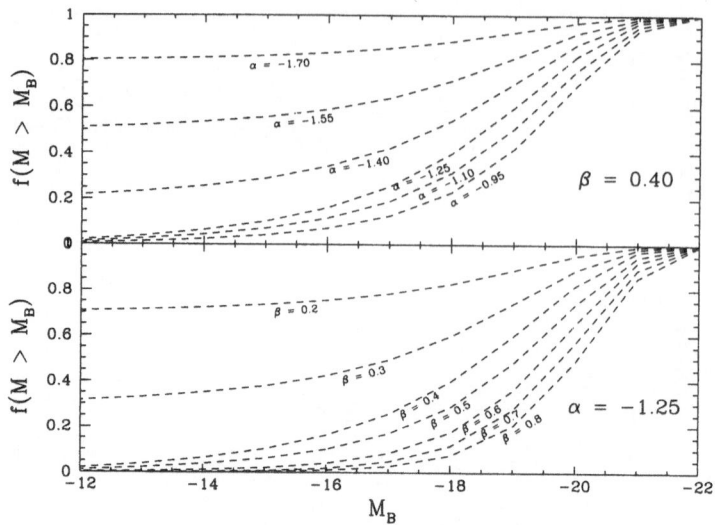

Figure 2. The fraction of quasar absorption caused by galaxies fainter than M_B as a function of α, the faint end slope of the luminosity function, and β, the scaling between galaxy size and luminosity.

dwarfs dominate the cross section if $\alpha + 2\beta < -1$. Since the high column density Mg II quasar absorbers are associated with luminous galaxies, it is likely that the more copious Lyman-α "forest" absorbers are traced by dwarf and LSB galaxies. At the high column density end, the limits on the space density of massive gas disks like Malin 1 (Briggs 1990), corresponding to $dN/dz < 0.019$, is consistent with the gas cross section of all normal galaxies ($dN/dz = 0.015 \pm 0.004$), and with the upper bound on damped Lyman-α absorbers at low redshift ($dN/dz < 0.05$).

References

Bingelli, B., Sandage, A., & Tammann, G.A. 1985, AJ, 90, 1681

Briggs, F.G. 1990, AJ, 100, 999

Bristow, P.D., & Phillips, S. 1994, MNRAS, 267, 13

Burstein, D., Willick, J. & Courteau, S. 1995, The Opacity of Spiral Disks, ed. J. Davies & D. Burstein (Kluwer: Dordrecht), p.73

Disney, M.J., & Phillips, S. 1983, MNRAS, 205, 1253

Heisler, J., & Ostriker, J.P. 1988, ApJ, 332, 543

Im, M., Ratnatunga, K.U., Griffiths, R.E., & Casertano, S. 1995, ApJLett, 445, L15

Impey, C.D., & Bothun, G.D. 1989, ApJ, 341, 89

Impey, C.D., Sprayberry, D., Irwin, M.J., & Bothun, G.D. 1996, ApJ, in press

Marzke, R.O., Huchra, J.P., & Geller, M.J. 1994, ApJ, 428, 43

Sprayberry, D., Impey, C.D., & Irwin, M.J. 1996, ApJ, in press

Valentijn, E.A. 1994, MNRAS, 266, 614

Walker, T.P., Steigman, G., Schramm, D., Olive, K., & Kang, H., 1991, ApJ, 376, 51

THE MORPHOLOGY OF MASSIVE, GAS-RICH LOW SURFACE BRIGHTNESS GALAXIES

PATRICIA M. KNEZEK
The Observatories of the Carnegie Institution of Washington
Las Campanas Observatory, Casilla 601, La Serena, Chile

1. Introduction

One of the more surprising discoveries in the recent work on low surface brightness galaxies (LSBs) is that a significant number are massive and possess significant quantities of atomic gas. Disk galaxies with sizes approaching that of Malin 1 ($M_{HI} \sim 2 \times 10^{11}$ $M_\odot$; D $\sim$ 147 kpc, $H_0 = 75$ km s^{-1} Mpc^{-1}; Impey and Bothun 1989) have been identified through the use of the Palomar Sky Surveys, with follow-up studies in HI. While many of these galaxies have large amounts of atomic gas and unusually blue disk colors, they have only weak regions of Hα emission, indicating little ongoing massive star formation, and their low surface brightnesses suggest extremely low stellar surface densities. The blue disk colors have ruled out the hypothesis that all LSBs are faded galaxies, but the difficulty in disentangling the difference between low metallicity and young stellar populations based on broadband optical colors has hindered understanding their stellar evolutionary history. It has been suggested (McGaugh 1992) that some LSBs may, in fact, be undergoing their first episodes of star formation since there is little evidence of a difference in the distribution of light in $UBVRI$ images. The LSBs in his sample are largely *dwarf* systems, however, and the situation is not necessarily comparable for more massive LSBs.

2. The Sample and Observations

The primary sample consists of galaxies identified from the *Uppsala General Catalog of Galaxies* (Nilson 1973) based on the first Palomar Observatory Sky Survey, and supplemented by galaxies from the second Palomar Sky Survey (Schombert and Bothun 1988; Schombert *et al.* 1992). These

D. L. Block and J. M. Greenberg (eds.), New Extragalactic Perspectives in the New South Africa, 509–512.
© 1996 *Kluwer Academic Publishers.*

galaxies are selected to be gas-rich ($M_{HI} \geq 5 \times 10^{10} M_{\odot}$) and of similar size
to giant spiral galaxies ($D_{25} \geq 25$kpc). No morphological criteria is applied
other than the galaxies must be identified as disks. Atomic hydrogen data
is from Schneider et al. (1990, 1992). Galaxies are separated into "high"
and "low" surface brightness samples based on their mean blue surface
brightnesses, which are initially estimated from published blue magnitues
and sizes from either Nilson (1973) or Schombert et al. (1992), then refined using our own data. Those galaxies with $\mu_B < 24.5$ mag arcsec^{-2}
are designated giant high surface brightness galaxies (HSBGs), and those
with $\mu_B > 24.5$ mag arcsec^{-2} are designated giant low surface brightness
galaxies (LSBGs). This is a mean blue surface brightness measured within
$\mu_B = 26.0$ mag arcsec^{-2} from our optical data. These data have been corrected for galactic extinction using an interpolation program provided by
Burstein and Heiles. No intrinsic extinction correction is applied.

The optical data were acquired using the 2.1 m telescope at San Pedro
Martir Observatory, in Ensenada, Mexico, the 0.9 m telescope at Kitt Peak
National Observatory, and the 1.3 m and 2.4 m telescopes at Michigan-
Dartmouth-MIT Observatory, using a variety of CCDs. The near-infrared
imaging has been done on the 1.3 m and 2.4 m telescopes at Michigan-
Dartmouth-MIT Observatory using the University of Massachusetts 256x256
NICMOS array. All data have been reduced using IRAF.

3. Discussion

Table 1 presents the mean global properties of the sample of gas-rich, massive galaxies. Mean values and their standard deviations are listed for the
giant low surface brightness galaxies (LSBGs), giant high surface brightness galaxies (HSBGs), and all the galaxies in the sample. The number
of galaxies used to calculate the mean is listed in parentheses. Dynamical
masses are calculated for those galaxies with $i > 17.5$ deg. Disk colors are
calculated between $23.0 < \mu(R) < 25.0$.

Both high and low surface brightness galaxies in this sample have comparable total gas masses, while LSBGs have, on average, a factor of five
lower total blue magnitude within $\mu_B = 26.0$ mag arcsec^{-2}. Correspondingly, they have higher gas mass to luminosity ratios (2.28 vs. 0.56 for
HSBGs). While there is no correlation between disk central surface brightness and absolute magnitude, the redder galaxies do appear to be the ones
with the longest disk scale lengths. There may exist a population of very
low surface brightness, red galaxies with long disk scale lengths, since there
are no galaxies with high central surface brightnesses and long disk scale
lengths. This may provide information about the underlying physics of disk
formation. Understanding these populations is important in order to deter-

TABLE 1. Mean Global Properties

	log M_{HI} / log $M_\odot$	log L_B / log $L_\odot$	$\frac{M_{HI}}{L_B}$ $\frac{M_\odot}{L_\odot}$	$\frac{M_{DYN}}{L_B}$ $\frac{M_\odot}{L_\odot}$	$\frac{M_{HI}}{\pi R^2_{\mu(B26)}}$ $\frac{M_\odot}{pc^2}$	$\frac{M_{HI}}{M_{DYN}}$	DISK $(B-R)$
LSBGs	9.88	9.77	2.28	67.1	13.9	0.16	0.817
	±0.29	±0.52	±2.98	±188	±20.8	±0.35	±0.222
(npts)	(49)	(49)	(49)	(40)	(49)	(40)	(49)
HSBGs	9.95	10.51	0.56	16.6	7.91	0.068	0.937
	±0.36	±0.43	±0.75	±27.9	±8.56	±0.052	±0.237
(npts)	(27)	(27)	(27)	(23)	(27)	(23)	(27)
All	9.91	10.03	1.67	48.7	11.8	0.124	0.860
	±0.31	±0.61	±2.56	±152	±17.5	±0.283	±0.232
(npts)	(76)	(76)	(76)	(63)	(76)	(63)	(76)

mine their affect on the galaxy mass and luminosity functions, since many of these systems are not dwarfs in any sense of the word.

The first thing that is apparent when examining the optical and near-infrared images of LSBGs is their morphological variety. Their morphology spans the entire Hubble sequence for disk galaxies, with about 30% of the sample possessing some sort of bar. This percentage is lower than the generally quoted 50% barred disk galaxies, although the number statistics are small. About 42% of the comparison sample of HSBGs appear to have bars. The difference may be real, however, as the multi-color imaging shows little evidence of significant quantities of obscuring dust in the LSBGs.

Our broadband optical ($BVRI$) and near-infrared (JH) imaging indicates that most of these systems *have* undergone previous episodes of star formation. Clear variations are apparent in the morphological appearance of most LSBGs and HSBGs from B to I or B to H. There are exceptions among the latest types, such as UGC 10313, however, where at least from B to I there is very little change. Optically, we find that the gas-rich galaxies have blue disks on average, whether they are high or low surface brightness systems, and there is a subset of the LSBGs which have *very* blue disks. These systems are well-fit by a simple exponential law, with $< (B - R) >= 0.749 \pm 0.166$. Many of the bluest systems, which we designate "pure disks", are too blue to be expained by moderately young, metal poor stellar populations based on models by Worthey, Faber, and Gonzalez (1992). It is these "pure disk" systems which show the least variation in the different bandpasses. Perhaps these galaxies are undergoing their first episodes of star formation.

4. Conclusions

Preliminary results of our observations indicate that:

• LSBGs have optical morphologies which span the entire range of Hubble disk types. While most galaxies appear to have undergone earlier episodes of star formation, multi-color imaging of some "pure disk" systems are consistent with no previous star formation episodes.

• Based on R images, roughly 30% of LSBGs appear barred, while 42% of HSBGs have bars. The percentage for HSBGs is similar to numbers found in earlier studies of disk galaxies, which suggest that approximately half of disk galaxies are barred. The LSBGs seem to fall below this number, especially since there is little evidence of significant quantities of dust hiding the presence of bars.

• There may be evidence for a very weak correlation between disk color and disk scale length, in the sense that galaxies with the longest disk scale lengths have redder colors.

• There is no correlation between the total luminosity of a galaxy and its disk central surface brightness. There does appear to be a relation between the disk central surface brightness of a galaxy and its disk scale length, in the sense that there are no galaxies with very high central surface brightnesses and long disk scale lengths. This is not easily explained as a selection effect, and may be a clue to the underlying physical processes of disk formation. There may, in fact, be a population of galaxies with very low surface brightnesses, red colors, and long disk scale lengths.

• Gas rich, massive disk galaxies have very blue disk colors on average. Although too little data in the near-infrared exists to more tightly constrain their underlying stellar populations, many of these disk systems have colors which are too blue to be explained by *both* moderately young stellar populations and poor metallicites. These galaxies may be undergoing their first star formation episodes.

References

Impey, C. and Bothun, G. 1989, *Ap. J.*, **341**, 89.
McGaugh, S. S. 1992, Ph. D. Thesis, University of Michigan.
Nilson, P. N. 1973, *Uppsala General Catalog of Galaxies, Uppsala Astron. Obs.*, **6**.
Schneider, S. E., Thuan, T. X., Magri, C., and Wadiak, J. E. 1990, *Ap. J. Supp.*, **72**, 245.
Schneider, S. E., Thuan, T. X., Magnum, J. G., and Miller, J. 1992, *Ap. J. Supp.*, **81**, 5.
Schombert, J. M. and Bothun, G. D. 1988, *A. J.*, **95**, 1389.
Schombert, J. M., Bothun, G. D., Schneider, S. E., and McGaugh, S. S. 1992, *A. J.*, **103**, 1107.
Worthey, G., Faber, S. M., and Gonzalez, J. J. 1992, *Ap. J.*, **398**, 69.

38 MICRON IMAGES OF GALAXIES:

THE INFRARED LUMINOSITY

G.J. STACEY, T.L. HAYWARD, G.E. GULL AND H. LATVAKOSKI
Cornell University
Ithaca, New York, USA

Abstract.
We report the first extragalactic observations with our new far-infrared imaging spectrometer/spectrophotometer, KWIC. We obtained 37 μm images of M82 and Arp 299, and both a 31.5 μm and 37.7 μm image of the 30 Doradus star formation region in the LMC. The raw spatial resolution of these images is $\sim$ 8.5″. The M82 data set is of sufficient signal to noise to permit a maximum likelihood image deconvolution to an effective beam of $\sim$ 5″. For these regions of very high star formation activity our far-IR images provide an extinction free probe of the source luminosity.

1. Introduction

Much of the luminosity in spiral galaxies including the Milky Way emerges at wavelengths longer than about 30 microns, in the far-infrared (FIR). This FIR radiation is "down converted" optical or UV radiation: dust in the interstellar medium absorbs optical and UV radiation, heats up, and reradiates the energy in the far-IR. Since the physical size of dust grains is small ($\sim$ 0.1 μm), the emergent far-IR radiation is not affected much by extinction, and therefore escapes the galaxy, cooling the dust. For most galaxies, the source of dust heating radiation is a combination of flux from regions of recent massive star formation, and the diffuse interstellar radiation field (Helou, 1986). Active galaxies may also have a significant contribution from embedded active galactic nuclei.

Our 37 μm image of the Orion A star formation region (Stacey *et al.* 1995) demonstrated that the 37 μm flux arises from the warm photodissociation region (PDR) at the interface between the Orion HII region and the

D. L. Block and J. M. Greenberg (eds.), New Extragalactic Perspectives in the New South Africa, 513–520.
© *1996 Kluwer Academic Publishers.*

parent molecular cloud. The 37 μm flux traces the same dust as 60 and 100 μm studies: the observed flux matches the longer wavelength color temperature, and optical depth maps very well. Therefore, at least for dusty star formation regions, the 37 μm flux traces the luminosity of the source. We show below that this holds for starburst galaxies as well. This is expected. It is easy to show the 37 μm flux is directly proportional to luminosity for star formation regions. For example, at the distance to Orion (480 pc), $L_{FIR}(L_\odot) = F_{37}$ (Jy) $\times$ 1.2 $\pm$ 0.5 for 50 $\leq T_{dust} \leq$ 200 K (Stacey et al. 1995). The advantage to working at 37 μm rather than longer wavelengths is the smaller beam size and very much larger format arrays available at 37 μm.

2. Instrumentation and Observations

2.1. THE KUIPER WIDEFIELD INFRARED CAMERA

The Kuiper Widefield Infrared Camera (KWIC, Stacey et al. 1993) is an imaging spectrometer/imaging spectrophotometer designed for use between 18 and 40 μm on the Kuiper Airborne Observatory (KAO). KWIC uses a 128 $\times$ 128 element Si:Sb BIB detector developed by Rockwell International for the Space Infrared Telescope Facility (SIRTF) project as its detective device. We chose a 2.73″ square pixel format with KWIC so as to oversample the diffraction limited beam on the KAO at 37 microns ($\lambda/D \sim 8''$). Even so, the large format array yields a very large field of view for KWIC: 5.8′ $\times$ 5.8′. There are two modes of operation for KWIC: imaging photometer mode (R = $\lambda/\Delta\lambda \sim$ 30 to 200), and imaging spectrometer mode (R $\sim$ 2000 to 10,000). The spectral resolution is obtained with fully tunable cryogenic scanning Fabry-Perot interferometers (FPIs). A single FPI is used for imaging photometer mode, and a second FPI is inserted into the beam for imaging spectrometer mode. These modes are rapidly interchangeable in flight. KWIC has flown a total of five flights on the KAO: two in February 1994, and three in May 1995.

2.2. OBSERVATIONS

We obtained continuum images of M82, Arp 299, and the 30 Doradus region of the LMC with KWIC. For these observations, we fixed the Fabry-Perot at a high telluric transmission wavelength, and used standard chopping and beam switching techniques. In 1994, we observed M82 and Arp 299 at a center wavelength of 37 μm, with a bandwidth (BW) of 0.5 μm, and integration time of 30 and 50 minutes respectively. In 1995, we observed 30 Doradus for 45.5 minutes at 37.7 μm (BW = 0.9 μm) and 19 minutes at 31.5 μm (BW = 1.1 μm). The data were calibrated with respect to the

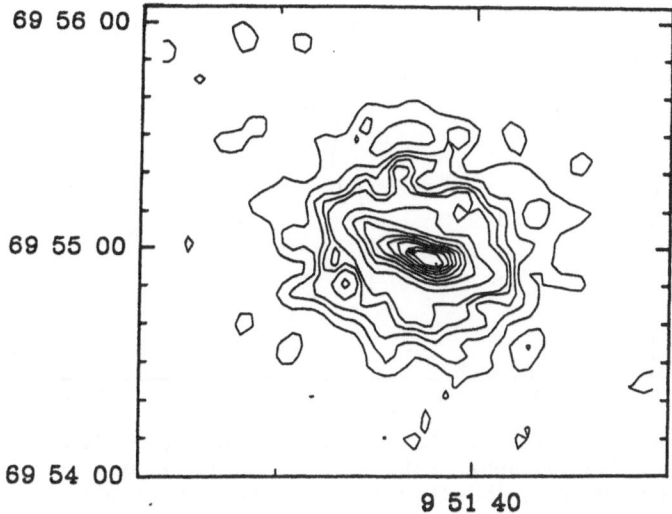

Figure 1. Maximum likelihood deconvolution of our 37 μm image of M82. Beam size is $\sim 5''$. Contours run from 2,4,6,8,10,20...100% of the peak flux: 95 Jy/beam. Noise (1 σ) ~ 0.5 Jy/beam. The dot represents the dynamical center of the galaxy.

integrated IRAS flux of M82, and Mars for the 1994 and 1995 observations respectively. The system noise equivalent flux density was about 20 Jy $Hz^{-1/2}$ for all of the observations. The beam size is 7.5'' and 8.5'' at 31.5 and 38 μm respectively. For all observations, we dithered the array to eliminate the effects of bad pixels (less than 100 on the array), and flat fielded by dividing by an image of our blackbody calibration source.

3. Results and Discussion

3.1. M82

M82 is long recognized as having a peculiar morphology including strong dust lanes occulting what appears to be the plane of the galaxy and filamentary structures extending well above the plane. This disturbed morphology is likely associated with the strong starburst that occurred in the inner kiloparsec about 10^8 years ago, perhaps triggered by a tidal interaction with M81 (Reike *et al.* 1980). The FIR luminosity in the inner kiloparsec is $\sim 3 \times 10^{10} L_\odot$ - more than the entire FIR luminosity of the Milky Way. M82 is the closest example (d ~ 3.3 Mpc) we have of a starburst galaxy, so that it is an intensely studied galaxy.

Our raw 37 μm image of M82 clearly resolves the inner starburst regions with an apparent width (including beam broadening) of 36'' and 14'' FWHM for the major and minor axis respectively. The peak flux in the raw

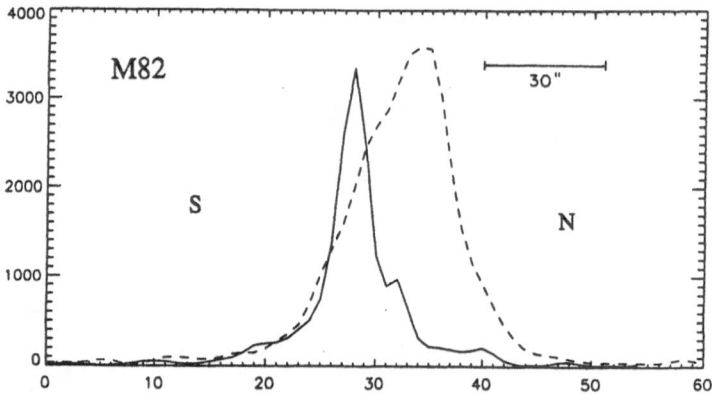

Figure 2. Cuts across the major (dotted line) and minor (solid line) axis of M82. The two cuts were overlaid to emphasize the "plateau" of emission at large scale heights. The y-axis is in arbitrary units, while the x-axis is in pixel units (2.73″).

image is 121 Jy/beam. Figure 1 is a maximum likelihood deconvolution of this image. The deconvolved image indicates an intrinsic source size of 23″ × 8.3″ (370 × 130 pc) so that it lies *within* the molecular ring as traced in its CO(1 - 0) line emission, suggesting the starburst has fragmented the local molecular ISM into small, mostly photodissociated cloudlets.

We find ∼ 75% of the 37 μm flux arises from these central starburst regions. The mid-IR work of Telesco *et al.* (1991) clearly show two peaks of emission on opposite sides of the dynamical center of M82. However, there is just one peak apparent at 37 μm ∼ 7″ to the southwest of the nucleus. This peak lies near a particularly intense region of star formation activity containing a young supernovae remnant, and an intense recombination line region. Many of the results above are in very good agreement with the tracking and scanning results of Joy *et al.* (1987). We find the run of flux from 10 to 100 μm in this region is well fit by a two component dust model: $T_{warm} \sim 60$ K, $\tau_{warm} \sim 8 \times 10^{-2}$, and $T_{hot} \sim 170$ K, $\tau_{hot} \sim 7 \times 10^{-5}$. The 37 μm optical depth in the warm component corresponds to $A_V \sim 10$. At 37 μm, about 90% of the flux comes from the warm component. Most of the FIR flux comes from the warm dust, so that, as for Orion, the 37 μm image is tracing luminosity.

An intriguing new result is revealed through close inspection of the minor axis profile. Figure 2 displays a cut across the minor axis, superposed on a similar cut across the major axis. There is a weak (∼ 2% of the peak), but significant "plateau" of 37 μm flux both north and south of the galactic plane. The plateau is well correlated with the extended emission seen at 10.8 μm by Telesco *et al.* (1991). The dust emission model outlined above works very well for this extraplanar emission, requiring only a smaller optical

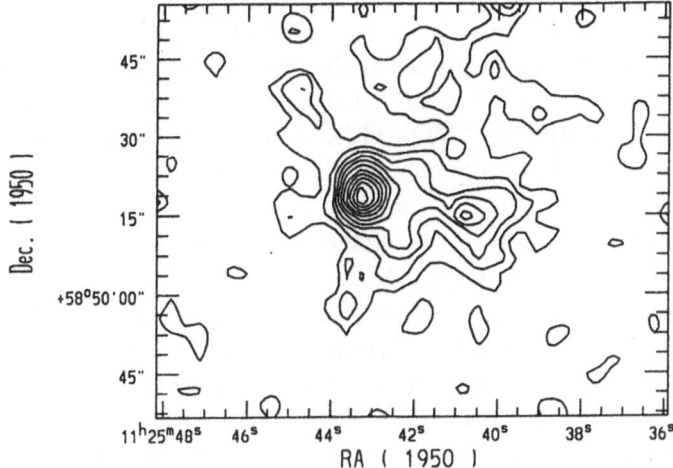

Figure 3. 37 μm image (8.5″ beam) of the Arp 299 system. The contour levels are 1.5 Jy/beam beginning at 1.5 Jy/beam (3 σ). A is the point source to the east (17 Jy), and B is the weaker source (8 Jy) to the west. C is not apparent, but lies ∼ 8″ to the north of B.

depth: τ_{warm} ∼ 1.6×10^{-3}, or A_V ∼0.2. Our data therefore show there is substantial warm dust at large (∼ 300 pc) scale heights, presumably heated by the UV from the starburst escaping along the minor axis. There is large scale extraplanar [CII] (158 μm) emission in the filaments as well (Stacey et al. 1991). The [CII] line cools the gas, and the 37 μm continuum image traces the dust that is necessary to provide a source for the photoelectrons to efficiently heat the gas.

3.2. ARP 299

Arp 299 is an extremely IR luminous (L_{FIR} ∼ 8×10^{11} L$_\odot$) interacting galaxy pair. Arp 299 has been well explored in the mid and far-IR (cf. Wynn-Williams et al. 1991, Joy et al. 1989). Our high spatial resolution, two dimensional image (Figure 3) resolves the system in its two primary sources: A, the nucleus of IC 694, and B, the nucleus of NGC 3690. Source C, the overlap region, is not apparent as an independent source. It lies on a ridge adjoining B. These three sources account for 46%, 22%, and ∼ 8% of the 37 μm flux respectively. The remaining 24% is "diffuse". Comparing our data to the 50 and 100 μm scans of Joy et al. 1989, and the mid-IR fluxes from Wynn-Williams et al. (1991), we again obtain a fairly good fit to the run from 10 to 100 μm data with a two component dust model. For component A: T_{warm} ∼ 46 K, τ_{warm} ∼ 5×10^{-2}, and T_{hot} ∼ 140 K, τ_{hot} ∼ 6×10^{-5}. For components B and C the model is similar except with smaller optical depths. Since the FIR flux is mostly in the warm component, and 70% of the 37 μm flux comes from this component, we again conclude that the 37 μm flux is tracing luminosity. Therefore, the FIR luminosities of the various components are: 3.7×10^{11}, 1.8×10^{11}, 0.65×10^{11}, and 1.9×10^{11} $L_\odot$ for A (IC 694), B(NGC 3690), C (overlap), and the diffuse components respectively. The FIR luminosity of the IC 694 nucleus is *ten times* the luminosity of M82! The luminosity of C is about twice that of M82 making it one of the most luminous non-nuclear starbursts known.

Sargent and Scoville (1991) take the ratio of the FIR luminosity to the molecular gas mass in an attempt to constrain the degree that an AGN contributes to the observed FIR luminosity for the nuclear components of the Arp 299 system. Following their arguments, we estimate less than 50% of the observed FIR luminosity of IC 694 can come from a nuclear starburst. The same line of reasoning would indicate an even smaller fraction for the NGC 3690 nucleus.

3.3. 30 DORADUS

We obtained both 31.5 and 38 μm continuum images of the 30 Doradus star formation region in the LMC. This region is of particular interest, as it occurs in a low metallicity dwarf galaxy, yet is the most luminous star formation region in the local group. These data were only recently properly reduced, so that we can give only a preliminary analysis here. However, apparent in both maps is the banana shaped structure representing the photodissociated surface of the parent molecular cloud to the NW. This interface is presented to us roughly edge-on. The KWIC FIR continuum images peak between the HII region, as traced by its near IR hydrogen recombination line emission, and the molecular cloud traced by its CO(1 - 0) line emission. The FIR emission is spatially coincident with the [CII] (158 μm) fine structure line distribution. The [CII] line traces the photodissociated surfaces of molecular clouds, implying that the FIR continuum traces the structure of PDRs as well. The FIR [CII] and [OI] fine structure line work has shown that low metallicity can have a large effect on the structure of molecular clouds (Stacey *et al.* 1991, Poglitsch *et al.* 1995). For example, the CO emitting cores of such clouds is dramatically reduced relative to the photodissociated skins of the clouds. If the source is nearly edge-on, then the observed width of the FIR emission indicates relatively large PDRs, in agreement with these results.

We have used the 31.5 and 38 μm fluxes to derive a color temperature map of the region. The peak color temperature is 130 K, in fairly good agreement with the previous large beam work of Werner *et al.* (1978). If the source of the heating for the ridge is the R136 complex at its projected

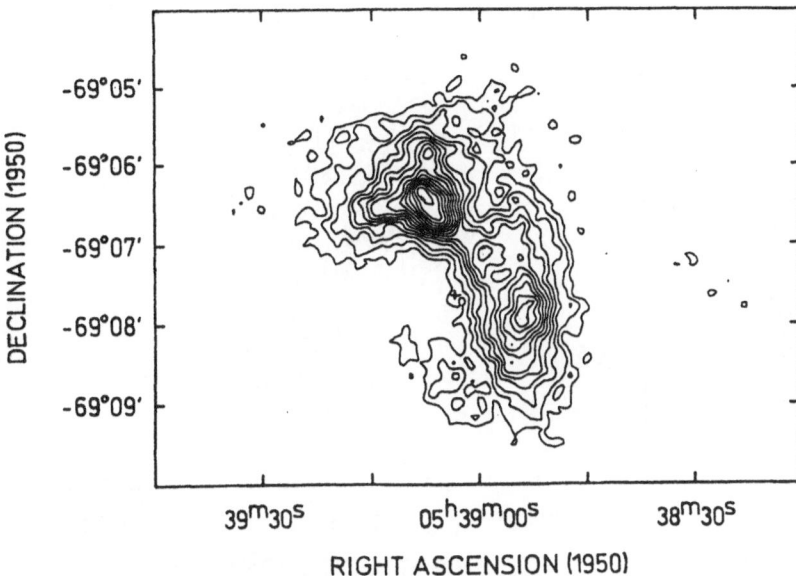

Figure 4. 38 μm image (8.5″ beam) of the 30 Doradus region of the LMC. The contour levels are spaced by 1.8 Jy/beam (4 σ), beginning with 1.8 Jy/beam and peaking at 28 Jy/beam. The cross marks the position of the exciting R136 star cluster

distance from the FIR peaks of 18 pc, then the observed color temperature is consistent with very small grains ($\sim$ 0.01 μm) as the source for the FIR emission. Curiously, however, our preliminary color temperature map shows little evidence for a color gradient from the R136 complex as one would expect for a centrally heated source. Perhaps the beam ($\sim$ 2 pc) is too large to reveal this gradient, or the source is not very close to edge on. We are also working to verify the registration between the 31.5 and 38 μm images.

4. Acknowledgments

We thank the staff and crew of the Kuiper Airborne Observatory for their excellent support, especially during the observatory's challenging final year of operations. We also thank Liang Peng and Kristin Nelson for their very significant contributions to the project. This work was supported by NASA grant NAG2-800 and an NSF graduate fellowship.

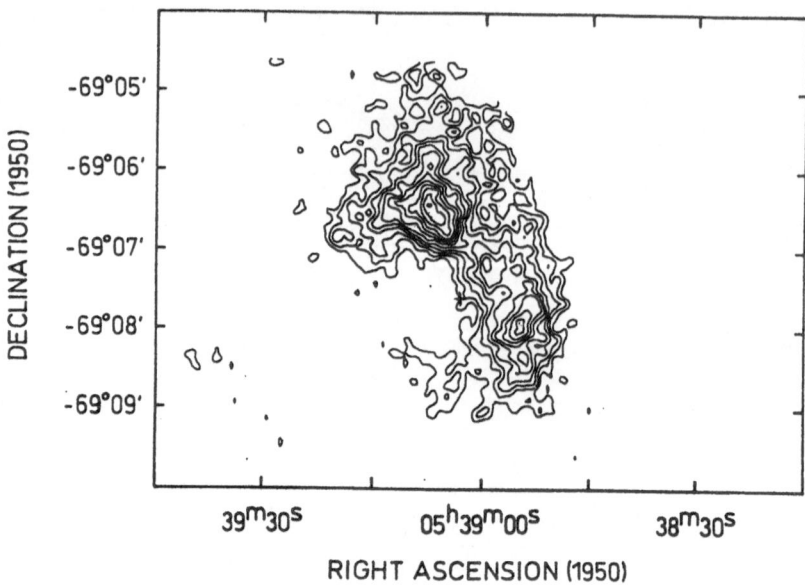

Figure 5.   31.5 μm image (7.5″ beam) of the 30 Doradus region of the LMC. The contour levels are spaced by 1.8 Jy/beam (4 σ), beginning with 1.8 Jy/beam and peaking at 20 Jy/beam. The cross marks the position of the exciting R136 star cluster.

5. References

References

Helou, G. 1986, *Ap.J.*, **311**,L33

Joy, M.J., Lester, D.F., and Harvey, P.M. 1987, *Ap.J.*, **319**,314

Joy, M.J., Lester, D.F., Harvey, P.M., Telesco, C.M., Decher, R., Rickard, L.J. and Bushouse, H. 1989, *Ap.J.*, **339**,100

Poglitsch, A., Krabbe, A., Madden, S.C., Nikola, T., Geis, N., Johansson, L.E.B., Stacey, G.J., and Sternberg, A. 1995, *Ap.J.*, **454**,293

Rieke, G.H., Lebofsky, M.J., Thompson, R.I., Low, F.J., and Tokunaga, A.T. 1980 *Ap.J.*, **238**,24

Sargent, A., and Scoville, N. 1991, *Ap.J.*, **366**,L1

Stacey, G.J., Geis, N., Genzel, R., Lugten, J.B., Poglitsch, A., Sternberg, A., and Townes, C.H. 1991, *Ap.J.*, **373**, 423

Stacey, G.J., Hayward, T.L., Latvakoski, H. and Peng, L. 1995, *Airborne Astronomy Symposium on the Galactic Ecosystem: From Gas to Stars to Dust*, ed. M.R. Haas, J.A. Davidson, and E.F. Erickson, (San Francisco: ASP), pp. 215.

Stacey, G.J., Hayward, T.L., Latvakoskyi, H., and Gull, G.E. 1993, *SPIE Proceedings*, **1946**, 238

Telesco, C.M., Campins, H., Joy, M., Dietz, K., and Decher, R. 1991, *Ap.J.*, **369**, 135

Werner, M.W., Becklin, E.E., Gatley, I., Ellis, M.J., Hyland, A.R., Robinson, G. , and Thomas, J.A. 1978, *Mon. Not. R. Astr. Soc.*, **184**, 365

Wynn-Williams, C.G., Eales, S.A., Becklin, E.E., Hodapp, K.-W., Joseph, R.D., Mclean, I.S., Simons, D.A., and Wright, G.S. 1991, *Ap.J.*, **377**, 426

DUST IN EARLY-TYPE GALAXIES: NEW RESULTS ON THE WAVELENGTH DEPENDENCE OF THE EXTINCTION

N. BROSCH

School of Physics and Astronomy, Beverly and Raymond Sackler Faculty of Exact Sciences
Tel Aviv University, Tel Aviv 69978, Israel

The determination of the total-to-selective extinction ratio $R_V = \frac{A(V)}{E(B-V)}$ is perhaps the most significant problem associated with the extinction curve (Lynds & Wickramasinghe 1968). Johnson (1968) measured R_V in many regions of the Milky Way and found values reaching as high as 6.3±0.4. In the late 1970s a few attempts were made to define an average law for the wavelength dependence of the extinction in the Galaxy (*e.g.*, Savage & Mathis 1979 [SM79], Seaton 1979). The extinction law differs significantly from one direction to another, mainly in the UV region (Fitzpatrick & Massa 1990). The deviations from an average relation appear minor in the optical and near-IR domains. Various ways to parametrize the extinction law were proposed, culminating with Cardelli, Clayton & Mathis (1989) where the shape of the law is determined by R_V. This parameter ranges from 2.85 in diffuse medium to 5.6 in dense clouds (O'Donnell 1994), while in the general ISM it is 3.15 (SM79).

It is interesting to establish whether the wavelength dependence of the extinction in other galaxies is similar to that in the Milky Way. In particular, I address here the question of the 'universality' of R_V. First steps in determining extragalactic R_V's resulted in widely different values. In M31 R_V is between 2.5 and 7, in SMC it is 2.72±0.18, and in LMC it reaches 3.41±0.15 (Lequeux 1988). The trend among the SMC, LMC and Milky Way has been interpreted by Lequeux as linked with metallicity: R_V has a maximum for $Z_\odot/2$, and decreases for lower and higher Z.

The determination of the extinction law in a galaxy requires either its separation into individual stars, in order to apply the pair method between similar stars with and without of extinction, or the development of a similar method for entire galaxies. The latter is done by comparing extinguished

D. L. Block and J. M. Greenberg (eds.), New Extragalactic Perspectives in the New South Africa, 521–522.
© *1996 Kluwer Academic Publishers.*

parts of a galaxy with unextinguished parts. I collected 22 values of R_V in early-type galaxies with dust lanes from the literature; Goudfrooij et al (1994) provided about half the values discussed here.

For most galaxies R_V <3.15 and the median value for 17 galaxies (after rejecting four extremal values) is 2.74. Therefore, the extinction in most E/S0 galaxies with dark lanes behaves more like dust in the diffuse medium than in dark, dense clouds. Goudfrooij et al (1994) claim that smaller values of R_V are encountered in galaxies with "smooth, regularly distributed dust lanes". However, it is clear that some objects with small R_V do not have smooth dust lanes at all (*e.g.*, N5128).

The value of R_V is an ∼linear indicator of the typical size of dust grains. Goudfrooij et al (1994) attributed smaller values of R_V to the activation of dust-destroying mechanisms in galaxies with smooth dust lanes. Another possibility is that in some dust lanes the conditions are **not favorable** for the coagulation of small dust grains into big grains and the grains remain small. In galaxies with smooth and regular dust lanes the conditions may not be suitable for grain growth and the grains, just as in the diffuse ISM, stay small. Only in dense and turbulent regions may the grains be able to grow and increase R_V.

The case of NGC 2685 demonstrates how difficult the derivation of the extinction law is. Reflection of the light distribution on the major axis about the center of the galaxy allows the determination of the wavelength dependence of the extinction at 32 spectral bands. It is necessary to subtract the light contribution from stars within the lanes (∼2% of the peak light distribution from the galaxy, measured from broad-band images). At ∼6" from the peak the law is ∼similar to that of SM79 with R_V ≈3.2. It is possible that the dust grains are larger than in the Milky Way, as the extinction law appears to saturate at short $\lambda\lambda$.

References

Cardelli, J.A., Clayton, G.C. & Mathis, J.S. 1989 ApJ **345**, 245.

Fitzpatrick, E.L. & Massa, D. 1986 ApJ **307**, 286.

Goudfrooij, P., et al. 1994 MNRAS **271**, 833.

Johnson, H.L. 1968 *Nebulae and Interstellar Matter*, B.M. Middlehurst and L.H. Aller (eds.), The University of Chicago Press, p. 167.

Lequeux, J. 1988 in *Dust in the Universe* (M.E. Bailey & D.A. Williams, eds.) Cambridge University Press, p. 449.

Lynds, B.T. & Wickramasinghe, N.C. 1968 Ann. Rev. A&A **6**, 215.

O'Donnell 1994, J.E. 1994 ApJ **422**, 158.

Savage, B.D. & Mathis, J.S. 1979 Ann. Rev. A&A **17**, 73.

Seaton, M.J. 1979 MNRAS **187**, 73.

NEW RADIATIVE TRANSFER MODELS OF DISC GALAXIES AND THE SCALE-LENGTH TEST FOR DUST DIAGNOSTICS

R.L.M. CORRADI, J.E. BECKMAN AND M.S. DEL RIO
Instituto de Astrofísica de Canarias, Tenerife, Spain

A. DI BARTOLOMEO
Dipartimento di Astronomia, Università di Padova, Italy

AND

E. SIMONNEAU
Institute d'Astrophysique de Paris, France

We present a set of *face–on* models of the radiative transfer (RT) of stellar light in exponential discs, taking into account absorption and *multiple scattering* (diffusion) by dust. At any point in the exponential disc, a *local* plane–parallel geometry is assumed. The RT equation is then solved using a moment method, with a high order closure relation for the moment system in order to take into account the anisotropy of the angular phase function for dust scattering. The moment system is solved using an implicit method which assures unconditional stability. We are extending the models to a 2-D geometry by means of a perturbation method (Simonneau et al., 1996, in preparation) in order to test the limits of the geometrical approximation employed. For the dust parameters, we have adopted the graphite-silicate models by Maccioni & Perinotto (1994) and Di Bartolomeo et al. (1995).

Special attention has been directed towards the effects on the observable photometric properties of different distributions of stars and dust perpendicular to the galactic plane. Models with a constant star to dust scale–height ratio z_s/z_d, as well as more realistic mixtures of stars representing the arm and interarm regions of the Galaxy, with *wavelength–dependent* scale–height ratios, are presented.

Face–on luminosity profiles, dust–induced colour excesses profiles, and observable scale–lengths are computed for various photometric bands (UB-VRIK) and for a wide range of on–axis optical thicknesses τ_V (from 1 to 50). The full set of models is presented in Corradi et al. (1996).

These models have been used to investigate the dust content of a sample of face–on spiral galaxies. The illustrative cases of NGC 6384 (Sbc) and

D. L. Block and J. M. Greenberg (eds.), New Extragalactic Perspectives in the New South Africa, 523–524.
© *1996 Kluwer Academic Publishers.*

NGC 157 (Sc) are discussed. The observed UBVRI luminosity profiles of NGC 6384 are reproduced by models with an on–axis optical thickness τ_V slightly larger than 4. This would imply a total extinction for the galaxy, within 3 disc scale–lengths and in the V band, less than 0.2 mag. The disc profiles of NGC 157 show a notable flattening in the inner regions. These are nicely fitted, in all bands, by models with $\tau_V = 20$, corresponding to a total extinction of ~0.8 mag.

Differential extinction causes the observed scale–length to change with passband. The measure of the variation of the observed scale–length with wavelength is therefore a potentially useful diagnostic (the so called *scale–length test*) of the total dust content of a disc galaxy. We have independently applied the method to the arm and the interarm regions of our sample of objects, as well as to their discs as a whole (Beckman et al., 1996). The results are consistent with those obtained by fitting the luminosity profiles of the galaxies, show that the arms exhibit greater optical depth in dust than the interarm zones, and support the conclusion that the internal extinction in spiral galaxies is low to moderate.

We stress the following inferences:
– In the optical range, models which include multiple scattering need roughly *twice* as much dust as models where scattering is neglected, in order to produce the same luminosity, colour, and scale–length deviations from the dust–free case.
– Changes in the adopted dust parameters can produce significant variations (several tenths of a magnitude) in the predicted colour excesses.
– Owing to the strong dependence of reddening on the geometrical distributions of stars and dust, no *standard* observational extinction law can be expected for external galaxies, even if the physical properties of the dust are invariant.
– Large on–axis optical thicknesses do not necessarily imply a large total extinction for the disc, in particular when scattering is included, because the local dust opacity decreases exponentially with radius. In the V band, in order to have a total extinction over the whole disc larger than 1 mag $\tau_V \geq 30$ is needed.

References

Beckman, J.E., Peletier, R.F., Knapen, J.H, Corradi, R.L.M., Gentet, L.J.: 1996, ApJ submitted
Corradi, R.L.M., Beckman, J.E., Simonneau, E.: 1996, MNRAS submitted
Di Bartolomeo, A., Barbaro, R., Perinotto, M.: 1995, MNRAS, 277, 1279
Maccioni, A., Perinotto, M.: 1994, A&A 285, 241

THE DISTRIBUTION OF COLD DUST WITHIN THE GALAXY

J. I. DAVIES, M.TREWHELLA AND H. JONES
University of Wales College of Cardiff,
Dept. of Physics and Astronomy, PO Box 913, Cardiff, UK

1. Introduction

Radiative transfer models of galaxies using exponentially distributed stars and dust have been used extensively to interpret surface brightness inclination tests, galaxy colour gradients, localised extinction and the far infrared (FIR) emission. The Standard Model (SM) used has been obtained mainly from observations of our Galaxy (extinction measures to stars and IRAS). Within the SM the generally excepted values for the vertical z and radial r dust scalelengths are about 0.13 and 3.5 *kpc* respectively. In this paper we describe how we have tested the SM using the emission from cold dust as measured by COBE (DIRBE). We have created all sky 140 and 240μ surface brightness maps from the DIRBE annual averaged sky maps. For references to the work described in this paper please consult 'The Opacity of Spiral Disks', Proc. NATO ARW Cardiff July 1994, Ed. J. Davies and D. Burstein.

2. The model

We have used a model in which the dust density and temperature are distributed exponentially in both the r and z directions. The dust is assumed to radiate as a grey body ($\beta = 2$) and we simulate the DIRBE observations as viewed from the Earth. We assume that on average the dust has the same emission properties at all points in the galaxy, leaving the emission defined only by the dust density and temperature. The emission is normalised using the observed emission along the Galactic plane. With this defined we can adjust the dust temperature distribution until it fits that of the observed colour temperature in both the r and z directions. We now have a model of the far infrared radiation from the Galaxy that is fully defined by the dust

525

D. L. Block and J. M. Greenberg (eds.), New Extragalactic Perspectives in the New South Africa, 525–526.
© *1996 Kluwer Academic Publishers.*

density distribution taken from extinction measures to individual stars and the dust temperature derived from the DIRBE observations.

3. The emission at 140 and 240μ

When we compare with the observational data we find that there is far more emission at high Galactic latitudes than can be accounted for by the standard model (too low by about a factor of 5), see fig 1. Two possible explanation are that there is much more cold dust at high galactic latitudes than has previously been thought or that there is a significant Extra-Galactic Background (EGB). We have discounted the EGB after considering the correlation of the FIR intensity with the 21cm column density at high Galactic latitudes. Extrapolating to zero 21cm column density we calculated values for the EB of about 1 and 0.5 MJ sr^{-1} at 140 and 240μ respectively. This still leaves about a factor of 4 too much emission at high Galactic latitudes.

4. Conclusions

We conclude that there is cold dust much higher above the plane of the Galaxy than previously thought. We calculate that the exponential scale height of this cold dust is about 0.4 *kpc*. As edge-on galaxies show a well defined thin dust lane within the galactic plane we hypothesise that there is a dust component 'clumped' within the plane that is responsible for the large extinction values. Above the plane there is an extended component that is less easy to detect by its extinction because it is more smoothly distributed. It is straight forward to show that once expelled, for example by SN explosions, the dust can be held above the galactic plane by radiation pressure.

Fig. 1 - The FIR emission above the Galactic plane at 140μ. Solid line is DIRBE data, dashed line is the Sm, dot-dash line is the best fit model.

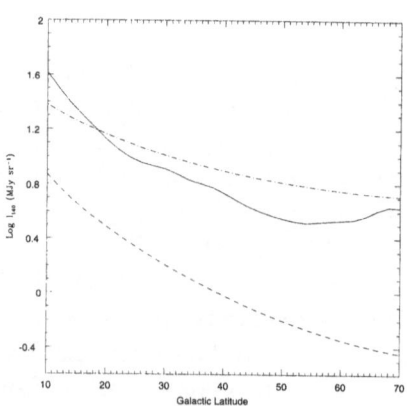

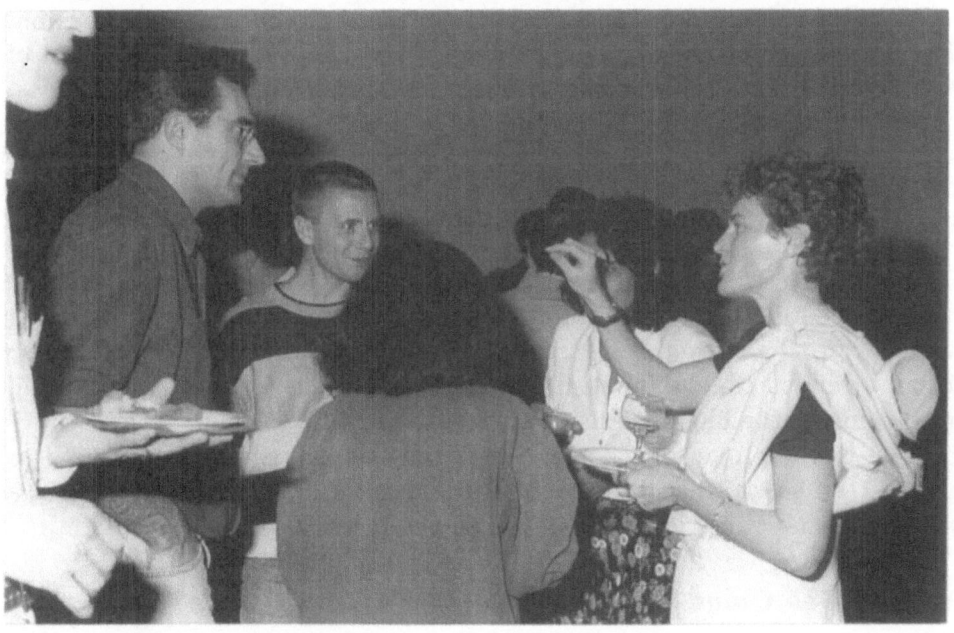

DETECTION OF EXTENDED DUST IN A RADIO-QUIET GALA
AT REDSHIFT 2.38

PAUL J. FRANCIS
University of Melbourne, School of Physics
Parkville, Victoria 3052, Australia

AND

BRUCE E. WOODGATE
NASA Goddard Space-Flight Center
Code 681, Greenbelt MD 20771, USA

Abstract.
We have discovered a compact group of galaxies at redshift 2.38, probably representing a proto-elliptical galaxy in the act of bottom-up formation.
The group is embedded in an extended (> 100 kpc radius) lumpy cloud of
gas. This gas cloud is dusty, even at large distances from any of the galaxies.

1. Background

During a search for protogalaxies, we found a group of elliptical galaxies at
redshift 2.38. The group is remarkable for its compactness ($\sim$ 5 galaxies,
each of $\sim 10^{11}$ solar masses, in a volume of diameter $\sim$ 100 kpc), and also
for an extended cloud of neutral hydrogen that seems to envelop it, causing
absorption in the line of sight to a background QSO 100 kpc away.

The merger time for galaxies this large and this close together is less
than a Gyr, so we speculate that these galaxies will have merged to form a
single giant elliptical galaxy long before the present.

2. The Gas Cloud

One of the galaxies in the group, B1, seems to contain a concealed radio-
quiet AGN. This AGN is photoionising a broad cone of the surrounding
gaseous envelope, triggering Ly-α and C IV emission. The Ly-α emission
extends out to more than 50 kpc from the galaxy, with an irregular lumpy

528

D. L. Block and J. M. Greenberg (eds.), *New Extragalactic Perspectives in the New South Africa*, 528–529.

morphology suggesting that the gaseous envelope is extremely inhomogeneous. Further evidence for this inhomogeneity comes from the absorption in the spectrum of the background QSO, which comes in three components.

3. Evidence for Dust

We have imaged the photoionised gas around B1 both in Ly-α (using the CTIO 4-m telescope) and in H-α (using an IR camera on the Siding Spring 2.3-m telescope). The ratio of Ly-α to H-α is ~ 0.8; well below the predictions of case-B recombination (~ 8). The ratio seems to remain low throughout the ionisation cone; even far from the galaxy B1.

We interpret this low radio of Ly-α to H-α as evidence for dust in the gaseous envelope. An extinction of $E(B - V) \sim 1$ mag. is required if the dust is in a foreground screen; if it is mixed in with the emission region, far less may be required due to resonant scattering of Ly-α photons.

4. Discussion

What is dust doing so far from a galaxy in the early universe? It is unlikely to have come from the galaxies themselves; the Hubble time at these redshifts is only $\sim$1–2 Gyr, and the dust would have to have travelled at in excess of 40km s^{-1} to have spread throughout the gaseous envelope in this time.

One possibility is star formation distributed throughout the 100 kpc gaseous envelope surrounding the proto-giant elliptical. Perhaps population 3 stars are seeding the gas cloud with heavy elements. Alternatively, compact galaxies below our flux limit may be scattered throughout the gas cloud. We have HST time to find out in Cycle 6.

The presence of distributed dust associated with a QSO absorption-line system also confirms Fall & Pei's (1993) contention that dust may be seriously biassing our view of the early universe.

5. References

More details can be found in Francis et al. (1996)

References

Fall, S. M., & Pei, Y. C. (1993), *ApJ*, **402**, 479
Francis, P. J. et al. (1996), *ApJ* in press (to appear Feb 1st)

EXTINCTION IN THE DIRECTION OF THE END-OF-BAR STAR FORMATION REGION

T. J. MAHONEY, P. L. HAMMERSLEY, M. LÓPEZ-CORREDOIRA
F. GARZÓN AND X. CALBET
Instituto de Astrofísica de Canarias,
E-38200 La Laguna, Tenerife, Spain

Abstract. VRI CCD photometry of selected Two Micron Galactic Survey fields towards the end-of-bar star formation region between $l = 18°, b = 0°$ and $l = 27°, b = 0°$ has been carried out at the 1.0-m Jakobus Kapteyn Telescope on La Palma. The sources appear to be mainly G/K dwarfs at a heliocentric distance of typically 1.5 kpc with a total visible extinction (A_V) of between 2.4 and and 4.8 mag

In July/August 1993 VRI CCD images were taken of 160 selected regions in the inner Galaxy covered by the Two Micron Galactic Survey (TMGS, see Garzón *et al.* 1993 with the aim of calibrating UKST photographic plates measured on COSMOS. These images, however, can be directly used in conjunction with the TMGS or in their own right. In particular, the images of the regions in or very close to the Galactic plane can be compared with others well away from the plane to estimate the amount of visible extinction in the direction of TMGS scans.

Field 117 has Galactic coordinates $l = 77.19°, b = -16.05°$; hence this region was used as a comparison field of low extinction. The first field, SFA1, is located at $l = 19.9°, b = 0.2°$ and the second field, SFB1 is at $l = 27.0°, b = -0.1°$ (i.e. both extremes of the star formation region). The $V, (V - R)$ HR diagram and $(V - R), (R - I)$ colour-colour plot for region SFB1 are shown in Fig. 1.a and 1.b, respectively.

If we assume low extinction in Field 117, then from the colours the bulk of the stars are G- to K-type dwarfs If it is also assumed that the stellar distribution is the same towards SFB1, then it is reasonable to conclude that the shift upwards and towards the right in the CC plot for the SFB1 stars is caused by extinction. In SFA1 a typical source has an apparent magnitude

D. L. Block and J. M. Greenberg (eds.), New Extragalactic Perspectives in the New South Africa, 530–531.
© 1996 *Kluwer Academic Publishers.*

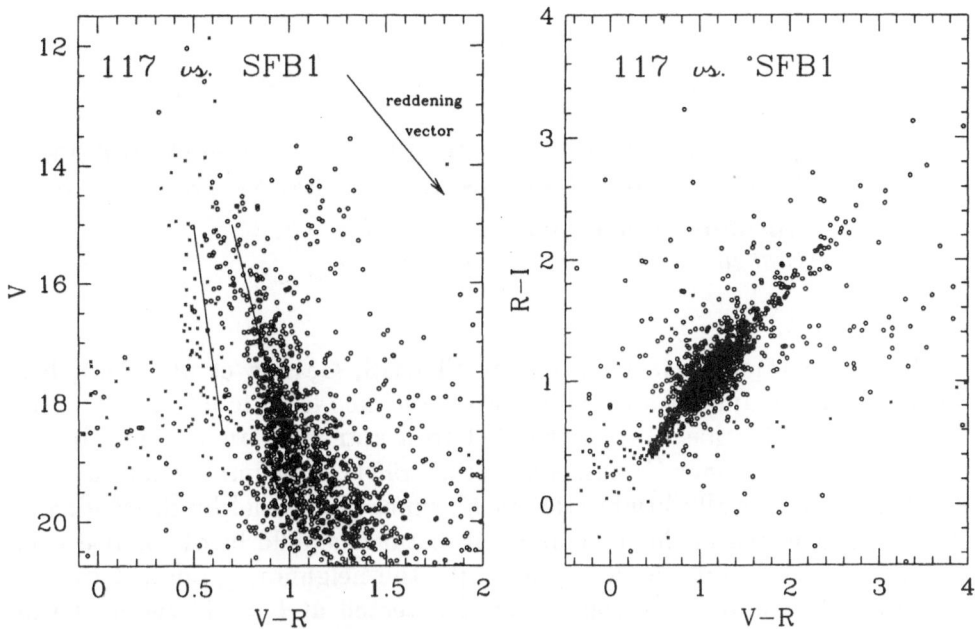

Figure 1. Left: V–$(V - R)$ HR diagram for the sources in Field 117 (crosses) and SFB1 (open circles). Right: $(V - R)$–$(R - I)$ plot for the same fields.

of $+19.5$. Using Van de Hulst curve 15 ($A_V = 1.0, A_R = 0.74, A_I = 0.47$) gives $E(R-I) = 0.27A_V$. For Field 117 and SFA1 the shift gives an $E(R-I)$ of between 0.6 and 1.2 approximately. Hence A_V lies between 2.4 and 4.8 mag. Although the density of late dwarfs is higher, the G dwarfs, at 2 mag brighter than K dwarfs, are seen much further out and hence make up the larger part of the sources seen. A typical G5 dwarf has an absolute V magnitude, M_V of about $+5.15$ mag (see Wainscoat *et al.* 1992). Hence, the standard formula $\log(r) = (m - M + 5 - A_V)/5$ gives a distance range for the SFB1 sources of typically 1.5 kpc.

References

Garzón F., Hammersley P. L., Mahoney T., Calbet X., Selby M. J., Hepburn I. D., 1993, MNRAS, **264**, 773

Hammersley P. L., Garzón F., Mahoney T., Calbet X., 1994, MNRAS, **269**, 753

Wainscoat R. J., Cohen M., Volk K., Walker H. J., Schwartz D. E., 1992, APJS, **83**, 111

SAMPLING THE BAR POPULATION

X. CALBET, M. LOPEZ-CORREDOIRA, P.L. HAMMERSLEY,
T. MAHONEY, F. GARZÓN AND J.E. BECKMAN

Instituto de Astrofísica de Canarias,
E-38200 La Laguna, Tenerife, Spain

The Two Micron Galactic Survey (TMGS, Garzón et al. 1993), has detected a large amount of bright sources on the Galactic plane at $l = 27°$ (Fig. 1 left). This has been attributed to a star formation region at the end of the Galactic bar (Hammersley et al. 1994). This implies that a large number of intrinsically bright stars in K, super-giants and bright M giants, should be expected in this region. Their distance should be of about 6 kpc. Comparing the counts in this region with the neighboring ones, it can be seen that the number of bright sources detected at $l = 27°$ should be of about 80-90% of the total. To confirm this the spectra of 51 bright TMGS sources were obtained.

The objects were observed at the 2.5-m INT (La Palma) with the IDS spectrograph in the 775 nm-950 nm range during the summer of 1995. The objects were TMGS selected bright K sources ($4.2\,\text{mag} < m_K < 4.8\,\text{mag}$) at $l = 27°$ on the Galactic plane.

The TMGS sources have a negligible error in position in R.A., but its error in δ is about $10''$. Doing spectroscopy, in the I band, of these frequently very red objects in these crowded fields can be a difficult task. Because of this, some of the originally selected sources were so faint in I that they could not .be observed; others were almost certainly misidentified. Despite this, the final results shown here are satisfactory.

All the stars were classified in spectral type. Their luminosity classes were derived from their Ca II triplet (849.8, 854.3, 866.2 nm) following Díaz et al. (1989). Some of these spectra are shown in Fig. 2. Their absolute magnitudes were obtained from their classification using the values shown in Wainscoat et al. (1992). The histogram of all the observed stars in absolute K magnitudes is shown in Fig. 1 right. Most of the stars are intrinsically very luminous in K as expected.

D. L. Block and J. M. Greenberg (eds.), New Extragalactic Perspectives in the New South Africa, 532–533.
© *1996 Kluwer Academic Publishers.*

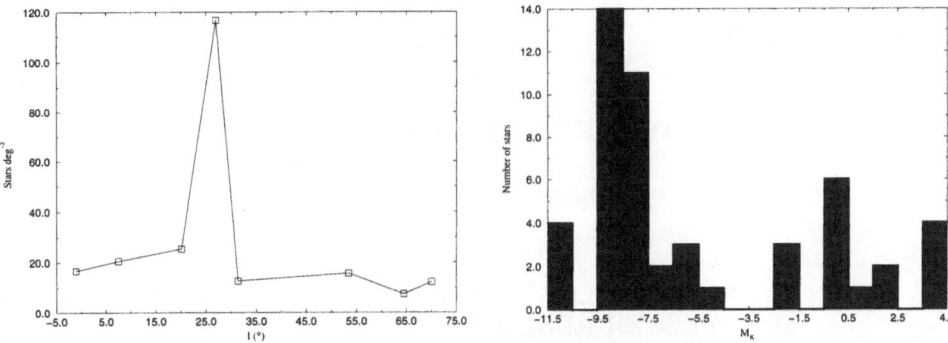

Figure 1. Left: The number of stars per square degree between $m_K = 4.2$ mag and $m_K = 4.8$ mag are plotted versus Galactic longitude on the plane ($b = 0°$). Notice the huge peak of bright stars at $l = 27°$ which corresponds to the end of the bar. Right: Histogram of the absolute K magnitude for all observed stars. Notice the relative large amount of intrinsically very bright sources.

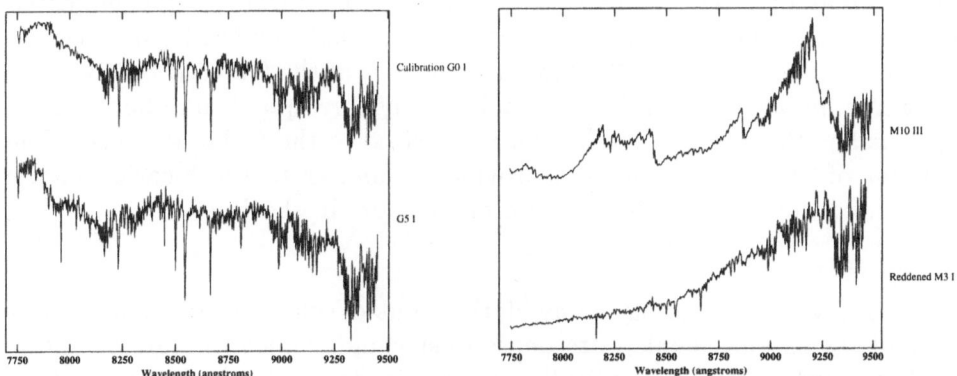

Figure 2. Selected spectra. Top left: Known G0 super-giant used as a calibration and shown for comparison. Bottom left: An observed TMGS source which has been classified as a G5 super-giant. Top right: An extremely late TMGS source, an M10 giant. Bottom right: A highly reddened TMGS source, an M3 super-giant.

References

Díaz A. I., Terlevich E., Terlevich R., 1989, MNRAS, 239, 325

Garzón F., Hammersley P. L., Mahoney T., Calbet X., Selby M. J., Hepburn I. D., 1993, MNRAS, 264, 773

Hammersley P. L., Garzón F., Mahoney T., Calbet X., 1994, MNRAS, 269, 753

Wainscoat R. J., Cohen M., Volk K., Walker H. J., Schwartz D. E., 1992, APJS, 83, 111

CHASING SHADOWS IN THE TMGS

P.L. HAMMERSLEY, T. MAHONEY, F. GARZÓN, X. CALBET AN]
M. LOPEZ-CORREDOIRA
Instituto de Astrofísica de Canarias,
E-38200 La Laguna, Tenerife, Spain

Abstract. Dust clouds in the Two Micron Galactic Survey (TMGS) show up as areas of lower than expected star counts. However, the increased penetration of the infrared wavelengths means that only major dust clouds produce noticeable effect. Within a few degrees of the Galactic centre there are a number of very dense dust clouds. Many correspond to radio features, such as Sagitarius C, so are known to be close to the Galactic centre. The amount of absorption not only provides a pointer to the location of the dust cloud but also the distribution of the stars in the Bulge.

The TMGS is finding many dark nebulae silhouetted against background stars. The nebulae are dense dust clouds which obscure the stars behind them, and so dramatically reduce the observed star density. Whilst in the visible only a relatively small amount of dust is required to produce a dark nebula, at 2.2μm very dense dust clouds are required (as in the cold cores of giant molecular clouds). Figure 1 shows the cumulative star counts to mk=+6 and mk=+7 towards $l=-0.3°$ and for comparison, at $l=7°$. Both traces show the star counts normalised to a square degree but that at $l=-0.3°$ has an effective spatial resolution of 0.25o so the effect of the extinction can be more clearly seen.

The majority of the TMGS sources in the direction of the Galatcic Centre are within about 350 pc of the Galactic Center (Garzón et al 1993). So all the dust producing the very obscured regions at $l=-0.3°$ must lie between us and these stars. Whilst it would be expected that the most obsure regions would be due to local dust this appears not to be the case. The darkest features, which can loose up to 80% of the expected sources, all coincide with features in the 10 GHz maps of Handa et al. (1984) which are known to be near the galactic centre. These include Sgr C and the North-

D. L. Block and J. M. Greenberg (eds.), New Extragalactic Perspectives in the New South Africa, 534–535.
© *1996 Kluwer Academic Publishers.*

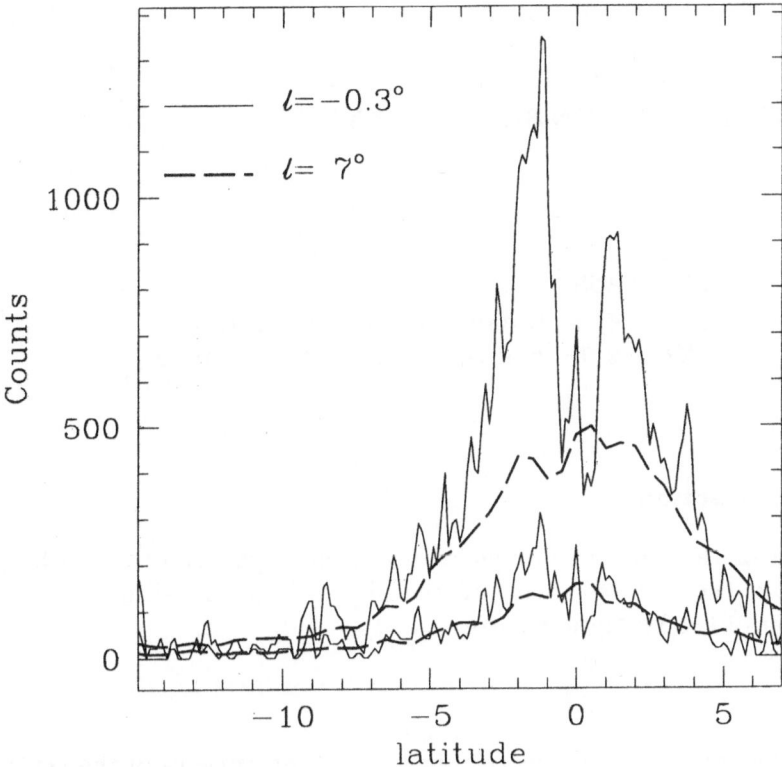

Figure 1. Star counts to $l=-0.3°$ and $l=7°$ to a K limiting magnitude of +6 (lower traces) and +7 (upper traces).

western Galactic Lobe (Sofue et al. 1986). This latter feature is of interest as whilst in the 10GHz maps it is clearly a U shaped feature arching from near the GC to $l=-0.5°$ it is only seen in extinction in the TMGS data and DIRBE 2.2μm data as half an arch near $l=-0.5°$. This suggests that the part of the lobe near $l=-0°$ is on the far side of the GC. Whilst that near $l=-0.5°$ is on the near side.

References

Garzón F., Hammersley P. L., Mahoney T., Calbet X., Selby M. J., Hepburn I. D., 1993, MNRAS, 264, 773
Handa, T., Sofue, Y., Nakai N., Hirabayashi H. & Inoue M., 1987 PSAPJ, 39, 709
Sofue, Y. & Handa, T., 1984, Nature, 310, 568

IMAGING THE INNER GALAXY

I.S. GLASS

South African Astronomical Observatory
PO Box 9, Observatory 7935, South Africa

1. Introduction

Imaging of the stellar component of the inner galaxy can only be accomplished in the infrared owing to the high column densities of dust in the direction of the Galactic Centre.

2. 1° × 2° Colour Map

SAAO has been involved in two major imaging surveys of the central portion of the galaxy. The first covered a region of 1 × 2 deg², including the Centre, at JHK (1.25 μm, 1.65 μm and 2.2 μm). Images in each colour were presented by Glass, Catchpole & Whitelock (1987) and a quantitative analysis of the data was presented by Catchpole, Whitelock & Glass (1990).

The three single-colour images have recently been combined by R.M. Catchpole into a three-colour Red, Green and Blue picture, corresponding to K,H and J (see colour plate 14 in these *Proceedings*). Whenever the intensity exceeded a preset maximum level, all the data for that pixel were adjusted to preserve the correct colour. For this reason, much of the detail of the central cluster has been lost. A linear scaling was used and the final image was made using a dye sublimation printer.

As pointed out by Glass *et al.*, the image contains many dark patches due to the exceedingly dense cores of the intervening interstellar dust clouds. It is estimated that complete blackness on this image corresponds to $A_V > 60$ mag. Apart from extended cores, the map shows evidence for dense filamentary structures. The colours of the background stars become increasingly red as the cores are approached. Most of the dark cores shown are clearly in the foreground, since their outlines correlate well with the positions of low-velocity features in the ^{13}CO radio maps.

D. L. Block and J. M. Greenberg (eds.), New Extragalactic Perspectives in the New South Africa, 536–537.
© 1996 Kluwer Academic Publishers.

3. Detailed Survey of Central 24 × 24 Arcmin²

The second survey project, which is ongoing, is to map the central $24' \times 24'$ of the Galaxy repeatedly, over a period of 3 years. The repetitive survey is carried out at K, while extra observations are made at J and H to enable the construction of colour-magnitude diagrams.

The PANIC (PtSi Astronomical Near-Infrared Camera) camera that we used is based on a chip of effectively 1040 × 520 pixels and has a 5 × 5 arcmin² field when attached to the 0.75m telescope at Sutherland. It was constructed as a joint project of the National Astronomical Observatory of Japan, the Institute of Astronomy, University of Tokyo, and the SAAO.

The survey consists of 25 slightly overlapping sub-fields, each of which has been observed 8 or more times so far. Following preliminary processing of the images, several thousand objects are measured from each one using DoPHOT. The output lists for each field are cross-correlated with a fiducial image, the photometry is placed on a uniform instrumental scale and the data for each individual object are tabulated in order of time. At present, variable stars are found by plotting the data for the brightest 200 stars in each sub-field and examining them by eye.

It is estimated from this work that the field contains about 300 long-period variables. Their positions have been found by relating objects appearing on both the infrared maps and the visible sky surveys to the stars of the HST Guide Star Catalog. Several of the OH/IR stars known from radio work have been recovered and the positional agreement is excellent.

The K, H-K diagram of each sub-field is obtained by cross-correlating the H image of the field with the fiducial field and forming H-K colours for all the stars detected at both wavelengths. From the H-K data it will be possible to construct detailed maps of the reddening in the whole area, using the fact that stars at the upper end of the AGB define a narrow locus in the colour-magnitude diagram (Frogel & Whitford, 1987). Most of the variable stars that we have found occupy the top right of the colour-magnitude diagram as in less obscured fields such as Sgr I (Glass, 1994).

I thank Dr R.M. Catchpole for allowing me to display the colour image, and Dr Y. Nakada, S. Matsumoto (Tokyo), Dr K. Sekiguchi (NOAJ), T. Ono (Nishi-Harima Astr. Obsy.) and B.S. Carter (Carter Obsy., Wellington) for their help and participation in the PANIC work.

References

Catchpole, R.M., Whitelock, P.A. and Glass, I.S. (1990) *MNRAS*, **247**, 479

Frogel, J.A. and Whitford, A.E. (1987) *ApJ* **320**, 199

Glass, I.S. (1994) in *The Nuclei of Normal Galaxies* eds. Genzel, R. and Harris, A.I., Kluwer, Dordrecht Kluwer, Dordrecht.

Glass, I.S., Catchpole, R.M. and Whitelock, P.A. (1987) *MNRAS*, **227**, 373

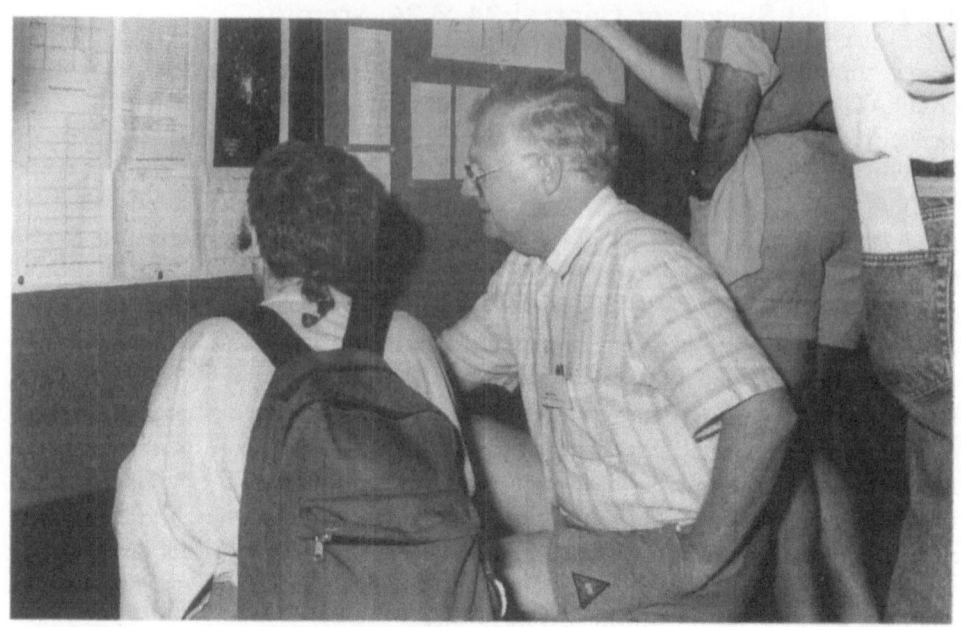

FABRY-PEROT IMAGING OF DUST GRAIN PROCESSING IN M82

M. A. GREENHOUSE, S. SATYAPAL AND H. A. SMITH

National Air and Space Museum, Smithsonian Institution, Washington DC 20560

C. E. WOODWARD

Department of Physics and Astronomy, University of Wyoming, Laramie, WY 82071

J. FISCHER

Remote Sensing Division, Naval Research Laboratory, Washington, DC 20375

AND

W. J. FORREST, J. PIPHER AND D. M. WATSON

Department of Physics and Astronomy, University of Rochester, Rochester, NY 14627

1. Introduction

Iron is an abundant refractory element; however, it is highly depleted from the gas phase of the interstellar medium of galaxies as a result of condensation into dust grains. Forbidden line emission from low ionization states of iron is greatly enhanced behind hydrodynamic shock fronts where sputtering processes volatilize the grains such that the gas phase abundance of iron can reach near-solar values. This enhancement can provide a new sensitive measure of supernova activity in galaxies, and the rate at which condensed solids are returned to the gas phase of the ISM.

Here we report the first [Fe II] Fabry Perot images of a galaxy. These data were obtained with the NASM/NRL Fabry-Perot system and the University of Rochester Third Generation Camera at the Wyoming Infrared Observatory 2.34 m telescope.

D. L. Block and J. M. Greenberg (eds.), New Extragalactic Perspectives in the New South Africa, 539–541.
© *1996 Kluwer Academic Publishers.*

2. Discussion

Two categories of [Fe II] emission are seen: a distributed background emission extending throughout the nuclear region of the galaxy, and 6 compact sources south west of the nucleus. We find excellent spatial correspondence between unresolved [Fe II] compact sources and K-band sources K5 and K6 mapped by Dietz et al. (1986). We find that K5 and K6 exhibit a [Fe II]1.644/H5-3 ratio that is roughly a factor of 5 higher than any other [Fe II] source in M82.

We have used the infrared extinction map produced by Satyapal et al. (1995) to correct our images for foreground extinction. We find that the [Fe II] sources are intrinsically compact, and do not result from holes in a foreground extinction screen. We note that the larger [Fe II] sources are similar in size to the Cygnus Loop (diameter ~ 40 pc, age $\sim 5 \times 10^4$ yr). In contrast, the most compact [Fe II] sources are at least a factor of 2 smaller, and as a consequence, much younger than the 40 pc sources.

The iron depletion we derive for these sources is an order of magnitude lower than typical interstellar values along reddened lines of sight in the Galactic ISM, and is roughly equivalent to model predictions of iron depletion behind moderate density 100 km s^{-1} shocks consistent with the idea that the [Fe II] morphology in M82 traces regions of grain destruction (Greenhouse et al. 1991).

We find that the distributed [Fe II] emission is extended to the south relative to H7-4 in the general direction of the galaxy's minor axis. This extended emission accounts for 90% of M82's [Fe II] luminosity. The absence of [Fe II] emission to the north, suggests that this emission is not tracing "starburst superwinds". An alternate interpretation is that the starburst region in M82 is propagating to the north (in projection) leaving a low depletion ISM in its wake. This interpretation is consistent with the collocation of [Fe II] and HII emission regions along the northern edge of the [Fe II] emission region.

Although several of the compact [Fe II] sources are in close proximity to one or more radio continuum sources, these new high spatial resolution data reveal no strong spatial correlation between the [Fe II] and radio continuum sources mapped by 1985 or Muxlow et al. 1994. The lack of spatial correlation between the [Fe II] and radio point sources and the size of the spatially resolved [Fe II] sources suggests that these components are tracing separate populations of supernova remnants that differ significantly in age.

We find that the [Fe II] sources are typically a factor of 20 larger than the 6 cm sources. However, we note that the size contrast between the [Fe II] and 6 cm remnants in M82 is similar to that exhibited by the Cygnus Loop (age 5000 yr) and Cas A (age 300 yr). A natural interpretation for the compact [Fe II] sources is that they trace a population of supernova remnants that are at least an order of magnitude older than those observed in the radio continuum. We note that the size distribution of the [Fe II] sources spans at least a factor of 2. Assuming that these remnants are in radiative expansion, they must span a factor of 10-20 in age. Hence these data suggest that [Fe II] remains an important coolant in supernova remnants for at least 10^3 years after outburst.

We find a strong correlation between the extinction corrected J-H color and the extinction corrected [Fe II]/H7-4 flux ratio within the central 500 pc of M82. In addition to correcting the near-infrared continuum for extinction, the free-free and free-bound components to the emission have been removed. Since the J-H color is only weakly effected by dust emission, the corrected J-H color can be ascribed to stellar photospheric emission alone. We find that the [Fe II]/Brγ line ratio throughout M82 correlates strongly with the age of the starburst as reflected by the color of photospheric emission from galaxy's stars. This correlation suggests that the [Fe II] emission regions in M82 are co-located with the post main sequence stellar population.

We conclude that these [Fe II] sources trace a population of supernova remnants in M82 that are substantially older than those revealed previously on 6 cm radiographs. We find that M82 contains a distributed [Fe II] emission component that is extended to the south along the minor axis, and accounts for 90% of the galaxy's [Fe II] luminosity. This emission may trace the wake of a spatially propagating starburst region. The gas phase iron depletion in M82 is an order of magnitude lower than typical Galactic interstellar values. In addition, we find that the [Fe II]/Brγ line ratio throughout M82 correlates strongly with the age of the starburst as reflected by photospheric emission from the galaxy's stars.

This work was supported by the Smithsonian Institution Scholarly Studies Program, NASA grant NAG1711, and NSF grant AST93-57392.

References

Dietz, R. D., Smith, J., Hackwell, J. A., Gehrz, R. D., and Grasdalen, G. L. 1986, AJ, 91, 758.

Greenhouse, M. A., Woodward, C. E., Thronson, H. A., Rudy, R. J., Rossano, G. S., and Erwin, P. 1991, ApJ, 383, 164.

Muxlow, T. W. B., Pedlar, A., Wilkinson, P. N., Axon, D. J., Sanders, E. M., and de Bruyn, A. G. 1994, MNRAS, 266, 455.

Satyapal, S., Watson, D. M., Pipher, J. L., Forrest, W. J., Coppenbarger, D., Raines, S. N., Libonate, S., Piche, F., Greenhouse, M. A., Smith, H. A., Thompson, K. L., Fischer, J., Woodward, C. E., and Hodge, T. 1995, ApJ, 448, 611.

THE HI CONTENT OF A COMPLETE SAMPLE OF SA GALAXI

W.K. HUCHTMEIER
Max-Planck-Institut für Radioastronomie
Auf dem Hügel 69, D-53121 Bonn, Germany

We present HI observations of a complete sample of Sa galaxies, i.e. of all Sa galaxies in the Revised Shapeley-Ames (Sandage and Tammann 1987, RSA) catalog. New HI observations were performed with the 100-m radiotelescope at Effelsberg which has a half power beam width of 9.3 arcmin at the wavelength of 21-cm. From a total of 120 Sa galaxies in the RSA 105 are detected in HI including observations published so far.

Correlations between global galaxian parameters are present as for other disk galaxies taking into account that the relative HI-content of Sa galaxies is lower than for later-type galaxies.

In general the relative HI content, i.e. the HI mass per luminosity (M_{HI}/L_B), increases towards late-type galaxies. This is the case even within the morphological type Sa, where M_{HI}/L_B increases with Sandage's (Hogg, Roberts, Sandage 1993) subtype.

Compared to other disk galaxies small (less luminous) Sa galaxies seem to be strongly HI-deficient. Some of these systems are located in the Virgo cluster or in the Coma I cloud. Is this effect connected with the mass loss of galaxies in clusters by the interstellar matter or tidal interactions that produces the well-known HI-deficiency, makes galaxies shrink and convert them into earlier type ones (spiral galaxies in the Virgo cluster being redder than field galaxies by nearly one morphological type, Kennicutt 1983)?

References

Hogg, D., Roberts, M.S., Sandage, A. (1993) AJ 106, 907
Kennicutt, R.C. (1983) AJ 88, 483
Sandage, A., Tammann, G.A. (1987) *A Revised Shapeley-Ames Catalog of Bright Galaxies* The Carnegie Institution, Washington D.C. [RSA]

D. L. Block and J. M. Greenberg (eds.), New Extragalactic Perspectives in the New South Africa, 542.
© 1996 *Kluwer Academic Publishers.*

MASSIVE GALACTIC HALOS AND DARK MATTER: UNUSUA

PROPERTIES OF NEUTRINOS IN THE DE SITTER UNIVERSE

NAIL H. IBRAGIMOV
*Department of Computational & Applied Mathematics,
University of the Witwatersrand, P.O. Wits 2050
Johannesburg, South Africa*

Abstract. A theoretical conclusion of the present work based on symmetry principles is that, in the de Sitter universe with a small constant curvature $K \neq 0$, a massless neutrino splits into two massive neutrinos. The unusual properties of neutrinos combined with the neutrino-as-dark matter hypothesis may play a significant role for understanding the nature of the material of dark halos surrounding galaxies.

1. Approximate representation of the de Sitter group

De Sitter space-time has the metric form $ds^2 = -(1 + \varepsilon |x|^2)^{-2} dx_\mu dx^\mu$, where $|x|^2 = x_\mu x^\mu$, $x_\mu = x^\mu$, $\mu = 1, \dots, 4$, and $\varepsilon = K/4$. Here K is the constant curvature of the de Sitter universe.

The *de Sitter group* (i.e., the group of isometric motions in the de Sitter space-time) differs from the *Poincaré group* in that the usual translations of space-time coordinates x^μ are replaced by more complicated transformations, the so-called "generalized translations". Since ε is small (according to cosmological data, $K \sim 10^{-54}$ cm^{-2}), one can use an approximate expression of these transformations in the framework of the theory of approximate groups. Then the result is rather simple and the generalized translation, e.g. along the x^1 axis is written:

$$\bar{x}^1 \approx x^1 + a + \varepsilon([(x^1)^2 - (x^2)^2 - (x^3)^2 - (x^4)^2]a + x^1 a^2 + a^3/3),$$
$$\bar{x}^j \approx x^j + \varepsilon(2ax^1 + a^2)x^j, \quad j = 2, 3, 4,$$

where a is a group parameter. The similar generalized translations along the other axes together with the six independent rotations of the variables x comprise the *approximate representation of the de Sitter group*.

D. L. Block and J. M. Greenberg (eds.), New Extragalactic Perspectives in the New South Africa, 544–546.
© 1996 *Kluwer Academic Publishers.*

2. Dirac equations

The Dirac equation in the de Sitter space-time (Dirac, 1935), in the case of zero mass (neutrinos) can be written in the linear approximation with respect to K as follows (Ibragimov, 1993):

$$\gamma^\mu \frac{\partial \phi}{\partial x^\mu} - 3\varepsilon(x \cdot \gamma)\phi = 0 \qquad (1)$$

with γ^μ the usual Dirac matrices in the Minkowski space-time, ϕ a four-dimensional complex vector, and $(x \cdot \gamma) = x_\mu \gamma^\mu$. Eq. (1) is reduced to the Dirac equation in the Minkowski space-time,

$$\gamma^\mu \frac{\partial \psi}{\partial x^\mu} = 0, \qquad (2)$$

by the substitution (in the first order of precision with respect to ε):

$$\psi = \phi - \frac{3}{2}\varepsilon|x|^2\phi. \qquad (3)$$

Eq. (2) is invariant under the transformation of the variables x:

$$y^\mu = ix^\mu, \quad i = \sqrt{-1}. \qquad (4)$$

By setting

$$\phi(x) = \varphi(y), \qquad (5)$$

we rewrite (3) in the form

$$\psi = \varphi + \frac{3}{2}\varepsilon|y|^2\varphi. \qquad (6)$$

The invariance of Eq. (2) under the transformation (4) means that

$$\gamma^\mu \frac{\partial \psi}{\partial y^\mu} = 0. \qquad (7)$$

Substitution of (6) into Eq. (7) yields:

$$\gamma^\mu \frac{\partial \varphi}{\partial y^\mu} + \varepsilon(y \cdot \gamma)\varphi = 0. \qquad (8)$$

Eq. (8) coincides with the Dirac equation (1) in the de Sitter space-time with the curvature $(-K)$.

The system of equations (1), (8) inherits all symmetries of the usual Dirac equation (2). Namely, the system (1), (8) is invariant under the approximate representation of the de Sitter group and under the transformation (4) - (5). Moreover, it is conformally invariant in the first order of accuracy (Ibragimov, 1993).

3. Discussion

The Dirac equations in the de Sitter universe have the following peculiarities due to curvature.

1. Eq. (1) can be regarded as a Dirac equation $\gamma^\mu \dfrac{\partial \phi}{\partial x^\mu} + m\phi = 0$ with the variable matrix valued "effective" mass $m = -3\varepsilon(x \cdot \gamma)$. Then, in the framework of the usual relativistic theory, neutrinos will have *small but nonzero* mass. It follows from the conformal invariance of Eqs. (1), (8) that these "massive" neutrinos satisfy the *Huygens principle* (Ibragimov, 1993) and move with the velocity of light.

2. Since Eqs. (2) and (7) coincide, there is no preference between two transformations (3) and (6), and hence between Eqs. (1) and (8). Consequently, a massless neutrino given by Eq. (2) splits into two distinct "massive" neutrinos described by the equations (1) and (8). These two particles are distinctly different if and only if $K \neq 0$. One of them, namely given by the equation (1), can be regarded as a *proper neutrino* and the other given by the equation (8) as an *antineutrino*. They have the "effective" masses $m_1 = -3\varepsilon(x \cdot \gamma)$ and $m_2 = 3\varepsilon(x \cdot \gamma)$, respectively.

3. We summarize. In the de Sitter universe with a small curvature $K \neq 0$, a neutrino is a compound particle *neutrino–antineutrino* with the total mass $m = m_1 + m_2 = 0$. It is natural to assume that only the first component of the compound is observable and is perceived as a massive neutrino. The counterpart to the neutrino provides the validity of the zero-mass-neutrino model and has the real nature in the *antiuniverse* with the curvature $(-K)$.

Acknowledgements

I am grateful to D.L. Block for valuable discussions. I also acknowledge the support from FRD research grant.

References

P. A. M. Dirac (1935) The electron wave equation in the de-Sitter space, *Ann. Math.* **Vol. 36, No. 3**, pp. 657–669.

N. H. Ibragimov (1993) Seven miniatures on group analysis, *Differential Equations* **Vol. 29, No. 10**, pp. 1511–1520. See also: N.H. Ibragimov, Applications to Celestial Mechanics and Astrophysics, Chapter 5 in *CRC Handbook of Lie Group Analysis of Differential Equations* **Vol. 2**, Ed. N.H. Ibragimov, CRC Press, Boca Raton–Ann Arbor–London–Tokyo, 1995.

DUST IN A RADIO GALAXY AT Z=4.25

ROB J. IVISON
Royal Observatory, Blackford Hill, Edinburgh EH9 3HJ

AND

ELESE N. ARCHIBALD
Institute for Astronomy, University of Edinburgh EH9 3HJ

Abstract. We describe rest-frame far-IR continuum measurements of 8C 1435+635, a radio galaxy at $z = 4.25$, obtained using the IRAM 30-m MRT and the 15-m JCMT. If the emission is thermal, then the derived dust mass lies in the range, $2 \times 10^9 < M_d < 8 \times 10^7\,M_\odot$ for $20 < T_d < 100\,K$, or $M_d \sim 1.6 \times 10^8\,M_\odot$ for $T_d = 60\,K$, similar to that derived for 4C41.17, suggesting a molecular gas mass of between 4×10^{10} and $9 \times 10^{11}\,M_\odot$. The upper limit obtained at $150\,\mu m$ (rest frame) limits T_{dust} to $< 100\,K$.

1. Introduction

There are approximately 50 known objects with $z \sim 4$, and it is natural to use them as observational probes of the early stages of galaxy formation. These young radio galaxies and QSOs are likely to be the sites of active star formation, and thus an obvious place to search for thermal emission from dust (see e.g. Dunlop et al. 1994; Chini & Krügel 1994) and line emission from CO.

During 1995 February, we used the IRAM 30-m MRT to observe 8 RQQs and a RG which satisfy $3.7 < z < 4.3$; we detected relatively strong continuum emission at 1.25 mm from 8C1435+635 (Lacy et al. 1994), and from PC2047+0123, a RQQ at $z = 3.80$. Both were as bright at 1.25 mm as 4C41.17, which was detected using both JCMT and IRAM. An extrapolation of the steepening centimetric radio spectrum of 8C1435+635 can account for less than 1% of the observed 1.25-mm flux, indicating that the emission is probably from dust.

Here, we describe follow-up measurements made during 1995 December using UKT14 on the JCMT.

547

D. L. Block and J. M. Greenberg (eds.), New Extragalactic Perspectives in the New South Africa, 547–548.
© *1996 Kluwer Academic Publishers.*

TABLE 1. Photometry.

Date (1995)	Telescope /Bolometer	λ_0 /μm	Flux Density /mJy	Note
Feb 17–22	IRAM, 7-channel	238	2.57 ± 0.42	Ivison 1995
Dec 24	JCMT, UKT14	152	$3\sigma < 26$	this paper

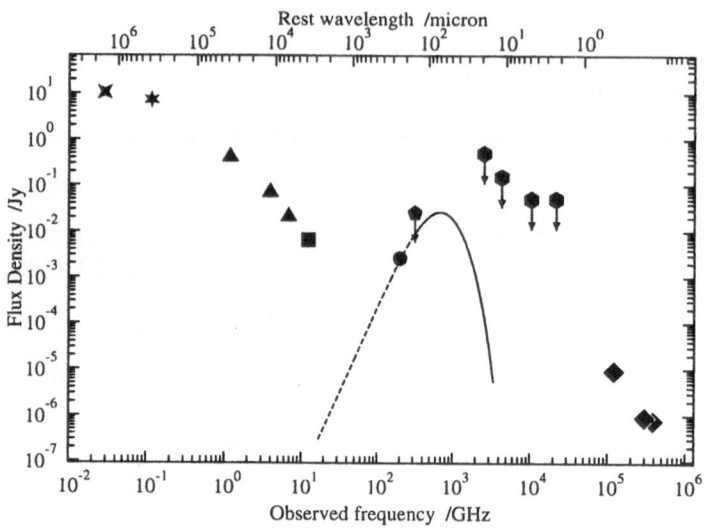

Figure 1.   SED of 8C1435+635. Data are from Ivison (1995). A 60-K greybody is illustrated. The 0.8-mm datum limits T_{dust} to < 100 K.

2. New Measurements

Data were obtained during 1995 December using the UKT14 bolometer on the 15-m JCMT, Mauna Kea. We concentrated on obtaining data at 0.8 mm, performing a total of 260×20-sec ON–OFF scans with our 16.8-arcsec FWHM near-Gaussian beam. Sky emission was subtracted by chopping the secondary mirror in azimuth, at a frequency of 7.8 Hz, with a throw of 60 arcsec. The weather was as good as it had been for some considerable time, but it was still not particularly useful: the opacity at 1.3 mm was around 0.07–0.08.

References

Chini, R. and Krügel, E. (1994) *A&A*, **288**, L33
Dunlop, J.S., Hughes, D.H., Rawlings, S., Eales, S. and Ward, M. (1994) *Nat*, **370**, 347
Ivison, R.J. (1995) *MNRAS*, **275**, L33
Lacy, M. et al. (1994) *MNRAS*, **271**, 504

VOYAGER OBSERVATIONS OF DUST SCATTERED STARLIG

J. MURTHY, D. T. HALL AND R. C. HENRY
Department of Physics and Astronomy
The Johns Hopkins University
Baltimore, MD 21218-2695, USA

AND

J. B. HOLBERG
Lunar and Planetary Laboratory
The University of Arizona,
Tucson, AZ 85721

1. Introduction

There are many different processes contributing to the diffuse ultraviolet radiation field: including heliospheric, Galactic, and extragalactic sources. Unfortunately, there are very few reliable observations of the ultraviolet background (see Murthy and Henry 1995), both due to the faintness of the signal and the need for space-based instrumentation. In this context, there has been a surprising neglect of the Voyager ultraviolet spectrometer (UVS) archives. Comprising more than 17 years of data from the long stretches when the two spacecraft were between their highly successful planetary encounters, these archives contain observations of a wide variety of astrophysical targets in the spectral region between 550 and 1700 Å. In the best cases, limits of about 100 photons cm^{-2} s^{-1} sr^{-1} Å^{-1} (Murthy, J. et al. 1991) can be set on the diffuse radiation field - sufficient to significantly constrain the various components of the UV background.

The two Voyager ultraviolet spectrometers are identical Wadsworth mounted objective grating spectrometers which cover the spectral range between 500 and 1700 Åover a field of view of 0.1 x 0.87. The spectral resolution of the instruments is 38 Åfor aperture filling diffuse sources and 18 Åfor point sources. The detectors are dual microchannel plates read out onto a linear self-scanned array of aluminum anodes, 126 of which are active for a bin size of 9.26 Å. The UVS is most sensitive at wavelengths below

549

D. L. Block and J. M. Greenberg (eds.), New Extragalactic Perspectives in the New South Africa, 549–550.
© 1996 *Kluwer Academic Publishers.*

1200 Å, with a rapidly declining response at longer wavelengths where a coating was applied to the bare microchannel plate. Out of the more than 8000 observations in the archives, we have identified 1414 as being potentially useful measurements of the diffuse radiation field.

2. Data Analysis

There are three independent constituents of the typical UVS spectrum: dark noise from the spacecraft's radioisotope thermoelectric generator (RTG); emission lines from the interplanetary medium; and a cosmic signal (which we define to be all radiation originating beyond the Solar System). We performed a χ^2 simultaneous fit of models for all three components to the observed spectrum. Here, we will describe only our model for the dust scattered component which divides the line of sight into a number of cells, calculates the local radiation field at each cell and then calculates the amount scattered to the observer with a Henyey-Greenstein scattering function (Henyey and Greenstein 1941). The amount of dust is scaled from H I 21 cm maps. We also tested for the presence of H_2 fluorescence using model spectra calculated from the relative emission strengths of Sternberg (1989) and for emission lines at the positions of C III (977 Å) and O VI (1032/1038 Å). It is important to state that we do not yet make any claims about the origin of the cosmic background: we are simply using an empirical model.

3. Results and Conclusions

Our preliminary results are now available as a table from any of the authors of this work and will shortly be placed in a location easily accessible over the network. There are a total of 1415 potential observations of the diffuse radiation field. Many of these are null results, with upper limits on the order of 100 photons cm^{-2} s^{-1} sr^{-1} $Å^{-1}$; however, we do also detect signals of several thousand photons cm^{-2} s^{-1} sr^{-1} $Å^{-1}$ at other locations. We are currently in the process of examining each observation in excruciating detail to rule out competing sources of emission. At this time oour results are in the nature of upper limits: we place an upper limit of 80 photons cm^{-2} s^{-1} sr^{-1} $Å^{-1}$ on the amount of extragalactic radiation at wavelengths below 1200 Å.

References

Henry, R. C. 1991, *ARA&A*, **29**, 89
Henyey, L. G., and Greenstein, J. L. 1941, *ApJ*, **93**, 70
Murthy, J. and Henry, R. C. 1995, *ApJ*, **448**, 847 Murthy, J., Henry, R. C., and Holberg, J. B. 1991, *ApJ*, **383**, 198

DUST AND STARS IN THE NUCLEAR REGION OF M51

N. PANAGIA[1,2], A. CAPETTI[1], S. SCUDERI[1,2,3],
H. LAMERS[1,4] & R.P. KIRSHNER[5]

[1] *STScI, Baltimore, MD 21218, USA.*
[2] *On assignment from Space Science Department of ESA.*
[3] *Osservatorio Astrofisico di Catania, Italy.*
[4] *Astronomy Department, University of Utrecht, Holland.*
[5] *Harvard-Smithsonian CfA, Cambridge, MA 02138, USA.*

1. Observations and Results

The observations were made with the Hubble Space Telescope (*HST*) [1] as part of the SINS (*Supernova INtensive Study*) project, to determine accurate UV (early epoch) and optical (late epoch) magnitudes of SNIc 1994I, which occurred about 15″ offset from M51 nucleus. In particular, we have *HST-WFPC2* images obtained on 1994 May 12, with the F255W and F336W filters (in these images the nucleus falls in WF2), and four more taken on 1995 January 15 with the F439W, F555W, F675W, and F814W filters (the nucleus is in PC1).

Our analysis is based on comparisons of color-magnitude and color-color diagrams to determine the intrinsic colors of the stellar populations and, from there, the distribution and the basic properties of the extinction.

We find that the galactic extinction law provides a good match to our observations to within 10%. The extinction in the nuclear core is is patchy (see Fig. 1) and is maximum on the Y-shaped dark lane ($A_V(max) \simeq$ 2.1 *mag*). Adopting a visual extinction efficiency $Q_V \simeq 1$ per grain, an average grain size of $0.15\mu m$ and a grain material density of 2.3 $g\ cm^{-3}$, the total mass in dust grains over the nuclear core (a region 0.83″ in radius) is about 330 $M_\odot$, which is about 10^{-5} the mass in stars (see below). Such a low amount of dust (approximately 100 times lower than in the solar

---

[1]The NASA/ESA *Hubble Space Telescope* is operated by AURA, Inc., under NASA contract NAS 5-26555.

D. L. Block and J. M. Greenberg (eds.), New Extragalactic Perspectives in the New South Africa, 552–553.
© *1996 Kluwer Academic Publishers.*

neighborhood) indicates that dust destruction processes have prevailed over formation processes during the lifetime of the nuclear core.

The stellar population in the core (less than 2″ diameter) is much bluer than that in the surrounding bulge (an oval region of 10-15″ size). The spectral distributions, as given by broad band photometry, indicate that while the bulge population is composed by stars of spectral type more advanced than about G0 and, therefore, older than 8 $Gyrs$, the earliest type stars in the core are about A0, implying a mass at turnoff of $\sim$ 3.4 $M_\odot$ and an age of $\sim$ 400 $Myrs$.

The total luminosity inside a 0.83″ radius (i.e. 30 pc for an adopted distance of 7.5 Mpc), is $L_{tot} \simeq 8 \times 10^7$ $L_\odot$. Adopting a Salpeter IMF, an upper mass of 3.4 $M_\odot$ and a lower cutoff mass of 0.1 $M_\odot$, the total stellar mass in the core is $M_{tot} \simeq 4 \times 10^7$ $M_\odot$. This corresponds to an average stellar density of 350 $M_\odot$ pc^{-2} which is about 3000 higher than in the solar neighborhood. After allowance for extinction, the starlight distribution in the inner nucleus appears to be almost perfectly circular and displays a distinct light peak in the middle (right panel: Fig. 1). We believe that the central point like source is the very nucleus of M51, which has a size smaller than 2pc and a luminosity of about 10^6 $L_\odot$.

The nuclear core radial profile is very well fitted by an exponential law $I(r) \propto exp(-r/25pc)$, suggesting that the stars are still in a non-relaxed status. The occurrence of a "middle age" starburst about 400 million years ago is the possible effect of the last "close encounter" of M51 with its dwarf companion NGC 5195.

Full account of this work will be presented in a paper (Capetti et al. 1996) which is going to be submitted to the *Astrophysical Journal* in early Spring 1996. See also color plate 15 (this volume).

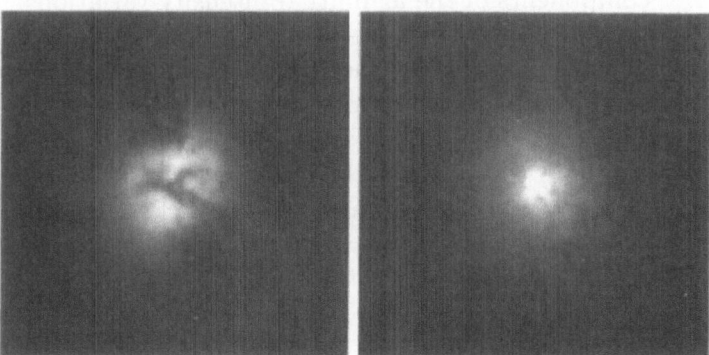

Figure 1 - True color image of the inner core (the central 4.6″ × 4.6″ area) of M51 obtained combining the B, V and R images. The left panel displays the observed image, and the right panel the reddening corrected one. In both panels North is to upper left corner.

GASEOUS AND STELLAR RESPONSES TO SPIRAL PERTURBATIONS DETECTED IN THE NIR

P.A. PATSIS
Max-Planck Institut für Astronomie
Königstuhl 17, D-69117 Heidelberg

AND

P. GROSBØL
ESO
Karl Schwarzschild Str. 2, D-85748 Garching bei München

1. Introduction

Among the conclusions of a comparative study on the morphology of 5 Sbc→Sc normal spiral galaxies in K′, B and V bands by Grosbøl & Patsis (1996) were the following: **1.** The spiral perturbation is marginally non-linear. **2.** Galaxies with a multi-armed morphology in B have a grand-design structure in K′. Also strong interarm features in B dim or even almost vanish in K′. **3.** In many cases an important m=1 term is detected in the Fourier components of the azimuthal intensity variations. The found differences in the morphology in near-infrared (NIR) and optical bands are in agreement with observations of other galaxies of the same type (Block et al. 1994). It is known, that observations in K′ refer mainly to the old disk population, while in B and V one observes young objects and gas. In this paper we investigate in what degree the observed differences correspond to differences in the responses of stellar and gaseous disks to an imposed spiral potential.

2. Stellar vs. Gaseous disks

The potential imposed is described in Contopoulos & Grosbøl (1988). The method for obtaining density maps of the stellar disks is also described in that paper. For the response of the gaseous disks we use SPH. In the present calculations we have added an m=1 component in phase with the main

D. L. Block and J. M. Greenberg (eds.), New Extragalactic Perspectives in the New South Africa, 554–555.

Figure 1. The gray-scale representation of the SPH response (after 6 pattern rotations) in the case described. The minima of the imposed potential are also given

m=2 spiral pattern. The amplitude of the spiral is such as to give a relative radial force perturbation of the order 3-5% of the axisymmetric force. We compare mainly the area between the 2/1 resonance and corotation, since we find, that the observed morphology in late type spirals can be modeled satisfactory only with models putting the end of the symmetric part of the arms at the 4/1 resonance. However, our gaseous models in many cases extend to the -2/1 resonance. In the example given here we deal with a spiral galaxy having a maximum rotational velocity 208 km/s, while the rest of the parameters describing its rotation curve are: $f_b = 0.02$, $\epsilon_b = 0.5 \ kpc^{-1}$, $\epsilon_d = 0.2 \ kpc^{-1}$. The imposed trailing spiral has a pitch angle 24° and an inverse scale length $\epsilon_s = 0.2 \ kpc^{-1}$. We find that:

- The pitch angle of the response spiral in the gaseous disks is smaller, i.e. we have a tighter spiral, than the stellar one. The latter remains almost logarithmic, as the imposed, until the 4/1 radius. The phase differences are of the order of 1-2° degrees and are larger close to the end of the symmetric part of the arms (Fig. 1)
- In the response of the gaseous models we obtain strong interarm features, spurs, broken arms and off-phase extensions beyond the end of the symmetric part of the spirals. These correspond to local density maxima of the stellar models much weaker than the spiral arms.
- The inclusion of the m=1 component in the models is necessary for obtaining a gaseous response resembling the morphology of grand design normal spirals in the optical.

References

Block, D.L., Bertin, G., Stockton, A., Grosbøl, P., Moorwood, A.F.M., Peletier, R.F. (1994) 2.1 μm images of the evolved stellar disk and the morphological classification of spiral galaxies *A&A* **288**, pp. 365-382

Contopoulos, G., Grosbøl, P. (1988) Stellar dynamics of spiral galaxies: self-consistent models *A&A* **197**, pp. 83-90

Grosbøl, P., Patsis, P.A. (1995) Disk Dynamics based on near-IR photometry, in *Spiral Galaxies in NIR, ESO/MPA Workshop*, **in press**

ATTENUATION AND STELLAR POPULATION GRADIENTS IN SPIRAL GALAXIES

Technique for Inferring E(B-K)

ROGER B. ROUSE
Arizona State University
Tempe, Arizona, U.S.A.

The reddening, E(B-K), in three spiral galaxies is inferred from observations and stellar population models. Results indicate that reddening by dust can decrease outward by up to 2 mags in B-K. An HST F606W image of NGC 4654 illustrates such a gradient (Figure 1).

This technique was derived from a continuing effort to separate the effects of age and metallicity on the integrated light of unresolved stellar populations. One would like to know how dust comes into play. The technique requires measuring a spectral index and a color at the same position in a galaxy. Longslit spectroscopy that also resolves the galaxy spatially and imagery in B & K are used. The stellar population models are those of Worthey (1993). Each model is for a single age and metal abundance. Ages range from 1.5 to 17 Gyrs. Metallicity ranges from -2.0 to 0.5.

One dimensional spectra are extracted from a longslit image using software apertures whose sizes are determined by the shape of the light profile in the spatial direction and the desire for sufficient S/N. The index Mgb is used because it is isolated from emission features. To obtain a color for a given value of Mgb a software aperture representing the longslit is placed on the B image using the galaxy center and a source in the disk (see Figure 2a). Then the flux is extracted resulting in a 1-d light profile that matches the longslit profile. A transformation from B is used to position the slit in K. The B and K light profiles are aligned with the longslit light profile (see Figure 2b). Then the apertures used to extracted 1-d spectra can be placed directly on the B & K light profiles. In each aperture B and K magnitudes are measured. Thus for each Mgb value a B-K color is measured, and each pair corresponds to one position in the galaxy.

The models define a narrow relationship in the log(Mgb) vs B-K plane. When overplotted the observed values form a nearly horizontal line to the right of the models (see Figure 2c). The models do *not* include dust or

D. L. Block and J. M. Greenberg (eds.), New Extragalactic Perspectives in the New South Africa, 556–558.

gas. Hence they represent unreddened stellar populations. Since the model relationship is due to variations in metallicity and age the observed points must be displaced because of the reddening effect of dust in the galaxy. A reddening is determined for each point along the slit by taking the difference in the model color and the observed color.

This inferred reddening is plotted against distance from galaxy center (see Figure 2d). The inner points are clearly redder than the outer points. This is *not* due to a change in age or metal content because such variations would cause the observed points to fall close to the model relationship, resulting in zero inferred reddening. The pattern of obscuration revealed in the HST F606W image of NGC 4654 confirms this conclusion (Figure 1).

Future work will include VRIJH imagery and other spectral indices for ten spiral galaxies. Also, estimates of the attenuation will be made using models that include dust, gas, and various geometries.

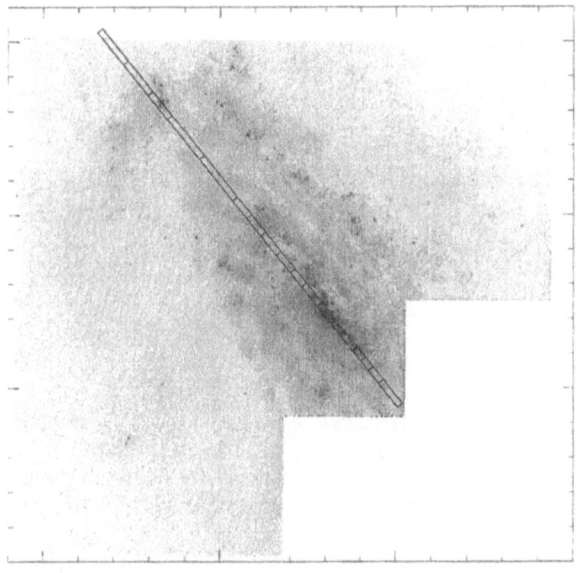

Figure 1.   An HST F606W image of NGC 4654

References

Worthey, G. 1993, Ph.D. thesis, U.C. Santa Cruz.

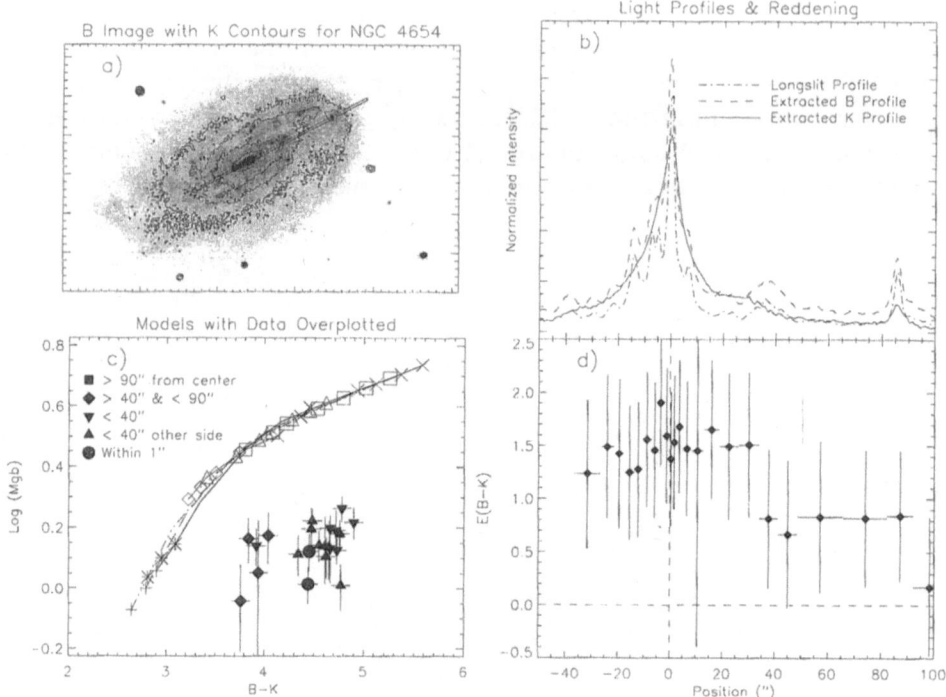

Figure 2. a) The slit position. B image with K contours b) B & K extracted profiles aligned with B-convolved longslit profile c) Models represented by lines with symbols, including all ages & metallicities. Data represented by filled symbols with error bars. d) Reddening decreases outward.

FIR AND CII EMISSIONS IN NGC6946.

A radiative transfer simulation of the penetration of UV photons in the interstellar medium of a galaxy.

SYLVAIN SAUTY, MARYVONNE GERIN, FABIENNE CASOLI
DEMIRM Dept
Observatoire de Paris - 61, Avenue de l'Observatoire-
75014 PARIS- France

1. Introduction

We present a first attempt to obtain a global and coherent model of the large scale emission of the gas tracers in a spiral galaxy. The purpose of the simulation is to use the knowledge of the basic processes in interstellar clouds and to apply it at a global scale. We have chosen the nearby spiral galaxy NGC 6946 as a suitable target since it has been mapped at many wavelengths, is close to face-on and presents a strong star formation activity.

2. The model.

The basis of the model is the cloud collisions code originally proposed by Combes & Gerin (1985) . We launch 20 000 clouds in the NGC6946 gravitational potential well deduced from a R band image, and adjust the pattern speed of the spiral structure to fit the distribution of molecular gas. Then OB associations are formed in these clouds according to an IMF with a slope of -2 and rapidly emerge from their parent clouds due ti their peculiar relative velocity of about 10 km^{-1}. The UV field in the model results only from the radiation of these OB stars associations and its calculation is the key part of the code. We use a Monte-Carlo approach to estimate the UV field (912-1100 Å and 912-2000 Å bands), launching 100 photons over 4 π steradians from each OB association and following their path in the galactic disc over 1 kpc. A 3D grid (1 pc resolution) allows us to track the photons's path in the diffuse gas before their absorption by molecular clouds to taking the scattering into account. We use the Milky Way extinction curve of Fitzpatrick & Massa (1988). Once a map of the local UV radiation field is created, we deduce the incident radiation field on the

D. L. Block and J. M. Greenberg (eds.), New Extragalactic Perspectives in the New South Africa, 559–560.
© 1996 *Kluwer Academic Publishers.*

gas clouds and the associated C^+ and FIR emissions, using the chemical code of Le Bourlot et al (1993) for the C^+ emission, and the dust model by Désert et al (1990), to predict the FIR emission of the clouds.

3. Preliminary results.

The model has a total continuum Lyman flux is 9.10^{53} s^{-1}, or 3 times the Galactic one, and a massive star formation rate of 2 $M_\odot$ yr^{-1}. The predicted luminosities (in U band and at 2000 Å) of the young stellar population are in agreement within 10 % with the observations, but the $H\alpha$ luminosity remains higher than the observed one by a factor 3-4 depending on the stellar models we use. From our present tests, we obtain relative contributions of OB stars to the global emissions as:

Tracer	molecular clouds	diffuse atomic phase
FIR	60-80%	1%
CII	10-30%	60%

the evaluation for molecular clouds depending strongly on the averaging or not (hot spot assumption) of the UV field at their surfaces. We deduce a global UV opacity, defined as $\tau_\nu = -\ln(N_{escaped}/N_{launched})$, in the range 0.8 at $\lambda = 930$ Å, 0.7 at $\lambda = 1100$ Å.

4. Discussion.

This study indicates that with simple assumptions about the massive star formation process and the structure of the galactic medium, we are able to reproduce qualitatively the observed features of the star formation tracers. The knowledge of the cloudy nature of the ISM is important to find out how far UV photons can travel away from OB associations, and the scaling law (mass/surface) appears as one of the major parameter, because it determines both the obscuration and the size of the emitting regions. It clearly appears that OB stars are not responsible for the whole heating of the dust, and we intend to include in the future the radiation field of the old stellar population to take into account the less energetic photons.

5. References

- Combes, F., Gerin, M., 1985, A& A, 150, 327-338
- Désert, F.X., Boulanger, F., Puget,J.L., 1990, A & A, 237, 215
- Fitzpatrick, Massa, 1988, ApJ, 328, 734
- LeBourlot, J., Pineau des Forêts, G., Roueff, E., 1993, A& A, 267, 233-254

THE COMPOSITION OF INTERSTELLAR DUST

ULYSSES J. SOFIA
National Research Council, NASA/GSFC
Code 681, Greenbelt, MD 20771 USA

Abstract. The majority of interstellar grain models are based on a dust population composed of carbon and silicate grains. Abundance studies using high-resolution absorption line spectra, however, suggest that other grain types, perhaps oxides or metal grains, are prevalent in interstellar clouds. The data further suggest that not enough carbon is depleted from the gas-phase interstellar medium to account for the carbon-based grains often used to explain the measured optical and UV dust opacity. Silicon atoms are likely to be more populous in grain mantles than in silicate grain cores.

1. Introduction

The majority of interstellar grain models are based on dust populations composed primarily or completely of silicates and carbon grains. For instance many models still rely on the Mathis, Rumpl & Nordsieck (1977; MRN) composition for dust which well reproduces the avaerage Galactic extinction curve.

Direct evidence of which elements likely compose interstellar dust can be obtained through interstellar gas abundance studies. *Copernicus* and IUE (Jenkins 1987, 1989) greatly contributed to our knowledge of grain compositions. Recently, higher-resolution data from the Hubble Space Telescope (HST) combined with improved atomic transition constants have placed tighter constraints on dust characteristics (Sofia, Savage & Cardelli 1993, Spitzer & Fitzpatrick 1993, Sofia, Cardelli & Savage 1994).

2. Interstellar Abundances

The Goddard High Resolution spectrograph on the HST has the ability to observe measurable transition lines of O, Si, Mg, and Fe in the interstellar

561

D. L. Block and J. M. Greenberg (eds.), New Extragalactic Perspectives in the New South Africa, 561–562.

medium (ISM), and is the first instrument to produce reliable measurements of interstellar carbon (Cardelli et al. 1991). These are the five elements other than hydrogen which contribute the largest number of atoms to dust grains.

Sofia et al. (1994) found from UV observations of O, Si, Mg, and Fe that as little as 35% of the silicon available in the ISM may be contained in dust grain cores. This contradicts evidence from observations of the 9.7μm feature which suggests that all of the silicon in the solar neighborhood should be incorporated into dust (Whittet 1992). The inconsistency likely results from the fact that the observations sample different conditions in the ISM. Sofia et al. suggest that the silicon atoms which are not in the grain core may reside in more tenuous dust mantles. They further suggest that more than 50% of the Mg and Fe atoms in dust cores are in metal and oxide grains rather than in silicates.

Five measurements of interstellar carbon abundances are reported by Cardelli et al. (1996). They find that the level of carbon depleted from the gas-phase does not vary in their sample, and is consistent with the 2175Å bump being formed from carbon-based grains, either PAH or graphite. The carbon grains which are often invoked to account for the extinction other than the bump (e.g. Mathis and Whiffen 1989) cannot be explained with the measured carbon depletions. There is simply not enough carbon "missing" from the gas phase to form the necessary carbon grains.

References

Cardelli, J.A., Meyer, D.M., Jura, M. and Savage, B.D. (1996) The Abundance of Interstellar Carbon, *ApJ*, in press.

Cardelli, J.A., Savage, B.D., Bruhweiler, F.C., Smith, A.M., Ebbets, D.C., Semback, K.R. and Sofia, U.J. (1991) Elemental Abundances in the Diffuse Clouds Toward ξ Persei, *ApJ*, **Vol. no. 377**, pp. L57–L60

Jenkins, E.B. (1987) Element Abundances in the Interstellar Atomic Material, in *Interstellar Processes*, eds. D.J. Hollenbach and H.A. Thronson, Jr., Reidel, Dordrecht, pp. 533–559

Jenkins, E.B. (1989) Insights on Dust Grain Formation, in *IAU Symp 135, Interstellar Dust*, eds. L.J. Allamandola and A.G.G.M. Tielens, Kluwer, Dordrecht, pp. 23–36

Mathis, J.S., Rumpl, W. and Nordsieck, K.H. (1977) The Size Distribution of Interstellar Grains, *ApJ*, **Vol. no. 217**, pp. 425–433

Mathis, J.S. and Whiffen, G. (1989) Composite Interstellar Grains, *ApJ*, **Vol. no. 341**, pp. 808–822

Sofia, U.J., Cardelli, J.A. and Savage, B.D. (1994) The Abundant Elements in Interstellar Dust, *ApJ*, **Vol. no. 430**, pp. 650–666

Sofia, U.J., Savage, B.D. and Cardelli, J.A. (1993) High Resolution Ultraviolet Observations of the Interstellar Diffuse Clouds Toward μ Columbae, *ApJ*, **Vol. no. 413**, pp. 251–267

Spitzer, L. and Fitzpatrick, E.L. (1993) Composition of Interstellar Clouds in the Disk and Halo. I. HD93521, *ApJ*, **Vol. no. 409**, pp. 299–318

Whittet, D.C.B. (1992) *Dust in the Galactic Environment*. Institute of Physics Publishing, Bristol.

DUST IN RADIO GALAXIES: CLUES TO ACTIVITY

W.B. SPARKS, S. DE KOFF, S. BAUM, D. GOLOMBEK,

F. MACCHETTO, J. BIRETTA
Space Telescope Science Institute
Baltimore, MD 21218, USA

G. MILEY
Sterrewacht, Leiden, The Netherlands

AND

P. MCCARTHY
Carnegie Institute, Pasadena, USA

1. HST Snapshots of 3CR radio galaxies

HST WFPC2 images of over two hundred 3CR extragalactic radio sources have been obtained. We have used these images to show the relationships between radio and optical jets and dust disks in unprecedented detail and with unprecedented clarity. We find a variety of morphological characteristics amongst the dust features, including dust disks that are perpendicular to jets, others that are not, and dust features that are less symmetric in appearance. The dust morphology in turn offers clues to the origin of the dust and the origin of nuclear activity. We investigate relationships between the dust morphology and characteristics of the associated radio sources and host galaxies in order to seek information on the physical connection between the two phenomena.

2. Results

We find dust (lanes, patches, wisps) in radio galaxies out to a redshift of ~ 0.5. Roughly 40% of the 3CR galaxies in the range $0.0 < z < 0.5$ show obvious signs of dust. There is no evidence of any change in this fraction between the low $0.0 < z < 0.1$ and intermediate redshifts $0.1 < z < 0.5$.

564

D. L. Block and J. M. Greenberg (eds.), New Extragalactic Perspectives in the New South Africa, 564–565.
© *1996 Kluwer Academic Publishers.*

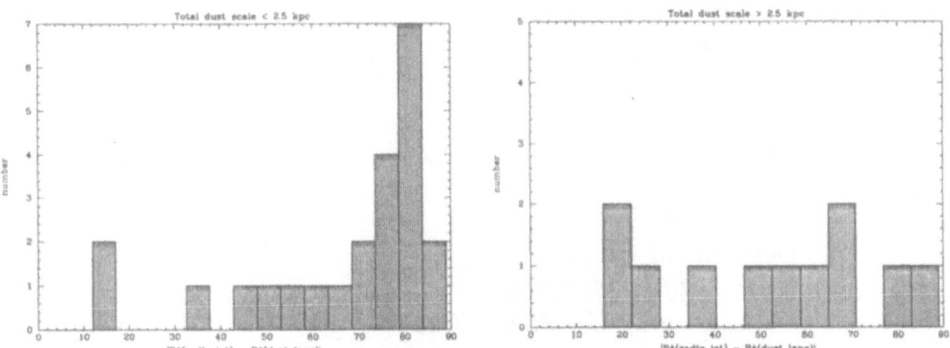

Figure 1.   Comparison of radio to dust lane axes for compact and extended dust lanes.

We compare the dust 'disk' orientation, taken to be the position angle of the dust closest to the nucleus, and the radio source axis. There is a strong tendency for the dust major axis to be perpendicular to the radio source axis, confirming the result of Kotanyi and Ekers (1979) which gives credence to the idea that gas from kpc scales ultimately feeds the central engine and affects its collimation axis. Note, however there is a surprising dispersion in the distribution of position angle offsets and there are even sources whose jets apparently issue directly into the dust disk.

In addition, we investigated whether the precision of the perpendicularity of the dust lanes to the radio axis correlates with other properties of the host galaxies. We found no significant differences between FR-I and FR-II sources, strong and weak radio power, and low or intermediate redshifts. On the other hand, there is a significant correlation with the total extent of the dust features visible in the sense that (i) dust and radio axes are perpendicular in sources where the dust extent is less than 2.5 kpc, but (ii) dust and radio axes are randomly oriented in sources where the dust extent is greater than 2.5 kpc, see Figure 1.

This supports an evolutionary scenario in which gas and dust is captured in an encounter and then settles into its preferred orbits and sinks into the inner few kpc of the galaxy where is fuels the central engine. In the early stages, when the gas and dust has not yet fully settled, the large scale dust does not appear perpendicular to the radio axis.

References

Kotanyi, C.G., and Ekers, R.D. (1979) Astron. Astrophys., **73**, L1.

DEFINING SPIRAL STRUCTURE
IN THE FLOCCULENT GALAXY NGC 5055

M.D. THORNLEY
Department of Astronomy
University of Maryland
College Park, MD 20742-2421

1. Introduction

Flocculent galaxies are important counterexamples to grand design spirals: they exhibit weak, discontinuous spiral structure at optical wavelengths, suggesting the absence of a spiral density wave (e.g., Elmegreen & Elmegreen 1985; Elmegreen & Thomasson 1993; Bertin 1993).

Here we present the results of a search for spiral structure in K' (2.1 μm) emission in a sample of nearby flocculents. In addition, modeling of the dust extinction in NGC 5055 through the use of B, V, R, I, J, and K' optical-NIR colors suggests that the asymmetric distribution of dust extinction in optical bands may explain why an inner two-arm structure in NGC 5055 was not detected previously. Thus flocculent galaxies may indeed have contributions from large-scale dynamics, instead of being affected only by local kinematics.

2. Uncovering Structure with Near Infrared Imaging

Near infrared (NIR) broadband images can be used to place limits on the level of density wave activity in flocculent galaxies by tracing structure in the old stellar population. Low-level, non-axisymmetric emission is detected in images of K' (2.1 μm) broadband emission of four nearby flocculent spirals (NGC 2403, NGC 3521, NGC 4414, and NGC 5055).

To better distinguish weak non-axisymmetric structures, we have modeled the axisymmetric distribution with a bulge and an exponential disk. These models were fit to azimuthally averaged radial profiles of the K' emission and subtracted from the original images. Three of the four sample

D. L. Block and J. M. Greenberg (eds.), New Extragalactic Perspectives in the New South Africa, 566–567.
© *1996 Kluwer Academic Publishers.*

galaxies (NGC 2403, NGC 3521, and NGC 5055) show evidence of two-arm spiral structure, while NGC 4414 shows kiloparsec-scale features in a less symmetric pattern.

3. Spiral Structure in NGC 5055

The strongest structure is detected in the prototypical flocculent NGC 5055 (Arm Class 3, Elmegreen & Elmegreen 1982, 1987). The map of residual K' emission shows symmetric, two-arm spiral structure at low levels. These arms extend to a radius of approximately 4 kpc, and then show a distinct drop in brightness. The brightness contrast between arm and interarm regions is approximately 1.25, similar to the arm-interarm contrast measured in M33 (Regan & Vogel 1994).

We have recently begun to model the variations in dust extinction over the observed galaxy disk, using optical-NIR colors (BVRIJK'). The model solves for visual optical depth and the ratio of the vertical scale heights for gas and dust; it uses Galactic values for dust opacities and albedos, and accounts for isotropic multiple scattering. For a full description of the model, see Regan, Vogel, & Teuben (1995).

The distribution of extinction in NGC 5055 is very asymmetric: optical depths in the northern half of the galaxy are approximately 0.5, increasing to 4-5 against the southern half of the galaxy. These results suggest a dust distribution which is not well mixed with the disk. The degree of asymmetry is similar to that seen in NGC 4826, for which Block et al. (1994b, see also Witt et al. 1994) have modeled the extinction source as an intervening dust screen. This highly asymmetric distribution has masked the presence of inner two-arm structure in the stellar disk at shorter wavelengths. These results serve to emphasize the value of NIR observations for placing constraints on the distribution of dust in galaxies, as well as uncovering the true structure of the stellar mass distribution (see also Block et al. 1994a).

References

Bertin, G. (1993), *PASP*, **105**, 640.
Block, D. L., Bertin, G., Stockton, A., Grosbøl, P., & Moorwood, A.F.M., & Peletier, R.F. (1994), *A&A*, **288**, 365 (a).
Block, D. L., Witt, A.N., Grosbøl, P., Stockton, A., & Moneti, A. (1994), *A&A*, **288**, 383 (b).
Elmegreen, D.M. & Elmegreen, B.G. (1982), *MNRAS*, **201**,1021.
Elmegreen, D.M. & Elmegreen, B.G. (1987), *ApJ*, **314**, 3.
Elmegreen, B.G. & Elmegreen, D.M. (1985), *ApJ*, **288**, 438.
Elmegreen, B.G. & Thomasson, M. (1993), *A&A*, **272**, 37.
Regan, M.W. & Vogel, S.N. (1994), *ApJ*, **434**, 536.
Regan, M.W., Vogel, S.N., & Teuben (1995), *ApJ*, **449**, 576.
Witt, A.N., Lindell, R.S., Block, D.L., & Evans R. (1994), *ApJ*, **427**, 227.

WHERE DOES DUST ABSORB MOST LIGHT?

Are We Working On The Wrong Wavelength?

M. TREWHELLA, J.I. DAVIES AND M.J. DISNEY
Department of Physics and Astronomy
University of Wales College of Cardiff
PO Box 913
Cardiff CF2 3YB
United Kingdom

1. Introduction

One commonly used method of calculating optical depth is to conduct an energy balance. As starlight is absorbed in the optical and reradiated in the far infrared (FIR), the optical depth can be deduced from the relative amounts of energy being emitted at these wavelengths [1]. In this paper we show that absorption from the NIR bands, particularly J, dominate the FIR output and that absorption from the U, B and V bands contribute a rather insignificant fraction.

2. Absorption Calculations

Our models [2] take an observed SED [3] and reconstruct what it would look like had it not been attenuated by dust. The difference between these SEDs is the contribution to the FIR. Table 1 shows the relative contributions to the FIR from all the bands for a sandwich model. The optical depths are scaled relative to the B-band by assuming that $\tau_\lambda \propto 1/\lambda$. The relative contribution from each band remains almost constant as τ_B increases, with the J-band always contributing the highest fraction ($\sim 20\%$).

3. The Explanation

Figure 1 is a plot of the average flux density across a band against wavelength. The area of each box is equal to the total energy given out. We have plotted both the observed (solid) and the intrinsic (dashed) spectrum. The area between the intrinsic and observed spectrum is largest for the J-band (20.9%), even though the J-band itself only loses 12% of its energy output. This can be compared with the B-band which loses 34% of its output but only contributes 9.6% of the FIR.

568

D. L. Block and J. M. Greenberg (eds.), New Extragalactic Perspectives in the New South Africa, 568–569.
© *1996 Kluwer Academic Publishers.*

TABLE 1. % age contribution to the FIR, $\zeta = 0.5$

| Optical | Band | | | | | | | | |
depth	UV	U	B	V	R	I	J	H	K
$\tau_B = 0.1$	8.4	5.4	9.5	9.0	16.8	10.7	20.9	11.6	7.3
$\tau_B = 1.0$	8.1	5.4	9.6	9.1	16.9	10.7	20.9	11.5	7.2
$\tau_B = 10.0$	3.1	3.1	6.4	6.9	15.8	11.6	26.0	16.0	10.6

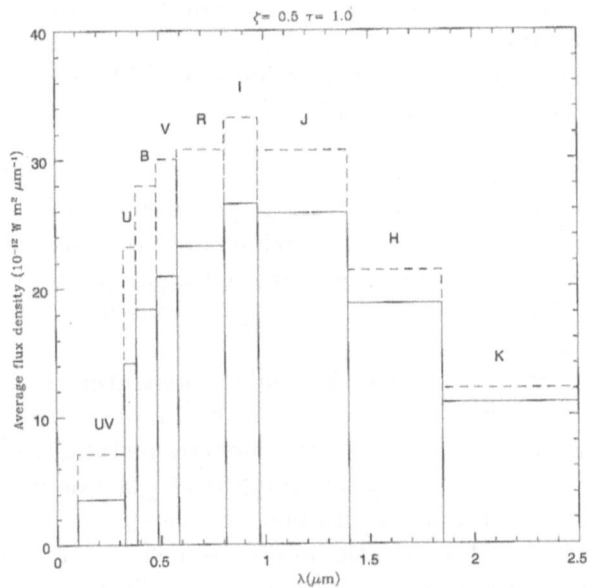

Figure 1. The Spectral Energy Distributions

4. Conclusion

Due to the large population of evolved stars in disk galaxies, a significant fraction of the stellar light is emitted in the NIR. Although the extinction is small in the NIR, the shear size of the NIR stellar luminosity means that the majority of the FIR radiation is reprocessed from the NIR.

References

1. *The Opacity of Spiral Disks*, (1994), proc. NATO ARW Cardiff July 1994, Ed. Davies J.I. and Burstein D.
2. Trewhella M., Davies J.I., Jones H. and Disney M.J. (1996) The Average Optical Depth of Disk Galaxies, *MNRAS*, submitted
3. Evans Rh. (1994) *Dust in Spiral Galaxies*, PhD. Thesis, University of Wales.

ON THE ANGULAR-MOMENTUM TRANSPORT IN SPIRAL AND BARRED GALAXIES

S. VON LINDEN[1], H. LESCH[2], F. COMBES[3]
1) Landessternwarte, Königstuhl, D-69117 Heidelberg, Germany
2) Universitätssternwarte, Scheinerstr. 1, D-81679 München, Germany
3) DEMIRM, Observatoire de Paris, 61 Av. de l'Observatoire, F-75014 Paris, France

The large-scale dynamics and evolution of disk galaxies is controlled by the angular-momentum transport provided by non-axisymmetric perturbances through their gravity torques. To continuously maintain such gravitational instabilities, the presence of the gas component and its dissipative character is essential.

The resulting re-distribution of angular momentum is equivalent to a "viscosity" (dubbed gravitational viscosity by Lin & Pringle 1987).

The coupling between gas and stars favours instabilities. By using 2D and 3D N-body simulations of a self-gravitating galactic disk composed of stars (38 000 particles) and gas (19 000 particles), we investigate quantitatively the efficiency of gravitationally driven viscosity and its relevance for large-scale mass accretion in disk galaxies. Collision between molecular clouds are highly inelastic, and to simulate the elasticity in our particle model, we adopt the same scheme and local processes as described by Casoli & Combes (1982) and Combes & Gerin (1985). The cloud ensemble is represented with its mass spectrum where giant molecular clouds (M $> 2 \cdot 10^5$ M$_\odot$) have a finite lifetime. After $4 \cdot 10^7$ years cloudlets are reinjected in the interstellar space with a steep power-law spectrum and a 2D random-velocity distribution.

The transfer of angular momentum outwards is accompanied by a gas inflow, i.e. the surface density Σ changes with time and radius. We understand our simulation as an experiment by which we can study the elementary processes in a disk galaxy. In our poster we present models with different input parameter for the disk, bulge and halo masses. We present as an example a disk dominated model (model D), a bulge dominated model (model B) and a model which only develop axisymmetric structure (model

D. L. Block and J. M. Greenberg (eds.), New Extragalactic Perspectives in the New South Africa, 570–571.
© 1996 Kluwer Academic Publishers.

A). The most unstable model D reveals globally the highest gravitational viscosity. The simulation exhibit very efficient angular momentum transport, driven by a strong and massive bar, revealed by considerable changes in the surface mass density.

The disk of model B is dominated by spiral structure and remains in average on a less active level than the barred galaxy in model D. To compare non-axisymmetric perturbances (bars, spiral arms) with purely axisymmetric disks we use simulation A.

We run our simulation $t = 2.4 \cdot 10^{10}$ years. For the different models we calculated the radial velocity for the gas disk. The conservation of mass and angular momentum together with the angular torque allows to calculate a viscosity $\nu(r, t)$. This deduced 'gravitational viscosity' does not correspond to a classical gaseous viscosity, but is a measure of the efficiency with which gravity torques transfer angular momentum.

In comparing the models we can remark that the gas mass is much more effectively accreted towards the central part in the disk dominated model (model D).

The viscosity depends essentially on the structural changes in the disk. For example the viscosity decreases inside the corotation radius, when the gas is depleted at this radius during these timesteps. The radial gas flows are then stopped, which translates into a smaller viscosity.

By comparing the time-averaged values for the three models, over $2.4 \cdot 10^{10}$ years we get: $1.80 \cdot 10^{28}$ cm s^{-1} (model D), $0.75 \cdot 10^{28}$ cm s^{-1} (model B), $0.25 \cdot 10^{28}$ cm s^{-1} (model A).

We show through N-body simulations that the galactic disk evolution time-scale is much shorter than the Hubble time, especially in the case of a high gas-mass fraction. The gas cooling maintains dynamical instabilities in the disk, with their corresponding gravity torques, and angular momentum transfer. The consequences are high central mass concentrations, that can react back on the stability, and lead to bar destruction or fading.

References

Casoli F., Combes F., 1982, A&A, **110**, 287
Combes F., Gerin M., 1985, A&A, **150**, 327
Lin D.N.C., Pringle J.E., 1987, MNRAS, **225**, 607

DUST, STARS AND NUCLEAR ACTIVITY IN THE CENTRES OF SBC GALAXIES

R.D. WOLSTENCROFT[1] & R.D. DAVIES[2]
[1] *Royal Observatory, Blackford Hill, Edinburgh, UK*
[2] *Jodrell Bank, University of Manchester, Macclesfield, UK*

1. Introduction

Emission from the central few Kpc of a normal Sbc spiral comes from a population of older bulge stars and one of younger disc stars, both modified by dust extinction. In some apparently normal galaxies a compact nuclear source may lurk whose presence may be detected in the radio continuum. In a 20cm VLA study(14 arcsec resolution) of 88 Sbc galaxies,Hummel et al. (1985) detected 36 with a flux density >5mJy. Of these Vila et al. (1990) detected 28 at a resolution of 1-2 arcsec: 21 contain a compact or slightly extended central component and 7 an extended structure with no discernible central component. The majority of the 28 sources have spectral indices consistent with synchrotron emission.In the absence of VLBI studies it is uncertain whether the 21 compact sources are truly compact and Seyfert-like or just compact starburst regions.In the sample of 28,only NGC 6814 and possibly NGC 4258 are known Seyferts.

2. NEAR-INFRARED STUDIES

To tackle the question of the nature and incidence of sources at the centres of nominally normal Sbc galaxies we have carried out JHKL' photometry and imaging of the 36 galaxies detected by Hummel et al(1985),using UKT9 and IRCAM on UKIRT. At the 22Mpc median distance of our sample,our apertures of 5.0,7.8,12.4 and 19.6 arcsec defined regions from 0.5 to 2 Kpc across. Errors in the colours were mostly <0.03 mag in J-H and H-K and <0.12mag in K-L'.The colours were corrected for Galactic absorption and redshift. Colours in the 5 arcsec aperture define a relatively narrow sequence in (J-H,H-K)(H-K,J-K) and (H-K,K-L') diagrams. At the red end of these sequences we find the Seyfert NGC 7314 with J-H=1.031(.040),

D. L. Block and J. M. Greenberg (eds.), New Extragalactic Perspectives in the New South Africa, 573–574.
© *1996 Kluwer Academic Publishers.*

H-K=0.822(.028) and K-L'=1.551(.035) and at the blue end a tight cluster of points, eg NGC 6384 with J-H=0.691(.019),H-K=0.212(.011) and K-L'=0.202(.060),typical of elliptical galaxies.To understand these sequences we need to explain how the colours of the actual stellar population are combined with other possible emission components such as HII region emission and 'hot dust' emission together with dust extinction in each galaxy.

The high photometric accuracy allows us to study the colour gradients especially at JHK. In J-K the most common colour pattern is a blue centre(1),a red ring(2) and a blue area beyond(3) where the numbers refer to the 5.0,7.8 and 12.4 arcsec apertures:the average J-K colour differences for the 9 galaxies in this class are 2/1=+0.070 and 3/2=-0.065. One of our 9 galaxies,NGC 4321,is common to the list of Shaw et al(1995) who find 4 galaxies with red CNR's. Other J-K colour patterns,apart from galaxies with no gradients(6 galaxies),are red centres with steep gradients into the blue (2 Seyferts). The possible connection between CNR's and the inner Lindblad resonance has been discussed by several authors (see eg Telesco et al,1993). If these CNR's are relics of the more prominent and presumably young CNR's,eg that in NGC 1097 (Wolstencroft et al,1983), then it remains to be explained why these rings are red (relative to the bulge).

The VLA observations show the presence of excess synchrotron emission spatially coincident with the near-IR CNR's,good examples being NGC 4321,4527,4536 and 6907. Furthermore those galaxies in the sample with a higher central radio concentration have redder NIR colours. This strong radio-NIR correlation suggests a model in which these redder colours are produced by small grains heated by UV photons generated in star forming regions to temperatures in the range 1000 to 1500K. In some objects such as the known Seyferts there may also be a power law contribution to the reddening: it follows a locus in the colour-colour diagrams very similar to hot dust. The recent COBE/DIRBE study of the Orion nebula complex by Wall et al(1996) has illustrated the potential importance of the NIR as well as the FIR emission from regions such as Orion when interpreting the colours of external galaxies.

References

Hummel, E., Pedlar, A., van der Hulst, J.M. & Davies, R.D., 1985, *Astron. Astrophys. Suppl.* **60,** 293

Shaw, M., Axon,D., Probst,R., & Gatley, I., 1995, *Mon Not R astr Soc* **274,** 369

Telesco, C.M., Dressel, L.L., & Wolstencroft, R.D., 1993, Astrophys J **414,** 120

Vila, M.B., Pedlar, A., Davies, R.D., Hummel, E., & Axon, D., 1990, Mon Not R astr Soc **242,** 379

Wall, W.F., et al, 1996, Astrophys J **456,** 566

Wolstencroft, R.D., Perley, R.A., & Tully, R.B., 1983, Mon Not R astr Soc **207,** 889

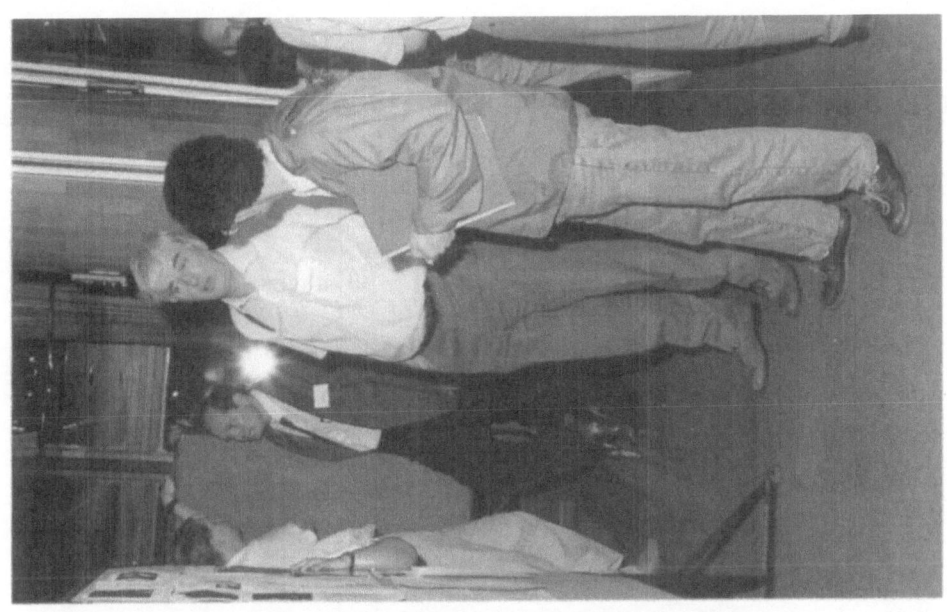

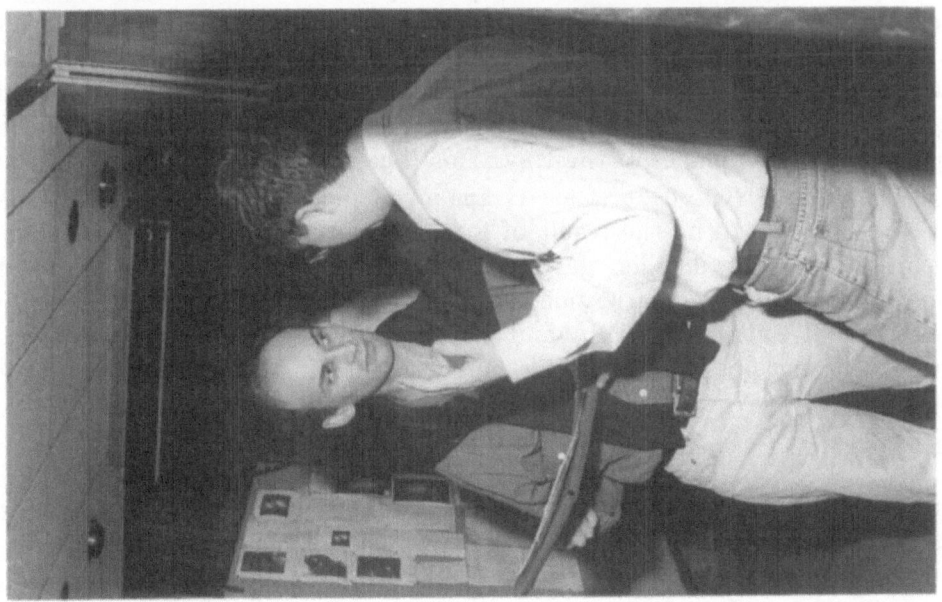

THE POLARIZING PROPERTIES OF DUST IN NGC 5128

S.M.SCARROTT
University of Durham, Durham DH1 3LE, UK.

AND

RAMON D. WOLSTENCROFT
Royal Observatory Edinburgh EH9 3HJ,Scotland.

1. Introduction

It is well established from radio and optical polarization measurements that relatively quiescent spiral galaxies possess large-scale magnetic fields which are believed to be generated by some form of dynamo action. In addition, several dust lane galaxies of quite different morphologies also show optical polarization that has been interpreted in terms of large-scale magnetic fields in the equatorial planes of these systems (Scarrott 1996).

The galaxy NGC 5128 (Cen A) has a conspicuous equatorial dust lane that is believed to be the result of a recent merger between a large elliptical galaxy and a smaller, gas-rich spiral galaxy. We have made detailed multi-colour optical polarization maps of NGC 5128 using the SAAO 1-m telescope at Sutherland, RSA and in this contribution we interpret these results in terms of a galactic-scale magnetic field in this galaxy.

2. Dust and the magnetic field in NGC 5128

In the figure we show an optical polarization map superposed on a greyscale intensity image of NGC 5128. At the extremities of the dust lane the polarization orientation is perpendicular to the dust lane and in this situation we believe that the polarization results from the scattering of light by dust in the lane due to the anisotropic illumination provided by the inner part of the elliptical component of the galaxy.

In the central regions of the dust lane the polarization is oriented parallel to the equatorial plane and the wavelength dependence of the polarization (Gledhill et al. 1996) agrees with the Serkowski relationship which was em-

D. L. Block and J. M. Greenberg (eds.), New Extragalactic Perspectives in the New South Africa, 576–577.
© 1996 *Kluwer Academic Publishers.*

Figure 1. A polarization map of NGC 5128 [S0+Spec] in the V waveband.

pirically derived for dichroic extinction by non-spherical paramagnetic dust grains partially aligned by the magnetic field in our Galaxy. This agreement suggests that the observed polarization maps out the magnetic field in the central regions of the galaxy – the field is parallel and confined to the equatorial plane of the dust lane. Many dust lane galaxies of quite different ilk (e.g. M104, IC 4329A & NGC2146) have similar polarization properties, including a similar wavelength dependence, which suggests magnetic fields are not an infrequent occurrence in these systems.

It is currently accepted that galactic magnetic fields in spiral galaxies are generated by dynamo action on time scales $\approx 10^9$ yr; however, it has been proposed that the dynamo action may be suppressed in strong tidal/merger interactions between galaxies (Lesch & Harnett 1993). Several dust lane galaxies fall into this category with the interactions taking place far more recently ($\approx 10^8$ yr ago in the case of NGC 5128) than the magnetic field generation times. If the dynamo action has been suppressed in NGC 5128 then we may be sensing the decaying remnant of the original field in the spiral galaxy or, there is a mechanism that regenerates the field at a much faster rate than anticipated after the interaction.

References

Gledhill,T.M., Foley,N.B., Scarrott,S.M. and Wolstencroft,R.D. (1996) *MNRAS*, in preparation.
Lesch,H. and Harnett,J. (1993) *A&A*, **268**, 58.
Scarrott,S.M. (1996) *QJRAS*, in preparation.

ASTRONOMY AT THE UNIVERSITY OF THE WITWATERSRAND

GEORGE C. BURIC
Department of Computational and Applied Mathematics,
University of the Witwatersrand, Johannesburg

For colleagues who could not personally travel to the host University for our Conference, a few words about Astronomy at the University of the Witwatersrand may be in order.

While the Department hosting this Conference is now officially known as the Department of Computational & Applied Mathematics, it was known – before the appearance of Workstations/desktops – as the *Department of Applied Mathematics & Astronomy*. One of our mentors, the late Professor A.E. H. Bleksley, wrote his DSc (on variable stars) here; Sir Arthur Eddington acted as his external examiner. Astronomy courses are offered at an introductory first year level by David Block and myself, while advanced astrophysics topics are offered in the Computational and Applied Mathematics Honours and Masters programs. Up to 40 astronomy students pass through our hands each year; students are offered live *hands-on* experience in photometry and spectroscopy, for example, in our Astrophysics Laboratory. Lectures to gifted school children and to the general public continue to generate much interest (eg. Figure 1). During the Conference, a public evening lecture entitled *Through the Eyes of the Hubble Space Telescope* by Duccio Macchetto was sold out; scores of public members had to be turned away.

For the past two decades, astronomical research in this Department has largely focussed on the morphology of spiral galaxies. Research bearing the Department's name has been featured on the cover of *Nature* (fig. 2). Photomicrographic tubes attached to Zeiss microscopes are available for detailed analyses of plates/negatives secured abroad.

Astronomy colleagues who studied physics and/or applied mathematics as students at this University have included Bernard Fanaroff (who co-developed the Fanaroff-Riley classes), Luis Balona, John Hutchings, Saul Teukolsky and David Block.

D. L. Block and J. M. Greenberg (eds.), New Extragalactic Perspectives in the New South Africa, 578–582.
© *1996 Kluwer Academic Publishers.*

Figure 1. A recent public astronomy lecture in Johannesburg entitled *'Our Universe: Accident or Design?'* [ref 1] drew a crowd of ten thousand. Our astronomy contingent continues to be actively engaged in such public outreaches. Photo: Ron Steele.

Figure 2. In the period 1990-1996, three astronomy letters to Nature bearing the University's name have been accepted. One involved the discovery of a family of cometary globules pointing to the CO peak in the Rosette GMC; another on the co-existence of two morphologies in the grand design spiral NGC 309 (in collaboration with R. Wainscoat) and a third paper on the discovery of dark infrared spirals in NGC 2841 (in collaboration with B. Elmegreen and R. Wainscoat). Courtesy: Nature.

It is perhaps not well known that the Yale Observatory was set up on the campus of the University of the Witwatersrand (Moore and Collins [2]). By 1925 all was ready: the Observatory was firstly operated solely by Yale University, but later (from January 1946) by the Universities of Yale and Columbia. The telescope was a 26-inch refractor; observer Dr H.D. Alden made a major contribution to the Yale Catalogue. Alden retired in 1945, and was succeeded by an equally energetic observer, C. Jackson. In 1952 the Yale Observatory was finally dismantled, and today only the building remains. Forty-four years later, it is a tremendous honour to have so many astronomers back on the campus again (Fig 3), including delegate Michael Rich from Columbia University.

The Yale Observatory building is within walking distance of the University's Planetarium, which boasts the largest Zeiss projector on the African

Figure 3. Our SOC elected review speakers and 'Eyes to the Future' Panelists sign their names at the University of the Witwatersrand: John C. Mather, J. Mayo Greenberg, Giuseppe Bertin, Rolf Chini, Paul W. Hodge, Eugene J. deGeus, James Lequeux, Richard Griffiths, Ronald J. Allen, Stephane Courteau, Francoise Combes, Bruce Elmegreen, Jay A. Frogel, Richard Brent Tully, Nicholas Devereux and David L. Block

continent. The seating capacity of our Planetarium is four hundred.

References

1. Block, David: 1992, *"Our Universe: Accident or Design?"* (Starwatch, Johannesburg).

2. Moore, Patrick and Collins, Pete: 1977, *"The Astronomy of Southern Africa"* (Howard Timmins, Cape Town), 111.

The Great Hall at the University of the Witwatersrand, Johannesburg. Photo: L. Greenberg.

EYES TO THE FUTURE

An Interactive Panel-Audience Discussion

RON ALLEN
Space Telescope Science Institute, Baltimore

BRUCE ELMEGREEN
IBM Watson Research Center, NY

PAUL HODGE
Dept of Astronomy,
University of Washington

AND

JOHN MATHER
NASA Goddard Space Flight Center

Editorial Comments: The final slot of the Conference Program was the very lively 'Eyes to the Future' session. Our four panelists were seated in the front of the lecture theatre, facing the audience. A television crew hired by the University's TV unit recorded the entire session for us; the TV tape was then transcribed into LATEX. Appearing in square brackets [...] below are occasional editorial comments – added at times for clarity and continuity of speech.

1. The Discussion Begins...

B. Elmegreen: [To Mayo Greenberg] [Referring to a transparency with a list of Mayo Greenberg's 'firsts'] ... Absolutely true, isn't this? That the temperature fluctuations and [other important properties of] dust grains were predicted at a very early time. What I didn't realise was that in an early Mayo Greenberg paper, spiral density waves (with age gradients and the whole triggering scenario) were also outlined well before the Roberts' paper became popular. In the mid-sixties, Mayo, you had a burst of very far-sighted research, which is being appreciated fully now thirty years later. Of course you also predicted that [the nucleus of] comet Halley would be black – and it was. So, with proper recognition of this, we'll now move

585

D. L. Block and J. M. Greenberg (eds.), New Extragalactic Perspectives in the New South Africa, 585–610.
© *1996 Kluwer Academic Publishers.*

on to the main presentations of this panel-audience discussion *Eyes to the Future*.

Mather: One thing that I found out in coming to the conference was 'Why did I come?' The original reason that I came was because David Block invited me to come and so I said 'OK I'm coming'. Now, as I've been here for all this time, I found out why it was so important to learn about dust, because it has affected our lives in many, many ways. Obviously you have all heard about these things, so none of this is a surprise, but I just wanted to point out that all of these things are really, really important to all of astronomy. The dust is where we find a lot of the results of nucleosynthesis, especially the dust that is not local to us. Now when we can study it far away, the dust made up the planets, made up ourselves. It hides from view almost everything that we would really like to know about, how things started, stars being born, the middles of galaxies, active galactic nuclei, the middles of quasars. Even intergalactic absorption may be important. So, dust gets in the way. Dust is a big problem for some people, as well as of great interest for other people. I think what we have heard shows that we now have many approaches for dealing with this, both mathematically and observationally. The dust traces the gas; helps us classify galaxies; controls star formation – it's a chemical factory, something that seems to be a pretty general process, that you can take ordinary chemicals (as Mayo explained to us) and hit them with ultra-violet and make tiny brown blobs. It traces the magnetic field ... we didn't hear a lot about that but it does get polarized in places. It might hide or be some of the dark matter. Finally, from my perspective, it's very important because dark, cold dust in our own galaxy makes it difficult for us to say what is the cosmic infra-red background at long wavelengths – which is what I set out to do with COBE a long time ago. We have heard about many different techniques for studying this dust (pretty much all optical techniques) because that's what we have to go on, but in principle, one can use cosmic ray information a little bit as well. Ranging from the far UV where you can get all the abundances, to the ultra-violet where you can study the particle sizes and types .. into the visible, where you can study the diffuse bands and atoms and molecules, and study the depletion of elements onto grains .. into the infra-red, where you can see the solid state bands that tell you about the actual content of the grains. Then, if you go to really, really, really long wavelengths – to where you can't see at all – you can see hydrogen ultra-fine structure. We have to do that experiment. Optical scattering gives us many ways to learn about dust. Again the particle size, shape and type, and the magnetic field influence the scattering, and in addition you can see around corners. You can see down into the nucleus of that quasar and see what it is really doing, even though you can't see it directly, and we study its emission; we have colour temperatures; there is a great deal that we can learn about the dust. It's not as helpful as we might have anticipated to study the far infrared emission because the spectra have turned out to be so ordinarily featureless. There's nothing with a sharp bump in it anywhere that says *'I'm a whatchamacallit grain'* but we do know, that at least, it's not a black body, and it's not a simple curve, so sometimes we can tell something about this.

There are some image processing techniques that have become very useful to us, as people have gotten multiband pictures, including infra-red, where you can see through the dust enough to get the dust constants and remove the obscuring effects. When you can map that dust and when you can correct for it, that changes the whole view of the galaxy. And in the lab there are a lot of things that we have heard about: (i) you can simulate the formation of the grains, (ii) you can measure the optical properties, (iii) you can look at the optical structures that they make, and (iv) you can even look inside the grains and see those little atoms of xenon and neon and so forth that tell you the individual history of every grain. So, there's a lot to know, and I was very impressed and very pleased to hear the experts on these things discuss how they found out. So, let me pass on to our next speakers.

B. Elmegreen: Now, what we've done here (referring to a transparency with a brief outline) is, collectively, assess the main categories of the presentations, and we've divided them into four main categories - (i) morphology (ii) dust properties (iii) dust distribution, and (iv) dark matter. What we would *like* to do - what we wanted to do - is have *you talk to us!* This may be very difficult but we'll try it. We'll first make everyone happy by putting their names on transparencies, so you know that you are all recognised, and we'll highlight several key points that came out of the talks. We would like to use these key points to begin discussions as part of this final session because we think that is most useful – for you to hear from each other and to get a flavour for what we are all thinking.

In the first broad topic of morphology, there are three issues:

(a) *New morphologies* - this was one of David Block's introductory points ... that by going into the infra-red, the morphology of galaxies changes. Now, people have been doing this in steps for over twenty years: first when B and I plates became available, the structure looked different. Now, in the K, that sequence of looking different is continuing. What we have learned is that in the K-band, galaxy discs look much more simple because all we tend to see are the rather smooth density enhancements from the *stars* which have high velocity dispersions. We think that the whole morphology and classification scheme of galaxies, in which the Hubble type (bulge to disk ratio, for example) or even rotation speed has been tied in to the tightness of spiral arms, the star formation structure and other population II aspects – this whole classification system needs to be revised.

(b) *Small scale structure* - We learned that we should look at the K-band (or even further into the IR if possible) to understand stellar dynamics, and then subtract this smooth structure away from the optical images to understand the gas, small scale processes, star formation and dust. I think this was an important new direction that should come out of this meeting.

(c) Also, we saw some beautiful pictures of galaxies in the deep field [Hubble deep field (HDF) pictures] and it's clear even though we may look at a galaxy locally in the V-band and see the same processes in K-band at high red shift, many of those HDF galaxies look *quite different* from local V-band galaxies, maybe because they are interacting more, maybe because there is more fluff and indescribable stuff around. But it's clear also that as part of this morphology issue, we should be able to make the classification system time dependent. And that will, I believe, be one of the major directions that deep space imaging from Hubble will give to us even if we don't have red shifts of the galaxies at the present time.

So, these are three points that came out of the morphology part of this meeting. I'd like to hear from you now, and we're prepared to wait a long time ... so let's see those hands come up. How important is this morphology issue? How important ... how revealing is it to see something different in the K-band? Is that obvious? Was that unexpected? Let's get some comments.

Disney: Well, I think you're absolutely right about that. What would worry me is that everybody will go off now and write their own morphology and we'll have 500 more different K-band morphological categories. I don't know how to solve it; perhaps several people at this conference should go away and agree to do it. At least to have some starting places for it – otherwise it might be as haphazard as it's been in the past in the optical.

Audience member: Is there a role for the IAU here?

B. Elmegreen: Is there a role for the IAU? We first need Jay Frogel to finish his catalogue – and he'll probably have the first crack at a morphology. But that's quite right, there is likely to be some diversity.

Debra Elmegreen: It's been a surprise to me at this meeting to see posters like Michele Thornley's ... where, what I always considered a classic flocculent galaxy turns out have to have this really pretty two arm structure in K-band ... also David Block's NGC 309 is a 3-armed galaxy yet very strongly m=2 in the K-band. However, we still have a number of flocculent galaxies that are still flocculent in the K-band and I am startled to find out that there are indeed very different types of flocculent galaxies. I think that is interesting - a density wave hiding in there for some but not others – I would like to know the difference.

B Elmegreen: Debbie tells me: *'Well, that's why I made Flocculent Class 2 and Flocculent Class 3!'* Flocculent Class 2 have HII regions that are very patchy of course, but they are often strung out along arms ... and those tend to be the same arms we're seeing very clearly in the K-band. Flocculent Class 3 is sort of more random.

D Elmegreen: [Laughter] No, the other way round, but...

B Elmegreen: Yes, 3 is more regular - indeed, the other way around! But it's clear, really in the K-band, where things look quite different.

Bosma: .... If you just look at the maximum rotation speed: Flocculent Class 3 are fast rotators while Flocculent Class 1 are slow rotators. So, the regularity in Flocculent Class 3 might be seen now in the K-band; it might indeed be that Class which has these two arm structures in the infra-red.

Pfenniger: One should get as *primary criteria* the ones that are the closest from the main physical factors, that is (i) the mass and (ii) the velocity. The K photometry seems the best for deriving the stellar mass, and the HI 21 cm for deriving the rotation velocity.

B Elmegreen: Right ... and somehow central concentration.

Pfenniger: Yes, right. But additional parameters should be considered as secondary.

Rich: I would caution people not to fit (or at least to take with a grain of salt) these fits of spiral profiles into $r^{1/4}$ and an exponential disc. There may be much more continuity between bulge and disc (in metallicity, age, and kinematics) than is generally appreciated. The ability to extrapolate the exponential disc into the nucleus does not necessarily imply its existence there. So while the bulge/disc decomposition fits the photometric profile, it does not guarantee that it actually translates into a physical separation of stellar populations.

Audience member: I'm not sure I understood you, Mike.

Another Audience member: ...but I didn't say I agree with you....

Courteau: I was just confused, because I was wondering what you felt about the age of bulges, for example. If you thought that bulges were close to ellipticals and you were basically saying ... that fitting as we do, exponentials in the core ... shouldn't be taken too much at face value as evidence say, for secular evolution... is that what you were saying?

Rich: No, actually not. I think that there may be a range of possibilities – what I was saying might actually be in favour of secular evolution. When you fit profiles (deconvolve disc from bulge), you mentally think of them as being separate entities often. I was just struck at seeing these dust lanes coming out at M31, thinking ... well, goodness ... you know – perhaps these bright stars have to do with a *disc* that's more prominent than Kent's decomposition might otherwise indicate.

Courteau: Just to follow up on your point: I think you're right just to say that one shouldn't think that [for example, in my presentation] ... the fact that we see these profiles that supports the mode of secular evolution: this just doesn't mean that secular evolution is necessarily at work. It's appealing evidence but it's not necessary evidence. I think what we really need are abundances; that's the real way to pin down any formation scenario between the disc and the bulge.

Emsellem: I would like to make a comment on this decomposition problem because I think we have to remember that we work with *projected quantities* and in fact it's amazing to see that from a V-band image ... and very simple models... one can reproduce the dynamics extremely well from the parsec scale up to kiloparsec scale. But the problem is then that if you slightly change the model ... When you look at the dynamics in space – the real true dynamics of the stars – it could be completely different ... from the interpretation of those dynamics you can then integrate it in ... you know, secular evolution or whatever ... in very different ways. So the problem [really] is: when you use these projected quantities – not to go too far with the interpretation. Look at the true physics.

Block: Bruce, I think the key point when one starts classifying galaxies in the near infra-red is that one must be very careful to make the classification as *simple* as possible. As we have mentioned, bars or small ovals occur rather often; furthermore, m=1 or m=2 components dominate. Higher m-modes do not dominate. I think it should be possible, given the large number of K-images, to develop a scheme which is reasonably simple ... one should not complexify it by developing too many parameters. I think that is the beauty of the K-band ... stellar mass distributions are just so beautiful ... usually so very regular.

Davies: Can I ask those who are better than I, whether they think that this is just a changing morphology with different wavelengths or whether this is an evolution in morphology? When you see different stellar populations with different wavelengths: is that reflecting how the gas mass changes with time?

B Elmegreen: The *current* distribution from old stars strongly affects the *current* distribution from gas. So, when the old stars were born, they wouldn't have had the spiral we see now. We're seeing *current* dynamics in a range of populations.

Davies: As each structure emerges in the K band... is that a reflection of what's happened in the past? Has it changed and evolved... so that the morphology changes as a function of time?

B Elmegreen: The time scale for these spirals (both for wrapping up and for the evolution of a mode) tends to be shorter than the age of the stars seen at K – which is about the age of the whole disc. So, the old stars were probably doing something different when they formed. A long time ago they had a lower velocity dispersion so they looked much more like the stars in blue photographs now.

Greenberg: I have what is probably a very naive question or point. Not all spirals are the same size. Secular evolution must depend on the *initial conditions* and it seems to be if you take a significant difference in size, things are going to evolve very differently, or could evolve very differently. I don't know, but I don't remember hearing differences in sizes of spirals coming up in the discussions.

B Elmegreen: You mean of the whole galaxy?

Greenberg: Of the whole galaxy, that's right. The total mass should make a difference in time scales and in the way things evolve and the way waves propagate and so on.

B Elmegreen: Well, the whole luminosity classification system essentially makes a connection between morphology and size ... and David Block has studied images as a function of size a long time ago. I believe you are absolutely right. As you go to smaller and smaller sizes, you get less and less importance from the rotation and more from turbulence and it's harder to generate smooth symmetric structures. And you also lose bulges ... which means you lose the inner Lindblad resonance, depending on the rotation curve. So, yes, there are variations with size.

Courteau: Mayo, in my presentation we showed a histogram of the distribution of scales ... of sizes of galaxies. I tried to convince you that I was covering pretty much all the typical sizes of spiral galaxies.

Impey: This is just a general point. The people who do the detailed kind of morphology work that we've been hearing about, on nearby galaxies, could actually do a great service for the people who are doing observational cosmology because we see the fine details of the structure and the bars and the discs and the cores and so on ... but when you come to the Sloan survey for which we will have digital images in five bands for 50 million galaxies ... when you come to the full database with HST, where the detailed information is not very high ... it would be very useful if the people who do morphology, where they have the full information, could evolve a sort of morphological classification scheme that was maybe more amenable to machine measurement – something based on moments, or asymmetry, smooth and condensed components, whatever. Then, they could correlate that with the detailed knowledge of the properties and give that as a tool to the people who will find vast numbers of galaxies at very faint levels where they don't have that information.

D Elmegreen: .... I'm aware of Barry Madore's efforts... at all wavelengths ... to have a universal classification... and I remember the Space Telescope morphology meeting a couple of years ago. Does anybody remember the names of a couple of other groups of people who were trying to do just that?

Griffiths: Well, yeah, they were doing that on neural network schemes – but the only thing that's been used since then is the fact that the compactness parameter may be related to asymmetry. There is in fact no scheme that's been adopted from that meeting ...held, as you say, two years ago.

B Elmegreen: One way to look at this is that morphology is useful only to humans. We're really after the *physics*. Morphological classification systems are sometimes very personal; they help us categorise things which probably are a continuum. They reflect more the way we *think* than the way physics is operating. If you go to quantitative systems, such as the simple system Dan suggested: central concentration... mass.. the basic quantities, and maybe a ratio of Fourier harmonics or something for spiral structure, then, you can get a little closer to the physics. But every effort I have seen to do this ends up ignoring 90% of what's interesting in the galaxy, as processed by our brains, and has made galaxies look the same numerically which are very different in reality. So, it's quite a difficult effort to make a physical classification system that is both quantitatively useful and comprehensive. There's something to be learned by actually looking at things, and we should feel free to modify our morphological classification systems based on what we see as we learn more. For example, we shouldn't force the deep space galaxies to fit into schemes we've invented for local galaxies. There's a bias one introduces in doing that. Morphological systems can change with time. The newer systems may have unknown connections to the physics, but still they are very useful in helping us know which directions to go to follow the physics.

Knezek: I just wanted to point out ...that [while] I agree that size in terms of spirals is very important, we shouldn't discount environmental effects. It's very important to know [what] the environment [of the] galaxy is. That can really make a difference in terms of how much star formation – and what sort of star formation – that disc undergoes.

B Elmegreen: Yes, thanks. Perhaps we can now move on to our next topic which is dust. Do you want to discuss modelling of dusty galaxies? Paul?

Hodge: An interesting question, and one that wasn't answered often, is why *is* the dust there? Spirals are sort of opaque in their centres and transparent in their outer parts, according to several studies. I think what I'd like to do now is stop and see what your reaction is to that claim.

Tully: Are we going to vote?

Hodge: That would be quite valuable! In the meantime, do you have a comment to add before we vote? ... Yes?

Xu: I saw one galaxy which is not opaque in its centre and that is M31.

Hodge: Yeah, that's right. I've seen many galaxies myself that aren't opaque in the centre ; in fact, it's hard to find one in the Local Group that *is* opaque in the centre. Our view of galaxies is often slanted towards the big famous giants.

Audience member: When you make a statement like that you have to be a little careful and specify whether you're referring to something that's edge-on or face-on.

Disney: I'm the little guy at the back here! ... Just a point here that was emphasised at the Cardiff meeting ... this discussion should certainly also take into account *luminosity*. I mean: a low-luminosity galaxy without much mass in it, may not have much interstellar medium and therefore ... may not be very opaque. You take a very luminous galaxy with a lot of interstellar medium, then of course .... so I think that's the extra thing you have to take into account.

Hodge: We've also seen evidence that the amount of dust, particularly the dust to gas ratio, may be different with different metal abundances, so metal abundances could also enter into it.

Can we go on to the next item? Dust is common in elliptical and S0 galaxies. It shows up as filaments, discs and rings. This is another example of how Baade was wrong. He said elliptical galaxies have no dust and they have no gas and they're all very old. That's three things that defined elliptical galaxies in Baade's view and they're all three of course wrong, as we now know. There were some very spectacular examples of that presented at this meeting. Does anybody have comments about that?

From the audience: ....[Unclear. Pertaining as to whether the dust in ellipticals is of external or internal origin]

Hodge: There were at least two claims, I think, that the dust in ellipticals was probably *external* and yet that wasn't discussed very much by others... so that's a really good question to throw out to the audience. Where *did* this dust come from? Why is it there? And there is the other question ... of why isn't there a lot *more* dust in those objects that have been processed so thoroughly into stars?

Goudfrooij: I think it's hard to imagine that this dust is generally from inside because... from a kinematic point of view: if you look into the centre, where the settling times are probably very short for dust lanes to form... the position angle of these dust lanes are *randomly* distributed with respect to the kinematic angle of the stellar body. I don't see how you can do this with an internal origin.

Impey: I guess I was left a little confused on the issue. [Especially for] the biggest ellipticals ... we know what the stellar populations are ... the models for the mass loss rates are pretty well determined... Is there a problem in that the material we expect to be there [is no longer there] ...

[should it have been] retained in the galaxy? Do we see it at the level we expect to see it [or] has it been removed somehow? In general, what's the answer to that question?

Hodge: Good question. Does anybody know the answer?

B Elmegreen: (To Impey) Right! You are asking if you just put a dust grain in an elliptical galaxy: will it go down or up, in time? It could do either. Radiation pressure could drive it out, or hot gas could drive it out, or it could fall. (To the audience): Is there an answer?

Goudfrooij: If you see hot gas even around an isolated elliptical, that means that the potential well must be very deep - in that case it would fall back after having been removed at some point... but the present models show that for the moderately luminous ellipticals, most gas is actually flowing *out* presently ... so I think it depends on the precise potential of these galaxies. I don't have a definite answer here, but ....

Impey: A related issue: I worry about AGN and so on. Can you generate enough gas in the central regions to fuel radio galaxies at the space densities and levels of radio activity that we see?

Sparks: [In our poster at this Conference] .... dust lanes in those radio elliptical galaxies [that we showed] were common, [but] there was too much dust to be explained by these internal processes. They do appear to be physically related to the radio source, though. If you go to more smaller scales, and more regular morphologies, they become perpendicular to the radio emission, so I think it all hangs together at this time ... but it definitely favours an *external* origin, just simply from morphological and angular momentum arguments.

Frogel: We had a similar discussion during one of the regular sessions the other day. Afterwards, I was trying to understand and fill in the holes in my ignorance ... just talking to Ray White ... I don't see him here today ... I asked him [why] there is far more dust in ellipticals than you'd expect just from stellar evolution. So then I asked: Well, where does it [the dust] come from? His explanation, which I don't think I now believe anymore, was that it comes from all these little companion galaxies that are [going] around the big galaxies, or [from] stars [that] are evolving and the dust just falls in. But it seems to me that even if you are a factor of two or three above what you expect from stellar evolution, these little companions just can't do that. So I ask... I propose a general question ... Where does the dust come from? So where does it come from?

Tully: It's not that much dust is it?

Anon: No, it's not much at all.

Anon: Well, how much is it?

Goudfrooij: Well, we're talking about 10^4 solar masses.

Frogel: Yeah, but how much more than you'd expect from just evolution within the galaxy itself?

Goudfrooij: Well, it's (unclear, shoulder shrugging) sometimes ten times higher because this ....

Frogel: That's enormous.

Goudfrooij: It's enormous, yes.

Frogel: So where does it come from?

Goudfrooij: I said ... from dwarf... (shoulder shrugging) ... well... small galaxies.

Frogel: I don't believe it ... because a dwarf isn't going to produce that. First of all a dwarf is going to be metal poor.

Goudfrooij: Yeah, that's a good point.

Frogel: Secondly, a typical dwarf is a tenth or a hundredth the mass of a giant galaxy, so you're talking about ... you know... what [amounts to a] magnitude of more dwarfs?

[Vigorous inaudible discussion, as at a soccer match. Several delegates speak simultaneously].

Goudfrooij: A small dwarf has clearly enough interstellar matter to provide these amounts.

Frogel: But ... I don't understand where it comes from in the dwarf then ... because you have far more stars in the elliptical galaxy than you have in a dwarf. Metal-rich stars produce significantly more dust over their lifetimes than will metal-poor stars so that per unit mass I would expect more dust in a giant E galaxy than in a dwarf.

Goudfrooij: They do not destroy it.

[A pause]

Goudfrooij: You're in a very different physical environment. The elliptical galaxies are supposed to have these X-ray-emitting hot coronas ... and so the calculation that everybody does is simply ... the dust destruction time times the injection rate ... [so] you get a small number. It's consistent at the low end with the amount of dust that's seen but probably not at the higher.

Frogel to Goudfrooij: Are people taking into account the fact that on average dwarf galaxies have lower mean metallicities than do luminous galaxies with the consequences that I alluded to above?

Goudfrooij: That's a difficult point. If you look in... as Bruce showed in his talk, there are dwarf galaxies actually produced in tidal tails around merging galaxies and recent measurements of the abundance in HII regions in those dwarfs, show that there can be solar metallicity there; much higher than in isolated dwarf galaxies. These might be the galaxies that we're talking about. They just might be!

Hodge: We're developing a new field of research here. If you can sprinkle the universe with new dwarf galaxies that are enriched in heavy elements by having these tidal encounters, maybe you can have an equilibrium situation. Was there one more question on that?

Regan: ... small amounts of dust are common, not dust is common. Without modern image processing techniques, you would never have seen this dust.

Hodge: This item [pointing to an item on the overhead, pertaining to cold dust and cold gas in the outer parts of galaxies] was a very new result, at least for M31 ... a very exciting one to me [Cuillandre et al., quoted by Lequeux and Guelin in this volume]. I think this is an area where we need to move into a great deal more research and see how in general the outer parts of galaxies can have so much dust. This was a very beautiful example of what we usually mean by modelling and of course we discussed the dust temperature varying with environment and bringing up the dust temperature leads into the next item.

Greenberg: Could I make one comment about the last thing? Dust temperature depends on the environment but it also depends on the dust. The properties of the dust are not invariant as you go around the Milky Way. As a matter of fact, just because the environment is different the dust properties are different, the optical properties can change... so that I think what we really have to do, is to try to fold in the variations in optical properties. For example, I was not completely aware of the fact that the cosmic ray density increases by six-fold or five-fold when you get about five kiloparsecs [from the galactic center], as compared to where we are now. That's going to make a big difference in the chemistry and the evolution of the grains and their properties, which means their optical properties ... which in turn apparently determines their temperature ... So, I think what we have to do is keep in mind it's not just the radiation field that determines the temperature, but the properties of the dust grain.

Hodge: Very good point.

B Elmegreen: Dust properties, our third topic: Now we're targeting six o'clock as a finish time ... and I think we can make that ... so, we have a good ten minutes to talk about this. You've seen discussions of temperature, composition and relatively little discussion of optical properties, in particular the high scattering albedo at K. In *temperature*, what David Block and collaborators said several years ago is right: IRAS missed a lot of dust – cold and very cold dust. Now there are various ways to see this dust by looking further in the infra-red, or by doing absorption studies. There was a missing factor of ten in dust mass, so now the amount of dust has been brought up to the typical solar abundance. We've learned that temperature is a delicate subject... because it varies with size, composition, and time; I can't add that much more to the discussion we've already had. There's an enormous amount that we've seen in the discussions which indicate how difficult it is to take a model from grain physics and convert that into an observed spectrum. There seems to be a lot of ambiguity in that conversion.

Now, *composition*: we've heard about the importance of organic refractories. We would like to get to the point where we can understand the entire evolutionary cycle of dust; perhaps, after capturing some of these pieces by [space]craft ... interplanetary dust, we can pin down the various types of dust. I would like to open this up to some discussion which could take about ten minutes and we're still on schedule.

Frogel: I recall hearing a year or two ago that when the old solar panels from the Space Telescope were brought down you could see all the holes from micrometeorites, you could study the distribution in sizes. [The next sentence was said in a softer tone of voice]. Does anybody know if this has been done?

B Elmegreen: If that's been done? You're asking if that's been done?

Cruikshank: I'm not specifically familiar with that, Jay. There was the Long Duration Exposure Facility [LDEF] that was retrieved in 1990. It had been in space for nearly six years and the record of impacts on that spacecraft made it possible to derive the mass influx of microscopic particles to the earth, which was nearly 40 000 metric tons per year. These were particles that were in the mass range E-9 to E-4 grams, and they were sufficient to make visible craters. People have searched in the vicinity in these little craters on the LDEF panels and found, among other things, fullerenes and other debris from the impacted body. I believe that C-60 has been found in material. To my knowledge, that's the only evidence for C-60 in meteoritic material, but I may not be current on that. There is potentially a lot of information in the LDEF panels, but it's rather limited in terms of actual physics of the interstellar dust. [For a discussion of fullerenes and the C-60 molecule, the reader is referred to, for example, *The Fullerenes: New Horizons for the Chemistry, Physics and Astrophysics of Carbon*, ed. H.W. Kroto and D.R.M. Walton, Cambridge University Press, 1993].

Disney: I was just going to comment that we are all of us dependent on a very tiny number ... ... all the astronomers are really dependent on a tiny number of people like Walt Duley and Mayo Greenberg, in knowing what the properties of the dust are. I was very interested ... (it was probably the biggest controversy of the many we had here: see Figure 1) ... whether grains can really get down to 2.7 degrees ... and I can remember several people changing their minds at least three times ..... So the question I'd like to ask the people who know about grains: Are we in for any surprises? Is it a field that some more people ought to go into? Or is it just a few details left

Figure 1. Can dust grains really get down to 2.7 degrees ... and even colder? Walt Duley (left) and Mayo Greenberg (right) deliberate. Mike Disney reminded us: "I can remember several people changing their minds at least three times..." Naomi Greenberg looks on.

to be worked out because there's a hell of a lot of other questions depend on the answer to that.

B Elmegreen (pointing to Mayo): There's only one person who can answer that!

Greenberg: Not just one ... but nevertheless, I would say it's more than just details at this point. ... I made a statement the other day ... I said if I were to go to heaven, my desire would be to go to a comet. Not just into outer space, but to go to a comet. Because from my point of view, comets really do contain a lot of interstellar dust just as it was in the sub-micron size. The right morphology, the right chemistry, and if I could do that, that would be fine. However, barring that, there are still many things that can be done and there are many groups working on them. There are laboratory approaches to understand how dust may be made; how it evolves. There are groups in California, in France, in Japan, in Italy and, of course, there's the group in Leiden ... and when ISO comes back [with data] we're going to have an enormous amount of additional information on at least some of the evolutionary properties of the mantles of dust as well as the basic properties of dust. So, I think what we're going to have to do is to wait a little bit. When that happens maybe we're going to have to change some of our basic ideas, maybe we'll simply have to elaborate on our basic ideas. There are conflicting theories of where the dust comes from, and how it evolves. I don't want to get involved in that. There are quite a few people working with different aspects of dust and so all I can say is ... let's wait a little while till ISO sends back some nice stories for us to work on.

B Elmegreen: Ron [Allen] has a beautiful slide of the Eagle Nebula which shows dust. (Excitement mounts. Lights darken, but there is no slide. Bruce to projectionist: Is anyone back there?) Ah.. the slide. [oohs...aughs...and excitement from the audience mounts, as the HST slide of the Eagle Nebula is shown].

Allen: This object is not in the central regions of any galaxy. It is about 6 or 7 kiloparsecs from the centre of our Galaxy ... and I think no one would argue that there are parts in it that are optically thick.

Greenberg: It looks just like cumulo-nimbus.

Allen: I wanted to [show this] very intricate structure which is available now; we don't have to wait for ISO on this one. There are regions in here that must be extremely optically thick; with τ of ... I don't know ... a hundred or more, perhaps, even though the physical thickness of these columns is only ~ 0.05 pc. [Turning to Greenberg] First of all, whatever is inside this cloud, are probably your grains covered in ice. [Addressing audience again] It is probably very cold in there; and [we can see] activity on the surfaces of these clouds because there are several hot stars up above this [out of the picture] that are flooding these surfaces with UV and boiling that gas off. You can just see the tops of these columns just evaporating off ... and exposing these things that someone called EGGS [Evaporating Gaseous Globules]. This whole thing is a classic example of a PDR, a Photo-Dissociation Region, with molecules and atoms and ions boiling off the surfaces of these very dense molecular clouds. The grains that came off with this hot gas will probably also have that blobby yellow stuff on the outside. [To Greenberg:] What did you call it? [Mayo:] Organic refractory stuff. [Allen continues:] And you know we don't have to wait for ISO, we can try to study it in objects like this.

Greenberg: Can we study it spectroscopically? Do we have the infrared [instrumentation] to do that at this resolution?

Allen: Well, the problem for the infrared is going to be background and foreground confusion along the line of sight. You're going to have to sort it out spectroscopically, using the velocities I presume. This picture of course is actually done in the optical light of three ions; the red is SII, the green is Hα, and the blue is OIII.

Greenberg: I have seen this and I haven't talked to anybody about it, maybe... I would like to talk with you about this after the meeting.

Allen: Well, I'm not the person who knows all that much about it. There are others who do, but here is a beautiful laboratory to work on this problem.

B Elmegreen: Any other comments on dust properties? Several of the main categories that you heard discussed include dust at high z which we just heard about today and the fact that it is there: [clearly], there must have been processing and heavy element formation even at high z. Perhaps it is quite important for extinguishing the light in quasars. HST is another main category which really stands by itself ... and any conference I've been to, in the last year or two, the HST talk is really so spectacular it's almost... a factor 10 away from where everyone else is. So I put these together [on this transparency] with mere exclamation points and perhaps we can talk more about that. We heard a very interesting talk about star formation and what the density profile tells us about models. So let me open this up for some discussion. We just finished a long discussion on dust at high z, but perhaps there is more on any of these topics.

Mather: I'd like to make a brief appeal that since NASA is now considering building the Next Generation Space Telescope ... anybody that has strong opinions that they want to tell us about ... should do so! We are at the beginning of thinking about it; so all comments are very helpful, especially now.

B Elmegreen: How big will the mirror be?

Mather: Good question. Bigger than Hubble. A lot bigger than Hubble.

B Elmegreen: So, it won't go up in the shuttle ... ten years away...

Mather: It may have to be folded.

Tully: Are we ready for interferometers?

Mather: NASA is certainly thinking about that also. Primarily for astrometry and for imaging... looking for planets. There is big excitement about that.

Disney: How far out could you possibly put it...?

Mather: You can put it fairly readily at the Lagrange point L2, which is in line with the sun and the earth. It is about 1.5 million kilometres away. So, the engineering is not too difficult for that.

Tully: What length baselines would you have for interferometers...?

Mather: I think that is a different topic. If you want to use the Next Generation Space Telescope as an interferometer, we could, for instance, put Michelson flats out on outriggers in front of it. I should say also that we want to make this cold so we can see the near infrared at least out to five microns and preferably a lot further.

Allen: The Astrometric Interferometer Mission that is being thought about now would have a maximum baseline of $\sim$ 10 metres. The elements of the interferometer are small, 0.3 m in diameter. That instrument is designed to measure star positions very accurately over wide angles in the sky, for positions and proper motions. It would have a modest imaging capability over a field of view of about a third of an arc second. Within that field it could do imaging with a resolution of about 10 milli-arc seconds at $\lambda = 500$ nm.

Tully: What stops us from going after real imaging capability within the interferometers in space?

Allen: Courage. Money.

Mather: People are just beginning to succeed with that on the ground. When we have a bit more experience it will become much more popular, I think.

Rich: ... [Unclear] Who do we contact at this point regarding the Next Generation Space Telescope?...

Mather: There are two leading scientists. I'm a study scientist at NASA Goddard; Peter Stockman at the Space Telescope Institute is leading their effort. So, do contact both of us.

B Elmegreen: I remember that Ron [Allen] had an interesting point about this [Hubble] deep field of galaxies. [If the images were] slightly deeper, they'll start to overlap and give us confusion. Would you elaborate on that?

Allen: Well: that's it, you've said it! There'll be a problem, because we have almost filled the whole 2D space with galaxies as it is ... and, if there is dust in those things we're not going see much further.

B Elmegreen: So ... the point is, we may be confusion limited before we reach the galaxy formation era.

Allen: We could be at the point where the dust curtain is going to drop – and that is it. Mike Fall showed some of that with the turnover.

Fall: It looks to me like the covering factor of galaxies in the HDF is about 10%. Now, the covering factor of high column density gas goes up with redshift roughly like $(1 + z)^{1.5}$ and reaches unity at $z \approx 4$. So if you assume that these covering factors are roughly equal, you can work backward to answer the question: How deep is the HDF? The answer turns out to be $z \approx 1$ (very roughly). This is not deep enough to cause the sort of dust screen I was talking about earlier. By the way, this argument doesn't mean you won't find some galaxies in the HDF at much higher redshifts. Very bright galaxies might even be seen out to $z \gtrsim 3$. But the 'characteristic' depth of the HDF, if you estimate it from the covering factor, is much shallower than can be probed by quasar absorption-line studies.

Lequeux: A way of getting rid of this problem is to use a large mm interferometer. With such an instrument you will see the thermal emission of dust shifted to the millimetre range. You can see it [to any] redshift because there is nothing like an absorbent screen in front of that. You will be limited by confusion but not by absorption.

Allen, to Fall:: [Inaudible]... What size where you using for the potential obscuring sizes of galaxies?

Fall: You don't need to assume a size, only a covering factor. And that is given to you by quasar absorption-line surveys. For such estimates, I took an HI column density threshold of $N_{HI} > 10^{20}$. I just assumed that the visible regions of galaxies correspond roughly to this threshold in HI column density. This is probably correct at the $\pm 50\%$ level.

Impey: I think the confusion limit issue is not really a problem for three reasons. First of all, the MDS is showing clearly that most of the excess faint galaxies (the blue ones and so on) are at very small scale lengths and small angular scale lengths. Second, the number counts of galaxies down towards this thirtieth magnitude level, which is the current limit, are flattening off quite substantially. And third, if you match the surface density of galaxies down to these incredible magnitude limits with the volume of the universe and the local luminosity function, then in an inflationary universe you'd see all the way down to luminosity at -14. Well, obviously galaxies evolve; but either way there is not enough volume there, for there to be enough surprises for the confusion limit to suddenly leap up and bite us with another couple of magnitudes.

Frogel: Ron, has any thought been given to repeating the deep field next year to increase signal to noise [by a factor of] two ... and to look for secondary variations? Also, my understanding is that these exposures were over a period of 10 days. Has anyone looked for supernovae?

Allen: [Unclear]... As of the day I left the Institute to come here, I hadn't heard anything startling about possible supernovae, but there is an approved Cycle 6 program to repeat the HDF with shorter exposures in order to look for them. There is a lot more to be done on the HDF. As a matter of fact: you can do it! The data is available to you. It's in the public domain.

Frogel: Well, is there any thought of repeating it next year?

Allen: Yes, there are thoughts of repeating it, and also of doing another HDF in the southern hemisphere... but I can tell you there are no specific plans yet. Don't forget that the initiation of this was not an easy process; a lot of telescope time went into this ... and that is a very valuable commodity. Bob [Williams] convened a committee, as you know, to think about this ... to decide what to do with his discretionary time. And I'm sure that there will be worries again, that there will be a lot of time going into follow-up projects. But if the community thinks this is important enough, it will be done I'm sure. It is a good suggestion that we should start thinking about consulting the community again in the course of this year.

Davies: With cosmological SB dimming, we're probably only seeing the very high surface brightness galaxies in an image like that [the HDF] and

so there could be lots and lost of low surface brightness galaxies which...
[unclear].

Impey: The number of counts are relatively shallow down to the magnitude limit of the Hubble field... Given that you're looking at an integrated number counts, something that would be very unusual in some integrated volume of that pencil beam [would be for] the number counts to steepen sharply over the next couple of magnitudes fainter ... I mean it would certainly have to be unphysical for that to happen. I agree that you'll start to pick up lower and lower intrinsic surface brightness objects, but as to whether it hits you with a confusion limit if you go a factor 10 deeper ... it is not obvious that you would.

B Elmegreen: I'd like to move on to the last topic before Ron speaks about molecules. That is dark matter. I know we can talk about this for a long time. But we still have these questions - where is it?; what is it? ... and the very intriguing possibility - that some of it is cold gas. The particularly good models these days are the ones that may be difficult, but not impossible, to observe. Are there any comments on this?

Allen: Well, we know pretty accurately what it is - its physical properties; its spatial distribution... its everything we can't observe. It's very intriguing. To be constructive, I'd say I think that [one of the things] that really are worth looking for that came out of Francoise [Combes'] discussion to think about - [are] the dimers. (To Combes): You mentioned the H_2-H_2 dimer, but there are other dimers - like the CO-H_2 dimer, which would be another possibility. If we find dimers at all, it is really strong evidence for a low temperature gas phase material. So that would be something well worth looking for. I think the idea that we might find low frequency lines is going to be useful impetus for a lot of low frequency projects. I like that a lot.

Lequeux: Another possibility which is not that different from this H_2 blob is planets or brown dwarfs. After all, you cannot place those fellows in a thick disc which is thickening to the outside of the galaxy ... and it will satisfy those old constraints. That is an alternative to that model ... and it is just the same thing in fact. It just collapses into dense material and that is it.

Tully: You have to keep in mind though that, whatever the stuff is, it is probably non dissipational. It formed very, very early and has not dissipated energy since. So, it is a totally different population than stars, for example. You have to have a totally different formation scenario. That is why I really don't buy this idea that the dark matter is in brown dwarfs because they would have had to form before the galaxies formed.

B. Elmegreen: Well... maybe there are several types of dark matter - the kind connected with rotation curves and the kind that closes the universe.

Pfenniger: It is certainly fundamental to understand well all the properties of hydrogen. Let us contemplate the sublimation curve of solid molecular hydrogen [the transparency of the curve representing Eq. [44] in Pfenniger & Combes 1994, A&A 285, 94, is shown]. Between 2K and 13K it covers almost 19 orders of magnitudes in pressure, from about $4 \cdot 10^1$ K cm^{-3} to $3 \cdot 10^{20}$ Kcm^{-3}, thus including comfortably the whole ISM pressure range. My interrogation is, since we have heard of temperature fluctuations of the dust grains below the average temperature, even below 3K, whether anybody would find unreasonable to expect in several cases condensation of H_2?

Greenberg: My recollection is, and I can't recall any of the details ... that I once published a paper which I, if I remember correctly, estimated that it was possible that if we could get grains... really cold grains ... down to four degrees, we could conceive of having solid hydrogen. In those days, it was impossible to think of grains going down to four degrees. So, it was a negative paper because I was really trying to say that somebody else was wrong when he said there was solid hydrogen in space. However, if we really could get to these regions where the temperature of the big grains is four degrees (I would have to take a look again at the numbers; I think I used good numbers for the solid hydrogen sublimation), that maybe four degrees might be adequate. I don't know... I [would have to] look at the numbers again. But it seems to me solid hydrogen is not out of the question. [When I got home I looked at the numbers on the vapor pressure of solid hydrogen in Greenberg and de Jong, *Nature*, volume 224, p 251. It turns out that even if the dust could be maintained at temperature $T_d = 4K$, the vapor pressure of solid hydrogen exceeds the ambient hydrogen pressure in a 10K gas unless $n_H > 5 \times 10^{10}$ cm^{-3}. *However*, at this density the gas collisions with the dust would deposit enough energy to raise the dust temperature well above 4K, so that I have to conclude that solid hydrogen can *not* exist as a dust mantle condensate.]

Cruikshank: Solid hydrogen has been detected in the interstellar medium reported in a paper by Sandford, Allamandola, and Geballe [Science 262, 400-402, 1993]. I believe it is in the W33A source. This is a weak absorption at 2.4 microns (4141 wavenumbers), but the detection is quite firm.

Greenberg: That's not solid hydrogen condensing on grains and making a nice solid hydrogen grain. That's what we really are looking for here... to really get that hydrogen on the grain, and that's a lot. I'm not saying that what they have isn't right; I'm just saying it isn't this.

B Elmegreen: Is there a region where grains are supposed to be this cold?

Greenberg: No, they're not that cold.

B Elmegreen: So this is a controversy still?

Pendleton: The source toward which Sandford, Allamandola, and Geball found the weak 2.4 μm absorption was WL5, an infrared source in the Rho Ophiucus cloud. They attribute the feature to frozen in H_2O rich interstellar ices. This part of the Rho Oph cloud is known to contain a lot of ice. Incidentally, Sandford and Allamandola [Astrophys. J. 409, L65-L68, 1993] predicted that this H_2 absorption should be present in sources as warm as about 70K, and might be detectable in W33A.

Greenberg: .... what was that again? W33A, it's a hot object. W33A shows a lot of heated grains, it can't possibly be that cold at W33A.

Pendleton: But there were other ice features associated with it, and it was typical of icy, dust-embedded infrared sightlines.

Allen: Well, very quickly on molecules, there were only a few of the participants who spoke about that subject specifically. I just tried to glean the important points that came to my ken when I wasn't suffering so much from jet lag that I was sleeping in the audience! [Editorial note: We never did catch Ron Allen napping!] So the X-factor is an issue that's troubled a number of us ... how big can it be in the outer parts and how small in the inner parts? How much of the inner discs of galaxies like M31 and NGC 7331 and M81 could be full of this kind of gas – and, if it's there: why doesn't it form stars abundantly? Why are we not seeing high mass stars in these regions if there is lots of gas there? That's a conundrum, I don't think we understand. We talked also about the CO and HCO+ absorption lines and the relevance of that to the chemistry and to the dynamics of the interstellar medium. The differences between molecular absorption spectra and the HI absorption spectra along the same lines of sight are very interesting I think, although we didn't go into that in great detail. As far as the real content of galaxies in H_2 is concerned, it looks like there's a whole subject here for other meetings, but what do you think? Anybody have any additional comments on things that I might have forgotten?

Tully: I'm finding an inconsistency. When we discovered that early-type spirals have HI holes in the centres, there was the comfortable coincidence that these galaxies have predominantly old populations near the centres. Now, if people are really trying to suggest that there is a lot of molecular gas in those holes, there is no deficiency of star-forming material after all. Is this inconsistent with the picture of low star formation in these regions?

Devereux: Also, those galaxies that you mentioned, they have these bright H-alpha spirals. If you work out how many ionising photons you need to ionise them, it's equivalent to about 10 000 O-9 stars. I'd be very surprised if any molecular gas were to survive there.

Allen: What about this very cold material that Francoise brought up. Do you have any additional remarks you'd like to make about that?

de Geus: Yes, we tried to find different ways of looking for molecular gas in far outer parts of external galaxies ... the parts within the extended HI disks, but well outside the optical discs of these galaxies... If the molecular gas is truly cold, of course, you have no chance of finding it in emission; finding it in absorption is very difficult, because you have a lack of background sources. Neil Tyson and I came up with an interesting thought ... if there is any kind of star formation which is associated with molecular gas in the far outer parts of external galaxies [and then] at some point, supernovae are going to come along ... type II supernovae ... they can *betray the presence* of star formation ... and molecular gas in external galaxies outside the optical discs. There is one very good example - NGC4699: an optical galaxy which has a huge extended HI disc. There [was a] type II supernova that went off in (I think it [was a] 1989 supernova...); we used the IRAM-30 meter and we looked for eight hours at the location of the supernova. We had a very marginal 4 sigma signal of detection of carbon monoxide and it was really too marginal to go out and say this is a real detection but I'm afraid we're never going to get more observing time to confirm... it is right on the limit of being [there] ...

Combes: Do you mean the supernova is enough to produce heavy elements and enrich the molecular clouds so that we can detect CO emission, or does the star formation produce only the heat for excitation?

de Geus: No, no, I think the supernova ... [indeed] the stars, formed from the molecular gas and the moot question in my mind is: Does the molecular gas exist before it has, for some reason, started forming stars? Or does the molecular gas condense from the HI and *immediately* starts forming stars so [that] there is no time difference between the existence of the molecular cloud and the formation of stars?

Allen: OK, any other comments about molecules? Then, I wanted to sort of put up a little philosophical summary, if you like, of some of the things that have struck me in terms of changing perceptions. That was, after all, one of the titles of our conference. For me, personally, I tried to think of historical precedents for this change-in-perception issue, and I can find some parallel with our changing view of the Universe in the early part of this century. Our view in the early 1900's was that the Universe was really a rather restricted area of space as we now know it ... with stars sort of distributed in an unusually strange way, in a flattened system with the sun nearly at the center and with a band of "no stars" along the mid-plane. This was Kapteyn's Universe [pointing to a sketch on a transparency: see Figure 2]. We now know of course the reason for that band in the middle was *dust*, and even Kapteyn worried repeatedly about that, but amazingly enough the pundits of the day could not find any convincing evidence for it in the data.

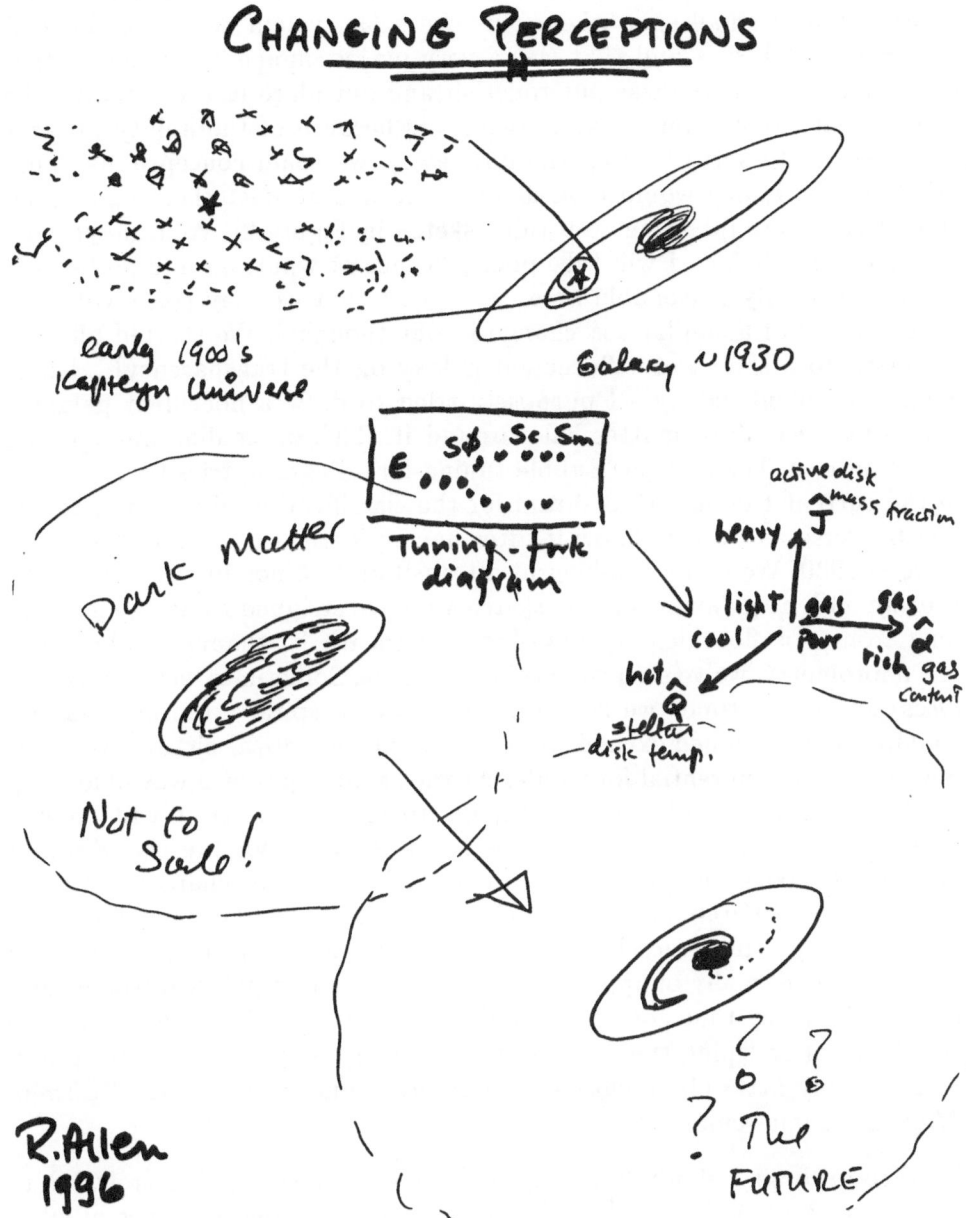

Figure 2. Changing Perceptions as sketched by Ron Allen, 1996

It took about fifteen, twenty years or so after the publication of the Kapteyn Universe, before dust was worked properly into the astronomers' points of view or psyche. Then we have the galaxy as we know it now –

[again pointing to another sketch in Figure 2] with Kapteyn's Universe kind of stuck into a small spot where our own solar system now is, embedded in our own Galaxy – and that was a very major change in our perception of the way the Galaxy was put together and our place in the Galaxy. The reason that came about, that change, was because of the ability at least conceptually to sort of sweep the dust away, we could conceptually think of what the Galaxy would look like if there was no dust there and that's this picture here [pointing to a third sketch in Figure 2]. Well, maybe it's going a little bit far - I will take poetic licence or whatever and go beyond what is probably reasonable here, to suggest that we may conceivably be on the verge of a similar sea change in our thoughts. We started off here [pointing to a sketch of a flocculent galaxy on the transparency] ...here's a nice flocculent galaxy - I purposely tried to draw a flocculent galaxy - with this great dark matter halo around it. This other diagram up here [pointing to a sketch of the Hubble tuning-fork diagram] tries to remind us of the current tuning-fork diagram for the classification of galaxies; this is our Universe as we now know it after having learned a lot about galaxies since $\sim$ 1930. We're now looking at a transition to a possible change in the way we look at galaxies. In this sketch we have stripped away the gas and dust from this flocculent galaxy and underneath it, thanks to the infrared photometry, we've seen something different. Sometimes we see formless discs, and other times we see discs that have a spiral structure that we couldn't have dreamed existed from looking at the optical picture. We have now up here the potential for a different way, a more physical way of looking at the classification of galaxies, thanks to this. Whether it will be just exactly this picture [pointing to the 3-D framework for the classification of spiral galaxies on the basis of their intrinsic modal characteristics as developed in IAU 146, p.93 and applied in A & A, 288, 366] which Giuseppe Bertin and his colleagues have worked out, I'm not sure. But we've got a possibility here really of applying the morphology to a physical framework, perhaps in a way that none of us could have dreamed of before we had the capability of sweeping the dust away from the galaxy in a figurative sense. So that's my little bit of closing philosophy. I understand that *Chairman Mayo* has some remarks.

Greenberg: I don't know if that's a good introduction [referral to 'Chairman Ma(y)o'] or not! All I would like to say along the lines of what Ron said is that – in the beginning there was [assumed to be] no dust and we had one perception. The next stage was that we had dust but that dust had just better be exactly the same everywhere – otherwise it was going to be a terrible nuisance. So we had a *uniform* extinction law and that's exactly what people liked. At the first astronomical meeting I ever went to, I had a suggestion about why the extinction law could be *different* in different regions of the Milky Way along different longitudes, depending on alignment. Somebody got angry with me because I was disrupting this nice way of reducing what you see when you get a reddening and therefore you know exactly what the extinction is and you know the distance. Well, we've gone so far beyond that I'm tempted to say that this is like 1,2,3, infinity – because now the role of dust in galaxies, in chemistry, probably in solar systems, comets, the origins of life – has become a very active field. But, and I'm just going to put in one piece of commercial - the curious thing is that - somebody as a matter of fact made a remark: Why aren't there more people working on dust? [To a member of the audience] Didn't you say that? Somebody said it. And the answer to that is that ... historically, astronomers, from the very beginning, have been *afraid* of dust. For example, what is dust doing in molecular clouds? All of the chemical work that was done on molecules in clouds was done with ion molecule chemistry, absolutely ignoring the fact that there was dust there. They ignored it from so many points of view it's impossible for me to add them all up. All I can say is that we now know that dust does play an active as well as a passive role in determining the chemical evolution of clouds. So the next step, I can say, is this. I would only hope that as a result of meetings like this, that the astronomical community would indeed recognise that dust is really an important subject, which is something which merits a little bit more support than it's had because, really, for many years, I was very lonely in this field. I started work at Leiden and practically nobody paid any attention to it. The chemical evolution of dust was something astronomers really didn't care about. When it was related to galaxies ... that was way beyond our way of looking at it. So all I can say is that dust, I think we now recognise, plays a role not only as a tracer of what goes on in space ... a corrector for making different modifications in our ideas of the pictures [morphology] of galaxies ...[but] also actively contributing to the chemical evolution of molecular clouds ... and in that sense, we do welcome more people to work in the field. It may become more competitive but it may be also more fruitful for everybody else in astronomy. Thank you.

[Applause]

Allen: We're finished.

B Elmegreen: David, did you have something to say?

Block: In conclusion, I'd like to thank everybody for putting in a tremendous amount of very hard work ... which has extended over the past fifteen months. I'd like to pay special tribute to the SOC; to the LOC; and to all student helpers. A special word of thanks to Ebrahim Momoniat, who has been working with me for the past fifteen months. I wish to thank every participant for joining us here in Johannesburg in the *New South Africa*; your participation has been invaluable. I know it has taken a lot of effort – and time – to hand in your camera-ready copies at Conference registration. We trust that the Proceedings will reflect the true spirit of the meeting. I do want to thank each delegate... every speaker ... the Chairpersons ... and our panelists this afternoon: Ron Allen, Bruce Elmegreen, Paul Hodge and John Mather. I must also add that the posters on display are magnificent. Everyone has worked very hard. Our 'poster judges' Mike Disney and Debra Elmegreen have urged that *all* posters should and must be published in the *Proceedings*. We are delighted to report that Eugene deGeus, on behalf of KLUWER, has agreed to this request.

My first contact with Anglo American and de Beers was, I believe, probably in October of '94; all of us are indebted to them for being our Principal Sponsor for this Event. I trust that everyone has really enjoyed this meeting ... and I trust that we can enjoy *New Perspectives in the New South Africa (Number II)* in possibly three year's time from now [1999/2000].... Judging from the tremendous enthusiasm that we received from delegates, Number II will indeed be a reality. To everyone, a heartfelt thank you.

[Applause]

Tully: We want to thank David Block for inviting us down here to participate in this wonderful meeting. Thanks, David.

[Applause]

EACH CHAIRPERSON REFLECTS ON THEIR SESSION...

Editorial comment: Some of the Chairpersons have provided us with a summary of the *papers* presented in their session, while others have given a summary of the *discussion* following the papers. We trust that readers will find the diversity of approach adopted by the different Chairpersons, useful.

Next, a special vote of thanks, from DLB, to the Chairpersons:

"There are a few occassions in one's life when time momentarily stands still. When you, Bruce and Mayo, presented me, on behalf of the Chairpersons and Panelists, with the magnificent wooden carving of a man with spear, and with 'The Hippo', time stood still. To each Chairperson and Panellist: I thank you."

1. Summary of Monday Morning Session, Jan 22 1996: Chair– J.M. Greenberg

The symposium "New Extragalactic Perspectives in the New South Africa" was appropriately initiated with the talk by David Block, the organizer and prime mover of what is turning out to be one of the most active fields in galactic astrophysics. He summarized his new methodology to probe the presence of dark arms in spiral galaxies where it is possible using *near infrared images without* the need for *far infrared* techniques to prove the existence of cold dust in these interarm regions. A point of the discussion raised by Greenberg was that in the radiative transfer calculations leading to this result the derived value of the dust albedo in the near infrared of $\alpha_{K'} = 0.9$ was far too high to be compatable with current interstellar grain models which predict a value $\simeq 0.25$. This question deserves deeper consideration because it seems to be explainable only by the presence of near infrared *emitting* sources. Nevertheless the pioneering work of David Block *is* consistent with temperature calculations for these same dust models (see Greenberg and Li earlier) so that his technique has provided a key to accounting for 90% of the dust which had been unobservable by the hitherto major IRAS source of data in the "far infrared". The major conclusions of this presentation were not questioned but rather raised new subjects demanding further research on galactic spirals by higher resolution submm techniques.

D. L. Block and J. M. Greenberg (eds.), New Extragalactic Perspectives in the New South Africa, 612–625.
© 1996 *Kluwer Academic Publishers.*

Mike Disney raised some deep questions about the properties of interstellar dust - or as he more appropriately called it interstellar smoke - and its consequences on observations of distant galaxies. He suggested that the whole question of the amount of dust obscuration in distant galaxies remains an open question. What about the dust in distant galaxies? What are the age effects? Are our theoretical calculations on simplified models of dust material and shape properties adequate to provide correct answers to the question of far infrared and submm emission? He suggested that there may exist significant amounts of *very* cold (<10 K) dust. In fact this turns out to be a reasonable conjecture (see Greenberg and Li, this volume).

The suggestion raised by Walt Duley in his paper on temperature fluctuations of small dust grains stimulated a heated discussion about his proposition that a sufficiently small particle spends a considerable fraction of the time at temperatures *below* that of the cosmic black body (CBB) radiation field at 2.7 K. This is a clearly defined mathematical result because of the long time interval between photon absorptions. During the discussion period it had not yet been resolved whether temperatures below the CBB temperature were consistent with thermodynamics but the fundamental aspect of the problem was clearly fascinating.

The Cosmic Background Explorer (COBE) probed not only the microwave background 2.7 K radiation but also measured the radiation emitted by the interstellar dust even down to 700 μm. In fact, it was pointed out by John Mather that the presence of very cold dust as a major fraction of the dust was clearly confirmed by the Far Infrared Absolute Spectrometer (FIRAS). This supported both the galactic observations of Block as well as the theoretical dust temperature calculations for the tenth micron dust component. The presence of both hotter and colder dust was discussed and how to relate these to suggested dust populations of smaller particles is yet to be fully explained.

Keeping the focus of his paper on the cold dust, Ron Allen showed how the so called X factor, used by astronomers to deduce the density of clouds where the dust extinction is not directly observable; namely, the ratio of the CO intensity to the hydrogen abundance, is influenced by the presence of very cold gas which makes the CO emission very faint. This cold gas is incapable of providing the CO excitation normally assumed and, in fact, Ron Allen gave as an example where the amount of cold gas in the inner disk of M31 was such that the mass surface density inferred from the virial theorem implied a value of $X \sim 50$ times (!) higher than expected from solar neighborhood results in our galaxy. He emphasized that trying to infer an extinction from sequentially applying the X factor and the "standard" dust to gas (hydrogen) ratio should be seriously reexamined because extinction of the cold molecular gas even in the neighborhood of even the UV sources may be significantly higher than indicated by the observable CO intensity.

The γ-ray spectral cut-offs produced in supernova remnants were interpreted by de Jager as indicating the presence of some dust at temperatures 25 K which, although cold, is significantly warmer than the average galactic dust temperature not directly in the vicinity of hot young stars.

2. Summary of Monday Afternoon Session, Jan 22 1996: Chair– R Griffiths

Jay Frogel started the Monday afternoon session with a review of optical and infrared images of galaxies, with an emphasis on the OSU survey in BVRJHK—the extinction at K being one-tenth that at V, and one-sixth of that in I. The ages and metallicity effects in stellar populations were illustrated using multicolor optical/infrared images and comparisons were made between maps of isophotes and isochromes. The colorful effects of stellar populations and dust were illustrated in a series of beautiful images.

Comment from N Brosch to J Frogel:

There is a problem in modeling stellar populations with broad-band colors only. We found that for Virgo BCDs we cannot distinguish between different evolutionary models using broad-band colors only, even when we included a band at 1600Å. The degeneracy was removed only when we included $H\alpha$.

Brent Tully showed data in BRIK and HI of a complete cluster survey. The faint-end slope to the dwarf galaxy luminosity function was found to be -1.0 over the luminosity range -20 to -16, with the implication that total cluster masses are not affected much by dwarfs. The bifurcated distribution of surface brightness was discussed in terms of a *'valley'* at the Scd/Sd classification. An important implication of the work of Tully *et al.* is that, at low luminosity, the HI-rich population of dwarfs is flat or declining. This may be in contrast with the steep L.F. of the dwarf field galaxies.

Michael Regan reported on a low value for the X-factor in NGC1530 and the spiral arms of NGC1068. During the discussion, it was pointed out that central regions of galaxies may well have a low X-factor, with a radial gradient. There is a clear need for high resolution dust maps and high dynamic range in CO mapping. This needs to be calibrated using local group GMCs.

Cong Xu commented on the importance of the inclusion of vacuum UV data when calculating the heating of dust, since the UV dominates the heating. Xu presented his 'frequency convertor' model as a useful tool in this calculation.

Darren de Poy showed that in the disks of normal galaxies, the contribution of a supergiant population to the total light at K, does not exceed the 20% level.

Ray White discussed the cases of galaxies overlapping in projection, and the resulting extinction. He showed that interarm regions are essentially transparent. High spatial resolution images from HST will shortly be used to provide further constraints on dust extinction.

3. Summary of Tuesday Morning Session, Jan 23 1996: Chair–P. Hodge

A highlight of the session (and of the meeting) was Mayo Greenberg's complete short course on dust in galaxies. Much of the current knowledge that many of us have acquired from laborious observational efforts was known long ago to Mayo from theoretical considerations and laboratory experiments. The formation and secular evolution of dust grains, their sizes and optical properties, and the consequent process by which the grains absorb and re-radiate ambient radiation and are heated to observed temperatures, were all described in elegant and enthusiastic detail.

Kris Sellgren reviewed our recent knowledge of dust grains and large molecules as gained from emission in the 1 to 25 micron wavelength range. Although there is still more to learn, especially about the sources of the emission features, much progress has been made. A significant fact is that, while features are found in long wavelength spectra of normal starforming galaxies, they are absent for AGN's, apparently because the large molecules have been destroyed.

Turning to shorter wavelengths, the session next explored the value of ultraviolet observations to discern the amount and nature of dust in galaxies. Using IUE spectra and UV photometry for a large sample of stars, Brosch mapped the UV extinction in the solar neighborhood, including its three-dimensional distribution. George Carruthers continued this discussion with some spectacular results from recent ultraviolet detectors in space and promised more, from launches scheduled soon.

The late Kuiper Airborne Observatory, recently retired from service, has provided many important results relevant to this conference. Beverly Smith and Gordon Stacey reviewed data from bulges and central areas of galaxies and from wide-field observations, respectively. Smith's results showed that dust heating in spiral bulges is common and that generally, the $100\mu m$ data is still too short a wavelength to detect the cold dust (a general theme of this conference). Stacey showed beautiful data for two active galaxies, M82 and Arp 299, which have important structure in multiple dust temperature components.

4. Summary of Tuesday Afternoon Session, Jan 23 1996: Chair– R.B. Tully

This session began on the theme of dust. The first two talks were about diagnostics of the properties of the dust in galaxies. In later talks the discussion moved on to models of specific systems.

Geoffrey Clayton was the first speaker and he gave an intercomparison of the ultraviolet reddening properties in the four galaxies with information: the Milky Way, the Large and Small Magellanic Clouds, and M31. The addition of M31 to the list of objects now studied in the ultraviolet has somewhat confused the situation. With just the MW, LMC and SMC, there was the appearance of a trend. The less metal-rich systems in that trio have successively less of a bump in $E(\lambda - V)$ *vs* λ distributions at 2175Å and turn up more sharply at short wavelengths. Along we come with M31 which has heavy element abundances comparable or greater than that for the Milky Way. The general shape of the extinction curve with wavelength is similar between M31 and the MW but the bump in M31 seems less pronounced. Uncertainties are large in this case. Could the "hump" grain populations really be different?

Patrick Boissé was concerned with distinctions between clumpy and uniform dust distributions. He pointed out differences that arise in effective optical thicknesses, influences of scattering – and mixing between dust and stars. Hmmm. He reports the 2175Å feature could be reduced with greater clumping. Is that relevant to Clayton and M31? He argues that complex filamentary structure occurs on scales of 1-100pc but that fluctuations are not important on scales smaller than 0.1pc.

Both Boissé and the fourth speaker, Eric Emsellem, used BVRI images of individual galaxies to determine parameters of the dust. Emsellem studied the Sombrero galaxy and concluded that a dust ring accummulated at the Outer Lindblad Resonance.

The third speaker, Nick Kylafis, studied the dust content of an edge-on galaxy through use of BVI images. He solved for the 10 parameters of a model with bulge –stellar disk – and gas disk components. He found that the radial scale lengths of stars and gas are similar and that the galaxy would be transparent if viewed face-on.

Ian Glass was the last speaker before coffee and he turned the discussion to infrared observations of the circumnuclear region of an active galaxy, NGC1068. It was discovered that there has been a gradual increase in intensity of this region over the last $\sim$ 15 years, with progressively larger amplitudes at longer wavelengths. Fluxes at K and L have approximately doubled over this time and the interval implies a dimension of $\sim$ 5 pc for the variable region.

Richard Griffiths was the sole speaker after coffee and he talked about the faint galaxy excess counts that have been coming out of HST observations, particularly with the Medium Deep Survey. He discussed both eyeball and machine compactness/asymmetry morphological characterizations. Some preference was indicated for the fading dwarf model. He mentioned issues like evolution of the brightest ellipticals, merger rates and constraints on cosmological parameters but I will not hold him to account.

5. Summary of Wednesday Morning Session, Jan 24 1996: Chair– B. Elmegreen

This session concentrated on the morphology of disk galaxies. The first speaker, Dr. Bertin, emphasized the implications of the modal theory of spiral structure in combined gas+star disks. K band observations have shown the symmetric types of spirals that are expected to result from wave modes in old stellar disks. Bertin suggested that flocculent galaxies might have their expected high stellar dispersions confirmed by observations of a difference between the stellar and gaseous rotation speeds. Comments after the talk included the suggestion (M. Fall) that perhaps the velocity dispersion could be measured directly for face-on galaxies (response: the in-plane component alone is important), and an inquiry (N. Brosch) about the importance of galaxy interactions in generating spirals.

The second talk, by Dr. Gonzalez, pointed out the existence of color gradients perpendicular to spiral arms, using a reddening-free index. Such gradients have been sought for a long time, but previous claims were questioned because of extinction and other effects. Dr. Gonzalez converted the observed color gradient into a star formation age gradient and derived the galaxy pattern speed. A comment after the talk (B. Elmegreen) suggested that color gradients might be expected even without star formation triggering because the gas and young star response to the spiral tends to peak along the inner edge of the old stellar response. Brosch also noted that the region with the color gradient had no $H\alpha$, whereas aging stars should include AB stars too, and these would make $H\alpha$. He suggested that diffuse $H\alpha$ would influence the reddening-free parameter as well.

The talk by Dr. Grosbøl considered non-linear spiral arm shapes using Fourier transform techniques, which predict variations in the ratio of the m=4 and m=2 components with galactic radius. Dr. Grosbøl also noted that I band spirals tend to be more tightly wrapped than K band spirals in the same galaxy.

Dr. Courteau discussed radial light profiles in galaxies and questioned the generality of the de Vaucouleurs form for bulges. Double exponential fits were equally good or better in many cases. Discussion followed about the apparent similarities between bulges and elliptical galaxies and what this means if bulges in spirals can be fitted to exponentials.

Dr. Calzetti modelled the dust distribution in starburst galaxies, using the ratios of H lines to determine the color excesses. She suggested that the bursts appear as if they were in the foreground, with stars and gas not co-spatial. A comment afterward suggested that FIR data could be used to help determine the relative position of dust+stars.

6. The Wednesday Afternoon Session, Jan 24 1996: Chair – J. Lequeux

[Editorial Comment: Professor Lequeux regrets that he did not take enough notes at the time to send in the summary for his Wednesday afternoon session, which he Chaired].

7. Summary of Thursday Morning Session, Jan 25 1996: Chair– N. Brosch

This session brought together various aspects of dust, as manifested by near-infrared (NIR) to far-IR (FIR) photometry, mapping, spectroscopy and their correlation with other observables, such as $H\alpha$ emission.

No real controversy was evident when Rolf Chini showed that many galaxies have sizable "cold" dust components. His mm measurements demonstrated that the dust temperature changes among different objects: whereas in normal galaxies, the cold dust is really cold ($\sim 15K$), it gets warmer the more active a galaxy is. Amazingly, in high redshift galaxies this dust is hottest (60K to 100K). Chini concluded that the "best" and only indicator of activity is the ratio of FIR luminosity to gas mass.

Andreani used mm observations to indicate that all the sub-mm background could be explained as emission from galaxies. However, there was some objection to her deconvolution of the spectral energy distributions, which used the IRAS values and the single point in the mm range to deduce up to three black-body components of the FIR emission. Ward-Thompson, in particular, remarked that the de-composition is far from unique. Andreani agreed that this is so, but she presented the results as "indicative".

Nikolaus Neininger showed beautiful images at 1.2 mm, obtained on the IRAM 30m telescope with a multi-beam bolometer, of edge-on spirals. His results showed that the dust follows faithfully the HI distribution, including the disk warp in NGC 4565. Interestingly, the cold dust detected at 1.2 mm does not always coincide with the CO emission. In N4565 the dust distribution is much wider (spatially) than that of the CO, and whereas the dust temperature at the center of this galaxy is 18K, it is only 15K at the peak of the HI distribution. This work pointed out vividly some of the difficulties inherent in the assumption that the CO traces well the dust.

Barbara Cunow showed us that galaxy disks are probably optically-thick near their centers, but probably optically-thin elsewhere. This refers to optical thickness at V. She used surface photometry of plate material in four bands and model fits, which consist of a simple galaxy (disk and bulge, disk of dust) and an assumed Galactic extinction law.

After the coffee break, Nick Devereux tried to convince us that the FIR emission in spiral galaxies is mostly dust heating by early-type stars. He showed that $H\alpha$ images, convolved to the HIRES IRAS angular resolution, appear very similar to the FIR images. The audience had various objections to this proposition. For example, Paul Goudfrooij asserted that, by mixing the $60\mu m$ emission with higher weight (2.35x) than the $100\mu m$ emission, Devereux is automatically biasing his FIR images to hotter dust, which is found near the OB stars. George Carruthers pointed out that both in our Galaxy and in the Magellanic Clouds there are O and B stars devoid of obvious HII regions. Thus, while a correlation between $L(H\alpha)$ and $L(FIR)$ appears to be established for galaxies, the correlation between $L(UV)$ and $L(FIR)$ is less well established. He proposed to measure extinction to O and B stars without HII regions and compare it with the FIR emission from the same locations to derive the distance of the dust from the star.

Yvonne Pendleton showed comparisons of near-IR spectra with laboratory analogs. She concentrated mostly on the $3.4\mu m$ absorption feature, which appear ubiquitously in the Galaxy in the diffuse ISM. Some matches with the EUREKA long-exposure samples were noted, but to accept these as identifications does require confirmatory observations in the $5-10\mu m$ region. This would help in understanding the rôle of UV processing of mantles. An indication that the core (Si)-mantle (organic) model of Greenberg for dust grains may be correct comes from the correlation of C-H and Si$-$0, which both increase toward the Galactic Center.

Gillian Wright showed us beautiful NIR spectra of active galactic nuclei (AGNs) which all show the $3.4\mu m$ absorption feature to different degrees. In particular IRAS 08572+3915 has the feature 4 × deeper than that to the Galactic Center, and even the sub-structure of the band appears to reproduce well. This indicates that the CH_2-CH_3 ratio is similar to that of the Milky-Way.

Despite the beautiful absorption and the existence of a dust lane just over the nucleus of this IRAS galaxy, the location of the absorbing matter is not yet clear. Is it in the thick accretion disk around the "central engine", in the "narrow-line region"? Duccio Macchetto does not think so. The identification of this location will help in identifying the accreted matter onto the giant black hole: stars (giants, supergiants) or molecular clouds. Among the galaxies studied, no correlation is evident between the $3.4\mu m$ and the $9.7\mu m$ features, contrary to the Galactic ratio of ~ 3.

8. Summary of Thursday Afternoon Discussion, Jan 25 1996: Chair–H. Butcher

Discussion following Fairall paper

QUESTIONS FROM THE FLOOR, ANSWERS BY FAIRALL

Q. Have you tried re-calculating the optical dipole with the galaxies corrected for extinction?

A. Not as yet—although as yet, we have only made a general correction for the A3627 galaxies and not for all the other galaxies partially observed by the Milky Way.

Q. Have you taken any deep photos of the cluster? Is there any evidence of lensing?

A. No, we have only the sky survey to work on, but we have started a program of CCD imaging of individual galaxies in the cluster. But, given the extinction, detection of lensing images would be difficult.

Q. If the cluster coincides with the Great Attractor, do you see any infall from the far side?

A. The cluster is very close to the predicted direction and redshift of the Great Attractor, but has insufficient mass to be the Great Attractor itself. However, as yet we are only measuring the core of the cluster itself—without independent distance estimates, we cannot as yet say whether we see any far side infall.

Q. Would it not have been better to use a near-IR sky survey when searching for these galaxies?

A. I understand that Reneé Kraan-Korteweg did explore this, but still found the IIIa-J survey better.

Q. Is there a cD galaxy in your cluster?

A. One of the central giant ellipticals is a well-known radio source; it could well be a cD galaxy. The extinction of course makes it difficult to see the extended envelope.

Comments by Ron Allen following the talk by De Geus:

1 *On the CO absorption measurements in the Galaxy:*

We were aware of the limitations of single dishes compared to interferometers for measuring absorption, and in fact we checked the results on 2013+370 (Lequeux, Allen, Guilloteau 1993, A&A 280, L23) using the interferometer at Plateau de Bure before publishing. Your observations of 0727 show we should have checked that one too! However, your data on that source were taken about 1.5 years later, and there is the real possibility of time variations. Frail and his collaborators have detected very small-scale structure in the ISM from time variations in the HI absorption spectra of pulsars. The variations in $\tau(\mathrm{HI})$ are as much as 30%, and I would expect the variations in $\tau(\mathrm{CO})$ to be even larger. In any case, as Dr. Lequeux has already mentioned, our interpretation of the CO and HCO+ absorption has changed and we no longer think this is good evidence for cold gas in the outer Galaxy.

2 *On cold gas in the inner disk of M31:*

I agree that the medium in these dust clouds will be clumpy. However, owing to the low UV flux in these parts of M31, we think the gas is dominated by low-density extended envelopes surrounding the clumps. In that case the CO lines will be optically thick with a high filling factor. This is consistent with our data and our calculations. In environments with higher UV fluxes the low density gas is dissociated leaving the higher-density cores exposed.

The models we have used are certainly not unique, but we have tried to be realistic and put in reasonable estimates for the sources of heat, i.e. UV photons and Cosmic Rays. It is true that low values of heat input of course lead to low temperatures. But if a high-temperature model is to be reasonable, then one has first to identify the heat source!

Question from Audience to Goudfrooij:

Your inference that there is a uniformly distributed component of the dust is based on the ratio

$$\frac{\mathrm{M(IRAS)}}{\mathrm{M(B-V)}} = 10$$

How can you get the mass from the (B-V) colour when you observe dust in the centre and you have a long column of stars in front of it? This is a reversed screen model, where all of the stars are in front of the dust.

Answer by Goudfrooij:

The dust mass of dust lanes from their colour excess is measured by assuming that $\sim 50\%$ of the stars in the galaxy in question is in front of the dust, at any line-of-sight. This will slightly under-estimate dust masses of very extended dust lanes, but anyway the uncertainty is only 50% at most.

9. Summary of the Friday Morning Session, Jan 26 1996: Chair– F. Macchetto

The session was opened with a review talk by Lequeux, who discussed the "Dust and Gas in the Outer Parts of the Galaxies". He reviewed the evidence for the existence of dust in the outer disks and the halos of spiral and irregular galaxies through different methods: direct detection of the thermal emission or indirectly through extinction of a background source. He described the current (KAO, mm wave) and future prospects (ISO, SOFIA, FIRST) for these studies. He discussed the statistical studies, based on the change of colours with galaxies' inclination to the line-of-sight and the somewhat still controversial results of the detection of opaque inner disks. He discussed in detail the results for M31, where they find a large dust/gas ratio and evidence for star formation (B stars) 116 arc min away from the center of the galaxy. He stated that in the SMC there is no clear evidence for dust detection, and there are controversial results for the halos of other galaxies (NGC 2835, NGC 3521) for which Lequeux would put upper limits of 3×10^8 solar masses (in the dust components). His main conclusion is that there is good evidence for the existence of dust in the disks and weak (or negative!) for the halos of spiral galaxies.

Schaefer discussed the FIRAS-COBE continuum spectrum and showed that a two component model is necessary to fit the observations. After fitting a dust model, he fits the remaining spectrum with molecular H at 12 degrees K, and finds agreement over 3 orders of magnitude. By correlating molecular and atomic H, he derives a density of 4.5×10^{18} cm^{-3}: which he claims can easily be *'buried'* in fractal structures in gravitationally bound fragments.

Pfenninger proposed a clumpy model for the structure of the ISM. He discussed the physical state of the ISM, showed that it is not in thermal equilibrium, reminded us that gravity plays an important role and that different scales are at work simultaneously. Current models for the ISM are based on the study of individual "clumps", and he proposed a description of the cold ISM based on a scale invariant geometry or fractal hierarchical structure, where clumps are, themselves, made-up of 5-10 subclumps over a range of 10^4 in scale.

Carignan explored the neutral hydrogen distribution. He reminded us that the dark-matter does not correlate with the luminous stellar distribution as determined from the optical rotation curves. He described a number of recent results where the dark-matter has been modeled from the 21cm observations and found that a typical ratio for dark-to-luminous matter is 35±20 for a sample of 20 galaxies.

A question from Rich to Carignan: "The dwarf spheroidal galaxies have some of the strongest evidence for dark matter, yet in general have no HI detection. This is especially true for Draco and Ursa Minor, which have some of the best established dark matter detections. If dark matter is some component of the ISM, and uniformly related to HI, why does this correlation not extend to the dwarf Spheroidal range?"

Response by Carignan: "What you mention for the DSph also apply to ellipticals which have very little or no gas. This doesn't mean that ellipticals have no dark matter. The correlation I mentioned in my talk only shows that the HI gas and the dark matter appear to be distributed similarly. This doesn't imply necessarily that dark matter is some component of the ISM, as you put it."

Combes described a candidate for the baryonic component of the dark matter of the Universe, namely extremely cold molecular gas in quasi-isothermal equilibrium with the cosmic-background radiation. This gas is "hidden" in a fractal structure, whose behavior is typically non-collisional at large scales, while it becomes dissipative at small scales. She described possible ways of detecting the existence of this gas, eg through absorption at UV wavelengths using QSO's as background souces. She showed that the distribution of gamma-ray emission is compatible with an extended distribution of the gas. She concluded by stating that "dark" baryons are required by current observations and that in her view the best candidates are not MACHOS, but a diffuse cold H component.

B. Elmegreen derived very interesting physical constraints on the surface density of the proposed diffuse component of dark-matter. He used a stability analysis to put limits on the values of the surface density of gas that would reproduce the observed rotation curves without introducing observable instabilities. Amongst his conclusions is that dark matter should exist in dense clumps.

Graham discussed the detection of dust and ice-band features around newly formed stars. These observations are difficult in that the different dust zones must be located before the dust disperses. He discussed IR spectroscopic observations of young stars and Herbig-Haro objects.

Cruikshank showed us that the interstellar dust can be detected directly as it passes through our Solar System and the primordial component can be detected in the meteorites and fragments of comets. He discussed detection of dust particles by the Ulysess and Galileo spacecrafts which show that these particles are coupled to the neutral interstellar gas flowing through the Solar System. He also showed the presence of organic compounds in meteorites. Of particular interest was the description of the sources of the different compounds: SiC and graphite in red-giants and the pulse phase of AGB, diamonds in SN explosions, etc.

Ward-Thompson discussed mm and sub-mm observations of pre-stellar cores. He showed that cores within a gravitationally bound molecular cloud have lifetimes of order 10^6 years, whereas the free-fall time is only 10^4 years. As a consequence, these cores cannot be in free-fall. His measurements show that the radial temperature profiles are flat and consistent with external heating. He proposed an ambipolar diffusion model and a magnetically supported pre-stellar core.

A long and very interesting discussion followed these excellent presentations.

My quote for the morning:

"Diamonds are girls' best friends ... are made in supernovae explosions ... and are found in South Africa!". Wow!

10. Summary of the Discussions in the Friday Afternoon Session, Jan 26 1996: Chair–A.P. Fairall

A lively discussion followed the Friday afternoon session.

Chris Impey asked Paul Francis: "Considering that flat-spectrum radio quasars have stronger synchrotron emission at optical and infrared wavengths compared to optically selected quasars – how could he then be sure that the redder B-K colours and the larger scatter were not due to nonthermal emission and variability?" Francis' reply was that, if the redness and scatter were caused by a strong synchrotron, one would expect it to dilute the emission lines – and no such dilution is seen. Also, the BL Lacs in his sample have colours identical to the rest of the sample.

Paul Goudfrooij asked Andrea Cimatti if he could say anything about dust distribution: whether the (patchy) light that is seen along the radio axis was indeed due to dust scattering. Cimatti's reply was that the polarized light extended over the whole of the high-redshift radio galaxy, thereby suggesting that the scattering particles are distributed and diffused in the ISM. However, he emphasised that one does not know the relative spatial distributions of dust and electrons.

Chris Impey was asked: "for how many galaxies in his sample did dust extinction in disks play an important part in selection?" Impey, however, was not convinced that disks existed in most of the low surface brightness galaxies. The radial profiles are generally exponential, but disk morphologies are not usually seen.

Andrea Cimatti questioned Paul Francis as to whether line emission in the K-band could be responsible for the red colours. Francis replied that it would only contribute a few tenths of a magnitude of scatter, which would be negligible for his purpose. Cimatti also asked how the redshift distribution of the Parkes and LBQ sample compared. Francis said they were identical.

Brent Tully remarked to Chris Impey that he does not believe that low surface brightness galaxies make a significant contribution to the baryonic content of the universe. Otherwise, one would expect to find much more of them in HI. Impey pointed out there was a big scatter in the HI luminosity function. Also Mike Disney disagreed with Tully, saying radio telescopes were not sensitive to low column densities, and 24-hr integration times might be needed. What the radio telescopes detect so far can easily be seen on the sky surveys. However, Patricia Knezek remarked that an HI survey (by John Spitzak, University of Massachusetts) to 7000 km/s was finding one catalogued galaxy for every catalogued HI galaxy *as well* as a population of nearby very low mass HI objects.

There were still many raised hands when discussion time ran out. The session chairman apologises that a question asked by George Carruthers to Patricia Knezek, and one to Michael Fall (on reducing the submillimetre background) are not reflected in the text above, as inadequate detail was recorded at the time.

THE RIDDLE...

At the Conference finale: the Braaivleis

BRUCE ELMEGREEN
IBM Watson Research Center, NY

I have a riddle:

What is it that has enormous mass, is always hidden, and is well studied in South Africa?

Cold dust, of course! Unless it is the hippo....

Figure 1. The Hippo: Presented to the SOC Chair by Bruce Elmegreen and Mayo Greenberg, on behalf of the Session Chairpersons and 'Eyes to the Future' Panelists.

626

D. L. Block and J. M. Greenberg (eds.), New Extragalactic Perspectives in the New South Africa, 626.
© 1996 *Kluwer Academic Publishers.*

'WHERE DUST IS STARS'

DAVID L. BLOCK
University of the Witwatersrand

It is fitting to conclude this volume never forgetting that sense of wonder, and that sense of awe, which observers feel time and time again when, perhaps on clear but cold nights, they step out onto a catwalk ... or step outside the Observatory building ... to gaze upon the Milky Way.

I quote from E.E. Barnard himself [1]:

"My telescope constantly swung back to the Milky Way, again to gaze on the 'broad and ample road where dust is stars.' So enraptured was I with these glimpses of the Creator's works that I heeded not the cold nor the loneliness of the night. And when the approaching dawn began to whiten the eastern skies, I sought out the great planet Jupiter, then only emerging from the solar rays, and beheld with rapture his four bright moons and vast belt system. But when the dawn had paled each stellar fire the coldness of the night impressed itself upon me and I retired from the field of glory."

References

[1] Quoted from *The Immortal Fire Within: The Life and Work of Edward Emerson Barnard* by William Sheehan: Cambridge University Press, 1995.

D. L. Block and J. M. Greenberg (eds.), New Extragalactic Perspectives in the New South Africa, 627–628.
© 1996 *Kluwer Academic Publishers.*

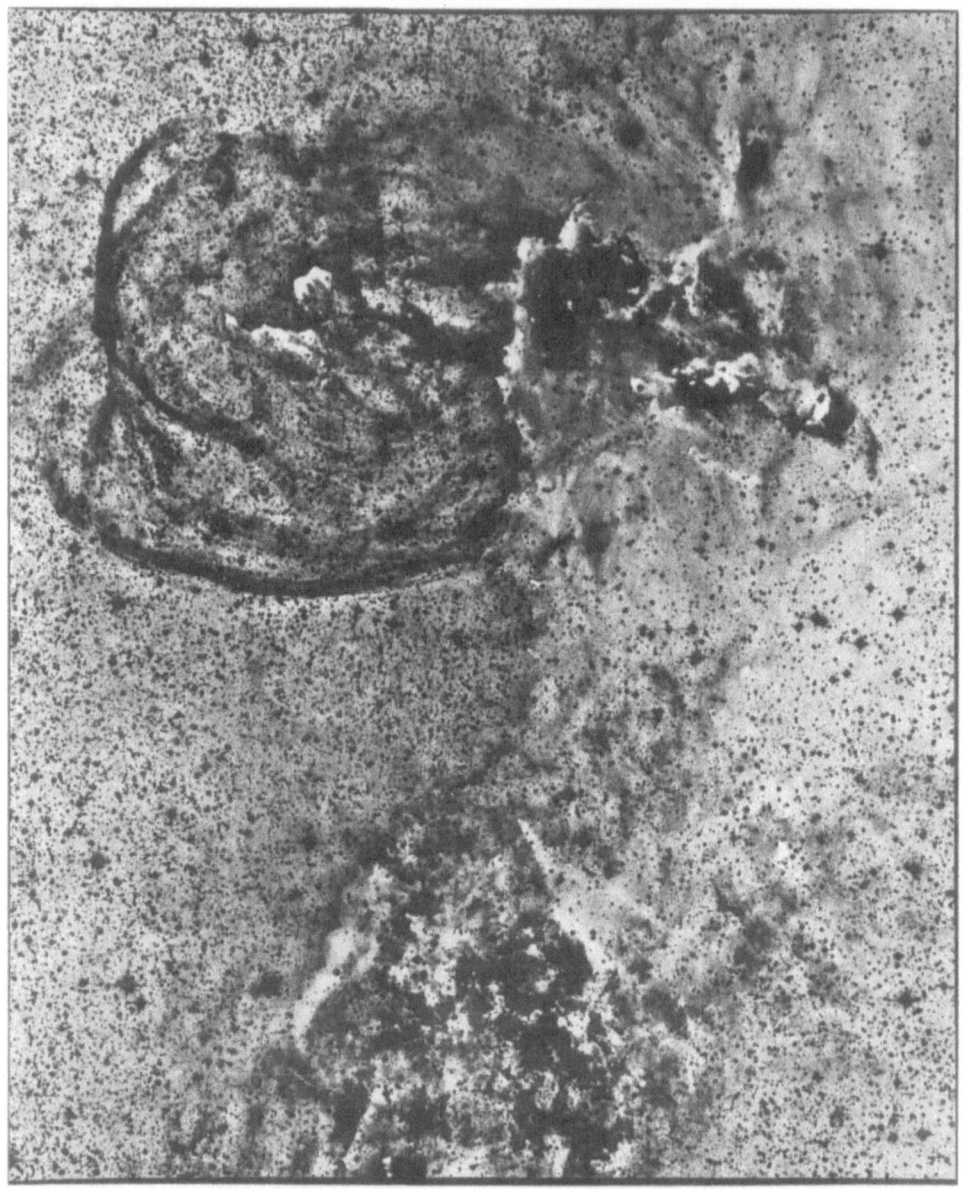

Figure 1. 'That field of glory: the Milky Way'. Areas spawning optically thick domains of dust and gas in our Milky Way, photographed from La Silla, with the ESO Schmidt telescope. Observers: D. Block and G. Pizzarro.

OUR LIST OF PARTICIPANTS

DK AITKEN
  5 Dewsbury Cottages, York YO1 1HB, UK.

L ALBERTS
  P.O. Box 35705, Menlo Park 0102, Gauteng, South Africa.

RJ ALLEN
  Space Telescope Science Institute, 3700 San Martin Drive, Baltimore, MD 21218, USA.

PM ANDREANI
  Dipartimento di Astronomia, Universita' Padova, Vicolo dell'Osservatorio 5, I-35142 Padova, Italy.

E ATHANASSOULA
  Observatoire de Marseille, 2 Place le Verrier, 13248 Marseille Cedex 04, France.

B BARSELLA
  Pisa University, Physics Dept., Piazza Torricelli 2, I-56126, Italy.

G BERTIN
  Scuola Normale Superiore, Piazza dei Cavalieri, I-56126 Pisa, Italy.

DL BLOCK
  Dept. of Computational & Applied Mathematics, University Witwatersrand, Private Bag 3, 2050 Wits, South Africa.

P BOISSÉ
  Ecole Normale Superieure/Radioastronomie, 24 rue Lhomond, 75231 Paris Cedex 05, France.

A BOSMA
  Observatoire de Marseille, 2 Place le Verrier, 13248 Marseille Cedex 04, France.

N BROSCH
  School of Physics & Astronomy, Beverly & Raymond Sackler Faculty of Exact Sciences, Tel Aviv University, Tel Aviv 69978, Israel.

GC BURIC
  Dept. of Computational & Applied Mathematics, University Witwatersrand, Private Bag 3, 2050 Wits, South Africa.

HR BUTCHER
  Netherlands Foundation for Research in Astronomy, P O Box 2, 7990 AA Dwingeloo, The Netherlands.

D CALZETTI
  Space Telescope Science Institute, 3700 San Martin Drive, Baltimore, MD 21218, USA.

C CARIGNAN
Université de Montréal, Department de physique, C.P. 6128, Succ. "Centre-Ville", Montréal, Québec H3C 3J7, Canada.

GR CARRUTHERS
E.O. Hulburt Center for Space Research, US Naval Research Laboratory, Code 7609, Washington DC 20375-5320, USA.

R CHINI
MPI für Radioastronomie, Auf dem Hügel 69, D-53121 Bonn, Germany.

A CIMATTI
Osservatorio Astrofisico di Arcetri, Largo E. Fermi 5, I-50125 Firenze, Italy.

GC CLAYTON
Center for Astrophysics & Space Astronomy, University of Colorado, Campus Box 389, Boulder, CO 80309, USA.

F COMBES
DEMIRM, Observatoire de Paris, 61 Av. de l'Observatoire, F-75014 Paris, France.

RLM CORRADI
Instituto de Astrofisica de Canarias, Via Lactea S/N, 38200 La Laguna, Tenerife, Espanã (Spain).

S COURTEAU
NOAO/KPNO, 950 N. Cherry Ave., Tucson, AZ 85726, USA.

DP CRUIKSHANK
NASA Ames Research Center, MS 245-6, Moffett Field, CA 94035-1000, USA.

I CRUZ-GONZALEZ
UNAM, Instituto de Astronomia, Apartado Postal 70-264, Ciudad Universitaria, Mexico D.F. 04510, Mexico.

B CUNOW
Dept Mathematics, Appl. Mathematics & Astronomy, University of South Africa, PO Box 392, Pretoria 0001, South Africa.

JI DAVIES
University of Wales College of Cardiff, Dept of Physics & Astronomy, P O Box 913, Cardiff CF2 3YB, UK.

EJ DE GEUS
Kluwer Academic Publishers, Spuiboulevard 50, 3300 AA Dordrecht, The Netherlands.

OC DE JAGER
Space Research Unit, PU vir CHO, 2520 Potchefstroom, South Africa.

DL DEPOY
Astronomy Dept, Ohio State University, 174 W. 18th Avenue, Columbus, OH 43214, USA.

N DEVEREUX
New Mexico State University, Astronomy Dept., Box 30001/Dept 4500, Las Cruces, NM 88003, USA.

M DISNEY
University of Wales College of Cardiff, Dept of Physics & Astronomy, PO Box 913, Cardiff CF2 3YB, UK.

WW DULEY
Physics Dept., University Waterloo, Waterloo, Ontario N2L 3G1, Canada.

BG ELMEGREEN
IBM Research Division, T.J. Watson Research Center, P O Box 218, Yorktown Heights, NY 10598, USA.

DM ELMEGREEN
Vassar College, Dept Physics & Astronomy, Poughkeepsie, NY 12601, USA.

E EMSELLEM
European Southern Observatory, Karl-Schwarzschild Straße 2, D-85748 Garching, Germany.

AP FAIRALL
Dept. of Astronomy, Univ. of Cape Town, Private Bag, Rondebosch 7700, South Africa.

M FALL
STSCI, 3700 San Martin Drive, Baltimore, MD 21218, USA.

MW FEAST
Dept of Astronomy, Univ of Cape Town, Private Bag, Rondebosch 7700, South Africa.

DCD FERREIRA
Euginia Villa 203, Johnston Street 80, Sunnyside 0002, Gauteng, South Africa.

MM FITZGERALD
The Planetarium, University Witwatersrand, Private Bag 3, Wits 2050, South Africa.

DA FORBES
Lick Observatory, University of California, Santa Cruz, CA 95064, USA.

PJ FRANCIS
University of Melbourne, School of Physics, Parkville, Victoria 3052, Australia.

JA FROGEL
Dept of Astronomy, The Ohio State University, 174 West 18th Ave., Columbus, Ohio 43210, USA.

F GARZÓN
Instituto de Astrofísica de Canarias, C/via Láctea S/N, E-38200 La Laguna, Tenerife, Espanã (Spain).

T GEBBIE
Dept Applied Maths, University of Cape Town, Private Bag, Rondebosch 7700, South Africa.

IS GLASS
South African Astronomical Observatory, P O Box 9, Observatory 7935, South Africa.

RA GONZÁLEZ
Astronomy Dept., University of California, Berkeley, CA 94720, USA.

P GOUDFROOIJ
European Southern Observatory, Karl-Schwarzschild-Straße 2, D-85748 Garching, Germany.

JA GRAHAM
Dept of Terrestrial Magnetism, Carnegie Institution of Washington, 5241 Broad Branch Rd NW, Washington DC 20015, USA.

JM GREENBERG
Laboratory Astrophysics, University of Leiden, Postbus 9504, 2300 RA Leiden, The Netherlands.

632

M A GREENHOUSE
National Air & Space Museum, Smithsonian Institution, MRC 321, Washington DC 20560, USA.

RE GRIFFITHS
Dept. of Physics & Astronomy, Johns Hopkins University, 3400 Charles St. N, MD 21218, USA.

P GROSBØL
European Southern Observatory, Karl-Schwarzschild-Straße 2, D-85748 Garching, Germany.

P HODGE
Astronomy Dept., University of Washington, Box 351580, Seattle, WA 98195-1580, USA.

WK HUCHTMEIER
Max-Planck-Institut für Radioastronomie, auf dem Hügel 69, D-53121 Bonn, Germany.

NH IBRAGIMOV
Dept Computational & Applied Maths, University Witwatersrand, Private Bag 3, WITS 2050, South Africa.

CD IMPEY
Steward Observatory, University of Arizona, Tucson, Arizona 85721, USA.

RJ IVISON
Royal Observatory, Blackford Hill, Edinburgh EH9 3HJ, Scotland.

A KASHLINSKY
NORDITA, Blegdamsvej 17, DK 2100 Københavnø, Denmark.

PM KNEZEK
OCIW/Las Campanas Observatory, Casilla 601, La Serena, Chile.

LE KUCHINSKI
Dept of Astronomy, Ohio State University, 5036 Smith Laboratory, 174 W. 18th Ave., Columbus, OH 43210-1106, USA.

ND KYLAFIS
Physics Dept., University of Crete, P O Box 2208, 714 09 Heraklion, Crete, Greece.

L LEEUW
Astronomy Dept, University Cape Town, Private Bag, Rondebosch 7700, South Africa.

J LEQUEUX
DEMIRM, Observatoire de Paris, 61 Ave. de l'Observatoire, F-75014 Paris, France.

FD MACCHETTO
Space Telescope Science Institute, 3700 San Martin Drive, Baltimore, MD 21218, USA.

JC MATHER
Infrared Astrophysics Branch, Code 685, NASA GSFC, Greenbelt, MD 20771, USA.

R MEDUPE
Astronomy Dept, University Cape Town, Private Bag, Rondebosch 7700, South Africa.

E MOMONIAT
Dept. of Computational & Applied Mathematics, University Witwatersrand, Private Bag 3, 2050 WITS, South Africa.

J MURTHY
Dept of Physics & Astronomy, The Johns Hopkins University, Baltimore, MD 21218-2695, USA.

N NEININGER
MPIfR, Auf dem Hügel 69, D-53121 Bonn, Germany.

PA PATSIS
Max-Planck Institut für Astronomie, Königstuhl 17, D-69117 Heidelberg, Germany.

YJ PENDLETON
NASA Ames Research Center, MS 245-3, Moffett Field, CA 94035-1000, USA.

D PFENNIGER
Geneva Observatory, University of Geneva, CH-1290 Sauverny, Switzerland.

MW REGAN
Dept of Astronomy, University of Maryland, College Park, MD 20742-2421, USA.

RM RICH
Columbia University, 538 W. 120th St., NY, NY 10027, USA.

RB ROUSE
Dept. of Physics & Astronomy, Arizona State University, 1011 E. Orange Apt. 66, Tempe, AZ 85281, USA.

S SAUTY
DEMIRM, Observatoire de Paris-61, Avenue de l'Observatoire, F-75014 Paris, France.

M SAUVAGE
Service d'Astrophysique, DSM/DAPNIA/SAP, C.E. Saclay, 91191 Gif sur Yvette CEDEX, France.

JH SCHAEFER
Max-Planck-Institut für Astrophysik, Karl-Schwarzschild-Str. 1, 85740 Garching, Germany.

J SCHOCHOT
Dept. of Information Systems, University Witwatersrand, Private Bag 3, Wits 2050, South Africa.

K SELLGREN
Astronomy Dept., Ohio State University, 174 W. 18th Ave., Columbus, OH 43210, USA.

BJ SMITH
IPAC/Caltech, MS 100-22 Pasadena, CA 91125, USA.

UJ SOFIA
National Research Council, NASA/GSFC, Code 681, Greenbelt, MD 20771, USA.

WB SPARKS
Space Telescope Science Institute, 3700 San Martin Drive, Baltimore, MD 21218, USA.

GJ STACEY
Astronomy Dept., Cornell University, Ithaca, NY 14850, USA.

TY STEIMAN-CAMERON
NASA Ames Research Center, MS 245-3, Moffett Field, CA 94035-1000, USA.

RS STOBIE
South African Astronomical Observatory, P O Box 9, Observatory 7935, SA.

MD THORNLEY
Dept of Astronomy, University of Maryland, College Park, MD 20742-2421, USA.

QD Tran
Service d'Astrophysique, DSM/DAPNIA/SAP, C.E. Saclay, 91191 Gif sur Yvette CEDEX, France.

M Trewhella
Dept of Physics & Astronomy, University of Wales College of Cardiff, P O Box 913, Cardiff CF2 3YB, UK.

RB Tully
Institute for Astronomy, University of Hawaii, 2680 Woodlawn Drive, Honolulu, HI 96822, USA.

F Vawda
Dept Computational & Applied Maths, University Witwatersrand, Private Bag 3, WITS 2050, South Africa.

S von Linden
Landessternwarte, Königstuhl, D-69117 Heidelberg, Germany.

D Ward-Thompson
Royal Observatory, Blackford Hill, Edinburgh EH9 3HJ, Scotland.

RE White III
University of Alabama, Dept. of Physics & Astronomy, Tuscaloosa, AL 35487-0324, USA.

PA Whitelock
South African Astronomical Observatory, P.O. Box 9, Observatory 7935, South Africa.

RD Wolstencroft
Royal Observatory, Blackford Hill, Edinburgh EH9 3HJ, Scotland.

J Wolterbeek
Dept Mathematics, Appl. Mathematics & Astronomy, University of South Africa, P O Box 392, Pretoria 0001, South Africa.

GS Wright
Joint Astronomy Center, 660 North Aohoku Place, Hilo, HI 96720, USA.

C Xu
Max-Planck-Institut für Kernphysik, Postfach 103980, Saupfercheckweg 1, D-69117 Heidelberg, Germany.

THE COLOUR PLATE SECTION

Plate 1: The colour cover for our Conference programme was designed by Jennifer van Schoor. The magnificent stained glass window featured on the cover overlooks the entrance lobby of Anglo American's new Head Office building in Johannesburg. In developing the design for the window, Cape-based artist, Karl-Heinz Wilhelm, set out to symbolise *"... the vast and diversified nature of Anglo American, which represents a veritable galaxy of mining and other activities."* The Principal Sponsor for the Conference was the Chairman's Fund of Anglo American and its sister company De Beers.

Plate 2a and 2b: K' image of NGC 2841, secured at Mauna Kea by R. Wainscoat, plotted in the top frame (Plate 2a) as a greyscale for log(intensity), and in the bottom frame (Plate 2b) as false-colour and 3D-projection, also for log(intensity). In Plate 2b, made using the IBM Data Explorer with the help of Dr C. Pickover, we have placed our artificial light source to illuminate the N-NE sector of the disk to show the dark IR spirals (arrowed) clearly. The bulge is clipped in the center. Reproduced courtesy B.G. Elmegreen and *Nature*. See Block, page 1.

Plate 3: Plate 3a (top): Deprojected and enhanced image of NGC 2841 reveals remarkable long gas+dust spirals at K' spanning azimuthal angles of $\sim 160°$. Enhancement with the IBM Image Access Executive in this rgb image is by subtraction of the average radial profile, boxcar averaged over 100 pixels, and normalized to a constant rms at each radius. The white lines on the left and right are 100 pixels (70'') and $R_{25} = 4.1'$ respectively. The northeast minor axis is up. Light and dark spirals that open up in a clockwise sense are evident. Plate 3b (bottom): V and K' band enhanced images of log(radius) versus deprojected azimuthal angle (shown with two full cycles on the horizontal axis, where the ticmarks are spaced by 180°; the vertical ticmarks are 0.2 to 0.7 R_{25}). Clockwise in the top image is leftward in the bottom image. Reproduced courtesy B.G.Elmegreen. See Block, page 1.

Plate 4: Arcsecond spatial resolution of widely distributed cold dust in the spiral galaxy NGC 4736. A B-K' colour map, wherein positive enhancements in B-K' are colour coded black. Optical and NIR colour excesses are sensitive to dust grains of all temperatures. The bulk of the dust grains in NGC 4736 are at low enough temperatures to have been undetected by IRAS. Typical V optical depths in the interarm regions are of order 0.75. The HII regions are colour coded blue; two of them, A and B, permit pixel registration between optical CCD and near-infrared NICMOS images. K' image of NGC 4736 secured at Mauna Kea by Dr Alan Stockton; B image courtesy Dr P. Mulder. Generation of colour matrix (c_{ij}) to specifically highlight the dust was facilitated on a SUN Workstation by the author after selection of appropriate transform functions, and the final image was produced via the CIBA process. See Block, page 1.

Plate 5: To construct this colour image of NGC4038/9 and the following one of NGC 2442, uniformly scaled and smoothed images obtained through H, R, and

B filters by Dr. K. Sellgren and G. Tiede on the 1.5 m at CTIO were assigned to the red, green, and blue guns, respectively of a colour digital camera. Note the very striking difference in stellar content of the two galaxies. For further details, please see the captions to the grey scale images of these galaxies in the article by Frogel et al. page 65.

Plate 6: For details of this three colour image of NGC 2442, see the caption for Figure 1 in the paper by Frogel et. al, page 65. Note the extreme blueness of the arms and the dust "pennants" trailing away from the linear feature.

Plate 7: Deprojected, face-on maps of (V-K') and intensity variation relative to the axisymmetric background in K' for (a) NGC 3223, (b) NGC 5085 and (c) NGC 5861. See Grosbøl and Patsis, page 251.

Plate 8a (top): M31 WFPC2 image formed by dividing V by I. The incomplete quadrant contains the PC image, which samples at twice the WF resolution and is centered on the nucleus; the entire mosaic is 160 arcsec on a side. All of the major dust features were first imaged by Johnson & Hanna (ApJ, vol. 174, L71: 1972) and cataloged by Hodge (AJ, vol. 85, 376: 1980). Some patches have $A_V > 0.5$ mag and we suspect that there are some unresolved clumps of dust. N is indicated by the arrow. See Rich et. al. page 325.

Plate 8b(bottom): M32 WFPC2 image formed by dividing V by I, as in plate 8a above. Notice that no dust is visible. The nucleus is centered on the incomplete PC quadrant, as in plate 8a. See Rich et. al. page 325.

Plate 9: Dividing V by F1042M ($1\mu m$) we search for dust in the near nuclear environments. Each postage stamp is a 100×100 pixel subraster of the PC (at 0.046 arcsec per pixel) and therefore each side is 4.6 arcsec. From left to right: division image, V, and F1042M. From top to bottom: the double nucleus of M31, M32, M33, and the nucleus of the brightest globular cluster in M31, G1. All of these nuclei contain unresolved point sources, but in no cases (even in M31) does dust affect the morphology. See Rich et. al. page 325.

Plate 10: Colour maps of the M31 nucleus from restored pre-refurbishment images. The maps in the triangular region on the upper right are made by dividing the images corresponding to each "matrix element". The wavelength of the images correspond to 6220Å for example, for F622W, etc. The far right hand row of residuals is spurious because the F1042M (1 micron) images could not be restored. On the far left, a far UV (F175W=1750Å) image obtained with the faint object camera has been divided by F785LP. This is the only image with a believable residual and it confirms King's finding that the two peaks P1 and P2 reverse their brightness in the UV. This image shows that normal dust is not responsible for the double morphology. See Rich et. al. page 325.

Plate 11a and b: Plate 11a (top) shows a λ 1.2mm continuum emission map secured using the IRAM 30-m telescope, of the edge-on spiral galaxy NGC 4565. The dust emission not only shows a central peak and an inner ring like the CO, but also, like the HI, a weaker, extended plateau. Comparable to the neutral hydrogen, the 1.2 mm contours are warped near the NW edge of the galaxy. See Neininger and Guelin, page 349.

Plate 11b: R-band image of the giant low surface brightness galaxy Malin 3, with inset of M51 at same scale. Image is 140×140 kpc (Hubble constant=100). See Impey, page 505.

Plate 12: Images of 8 luminous elliptical galaxies. For each galaxy, the $B - I$ colour index distribution is shown in the upper image and the distribution of $H\alpha +$ [$N\ II$]-emitting gas is shown in the lower image. In the upper images, the colour white indicates the reddest regions and purple the bluest. The galaxies are (top left to bottom right): *NGC 3962, NGC 4125, NGC 4278, NGC 4589, NGC 4696, NGC 5018, NGC 5044,* and *IC 1459*. For each galaxy, the B-band isophotal contour at half a de Vaucouleurs effective radius (R_e) is superposed (except for *NGC 4696* and *NGC 5044*, where the contour at $R_e/4$ is superposed). Dust (indicated by reddened structures with a morphology different from the generally elliptical isophotes of stellar galaxy light) is found to be present in about 40% of luminous ellipticals. As a rule, these dust features are associated with ionized gas (and vice versa), and may be relics of accreted galaxies. Images obtained with the 1.54-m Danish telescope at the European Southern Observatory and the 1.0-m Jacobus Kapteyn Telescope at La Palma Observatory. The images are typically 1 arcmin wide. See Goudfrooij, page 400.

Plate 13: Fractal cloud models of dimension $D = 1.7$, 2.0, 2.5 and 3.0, from bottom left clockwise to bottom right. At each level of the fractal the number N of sublevels is calculated by a Poisson distribution of mean 8. The length ratio x between each adjacent level is given by $x = N^{1/D}$. The position of each sub-clump is distributed according to a truncated isothermal law. Five levels are calculated; the total number of clouds at the smallest scale is about 42 000. See Pfenniger, page 439.

Plate 14: The South African Astronomical Observatory (SAAO) has been involved in two major imaging surveys of the Inner Galaxy. In this photograph, three single-colour images are combined by R.M. Catchpole into a three-colour red, green and blue picture, corresponding to K, H and J filters. It is estimated that complete blackness on this image corresponds to a visual extinction exceeding 60 magnitudes. See Glass, page 536.

Plate 15: True colour image of the inner core (the central 4.6″ × 4.6″ area) of M51, obtained by combining B, V and R images. The maximum visual extinction in the Y-shaped dark lane is ∼ 2.1 mag. Dust destruction processes have prevailed over formation processes during the lifetime of the nuclear core. See Panagia et. al., page 552.

Plate 16: The Continent of Africa beckons: photographed in jeep are family Macchetto, Griffiths, Block and Greenberg, heading for the bush at the *Mala Mala* Game Reserve – with armed ranger and professional tracker; an encounter with buffalo, one of the *Big Five*; stuck in knee-deep mud; *Mala Mala* hats provide welcome shade under the African sun to Mayo, David and son Aaron.

[Editorial note: Below each plate to follow, we refer to the first author of the *paper* in which the plate is discussed].

638

JANUARY 22-26. 1996

NEW EXTRAGALACTIC PERSPECTIVES IN THE NEW SOUTH AFRICA

Changing Perceptions of the Morphology,
Dust Content and Dust-Gas Ratios in Galaxies

The Conference Programme Cover: Colour Plate 1

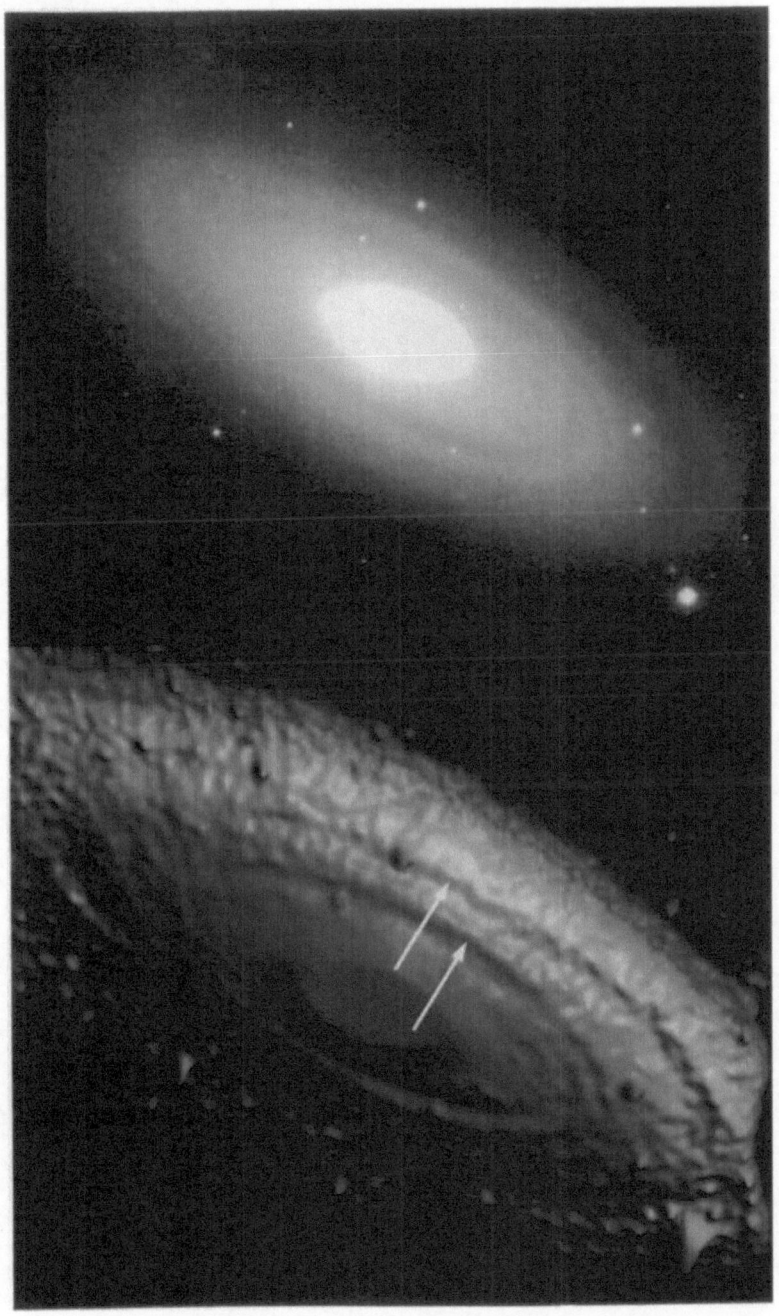

Block: Colour Plate 2a (top) and 2b (bottom)

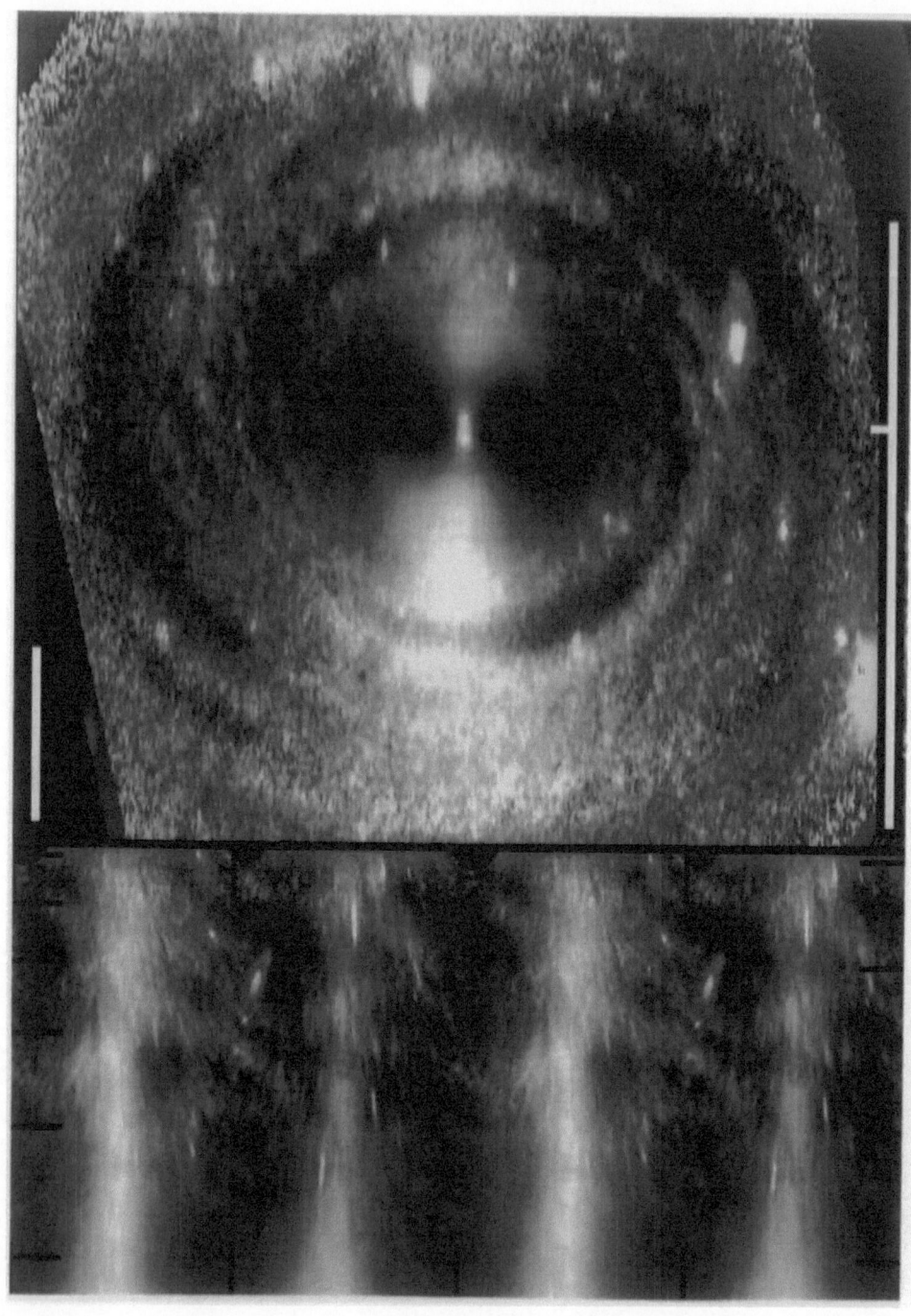

Block: Colour Plate 3a (top) and 3b (bottom)

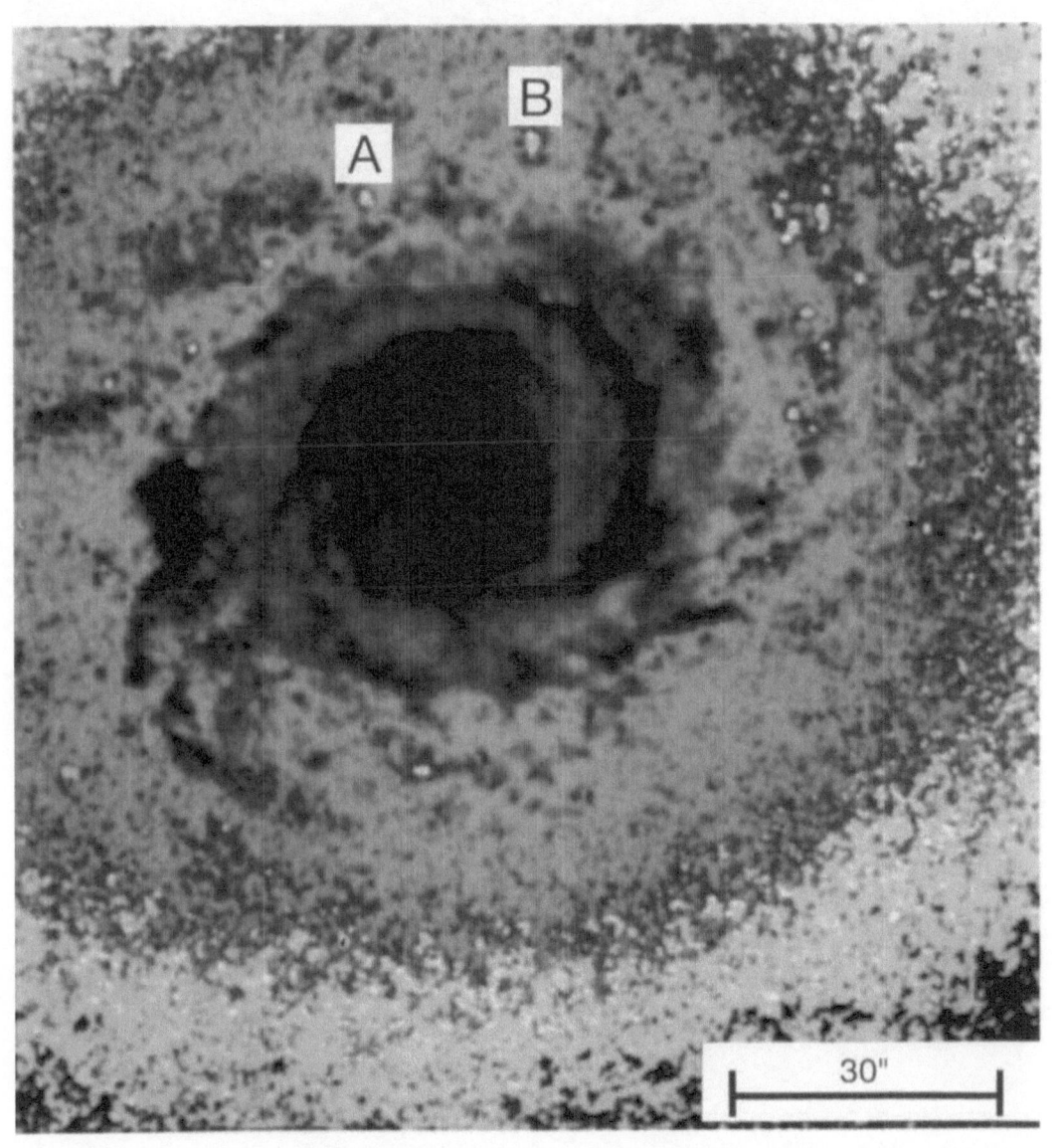

Block: Colour Plate 4

642

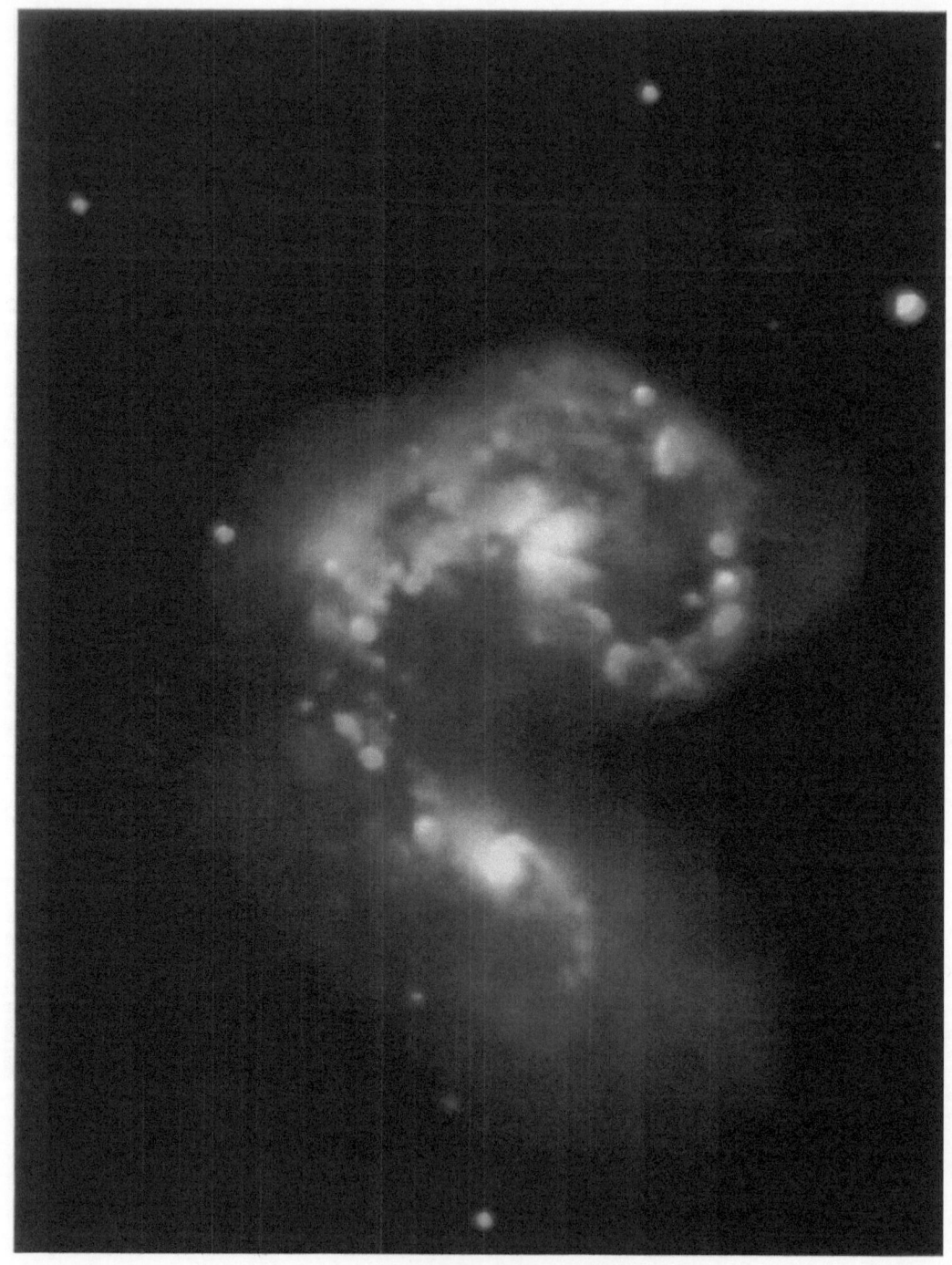

Frogel: Colour Plate 5

Frogel: Colour Plate 6

644

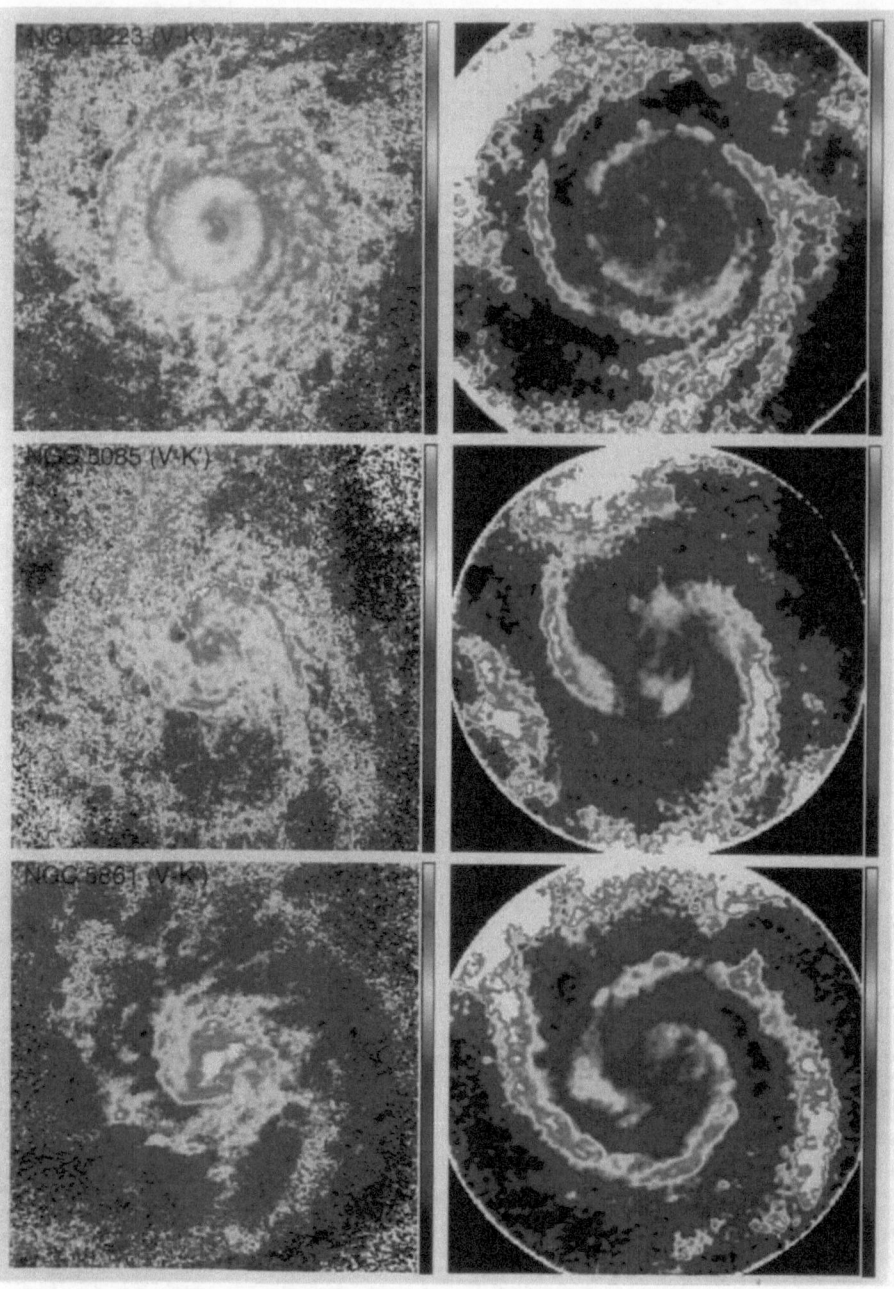

Grosbøl and Patsis: Colour Plate 7

Rich et. al.: Colour Plate 8a (top) and 8b (bottom)

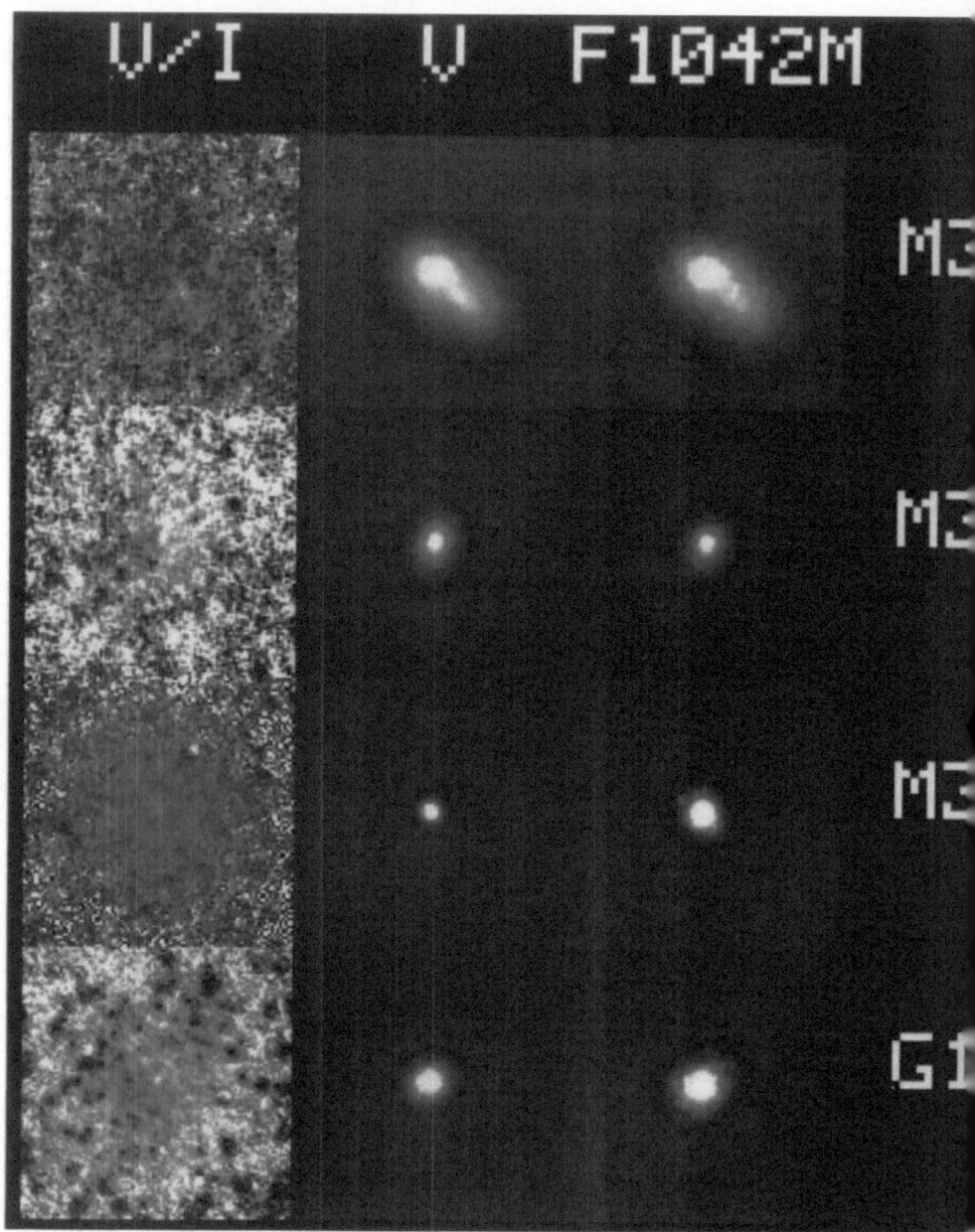

Rich et. al.: Colour Plate 9

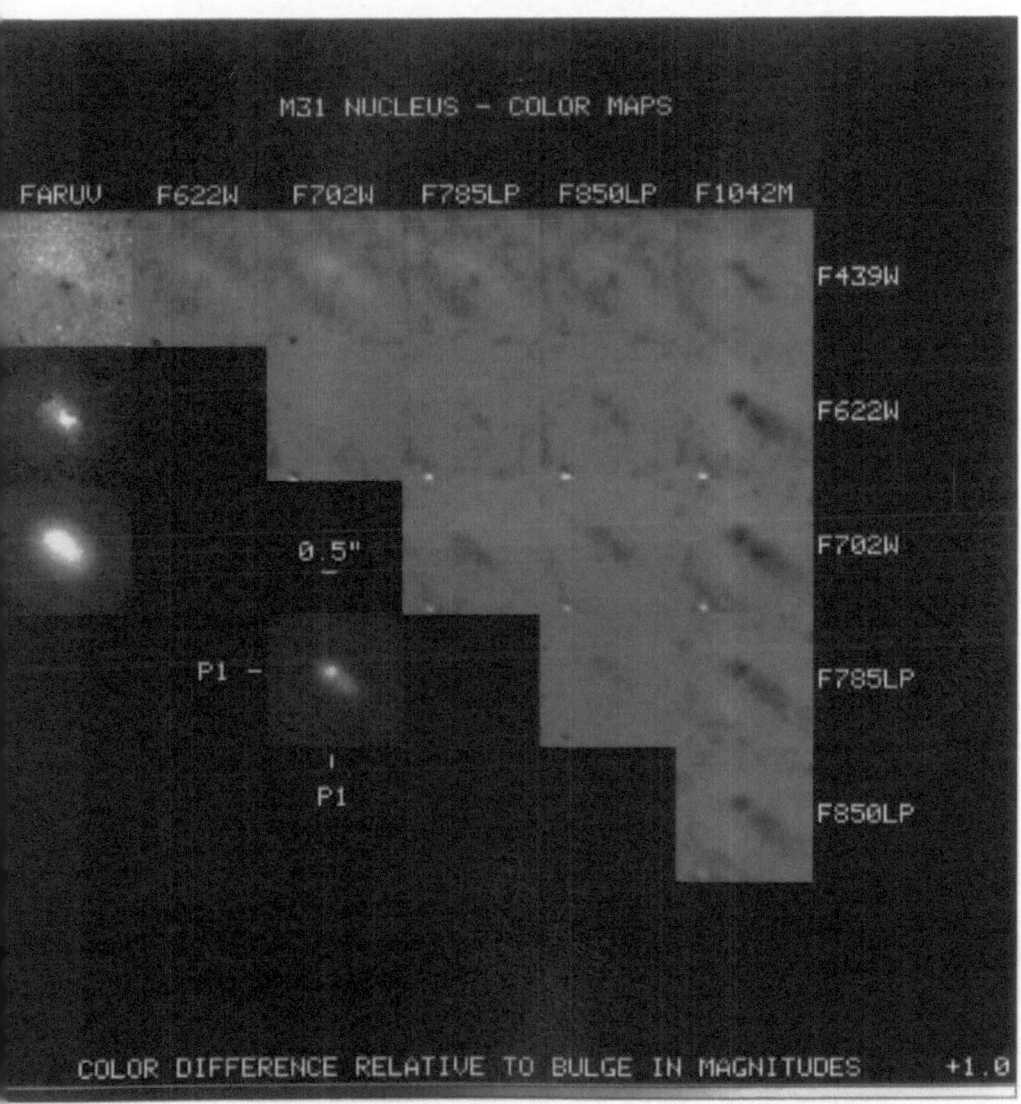

Rich et. al.: Colour Plate 10

648

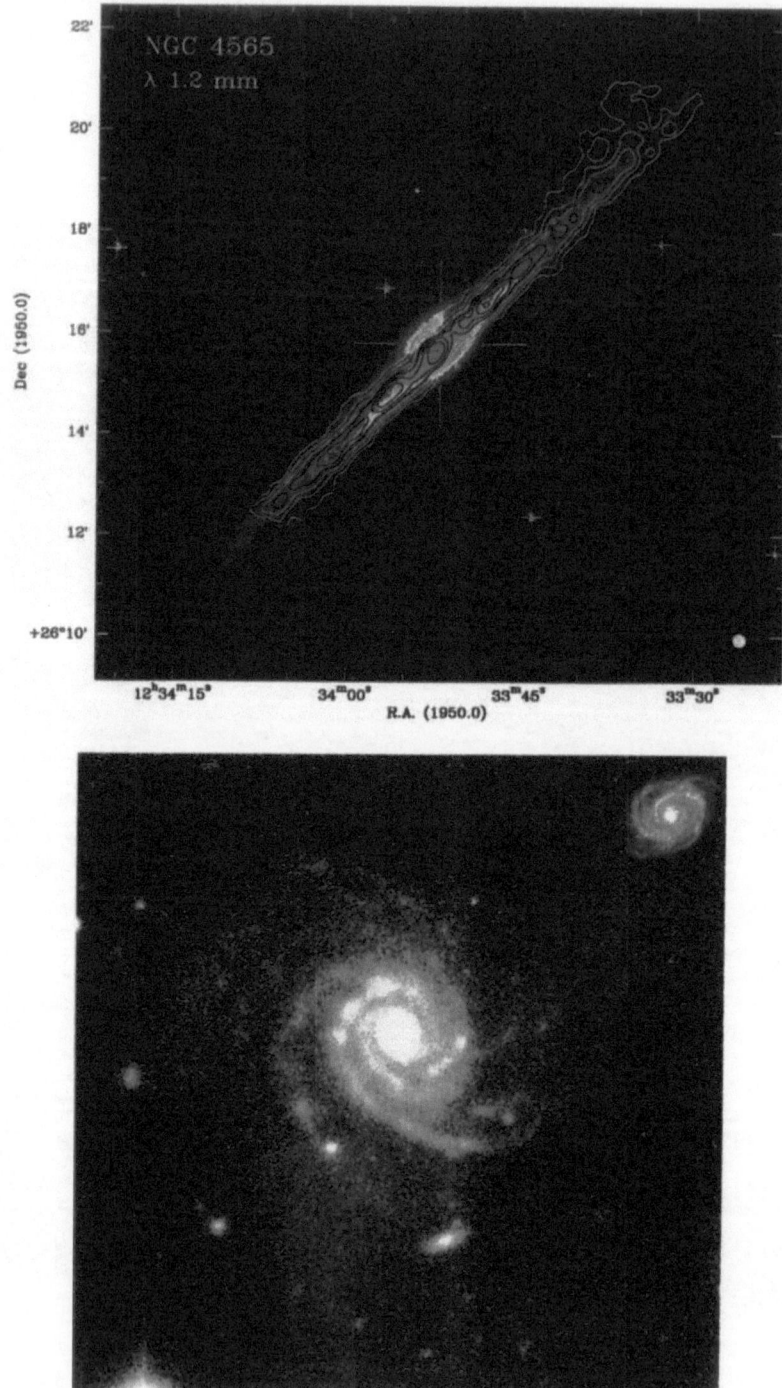

Neininger and Guelin: Colour Plate 11a (top). Impey: Colour Plate 11b (bottom)

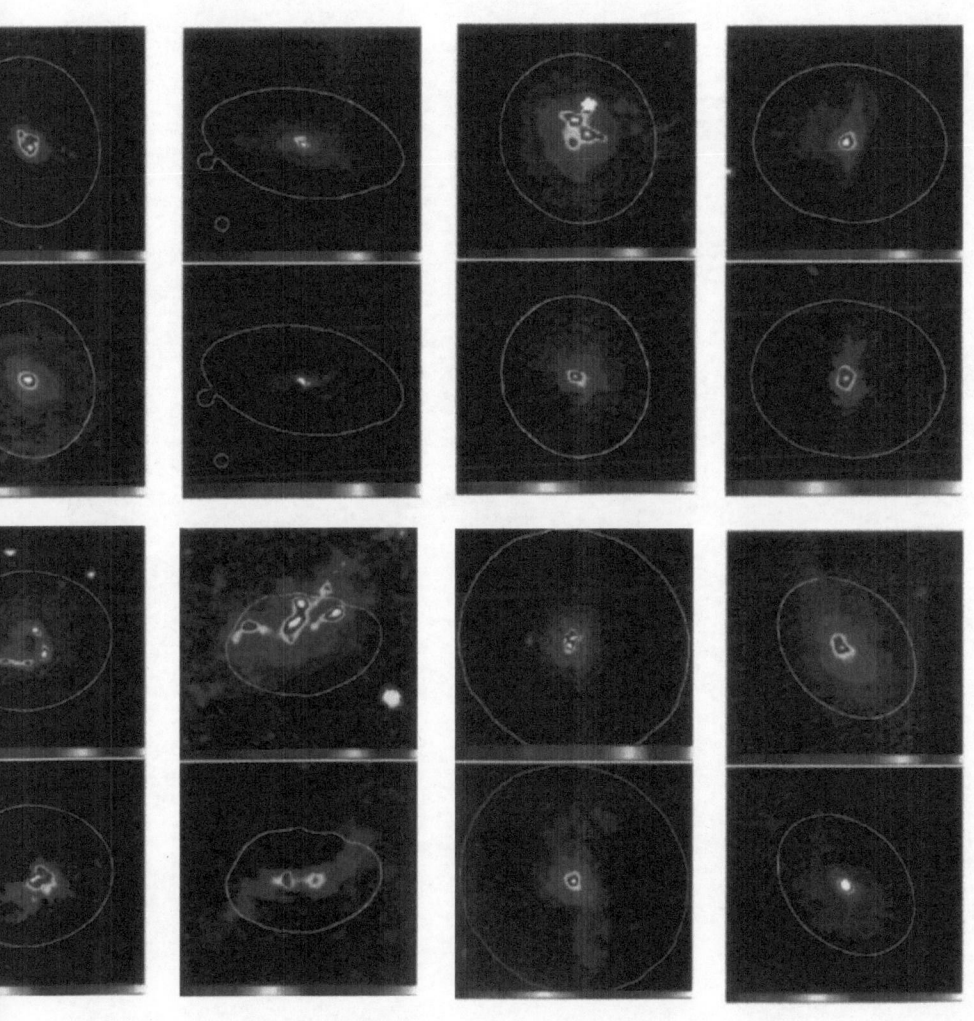

Goudfrooij: Colour Plate 12

Pfenniger: Colour Plate 13

Glass: Colour Plate 14

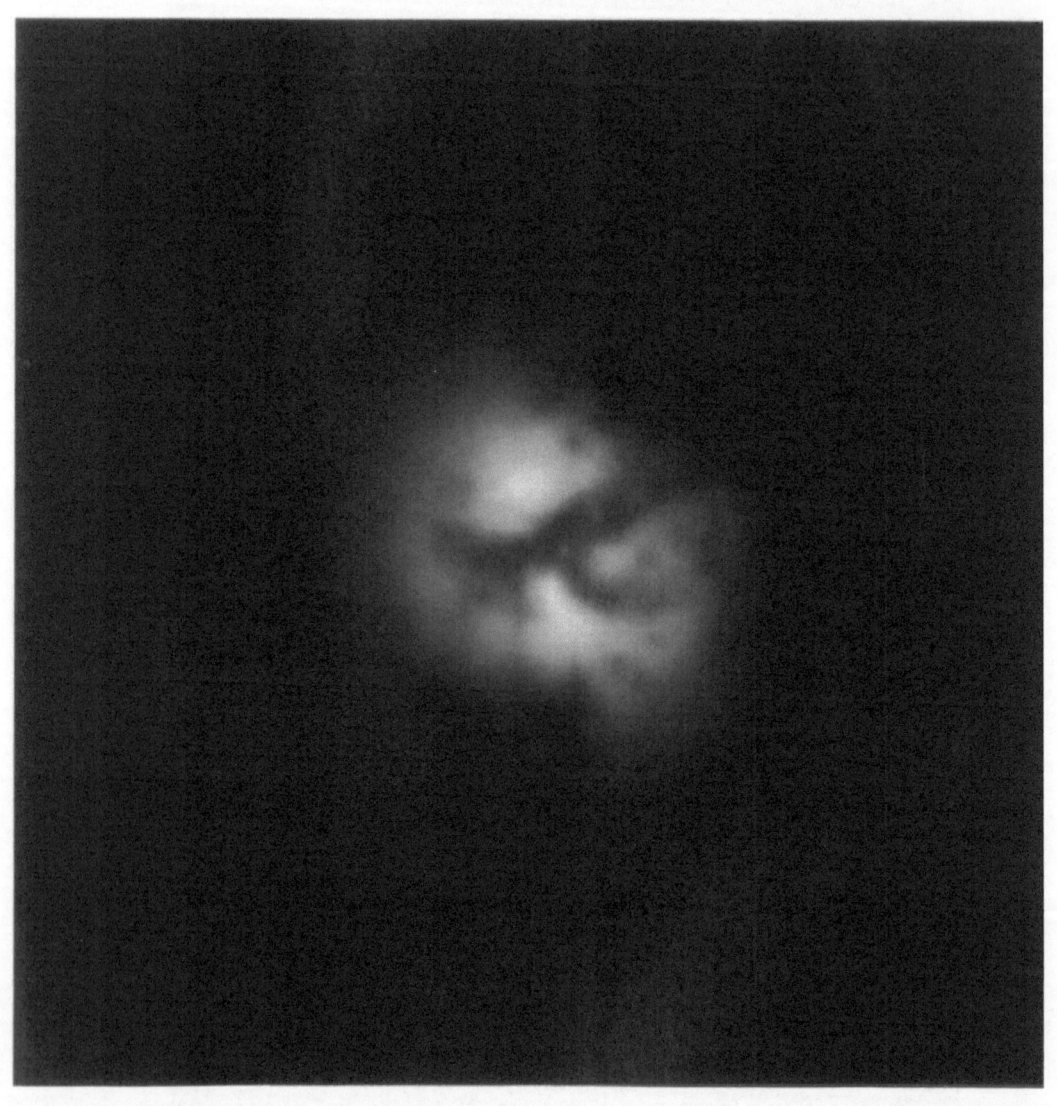

Panagia et. al.: Colour Plate 15

The Continent of Africa beckons: Colour Plate 16